VOLUME **4**
1700-1799

Science
and
Its
Times

Understanding the

Social Significance of

Scientific Discovery

Science and Its Times

Understanding the
Social Significance of
Scientific Discovery

Neil Schlager, Editor
Josh Lauer, Associate Editor

Produced by Schlager Information Group

Detroit
New York
San Francisco
London
Boston
Woodbridge, CT

Science and Its Times

VOLUME 4

1700-1799

NEIL SCHLAGER, *Editor*
JOSH LAUER, *Associate Editor*

GALE GROUP STAFF

Amy Loerch Strumolo, *Project Coordinator*
Christine B. Jeryan, *Contributing Editor*

Mary K. Fyke, *Editorial Technical Specialist*

Maria Franklin, *Permissions Manager*
Margaret A. Chamberlain, *Permissions Specialist*
Shalice Shah-Caldwell, *Permissions Associate*

Mary Beth Trimper, *Production Director*
Evi Seoud, *Assistant Production Manager*
Wendy Blurton, *Senior Buyer*

Cynthia D. Baldwin, *Product Design Manager*
Tracey Rowens, *Senior Art Director*
Barbara Yarrow, *Imaging and Multimedia Content Manager*
Randy Bassett, *Image Database Supervisor*
Robyn Young, *Senior Editor, Imaging and Multimedia Content*
Pamela A. Reed, *Imaging and Multimedia Content Coordinator*
Leitha Etheridge-Sims, *Image Cataloger*

Contents

Exploration and Discovery

Life Sciences and Medicine

Contents

1700-1799

Mathematics

Physical Sciences

Preface

The interaction of science and society is increasingly a focal point of high school studies, and with good reason: by exploring the achievements of science within their historical context, students can better understand a given event, era, or culture. This cross-disciplinary approach to science is at the heart of *Science and Its Times*.

Readers of *Science and Its Times* will find a comprehensive treatment of the history of science, including specific events, issues, and trends through history as well as the scientists who set in motion—or who were influenced by—those events. From the ancient world's invention of the plowshare and development of seafaring vessels; to the Renaissance-era conflict between the Catholic Church and scientists advocating a sun-centered solar system; to the development of modern surgery in the nineteenth century; and to the mass migration of European scientists to the United States as a result of Adolf Hitler's Nazi regime in Germany during the 1930s and 1940s, science's involvement in human progress—and sometimes brutality—is indisputable.

While science has had an enormous impact on society, that impact has often worked in the opposite direction, with social norms greatly influencing the course of scientific achievement through the ages. In the same way, just as history can not be viewed as an unbroken line of ever-expanding progress, neither can science be seen as a string of ever-more amazing triumphs. *Science and Its Times* aims to present the history of science within its historical context—a context marked not only by genius and stunning invention but also by war, disease, bigotry, and persecution.

Format of the Series

Science and Its Times is divided into seven volumes, each covering a distinct time period:

Volume 1: 2000 B.C-699 A.D.

Volume 2: 700-1449

Volume 3: 1450-1699

Volume 4: 1700-1799

Volume 5: 1800-1899

Volume 6: 1900-1949

Volume 7: 1950-present

Dividing the history of science according to such strict chronological subsets has its own drawbacks. Many scientific events—and scientists themselves—overlap two different time periods. Also, throughout history it has been common for the impact of a certain scientific advancement to fall much later than the advancement itself. Readers looking for information about a topic should begin their search by checking the index at the back of each volume. Readers perusing more than one volume may find the same scientist featured in two different volumes.

Readers should also be aware that many scientists worked in more than one discipline during their lives. In such cases, scientists may be featured in two different chapters in the same volume. To facilitate searches for a specific person or subject, main entries on a given person or subject are indicated by bold-faced page numbers in the index.

Within each volume, material is divided into chapters according to subject area. For volumes 5, 6, and 7, these areas are: Exploration and Discovery, Life Sciences, Mathematics, Medicine, Physical Sciences, and Technology and Invention. For volumes 1, 2, 3, and 4, readers will find that the Life Sciences and Medicine chapters have been combined into a single section, reflecting the historical union of these disciplines before 1800.

Arrangement of Volume 4: 1700-1799

Volume 4 begins with two notable sections in the frontmatter: a general introduction to science and society during the period, and a general chronology that presents key scientific events during the period alongside key world historical events.

The volume is then organized into five chapters, corresponding to the five subject areas listed above in "Format of the Series." Within each chapter, readers will find the following entry types:

Chronology of Key Events: Notable events in the subject area during the period are featured in this section.

Overview: This essay provides an overview of important trends, issues, and scientists in the subject area during the period.

Topical Essays: Ranging between 1,500 and 2,000 words, these essays discuss notable events, issues, and trends in a given subject area. Each essay includes a Further Reading section that points users to additional sources of information on the topic, including books, articles, and web sites.

Biographical Sketches: Key scientists during the era are featured in entries ranging between 500 and 1,000 words in length.

Biographical Mentions: Additional brief biographical entries on notable scientists during the era.

Bibliography of Primary Source Documents: These annotated bibliographic listings feature key books and articles pertaining to the subject area.

Following the final chapter are two additional sections: a general bibliography of sources related to the history of science, and a general subject index. Readers are urged to make heavy use of the index, because many scientists and topics are discussed in several different entries.

A note should be made about the arrangement of individual entries within each chapter: while the long and short biographical sketches are arranged alphabetically according to the scientist's surname, the topical essays lend themselves to no such easy arrangement. Again, readers looking for a specific topic should consult the index. Readers wanting to browse the list of essays in a given subject area can refer to the table of contents in the book's frontmatter.

Additional Features

Throughout each volume readers will find sidebars whose purpose is to feature interesting events or issues that otherwise might be overlooked. These sidebars add an engaging element to the more straightforward presentation of science and its times in the rest of the entries. In addition, each volume contains photographs, illustrations, and maps scattered throughout the chapters.

Comments and Suggestions

Your comments on this series and suggestions for future editions are welcome. Please write: The Editor, *Science and Its Times,* Gale Group, 27500 Drake Road, Farmington Hills, MI 48331.

Advisory Board

Amir Alexander
Research Fellow
Center for 17th and 18th Century Studies
UCLA

Amy Sue Bix
Associate Professor of History
Iowa State University

Elizabeth Fee
Chief, History of Medicine Division
National Library of Medicine

Lois N. Magner
Professor Emerita
Purdue University

Henry Petroski
A.S. Vesic Professor of Civil Engineering and
*   Professor of History*
Duke University

F. Jamil Ragep
Associate Professor of the History of Science
University of Oklahoma

David L. Roberts
Post–Doctoral Fellow, National Academy of
*   Education*

Morton L. Schagrin
Emeritus Professor of Philosophy and History of
*   Science*
SUNY College at Fredonia

Hilda K. Weisburg
Library Media Specialist
Morristown High School, Morristown, NJ

Contributors

Amy Ackerberg-Hastings
Independent Scholar

Peter J. Andrews
Freelance Writer

Kenneth E. Barber
Professor of Biology
Western Oklahoma State College

Charles Boewe
Freelance Biographer

Kristy Wilson Bowers
Lecturer in History
Kapiolani Community College, University of
Hawaii

Sherri Chasin Calvo
Freelance Writer

Thomas Drucker
Graduate Student, Department of Philosophy
University of Wisconsin

H. J. Eisenman
Professor of History
University of Missouri–Rolla

Ellen Elghobashi
Freelance Writer

Loren Butler Feffer
Independent Scholar

Keith Ferrell
Freelance Writer

Randolph Fillmore
Freelance Science Writer

Richard Fitzgerald
Freelance Writer

Maura C. Flannery
Professor of Biology
St. John's University, New York

Donald R. Franceschetti
Distinguished Service Professor of Physics and
Chemistry
The University of Memphis

Jean-François Gauvin
Historian of Science
Musée Stewart au Fort de l'Île Sainte-Hélène,
Montréal

Phillip H. Gochenour
Freelance Editor and Writer

Brook Ellen Hall
Professor of Biology
California State University at Sacramento

Diane K. Hawkins
Head, Reference Services—Health Sciences Library
SUNY Upstate Medical University

Robert Hendrick
Professor of History
St. John's University, New York

James J. Hoffmann
Diablo Valley College

Leslie Hutchinson
Freelance Writer

P. Andrew Karam
Environmental Medicine Department
University of Rochester

Evelyn B. Kelly
Professor of Education
Saint Leo University, Florida

Contributors

Judson Knight
Freelance Writer

Lyndall Landauer
Professor of History
Lake Tahoe Community College

Josh Lauer
Editor and Writer
President, Lauer InfoText Inc.

Brenda Wilmoth Lerner
Science Correspondent

K. Lee Lerner
Prof. Fellow (r), Science Research & Policy Institute
Advanced Physics, Chemistry and Mathematics,
*   Shaw School*

Eric v. d. Luft
Curator of Historical Collections
SUNY Upstate Medical University

Lois N. Magner
Professor Emerita
Purdue University

Marjorie C. Malley
Historian of Science

Amy Lewis Marquis
Freelance Writer

Ann T. Marsden
Writer

Kyla Maslaniec
Freelance Writer

William McPeak
Independent Scholar
Institute for Historical Study (San Francisco)

Lolly Merrell
Freelance Writer

Leslie Mertz
Biologist and Freelance Science Writer

Kelli Miller
Freelance Writer

J. William Moncrief
Professor of Chemistry
Lyon College

Stacey R. Murray
Freelance Writer

Stephen D. Norton
Committee on the History & Philosophy of Science
University of Maryland, College Park

Steve Ruskin
Freelance Writer

Elizabeth D. Schafer
Independent Scholar

Neil Schlager
Editor and Writer
President, Schlager Information Group

Keir B. Sterling
Historian, U.S. Army Combined Arms Support
*   Command*
Fort Lee, Virginia

George Suarez
Freelance Writer

Todd Timmons
Mathematics Department
Westark College

David Tulloch
Graduate Student
Victoria University of Wellington, New Zealand

Roger Turner
Brown University

A. Bowdoin Van Riper
Adjunct Professor of History
Southern Polytechnic State University

Stephanie Watson
Freelance Writer

Karol Kovalovich Weaver
Instructor, Department of History
Bloomsburg University

Giselle Weiss
Freelance Writer

A.J. Wright
Librarian
Department of Anesthesiology
School of Medicine
University of Alabama at Birmingham

Michael T. Yancey
Freelance Writer

Introduction: 1700–1799

The Age of Enlightenment Carries the Scientific Revolution Forward

During the sixteenth and seventeenth centuries the way educated people viewed the natural world and their relationship to it underwent a radical transformation. Known as the Scientific Revolution, this change was based on the work of such scientists and philosophers as Francis Bacon, Robert Boyle, Nicolas Copernicus, René Descartes, Galileo Galilei, William Harvey, Johannes Kepler, Gottfried Leibniz, and John Locke. It reached its crowning achievement with the publication of Isaac Newton's laws of motion in 1687.

As a result of the Scientific Revolution, by the beginning of the eighteenth century people had great confidence in the ability of reason to explain the natural world. They believed that scientific methods (such as those that had led to Newton's achievements in physics) could give rational explanations for all phenomena. Not only were Newton, Leibniz, and Locke still alive as the new century began, but Newton and Leibniz subsequently published major new works. They were joined by others who shared their faith in rational explanations, and who were attracted by the continuing success of this new empirical approach. Devoting themselves systematically to problems in science and technology, they critiqued, applied, and expanded this new way of thinking about the world and humanity's place in it. Knowledge expanded and practical applications of science grew at an unprecedented rate. Because of this intellectual ferment and the progress that came from it, the eighteenth century is known as the Age of Enlightenment.

Major discoveries about the composition of the physical world, made by men like Henry Cavendish, Joseph Priestley, and Joseph Black, were interpreted and synthesized into a theoretical framework by Antoine Lavoisier, who established chemistry as a distinct science. The modern science of biology began to take shape as new systems of nomenclature and classification were developed by Carolus Linnaeus. The organization of knowledge in both these fields facilitated learning and understanding.

In the life sciences, the century saw significant progress in the understanding of photosynthesis, plant hybridization, the role of nerves in muscle contractions, and the electrical basis of nervous impulses. The science of nutrition was launched by Rumford, and inoculation for the prevention of smallpox was developed by Edward Jenner.

Mathematics continued to play a significant role in the development of the physical sciences, and much progress came as scientists found mathematical expressions for much of the physical world. During the eighteenth century, Pierre Laplace and Joseph Lagrange made particularly significant contributions in statistics, probability theory, calculus, and analysis.

Similar strides were made in the physical sciences. The work of Joseph Black, Benjamin Thompson (Count Rumford), and others led to important progress in the understanding of heat and its transfer; Benjamin Franklin and Luigi Galvani provided an understanding of electricity; and Daniel Bernoulli laid the foundations of the science of hydrodynamics.

New Technology Leads to the Industrial Revolution

Applying this new scientific knowledge to technology led to new processes and inventions. Life was made easier by such inventions as the flushing toilet and bifocal spectacles. Benjamin

Franklin invented the kitchen stove, liberating women and servants from the difficulty of cooking on the open hearth.

The production of manufactured goods was greatly enhanced by the development of efficient steam power, the blast furnace, and the hydraulic press. Inventions like the flying shuttle, the spinning jenny, the power loom, and Eli Whitney's cotton gin improved textile weaving. These (and other) inventions led to a new industrial system in which work was concentrated into a single factory that employed many workers, replacing traditional cottage industries in which work was done by individuals in their homes. The Industrial Revolution had begun, and the way people lived and worked was changed forever.

During the eighteenth century, the invention of an accurate marine chronometer, development of navigational quadrants, and other new technology that aided navigation significantly increased exploration of the world and led to an expansion of worldwide trade. Of particular importance were the circumnavigations of the globe by Captain James Cook. On land, exploration of California, Alaska, and the African interior began during this time. The region beyond the Appalachian Mountains was opened for settlement through the efforts of pioneers such as Daniel Boone. Nor were the heavens neglected — hot air balloons were first used for human transportation during this time.

Adverse Effects of the Growth of Science and Technology

The extensive progress in science and technology during the Enlightenment created change that was sometimes painful. While expanding industrialization and trade enlarged the middle class of merchants and manufacturers and improved their living conditions, the change from piece-work manufacture in the home to factory production had negative consequences for many people. The shift spurred the development of industrial cities, whose rapid growth produced squalid slums with a variety of health and social problems.

Back on the farm, the development of scientific agriculture and better equipment reduced the need for farm workers ("freeing" them for factory work). Furthermore, the expanding wool market encouraged wealthy farmers to convert much of the open land previously used for growing crops into sheep pasture. Enclosure, as it was called, left many small farmers without

the means to survive, and their farms were effectively dissolved. Entire communities, such as the crofting (cooperative farming) towns of Scotland, were abandoned, and the population was forced to seek a means of livelihood elsewhere, often in the teeming industrial cities. Many emigrated to America, where they served as an impetus for westward expansion.

A particularly cruel effect of technological progress was the rise of slavery, especially in the American South. Ironically, in the closing decades of the eighteenth century, slavery was becoming an economic liability, especially in the Southern states. Slaves were expensive, and there were no crops that could overcome in profits what they cost to maintain, however poorly. Cotton was a labor-intensive crop, and deseeding it (except for Sea Island cotton, which could only be grown along the coast) was so time-consuming that it was hardly worth the effort to plant it. Eli Whitney's invention of the cotton gin in 1793 changed everything. Suddenly cotton seeds could be removed quickly and easily, and even the hard-to-seed inland varieties became highly profitable. Cotton became a cash crop almost overnight, when farmers realized they could sell to the British textile industry, whose demand for cotton to feed their mills was insatiable. This assured the South's reliance on slave labor for the next 60 years.

Religion was especially influenced by the developing scientific (and quasi-scientific) ideas of the Enlightenment. The church had dominated life in the West before the Scientific Revolution, yet its influence was gradually diminished by the emergence of science, with its belief that nature was both rational and understandable. Among intellectuals, there was a rise in deism, the belief that God created but then withdrew from the world, and in atheism, the denial of God's existence. Geology, which became a separate scientific discipline during the eighteenth century, resulted in widespread debate on the accuracy of the biblical creation story in Genesis. Although the Catholic Church and other denominations remained strong, a decreasing confidence in established church doctrine and a wish for a more individualized, less formal, religious expression grew. The founding of Methodism in England and the Great Awakening revival in America were direct results.

Science, Technology, and Politics

In the political arena, the eighteenth century was not a peaceful one. While monarchy remained the most widespread means of government in the West, the middle class began to demand freedom from arbitrary hereditary rule. This led to two major revolutions: the American colonists against England, resulting in the birth of a new self-governing nation; and the middle class and peasants of France against the king and the *ancien régime*, resulting in years of turmoil, and eventually the usurpation of power by Napoleon.

Because industrial and scientific growth was concentrated in France and England, these nations became the primary world powers. Their competition for territory, natural resources, and markets caused a number of costly and destructive wars, in which new technology increased casualties. These included the War of Spanish Succession, the Seven Years' War (fought as the French and Indian War in America), and the Napoleonic Wars. These political realities affected science directly. Governments financed the development of new technologies because of their potential contribution to the national economy and possible application to warfare, a practice that has remained in effect ever since.

Cultural Effects of Scientific Thought

The eighteenth century foreshadowed the profound effect that science and technology would have on all areas of human endeavor in succeeding centuries. Enlightenment philosophy was deeply influenced by the widespread confidence in new scientific ideas and the rationality of all things, including economic, social, and political matters. Philosophical systems developed either in support of these new ideas or in reaction to them. John Locke's ideas formed the philosophical bases of the American and French revolutions. The writings of François Voltaire, Jeremy Bentham, Immanuel Kant, David Hume, Jean-Jacques Rousseau, Edmund Burke, and George Berkeley were widely read in the eighteenth century and remain influential today. Their works are part of the foundation on which our understanding of truth rests, and they influence our understanding of ourselves and the world around us. In addition, the economic principles of Thomas Malthus and Adam Smith were instrumental in justifying laissez-faire policies and the spread of the Industrial Revolution.

Music, literature, and art also developed greatly during the Age of Enlightenment. Changes in these fields reflected a cultural response to the Scientific Revolution and the Enlightenment. Styles evolved from a concentration on mathematical form and precision (the Baroque era) through a nostalgic return to simpler ideals (the Classical period) into an idyllic focus on the natural world (the Romantic style). Some historians argue that music reached its highest point during the eighteenth century with the introduction of the concerto, symphony, sonata, and opera in their modern forms; the invention of the piano; and the compositions of Johann Sebastian Bach, Georg Friederich Handel, Antonio Vivaldi, Joseph Haydn, Wolfgang Amadeus Mozart, and Ludwig van Beethoven.

The principal visual artists of the eighteenth century portrayed people and the natural world realistically. They included Thomas Gainsborough and Joshua Reynolds in England, Jacques-Louis David (who portrayed French patriotic fervor) and Jean-Baptiste Chardin in France, Francisco Goya (whose works involve social criticism) in Spain, and Gilbert Stuart and John Singleton Copley in America.

Prose forms, especially the essay, were developed into effective means for influencing political thought and disseminating scientific and other ideas. Joseph Addison, Denis Diderot, and Jean-Jacques Rousseau were particularly successful essayists; Jonathan Swift and Françios Voltaire were especially adept with satire. The English novel flourished with such authors as Daniel Defoe and Henry Fielding. The Scotsman Robert Burns made colloquial poetry an acceptable form. A strong Romantic movement developed at the end of the century, especially in poetry. Its practitioners, who were strongly influenced by the scientific observations of the naturalists, included Friedrich Schiller, Johann Wolfgang von Goethe, William Wordsworth, William Blake, and Samuel Taylor Coleridge.

The effort to codify all knowledge scientifically also led to the publication of Samuel Johnson's *Dictionary* in England and *Encyclopédie* in France. Both were the first of their kind; the latter was regarded as the Bible of the Enlightenment. The first daily newspaper, the *Daily Courant,* debuted in England in 1702. By the end of the century, public newspapers were commonplace among in France and the American colonies as well. The Enlightenment ideal of knowledge and rational discovery was being disseminated to the people.

J. WILLIAM MONCRIEF

Chronology: 1700–1799

1700-21 Power struggles consume Europe, as the continent is embroiled in the War of the Spanish Succession (1700-1714) and the Great Northern War (1700-1721).

1701 Frederick III, elector of Brandenberg, establishes the kingdom of Prussia with himself as King Frederick I.

1707 England and Scotland form the United Kingdom of Great Britain, combining the Scottish cross of St. Andrew and the English cross of St. George to form the Union Jack.

1714 Gabriel Daniel Fahrenheit invents the first accurate thermometer, along with the scale which bears his name, in 1730; Réné Antoine Ferchault de Réaumur develops his own thermometer and scale; and in 1742 Anders Celsius introduces the centigrade scale later adopted internationally by scientists.

1728 Daniel Bernoulli, studying the mathematics of oscillations, is the first to suggest the usefulness of resolving a compound motion into motions of translation and rotation.

1735 Swedish botanist Carolus Linnaeus outlines his system for classifying living things, with a binomial nomenclature that includes generic and specific names.

1740-48 The War of the Austrian Succession, involving numerous European nations, results in the establishment of Prussia as a major European power.

1750s-70s The Enlightenment spawns great works, including the *Encyclopédie* of Diderot and d'Alembert (1751-72), Samuel Johnson's *Dictionary of the English Language* (1755), David Hume's *An Enquiry Concerning Human Understanding* (1758), Voltaire's *Candide* (1759), Rousseau's *Du contrat social* (1762), and Adam Smith's *Wealth of Nations* (1776).

1756-91 Meteoric career of Wolfgang Amadeus Mozart, who was born six years after Bach's death in 1750, and who in 1787 taught a young Ludwig van Beethoven.

1756-63 The first worldwide conflict of modern times, the Seven Years War, is fought in Europe, North America, and India, resulting in establishment of Britain as world's leading colonial power and Prussia as an up-and-coming force on the European continent.

1768 Lazzaro Spallanzani, who three years earlier had published data refuting the theory of spontaneous generation, concludes that boiling a sealed container prevents microorganisms from entering and spoiling its contents.

1768 Captain James Cook embarks on the first of three voyages over the next twelve years, which included expeditions to the Pacific, the ice fields of Antarctica, and the northern Pacific coasts of North America and Asia.

1769 James Watt obtains the first patent for his steam engine, which improves on ideas developed by Thomas Newcomen half a century earlier and is the first such engine to function as a "prime mover" rather than as a mere pump.

1774 Joseph Priestley, an English chemist, discovers oxygen.

1775-83 Britain's colonies in North America revolt against the mother country, declare independence (1776), and secure victory after an eight-year war.

1781 German-English astronomer William Herschel discovers Uranus, the first planet discovered in historic times.

1783 Jean François Pilatre de Rozier and the François Laurent, marquis d'Arlandes become the first human beings ever to fly, in a Montgolfier hot-air balloon.

1787 The United States Constitution is ratified, formally establishing the new nation.

1789 The French Revolution begins.

1792 France declares itself a republic; a year later, Louis XVI and Marie Antoinette are executed, and the Reign of Terror begins.

1793 Eli Whitney invents the first cotton gin, which greatly stimulates the U.S. cotton industry—and helps perpetuate slavery in the process.

1795 Gaspar Monge makes public his method for representing a solid in three-dimensional space on a two-dimensional plane—i.e., descriptive geometry, a military secret that had long been guarded by the French government.

1799 The corrupt Directory, which assumed power in 1795 and ended the Reign of Terror in France, is replaced by the Consulate under Napoleon Bonaparte.

Exploration and Discovery

Chronology

1741 Thirteen years after his first voyage of exploration in the sea channel between Siberia and Alaska that bears his name, Vitus Bering discovers the Alaskan mainland and Aleutian islands.

1743 French explorer Charles-Marie de La Condamine leads the first scientific exploration of the Amazon River.

1748 Excavation of Pompeii, destroyed by a volcano some 1,700 years before, begins.

1767 Samuel Wallis leads the first European expedition to Tahiti.

1768 Captain James Cook embarks on the first of three voyages over the next twelve years, which will include expeditions to the Pacific, the ice fields of Antarctica, and the northern Pacific coasts of North America and Asia.

1769 Junipero Serra explores California and establishes the first Spanish mission at San Diego; over the next seven years, he also establishes missions at Los Angeles and San Francisco.

1783 Jean François Pilatre de Rozier and the François Laurent, marquis d'Arlandes become the first human beings ever to fly, in a Montgolfier hot-air balloon.

1786 Michel-Gabriel Paccard and Jacques Balmat become the first to reach the summit of Mont Blanc.

1792 George Vancouver charts the Pacific coast of North America from California to Alaska.

1792 Pedro Vial establishes the Santa Fe Trail between Santa Fe, New Mexico, and St. Louis, Missouri.

1796 Mungo Park explores the African interior and is the first European to discover the Niger River.

1799 Soldiers in Napoleon's army occupying Egypt discover the Rosetta Stone, whose translation two decades later will unlock the mystery of ancient hieroglyphics.

Overview:
Exploration and Discovery 1700-1799

Background

Explorers throughout history have been driven by a desire for discovery that has incorporated a multitude of objectives both personal and nationalistic. From the conqueror to the adventurer, all types of explorers, both men and women, have traveled to the furthest corners of Earth. As man broadened his horizons, one of the strongest forces driving further exploration became the pursuit of trade, especially in luxury goods such as precious metals, jewels, furs, silk, aromatic scents, and spices. In the 1600s organizations such as the East India Company made historic ocean voyages to the Orient and South Pacific. Trade soon led to permanent trading posts and these in turn led to colonial occupation such as the colonies founded in North America. By the end of the seventeenth century explorers also began to venture forth for nobler motives—some as missionaries, others for the love of travel, as well as those interested in satisfying scientific curiosity.

In the eighteenth century explorers made great strides in compiling more accurate geographic and meteorological data and maps, and contributed to political history and expansion, diplomacy, and geography. Their expeditions also helped to dispel many myths and superstitions regarding the oceans and continents of Earth. Others traveled to expand the new sciences of mathematics, physics, and astronomy (all of which influenced navigation). Still others widened the knowledge of archaeology, geology, anthropology, ethnology, and other natural sciences.

European Nations Explore the Pacific Ocean

Much European exploration had concentrated in previous centuries on the Atlantic Ocean and the lands bordering its coastlines. In the 1700s Europe's nations began to survey, explore, and lay claim to lands along and islands in the Pacific Ocean, the largest of Earth's three great oceans. Prior to the eighteenth century the Pacific had been a vast sea to be crossed by circumnavigators and others seeking routes to the East or by men seeking undiscovered continents. Although the search for the elusive *terra australis*, a legendary southern continent filled with mythic,

fantasy creatures, still spurred voyages to the Pacific in the early and middle part of the century, more significant exploration was accomplished by those seeking to learn more about the great ocean itself.

The quest for *terra australis* led Dutch Admiral Jacob Roggeveen (1659-1729) to the first of many islands to be visited by Europeans. On Easter Day 1722 his voyage, sponsored by the West Indian Company, resulted in the discovery of Easter Island, and set the stage for future voyages to the South Pacific. Similarly, the British Admiralty selected Captain Samuel Wallis (1728-1795) for a voyage of exploration to the South Pacific in search of *terra australis*. Wallis instead discovered Tahiti, introducing the island to European society.

The North Pacific was also explored during this time. In 1728 and again in 1733-41, the Russian Navy sent Vitus Bering (1681-1741) on voyages to map large portions of Russia's coasts and northwestern North America. These voyages had a great impact on Russian trade in the area. The strait separating Asia from North America is named for Bering. Also exploring the North Pacific was British Captain George Vancouver (1757-1798), who surveyed and mapped the Pacific coast from Alaska to Monterrey between 1790-95. Vancouver's voyage, its critical survey, accurate soundings, and the coastal data returned had a tremendous impact on the expansion of British control of land and sea in the region.

The French, long embroiled with political and national concerns at home, made their first major ocean explorations in the 1700s—and the first French navigator to sail around the world on a voyage of discovery was Louis-Antoine de Bougainville (1729-1811), who spent a significant time in the South Pacific. In 1785 Jean-François de Galaup, comte de la Pérouse (1741-1788?), explored the North Pacific from China to Japan. On the return voyage via Australia, the comte and his crew were lost at sea. In an effort to discover La Pérouse's whereabouts, the French sent Antoine de Bruni (1739-1793) to the South Pacific, where he charted the Tasmanian and Australian coasts and many of the region's islands before dying of scurvy. De Bruni's accurate maps allowed France to lay claim to numerous islands he discovered—France soon expanded its territorial possessions to include many South Sea islands. His records of oceanographic and meteorological data were invaluable to future mariners and aided the French in planning trade routes and military objectives in the South Pacific.

The most significant Pacific explorer was Britain's Captain James Cook (1728-1779). His three major voyages of discovery to the Pacific yielded vital data for navigators, botanists, and naturalists, and the medical sciences (especially with regards to proper diet to prevent scurvy). On his first voyage, Cook made important celestial observations, circumnavigated New Zealand, and explored the east coast of Australia; on his second, he circumnavigated the globe from west to east, discovered New Caledonia, the South Sandwich Islands, and South Georgia, was the first to travel below the Antarctic Circle, and his voyage of over 60,000 miles (96,560 km) also proved that *terra australis* did not exist; on his third voyage, he proved the Northwest Passage was not a practical route from the west and discovered Hawaii, where he was killed during a second visit. The data collected on his voyages provided a more realistic map of the globe. Overall, as a single man, Cook had a tremendous impact on the world he lived in and helped shift the world's focus from exploration to development.

Europeans in North America and Africa

Although the Pacific was the region most visited by explorers in the 1700s, other expeditions were traveling to unexplored lands closer to Europe and its colonies. From 1731-43, Pierre Gaultier de La Vérendrye (1685-1749) led an expedition team that included his sons on an extensive journey that established numerous forts and trading posts throughout the northern half of North America and spurred the fur trade and Indian relations for his sponsors. (Others who influenced fur trade in the northernmost parts of North America included seafarers Bering and Vancouver.) From 1736-43 Frenchman Charles-Marie de La Condamine (1701-1774) led the first scientific exploration of the Amazon River. Other explorers blazed trails in North America that aided expansion to the West—such as Frenchman Pedro Vial (1746?-1814) who was hired by the Spanish governor of Santa Fe to establish trade routes from there to St. Louis, New Orleans, and San Antonio. (The 1803 signing of the Louisiana Purchase would give American settlers direct access to Vial's Santa Fe Trail.)

Trailblazing translated to rivers in Africa, where in 1772 British explorer James Bruce (1730-1794) became the first European to follow the Blue Nile to where it converged with the White Nile in Ethiopia. In 1795 Scotsman Mungo Park (1771-1806) located the Niger River and followed it over 1,000 miles (1,609 km) through the African interior. His adventures, which he published, and his description of Africa fueled Europe's interest in the continent. Another unique expedition was one sponsored by the Danish government to the Near East from 1761-67—the first European expedition to that area of the world. The only survivor, a German explorer and surveyor, was Carsten Niebuhr (1733-1815), who continued exploring even after the deaths of his companions, returning to publish several important reports as well as a three-volume set of notes from the expedition's naturalist. Even the possibility of death was not a deterrent for the most intrepid of adventurers, and many were to follow in Niebuhr's footsteps in the nineteenth century.

As European explorers traveled around the world, other discoveries and firsts were made closer to home. The year 1786 saw the birth of modern mountaineering as three men made momentous ascents of Mont Blanc in the Alps. The birth of modern archaeology can be traced to three separate events in the eighteenth century—the 1738-48 discovery and meticulous excavation of the cities of Herculaneum and Pompeii in Italy, covered since A.D. 79 by volcanic debris; the 1790 discovery and revolutionary interpretation of Stone Age tools on the lands of John Frere (1740-1807) in England; and the 1795 discovery of the Rosetta Stone in Egypt by French soldiers in Napoleon's army. All three events had lasting impact on the science of archaeology.

Conclusion

By the end of the eighteenth century superstition and hearsay about the world's lands and oceans were a thing of the past. In the 1800s men turned to science, and governments turned to colonial expansion. Historic ocean voyages, epic adventures, and exhaustive expeditions rapidly expanded national boundaries and imperial domains as well as scientific knowledge in the fields of botany, zoology, ornithology, marine biology, geology, and cultural anthropology. By the end of the nineteenth century few areas of the world remained undiscovered and unexplored by man.

ANN T. MARSDEN

Voyage Into Mystery:
The European Discovery of Easter Island

Overview

Dutch Admiral Jacob Roggeveen (1659-1729) made the first European discovery of Easter Island on Easter Day, April 5, 1722, and ended 1,400 years of isolation on the island. Triangular shaped, Easter Island or Rapa Nui as it is known locally, is located 2,300 miles (3,700 km) west of the Chilean coast in the South Pacific Ocean. Over 2,000 miles (3,200 km) from the nearest populated center, Rapa Nui is one of the most isolated settlements in the world. The island is small, only 60 square miles (155 sq km), and is barren except for the hardy grasses that grow there, but is noted because of the large mysterious statues or moai that dot the island. Although the discovery of this island was not considered important at the time, it has since attracted the attention of archaeologists and scientists from all over the world.

Background

It was largely the hunt for riches and commerce that led to the exploration of the South Pacific Ocean by Europeans. It was commonly believed there was a large super continent called *terra australis incognito* in the Southern Hemisphere, and many expeditions left for the Pacific in search of it. Vasco Núñez de Balboa (1475-1519) was the first European to sight the Pacific in 1513, and seven years later Ferdinand Magellan (1480?-1521) rounded South America and sailed across the Pacific Ocean. It was the Spanish, interested in trade, who led the initial explorations of the South Pacific from 1567-1606. The Dutch, who were excellent seamen, followed. Jakob Le Maire (1585-1616) was an entrepreneur who explored the Pacific in 1615 and 1616, followed by fellow Dutchman Abel Tasman (1603-1659) in 1642, who worked for the East Indian Company.

Roggeveen's voyage used the knowledge of the Dutchmen who preceded him, as well as that of Englishmen William Dampier (1652?-1715) and Edward Davis. In 1687 Dampier and Davis were in the Pacific in search of the southern continent and reported "seeing" a low sandy island, and Davis said he could make out the faint outline of mountains in the background. This was of particular interest to Roggeveen, who had inherited from his deceased father the rights to an expedition to the South Pacific with the West Indian Company. Retired from a position with the East Indian Company, Roggeveen renewed the proposal with the rival West Indian Company. Desirous of finding *terra australis* and aware of the accounts given by Dampier and Davis, the company approved the expedition and provided Roggeveen with three ships, the *Arend, African Galley,* and *Thienhoven.* Roggeveen and his crew of 233 departed from Holland on August 21, 1721.

Crossing the Atlantic, they touched briefly at the Falkland Islands, and sailed for Le Maire Strait and Cape Horn. It was a three-week passage to the Pacific during cold weather, which correctly convinced Roggeveen there was a large landmass in the polar region, but he thought it was part of *terra australis.* The next stop was the Juan Fernandez Islands off Chile, where Roggeveen was so enthralled that he planned to return and establish a settlement. From these islands, Roggeveen sailed west, looking for Dampier's island.

The crew aboard the *African Galley* was the first to see the what was subsequently named Paasch Eyland (Easter Island), on April 5, 1722. Excited, Roggeveen and his crew thought it could indicate the presence of the elusive southern continent. Staying offshore, they noticed smoke coming up from various parts of the island the next day. Roggeveen decided to send the well-armed *Arend* and *Thienhoven* closer to look for a suitable place to lay anchor. With bad weather on April 7, the ships were not able to drop anchor, but an islander did canoe out to visit one of the ships. The Dutch were amazed by the totally nude man who boarded their ship. He was described as being well-built and tall, with tattoos all over his body. The islander was equally amazed by the Dutch, and marveled at their well-built ship. The crew sent him back with two strings of blue beads, a small mirror, and a pair of scissors. Following this, Roggeveen brought his ships closer to the island and was disappointed to see it did not fit the description of Dampier's island. On April 8, all ships set anchor offshore, but the weather was still too bad to go ashore, and the following day more islanders came out to meet the Dutchmen. They

Giant statues at Easter Island. *(Susan D. Rock. Reproduced with permission.)*

too admired the Dutch ships, and were so bold they stole the hats right off the men's heads and dove back into the ocean. Roggeveen organized a shore party of 134 men on the same day. While cautious, the crew was curious about the island, as they could see on shore the huge megaliths that have made the island famous.

Impact

Rowing ashore on April 10, 1722, Roggeveen and his crew climbed over the rocks that covered the shoreline and began marching into the interior, but were deluged by a large gathering of islanders. As they were coming into formation, Jacob heard shots fired from the back. An islander had tried to grab a musket from one of the men, who in return struck him, while another islander grabbed at the coat of one of Roggeveen's men. In defense, the islanders picked up rocks and the nervous crewmembers shot at them. In the end, 10 to 12 of the islanders lay dead, while several more were injured. Settling down quickly, the islanders tried to restore the peace by presenting Roggeveen with large amounts of fruit and poultry. Relations remained friendly, and Roggeveen was shown around part of the island. He noticed about 20 well-made huts, and several poorly-made canoes. A lack of women and children were also noticed. Naturally, the statues were of great interest to Roggeveen, with some of them as a high as 30 feet (9.1 m), and carved in human form. It was difficult for the Dutch to understand how the statues could have been erected since there were no trees to provide poles for leverage. Roggeveen concluded, incorrectly, they were made of clay and surfaced in stone. Only remaining on the island one day, the three ships sailed eastward in search of Terra Australis, which Roggeveen was convinced must be close.

Roggeveen sailed on in search of the southern continent. In mid-May he came to the fringes of the Tuamotus Islands, were he lost 10 men in an altercation with local residents. His poor luck continued, when he lost one of his ships on the reefs that surround the Tuamotus. The men became discouraged after sailing in the Pacific for another month, still unable to find the continent. A meeting was held, and Roggeveen decided to sail west for the Dutch outposts in Batavia (Jakarta).

Enroute to the outposts, Roggeveen passed the island of Bora-Bora and then the Samoan Islands. Roggeveen and his expedition were the first Europeans to see the Samoan Islands, but only went ashore briefly to get fresh fruit and water. By this time the crew was ravaged by scurvy, which killed 140 crewmembers. Passing between the island groups of Tuvalu and Kiribati, they headed north of New Guinea and onto the Moluccas, which were part of the Dutch East Indies. They arrived at Batavia in September 1722, where the rival East Indian Company confiscated their ships claiming they were trespassing on their territory. Virtually taken prisoner, they were escorted back to the Netherlands by the company. Later, the East Indian Company was taken to court and ordered to pay restitution to Roggeveen and his crew.

It was nearly 50 years before the island was revisited by Europeans, then by the Spanish, led by Don Felipe Gonzalez, who arrived in 1770. They too noticed there were no women on the island and suspected the islanders had underground hiding places. It was also noticed by the Spanish that the moai were not made of clay, but of stone. Four years later the island was visited by Captain James Cook (1728-1779), who actually saw the entrances to the underground caves, but was not permitted access. It was on Cook's second voyage (1772-75) to the Pacific that he proved the southern continent Terra Australis did not exist. The French arrived at Easter Island in 1786 and confirmed the existence of the caves when they were escorted through the hidden caverns. But disaster struck in 1862, when the Peruvians conducted a major slave raid on Easter Island, taking more than 1,000 people. Later they were forced to return their captives to the island, but by then illness and disease had killed most of them. The survivors returned to the island only to spread smallpox to the remaining population, reducing it to just 111.

While European intrusion on the island had devastating effects, its ecology and civilization were already in a state of crisis when the Europeans arrived. Rapa Nui was once a sub-tropical island, thickly covered in palm trees and home to many different bird species. Polynesians, as it has been determined, probably first came to the island around A.D. 400. A rich and complex society developed and the population swelled to nearly 10,000. Rival clans developed and each built moai for political as well as religious reasons. A period of decline came in A.D. 1500, as the growing population put too much pressure on the island's ecosystem, and all the palms were cut down to move moai or to supply fuel for the islanders. As resources dwindled, wars followed and the population fell to approximately 2,000, while Easter Island was reduced rock and grass. After a long and colorful history Chile annexed Easter Island in 1888. The islanders became full Chilean citizens in 1965, when a civilian governor was appointed to the island.

Since Roggeveen did not find the southern continent, his sponsors considered his expedition a failure, though Roggeveen, along with explorers like Captain Cook, contributed greatly to European knowledge of the South Pacific. Roggeveen's findings inspired the imaginations of laypersons and scientists alike, and archaeologists have learned much about the lives and travels of ancient humans from Easter Island. The moai have long been the subject of fascination and controversy, with some even suggesting the giant megaliths were built by aliens. Easter Island's fate also serves as a reminder of Earth's fragility, and the responsibility we have to preserve and protect it for future generations.

KYLA MASLANIEC

Further Reading

Bohlander, Richard E., ed. *World Explorers and Discoverers.* New York: Macmillan, 1992.

Heyerdahl, Thor. *Aku-Aku: The Secret of Easter Island.* Chicago: Rand McNally, 1958.

Orliac, Catherine, with Michel Orliac. *Easter Island: Mystery of the Stone Giants.* Translated by Paul G. Bahn. New York: H. N. Abrams, 1995.

First Scientific Exploration of the Amazon River Led by Charles-Marie de La Condamine

Overview

In an expedition intended to take the most accurate measurements ever of Earth, a team of French scientists were given permission as the first foreigners to be allowed into the New World territories of the Spanish Empire for the purpose of conducting scientific research. At the end of years of work the expedition's leader, Charles-Marie de La Condamine (1701-1774), undertook the first scientific expedition down the length of the Amazon River from its headwaters in the Andes Mountains to its mouth on the Atlantic Ocean. Previous explorers to the area were military or government agents acting on behalf of the Spanish or Portuguese authority in the New World or clergy accompanying them. La Condamine's exploration, which occurred over the course of 4 months in 1743, focused on observing the river and its environment. The expedition would turn out to be the main achievement of the original mission.

Background

In the first half of the eighteenth century physicists, geographers, and astronomers had come to the conclusion that the various forces acting on Earth as it spun on its axis changed its shape from the perfect sphere it was long assumed to be. Two conflicting theories arose as to how that shape was imperfect. English physicist Sir Isaac Newton (1642-1727) calculated that the planet flattened out at the poles and bulged at the equator. A conflicting theory was put forward by two French astronomers, Giovanni Domenico Cassini (1625-1712) and his son Jacques (1677-1756), who made measurements demonstrating that Earth was elongated at the poles and drew in at the equator. The French Academy of Sciences decided to settle the matter by sending two expeditions out to make the same measurements where they would show the greatest difference. One team was sent to northern Scandinavia to make the measurements close to the North Pole. The other team, led by the mathematician Charles-Marie de La Condamine, would go to northern Peru, where the equator passed through the Andes Mountains in South America. Each expedition would take accurate measurements of the distance covered by one degree of latitude and compare the measurements back at the Academy of Sciences in Paris.

La Condamine's team was given unprecedented permission from the Spanish Crown to travel into its South American territories to conduct their research. In May 1735 they sailed from France to what is now Colombia, and from there traveled to the Isthmus of Panama, where they crossed overland to the Pacific Ocean. From there they sailed to the northern portion of the Peru Territory (now Ecuador) and ascended the mountains to the city of Quito, where they would make their measurements. Delays plagued the expedition: accusations of espionage, meddling colonial officials, disputes over the participation of Spanish scientists, and death

threats. In the middle of all their delays word arrived from France that the Arctic expedition had returned, and their data confirmed Newton's theory that Earth flattened at the poles. La Condamine's team continued its work, and in 1743 they made their last measurements. The expedition split up with only La Condamine making an immediate return to France.

La Condamine chose to make his way back to France by embarking on a mapping expedition down the Amazon River. He chose a route that began at the furthest navigable reaches of the Marañón river and proceeded through the dangerous pass at Pongo of Manseriche for the expressed purpose of seeing the pass. In June 1743 La Condamine and his native Andean guides left from the river port of Jáen in what is now northern Peru, about 100 miles (160.9 km) from the Pacific coast of South America. Traveling on a raft built by his guides, La Condamine had several close calls with not only his life but also the eight years of research and scientific instruments he was transporting back to France. However, during the expedition's arrival in South America La Condamine had been introduced to latex made from the sap of the rubber tree. Early on in the trip he was able to make rubber-treated sheets of cloth into waterproof bags that he used to protect his scientific instruments from the tropical moisture. After passing through the Pongo of Manseriche, where the river narrowed from 1,500 to 150 feet (457 to 45.7 m) across, La Condamine again almost lost his raft and work before emerging out of the mountains and onto the flat plain of the Amazon basin.

The raft arrived at a settlement on the river at Borja, where a priest provided him with a map of the area and accompanied him for the next portion of the voyage. At Borja the expedition changed from rafts to two large canoes, each 44 feet (13.4 m) long and 3 feet (0.9 m) across. Safer in the new canoes and with rowers paddling day and night, La Condamine took up the task of mapping and measuring the river. In late July the team reached the place where the large Ucayali River meets the Amazon and observed the Omaguas, a tribe first encountered by the missionary Padre Fritz years before. La Condamine noted the Omagua practice of placing the heads of newborn babies between wooden boards to squeeze them into a rounder shape and their cultivation of hallucinogenic seeds for ritual uses. By early August the expedition had entered Portuguese territory and the mission of São Paulo, where European influences were strong and La Condamine saw brick buildings and women wearing clothing imported from England. From there they continued downstream with more than 1,200 miles (1,931 km) ahead of them to the Atlantic. Below where the Rio Negro meets the Amazon, La Condamine observed the influences of the Atlantic tide on the river. The tides coming in from the ocean, still 700 miles (1,126 km) downriver, created two currents on the river, one at the surface and another in the opposite direction below. Along with the movements of the river, La Condamine recorded the animals he saw living along the river, including crocodiles, monkeys, vampire bats, anacondas, parrots, and frogs used by the river's inhabitants for their deadly poison.

On September 19, 1743, almost four months since setting out on the river, La Condamine reached the city of Grão Pará, now called Belém, near the mouth of the river. After several months in Grão Pará, La Condamine continued on to Cayenne, in what is now French Guyana, by way of the mouth of the river. He took another canoe with 22 rowers to explore Marajó Island at the very end of the river. Beyond the island the canoe crossed the river at it's widest point and reached the flatlands of Macapá, which he observed was at 3° north latitude and would have served just as well for the French Academy's expedition as the Peruvian Andes while being far more accessible. From Cayenne La Condamine was able to get a ship back to Europe, where he arrived at the French Academy of Sciences on February 23, 1745, almost 10 years after he had left.

Impact

La Condamine's voyage did not help settle the dispute over the shape of Earth. In the end he was in disagreement with members of the Academy over the meaning of his data. His precise calculations and mathematical corrections of the existing maps and measurements in South America improved navigation, and his explorations of the river's tributaries and islands made important corrections to the imperfect maps of the day. As a mathematician, his expedition down the Amazon ushered in a new era of scientific endeavor in the New World and helped to stimulate the scientific explorations of the nineteenth century.

Not least of La Condamine's observations of the natural life in the Amazon was his ingenious application of rubber as a waterproofing treatment for textiles. This would be the first of many applications of rubber that would provide a booming economy and great changes throughout the Amazon basin La Condamine observed. Some of his observations of the natural and cultural life of the region are still accurate today, but they are also notable as a record of what has changed in the last 250 years in what is now a threatened and contested part of the world. In 1743 he already observed the loss of native languages and beliefs to Spanish and Portuguese incursions into the indigenous culture. He noted insufficient efforts to protect the native Amazonians from the same European diseases that pose a threat to the few tribes that still avoid outside contact even today.

GEORGE SUAREZ

Further Reading

Palmatary, Helen Constance. *The River of the Amazons: Its Discovery and Early Exploration, 1500-1743*. New York: Carlton Press, 1965.

Smith, Anthony. *Explorers of the Amazon*. New York: Viking, 1990.

Encountering Tahiti: Samuel Wallis and the Voyage of the *Dolphin*

Overview

During the seventeenth century scientists made significant discoveries in the fields of mathematics, physics, and astronomy—fields necessary for the improvement of navigation. These advances led to the development of the chronometer (a timepiece used to determine longitude), modifi-

cations in ship design, and increased accuracy in navigation. The result was a blending of science, exploration, and economics that culminated in the Pacific explorations of the eighteenth century.

In 1766 the British Admiralty appointed Samuel Wallis (1728-1795) to command a voyage of exploration to the South Pacific, continuing the

An early photograph of a Tahitian man. *(Corbis Corporation. Reproduced with permission.)*

search for the elusive *terra australis*—the great southern continent and huge, theoretical landmass then thought to occupy much of the largely unexplored Southern Hemisphere. While Wallis did not find the continent of Australia, he did land in Tahiti, bringing this lush island's inhabitants perhaps their first contact with European society.

Background

By 1766 European explorers had searched for new lands for nearly 300 years, driven primarily by the desire for new trade routes or territory that might provide new wealth. These voyages revolutionized European understanding of world geography—discovering North and South America, charting the coasts of Asia and Africa, and dispelling myths about boiling temperatures

near the equator and ferocious sea monsters in distant parts of the ocean.

England, Spain, Portugal, and France had all established colonies in the New World, and voyages across the Atlantic and around the Cape of Good Hope to Asia and the East Indies had become relatively uneventful. But the Pacific Ocean remained largely unknown apart from a handful of Spanish settlements along the west coast of South America and ports on the mainland of Asia that were occasionally visited by trading ships. Spanish and Dutch explorers in the sixteenth and seventeenth centuries had stumbled across a few islands in the South Pacific and the coasts of Australia and South Island of New Zealand. But these discoveries were vaguely documented and woven into the myth and mystery of the times.

Lack of knowledge, however, did not mean a dearth of theories about what might exist on the fringes of the known world. The notion that a great continent existed at the bottom of the world dated back to ancient times, when the Greeks argued that a landmass around the South Pole must exist to balance the continents in the Northern Hemisphere. Most Europeans of the eighteenth century still believed a continent had to exist somewhere in the yet unexplored regions. Mapmakers and geographers of the time called it *terra australis incognita*—the unknown southern land, or *terra australis nondum cognita*—the southern land not yet known.

For more than a century, despite occasional voyages into the Pacific, England had concentrated its colonization efforts in North America. However, in the mid-1760s England decided to pursue new markets in the unexplored Pacific, hoping to find the lost southern continent. The first expedition sailed secretly in March 1764 under the command of John Byron (1723-1786). The *Dolphin* completed her clandestine voyage, and brought back useful information suggesting that *terra australis incognita* did indeed exist. With this information, the British government decided to send the *Dolphin* out again under the command of Captain Samuel Wallis.

Wallis sailed from Plymouth on August 22, 1766, accompanied by the *Swallow*, commanded by Philip Carteret (1733-1796). Wallis had orders to first establish a base in the Falkland Islands to give England a political advantage in their further exploration of the Pacific. Then, he was to continue sailing until he reached 100–120° longitude—the possible site of *terra australis*. If land was found, he was to return to England. If land was not found, the *Dolphin* was to continue exploring in search of islands and return to England by way of China and the East Indies.

The *Dolphin* was outfitted with a still for producing fresh water, a forge for ship repairs, and 3,000 (1,361 kg) pounds of "soup" (a concentrated syrup of oranges and lemons) for preventing scurvy. The crew of 150 included many seasoned and experienced men from the first voyage of the *Dolphin*, as well as the usual assortment of uneducated men from England's lower classes. One sailor named George Robinson kept a journal that is one of the best travel accounts kept by eighteenth-century explorers. The *Swallow*, however, was not as lucky. The ship was already 30 years old, and poorly provisioned, relying on the the *Dolphin* for most of her supplies. Accompanying these two ships was a store-vessel that was

to drop supplies at Port Egmont on the Falkland Islands before returning to England.

Three months after leaving England, the group reached the Atlantic entrance to the Strait of Magellan. During the next four months, fierce, frigid winds and rough seas slowed their pace. Both the *Dolphin* and the *Swallow* were in serious trouble many times. After struggling to stay together, the ships became separated in heavy fog just as they entered the Pacific. The *Dolphin*, a much faster and better built ship, sailed ahead while an easterly current sucked the slower *Swallow* back into the strait. The two ships did not meet again; the captains pursued the search for a southern continent independently. Wallis, instead of following the well-established sea route northward along the coast of Chile, headed northwestward, traversing a vast, little-known area of the South Pacific. Carteret, upset at Wallis for what he thought was a deliberate desertion, sailed farther to the south.

Six weeks into the Pacific, the *Dolphin* had encountered no signs of land, and the months aboard the ship with little in the way of fresh food or exercise began to take its toll on the crew. Wallis was unusual among ship captains in his attention to the health and well being of his men. However, in spite of the crew washing down the ship everyday with vinegar and airing the hammocks, the crew "looked very pale and Meager," and many were suffering from scurvy.

During the following weeks, islands were sighted. Many were inhospitable coral atolls or guarded by hostile natives. On occasion they were able to land and trade nails and beads for water and coconuts. The crew's desperation rose as their food and water dwindled. One day, however, they spied a mountain in the distance, shrouded in clouds. Beyond it, far to the south, rose what appeared to be a mountain range. The elated men believed they had reached the fabled *terra australis* and sailed on through the night thinking they had made the most important discovery since Christopher Columbus (c. 1451-1506) landed in the New World.

The next morning Wallis's crew sighted a massive mountain that looked to be 7,000 feet (2,133 m) high, stretching for several miles. The landscape was thickly covered with trees, contained brilliant green flora, and had many streams entering the ocean. Wallis assumed that the mountain was just the tip of the southern peninsula, or perhaps an island off the sought-after southern continent's coast. Recognizing the importance of this find, Wallis named it King

George's Island in honor of the British king. The island's native name was Tahiti.

As the *Dolphin* sailed closer, dozens of canoes paddled out from shore. Robinson noted, "they lookt at our ship with great astonishment holding a sort of council of war amongst them." The *Dolphin* crew showed their friendly intentions by dangling beads and trinkets over the side, while the islanders waved plantain branches. Nearly 800 men and 150 canoes lined up alongside the unfamiliar vessel. Finally, a party came aboard the ship. The encounter ended with Wallis firing a canon over the ship and the delegation jumping overboard.

After a few more skirmishes during the next few weeks, Wallis was eventually able to establish friendly relations with the islanders and anchor in one of the island's bays. Wallis himself was personally welcomed by Obera, the Queen of Tahiti, and from that point on a brisk trade ensued, exchanging nails, clothing, and pots and pans for fresh food and water. Although Wallis soon realized that Tahiti was probably not the southern continent, the crew spent another five weeks in Tahiti resting, repairing the ship, and planting a garden before continuing their voyage. The men, aided by Robinson's sensitivity to the local inhabitants, were able to leave a positive impression behind, an unusual occurrence for the times.

Impact

Having found a place of such beauty and abundance, Wallis decided to return to England to report his findings rather than continuing his search for *terra australis*. On the return trip the *Dolphin* made stops in present-day Tonga and Indonesia, and also located a group of islands west of Samoa and north of Fiji now known as the Wallis Archipelago. The *Dolphin* completed her circumnavigation by sailing around the Cape

of Good Hope, arriving back in England on May 20, 1768.

Wallis's news of Tahiti aroused the interest of the British government, which decided that the island would be a suitable spot to send scientists to observe the Transit of Venus, a major astronomical event whose measurements would help determine the distance from the Earth to the Sun. Later that year Captain James Cook (1728-1779) was sent to Tahiti with scientists and naturalists on the first of his three voyages in search of the elusive southern continent. In 1773 Wallis published his *Account of the Voyage Undertaken for Making Discoveries in the Southern Hemisphere*, detailing his circumnavigation of the world.

The *Swallow* also eventually made it to Tahiti, nearly four months behind the *Dolphin* and after an arduous journey filled with extreme hardship. The crew was assisted by the French, who repaired damage to the ship, gave them food and water, and offered navigational help. After landing in Tahiti and discovering that Wallis and his crew had already sailed back to England, Carteret and the *Swallow* did the same.

Although the voyages of the *Dolphin* and the *Swallow* failed to prove the existence of *terra australis*, their voyages opened a previously unknown part of the South Pacific. Thanks to improved navigational capability, the ships safely traveled thousands of square miles of open sea, and brought back to European society the news of previously unknown lands and peoples.

LESLIE HUTCHINSON

Further Reading

Beaglehole, J.C. *The Exploration of the Pacific*. Palo Alto, CA: Stanford University Press, 1934.

Sharp, Andrew. *Discovery of the Pacific Islands*. Oxford, England: Oxford University Press, 1960.

Withey, Lynne. *Voyages of Discovery*. New York: William Morrow and Company, 1987.

George Vancouver Charts the Pacific Coast of North America from California to Alaska

Overview

George Vancouver (1757-1798) was an English sea captain and member of several expeditions to the South Pacific and the coast of North America

in the late eighteenth century. He was instrumental in the gathering and dissemination of knowledge about the Pacific Coast of the continent of North America. On several visits to the area, he explored it, met with natives and the Spanish, and surveyed and mapped its features from Alaska to Monterey in California. His work helped to establish the claim of Great Britain to ownership of western Canada. He also studied and charted many Pacific Ocean islands, including Hawaii, and made allies of natives for England everywhere. He proved that a practical and usable Northwest Passage through North America did not exist.

Background

George Vancouver sailed on Captain James Cook's (1728-1779) second and third voyages of discovery to the Pacific Ocean from 1776 to 1780. He served on Royal Navy ships in the North Sea and the Caribbean Sea, learning the technical aspects of surveying, mapping, and charting these areas. By the time he became the leader of a new expedition in 1790, he had a great deal of experience and had visited many lands and seas.

In 1790 Vancouver obtained command of an expedition to the Pacific Ocean. His flagship was called *Discovery*, a newly built namesake of the ship Captain Cook had sailed to Hawaii a decade earlier. Vancouver's expedition was instructed to survey the western coast of North America, meet with Spanish representatives in Canada in order to formally receive property Spain had taken from the British, try again to locate a Northwest Passage, and attempt to complete Cook's survey of the Hawaiian Islands. *Discovery*, accompanied by a smaller ship called *Chatham*, sailed on April 11, 1791.

They arrived in Hawaii in the fall and spent the winter there. Then, in March 1792 *Discovery* and *Chatham* sailed north. After 2,000 miles (3,219 km), they reached Cape Mendocino on the northwest coast of North America, about 300 miles (483 km) north of San Francisco Bay. It was a good place to replenish the ship's fresh water, wood, and fresh stores. When these tasks were complete, they began to survey the rugged coastline moving northward toward Alaska. Their pattern of work was to move toward the land each morning, sailing close to shore, slowly and safely, making charts and drawings of the coast as they went. At night they would stand off shore a few miles to avoid contact with the rocks and reefs that are prevalent on this rugged coast.

About 100 miles (161 km) north of Cape Mendocino, Vancouver's ships entered a large waterway between two expanses of land called the Straits of Juan de Fuca. They found an an-

chorage, and Vancouver recorded their position and a description of the land around them. After careening and cleaning the hulls of both ships, replenishing their water, wood and whatever game they could find, the ships moved west into the strait and discovered a complex of bays and rivers that Vancouver named for one of his officers, Peter Puget. The company spent the next month surveying Puget Sound, which was difficult and time consuming because of its complex arrangements of bays, inlets, rivers, and islands. When the task was finished, they sailed back to the Straits of Juan de Fuca and turned north up another inland waterway east of a landmass, uncertain of where it would lead. When they reached its northern end, it was clear that the land to the west was a large island and not part of the continent. It was later named Vancouver Island in Vancouver's honor. On this voyage, Vancouver sighted the summit of a tall mountain and named it Mount Rainier after Peter Rainier, a fellow navigator. He also reiterated Great Britain's claim to the whole area.

In August 1792 Vancouver reached Nootka Sound on the west coast of Vancouver Island and met with the Spanish, who still disputed the ownership of the area with Great Britain. Vancouver was fully aware that England and Spain were at odds over who had the right to control and exploit this land. England had long ago declared that any country wishing to maintain a valid claim on any land in the world had to establish a colony there and to physically control the land before their ownership rights were recognized by the rest of Europe. Spain had claimed so much land on this coast that they had never been able to control, exploit, or even explore all of it. They did not agree with the informal rules the British had proclaimed, but were willing to negotiate, especially since they had lost their right to this northern part of the land in a struggle with England. Vancouver represented the British government in the negotiations. While there, he sent one of his ships to search for a group of islands that the Spanish called Los Majos, or Islas de Mesas. They were not where the Spanish said they were. It was clear that the Spanish had seen some islands, perhaps the Hawaiian Islands, but had the location wrong.

Negotiations with the Spanish came to a standstill until the arrival of further instructions from London and Madrid, so Vancouver sailed for California. On the way, he sent *Chatham* into the river, now called the Columbia River, to survey, as *Discovery* was too large. Both ships arrived in San Francisco Bay in November 1792.

They spent ten days in this area, which had been settled and claimed by the Spanish for 20 years. They received help, food, and friendship from Spanish authorities, missionaries, and citizens. Vancouver also sailed the 100 miles (161 km) south to Monterey Bay, surveying as he went. The rest of that winter of 1792-93 was spent in Hawaiian waters. The following summer, *Discovery* and *Chatham* sailed back to the west coast of America and completed the survey. They took a trip as far north as Cook Inlet in Alaska, searching for the west end of the fabled Northwest Passage. Vancouver came to the conclusion that this fabled passage did not exist.

Impact

Vancouver detailed his voyages in five books and was halfway through the sixth when he died in 1798. The importance of this work was clear, and his friends made sure that his *Voyage of Discovery to the North Pacific Ocean and Round the World in the years 1790-1795* was published. It appeared in 1798 complete with narrative, charts, maps, and drawings. It had a tremendous impact on the expansion of British control of land and sea, though at first it did not receive much attention because England and other countries in Europe were involved in a war with France and its leader Napoleon. After the war was over in the early nineteenth century, Vancouver's work was reprinted in three volumes and was circulated widely.

The accurate soundings and coastal information in Vancouver's book were invaluable. They included the location of dangerous rocks and off-shore islands, sandbars, entrances to bays, and the location of viable, usable harbors. This ground-breaking work gave the English a significant claim to the northwest coast, at least enough to satisfy the king and his countrymen. With its meticulous detail, the book enabled many other would-be explorers to sail to that far away coast to see how it could be exploited. Possibilities for expansion of the fur trade were clear for any who could make the enterprise work. British and American ship owners and traders turned their attention to it, as did the Russians, who established a settlement in Alaska in the 1790s. The Spanish were affected in that they could now see the extent and complexity of the land they had claimed for 200 years but had never been able to explore, much less control. At the time, the Americans could support little effort to obtain the area as they were still establishing their new country. They were in no position to fight the British or the Spanish for control of the

western coast of the continent, but they had always claimed it, and many had their eyes on it. Little American exploration was undertaken until the Lewis and Clark expedition reached the area in 1805. Puget Sound was also shown to have a tremendous potential for fishing and fur-trading. While Vancouver was not the first to explore the Columbia River, another English sailor, Captain Robert Gray on the ship *Columbia*, did the initial exploration of it. Vancouver later sent *Chatham* up the river to make a cursory survey of it, and the ship's findings were included in his book. This river, about 100 miles (161 km) south of Puget Sound is today an important waterway and is the dividing line between the states of Oregon and Washington.

In the nineteenth and early twentieth centuries, Vancouver's work was often overshadowed by the dramatic events that surrounded the voyages of Captain Cook, but modern historians have given him his proper place in the company of explorers. Vancouver's survey of the Pacific Coast was the most arduous undertaken to that time, the accuracy of his notations was remarkable, and his descriptions of the terrain were realistic and precise. One hundred years later, his charts were still the best and most trusted by those who sailed the west coast of North America, Canada and Alaska.

George Vancouver's contributions to the world cover several fields: exploration, political history, diplomacy, and geography. He explored some of the most inaccessible places in the Pacific Ocean. He settled problems with Spain and came close to obtaining control of the Hawaiian Islands for England. He made the shape of the west coast of the North American continent known, and his surveys were the only accurate information on the region for many years. Fifty years later, they were used in a dispute between the United States and Great Britain over claims to Oregon, Washington, and Western Canada.

LYNDALL LANDAUER

Further Reading

Anderson, Bern. *The Life and Voyages of Captain George W. Vancouver, Surveyor of the Sea.* Toronto: University of Toronto Press, 1960.

Batman, Richard. *The Outer Coast.* San Diego: Harcourt Brace Jovanovich, 1985.

Godwin, George. *Vancouver, A Life.* New York: D. Appleton & Co, 1931.

Vancouver, George. *Voyage of Discovery to the North Pacific Ocean and Round the World in the years 1790 to 1795.* 3 vols. London: G.G. & J. Robinson, 1798.

Pedro Vial Charts the Santa Fe Trail and Opens the Southwest to Exploration and Trade

Overview

By the mid-eighteenth century the Spanish inhabited most of the settlements in the American Southwest and had begun exploring routes east of them, looking for new trade opportunities and invading enemies. The French had already sent many representatives across the barren, Indian-protected West in search of gold, silver, and trade. At the same time, Americans who had settled the New England colonies were moving westward, looking for land, trading opportunities, and natural resources. No one, however, had established a mapped, straightforward trail from west to east. Aware of the encroaching Americans, the persistent French, and the need to access other cities more easily, Spain blazed the first established trade route. In 1786 the Spanish governor of New Mexico hired Pedro Vial (1746?-1814), a Frenchman known for his rapport with Indians and a resourceful explorer's spirit, to blaze routes to St Louis, New Orleans, and San Antonio from New Mexico. Immediately following Vial's successful journeys to each of these cities, by 1803 the Santa Fe Trail was thick with traders and eventually American settlers, heading to Santa Fe and beyond.

Background

By the sixteenth century the Spanish, in search of gold, riches, and new colonies, had ventured north from early settlements in Florida. They traveled through Missouri and westward and eventually penetrated New Mexico, Arizona, and California. By 1609 the Spanish had settled into Santa Fe, New Mexico, a remote outpost at the edge of the Sangre de Cristo mountains. The only main commercial route to Santa Fe was the Chihuahuan Trail, which connected to Chihuahua, Mexico, nearly 600 miles (965.6 km) south. Resources were scarce in Santa Fe, so the Spanish traded frequently with nearby American Indian tribes, such as the Pueblo, the Comanche, the Navajo, and the Apache. The tribes, however, attacked and raided the Spanish settlement. It was a hostile relationship, but the Spanish needed the Indians for their goods and their knowledge of the frontier.

With settlements in Louisiana, Illinois, and Missouri, the French had already explored the main trade routes of the Midwest: the Missouri, Ohio, and Red Rivers. Looking for rumored gold, silver, and turquoise mines and a route to the "western" (Pacific) sea and Asia, French frontiersman eventually traveled overland, and some made it as far as New Mexico. The Spanish, however, were not welcoming to these French travelers and frequently imprisoned them (shipping them as far as Mexico City) or simply killed them. French who sailed to the shores of Texas were fought back when attempting to move west. The Spanish refused to accept any foreign goods—except Indian goods—and punished any person with this "contraband" in their possession. Finally, in 1739 Paul and Pierre Mallet, two Frenchmen from Canada, traveled down the Missouri and across Nebraska, Kansas, and southwestern Colorado, and were received at Santa Fe safely. News of the unusual kindness of the Spanish toward the Mallets reached the French city of New Orleans, so more expeditions were launched toward Santa Fe.

The Spanish, however, were growing nervous. Frenchmen were venturing across Texas and New Mexico in search of horses, and the Spanish feared an invasion. All this stopped, however, after the French and Indian War (also called the Seven Years' War) in 1763. The Spanish assumed control of Louisiana by 1769 which provided an great incentive to set up a trade route to the prosperous east. It was much closer than Mexico, and the French threat seemed all but gone.

American settlement in the east, by 1776, was swelling. The Spanish were aware of "Kentuckians," as they called the early Americans, moving west. They saw Louisiana as a barrier to the spreading pioneers, and knew that establishing a trade route from Santa Fe was extremely important to do before the Americans could. Up to this point in history, several men—including some French, a few Spanish and several Indians over thousands of years—had traversed the plains from east and north to Santa Fe, but an established, mapped route did not exist. In 1786, the governor of New Mexico sent Pierre (Pedro) Vial on an expedition to blaze a trail from Santa Fe to San Antonio, in the hopes of connecting the isolated Santa Fe to the east.

Pedro Vial, born Pierre Vial and originally from Lyons, France, was an expert gunsmith

Travelers on the Santa Fe Trail, mid-1840s. *(Corbis Corporation. Reproduced with permission.)*

who had traveled throughout the plains frontier and Texas and became familiar with the ways and language of many of the Indian tribes along the route. For any European traveling through the wild country, attacks from Indians were the most dangerous challenge; these attacks had prevented any established trade route up to the time of Vial's departure. So in 1786 Juan Bautista de Anza, the governor of New Mexico, sent Vial out with one man and provisions to map a course from Santa Fe to San Antonio, Texas.

Vial, who is guessed to have been 40 years old, set out to travel through friendly Indian hamlets that he knew of, stopping along waterways and springs en route. He became very ill early in the journey, fell from his horse, and traveled 150 miles (241.4 km) out of the way to a Comanche Indian camp. For two months the Comanche cared for him and brought him back to health. He continued on, negotiating with other tribes along the way—Apache, Tawakoni, more Comanche—and getting advice on the route to San Antonio.

During this series of encounters, Vial was skillfully assuring the tribes that any Spanish traders they might encounter in the future would be friendly. This was an important achievement on Vial's journey, since the Spanish relationship with tribes in New Mexico histori-

cally had been violent. The guarantee of a safe passage through the eastern Indian country was as vital as mapping the terrain. Vial was so successful in this public relations effort that many Indians accompanied him along his journey. He reached San Antonio, turned around, and followed the exact same route back to Santa Fe, arriving May 26, 1787, with a crowd of Comanche at his side. Overall, the journey took nearly seven and a half months and covered some 1,157 miles (1,862 km).

Immediately upon Vial's return, the governor sent out another explorer, José Mares, to retrace most of Vial's steps (and take advantage of the Indian relationships Vial had established) but in a more efficient manner. The Spaniards wanted to know the fastest route possible, and Mares completed the round trip in 845 miles (1,360 km), nearly 125 miles (201 km) less than Vial.

Satisfied by Mares's efficient route, and still feeling the pressures of the Americans' feverish trading along the Mississippi and Missouri rivers, the Spanish felt they needed to press farther east. The governor of New Mexico, Fernando Concha, sent Vial back out, this time toward Natchitoches, a Louisiana outpost just north of New Orleans. Vial left on June 24, 1788, with several companions, including cavalrymen and local New Mexicans. He once again followed routes that had water, sat with Indian tribes, and smoked with them, offering them gifts. They skirted along the Red River, staying in Comanche, Wichita, and Taovaya villages and several well-established Tawakoni villages that became popular stopovers for future travelers.

Vial encountered herds of wild mustangs, deer, and wild boar, and he traversed canyons, rivers, and rocky passes. They reached Natchitoches in 663 miles (1,067 km) after 26 days, turned south through San Antonio, and arrived back in Santa Fe on June 24, 1789. Although many of the expedition party were ill, probably from malaria, they cut their return trip to 632 miles (1,017 km) and 23 days. Vial's companions included two literate diarists, who kept more detailed records of the tribes and terrain they encountered than Vial had kept on his first expedition to San Antonio. They also listed the supplies that were traded with the Indians en route. Some of these items included tobacco, petticoat cloth, beads, chairs, spurs, soap, and hair ribbons. And while the road from Santa Fe to the east was now open, the Spanish didn't immediately start trading. They were more concerned with the increasing exploration by the Americans.

The major towns east of Santa Fe included St. Louis, New Orleans, and San Antonio, all of which were connected in some way to immense river systems. Of these towns and their rivers, none was more important than St. Louis, where the Missouri River connected the French in Illinois with the Americans all along the land east of the Mississippi. Once again, Governor Concha rallied Vial for another explorative journey. He departed on May 21, 1792, out of the small village of Pecos, just outside of Santa Fe for St. Louis, with two men and four horses.

They followed along the Colorado River until they reached southern Colorado and the Arkansas River, and then turning north they followed the Canadian River. They were almost killed by a band of Kansas Indians until some of the tribe recognized Vial from his previous wanderings and spared his life. Stripped naked, they were held captive for several days. They set out across Kansas and Missouri until they reached the Missouri River and sailed into St. Louis on October 3, 1792. After presenting the governor with a letter from Concha, announcing the purpose of Vial's trip, he stayed until June 14, 1793. The trip took 82 days and covered 1,100 miles (1,770 km).

When Vial began his return, so many months later, he had resupplied with hats, handkerchiefs, razors, shirts, trousers, mosquito nets, gunflints, soap, mirrors, and bullets. They sailed upriver on a pirogue with four rowers, until they met a Pawnee village, where they stayed and traded for a couple weeks, establishing a solid relationship with the tribe. Vial returned to Santa Fe on November 16, 1793, completing what would be his last groundbreaking expedition.

Impact

Vial's expeditions signaled both an end and a beginning: the gradual end of Spanish authority in the United States and the beginning of American conquest of the area west of the Mississippi. In 1797, with the Spanish growing more wary of American and French explorers taking advantage of their new trade routes, Vial left for a Comanche village, where he lived until 1803. That same year, the Americans signed the Louisiana Purchase, which doubled the size of the existing United States territory and gave the Americans control of Louisiana and direct access to the Santa Fe Trail. Trappers and traders soon frequented the trail, heading to the West and Santa Fe, which had earned a reputation as a paradise of jewels and gold, where all nationalities mixed in a kind climate. Of course, the route was not without hazards, as not everyone had a friendly relationship with the tribes along the route.

In 1821, the year of Mexican independence from Spain, the restricted trade barriers set up by Spain disappeared, Spanish rule retreated, and Santa Fe became a western trading destination for the entire country. By 1822 William Buckle led the first wagon train from St. Louis along Vial's Santa Fe Trail, and kicked off a massive migration of American pioneers. These early traders carried goods on mules, covering about 12 miles (19.3 km) a day, making the typical trip in six to eight weeks. Kansas City was en route, and it soon flourished with the trail traffic. By 1860 more than 9,000 men and women, 28,000 oxen, and 3,000 wagons had traversed the trail. By 1866 this number had doubled. By the end of the nineteenth century, the railroad laid its tracks across the prairie and the trail opened up to everyone, not just hardy adventurers.

LOLLY MERRELL

Further Reading

Brandon, William. *Quiviria*. Athens, OH: Ohio University Press, 1990.

Cather, Willa. *Death Comes to the Archbishop*. New York: Vintage Books, 1990.

Loomis, Noel M. and Abraham Nasatir. *Pedro Vial and the Road to Santa Fe*. Norman, OK: University of Oklahoma Press, 1967.

Vitus Bering's Explorations of the Far Northern Pacific

Overview

On voyages in 1728 and again from 1733 until 1741, Vitus Bering (1681-1741) explored the northernmost reaches of the Pacific Ocean. On these voyages he discovered (rediscovered, actually) the Bering Strait, which separates Asia from North America. He also mapped large portions of Russia's north and east coasts as well as northwesternmost North America. Because of his work, Russia was able to lay claim to Alaska, a long stretch of the North American Pacific coast, and the fur trade encompassed by this area.

Background

European exploration of North America started on its east coast, primarily because the nations conducting this exploration were on the Atlantic Ocean and this was the easiest path for them to follow. Because of this, the western coast of North America was explored primarily by the Spanish (who had crossed Central America to the Pacific Ocean), but they did not generally explore north of San Francisco Bay. Voyages to the north were blocked by ice and poor weather, severely limiting exploration in this direction. As the eighteenth century opened, the coast of America's Northwest was completely unmapped and unexplored by Europeans.

Europeans were, however, interested in exploring this area. The major seafaring nations wanted an east-to-west sea passage across North America, which they hoped would enable them to find a shorter route to the Far East. Russia, however, was another story. While Britain, France, and Holland tried to push across the top of North America in search of the Northwest Passage, Russia wanted to open a Northeast Passage that would link the Russian West to its ports and potential trading partners in the East. So Russian explorers pushed towards the Pacific from the other direction.

In 1648, a Cossack sailor named Semyon Deshnyov sailed with seven ships through the Bering Strait from north to south, becoming the first European to do so. However, accounts of his voyage were lost for several decades, long enough for Bering to receive credit for this discovery.

More than seventy years later Peter the Great became intensely interested in finding a sea passage between Russia and North America. In 1725 he sent Vitus Bering, a Dane serving in the Russian Navy, on a voyage of discovery. Setting off from the Kamchatka Peninsula, Bering sailed north along Russia's east coast, mapping the shoreline as he sailed, and passing successfully into the Arctic Ocean. Although this proved conclusively that such a passage existed, it was exceedingly foggy and he never actually sighted North America. Because of this, there were those who still questioned his discovery. Not until 1732, when others in the Russian Navy made the same transit in good weather and sighted land to both east and west, was Bering's discovery finally accepted.

The next year, Catherine the Great gave Bering command of a more ambitious expedition, the Great Northern Expedition of 1733-1743. This expedition, one of the largest ever mounted, employed over 1,000 people in exploring and mapping northeast Russia, the American Northwest, parts of the Russian Arctic coast, and the northern Pacific coasts of Russia and North America. The expedition was both phenomenally expensive and successful, returning staggering amounts of scientific information, maps, and other useful knowledge.

In 1741, towards the end of the Great Northern Expedition, Bering and his crew became stranded for the winter on what is now called Bering Island, not far from the coast of Kamchatka. There, Bering and most of his crew died of scurvy, other diseases, and injury during the long Arctic winter. The expedition continued for another two years, returning with maps of the Russian Arctic and Pacific coasts, the

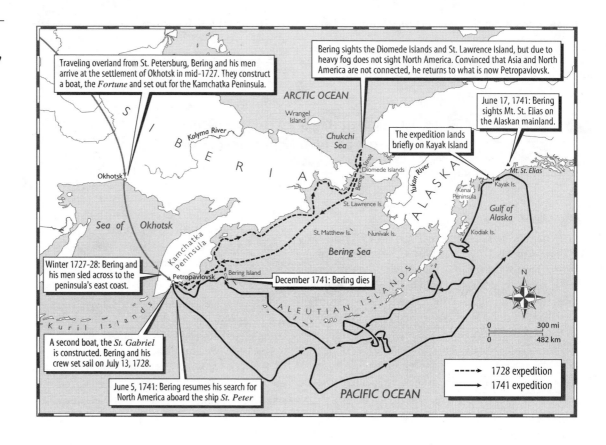

Traveling overland from St. Petersburg, Bering and his men arrive at the settlement of Okhotsk in mid-1727. They construct a boat, the *Fortune* and set out for the Kamchatka Peninsula.

Bering sights the Diomede Islands and St. Lawrence Island, but due to heavy fog does not sight North America. Convinced that Asia and North America are not connected, he returns to what is now Petropavlovsk.

June 17, 1741: Bering sights Mt. St. Elias on the Alaskan mainland.

The expedition lands briefly on Kayak Island

Winter 1727-28: Bering and his men sled across to the peninsula's east coast.

December 1741: Bering dies

A second boat, the *St. Gabriel* is constructed. Bering and his crew set sail on July 13, 1728.

June 5, 1741: Bering resumes his search for North America aboard the ship *St. Peter*

- - - → 1728 expedition
——→ 1741 expedition

The voyages of Vitus Jonassen Bering, 1728 and 1741. *(Gale Research.)*

Aleutian Islands, and much of the American Pacific coast.

Impact

The first and most fundamental result of Bering's journeys was Russia's formal claim to what is now Alaska and much of the Pacific Northwest. Because her ships reached these areas before any rival European powers, Russia was able to claim the land and all its wealth, establishing small colonies and stations on many islands and the Alaskan mainland. Here the Russians traded and trapped for valuable furs and skins, fished the waters, and hunted seals and whales.

Russia governed its North American territories for over a century. In 1867, however, U.S. Secretary of State William H. Seward bought them from cash-poor Russia for what is now recognized as a ridiculously low price: $7 million. At the time, the sale was publicly ridiculed and the territory was referred to as Seward's Folly or Seward's Icebox. It would be another three decades before this opinion changed, when gold was discovered.

Russia had settled colonies in Alaska to collect furs and fish, just as the British and French

had done in North America. This trade was lucrative and, in some years, perhaps a thousand ships would dock at various colonies to exchange supplies for trade goods. However, even in the late nineteenth century, nobody thought it possible that the frozen expanse held anything other than game or lumber.

Then the discovery of gold in Alaska sparked a gold rush, bringing thousands of people into the territory in just a few years. Over the next decades, other minerals were found as well, culminating in the discovery of huge petroleum reserves on the Alaskan north coast. Gold and oil made Seward's Folly a vital part of the American economy, while its location near Russia gave it exceptional strategic value.

Little of this came without conflict, however. At first, the conflict was a curious one: Russia wanted to sell Alaska badly, but the U.S. had no desire to purchase it. At this time, the conflict was between opposing sides in the U.S. Congress. When it was finally purchased, Congress then refused to appropriate money to pay the final bill for months. Then, feeling it a worthless land of polar bears and glaciers, Congress refused to maintain the new territory, letting it languish instead.

The first international dispute came in 1881, when the U.S., Canada, and Britain began wrangling over fishing and sealing rights in the Bering Sea. Claiming the entire sea as territorial waters, the U.S. tried to exclude other nations from hunting and fishing in them. An international arbitration body decided decided against the U.S. in 1893, ruling that the Bering Sea was open to all nations. In 1911 the U.S., Canada, and Japan signed a further agreement on seal hunting in the Bering Sea. Japan withdrew in 1941, leaving the U.S. and Canada with an interim agreement that was revised in 1956 to include the Soviet Union.

During World War II, the waters first explored by Vitus Bering proved to be of military importance. Trying to draw American attention away from Pearl Harbor, the Japanese Navy staged diversionary raids near the Aleutian Islands. These attacks continued throughout the war, and included a Japanese invasion of some islands. During the Cold War, the Aleutians and the Bering Sea were important for both American and Soviet navies. Nuclear submarines on both scientific research and military missions routinely passed through the Bering Straits to patrol beneath the Arctic icepack. In fact, at least one class of Soviet ballistic missile submarines was designed specifically to operate under the ice, breaking through if necessary to launch their missiles against NATO. These were pursued by U.S. attack submarines, although such pursuits were difficult in the noisy waters beneath the drift ice.

In the mid-1990s Russian nationalists began clamoring for the return of Alaska to Russia. In a state of shock after the collapse of the Soviet Union, Russia was increasingly seen as a poor and almost pathetic former power. This led to a nationalistic backlash as opportunistic politicians attempted to ignite nationalist passions in the hearts of Russians. Among the wild claims made was that Alaska was most properly a part of Russia because Russians had discovered and settled it first. Noting the low price of purchase, these politicians claimed that Alaska had been stolen from the Russian people and should be returned. Whether driven by political ambition, sincere nationalistic feeling, economic calculation, or a combination of the three, these claims obviously went nowhere.

Bering's initial voyages of discovery resulted in riches that he could not even have guessed when he first set forth to explore the Russian Far East. Although he did not realize it, his maps and discoveries in the Bering Sea, and along the Russian and American Pacific coasts, would lead to economic gain, military confrontation, political dispute, and more.

P. ANDREW KARAM

Further Reading

Books

Fisher, Raymond. *Bering's Voyages: Whither and Why.* 1977.

Michner, James. *Alaska.* New York: Random House, 1988.

Steller, Georg Wilhelm, and O.W. Frost. *Journal of a Voyage With Bering 1741-1742.* Stanford University Press, 1993.

Alexander Mackenzie Becomes the First European to Cross the Continent of North America at Its Widest Part

Overview

As an agent of the Northwest Company trading furs in western Canada, Sir Alexander Mackenzie (1755?-1820) became the first European to cross the continent of North America at its widest part, north of the Spanish territories in Mexico. He achieved this feat in two landmark expeditions: exploring the Mackenzie River to the Arctic Ocean at Beaufort Sea (1789) in Western Canada and traveling overland from Lake Athabasca to the Pacific Ocean over the Canadian Rockies (1793). While working towards expanding the reach of the Northwest Company, Mackenzie made new contact with tribes that had never before seen Europeans and created new trading routes for furs and other goods in western Canada.

Background

By the late eighteenth century, two royally chartered companies dominated the fur trade in

Canada: the Hudson's Bay Company, doing business from outposts along the shores of the bay, and the Northwest Company, which had established trading posts in the country's interior, where furs trapped by the Native American tribes were exchanged for manufactured European goods and rum. Complex trading relationships between the tribes of North America already existed, and independent traders and trappers had for years been infiltrating the cross-continental trade. The royally chartered companies sought to expand that trade by creating trading posts where the tribes could bring furs for exchange and that would serve as exchange centers in the company's trade network.

Alexander Mackenzie began working in the offices of the Northwest Company at the age of 16 in the last days of the American war for independence. In the Montreal and Detroit offices of the company, he learned the fur trade and eventually formed a venture that sent him out into the interior of the country to make contact with the Native American tribes that sold furs to the company.

Mackenzie established trading relationships with the tribes that lived in the Canadian interior and founded trading posts along the rivers that led to Hudson's Bay. These posts were able to intercept the river trade that otherwise would have gone to the shores of Hudson's Bay to be traded by the Hudson's Bay Company, the Northwest Company's chief competitor.

After gaining control over a large portion of Canada for trade, Mackenzie established a trading post on the shore of Lake Athabasca that would become the base for his later expeditions to the Arctic and Pacific Oceans. In the years before the first trip north, Mackenzie gathered essential information about the geography of northern Canada from the tribes that came to trade at the newly established Fort Chipewyan.

On June 3, 1789, Mackenzie departed from Fort Chipewyan in canoes, traveling across Lake Athabasca and up the Slave River to Great Slave Lake, where he had been told he would find a large river that would take him to the Pacific Ocean. The expedition consisted of two canoes, a group of trappers already familiar with the southern portion of their route, and a member of the Chipewyan tribe named Nestabeck whom Mackenzie called English Chief. (Almost 20 years earlier Nestabeck had accompanied the Chipewyan leader Matonabbee when he guided Samuel Hearne [1745-1792] on the overland trip from Hudson's Bay to the Arctic Ocean

along the Coppermine River.) In early June ice still blocked some waterways in northern Canada, and the expedition had difficulty finding the northbound outlet from Great Slave Lake. The expedition spent the final week of June looking through the lake's icy marshes for the source of the river known by the Cree as "swiftly flowing waters" and now known by Canadians as the Mackenzie River.

The river proved to be wide and navigable as the Native Americans had described. Traveling along the large river's current, the expedition was able to cover about 75 miles (121 km) each day. By the middle of July the expedition noticed that the river had taken a turn towards the northwest and that the Rocky Mountains still lay between them and the Pacific Ocean. As they continued north Mackenzie continued trading with Native American settlements he encountered, exchanging their furs for ironware, rum, and tobacco and hoping through these trades to learn more of the local geography. Mackenzie's trading gifts with newly encountered tribes contrasted with his management of his own expedition members, who by July 8th were threatening to abandon the demanding and relentless Mackenzie. Mackenzie was able to get his men to continue, and they canoed past abandoned Inuit settlements and whale carcasses along the river until they reached an island from which they could see nothing but ice-covered water for 6 miles (9.65 km). They reached the river's delta on the Arctic Ocean on the July 14, 1789. Navigational instruments clearly showed that they had not reached the Pacific. When a rising tide of saltwater swamped baggage they had left out of the canoes, the disappointed crew returned upriver, arriving back at Ft. Chipewyan on September 12, 1789.

On his next expedition during 1792-93, Mackenzie again tried to reach the Pacific Ocean. Taking a plan outlined by officials at the Northwest Company five years earlier, Mackenzie set out to make contact with the Russian, Spanish, and newly independent American traders conducting business on the Pacific coast of what is now British Columbia. Mackenzie was rumored to have wanted to proceed from the Pacific along the established Russian trading routes to Siberia. The trip would have been the realization of a 300-year-old goal of establishing a trading route to transport North American goods over the continent to the markets in Asia. Russian traders already had a firm presence that would extend as far south as California and were taking North American furs across the pacific to

A rock marked by Alexander Mackenzie near the Pacific Coast in Bella Coola, British Columbia. *(Corbis Corporation. Reproduced with permission.)*

the trading port of Kiatka. There the furs were sold to the caravans that carried them to Peking and from there throughout China. Rather than evict the Russian traders from the Pacific coast as the Spanish had tried to do, Mackenzie hoped that through the Russians he could link their trading network to Asia and create a series of outposts from London to Montreal through Fort Chipewyan to Kiatka and Peking.

On October 10, 1792, the expedition began canoeing west up the Peace River towards the Continental Divide. Stopping at existing trading posts along the way, they continued to trade rum and tobacco for furs and information. The trading presence of the Canadians to the east and the Russians on the coast to the west had strengthened contacts between tribes between the two economic powers. Before crossing the Continental Divide and beginning the descent to the ocean, the expedition was met by members of a western tribe that offered guidance in reaching the Pacific in hopes of establishing a trading relationship with the Northwest Company. Unlike the voyage to the Arctic Ocean, however, the explorers did not have the advantage of a large river to carry them all the way downstream. After spending the winter in the newly established Fort Fork, the expedition paddled upstream along the Peace and Parsnip Rivers at just the time of year when they were overflowing with melting snow from the mountains. After

crossing the Continental Divide the expedition wrongly thought they had reached the headwaters of the Columbia River. To avoid going as far south as the mouth of the Columbia River at the present boundary between Washington state and Oregon, the expedition decided instead to travel the final 150 miles (241 km) to the Pacific overland. The expedition arrived at the coast on July 20, 1793.

Impact

Once at the coast the expedition found no Russian or Spanish traders, but instead Native Americans who were hostile to the Europeans. Mackenzie established a defensive position on a small island off the coast where he continued trading with nonhostile natives who approached them. In the end Mackenzie's voyage to the Pacific did not establish a strong trading relationship with Russia or even a thoroughfare across the Rockies to transport good and settlers.

Mackenzie's explorations, however, did help to develop a trade network for the Northwest Company. His goals were financial rather than scientific as he sought to expand the Northwest Company's trading territory. The Company would continue to be a strong financial presence in northwestern Canada well into the twentieth century. That presence was due, in large part, to Mackenzie's efforts in expanding the company's presence in the Canadian interior, where it would become a force for economic development even after the decline of the fur trade.

GEORGE SUAREZ

Further Reading

Daniels, Roy. *Alexander Mackenzie and the North West.* London: Faber, 1969.

Gough, Barry. *First across the Continent: Sir Alexander Mackenzie.* Norman: University of Oklahoma Press, 1997.

Lamb, W. Kaye, ed. *The Journals and Letters of Sir Alexander Mackenzie.* Cambridge: Cambridge University Press, 1970.

The Explorations of
Pierre Gaultier de Varennes et de La Vérendrye

Overview

The period prior to the exploration of the Canadian West by Pierre Gaultier de Varennes et de La Vérendrye and his sons was full of turmoil between England and France. Often at war with one another, the competition to settle North America under their own flags became intense. England had many fur trading posts in Canada, and the French were eager for the profits of the fur trade as well. The French resolved to find a route from New France in eastern Canada to the fabled "Western Sea," thought to be an outlet to the Pacific Ocean, and to make their own mark by establishing a string of forts in the western Canadian wilderness. La Vérendrye's expeditions, though never reaching the Western Sea, were largely responsible for opening the central interior of Canada to French control.

Background

Pierre de La Vérendrye was born in 1685 in Trois Rivières, on the banks of the St. Lawrence River. Entering the French army in 1697, La Vérendrye began his military career at the age of 12, and served in Canada and France. After a period of peace between the English and French, La Vérendrye returned to Canada to take up the fur trade, and accepted command of the Lake Nipigon trading center in 1726. While at the fort he met a tribesman, Ochagach, from the Kaministikwia River. From him, La Vérendrye heard of a great lake and a river that flowed from it through a "treeless country where roam great herds of cattle," leading to a "great salt sea." His interest sharpened, La Vérendrye decided to explore this country and to find a route to the Western Sea for France.

After receiving reports from La Vérendrye, Governor Marquis de Beauharnois of New France interviewed him in 1730. La Vérendrye was appointed commissioner of a project to find the Western Sea, and received financial support from local merchants who wanted a share in the fur trade. On June 8, 1731, the expedition team of La Vérendrye, his cousin Christophe Dufrost, sieur de La Jemeraye, La Vérendrye's sons, Jean-Baptiste, Pierre, and François, and a crew of 50 men embarked for the west by canoe. Vast amounts of supplies, such as tobacco, kettles, axes, saws, powder, and muskets were needed to trade with the tribes of the west. The route was not an easy one, with rugged, difficult terrain.

The expedition canoed 210 miles (336 km) west on the Ottawa River, onto the Mattawa River and then the French River, averaging 20 miles (32 km) a day. Portages (land crossings) were many and very difficult because of the heavy load, as well as the constant annoyance of biting flies and mosquitoes. Passing by the islands of northern Lake Huron they arrived at Fort Michilimackinac between Lake Huron and Lake Michigan, and enjoyed a brief rest before moving on. After a month of storms and bad weather they reached the Grand Portage on August 26, 1731, 40 miles (64.4 km) southwest of Fort Kaministikwia, the present site of Thunder Bay, Ontario. A difficult series of portages stood between Lake Superior and Rainy Lake, where La Vérendrye wanted to establish his first fort. The men broke into open mutiny as some feared "evil spirits" and others did not want to continue so late in the season. La Vérendrye decided to winter at Kamistikwia with some of the men, while La Jemeraye pressed on to Rainy Lake with Jean-Baptiste and the remaining crew.

Impact

After canoeing 200 miles (320 km), Jean arrived at Rainy Lake. Fort St. Pierre, named after La Vérendrye, was built where the Rainy River leaves Rainy Lake towards the Lake of the Woods. La Vérendrye, with the rest of the crew and his two sons, resumed travel west on June 8, 1732, exactly one year after the expedition began, and arrived at the new fort one month later. A profitable fur trade had developed and the loyalties of the Cree, traditional enemies of the Sioux, were won with gifts of guns and shot. Going west, with an escort of 50 Cree and Monsoni canoes, the expedition reached The Lake of the Woods and followed along the present Canadian-American border. A peninsula that ran out into the lake was chosen for the sight of their second fort, named Fort St. Charles after Governor Beauharnois. The fort, located in present-day Minnesota, was a good location for trade, and La Vérendrye decided to winter at the fort to help establish the fur trade.

Early in April 1733 La Vérendrye sent a small party with La Jemeraye and Jean-Baptiste to explore the Maurepas (Winnipeg) River. La Jemeraye returned to the fort, where he later died, and Jean went on and established Fort Maurepas at the point where the Winnipeg River enters Lake Winnipeg. La Vérendrye remained at Fort St. Charles, where he continued trade with the Cree and Assiniboines. The two tribes asked La Vérendrye to allow his son, Jean-Baptiste, to join them in war against the Sioux. La Vérendrye agreed, failing to distance himself from the tribal conflicts of the Cree, Monsoni, Assiniboines and the Sioux.

After spending the winter of 1734-35 in Montreal, La Vérendrye returned to Fort St. Charles on Sept 6, 1735, and sent Pierre and François to Fort Maurepas on February 7, 1736, for supplies. Jean-Baptiste was sent to retrieve supplies from Fort Michilimackinac on June 5, 1736. Due to Sioux attacks in the area, La Vérendrye sent a party to retrace Jean's route to the fort. On June 22, 1736, La Vérendrye met with news of his son's death, plus 20 others with him. The Sioux had attacked, and the men were found carefully laid in a circle with their decapitated heads beside them, wrapped in beaver skins.

During the difficult winter of 1736-37 at Fort St. Charles, La Vérendrye learned about a tribe further in the west called the Mandans. They were reported to be white and described as living in forts and tilling the ground. La Vérendrye inquired about the great river the Mandan lived on (Missouri River) and wondered if it was the river of the west they were seeking. He was told it ran south and discharged into the Pacific, where there were towns of white men. By this time La Vérendrye knew that the river of the west described by Ochagach was only the Winnipeg River, and the great sea was probably Lake Winnipeg. La Vérendrye decided to make a voyage to the home of the Mandans. Going back to Montreal before he left for the west, La Vérendrye reached the forks of the Red and Assiniboine Rivers on September 24, 1738, where the city of Winnipeg, Manitoba, now stands. Despite low waters the crew continued up the Assiniboine a few days later, while La Vérendrye walked across the plains. With river levels getting even lower, La Vérendrye stopped on October 3, 1738, and erected Fort la Reine, now Portage la Prairie, Manitoba.

On October 18, 1738, La Vérendrye, with Assiniboine guides, left for the Mandans. They traveled towards the Pembina River, around the Turtle Mountains, and passed Bismarck, North Dakota. They met a Mandan chief on November 25, 1738, whose party took La Vérendrye to the first fort of the Mandan's on November 30, 1738, near a small river about 15 miles (24 km) northeast of the Missouri River. Further down river there were five more villages. The Mandans impressed La Vérendrye, as they lived in large comfortable homes arranged in neat little streets inside a well-built fort. However, La Vérendrye was disappointed that they were not white as had been reported. La Vérendrye remained on the treeless plain for much of the winter, and learned of other tribes even further to the west of the Mandans. Returning to Fort Reine on February 10, 1739, La Vérendrye was very ill and needed time to recuperate.

While at Fort Reine, La Vérendrye sent his youngest son, Louis Joseph, who joined the journey after the death of his eldest brother, to investigate the shores of the major prairie lakes and to select sites for forts. Louis traveled along Lake Manitoba to Lake Dauphin and marked out the site for Fort Dauphin. Going northwest to the mouth of the Saskatchewan River, he staked out a second site close to present-day Grand Rapids, and then proceeded northwest once again. Where the Carrot and Pasquia Rivers join the parent Saskatchewan River, Louis marked a spot for Fort Paskoyac, today known as The Pas, Manitoba, and then returned to Fort Reine for the winter of 1739-40.

The trip to the prairies was not profitable since there were not many furs to be traded, and the expedition proceeded under a cloud of debt. After going to Montreal to clear up some legal issues involving his debtors, La Vérendrye returned to Fort Reine on October 13, 1741. During this time, his son, Pierre, returned to the Mandan and Horse tribes in the west and was able to determine that the Missouri River was not the river of the west, as it flowed south and into the Gulf of Mexico, not the Western Sea. Pierre returned to Fort Reine with horses he received from the western tribes and was responsible for bringing the first horses to the Canadian Prairies.

Wintering at Fort Reine in 1741-42, La Vérendrye sent Louis and François for a third expedition to the Mandans, while he sent Pierre to establish Fort Dauphin. La Vérendrye's two sons returned to the Mandans on May 19, 1742, and explored the Cheyenne and Missouri Rivers and their tributaries. Escorted by the Bow Indians, whom they met during their explorations, the expedition reached the Rocky Mountains on January 12, 1743, but news of a rival tribe in the area forced the party to return to the Bow villages on February 9, 1743. A leaden plate, with an inscription, was buried by La Vérendrye's sons at Pierre, North Dakota, on their way back to Fort Reine, which was later unearthed by high school students in 1913. The expedition returned to Fort Reine on July 2, 1743.

Although the French government was unhappy with La Vérendrye's efforts to find the Western Sea, he and his expedition accomplished much. They had, between 1731 and 1747, covered almost 1,000 miles (1,600 km) and established a string of important trading posts. The forts he established set the foundation for the settling of the rugged Canadian West. The city of Winnipeg (Fort Rouge) is a thriving community that still serves as the gateway to the west, and is the capital city of Manitoba. Today, the Turtle Mountain Provincial Park, located on the Canadian-American border, acts as a reminder of the explorations of La Vérendrye and his men. In 1744 Pierre de La Vérendrye retired as Commandant of the Western Posts, but still planned to return west when he died in Montreal on December 5, 1749.

KYLA MASLANIEC

Further Reading

Bolander, Richard E., ed. *World Explorers and Discoverers.* New York: Macmillan, 1992.

Crouse, Nellis M. *La Vérendrye: Fur Trader and Explorer.* Toronto: Ryerson Press, 1956.

Kavanagh, Martin. *La Vérendrye: His Life and Times.* England: Fletcher & Son, 1967.

Long, Morden H. *Knights Errant of the Wilderness: Tales of the Explorers of the Great North-West.* Toronto: Macmillan, 1919.

Syme, Ronald. *Fur Trader of the North: The Story of Pierre de la Vérendrye.* New York: Morrow, 1973.

Captain Cook Discovers the Ends of the Earth

Overview

Almost half the globe was a mystery before James Cook began his voyages. By the time he was finished, it was clear that there were no large landmasses left to be discovered on the planet. He proved that the legendary southern continent, *terra australis*, did not exist and showed that there was no practical Northwest Passage from the Atlantic to the Pacific. Cook also discovered Hawaii, New Caledonia, and many other remote islands. He was the first recorded individual to travel below the Antarctic Circle. His voyages made major contributions to science, especially botany, and helped establish good nutrition as the way to prevent scurvy.

Background

All the inhabited continents had been discovered before Captain James Cook (1728-1779) began his three major voyages of discovery, but nobody knew this fact. Large portions of the Southern Hemisphere remained unexplored. Maps showed blank space or *terra australis*, a legendary southern continent filled with fantasy creatures.

Symmetry was one of the main arguments for the southern continent. Should not there be as much land in the Southern Hemisphere as in the Northern? The known landmasses of the Southern Hemisphere only amounted to about a quarter of those of the Northern. Wind patterns were also used to make the case, and it was even declared that more land was necessary to keep the Earth's motion in balance. By Cook's time, the imaginary *terra australis* had already shrunk considerably. For centuries, it was thought to connect Africa to Ceylon, but that idea was destroyed by Bartolomeu Dias's (1450?-1500) trip around the southern tip of Africa in 1488. Abel Tasman (1603-1659) sailed completely around Australia, proving it was not the great southern continent. He also showed that there was open ocean—and plenty of it—between Africa and Australia. Other voyages (by the French) found islands, providing hope that the South Pacific was the place to look.

The British, both a naval and commercial power, were especially keen to find the southern continent. They had the opportunity for surreptitiously undertaking a search thanks to an astronomical event. On June 3, 1769, there would be a transit of Venus across the Sun. This event was of both practical and academic interest. Observations of the transit could be used to get a better estimate of the mean distance from Earth to the Sun, and these estimates could be used for celestial navigation. The Royal Society, therefore, decided to sponsor an expedition to Tahiti to observe the transit. Alexander Dalrymple (1737-1808) might have been the logical choice to lead the expedition. He was a member of the Royal Society, a nobleman, and Britain's self-proclaimed expert on *terra australis*. But Cook, who was none of these things, was the society's choice. He was an experienced seaman who had a record of taking meticulous readings. He had done an expert survey of the difficult passages of the St. Lawrence River in 1766. He had also observed an eclipse of the Sun in Newfoundland and sent his calculations to the Royal Society.

Cook's first two decisions as head of the expedition were critical to the success of this voyage and later ones. First, the ship he selected was not a flashy warship, but a sturdy, flat-bottomed collier, a choice that may have saved him from disaster when he discovered the Great Barrier Reef and ran the ship aground. Second, he made the health of his crew an obsession, insisting on strict hygiene on his ship and introducing a variety of foods aimed at preventing disease, especially scurvy. As a result, there were relatively few fatalities—perhaps just one from scurvy—on Cook's three voyages during a total of almost eight years at sea. By comparison, only 55 out of the original 170 members of the crew of Vasco da Gama (1460-1524) safely returned

home from their voyage of a little over two years. Most of those lost had died of scurvy.

On Cook's first voyage (1768-1771), after the astronomical observations were complete, he set sail in search of *terra australis*. On the trip, he discovered the Admiralty Islands and the Society Islands. He also circumnavigated New Zealand and determined that it consisted of two islands. He explored the east coast of Australia, to the delight of Joseph Banks (1743-1820), the ship's botanist. Banks was a prominent and wealthy scientist, who, at his own expense, had brought along a stash of scientific equipment, an assistant, and four artists. In addition to collecting botanical specimens, he investigated the wildlife and later showed that almost all native mammals of Australia were marsupials. The presence of mammals that were more primitive than those on other continents became an early indicator of evolution.

The second voyage (1772-1775) involved a systematic search for the southern continent. Cook circumnavigated the globe, traveling west to east and venturing as far south as he could. Cook discovered New Caledonia, the South Sandwich Islands, and South Georgia. He was the first individual to travel below the Antarctic Circle, where he found unnavigable ice, harsh winds, and no evidence of *terra australis*. By traveling over 60,000 miles (100,000 km), Cook succeeded in proving a negative, and his reputation for precision and reliability was such that his evidence and conclusions were accepted as fact.

Cook's third voyage (1776-1779) took on a less ancient myth, that of the Northwest Passage. Columbus's idea of sailing west to reach Asia had not died, even though the Americas were in the way. Explorations to the south had shown the only way through was via Tierra del Fuego, at the southernmost tip of South America. This trip was about an 8,000-mile (13,000-km) detour, and it was hoped that a way through navigable rivers and inlets of North America would provide a more direct option. Fifty other explorers, including Henry Hudson (1565?-1611) and John Cabot (c. 1450-1499), had probed the North American continent from the east. Cook took on the project from the west. He went from Vancouver up along the coast of Alaska and continued north until he found himself once again in unnavigable, icy waters. He proved that there was no practical way through North America, although attempts continued into the next century, often with disastrous results. Cook found his own disaster on his return from the north. He

1700-1799

25

stopped at the Sandwich Islands (Hawaii), which he had discovered earlier. This second visit led to a quarrel and he was killed by the natives.

Impact

Cook's voyages essentially ended the age of exploration. While the Antarctic continent remained unexplored and many remote regions within the known continents still held secrets, there were no more undiscovered lands. Most people accepted Cook's conclusions about *terra australis* because he was universally respected as an honest, competent, and meticulous navigator. Cook's journals, however, did not discourage people from traveling to remote regions. His detailed descriptions, especially of the abundant wildlife, sent whalers, hunters, and traders into the regions he had explored. Cook, though, did shift the focus from exploration to development, and, by providing a more realistic map of the globe, he indirectly freed up funding for other ventures, including the digging of the Panama Canal.

The discoveries made on Cook's voyages had major geopolitical ramifications. Hawaii ultimately became a U.S. state. Cook also established claims to Australia and New Zealand that made them important British colonies and, later, nations with distinctive cultures and histories. While Abel Tasman's encounter with the Maori (1642) had been bloody, Cook dealt with them diplomatically and established friendly relations (1769). He led the way in opening Oceania to trade and European influence. In the following century, almost all the remaining inhabited regions were claimed by Germany, France, England, and the United States. Missionaries went and converted the local populations to Christianity. They also brought devastating diseases, including smallpox, tuberculosis, and measles. Cultural changes, including Western-style governments and modern technologies, took hold in these regions.

From the start, Cook's voyages were scientific. Astronomy may have been the primary purpose of the first voyage, but the most important contributions were to botany. A mother lode of new plant species was brought home by Joseph Banks. These species became the basis for Kew Gardens and, indeed, the establishment of botany as a separate scientific discipline. The practice of including a scientist on exploratory voyages became fashionable, most notably

Charles Darwin (1809-1882) on the *Beagle* and T. H. Huxley (1825-1895) on the *Rattlesnake*. The role of science officer also found a place in popular culture, reaching an apotheosis of sorts with Spock on *Star Trek*.

Banks went on to make scientific expeditions to Iceland and the Hebrides and to serve as president of the Royal Society for over 40 years. Of equal value were the contributions of Banks as a sponsor to rising stars of science, including Robert Brown (1773-1858), who discovered the cell nucleus and Brownian motion. Banks also became a strong advocate for the settlement of Australia and is sometimes called the "Father of Australia."

On his second voyage, Cook himself made a key contribution to science, for which he was honored with the Copley Medal. Cook wrote about his experiences in using different foods to prevent scurvy. Though it would be over a hundred years before the role of vitamin C in his success was understood, Cook provided a practical solution to scurvy that helped the British dominate the seas for the next century. He also helped provide a basis for the science of nutrition. Our understanding of the role of vitamins in preventing scurvy, rickets, beriberi, and pellagra is owed in part to the attention and diligence of Cook.

Cook's voyages led to an incident that was to capture the public imagination. William Bligh (1754-1817) was shipmaster under Cook. In 1787 Bligh was given command of his own ship, the *Bounty,* and received orders to transport breadfruit from Tahiti to the West Indies. Bligh's crew, led by Fletcher Christian, mutinied. Bligh, set adrift in a small craft with his officers, achieved one of the great feats of navigation, traveling over 4,000 miles (6,000 km) in an open boat to reach Timor in the East Indies. Bligh lived to testify against the mutineers. The story of the mutiny on the *Bounty* entered popular culture and Bligh's name became synonymous with a harsh taskmaster.

PETER J. ANDREWS

Further Reading

Asimov, Isaac. *Isaac Asimov's Biographical Encyclopedia of Science & Technology.* New York: Doubleday, 1976.

Boorstin, Daniel J. *The Discoverers.* New York: Vintage Books, 1985.

Hough, Richard. *Captain James Cook.* New York: W.W. Norton, 1997.

Samuel Hearne Is the First European to Reach the Arctic Ocean by Land Route

Overview

In the late 1760s reports from Native American traders at the Hudson's Bay Company's Prince of Wales Fort told of a large river leading to the Arctic Ocean with a wealthy copper mine at its mouth. The Hudson's Bay Company was encouraged by the idea of a mine next to a navigable river that could be used to transport its riches into the country's interior. Samuel Hearne (1745-1792) was sent on three attempts to discover the river and its mine, and in the course of his travels he mapped a large portion of the Canadian interior and closely observed the lives of the tribes living there. Most of all he proved the difficulty of long-distance travel overland through the Arctic.

Background

By the late 1760s the Hudson's Bay Company had established several forts along the bay's coast for trading and shipping furs and other goods extracted from northern Canada. An ultimate goal of the company was to explore the area for a Northwest Passage by which ships from England could sail to the trading ports of Asia. The company was one of two operating in Canada under exclusive royal charter to trade animal skins with the Native American tribes of Canada. The Hudson's Bay Company's presence consisted mostly of a string of forts along the shore of the bay that served as trading posts, where Native American trading parties could exchange their furs for manufactured goods from Europe. As of 1770 the company had not extended its presence into the interior of northern Canada.

In 1768 a group from the Chipewyan tribe of western Canada arrived at Prince of Wales Fort at the site of present-day Churchill, Manitoba, with samples of copper they claimed had come from a large and productive mine by a great river to the northwest. The commander of the fort assembled an expedition of three Englishmen led by Samuel Hearne to find the mine. The expedition was to travel with a group from the Chipewyan tribe.

On this first attempt in the winter of 1768, the two Englishmen under Hearne were unable to stand the rigors of traveling with the Chipewyan, and Hearne was forced to turn back

and join another group of Chipewyan traveling to Prince of Wales Fort. Hearne was barely able to get the two men back to the fort alive, but through the experience he began to learn how to plan for and survive the extended trip required to reach the elusive mine. On his next attempt Hearne took no other Europeans and greatly reduced the load he had to carry on the journey. During his first attempt to reach the Coppermine River, more Chipewyan had arrived at Prince of Wales Fort, and Hearne retained them as guides for his next attempt.

Hearne made his second attempt only three months after the first and traveled alone with another party of Chipewyan guides. After six months he lost his equipment, including the sextant he needed to map the route to the river, when a high wind blew down his tent. He was thus forced again to return to the Prince of Wales Fort. On his return to the Fort he was taken in by a Chipewyan leader named Matonabbee who offered to guide him to the Coppermine River on his next attempt. Despite the protests of the company's chief agent at Prince of Wales Fort, who preferred that Hearne travel with members of the tribe already familiar with the Company, Hearne set out again for the Coppermine River with Matonabbee and his party in early December of 1770, with winter approaching.

The party left Prince of Wales heading west to avoid traveling through harsher terrain during the first winter months of the journey, dragging their sleds behind them. In early February of 1771 the group crossed the Kazan River and walked over the frozen surface of Lake Kasba 300 miles (483 km) from Prince of Wales Fort and just north of what is now the border between Manitoba and Alberta. In April they reached Lake Clowey, an eastern arm of Great Slave Lake, from which they proceeded north towards the coast of the Arctic Ocean.

Along the route more Chipewyan traveling to the mouth of the Coppermine River continually joined the group. In this ever-expanding group Hearne was given the opportunity to observe the tribe's life from within. Hearne noted the habit of going completely without food for days at a time when it was not available and gorging themselves until they could not walk when it was. On the trail the party hunted for food including swans, geese, caribou, and musk oxen, all of which were eaten raw when wood was not available to build a fire for cooking. Hearne also witnessed the heavy workload given to the women of the tribe. Matonabbee attributed the failure of his two previous attempts to reach the river to the fact that he did not bring any women along to share the heavy work of the expedition. He explained the Chipewyan tradition of having the women of the group carry the vast majority of the load on the trail and making them the first to go hungry when food became scarce.

The cycles of feast and famine continued as the group traveled north. In June Hearne discovered that the Chipewyan were traveling as a war party planning a massacre of the Inuit settlement at the mouth of the river. Hearne objected to the plan and declared that he would not participate in the massacre except to defend himself if he were attacked. Leaders of the party were offended by his unwillingness to fight on their behalf and threatened to eject him from the traveling group, but Matonabbee intervened on Hearne's behalf and he was allowed to stay with the group. Later that month the group camped on Lake Peshew, where the women and children were to stay while the men formed an attack party and raided the Inuit settlement. On July 14, 1771, they reached the Coppermine River, and Hearne saw that the accounts given by the Chipewyan that came to Prince of Wales Fort exaggerated the size of the river. The Coppermine proved to be too shallow for boats to travel upstream for shipping to the Canadian interior as the Hudson's Bay Company had hoped.

The Chipewyan painted their shields and bodies in preparation for their attack on the Inuit. Hearne was still determined not to fight and was given a shield and weapon for his own protection. In the early morning hours of July 17, when the Arctic sun was at its lowest point sitting just on the horizon, the Chipewyan attacked the settlement and killed all but a few Inuit, who were barely able to escape from a nearby encampment. In the aftermath of the attack Hearne observed scattered whalebones and sealskins in the Inuit camp and concluded that they had reached the mouth of the Coppermine River at the Arctic Ocean. He then took possession of the coastline for the Hudson's Bay Company.

The search then began for the copper mine that the Chipewyan had described to the agents at Prince of Wales Fort years before. Like the story of the wide, navigable river, the rich copper mine also turned out to be an exaggeration.

After hours of searching Hearne only found one large lump of copper ore among the rocks along the river. Meanwhile the Chipewyan ransacked the Inuit settlement, destroying tents and pottery and collecting all the copper items for themselves. The group turned south to rejoin the main encampment and return to Prince of Wales Fort. At Great Slave Lake Hearne again lost his sextant, as he had on his second attempt to reach the Coppermine River, and was unable to map the rest of the return to Hudson's Bay.

Impact

Hearne returned to the fort on June 29, 1772. After a year and a half traveling through hostile environment he found neither large deposits of copper nor a new waterway to transport goods through Canada. But while he found no new resources in the extreme northwest of Canada, the trip did demonstrate to the Hudson's Bay Company that it would not be able to depend on having just a presence on the shores of Hudson's Bay. The company would have to build trading posts in the interior of Canada in order to extend its trade with the tribes there. Hearne himself was sent to command such a post when he was appointed to Cumberland House in what is now Saskatchewan. He remained there until 1775 when he was made governor of Prince of Wales Fort. At the end of the American Revolution the French invaded Hudson's Bay and threatened to attack the fort. Hearne surrendered and was taken prisoner by the French until the end of the war, when he returned to Prince of Wales Fort.

Hearne's overland trip through Canada was the last of its kind in which a lone European explorer would travel alone among a group of Native Americans overland through the Arctic. In the next 20 years Alexander Mackenzie (1755?-1820), working for the competing Northwest Company, would use the same strategy as Hearne in relying on the geographical expertise of the Native Americans to travel through the country and build a business network among the tribes of the interior. Mackenzie was able to use that method to build a vast trading network that stretched from Montreal to the Arctic Coast to the Pacific Ocean.

GEORGE SUAREZ

Further Reading

Speck, Gordon. *Samuel Hearne and the Northwest Passage.* Caldwell, Idaho: Caxton Printers, 1963.

"The Journeys of Samuel Hearne" http://web.idirect.com/ ~hland/sh/title.html

Mungo Park's African Adventures

Overview

In 1795, a young Scottish surgeon, Mungo Park (1771-1806), was hired by the African Association, a British organization that sponsored African exploration, to locate and, if possible, map a large river thought to flow in the African interior, now known to be the Niger River. Park found the Niger and explored parts of it before his untimely death in 1806 at the age of 35. Following his expeditions, several more years were to pass before the Niger would be completely mapped into the Gulf of Guinea. However, by locating the Niger and following it for over 1,000 miles (1,609 km) through the African interior, then writing about his adventures, Mungo Park helped increase knowledge of African geography and fueled Europe's interest in this large continent.

Background

By the end of the eighteenth century, vigorous exploration of the interiors of major continents was well underway. In North America, the eastern part of the continent was well known, and major portions west of the Mississippi had been explored by the Spanish and the French. South America had been explored by the Spanish, and much of Asia had been visited or described as well. The Australian interior remained a mystery, nor was anything known of the African interior. Of these, Africa was of far greater interest because of its animals, great lakes and rivers, natives, and jungle. It simply seemed more exotic, dangerous, and interesting than Australia. It was also more accessible, lying just a few thousand miles from Europe.

There were also many tales of Africa: travelers' stories dating back to the times of the ancient Greeks and Romans, biblical stories, and legends with unknown origins. Africa was a continent about which just enough was known to excite the imagination, but not enough to satisfy it. In addition, European governments saw Africa as a possible source of raw materials, new colonies, and economic or political gain. Since both European nations and the American colonies were also engaged in the slave trade, they paid frequent visits to Africa to purchase human cargo. Finally, scientists simply wanted to learn about the continent, its animals and inhabitants, and its geology and geography. In short, there were many people interested in Africa.

To help satisfy this knowledge, expeditions were launched by both public and private organizations. One of these, the African Association, commissioned a group in 1795 to locate a major river rumored to lie in the African interior. They chose a surgeon named Mungo Park to lead the expedition; three years earlier he'd been a medical officer on a ship in the East Indies trade, and had spent some time studying the flora and fauna of Sumatra.

Park began his journey by sailing to Gambia, on the Atlantic coast of northern Africa. From there, he made his way overland through deserts and over mountains, traversing country inhabited primarily by Muslims, reaching the Niger River after several months of hard travel. Falling seriously ill was only one of his adventures; he was also imprisoned, robbed, and threatened. Upon reaching the Niger, Park realized it would be suicidal to try to press on to Timbuktu as he had originally planned, so he returned to Gambia, returning to England in 1797.

Back in England, Park married, wrote a book, and became licensed in surgery. In 1805 he set out again on another expedition sponsored by the African Association, accompanied by nearly 40 men, trying again to map the course of the Niger. This time, after reaching the river, they built boats and sailed along it for over 1,000 miles (1,609 km), mapping its course as it flowed to the east and turned south. Disease, however, killed all but 11 of his expedition members, and the weakened party was never to reach the mouth of the Niger. They were killed in a battle with natives near the present city of Bussa in 1806.

Impact

Upon returning from his first trip to Africa, Park wasted little time in compiling an account of his adventures, *Travels in the Interior Districts of Africa* (1797), in which he described the African landscape, the people he met, and the difficulties he encountered. His book was a success because it detailed what he observed, what he survived, and the people he encountered. His honest descriptions set a standard for future travel writers to follow. This gave Europeans a glimpse of what Africa was really like. Park introduced them to a vast, unexplored continent with huge rivers, untamed reaches of land, a va-

A view of the Niger River near Timbuktu. *(Corbis Corporation. Reproduced with permission.)*

riety of human cultures and societies, and a great abundance of potential that was, from the standpoint of eighteenth-century Europe, completely unexploited.

After Park's disappearance public and political interest in Africa began to increase. He had proved that Africa could be explored, showing that it was possible to journey through unknown territory to a major African river, with few supplies and little help—but that doing so was dangerous business. More than 15 years would pass before the next major expedition left for Africa. (This is surprising when you consider that Africa, is, after all, geographically closer to Europe than either of the Americas or Asia. Yet, trade was established with India and China, colonies were established in both North and South America, and a struggling colony was present in Australia before African exploration was well underway.) Hugh Clapperton, Dixon Denham, and Walter Oudney led a three-year expedition for the British government (1822-1825) through Saharan and sub-Saharan Africa—and returned to England to tell about it. They were followed by many others in subsequent decades, culminating in the epic journeys of David Livingstone (from 1852 until his death in 1873).

Perhaps the most lasting effect of Park's travels, though, was their influence on European governments, which were at that time intent on building their empires and quelling domestic troubles. Africa seemed a way to move their competition away from Europe while at the same time opening markets for the flood of goods produced by the Industrial Revolution then sweeping through Europe.

Over the next century, Britain, France, Italy, Germany, Portugal, and Belgium all established (or tried to establish) colonies, trading outposts, or both in Africa. Although warfare between competing European powers rarely erupted, the natives often resisted European incursions. The African tribes, however, could neither coordinate their efforts nor overcome the technological advantage of European weapons. In every instance but one (Ethiopia, who defeated the Italians in 1896), they failed to resist the onslaught of European colonizers.

Despite subduing the native populations, most European nations failed to extract the same economic or political advantage from Africa that they did elsewhere. Although parts of Africa are rich in gold, diamonds, and other mineral wealth, this was not immediately apparent in the late eighteenth and nineteenth centuries. Parts of Africa are suitable for agriculture, but not to the extent that North America is. Tropical woods fetched a high price in European markets, but tropical diseases and often-hostile tribes made harvesting them difficult and expensive. Big game hunting was popular, but was too expensive for all but the very wealthy. For these and other reasons Africa failed to live up to Europeans' initial hopes or expectations, and very few African colonies became economic boons to their mother nations.

At the time of Park's journeys, however, this was all in the future. Upon his arrival, Park found a continent full of mystery and promise, ripe with expectation. This is what he reported in his book, and helped entice future explorers to follow the path he started.

P. ANDREW KARAM

Further Reading

Books

McLynn, Frank. *Hearts of Darkness: The European Exploration of Africa.* Carroll & Graf, 1993.

Parker, Geoffrey. *The World, An Illustrated History.* New York: Harper & Row, 1986.

James Bruce Explores the Blue Nile to Its Source and Rekindles Europeans' Fascination with the Nile

Overview

From ancient times, the existence and survival of Egypt has depended on the Nile River. About 4,000 miles (6,437 km) long, the Nile is the longest river in the world and consists of two main branches. The longer branch, often referred to as the White Nile, rises from the heart of central Africa and flows more than 3,000 miles (4,828 km) to the Sudan, where it is joined by the Blue Nile. At this junction, the two tributaries form the greater Nile, which then courses through Egypt and drains at the wide Nile Delta into the Mediterranean Sea.

The ancient Egyptians probably knew that the source of the Blue Nile was Lake Tana in Ethiopia, but the headwaters of the White Nile remained a mystery for centuries. In 457 B.C., the Greek historian Herodotus (c. 484-420 B.C.) attempted to locate the Nile's source and followed the river to Aswan, but he was unable to progress any farther. Six hundred years later, the great Greek geographer and astronomer Ptolemy (fl. A.D. 127-145) described the Nile as originating from two lakes near the Mountains of the Moon in central Africa and snaking northward to the Mediterranean Sea.

European interest in the Nile waned until the seventeenth century, when Portugal in particular became interested in Africa for both religious and commercial purposes. In 1618 a Jesuit priest, Father Pedro Paez (1564-1622), traveled from Arabia to Ethiopia; in the mountains south of Lake Tana he came upon some swampy soil and a spring that he correctly recognized as the source of the Blue Nile. His fellow missionary, Father Jeronymo Lobo, wrote about Paez's discovery; Sir Peter Wyche and Samuel Johnson of Britain later translated Father Lobo's work. A few Europeans continued to explore the Blue Nile, but no one reached its source again until 1770, when James Bruce (1730-1794) of Scotland declared that he was the first European to have discovered the primary source of the Nile. He was, of course, mistaken on both counts: he had only found the source of the Blue Nile, and he was not the first European to do so. Nonetheless, James Bruce's historical role is significant: his adventures rekindled Europe's fascination with the Nile and lured many nineteenth-century explorers to Africa in search of the river's mysterious source.

Background

James Bruce was born in 1730 in Scotland. He lost his mother when he was only three years old, and grew from a delicate child into a strong man more than 6 feet (1.83 m) tall, with red hair and a deep, booming voice. An aristocrat who was the heir to his family estates at Kinnaird, Scotland, Bruce was educated at Harrow and later at Edinburgh University. At his father's insistence, Bruce studied law and eventually joined the East India Company in London, where he met and married a wealthy young woman. Within nine months, however, his wife died, and at 24 years old, James Bruce was a lonely, embittered young man. He turned to travel and the study of languages, and in 1762 he accepted the post of British consul in Algiers, where he continued to study Oriental languages and medicine.

In Algiers the Barbary pirates and the cruel Ali Pasha made life miserable for Bruce, and finally, in 1765, he was allowed to leave his post. He traveled to various archaeological sites in Greece, Syria, and modern Lebanon before arriving in Cairo in 1768 with a young Italian artist, Luigi Balugani. In Egypt James Bruce began to pursue his dream of locating the source of the Nile River.

Erroneously convinced that the Blue Nile was the larger of the two branches of the Nile, Bruce traveled from Cairo up the Nile to Aswan. When further progress along this route proved impossible, he chose instead to approach Ethiopia, or Abyssinia as it was then called, from its coast. Traveling inland from the Red Sea, Bruce and his party reached Gondar, then the capital of Ethiopia, in February 1770 and immediately became embroiled in the savagely brutal tribal warfare of Ethiopia. Bruce's very survival is almost miraculous and undoubtedly due to his knowledge of the native language and to his medical expertise, especially in treating smallpox.

Determined to find the source of the Nile, Bruce hiked about 70 miles (112.6 km) into the mountains on the southern edge of Lake Tana. On November 4, 1770, he and his party reached a swampy area that he mistakenly thought was the source of the great Nile River. He then proclaimed himself to be the first European to reach the spring, pointedly ignoring Father Lobo's record of Pedro Paez's presence at the same place in 1618. James Bruce was fanatically anti-Catholic and scoffed at any claims made by Paez and Lobo. Ironically, his refusal to acknowledge Paez's discovery was later a factor that caused others to mock Bruce's own great accomplishments.

Bruce returned from his mountain trek to an ugly civil war in Gondar and was unable to leave Ethiopia. He remained there and devoted his energy to writing a history of the local kings and to assembling a collection of manuscripts and indigenous flora. He recorded in gruesome detail the bloody battles and the frequently licentious customs of the Christian people of Gondar before being allowed to leave Ethiopia in December 1771.

Bruce and his party chose to return to Cairo by land and reached the Moslem city of Sennar in April 1772, where he remained for four months before traveling to Khartoum, the point where the two Nile rivers join. No record exists, however, that Bruce even acknowledged what he must have so clearly observed at Khartoum: that the White Nile was obviously the larger of the two tributaries and that Lake Tana and the nearby spring could not, therefore, be the primary source of the Nile River.

Bruce was ill by this time with guinea worm, a flesh-eating parasite. Nevertheless, he pushed on and finally joined a caravan headed for Cairo. His party reached Aswan on November 28, 1772, and Cairo thirty days later. After two months, he left Egypt and sailed for Marseilles, France, where he remained for a month, seeking medical treatment for his leg, traveling and being received by King Louis XVI. From France he journeyed on to Italy, again in search of medical care. At last, in June 1774, he returned to London, but not to the reception he had eagerly anticipated. King George III as well as the educated elite rebuffed Bruce's account of his years in Ethiopia, often believing he had fabricated his fascinating tales of butchery and barbarism. Additionally, Bruce was condemned because he had repudiated the words of the Portuguese priest, Father Lobo, and continued to maintain that he was the first European to reach the Nile's source above Lake Tana.

Spurned by much of London society and intelligentsia, Bruce returned to his native Scotland, where he was more appropriately welcomed. He married Mary Dundas, who bore him several children but tragically died in 1788, 14 years after they had wed.

After his wife's death, Bruce yielded to friends' urgings to publish his journals, working in London with B. H. Latrobe for a year to transcribe his writings into five volumes. *Travels to Discover the Sources of the Nile, in the Years 1768, 1769, 1771, 1772, and 1773* was published in 1790, but Bruce refused to pay Latrobe for his efforts. Likewise, he refused to credit the Italian artist, Balugani, who had died in Ethiopia, for his many superb drawings. In fact, Bruce took personal credit for the Italian's drawings. The book's reception was dismal. Critics lampooned his stories of what he had observed in Ethiopia; most of the book was regarded as mere fiction and became the subject of many jokes.

Angry and dejected, James Bruce returned to his family in Scotland. At the age of 64, he accidentally fell on the staircase in his home and was knocked unconscious. He died the next day, on April 27, 1794.

Impact

Although James Bruce and his book were reviled in England, the French regarded his contribution as serious. Indeed, the work of James Bruce was most influential in leading the French to Africa, especially to the Nile. Napoleon's Egyptian campaign in 1798 resulted in the brief French occupation of that country, further exploration of the Nile, and the discovery of the Rosetta Stone.

The primary source of the Nile, however, remained a mystery until the mid-nineteenth century, when British explorers Sir Richard Burton (1821-1890) and Captain John Speke (1827-1864) set out from Zanzibar toward central Africa. Speke eventually located a large lake in 1858, which he named Lake Victoria in honor of the British sovereign, and realized it was the source of the Nile. In 1875 Sir Henry Morton Stanley (1841-1904) was the first to sail completely around Lake Victoria, the second largest freshwater lake in the world, and saw the snow-capped Ruwenzori mountain range, Ptolemy's "Mountains of the Moon."

Even though James Bruce had erred in thinking that an Ethiopian swamp high above sea level was the primary source of the Nile, and despite the fact that he inflated his role in the exploration of the Nile River, his contribution to the history of the continent of Africa is important. Not only was he the first European to follow the Blue Nile to where it converged with the White Nile, but his accounts of his travels and the years he spent in Ethiopia are still regarded today as epic. Not until 1960 were the headwaters of the Blue Nile fully charted, almost 200 years after James Bruce climbed more than 6,000 feet (1,829 m) to a swampy spot in the Ethiopian highlands.

ELLEN ELGHOBASHI

Further Reading

Morehead, Alan. *The Blue Nile*. New York: Harper & Row, 1962.

Reid, J.M. *Traveller Extraordinary*. New York: W.W. Norton & Co., Inc., 1968.

Udal, John O. *The Nile in Darkness: Conquest and Exploration 1504-1862*. Michael Russell, 1998.

The North Pacific Voyages of the Comte de La Pérouse

Overview

Between 1785 and 1788 Jean-François de Galaup, comte de La Pérouse (1741-1788?), sailed on a mission of exploration to discover islands and lands not yet found by France's European rivals. With two ships, *La Boussole* and *Astrolabe*, La Pérouse traveled to many parts of the Pacific, although he is best-known for his explorations in the North. In particular, he discovered the strait named for him that separates the islands of Sakhalin and Hokkaido, connecting the Sea of Okhotsk with the Sea of Japan. These and other discoveries helped complete this phase of Pacific Ocean exploration and established a French presence in the region.

Background

European exploration of the Pacific Ocean began with Vasco Núñez de Balboa's (1475-1519) first view of the ocean from the Isthmus of Panama in 1513. Six years later, when Ferdinand Magel-

lan (c. 1480-1521) set out to circumnavigate the globe, he became the first European to sail the largest ocean on Earth. One of Magellan's goals, like Christopher Columbus (1451-1506) before him, was to find a new route to the Orient, to open it for further trade with Europe.

Over the next two centuries, exploration in the Pacific was sporadic at best. While its periphery was visited and mapped to some extent, there were no real systematic efforts to document what lay beyond those areas thought to be commercially valuable or that held riches such as gold or spices. Even midway through the eighteenth century much of the Pacific and many of its shores were unknown.

The exploration that was done, and the few trading posts and colonies that were established in the Pacific were almost exclusively English, Dutch, or Spanish. Although the French boasted the largest nation in Europe, a strong navy, and a vigorous economy, France's concentration on European affairs blinded her to the possibilities that existed overseas. Finally, in the latter half of the eighteenth century, France awoke to the possibilities that existed in the Pacific.

During this time, voyages of exploration were dispatched to all parts of the Pacific, charged with finding lands not already claimed or visited by France's European rivals. Three of the best-known explorers were Antoine Raymond Joseph de Bruni, chevalier d'Entrecasteaux, (1739-1793), who visited Australia and neighboring regions; Louis-Antoine de Bougainville (1729-1811), who explored the islands around Indonesia and New Guinea; and La Pérouse.

Setting out from France in August 1785 for the Pacific, La Pérouse rounded Cape Horn on his way to Easter Island, the Sandwich Islands (now Hawaii), and up the coast of North America. His first goal, to find the long-sought Northwest Passage between the Atlantic and the Pacific, was futile, and he stopped his search when he reached the southern shore of Alaska. From there, he headed south, reaching as far as San Francisco before crossing the Pacific. He arrived in southern China in early 1787 and proceeded on to Manila, then owned by the Spanish. Stocked with water and food, he headed north along the Asian coast, reaching Japan in early summer.

La Pérouse reprovisioned in Japan, then headed north through the Sea of Japan. He first passed through the Tatar Strait, a long and narrow channel separating the island of Sakhalin from the Asian mainland, and also sailed through what is now known as the Straits of La Pérouse, which separate Sakhalin and Hokkaido. He stopped briefly in Petropavlosk on the Kamchatka Peninsula, where he dispatched his notes and journals back to France before heading for Botany Bay in New Holland (now Australia). When the ships stopped at the Navigators' Islands (now Samoa) en route, a dozen of La Pérouse's crew were killed by the natives. The expedition reached Botany Bay, from which they set forth again on March 10, 1788. Neither La Pérouse's nor his crew were seen again and, until 1826, nothing more was known of them.

Later explorers, both English and French, found that his ships had apparently been wrecked in the Solomon Islands, not far from New Guinea. Some crew members were killed by natives; others escaped and apparently perished at sea. Only a few artifacts from the ship and its crew were found.

Impact

The most obvious impact of La Pérouse's voyages was the return of scientific and geographic knowledge from a part of the world not yet visited by Europeans. Animal, plant, and geologic specimens were returned to France, along with detailed notes and journals about weather and sea conditions, maps of the coasts visited, and information about much of the Pacific Ocean, gathered over the better part of two years. By themselves, these records could only give a snapshot of, for example, climate and seas, but when combined with the discoveries of other expeditions, a much more complete and detailed picture was gradually assembled. Some of these observations were of scientific interest only. Others were more general, and of use to French commerce and the military, which began moving into the area in later decades.

La Pérouse's voyages also presaged, to some extent, later voyages to Japan aimed at opening trade with the isolated nation. Although La Pérouse did not attempt to force trade issues, his journals provided valuable information for later visits, including Matthew Perry's (1794-1858) 1852 visit that forced Japan to open trade with the West. In addition, the Straits of La Pérouse are a frequently used transit path for ships traveling to Japan, forming the conduit from the Sea of Okhotsk to the Sea of Japan. These straits are also militarily significant; American submarines used them during the Second World War in their attack on the Japanese Empire, and they are currently thought to be equally important for Russian ships and submarines.

Of more importance was La Pérouse's role in helping France gain a foothold in the Pacific. As mentioned above, despite having the strongest navy in Europe, France had previously concentrated on Continental European affairs, almost to the exclusion of overseas explorations. Had France given the same resources to exploration as did Britain, Holland, and Spain, she may well have emerged as the strongest colonial power by the end of the eighteenth century. Instead, as that century drew to a close, France had only her North American colonies in Canada and Louisiana, small outposts in South America and the Caribbean, and some scattered islands in the Indian Ocean. This changed in 1785, when King Louis XVI sent La Pérouse to the Pacific, to visit "all the lands that had escaped the vigilance of Cook (the great English explorer)."

With this charge, La Pérouse's voyages began an era of French efforts to gain a stake in the Pacific. Over the following decades, they ended up with outposts of limited commercial value, but of strategic importance. Scattered throughout the Pacific, French territories still exist, including the atolls where France performed atomic bomb testing as recently as the 1990s. Although her Pacific possessions never returned the same wealth as those of the Dutch or the British, France's islands spanned the ocean, giving her ships bases of operation and places to provision, shelter, and make repairs as they crossed the Pacific on voyages of exploration, commerce, and war.

Shortly after La Pérouse's death, however, events in France prevented follow-up on his and other explorers' successes. With the onset of the French Revolution in 1791, France descended into turmoil that was to last for several years. Shortly thereafter, Napoleon's military successes kept France's attention focused on matters close to home, and her overseas possessions were an afterthought. This helped ensure that France, despite her military and economic prowess, would never enjoy the same economic advantage from her overseas colonies as the English and Dutch.

La Pérouse set out to make scientific discoveries, to open trade routes, and to help France establish a presence in the Pacific Ocean. Although he failed to open any new trade routes, he met his other objectives, which helped bolster France's status as a great power. In addition, the Straits of La Pérouse are a frequently used waterway when not frozen, navigated by both military and commercial ships of many nations.

La Pérouse accomplished a great deal for France, although he died before being recognized for his achievements. As one of the first French explorers in the Pacific, he helped set high standards for those who followed. He helped France to establish herself as a player in the global game of territorial dominance, and helped France establish outposts of military (if not commercial) significance that are retained to this day.

P. ANDREW KARAM

Further Reading

Books

Bohlander, Richard, ed. *World Discoverers and Explorers.* New York: MacMillan Books, 1992.

Great Voyages of Discovery: Circumnavigators and Scientists. New York: Facts on File Publications, 1983.

Internet Sites

"Discoverers Web." http://www.win.tue.nl/~engels/discovery

Carsten Niebuhr Describes the Near East

Overview

The first European scientific expedition to the Near East, dispatched in 1761, was an unlucky one for its crew. One after another they died, until the surveyor, Carsten Niebuhr, was left alone. Niebuhr continued exploring and, upon his return, published several important reports. These included maps that were used for a century, cuneiform inscriptions copied from the ruins at Persepolis, and botanical data gathered by the expedition's lost naturalist.

Background

Carsten Niebuhr was born in Ludingworth, Hanover, Germany, on March 17, 1733, into a farming family. He was orphaned while still an infant, and the family farm was sold, the proceeds being split between Niebuhr and his several brothers and sisters. The boy was set to farming work, but no longer had a farm of his own. Although his prospects seemed quite limited, the resourceful Niebuhr managed to learn surveying and pick up a smattering of a number of scientific and technical trades. Eventually he was admitted to the University of Göttingen to study mathematics.

In 1760 King Frederick V of Denmark, at the urging of the Hebrew scholar Johann David Michaelis of Göttingen, authorized the first European scientific expedition to the Near East. The naturalist on the expedition was to be Pehr Forrskal, a gifted student of the founder of botanical classification, Carl Linnaeus (1707-1778). Forrskal, a Swede whose writings on civil liberty had been confiscated by Linnaeus himself and condemned from Swedish pulpits, jumped at the offer of a lengthy journey financed by the

Danish crown. The official in charge of organizing the expedition, Baron von Bernstorff, also had connections at the University of Göttingen, and Niebuhr was invited to participate as a surveyor and engineer.

The expedition was intended to explore the culture and history of the Near East, especially the background of the Bible, as well as the region's flora and fauna. To that end, the party included, besides Niebuhr and Forrskal, the Danish linguist and Orientalist Friedrich Christian von Haven and Christian Carl Kramer, a Danish physician and zoologist. Georg Baurenfeind, an artist from southern Germany, went along to assist in documenting the group's discoveries. A Swedish ex-soldier named Berggren was also included.

The explorers sailed from Copenhagen in January 4, 1761, on the Danish military vessel *Groenland*. The small expedition was loosely organized, with no official leader. Perhaps this affected the relationships between some of the team members, which were difficult almost from the start, although Niebuhr and Forrskal seemed to get along well. The weather too was stormy, impeding the ship's progress for months. The situation deteriorated to the extent that von Haven actually disembarked and traveled overland to the port of Marseilles, rejoining the group when the ship finally docked there in May.

After a stop at Malta, the travelers continued to Istanbul (then called Constantinople), where they boarded a Turkish ship headed for Egypt's port city of Alexandria. This leg of the voyage was much more pleasant, uneventful but for the opportunity to enjoy the company of a group of Turkish girls in the adjacent cabin. From Alexandria, they traveled up the Nile River to Cairo, and went on to Mt. Sinai and the Suez.

In October 1762, for their journey to the Muslim holy city of Mecca, they disguised themselves as pilgrims. From the Arabian port of Jidda, they traveled south along the eastern shore of the Red Sea in an open boat called a *tarrad*. Southern Arabia was completely unknown territory to Europeans at that time. An Englishman named John Jordain had visited the Turkish pasha there in 1609. In 1616 the Dutchman Pieter van der Broeke traveled to the Yemini capital of Sana and wrote admiringly of the great pillared mosque. But no group had previously been sent there for purposes of gathering extensive information.

The travelers made frequent landings until they reached the harbor of Luhayyah in northern Yemen on December 29, 1762. They stayed as the guests of the local emir, who became fascinated by their microscope. Next they journeyed over the coastal plains by donkey to Mocha in the southwest.

It was during this part of the trip, in 1763, that disaster began to strike. Niebuhr and von Haven both contracted malaria, and in May von Haven succumbed to the disease. Forrskal also became ill after a journey into the hills to collect herbs. His specimens were confiscated and destroyed by customs authorities in Mocha, and he died in July.

The remaining members of the expedition then traveled to Sana. The imam received them graciously, housing them in a villa and providing them with money and camels for their return to Mocha. The four travelers, all of them sick and feverish, had to be carried onto the English ship bound for Bombay, hoping to find a healthier climate in India. Baurenfeind died onboard on August 29, and Berggren the next day. Kramer followed on February 10, 1764 in Bombay. Niebuhr was the expedition's sole survivor.

Adopting native dress and diet, Niebuhr remained in India for 14 months. He embarked upon his long journey back to Europe with a visit to Muscat in southeastern Arabia, followed by a voyage on a small English warship to Persia. He traveled overland through Shiraz to Persepolis. There he copied many ancient cuneiform inscriptions from the ruins of the palace destroyed by Alexander the Great in 332 B.C. In doing so he provided a tremendous gift to scholars, who had been trying to decipher the cuneiform scripts with very few samples to work from.

Niebuhr went on to Babylon, Baghdad, Mosul, and Aleppo. From the Mediterranean coast he sailed to Cyprus before returning to visit Jerusalem in 1766. He then proceeded up the coast and over Turkey's Taurus mountains to Istanbul.

Impact

Upon his return to Copenhagen in 1767, Niebuhr conscientiously began working on an official report from the ill-fated expedition. The result still makes fascinating reading, being filled with colorful details and entertaining anecdotes. As one might expect, Niebuhr's writings also exhibit the ethnocentricity of an eighteenth-century traveler commenting upon the "ignorance and stupidity" of those unfamiliar with the European way of life.

The maps Niebuhr made remained in use for more than 100 years. In fact, his measurements of the Nile Delta were so exact that they were used in building the Suez Canal (1859-69). He was also scrupulous in preserving the work of the unfortunate Forrskal, compiling, editing, and publishing three volumes of his notes: *Descriptiones animalium, Flora aegyptiaco-arabica,* and *Icones rerum naturalium.*

Niebuhr was married in 1773; his only son, Barthold Georg Niebuhr, was to become an eminent historian. The family left Copenhagen in 1778, when Niebuhr was offered a position in the civil service in Holstein, and took up residence in Meldorf. Niebuhr died there on April 26, 1815, by which time many European soldiers, scientists, and adventurers, beginning with Napoleon, had established a presence in the Near East.

SHERRI CHASIN CALVO

Further Reading

Bidwell, Robin. *Travellers in Arabia*. London: Garnet Publishing, 1994.

Hansen, Thorkild. *Arabia Felix: The Danish Expedition of 1761-1767*. London: St. James, 1964.

Hepper, F. Nigel and I. Friis. *The Plants of Pehr Forrskal's Flora Aegyptiaco-Arabica*. Kew: Royal Botanical Gardens, 1994.

Niebuhr, Carsten. *Reisebeschreibung nach Arabien und andern umliegenden Ländern*. 1774. English translation as "Travels through Arabia and Other Countries of the East" by R. Heron, London, 1799. Facsimile edition by Garnet Publishing, London, 1994.

Antoine de Bruni
Charts the Tasmanian Coast

Overview

During the closing years of the eighteenth century, French exploration of the Pacific Ocean began in earnest. One of the most important French sea captains during this period was Antoine de Bruni, known as the Chevalier d'Entrecasteaux (1739-1793), whose voyages to Australia returned valuable knowledge of this continent and its neighboring islands. Bruni's voyages helped establish France as a serious presence in the Pacific and served to counterbalance the existing presence of the British, Dutch, and Spanish in these waters.

Background

European exploration of the Pacific Ocean began in 1519, when Ferdinand Magellan (1480?-1521) led his small fleet through the Straits of Magellan into the reaches of the Pacific. Although Magellan did not live to complete the voyage, the reports returned by his expedition whetted the appetites and imaginations of most major European powers. Sensing the opportunity to expand colonial and trade empires, Britain, Spain, and Holland rushed to explore the Pacific and its neighboring lands.

Oddly absent in this rush was France, who was concentrating on Continental European matters almost to the exclusion of overseas territories. With the exception of her American colonies in the Caribbean, Quebec, and Louisiana, France chose to forgo empire-building at first.

By the late eighteenth century, it became obvious that this policy was flawed. Spain returned untold amounts of gold, silver, and gems from her New World colonies, Britain's American colonies gave her a commercial and strategic foothold in the Western Hemisphere, and the Dutch colonies and trading centers in the East Indies were returning huge dividends to Holland. Following the resounding success of James Cook's (1728-1779) voyages of discovery, King Louis XVI realized that France must follow suit to remain a great power.

In November 1766, Louis-Antoine de Bougainville (1729-1811) led France's first expedition to the Pacific. A veteran of France's war against England in Canada, Bougainville had seen France lose many of her possessions to the English. Bougainville's ships passed into the Pacific in January 1768 and arrived at the eastern shore of Australia in June of that year, the first European to do so. He continued on and returned to France in 1769, the first French naval captain to successfully complete a circumnavigation.

Bougainville was followed by du Fresne, who, in 1772, made the second French landfall in Australia, on the island of Tasmania. He was followed by Yves-Joseph de Kerguelen-Tremarec, who set forth from France's base on the Ile de France (now called Mauritius) in search of the austral continent. Although Kerguelen did not find it, another ship in his squadron, separated in a storm, did proceed on to Australia, claiming possession of the west coast for the King of France. Unfortunately, Kerguelen died before this information could be transmitted to France, and the claim was not recognized.

The next French expedition into Australian waters was led by the Comte de la Perouse (aka Jean de Galaup, 1741-1788?) in 1785. After two years of exploration, la Perouse left Botany Bay in March 1788 and was lost at sea. La Perouse's disappearance was one of the factors leading to yet another French expedition to these waters, this time led by Antoine Raymond Joseph de Bruni. Bruni and his crew arrived in Tasmania (then known as Van Dieman's Land) in April 1792 and spent the next several months charting the Tasmanian and Australian coasts, while failing to find any sign of la Perouse.

During his search for la Perouse, Bruni discovered a number of islands and made significant progress in mapping the coasts of Tasmania, parts of Australia, and many of the region's islands. He did not, however, locate la Perouse, his crew, or their remains. As it turned out, Bruni was not only unsuccessful in locating any trace of la Perouse, but he was also to die of scurvy before completing his mission and returning to France.

Impact

Bruni's voyages were part of a larger picture of French explorations in the Pacific, and their impact must be viewed in that larger context. However, his work was also important in and of itself. For this reason, the impact of his work alone will be examined, as will how his work fit into the context of the times.

The most immediate impact of Bruni's expedition lay in his geographical discoveries. Bruni visited lands that had been previously unvisited by Europeans, or that were marginally known. By constructing accurate maps of their locations, he provided a service to the French government, which was interested in laying claim to these islands. In addition, accurate maps were of military importance, helping to plan both attacks and defense of colonial and military outposts. And, finally, by showing the locations of newly discovered islands, Bruni helped future mariners avoid unpleasant surprises.

In addition to constructing maps of new coastlines and lands, Bruni maintained records of weather, ocean currents, prevailing wind directions, and other oceanographic and meteorological information that was turned in to the French government when his expedition returned to France. This information was very important to future mariners because, when combined with similar records maintained by other captains, they helped in developing a comprehensive picture of ocean and weather conditions throughout the world. This information, in turn, could be used to help plan trade routes, travel times, locate military bases, and so forth. In those times, the oceans were the world's highways, and the winds were the engines that drove commercial and military vessels around the globe. Bruni and his fellow captains provided the French government with information, from which accurate maps were developed, showing French merchant and naval vessels how best to navigate the oceans. Such maps were closely held state secrets precisely because of the commercial and strategic advantage they could confer.

In the larger setting, Bruni and his fellow captains were charged with helping France make up for lost centuries of exploration. In mounting an all-out effort to explore uncharted waters, the French hoped to offset the advantage enjoyed by the other major powers. Bruni's part of this effort was to help explore the Pacific Ocean in the vicinity of Australia while searching for the missing la Perouse. Through his efforts, he was able to help France claim some important territories in this part of the world, adding to those already claimed by Bougainville, la Perouse, and others. Through these territorial possessions, which included much of Polynesia, Indochina, and many of the islands near Australia and New Guinea, France was able to exercise some degree of military prowess in this strategically important part of the world.

It is also important to recall that, at this time, the great European powers were in the process of becoming history's first truly global powers—that era's equivalent of today's superpowers. Britain, Holland, and Spain all had territorial possessions in the New World, the Indian Ocean, and in Southeast Asia. Only France, arguably the strongest European power, lacked such a far-flung empire. Bruni's voyage, among others, was her attempt to redress this imbalance of power and to return France to the upper echelons of global political importance.

The South Seas islands, including those in areas visited by Bruni, were of great interest to the French philosophers of the day, primarily because they provided an opportunity to see humanity unencumbered by the trappings of society and civilization. At this same time, Jean Jacques Rousseau had published his treatise on the inherent nobility of man. Arguing that man was inherently good and that it was only society that made men bad, Rousseau speculated about the "noble savage" that, in the absence of society, would reveal humanity's true nature. These philosophers saw the apparently simple cultures of the South Pacific with their seemingly simple lifestyles and free sexual activity as confirming their speculations.

Finally, these discoveries provided France the opportunity to establish military bases in precisely those areas from which they could prey on Dutch, Spanish, and British merchant shipping. Heavily laden merchant vessels bound for European ports sailed through a handful of choke points that could be watched by a relatively small number of naval vessels. Seizing these merchant ships on the high seas not only hurt France's enemies, but enriched France at the same time since the French government would sell the cargoes at market value.

In short, Bruni's mission, though it failed to accomplish the original goal of locating and (if necessary) rescuing la Perouse's expedition, was nonetheless successful in many respects. As an individual expedition, it returned a great deal of valuable information about the South Seas that proved valuable to the French government and to French military and commercial vessels. In addition, in conjunction with information returned by other captains, France was able to compile a better set of information about ocean and weather conditions throughout the world, giving her ships added advantage when sailing those waters. Finally, by claiming lands for France, Bruni and his compatriots helped France to establish herself as a global power as she now owned territories across the world and, from these territories, could challenge merchant and military vessels belonging to the other great powers.

P. ANDREW KARAM

Further Reading

Books
Allen, Oliver. *The Pacific Navigators.* Time-Life Books, 1980.

Other
"France's Role in Exploring Australia's Coastline." http://www.france.net.au/site/presse_info/af/expl.html.

Excavations at Pompeii and Herculaneum Mark the First Systematic Study in Archeology

Overview

When Mount Vesuvius erupted in August of A.D. 79, it destroyed Herculaneum and Pompeii, two lively, bustling Roman cities. The eruption also had the effect of perfectly preserving that moment of tragedy, the daily routines of life in classical Italy, and the structure and art of those ancient cities. Directed by King Charles III of Naples, the initial excavations of Pompeii and Herculaneum in the eighteenth century were the first large-scale archaeological projects in history. The revolutionary methods used during the excavation unearthed enormous historical, architectural, and artistic treasures and formed the basis of the meticulous approach used in today's archaeological techniques.

Background

The volcanic mountain of Vesuvius rises 4,190 feet (1,277 m) above a fertile valley near the Bay of Naples, once home to a bustling seaport of the first-century Roman Empire. The original city of Pompeii was 128 feet (39 m) above sea level, sitting on a prehistoric lava flow about 4.97 miles (8 km) from Mount Vesuvius. Herculaneum, another village in the region, was less than 4.35 miles (7 km) from the base of the volcano. In August of A.D 79. Vesuvius belched out a great plume of smoke, spewing pumice, black ash, stones, and deadly gases from its crater. The steam was estimated to be near 750° Fahrenheit, and with it were hot ash and lethal gasses traveling 100 mph (161 kph). The poisonous eruption swallowed two towns in its path: Pompeii, with nearly 20,000 people, and Herculaneum, 10 miles (16 km) west with approximately 5,000 citizens. The eruption, while swift in its destruction, layered Pompeii in a preserving material nearly 25 feet (7.62 m) deep. The first layer consisted of pumice and lava pebbles. It was followed by a deep layer of ash, and finally a layer of fertile soil collected on top of the ash. Herculaneum was buried even deeper—nearly 65 feet (19.8 m) in places—but frozen under a hardened rock layer of hot mud lava.

The cities were all but forgotten over the years, although the texts of some letters, recording the moment of the eruption, had lasted through the centuries. Probably influenced by a new interest in these letters and some allusions to Pompeii and Herculaneum in ancient art, memory of the towns surfaced in sixteenth-century Italian maps of the region. Count Muzio Tittavilla turned up pieces of ruined buildings and wall paintings while irrigating his land. He notified the architect Domenico Fontana of the finds, but no more investigation was made. Ancient historians, however, were intrigued by the finds and resurrected an interest in Pompeii and Herculaneum in the early seventeenth century.

In 1709 workmen clearing a well in the region encountered a wall of the ancient theater of Herculaneum. Prince D'Elboeuf of Austria, who had occupied land near the site, heard of the discovery, bought the land, and hired the diggers to continue, taking the marble decorations of the stage and many statures from the theater. More looting over the years followed, until King Charles III of Spain was alerted to a tablet discovered in the region. It read *Theatrum Herculanensi,* leaving no doubt that the ancient city of Herculaneum lay beneath the site.

Up to this point in history, archeology had been a historian's science. Typically, an interested antiquarian would discover a relic, then guess at an explanation for its use and existence. These guesses had formed myths and biblical tales, fueled philosophies, and the led to the creation of deities. The method was more than inexact as a recreation of history—it was frequently inaccurate and unreliable. In addition, most people drawn to excavations in the sixteenth century were seeking treasure, not historical reenactments. The excavation of Pompeii and Herculaneum, however, with the towns' perfect preservation, was the first to systematically record and uncover the treasure of history that lay beneath the volcanic rock and ash.

Impact

Vases, pieces of buildings, parts of statues, and fragments of artistic paintings were frequently discovered on the grounds near and around the site of Pompeii and Herculaneum. These relics were turned up during the plowing of fields, road construction, or digging wells. During the early 1700s Europe was entering a sort of classical revival in terms of its architecture, art, and images, so these finds could bring a hefty price at the right marketplace. Needless to say, any "excavation" in the early 1700s was more of a looting than a scientific endeavor. Several years after Prince D'Elboeuf plundered the land he purchased, King Charles III of Bourbon acquired the site at the skirt of Vesuvius and assigned his royal engineers to undertake a dig. In 1738 the workers discovered Herculaneum, and the immense excavation began.

Up until this point, even with the sporadic looting and treasure hunting, few people knew exactly where Pompeii and Herculaneum were located. This was partly because the eruption had dramatically changed the coastline, so any previous written geography no longer matched the 1738 landscape. The discovery of the theater at Herculaneum placed the city on a map 4.5 miles (7.24 km) southwest of Vesuvius. Now that Herculaneum was located, the search for the site of Pompeii was next. Local citizens were familiar with some remains of an ancient city on a spur of land near the sea and had referred to it casually as *la civita,* or "the city."

Years previous, in the sixteenth century, residents had found a tablet inscribed with *decurio Pompeiis,* but thought it referred to a Roman statesman. In 1748, however, the workers at the Herculaneum site, lured by the rumors that *la civita* was indeed Pompeii, began digging on the spur. Soon they uncovered wall paintings, a skeleton, and coins. The Spanish workers were ecstatic. The digging here was easier, because while Herculaneum was a solid mass of lava rock, Pompeii was a more shallow layered ground of soft soil and porous ash and pumice. It wasn't until 1763, however, when workers discovered an inscription that included the name of the city, that they were absolutely certain their dig site was Pompeii.

The goals during the first excavation were not entirely noble, however. King Charles was

An early photograph of the Tempio de Venere ruins in Pompeii. *(Corbis Coporation. Reproduced with permission.)*

more interested in adding to his personal collection of classical artifacts than analyzing history. Frescoes were removed from walls, statues and valuable artifacts were removed without recording their position in the city, and tunnels were blasted out with gunpowder to move things along more quickly. Temples and houses were ransacked. Foreign dignitaries were brought in as tourists to witness "miraculous" discoveries of treasures that had been removed, then "unearthed" again before the eyes of the impressed visitor.

In 1755, however, a collection of academics saw the growing problem in Herculaneum and Pompeii and formed a group called the Academy of Herculaneum to record in detail the findings at the site. Johann Winckelmann (1717-1768), a German antiquarian, was the first to begin a meticulous record of the objects, placing them in their found locations, in an attempt to understand the events that took down the city and analyze the culture of the ancient lives there. One of the greatest finds in this early, more academic stage of the project was the unearthing of a barracks that served the gladiators of the day. Diggers headed up the side of the barracks and encountered nearly 50 skeletons. The bodies were from all classes of people—evident from their dress and ornaments—who most likely huddled in the barracks to escape the invading

ash and mud. This find was one of the first to offer a snapshot of the panic of the eruption. It became one of hundreds of similar scenes that would define the tragedy of that day in Pompeii and Herculaneum.

The unique volcanic quality of the ground in this region of Italy had preserved many fragile structures, including wooden beams, furniture, clothing, and even food. From 1750 to 1764 the excavation was directed by Karl Weber, a German engineer. Weber diagrammed the architectural plans of the ruins while recording and documenting the precise location of all found artifacts and bodies. Famous sites, such as the Villa of the Papyri, which held a library of papyri, or philosophical writings, were uncovered during Weber's directorship. The sensational findings brought scholars, royalty, and famous writers, poets, and artists to the dig sites. The land surrounding Vesuvius, however, became France's possession in 1798 under Napoleon, and for many years the excavations were, once again, more for bounty than study. Finally, Italy took back her cities in 1860 and hired a man named Giuseppe Fiorelli (1823-1896) as director of excavations.

Fiorelli took the landmark excavating techniques that started in Pompeii and Herculaneum in 1738 and made them more scientific. He required that every piece of architecture be numbered, he organized the site into districts, and he created a method of duplicating the corpses preserved in Herculaneum's impossible lava rock. He filled the contours of hollow spaces with plaster of Paris, then chipped the stone away. The result was the re-creation of a human body as it died. The dramatic tragedy was relived in the images of struggling men, women, children, and animals, all going about their daily routines when the eruption occurred.

Fiorelli's meticulous technique has changed little today, and it built on the basic, although frequently misguided, approach to the first excavations at the cities of Vesuvius. The cities themselves are only two-thirds completely excavated, and new techniques to preserve the exact moment of their destruction are constantly improving.

LOLLY MERRELL

Further Reading

Books

Brilliant, Richard. *Pompeii* A.D. *79, The Treasure of Rediscovery*. New York: Clarkson N. Potter, Inc., 1979.

Grant, Michael. *Cities of Vesuvius*. New York: Macmillan Company, 1971.

Time Life Books. *Pompeii: The Vanished City*. Alexandria, VA: Time-Life Books, 1992.

Internet Sites

Pompeii Forum Project. http://jefferson.village.virginia.edu/pompeii/page-1.html.

The Birth of Alpinism

Overview

Modern mountaineering had its beginnings in the 1700s when humans began to take to the peaks for reasons of scientific discovery and adventure. The names of Michel-Gabriel Paccard (1757-1827), Jacques Balmat (1762-1834), and Horace-Bénédict de Saussure (1740-1799) and their initial climbs up the 15,771-ft (4,807 m) tall Mont Blanc, the highest peak in Europe, are often cited as the starting point for the present-day sport of mountaineering, or alpinism. Paccard and Balmat were the first to the top of the Mont Blanc in 1786. Saussure followed a year later.

Background

The interest in the natural world heightened in the 1700s as humans began to explore more and more of Earth, to find new plants and animals, and to wonder about the science behind their discoveries. One area of particular interest involved the mountains. At the time, few tall mountains had been climbed, not only because the climbs were difficult and dangerous, but because it was unknown whether humans could even survive the extreme conditions suspected at higher altitudes. The numerous legends of evil spirits who lived among the peaks also kept people from exploring the heights.

Human fascination eventually won out over the fears and potential dangers. As early as 1742 scientists began pointing to Mont Blanc as the highest peak in all of western Europe. While that designation brought attention to the mountains, it was not until 1760, when Swiss scientist and mountain explorer Horace Bénédict de Saussure offered a monetary prize to the first person who climbed it, that the name of Mont Blanc gained prominence. Saussure hoped to be the first to reach the pinnacle, but offered the prize with the intention of encouraging others to make the climb as well.

The challenge inspired many to attempt the summit, but no one was able to find a path to the top until Paccard and Balmat did in 1786. The 24-year-old Balmat, described variously as a peasant, geologist, gem cutter and fur trader, came across what he felt was a passable path to the summit in June 1786. His discovery was serendipitous: after partially ascending the mountain, he became lost overnight and nearly died from the cold, but stumbled upon a potentially successful route.

Paccard was a 29-year-old scientist and physician whose scientific interest in the natural world prompted his involvement in mountain exploration. Like Balmat, he was a resident of Chamonix, the town at the base of Mont Blanc. Paccard hired Balmat as a porter, and planned the ascent for August, two months after Balmat had nearly died on the mountainside.

The two men left town with only walking sticks on August 7. Following Balmat's path across snowy slopes and precipitous crevasses,

Mont Blanc. *(Corbis Corporation. Reproduced with permission.)*

they climbed the mountain. At least one observer from the town used a telescope to follow the men, recording their successful attainment of the apex at 6:23 P.M. on August 8. True to his scientific inclination, Paccard took various measurements on the summit before he and Balmat made the return trip. They were back in Chamonix by the following day. Paccard, who climbed the mountain to satisfy scientific curiosity rather than for monetary gain, sent Balmat to Saussure in Geneva to pick up and keep the prize money.

Despite the apparent generosity on the part of Paccard, Balmat spread tales that he did most of the work, reached the summit first, and then had to backtrack and literally drag Paccard on the last leg of the journey to the top. Paccard's reputation was also damaged by a jealous mountain-climber named Marc-Théodore Bourrit and by Alexandre Dumas, author of *The Three Musketeers,* who both published stories about Paccard based on Balmat's self-serving accounts. More than a century after the deaths of Paccard and Balmat historians found documents written by Balmat that gave Paccard due credit, noting that Paccard made the journey under his own power and inferring that Paccard may actually have been the first of the two men to reach the summit of Mont Blanc.

The successful ascent of Paccard and Balmat spurred Saussure to put together a large climbing team and attempt Mont Blanc the following year. When he reached the top, Saussure conducted numerous scientific studies during a four-hour period. A highly respected and wealthy scientist, Saussure garnered considerable public attention for his successful climb and his ensuing published chronicle of the journey.

Impact

As the first to reach the summit of Mont Blanc, Balmat and Paccard's ascent was significant. However, the true impact of their effort was realized through Saussure's climb a year later. Saussure's status as an aristocrat and respected scientist ensured that word would quickly spread about the conquest of Mont Blanc, and mountaineering soon became an exciting new sport.

Alpinism grew in popularity. People began to climb mountains not for scientific reasons but for the adventure of being the first to reach a new summit. Others did not care to be first, but instead sought merely to climb. By the mid-1800s French and Swiss guides were leading mountaineers by established routes to peaks throughout Europe. The sport of mountaineering received a monumental boost in 1865 when the alpine team of English artist Edward Whymper (1840-1911) made the first successful climb of the 14,692-ft (4,478 m) tall Matterhorn on July 14, 1865. After this great ascent in Europe, mountaineers began to look for new challenges on already climbed mountains by seeking more difficult lines of ascent. Saussure, Balmat and Paccard, for instance, had climbed the less severe slope of Mont Blanc's north face, but later climbers began to consider and attempt the steep and often ice-covered east, or "Brenva," face, and the even steeper south face with its 1,500-ft (457 m) tall pillars of rock.

Mountaineers also turned their sights on more exotic lands. From 1895 to 1900 climbers reached the highest summit of the South American Andes, the 22,834-ft (6,960 m) Aconcagua; the top of the 13,766-ft (4,196 m) Grand Teton in the Rocky Mountains; and the heights of Mount St. Elias, an 18,009-ft (5,489 m) mountain that borders Alaska and Canada. An American climbing team in 1913 triumphed over Mount McKinley, which, at 20,320 feet (6,194 m), is the tallest mountain in North American.

Many mountaineers became fixated with the soaring heights of the Himalayan chain in the early- to mid-1900s. In 1933 a Soviet group ascended a 24,590-ft (7,495 m) peak, and three years later an English team pressed to a 25,643-ft (7,816 m) peak in the Himalayas. Climbing continued—and altitudes increased—following World War II. By 1955 mountaineers had reached peaks of more than 28,000 feet (8,534 m). For many, the last remaining challenge was Mount Everest, the 29,028-ft (8,848 m) Himalayan behemoth. A New Zealand team, led by Edmund Hillary (1919-) and Tenzing Norgay (1914-1986), reached its summit on May 29, 1953.

Since its humble beginnings, when Paccard and Balmat struck out with walking sticks to climb the tallest peak in western Europe, mountaineering has gone through major changes. One of the first was the use of veteran guides to lead parties to the summits. Mountaineers also began to develop and use simple climbing aids, such as rope and ice picks. More advanced gear, including anchors, pitons, and specialized footgear, especially crampons, made it possible to climb formerly insurmountable obstacles, even sheer rock faces such as El Capitan in the North American Sierra Nevada mountain chain.

The earliest climbers made their ascent with simple means and for basic purposes: to learn whether humans could survive high altitudes and to gain a greater understanding of the workings of nature. They accomplished both, and along the way also established alpinism as a sport that now draws tens of thousands of enthusiasts every year.

LESLIE A. MERTZ

Further Reading

Books

Ardito, Stefano. *Mont Blanc: Discovery and Conquest of the Giant of the Alps.* Translated by A. B. Milan. Seattle, WA: Mountaineers, 1997.

Engel, C. *A History of Mountaineering in the Alps.* New York: Scribner, 1950.

Raymann, Arthur. *Evolution de l'alpinisme dans les Alpes françaises.* Genève: Slatkine, 1979.

Periodical Articles

Bernstein, R. "The French Celebrate a Summit of Their Own." *The New York Times,* section 1, 10 August 1986, p. 16.

Hyde, W. "The Ascent of Mont Blanc." *National Geographic* (August 1913): 861-942.

Meisler, S. "France Gives a Hero His Due for First Conquest of Europe's Highest Peak." *Los Angeles Times,* part 1, 9 August 1986, p. 6.

John Frere Discovers
Prehistoric Tools in England

Overview

John Frere (1740-1807) was an English landowner with a modest political career and enough of an interest in archaeology to join the London-based Society of Antiquaries. He discovered a group of chipped-flint objects in a brick-earth quarry near Hoxne in 1790, and described them in a June 1797 letter to the Society as "weapons of war, fabricated and used by a people who had not the use of metals." What made the tools remarkable, first to Frere and later to others, was that they lay beneath 12 feet (3.66 m) of undisturbed soil and gravel, below (and thus older than) a sand layer containing shells that appeared to be marine and the bones of a large, apparently extinct mammal. Frere con-

Prehistoric flint tools. *(Corbis Corporation. Reproduced with permission.)*

cluded that the tools and their unknown makers belonged to a time long before humans were thought to have existed. Frere's interpretation of the stone tools challenged current ideas about of the early history of the human race and set the stage for the evolution of a new understanding over the next 60 years.

Background

Frere's letter to the Society of Antiquaries, read before the society in 1797 and published in the 1800 issue of its journal, was hardly an intellectual bolt from the blue. It presented revolutionary interpretations but rooted them in well-established, widely accepted scientific principles.

Frere could, for example, assume that his readers would readily agree that the objects he found were stone tools, shaped by human hands. The decline of Platonic and Aristotelian ideas during the seventeenth century had put an end to the once-popular belief that such objects were formed, where they lay, by the "generative powers" of Earth itself. It made far more sense, within the mechanical view of nature popularized by René Descartes (1596-1650) and Isaac Newton (1642-1727), to conclude that objects that looked like stone axe heads were just that. The human origins of such objects were, by Frere's day, regarded as self-evident. Frere knew that he could display a picture

of one, call it a human artifact, and go on to more complex issues.

Frere could also assume his readers' acceptance of the idea that the oldest layers of sediment in the sequence he described would be at the bottom and the youngest at the top. That premise, too, was an idea from the late seventeenth century that scientists of the late eighteenth century regarded as axiomatic. First formulated by Danish clergyman-scientist Niels Stensen (1638-1686), the "law of superposition" was originally used as a tool for understanding the relative ages of rock formations. By Frere's time, it was also being used by paleontologists to determine the relative ages of fossils. Frere could, once again, allow his chain of reasoning to remain implicit, knowing that his audience would understand it. He could simply assert, without further explanation, that the tools must be older than the shells and bones in the bed of sand above them.

Frere could, finally, assume his readers' belief in an old Earth that had undergone both geological and biological changes since its origin. European scholars had, as late as the late seventeenth century, maintained that the human race was 6,000 years old (a figure deduced from Old Testament genealogies) and Earth (based on a literal reading of Genesis) only days older. This view of a young Earth created in essentially its modern form slowly crumbled, however, over the course of the eighteenth century. Geological and paleontological evidence for a long, eventful Earth history accumulated steadily, and liberal interpreters of scripture suggested that Genesis should be read as poetic metaphor rather than detailed reportage. The consensus that emerged by Frere's day gave Earth itself a long history in which its flora, fauna, and landscape had changed radically and perhaps repeatedly. It continued, however, to assign the human race an age of about 6,000 years. It was this durable belief that, Frere believed, the discoveries at Hoxne called into question.

Impact

Frere's letter sank without a trace in the sea of turn-of-the-century archaeological literature, generating little excitement either at its 1797 reading or its 1800 publication. The letter's long-term conceptual impact, on the other hand, was enormous. It introduced revolutionary ideas that were elaborated and reinforced by others over the next six decades until, in the early 1860s, they became the foundations of the new field of prehistoric archaeology.

Frere's letter made two substantive claims. The first, that the tools had been made by "a people who had not the use of metals," reflected the general consensus of eighteenth-century archaeologists that nobody capable of working metals would make tools of stone. The conceptual revolution that Frere set in motion lay in the second claim: that the tools and their makers belonged to "a very remote period indeed, even beyond that of the present world." Frere's readers would have understood "beyond ... the present world" to mean "before Earth's flora, fauna, and landscape looked the way they do now." That claim challenged the established archaeological ideas of the time in three important ways: methodological, chronological, and religious.

Eighteenth-century archaeology centered around the collection of beautiful, striking objects and the interpretation of written inscriptions. Archaeologists tended, as a result, to pay far closer attention to the remains of advanced urban civilizations (the Greeks, the Romans, and, beginning at the end of the eighteenth century, the Egyptians) than to those of less sophisticated "barbarian" societies. This preference was partly aesthetic and partly practical. Urban civilizations produced more interesting, more durable, and more attractive artifacts, but the fact that they were nearly always literate meant

that a wealth of information about their politics, economic trends, and religious beliefs could be found in their writings. Understanding anything about a society incapable of recording its thoughts in writing seemed, by comparison, a hopeless and unrewarding task.

Frere's analysis of the Hoxne site suggested a radically different approach to archaeology. Taking the illiteracy of the toolmakers as a given, he inferred what he could from the tools themselves and from their geological and paleontological context. The information he gleaned by these methods—a rough sense of the toolmakers' relative age and a few glimpses of their culture—were trivial compared to what a classicist could wring from a single Roman inscription. By doing it, however, Frere showed that archaeology could ally itself with geology and paleontology as well as the old classical disciplines of Greek, Latin, and ancient history. That new alliance would, in time, become a defining feature of prehistoric archaeology.

Eighteenth-century archaeologists' focus on literate urban civilizations meant that their study of Western Europe effectively began with the Romans. They acknowledged the existence of pre-Roman peoples but paid little attention to them. In Britain archaeologists' understanding of the pre-Roman "Celtic Period" was a patchwork quilt of disconnected facts: snatches of description from Roman writers' accounts of conquest, catalogs of burial mounds and artifacts, and speculations on mysterious stone structures. Information as basic as the extent and internal chronology of the period remained a mystery, and discussions of the Ancient Britons' lives and culture often owed as much to fantasy as to fact. Few archaeologists found this level of uncertainty troubling. Absent evidence to the contrary, most assumed that the Celtic Period had been a brief, unimportant prelude to the culturally diverse Roman, Saxon, and Norman eras that followed it.

Frere's conclusions about the Hoxne site suggested a very different picture. The stone tools he described belonged to a people far less advanced than those who fell before the Roman legions. Frere had, moreover, placed the tools and toolmakers from Hoxne much further back in time than the Celtic Period had ever been thought to extend. Frere, in his brief letter, implied that the pre-Roman history of Britain ("prehistory," a more concise and versatile term, would not be coined until 1851) was longer and more culturally diverse than his colleagues had supposed. This revised chronology encouraged—even demanded—closer and more rigorous study of Britain's pre-Roman inhabitants.

The religious implications of Frere's arguments were more abstract than the chronological and methodological ones but carried more cultural freight. The belief that Earth's flora, fauna, and landscape were created specifically for humans had a long history and a central place in Judeo-Christian thought. It reinforced ideas about human uniqueness, human dominion over nature, and the providential nature of God's design of the natural world. Its power was such that it was reinterpreted, rather than discarded, as scientists' ideas about Earth history changed. Seventeenth-century scientists argued that Earth was only days older than its human inhabitants because, created for human use, it served no purpose standing empty. Eighteenth-century scientists, who saw the human era as the last brief segment of Earth's long history, argued that the first humans did not appear until Earth had taken on its modern, "finished" form and was ready to receive them. Frere's placement of his toolmakers on an Earth not yet in its modern form (and so "unfinished") demanded, if taken seriously, that the venerable old idea be rethought once again.

John Frere did not single-handedly overturn the prevailing belief that humans were little more than 6,000 years old. Nor did he single-handedly create the discipline of prehistoric archaeology. He set both processes in motion, however, by approaching familiar data from a novel perspective. By doing so, he established a line of thought and investigation that reached a climax 60 years later. The belief that humans had lived "beyond ... the present world" was established beyond reasonable scientific doubt in 1859, and the new field of prehistoric archaeology came into its own by 1865. Both had their roots in Frere's willingness, in 1797, to reexamine old assumptions.

A. BOWDOIN VAN RIPER

Further Reading

Books

Bahn, Paul G., ed. *The Cambridge Illustrated History of Archaeology*. Cambridge: Cambridge University Press, 1996.

Bowler, Peter J. *The Norton History of the Environmental Sciences*. New York: Norton, 1993.

Grayson, Donald K. *The Establishment of Human Antiquity*. New York: Academic Press, 1983.

Heizer, Robert F., ed. *Man's Discovery of His Past: Literary Landmarks in Archaeology*. Englewood Cliffs, NJ: Prentice-Hall, 1962.

Piggott, Stuart. *Ancient Britons and the Antiquarian Imagination: Ideas from the Renaissance to the Regency*. London: Thames and Hudson, 1989.

Van Riper, A. Bowdoin. *Men Among the Mammoths: Victorian Science and the Discovery of Human Prehistory*. Chicago: University of Chicago Press, 1993.

Periodical Articles

Frere, John. "An Account of Flint Weapons Discovered at Hoxne, in Suffolk." *Archaeologia 13* (1800): 204-205 [reprinted in Grayson (1983), 55-56, and Heizer (1962), 70-71].

John Byron's Record-Setting Circumnavigation on the *Dolphin*

Overview

In 1764, John Byron (1723-1786) left England in command of a two-ship expedition to circle the globe. He returned slightly less than two years later, having set a record for the fastest circumnavigation to date, and the first commander to circle the globe without losing a ship. While Byron did not accomplish some of his goals of locating new territories for Britain (with the exception of laying claim to the Falkland Islands), he did help to set a standard for both speed and safety on such an epic voyage.

Background

Contrary to popular legend, the world had been known to be spherical since the time of the ancient Greeks. Christopher Columbus's (1451-1506) achievement lay not in "proving" the world to be round, but in being the first to try to exploit this fact in attempting a faster route to the Far East. In fact, Columbus erred greatly in his calculations and, were it not for the unexpected presence of the Americas, his crew would almost certainly have perished trying to reach the Orient.

In spite of the long-standing knowledge of the Earth's shape, nobody attempted to circumnavigate it until Ferdinand Magellan's (1480?-1521) lieutenant, Juan Sebastian d'Elcano (1476?-1526), returned in command of Magellan's remaining ship in 1522 (Magellan, of course, was killed during the voyage and failed to complete the journey). Following this feat, the next circumnavigation was accomplished by Sir Francis Drake (1540?-1596) almost 60 years later, and others followed.

Each time a circumnavigation was completed, the sponsoring nation gained territories, prestige, and possible military or trade advantages over its rivals. In the nearly continual state of war and conflict that characterized Europe in the sixteenth, seventeenth, and eighteenth centuries, such advantages were seen as vital to the national interests of the great powers of Britain, France, Holland, and Spain.

Of increasing importance during these centuries were the trade routes to the Orient. Spices, silk, porcelain, tea, and other commodities were the foundation of great wealth, and all nations were avidly seeking gold, silver, and gemstones to help fill their treasuries. Trade routes helped ships bring these good to the mother countries, and military outposts were needed as bases to help protect the cargo vessels, which made rich prizes for any attacking nation. Because of this, many voyages of exploration and discovery had multiple goals: to locate new trading partners, to gather information about wind, weather, and ocean conditions that could help ships to reach their destinations more quickly and safely, and to scout for likely military bases at which ships could be stationed to help escort merchant vessels safely to and from their destinations. It was on such a mission that the *Dolphin* departed England in July 1764.

The expedition, under the command of John Byron (grandfather to the poet), was charged with exploring the Pacific Ocean in order to help Britain gain and maintain an advantage over her Continental rivals. On his way across the Atlantic, Byron laid claim to the Falkland Islands for Britain, apparently unaware that they had already been claimed by the Frenchman Louis-Antoine de Bougainville (1729-1811). Rounding the tip of South America, he continued into the Pacific, but here he apparently either decided to disregard his orders or he was simply unlucky. In any event, he managed to cross the Pacific in the latitude of the trade winds, and did so without discovering any lands

of interest. However, his long, uninterrupted stretch of steady sailing gave him the fastest crossing of the Pacific to date. He then turned his attention to returning home by the fastest route, reaching England again only 22 months after his departure. A few months later, *Dolphin* was outfitted for another circumnavigation, becoming the first vessel to complete this arduous journey twice.

Impact

Previous circumnavigations had all taken longer, and had lost ships and men. In fact, both Magellan and Drake set forth on their circumnavigations only accidentally, when pursuit by enemy vessels made any other return home impractical or unsafe. By comparison, Byron set forth with the express intent of circling the globe. However, in addition to his record-setting pace, Byron's expedition was the first to complete this voyage without losing a ship. By comparison, Magellan left with five ships and 265 men and returned with 18 men manning a single ship. Perhaps the single worst trip to have been on, however, would have been the ill-fated expedition led by George Anson (1697-1762) between 1739 and 1744. Anson left port with a total of 1,939 men, only 500 of whom survived the five-year voyage.

Although Byron's voyage did not result in discovery of many new lands for England and was not nearly so successful as the later voyages of James Cook (1728-1779), it was not without accomplishment. In particular, England gained some national pride from hosting such a rapid and safe voyage of this magnitude, and the Royal Navy gained some degree of acclaim through this accomplishment. In addition, by sailing with the trade winds so rapidly across the largest ocean on Earth, Byron helped to show the value of these routes. The trade winds had been known for some time; what was not known was that they were so constant over this vast expanse of water. Finally, Byron set the standard for future circumnavigations.

In fact, Byron's mission was inspired by a French writer, who urged the French government to send expeditions to claim lands for French bases, to help protect French merchant ships, and to trade in the Orient. However, the French government failed to act on this advice, leaving these seas to the English. Although Byron discovered nothing of importance in his voyage, he gave Britain valuable insights into the rigors of long-duration sea journeys—insights that were to prove valuable in future expeditions.

In addition, by showing that a British ship could successfully circle the globe so rapidly, the navy and the nation gained a sense of pride in their accomplishment. The added fact that Byron did not lose a single vessel was even better. In a sense, after the *Dolphin* and *Tamar* returned to Britain, there was the realization that voyages of such length, while still hazardous, could now be treated as "just" another voyage, albeit one of great length. This helped to mark the end of the romance of seaborne exploration and the start of the practical business of far-flung commercial and naval enterprises.

Byron also helped increase knowledge of the trade winds in the Pacific. Trade winds, the steady winds that typically blow from east to west in low latitudes (i.e., near the equator), were the global superhighways in the age of sail. By descending to the latitudes in which they blew, a ship could sail for weeks across the Pacific without adjusting its sails or rudder significantly. With steady winds and a known distance to sail, voyages began to become more predictable, and sailing became somewhat more scientific.

This increase in scientific knowledge of the Pacific and its winds, in turn, helped the British to better understand the oceans upon which their nation depended for its commercial wealth and military strength. As with merchant vessels, it was a great help to a naval captain to know how long a voyage might last, and how to cut that time to the bare minimum necessary. With this knowledge, he could better plan his supply of food and water, for example, to ensure that his men did not starve while, at the same time, taking the greatest amount of cargo, guns, or passengers possible. Although much of this information became superfluous once the age of sail had passed, it was vitally important at the time, and failure to understand wind patterns or crossing times could (and did) result in the loss of ships, lives, and cargo on a regular basis.

The last legacy of Byron's journey was the standard he set for both speed and safety while circling the world. Future such expeditions would be judged by their speed and the number of ships lost as well as by the amount of information returned by the ship at voyage's end.

All in all, Byron's voyage was not successful in the manner planned when he departed. He claimed only minor territories for Britain, and discovered no new lands of any significance to add to Britain's overseas possessions. However, he did help to break new ground in finding a fast and relatively safe route across the expanse of the Pacific Ocean by making full use of the trade winds, and he helped set a high standard for future voyages.

P. ANDREW KARAM

Further Reading

Books

Wilson, Derek M. *The Circumnavigators*. Evans & Company, 1989.

The Origin of Human Flight

Overview

For centuries humans dreamed of flying. The ancient Greek myth of Icarus, killed attempting to fly when the wax on his artificial wings melted, was an early expression of this desire. This myth also reflected the realization that human flight would be difficult, perhaps impossible. Yet the dream persisted. Writers who speculated about future societies often included controlled human flight in their utopian fiction. Cyrano de Bergerac (1619-1655) described interplanetary journeys in several of his stories, while Jonathan Swift (1667-1745) made a huge flying island the focus of the third voyage in *Gulliver's Travels*.

Medieval and renaissance scholars Roger Bacon (1220?-1292) and Leonardo da Vinci (1452-1519) speculated on the possibility of human flight. Da Vinci even designed a heavier-than-air machine and a parachute. But human flight remained unachieved until 1783.

Background

By the seventeenth century, scientists realized the earth was surrounded by heavy gas. They assumed that this ocean of air could be navigated, just as ships sailed across water. Various devices were invented to achieve flight, such as artificial wings and crafts relying on muscle power. The repeated failures of these devices led to investigations of lighter-than-air balloons. This decision now seems mistaken, since heavier-than-air craft are clearly superior to lighter-than-air vessels in terms of carrying capacity and control. But the paucity of aerodynamic knowledge and the lack of an adequate engine gave eighteenth-century inventors no choice in the matter.

Three developments in the eighteenth century made human flight possible. The first was when the English scientist Henry Cavendish (1731-1810) discovered hydrogen in 1766. He found that iron placed in a dilute solution of sulfuric acid produced a gas fourteen times lighter than air. Because it was so combustible, it was called "inflammable air." One problem of using hydrogen in terms of flight was that it easily seeped through cloth and so apparently could not be contained in lightweight materials. It was also expensive to make on a large scale and it was very dangerous; a spark of static electricity could set it on fire.

The second eighteenth-century condition that led to human flight was not scientific, but rather social in nature. Middle-class men began taking a great interest in science. In part, this was a result of the growing application of science to the industries owned by the middle classes. Interest in science was also an indication of higher social status. Pursuing scientific experiments, even as a dilettante, was evidence that an individual possessed the leisure time and income necessary for such pursuits. It was no accident that the men most responsible for the early development of human flight were middle-class scientific amateurs.

A final factor leading to successful human flight was the influence of the enlightenment, an eighteenth-century intellectual movement based on the belief that if reason and scientific inquiry were applied to physical and social conditions, the laws governing those conditions could be discovered. Following these laws would lead to unlimited progress. This faith in progress was very strong in France, especially after the publication of Denis Diderot's (1713-1784) *Encyclopédie*, a summary of the scientific and technological advances of the century. It created an optimistic belief that the riddle of human flight could be solved.

Impact

The first major step toward human flight occurred on June 4, 1783, when Joseph (1740-1810) and Etienne (1745-1799) Montgolfier made a public scientific demonstration in the main square of Annonay, a small town in southern France near Lyon. They inflated a 35-foot (10.7-m) paper-covered cloth globe, which floated upward about 6,000 feet (1,829 m) and landed a mile and a half (2.4 km) away. Although it carried no passengers, the Montgolfiers had succeeded in achieving the first large-scale balloon voyage in history, bringing human flight much closer to reality.

The Montgolfier brothers were middle-class businessmen, not scientists. Their father was a wealthy paper manufacturer who supported his sons' scientific dabblings. They became interested in the problems of flight in 1782. Their experiments with silk and paper models led them to conclude that hydrogen could not be contained long enough to permit flight. But they did discover that heated air became sufficiently rarefied (less dense) to lift a balloon and it did not diffuse through its cover. They erroneously believed that the smoke of the fire, not the heated air, provided the lifting power and spent weeks experimenting with different types of fuel to get the "ideal smoke," eventually settling on a mixture of wet straw and wool. So two scientific amateurs had inadvertently provided a solution to the problem of lifting a heavy craft off the ground for an extended period of time.

The Montgolfiers immediately notified the scientific establishment of their success and Etienne went to Paris seeking a grant to cover their expenses. The family also hoped their fame would result in lucrative government contracts for their paper business. News of their success galvanized the capital. A popular science lecturer, Jacques-Alexandre-César Charles (1746-1823) decided to compete with the Montgolfiers in an effort to be the first to achieve human flight. Assisted by two clever instrument makers, the brothers A.J. and M.N. Robert, Charles set out to make a workable hydrogen balloon.

Charles and the Roberts succeeded in making enough hydrogen for an unmanned flight from Paris on August 27, 1783, although they needed half a ton of iron filings, a quarter of a ton of acid, and several days to do so. They also managed to make a leak-proof sack of taffeta covered with a rubberized paint to hold the gas. Charles's other contributions to the development of hydrogen balloons are equally important: the valve line to release gas allowing immediate descent (in a hot air balloon, descent occurs only when the air cools); the "appendix"

First human flight in a Montgolfier balloon. *(Corbis Corporation. Reproduced with permission.)*

(an opening through which gas expanding at higher altitudes can automatically escape before bursting the balloon); the use of bags of sand for ballast and of a grapnel (anchor) to aid in landing; the "nacelle" or gondola (a wicker basket suspended beneath the balloon).

A few weeks later, Etienne Montgolfier launched a hot air balloon from Versailles with Louis XVI and his court in attendance. It carried aloft a rooster, duck, and sheep and all three survived, making human flight the next logical step. There were now two successful techniques of ascending, one physical and one chemical: a balloon utilizing the Montgolfiers' method of relying on heated air would be called a *montgolfière*, while a balloon using a gas different from air would be labeled a *charlière*. In 1783 the Montgolfiers' method was the simplest and cheapest; Etienne only needed 90 pounds (40.8 kg) of straw and wool in the Versailles flight to lift a load of 1,000 pounds (454 kg). Its disadvantages for human flight were that a *montgolfière* could not rise as high as a hydrogen balloon, its fire needed constant tending, and it could not descend as quickly or safely.

Etienne won the race for the first human flight in an untethered, freed balloon. On November 21, 1783, a *montgolfière* carrying two men ascended from the Bois de Boulogne and flew over Paris for about 20 minutes. The two passengers were Jean François Pilâtre de Rozier (1757-1785) and François Laurent, marquis d'Arlandes (1742-1809). They were the first aeronauts, and the first humans to ride freely in the air. It was d'Arlandes's only ascent; Etienne Montgolfier himself never ascended in an untethered balloon. On December 1, 1783, Charles and the older Robert brother ascended from the Tuileries in Paris in a new *charlière*. By now a balloon mania had seized Paris and half the population of the city turned out for the flight. Charles and Robert flew for over 2 hours and covered 25 miles (40 km). After landing, Charles took off again in history's first solo flight. He reached 10,000 feet (3,048 m) when the cold drove him back to earth. Although this was Charles's only flight, he proved that there was a height barrier to balloon flights. He also showed that hydrogen balloons were far superior in range (both distance and height) to hot air balloons. In addition, landings usually destroyed *montgolfières*, while the smaller and more durable *charlières* could be used repeatedly. As soon as cheaper and quicker means to manufacture hydrogen were developed, the *montgolfières* became quite scarce.

Meanwhile, Joseph Montgolfier was in Lyon constructing a monster balloon to carry multiple passengers. On January 19, 1784, he and six other men made a successful ascent, his first and last. On June 4, 1784, a Madame Thible made the first free flight by a woman when she and a male companion ascended in a *montgolfière* from Lyon. Another aeronautic milestone was reached on January 7, 1785, when a Frenchman named Jean Pierre Blanchard (1753-1809) flew from England to France in history's first overseas flight. Pilâtre de Rozier tried to cross the English Channel in the other direction on June 15, 1785, but his balloon caught fire and he was killed. Thus, the first man to fly was also the first to be killed in an aerial accident.

After de Rozier's death, advances in ballooning slowed considerably. Although observers from a tethered balloon helped the French win the battle of Fleurus in 1794, it was not until the American Civil War that balloon observations took on military significance. Nor were any scientific advances achieved through flight in the next half-century. The inability to control the flight direction of a balloon reduced it to purely entertainment uses throughout the early nineteenth century. No fair or civic celebration was complete without a balloon ascent, with acrobats performing dangerous stunts high above the sensation-seeking public; a parachute descent usually ended the show. After the first night ascent in Paris on June 18, 1786, fireworks set off from balloons became common (and risky because of the hydrogen). Samuel Johnson's (1709-1784) remark about balloons remained accurate: "In amusement, mere amusement, I am afraid it must end, for I do not find that its course can be directed so that it shall serve any useful purpose in communication." Even famous nineteenth-century aeronauts such as Gaston Tissandier (1843-1899) realized the future of flight lay in the development of heavier-than-air machines.

ROBERT HENDRICK

Further Reading

Books

Becker, Beril. *Dreams and Realities of the Conquest of the Skies*. New York: Atheneum, 1967.

Gillispie, Charles Coulston. *The Montgolfier Brothers and the Invention of Aviation, 1783-1784*. Princeton: Princeton University Press, 1983.

Gough, J.B. "Jacques-Alexandre-César Charles." In *Dictionary of Scientific Biography*, Vol. III, edited by C.C. Gillispie. New York: Scribner's, 1971.

Rolt, L.T.C. *The Aeronauts: A History of Ballooning, 1783-1903*. New York: Walker, 1966.

Smeaton, W.A. "Etienne Jacques de and Michel Joseph de Montgolfier." In *Dictionary of Scientific Biography*., Vol. IX, edited by C.C. Gillispie. New York: Scribner's, 1974: 492-494.

The Rosetta Stone Is Discovered by Napoleonic Soldiers

Overview

The Rosetta Stone was discovered in 1799 by a member of Napoleon's Egyptian expeditionary force. The Stone is a stela fragment carved during the reign of Ptolemy V (205-180 B.C.) and is inscribed in two different languages with three different scripts—hieroglyphic, demotic, and Greek. The importance of this artifact as a potential key for deciphering hieroglyphics was immediately recognized and then confirmed when a translation of the Greek established that the other two scripts contained the same message. News of the discovery created a sensation, spawning renewed efforts at decipherment that culminated in the stunning success of Jean-François Champollion (1790-1832).

Background

Discovery of the Rosetta Stone removed one of the three principal impediments to progress in deciphering hieroglyphics—lack of a bilingual inscription. Others were the nonexistence of a large corpus of accurately copied inscriptions and the false belief that hieroglyphics were essentially symbolic. The orthography of hieroglyphics was partly responsible for the latter view, its pictorial nature helping to disguise the fact that it encodes the spoken language of ancient Egypt.

The first hieroglyphic inscriptions date to the beginnings of Pharaonic Egypt (c. 3300 B.C.). The script evolved from about 700 characters during the Old Kingdom (2705-2250 B.C.) to over 6000 during the Ptolemaic period (332-30 B.C.). The gradual infusion of Hellenism throughout the Ptolemaic period coupled with the Roman conquest (30 B.C.) and subsequent infusion of Christianity slowly eroded Pharaonic culture. Writing of the native language in hieroglyphic and hieractic (a simpler, cursive form of hieroglyphics) was gradually replaced until by 250 B.C. only demotic (a popular form of hieroglyphics) remained in general use. This lasted until the fifth century A.D., after which Coptic (a script composed of the Greek alphabet supplemented by seven demotic characters) enjoyed a period of ascendancy before the Arab conquest in 641. The last hieroglyphs were carved in 394, but by this time priestly secrecy and a heavy veneer of mysticism served to obscure the script's true meaning.

The European Renaissance witnessed a reawakening of interest in Egypt as classical and Greco-Roman works were rediscovered. Authors such as Herodotus, Strabo, and Plutarch had written of the esoteric nature of Egyptian knowledge, and Neoplatonists such as Plotinus, Porphyry, and Iamblichus developed these themes while focusing on the symbolic nature of hieroglyphics, maintaining that they recorded pure moral and philosophical ideas unfiltered by language. Greatly influenced by such works, Renaissance efforts at decipherment devolved into attempts to explain the mystical significance of hieroglyphics rather than reading them.

From the late seventeenth century on the number of scholars visiting Egypt gradually increased. The antiquities they collected and their writings greatly facilitated the study of ancient Egypt and hieroglyphics. Modern Egyptology, though, begins with Napoleon Bonaparte's (1769-1821) Egyptian Campaign (1798-1801). In addition to the expedition's political and military objectives, Napoleon wished to recover Egypt's lost wisdom. Consequently, over 150 scientists, scholars, and artists disembarked with the invasion fleet. After his victory at the battle of the Pyramids (1798), Napoleon established the Institut d'Egypte in Cairo, from where the French savants were to explore and report on all aspects of Egyptian culture. The culmination of their work was published in the monumental *Description de l'Egypte* (1809-22).

The expedition's most memorable discovery was made by the army in July 1799. During construction of Fort St. Julien at port el-Rashid (ancient Rosetta) in the Nile Delta, engineering officer Pierre-François Xavier Bouchard (1772-1832) uncovered an irregularly shaped, dark-gray slab that he immediately identified as a stela fragment inscribed with three different scripts. When translation of the Greek established that the other two scripts recorded the same message, news of what might be the key to unraveling the mystery of hieroglyphics quickly spread.

Impact

Many copies of the inscription and casts of the fragment were quickly made and distributed. Copies reached Paris in the fall of 1800, but when the French surrendered to the British in Egypt in 1801, the Rosetta Stone passed into British hands and was officially donated to the British Museum in June 1802.

That same year important contributions to unlocking the secret of hieroglyphics were made by the French orientalist Silvestre de Sacy (1758-1838) and the Swedish diplomat Johan David Åkerblad (1763-1819). Sacy decided to ignore the Rosetta Stone's hieroglyphic inscription and concentrate on the more complete demotic. He succeeded in locating some of the proper names, but his analysis of the individual signs and transliterations were incorrect. Åkerblad had more success. In an open letter to Sacy, he showed that proper names and foreign words were written phonetically in demotic, although he incorrectly assumed demotic to be primarily alphabetic.

Coptic scholarship also proved of fundamental importance in deciphering hieroglyphics. Pietro della Valle (1586-1652) and others had uncovered many Coptic manuscripts during the seventeenth century. Based on his study of them, Athanasius Kircher (1602-1680) had claimed Coptic was the language of ancient Egypt. In his letter to Sacy, Åkerblad identified important aspects of the Rosetta Stone's demotic and correlated them with their Coptic equivalents. In 1808 Sacy refined Kircher's claim by suggesting Coptic grammar preserved something of the grammar of hieroglyphics, a fact later exploited by Champollion. None of this, however, challenged the false belief that hieroglyphics were essentially symbolic.

The next significant step toward decipherment was taken by Sacy in 1811. In ideographic languages such as Chinese, there is a difficulty rendering proper names. One of Sacy's students first introduced the idea that foreign words and names were written phonetically in Chinese using standard characters appropriately marked to distinguish phonetic usage. Additionally, it previously had been suggested by Abbé Jean-Jacques Barthélemy (1716-1795) that the oval-shaped cartouches—essentially ovals with hieroglyphs inside them—in hieroglyphic texts might contain the names of kings or gods (1761). From these facts Sacy conjectured that cartouches encircled hieroglyphs that were employed phonetically. Though incorrect, Sacy's suggestion provided the framework under which all work on hieroglyphs proceeded over the next ten years and played a central role in Champollion's first solution.

Further steps towards decipherment were made by Thomas Young (1773-1829), who is best known for his work on the eye's physiology and wave theory of light. By 1816 he had proposed an alphabet that allowed him to decipher Ptolemy's name inside the single cartouche of the Rosetta Stone's hieroglyphic section. Unable to account for all the hieroglyphs in this cartouche, he concluded that only foreign names were written phonetically, with titles and epithets appearing symbolically. Young was also the first to express doubt over the purely ideographic nature of hieroglyphics, noting that with fewer than a thousand hieroglyphic characters typically in use there could be no simple one-to-one correspondence between symbols and the ideas or objects represented. He further realized that demotic was not primarily alphabetic—it contained ideograms as well as phonetic symbols. Despite these insights, Young continued to maintain hieroglyphics were primarily symbolic, with phonetic uses being ancillary.

The key to deciphering hieroglyphics was at last provided by the brilliant French linguist Jean-François Champollion. Highly precocious, his fascination with Egypt began at an early age when he heard stories of the Rosetta Stone's discovery. In 1806, at age 16, Champollion presented a paper before the Société des Sciences et Arts de Grenoble arguing that Coptic was the language of Ancient Egypt. He then went to Paris in 1807 to study Arabic with Sacy and to acquire fuller knowledge of other languages considered relevant for solving the puzzle of hieroglyphics.

Champollion's work proceeded slowly due to insufficient and inaccurately copied inscriptions. The need for an expanded corpus of texts and collateral evidence provided by miscellaneous antiquities was gradually met as volumes of the Description de l'Egypte appeared and new manuscripts and more accurate copies of monument inscriptions were brought back from Egypt. By 1821 Champollion's analysis of this material allowed him to firmly establish the distinction between hieroglyphics, hieratic, and demotic as well as compiling tables of their equivalent signs. He also demonstrated that hieratic and demotic were primarily ideographic, not alphabetic, and subsequently realized that the proper names and foreign words written phonetically in the Rosetta Stone's demotic could be used to help decipher phonetic hieroglyphs. On Friday, September 27, 1822, before the Académie des Incriptions et Belles Lettres in Paris, he announced his success in constructing a phonetic alphabet that allowed him to accurately read cartouches.

The Rosetta Stone. *(Corbis Corporation. Reproduced with permission.)*

As important as this was, especially for dating Ptolemaic and Roman ruins, the decisive step in decipherment was not taken until early the next year when Champollion realized pure hieroglyphics—those outside the cartouches—were primarily phonetic, not symbolic. This insight was motivated by a number of factors. First, his analysis of the Rosetta Stone revealed that just under 500 words of the Greek text were represented by 1,419 hieroglyphic sign-groups, indicating hieroglyphics could not be purely ideographic. Furthermore, the 1,419 sign-groups were constructed from just 66 distinct characters, suggesting the possibility of a phonetic script. Champollion also noted that the few dozen sign-groups from cartouches he had already deciphered constituted over two-thirds of all hieroglyphic inscriptions, providing more evidence that hieroglyphics were primarily phonetic.

Champollion's suspicions were confirmed when he successfully applied his phonetic alphabet to the Rosetta Stone inscription. However, understanding what he was deciphering was another matter, one that turned quite decisively on his knowledge of Coptic. As he proceeded he was able to identify recognizable Coptic words and elements of Coptic grammar. By comparison with Coptic he was then able to reconstruct the grammatical forms of hieroglyphics. After expounding the principles of hieroglyphics in *Précis du système hiéroglyphique* (1824), Champollion traveled to Turin's Museo

Egizio, which at the time possessed the world's most extensive collection of Egyptian antiquities, and Egypt to test his method. Further research has confirmed, refined, and extended Champollion's work, which serves as the cornerstone of modern Egyptology.

STEPHEN D. NORTON

Further Reading

Davies, W. V. *Reading the Past: Egyptian Hieroglyphs*. London: British Museum Publications, 1987.

Iverson, Erik. *The Myth of Egypt and Its Hieroglyphs in European Tradition*. Copenhagen: Gad, 1961.

Parkinson, Richard. *Cracking Codes: The Rosetta Stone and Decipherment*. Berkeley, CA: University of California Press, 1999.

Pope, Maurice. *The Story of Decipherment*. Revised ed. New York: Thames and Hudson, 1999.

Siliotti, Alberto. *The Discovery of Ancient Egypt*. Boston, MA: White Star Publishers, 1998.

Biographical Sketches

Sir Joseph Banks
1743-1820
English Naturalist

In 1768 Joseph Banks took part in the first expedition of Captain James Cook (1728-1779), from which he returned to England with a thousand new species of plants, a thousand more of birds and fish, and "insects innumerable." While still a young man, however, Banks ended his days as an explorer to become president of the Royal Society, and later a sponsor of expeditions to Africa, Australia, the Pacific, and the Arctic.

Born in London on February 13, 1743, Banks came from an exceedingly wealthy family with an enormous estate in Lincolnshire. He undertook his early education at Eton, and it was there, at age 14, that Banks, as he later recalled, discovered his calling in life. One summer evening, he was walking near the school when he suddenly became aware of the variety of flowers growing along the lane. Inspired to study nature, thereafter he learned as much as he could about botany. Upon arriving at Oxford University and discovering that it had no professor of botany, the young heir (his father had died when he was young) simply hired a

botany professor from Oxford's rival, Cambridge, and brought the instructor back to Oxford to teach him.

After graduating from Oxford in 1763 and gaining his full inheritance the following year when he turned 21, Banks became a fellow of the Royal Society in 1766. He went on his first voyage later that year, aboard the *Niger* to collect plant specimens in Newfoundland and Labrador. Returning to England, he soon learned about the opportunity of a lifetime: a chance to sail with Captain Cook as onboard naturalist. Banks would have to pay his own expenses, and those of his staff, but that was no problem for him, so in 1768 he left England with Cook aboard the *Endeavor*.

The crew had many adventures and misadventures along the way, and in Tahiti Banks allowed himself to be tattooed—one of the first Westerners to do so. He is also rumored to have engaged in amorous involvements with at least one of the beautiful Polynesian women he met. But he also found time for work, and as the ship sailed from Tahiti to New Zealand and Australia, he collected numerous specimens of animal and plant life. He returned to a hero's welcome, receiving praise from King George III (destined to

Joseph Banks. *(Archive Photos. Reproduced with permission.)*

Inland Parts of Africa" in 1788. Over the next 17 years, the African Association, as it was called, sent a series of failed expeditions to find the source of the Niger River in West Africa. Most notable among the many explorers sent out by the African Association was Mungo Park (1771-1806), who drowned while looking for the elusive source of the river.

In addition to his Africa endeavors, Banks sponsored an 1801 voyage to Australia by Matthew Flinders (1774-1814), and from 1817 became involved in renewed efforts to find a Northwest Passage between the Atlantic and Pacific oceans. At this point in his life, however, he was becoming increasingly infirm, having been confined to a wheelchair since 1804 due to gout. He died on June 19, 1820.

JUDSON KNIGHT

Vitus Jonassen Bering
1681-1741
Danish Explorer

One of the most celebrated endeavors of seventeenth- and eighteenth-century exploration was the search for the Northwest Passage, a route between Europe and Asia via the frozen seas north of Canada. Less famous was the quest for the Northeast Passage, or a means of navigating between the furthest eastern extremities of Asia and the western tip of North America. Perhaps the greatest figure in the search for the Northeast Passage was Danish navigator Vitus Bering. Sent on two expeditions by the Russian czars, he explored Russia's Far East and the offshore islands of Alaska, and proved the existence of a passage between Asia and North America.

Born in Horsens, Denmark, in 1681, Bering grew up around the sea, and as a young man joined the Dutch navy. At that time Holland had a vast international empire, and his work gave the young Bering an opportunity to see the East Indies (modern-day Indonesia). Eventually he joined the Russian navy, then a recent creation of Czar Peter the Great, and took part in the Great Northern War (1700-21) between Russia and Sweden. His bravery so impressed Peter that in 1725 the czar commissioned him to lead an eastern expedition.

Leaving the Russian capital of St. Petersburg, Bering and his crew traveled overland across Siberia, bringing with them the materials for building a boat. They arrived at the Sea of Okhotsk, which separates the Russian mainland

be a lifelong friend of Banks's) and Swedish naturalist Carl Linnaeus (1707-1778).

Ego seems to have gotten the better of Banks for a time, because only his hubris can explain why he failed to join Cook on the his second voyage in 1772. Banks insisted on bringing a staff of 15 people, including two horn players, which would require the building of extra cabins on deck. Cook ordered these cabins torn down because they would make the craft unseaworthy, yet Banks refused to reduce the numbers of his staff. So Cook left without him—a fact Banks rued for the rest of his life—and, except for a voyage to Iceland later that year, Bank's travelling days were over.

The second phase of his career began in 1778, when the 35-year-old Banks was elected president of the Royal Society, perhaps the most honored position in the scientific world at that time. In this capacity, he established a large library of travel books in the British Museum, a library still in existence more than two centuries later. He also influenced exploration in many ways, including his suggestion to Captain William Bligh (1754-1817) that he sail to Tahiti to collect breadfruit—as Bligh later did on his infamous *Bounty* voyage.

Banks was particularly interested in the exploration of Africa, to which end he formed "An Association for Promoting the Discovery of the

Vitus Jonassen Bering. *(The Granger Collection, Ltd. Reproduced with permission.)*

from the Kamchatka Peninsula, in 1727. There they built a boat and sailed across to Kamchatka. They then sledded across the peninsula to its east coast, where they built a second boat, the *Gabriel.*

In the summer of 1728, Bering sailed up Kamchatka's east coast and beyond, almost to the extreme northeastern tip of the Asian continent. To the east he saw a large island, which he named St. Lawrence (now part of Alaska), but he could go no further north due to the ice. Therefore he turned southward, spending the winter of 1728-29 in Kamchatka before ultimately making his way back to St. Petersburg.

At the royal court, Bering persuaded the Czarina Anna to commission a second voyage, and in 1733 he took charge of what was dubbed the "Great Northern Expedition." Due to a number of delays, however, it was only in 1740 that the expedition set sail from the east coast of Siberia. When the earlier expedition arrived at the Sea of Okhotsk, Bering had been 46 years old; now he was nearly 60.

This second expedition was much larger, with several scientists on board, and therefore Bering took two ships: the *St. Paul,* of which he was captain, and the *St. Peter,* with Alexei Ilyich Chirikov (1703-1748), who had sailed on the first voyage, at the helm. They waited out the winter of 1740-41 on the east coast of Kamchat-

ka, at a base they called Petropavlovsk after their two ships. Later a town would spring up on the site and become the largest city on Kamchatka.

The two ships finally sailed on June 5, 1741, and were quickly separated. Chirikov and his crew made it to North America and sent out two reconnaissance boats that never returned. Ravaged by scurvy, they limped back to Petropavlovsk a few months later. Bering and the rest of those aboard the *St. Paul* were less fortunate.

Sailing south and then east, on July 17 he caught sight of the American mainland at Mount St. Elias. Later he sighted many of the Aleutian Islands, but by then Bering and the others were suffering the effects of scurvy. Their ship wrecked on a barren island along the eastern coast of Kamchatka, where they suffered miserably during the winter that followed. Bering was among the casualties, dying on December 8, 1741. In the following August, the few survivors made it back to Petropavlovsk. Today the island where Bering died is known as Bering Island, and the passage between Siberia and Alaska is called the Bering Strait.

JUDSON KNIGHT

William Bligh
1754-1817
English Naval officer

Thanks to the book *Mutiny on the Bounty* (1932) by Charles Nordhoff and James Hall, as well as many motion pictures on the subject, Captain William Bligh remains a symbol of arrogant power. The incident on the H.M.S. *Bounty,* when Bligh's crew mutinied and cast him and his supporters adrift on a boat in the Pacific, was not his last experience on the wrong end of an insurrection. Later, as governor of New South Wales, Australia, Bligh's authoritarian style helped to spark another revolt.

Bligh was born on September 9, 1754, in Plymouth. The son of a customs officer, he began his career at sea early, going away as a cabin boy at age seven. He joined the Royal Navy at 16, and by 22 Bligh had command of a ship, the *Resolution,* during part of the third voyage of Captain James Cook (1728-1779). By the time he returned in 1780—Cook was killed along the way—Bligh was a captain of recognized talents.

During the 1780s, Bligh commanded a number of ships in Britain's merchant marine, and on several voyages his first mate was Fletcher Christian. The latter, an intensely sensitive man with a taste for scholarship rather than seamanship, had been forced by the loss of his family's fortunes to seek a naval career. Though he was inclined to smart at insults from superiors, Christian got along well with Bligh. In 1787 the British government needed someone to command the *Bounty* on a voyage to Tahiti for breadfruit trees, which would be transplanted to the West Indies to provide cheap food for slaves. Sir Joseph Banks (1743-1820) suggested to Bligh that he command the expedition, and Bligh accepted.

The voyage was fraught with disaster, not least because neither the ship nor its crew was suited to their demanding mission. Bligh had at least learned from Cook how to forestall scurvy by supplying the men with lemon juice, and even kept a fiddler aboard to play while the men danced—a much-needed form of exercise. Morale lifted when the crew reached Tahiti, but from Bligh's perspective this was no comfort. Seduced as much by the languid tropical ambience of the place as by the beauty of its women—many of whom cheerfully gave themselves to the sailors—the crew became even less reliable. Bligh struggled through the six weeks in Tahiti, writing down his observations on Tahitian culture before setting sail once again.

On the night of April 28, 1789, a group of mutineers broke into Bligh's cabin, held a cutlass to his throat, and took over the ship. Leading the group was Christian, who had become increasingly dissatisfied with Bligh. The latter had lately taken to criticizing his painfully sensitive first mate, and Christian could no longer stand it. He and his mutineers set Bligh adrift in an open boat with 18 men who chose to accompany him.

The reality of the mutiny was quite different from the Hollywood version. Christian proved an incapable leader, and was later killed by Tahitians. By that time the mutineers had landed on Pitcairn Island near Tahiti, taking with them a number of Tahitian women, as well as a few Tahitian men they intended to enslave. They had burned the *Bounty* as a means of preventing retreat, but by the time a naval expedition found them, all but one of the men had been killed by the would-be Tahitian slaves. The one remaining man, John Adams, had become a fervent Christian, and had organized the community—which included the women and a larger gathering of children, descendants of the mutineers—along religious lines. The Admiralty resolved to leave

Captain William Bligh. *(Corbis Corporation. Reproduced with permission.)*

Adams where he was, to provide an inspiration for South Sea missionaries.

As for Bligh and his loyalists, they underwent a grueling seven-week voyage of 3,618 mi (5,823 km) in the open boat before finally arriving, half-starving and half-crazed from thirst, on the Dutch colony of Timor in what is now Indonesia. Bligh was tried before a court martial and acquitted in 1790. Later, he went back to Tahiti to complete his original mission of transplanting the breadfruit trees.

Bligh commanded British vessels in action against Napoleon's navy at Camperdown and Copenhagen in the late 1790s and early 1800s, and faced another mutiny aboard the H.M.S. *Nore* in 1797. In 1801 he was elected a Fellow of the Royal Society, but in 1805 he was reprimanded by the Admiralty for his abusive treatment of a junior officer.

Later, in 1805, Banks recommended Bligh for the governorship of New South Wales, and Bligh arrived in the colony the following year. He set out to suppress the rum traffic and the trade monopoly enjoyed by officers of the New South Wales Corps, and soon found himself at loggerheads with a faction led by John Macarthur. Bligh had Macarthur tried for sedition, so the officers deposed Bligh in what came to be known as the Rum Rebellion.

Later, Major George Johnson, the man elected by the officers to replace Bligh, was dismissed from the service. Bligh himself was exonerated by default, though a number of figures in the British government criticized his harsh leadership style, which in their view had helped precipitate the revolt. He later became an admiral before retiring to Kent, and died in London on December 17, 1817.

JUDSON KNIGHT

Louis-Antoine de Bougainville
1729-1811
French Explorer

L eader of the first French circumnavigation of the globe (1767-69), Louis Antoine de Bougainville explored the South Atlantic and Polynesia before reaching the Great Barrier Reef around Australia. He discovered a number of the Solomon Islands, including one named for him, before making his way back to his point of origin. In the course of the voyage, he and his crew learned that, unbeknownst to them, their party included the first woman to circumnavigate the globe.

Born in Paris in 1729, Bougainville was raised as a member of the French nobility. He was trained for a career in the military, and obtained his first fighting experience in Quebec during the Seven Years War (1756-63), known in North America as the French and Indian War. After his superior, General Montcalm, was killed, Bougainville spent the remainder of the war fighting in Germany.

France lost most of its overseas empire in the war, and Bougainville was among the French patriots who resolved to win a new empire. In 1765 he claimed a group of islands off the coast of South America, dubbing them the Malouines after the French port of Saint-Malo. Spain claimed the islands too, however, and since King Louis XV wanted to maintain good relations with his allies in Madrid, he ordered the colony turned over to the Spaniards. The Spanish changed the French name slightly, calling the islands the Malvinas. Meanwhile, British forces were claiming another part of the islands—which they called the Falklands—for England. This conflict would ultimately lead to the Falklands War more than 210 years later in 1982.

Louis-Antoine de Bougainville. *(The Granger Collection, Ltd. Reproduced with permission.)*

Ordered to remove all French subjects from the Malvinas, Bougainville did so. After reaching Rio de Janeiro, however, he received a more intriguing command from the crown: he was to continue journeying around the world. With him was an astronomer named Pierre Antoine Véron, who would use newly developed technology to determine the correct longitude at any given spot. Taking two ships, the *Boudeuse* and the *Etoile,* the expedition sailed from Brazil late in 1767.

The ships took nearly eight weeks just to get through the perilous Straits of Magellan, but after many misadventures they finally found their way to Tahiti. They were only the second Europeans to arrive there, after a British expedition the year before, and like their counterparts they found it to be a veritable paradise. While in Tahiti, Bougainville learned that the ship's botanists' valet, Bare, was a woman who had disguised herself as a man in order to get on the ship and go where no woman had ever gone before.

From Tahiti, the expedition sailed to Samoa, and then to Vanuatu (formerly the New Hebrides), becoming the first Europeans to reach those islands since 1605. They nearly ran aground on the Great Barrier Reef and never reached Australia, which Bougainville incorrectly thought was connected to Vanuatu by a land bridge. Sailing eastward, they entered the Louisiade Archipelago, which Bougainville named, and the Solomons, many of whose islands he named. Arriving at the island of New Britain, Véron was able to make the first accurate calculation of the Pacific's width.

In the Moluccas, or Spice Islands, Bougainville carried out a mission of agricultural espionage ordered by the king, stealing clove and nutmeg plants and transporting them to Mauritius in the Indian Ocean, where they could be cultivated as a means of breaking Holland's monopoly on the spice trade. By the time they reached Mauritius in early 1769, the men needed to recover from scurvy, but eventually they were on their way again, and in a few months' time had reached Saint-Malo. Not only were they first Frenchmen to circle the globe—along with the first woman of any nationality—but their loss of only seven lives was a record at the time.

Bougainville went on to adventures in the American Revolution, serving aboard a French ship sent to assist the fledgling republic in its fight against Britain. He became a field marshal in 1780, and when the French Revolution came in 1789, he retired to occupy himself with writing scientific papers. Napoleon later granted a number of high honors to Bougainville, who died in 1811.

JUDSON KNIGHT

James Bruce
1730-1794
Scottish Explorer

When James Bruce returned to England in 1774 after years spent in the mysterious land of Ethiopia, his fantastic tales gained him a reputation as a liar. Yet modern research has confirmed the accuracy of Bruce's account, which he set down in writing in 1790.

Born in 1730 in Stirlingshire, Bruce was the son of an aristocratic family who owned an estate at Kinnaird House. His mother died when he was three years old, and his father later arranged for him to be taught by a private tutor in London. He went on to Harrow School and Edinburgh University, where his father expected him to study law; Bruce's interests, however, were in languages and art.

Bruce married in 1752, but his wife died in childbirth nine months later. To recover from his grief, he traveled around Europe, returning to London after the death of his father in 1758. During this time he became fascinated with

James Bruce. *(The Granger Collection, Ltd. Reproduced with permission.)*

Ethiopia, or Abyssinia as it was then called, and resolved to go there. In particular, he hoped to find the source of the Blue Nile, which together with the White Nile, a deeper river that flows northward from its source at Lake Victoria in Central Africa, forms the Nile proper at Khartoum in modern-day Sudan.

Appointed British consul-general to Algiers in 1762, Bruce served in that position for two years, then spent considerable time traveling throughout North Africa, devoting himself to learning languages. By 1768 he had wound up in Cairo, accompanied by an Italian assistant named Luigi Balugani. He ultimately made contact with the patriarch of Ethiopia, a leading church official in that land, which was Christianized during the fourth century A.D.

Taking with him a letter of introduction from the patriarch, Bruce endeavored by a number of means to reach Ethiopia. Stopped by war from going in via the Nile, he used a circuitous route that took him across the Red Sea to the Arabian Peninsula, and back again to the ancient city of Aksum. Finally, he arrived in Gondar, then the capital, in 1770, and ultimately met Ras Michael of Tigre, ruler of the Abyssinian Empire.

Ras Michael invited Bruce on an expedition up the Little Abbai River, thought to be the source of Blue Nile, but constant fighting between Ethiopia and its neighbors forced them to turn back. Late in 1770, Bruce and Balugani made a second expedition up the Little Abbai, reaching its source on November 4. Unbeknownst to them, the explorer Pedro Paez (1564-1622) had already been there in 1618.

In the months that followed, Bruce collected extensive knowledge about Ethiopia. Balugani died of dysentery, and Bruce left Ethiopia in December 1771. He spent time in Sennar, a Muslim town in what is now Sudan, studying the life of the people there before taking the Nile north to Cairo, where he arrived in 1773.

Bruce returned to a number of disappointments in Europe. Before leaving, he had become engaged to a woman, and later, thinking him long dead, she had married an Italian noble. In Paris, a cartographer informed him of Paez's earlier journey to the source of the Blue Nile; and finally, after an initially warm reception in England, the explorer soon gained a reputation as a spinner of tales.

Licking his wounds, Bruce retired to Kinnaird House, married a much younger woman, and lived happily until 1788, when she died suddenly. Once more seeking to assuage his loss, he set out to write of his adventures, and published a book in 1790. It was received with no more credulity than his earlier verbal accounts of his journeys—even though modern scholarship has confirmed most of his findings. Bruce died in 1794 after he fell down a flight of stairs on his way to escort a lady to her carriage.

JUDSON KNIGHT

John Byron
1723-1786
British Explorer and Naval Admiral

After the exploration of Australia and New Zealand by Dutchman Abel Tasman (c. 1603-1659), a feverish period of sea exploration began, setting the stage for a surge in European colonialism during the eighteenth and nineteenth centuries. Since Ferdinand Magellan's (1480?-1521) first circumnavigation in 1520, the globe had been circled by ship numerous times. However, the need for faster, easier trade routes and a curiosity about unknown lands fueled a new breed of seaman. Admiral John Byron of the British Navy, sent to find the elusive "southern continent," discovered the Falkland Islands as well as many other smaller islands in what would be the fastest circumnavigation at the time.

John Byron. *(National Maritime Museum. Reproduced with permission.)*

Born into a family of Navy men, Lord Byron began his naval career in 1731 at age nine, when he became a midshipman. Nine years later, and still a midshipman, Byron took up with the vessel *Wager*, a supply ship with a reputation as the worst boat in the Navy. When it rounded Cape Horn in 1741, *Wager* separated from the fleet and wrecked on the rocks of an island near the Strait of Magellan. The crew, no longer under the rules of the military, disbanded and fended for themselves. Byron, who later wrote a book about the perilous trip, lived with very little food, shelter, or clothing for 13 months. He then moved to the hostile, Spanish-occupied territory of South America until a French vessel rescued him and brought him back home, nearly five years after the shipwreck.

Upon his return in 1746, he rose through the ranks of the Navy rapidly. He commanded several ships and eventually entire fleets of frigates. In 1764 he was assigned to an older frigate ship, the *Dolphin*, that was rotten with worm holes through its hull. The Navy used an experimental copper sheathing to cover the hull and prevent further damage. Byron's assignment, under secret government orders, was to seek out new foreign trade markets, and if possible claim any undiscovered or unclaimed lands for the British. He "rediscovered" the Falkland Islands, off the tip of Patagonia, and claimed them as British.

Unfortunately, unbeknownst to Byron, the French had already claimed the island chain. For years the British squabbled with the French, and eventually the Spanish, over control of the Falklands, but Byron had provided one of the most accurate navigational maps of the chain. After touching the Falklands, however, Byron's journey became somewhat disappointing. He toured the south seas for a "southern continent" that the British hoped to find, but soon made a fast track to the South Pacific to "rediscover" the Solomon Islands, which were said to be rich with gold and silver. Byron missed the Solomon Islands as well as Tahiti, but he did manage to spot many smaller islands.

While many observers criticized Byron's journey as unsuccessful and Byron himself as a lazy explorer, his journey was the fastest circumnavigation up to that point in history. His trip also ignited British interest in the South Pacific, where English products were eventually traded with great success.

Byron was eventually assigned to Newfoundland, where he was appointed governor, and in 1778 he became a vice-admiral. While commanding a fleet of British ships sailing to America, he lived through one of the worst Atlantic Ocean gales in history, securing his nickname, "Foul-weather Jack." Byron died in 1786, but his legacy continues to live on through the literary work of his grandson, the poet Lord Byron, who based his work *Don Juan* and some of his other poetry on John Byron's harrowing sea travels.

LOLLY MERRELL

Jacques-Alexandre-César Charles
1746-1823
French Inventor and Scientist

Jacques-Alexandre-César Charles, with Nicolas Robert, ascended in the world's first hydrogen balloon in 1783. He was also a physicist and mathematician and is perhaps better known in this capacity as the person who developed Charles's law, which relates gas temperatures and pressures. In both roles, he made important scientific and technical contributions that have had lasting effects on both science and society.

Charles was born in 1746 in Beaugency, France. Very little is known about his childhood, but he began his professional life as a clerk in the French finance ministry. From there he turned increasingly to science, experimenting

with electricity at first. In fact, it may have been Charles's experiments with electricity that showed him how to pass an electrical current through water, separating it into its components of oxygen and hydrogen.

In September 1783 the Montgolfier brothers' first balloons ascended into the air, lifted by hot air. Not knowing that simply heating air could create lift, the Montgolfiers believed that a special gas was formed by burning straw, which they called "Montgolfier gas." Charles mistakenly thought that Montgolfier gas was hydrogen, and he hastened to duplicate their experiment by filling his own balloon with hydrogen. In November 1783 Charles and Nicolas Robert climbed into the balloon they had built and rose into the air as the first truly lighter-than-air flight began. Ascending to an altitude of over a mile (1.61 km), they drifted for several miles before setting down in a field, scaring the peasants who thought they were being attacked by a strange creature. The peasants "killed" the balloon by stabbing it, then dragged it away.

The major advantage of Charles's design was that, without a fire burning beneath the balloon, the risks of a blaze were greatly reduced. In fact, with the substitution of helium for hydrogen, today's balloons are very similar in design to Charles's.

Following his first flight, Charles made additional balloons, financed in part by charging admission to see the balloon fly. His ballooning experiences piqued his curiosity about heated gases, and he spent much of the rest of his life experimenting and formulating what is now known as "Charles's law." This states that, under conditions of constant pressure, a heated gas will expand in volume. This is also known as Gay-Lussac's law, in honor of its co-discoverer.

Over the following decades, Charles's law, Boyle's law, and others that describe the properties of gases under a variety of changing conditions were shown to be part of a more universal ideal gas law. The ideal gas law describes how, in a gas, pressure, temperature, volume, and the number of molecules of a particular gas all relate to each other under a variety of changing conditions. This law is used to predict the behavior of gases when compressed, heated, expanded, and the like. In fact, air conditioning, refrigeration, and similar industries absolutely depend on the proper application of these gas laws to help transfer heat from one point to another, cooling in the process.

In 1795, in honor of his many accomplishments, Charles was elected a member of the Académie des Sciences in France. Later, he became a professor of physics, continuing his work with the characteristics of gases under a variety of conditions. Interestingly, most of his scientific publications deal with mathematics instead of with those discoveries for which he is best known. Charles died in April 1823 in Paris.

P. ANDREW KARAM

Captain James Cook
1728-1779
English Explorer

James Cook was one of the foremost figures of the so-called "Age of Exploration." During his career, Cook circumnavigated the globe twice, and captained three voyages of discovery for England. He made significant contributions to the fields of surveying, cartography, advanced mathematics, astronomy, and navigation. The detailed records of his voyages and contacts with various native peoples are considered the first anthropological survey of the Pacific Islands, Australia, and New Zealand. Cook's voyages sparked European and American interest in Pacific colonization.

James Cook was born in Marton-in-Cleveland, Yorkshire, England. As a youth, he received a modest education, but was a dedicated self-study of mathematics, surveying, and cartography. Cook was apprenticed to a small shop owner, but later left to join a merchant collier fleet at Whitby. Cook earned his mate's certificate, but his merchant career was cut short by his decision to enlist with the Royal Navy in 1755 at the outbreak of the Seven Year's War (the American phase is known as the French and Indian War, 1756-1763).

Cook was sent to America in 1756 not only as a seaman, but also as a cartographer. His first charge was to conduct soundings and draw charts of the St. Lawrence River. Cook's charts were later used by British forces for their attack on Quebec. He was next named surveyor of Newfoundland and carried out that project until 1767. Cook's maps were so precise that many were used for a century.

As Cook gained renown for his cartography, he also submitted a paper to the Royal Society on astronomical observation and navigation. His work on determining location using

James Cook. *(Corbis Corporation. Reproduced with permission.)*

the moon commanded the attention of not only scholars but also the British government. In 1768, Cook was appointed to command an expedition to the Pacific—the first of three great voyages. The stated purpose of Cook's Pacific expedition was to observe the transit of Venus. At the completion of that task, Cook continued to record significant discoveries. In the South Pacific, he discovered and named the Society Islands. Cook then sailed to New Zealand, which he reported upon favorably as a potential site for British colonization despite the lack of domesticated animals. Venturing from New Zealand, Cook sailed to the eastern coast of Australia and charted the coastline before claiming the land for Britain. On the return voyage, Cook's crew was stricken with disease—a then common occurrence at sea. One-third of his crew died from malarial fever and dysentery.

Cook was scarcely back in Britain for year before he received his next appointment. He was granted two ships, the *Adventure* and the *Resolution*, and sent back to the Pacific to further complete the exploration of the Southern Hemisphere. Cook was charged with finding a southern continent that was thought to exist in the extreme South Pacific. The mysterious continent was supposed to be temperate with fertile land. Cook left Britain in 1772 and sailed for the extreme southern Atlantic. Pushing his way through freezing temperatures and ice flows, Cook sailed along the edge of Antarctica. The frozen Antarctic was certainly not the fabled southern continent. Cook's circumnavigation of the southern Pole put an end to the legend. Cook again stopped in New Zealand, this time introducing some European plants and domestic animals into the indigenous landscape. He discovered, charted, and named several more islands as he finished his journey.

On both voyages Cook made pioneering provisions for his crew. To avoid scurvy, which usually plagued sailors on long sea crossings, Cook made sure that his crews' quarters were clean and well ventilated. He also provided a diet that included lemons, cress, and sauerkraut. Although Cook lost 30 men on the first voyage, only one perished from disease on the second. None of the deaths were caused by scurvy.

Cook embarked on his third and final voyage in 1776, turning his efforts to the Pacific coast of North America in search of the fabled Northwest Passage. Enroute he discovered present-day Hawaii, which he dubbed the Sandwich Islands, in January 1778. Cook enjoyed a record of very amicable relationships with the native peoples he encountered on his expeditions, and his initial contact with in the Sandwich Islands was no exception. After a fortnight, the ships left Hawaii for the American Pacific Northwest, where Cook created detailed maps that were used in later explorations, including the Lewis and Clark expedition. However, Cook failed to locate the Columbia River and thought that Victoria Island was part of the mainland. Despite these flaws in his cartography, Cook's expedition, and his records of contact with various native peoples who possessed great natural resources, created a new interest in trade and settlement in the Pacific Northwest.

By January 1779, the ships were once again anchored in the Hawaiian Islands. This time, however, their reception was distinctly unfriendly. Cook decided to leave the islands within a month, but was forced back by a storm only a week later. The native population was by now openly hostile and when one of Cook's cutters (a small boat) was stolen, Cook took the tribal chief hostage to guarantee the boat's return. In the ensuing commotion, a shot was fired. The natives then attacked the sailors, killing Cook in the process. He was 51.

BRENDA WILMOTH LERNER

Johann Georg Gmelin
1709-1755
German Explorer and Botanist

During the period from 1733 to 1743, German botanist Johann Georg Gmelin explored a wide area of Siberia. These expeditions yielded numerous plant specimens, which he later described in his writings. Also significant was his identification, in 1735, of permafrost, a permanent frozen layer of earth that exists in northerly regions.

Born in 1709, Gmelin (pronounced gMAY-leen) became a professor of chemistry and natural history. Like many German-speaking scientists of the eighteenth century, he was drawn eastward by a job offer from the St. Petersburg Academy of Sciences in Russia, where he became a professor at the age of 22 in 1731.

Two years later, Gmelin began to travel much further east, as he embarked on what turned out to be a decade's worth of exploration in Siberia. His journeys took him through a number of towns and regions, including Tobolsk, Semipalatinsk, Tomsk, Krasnoyarsk, Irkutsk, and Yakutsk. Traveling eastward as far as the Lena River, he then began making his way back toward St. Petersburg, along the way collecting enormous numbers of specimens from the taiga and tundra.

Gmelin spent four more years at the St. Petersburg Academy, writing the vast work later published by the Academy in four volumes as *Flora sibirica* (1747-69). In 1749 he took a position as a professor at Tübingen in Germany, where he remained until his death in 1755. Gmelin published *Journey across Siberia,* his diaries of the ten-year expedition, in 1751. Translated into numerous languages, the book became a best-seller. Gmelin's nephew, Leopold Gmelin (1788-1853), later became a famous chemist and author of a definitive textbook in that discipline.

JUDSON KNIGHT

Sir William Hamilton
1730-1803
English Archaeologist and Geologist

During his many years as British envoy to the court of Naples (1764-1800), Sir William Hamilton conducted extensive archaeological investigations of Roman antiquities. Among the sites of his greatest activity were the volcanic remains around Vesuvius in Italy and Etna in Sicily.

Grandson to the third duke of Hamilton and son of Lord Archibald Hamilton, governor of Jamaica, young William Hamilton came from a distinguished line. He entered into military service at age 17, in 1747, but left the army after marrying a Welsh heiress in 1758.

Sent to Naples in 1764, Hamilton soon took an interest in archaeology and the natural sciences. He studied Vesuvius and Etna, and between 1772 and 1783 published several studies on earthquakes and volcanoes. Also significant were his excavations at Herculaneum and Pompeii, towns devastated by the volcanic eruption of Vesuvius in A.D. 79. Hamilton amassed an extensive collection of antiquities, which he sold to the British Museum in 1772; he was knighted during the same year.

Hamilton lost his wife a decade later, at which point he inherited her estate in Swansea. During the 1780s he continued his studies in Italy, but he was about to become embroiled in a set of personal intrigues as perplexing as anything in the archaeological record.

Hamilton was appointed privy councillor in 1791, the same year he married Lady Emma Hamilton. He was 61 years old, his new bride 26, and unbeknownst to him, Emma had been mistress to his nephew Charles Greville since 1786.

The couple spent a few relatively peaceful years, then in 1800 they went on a trip to Palermo, Sicily, with Lord Horatio Nelson. A year later, Emma gave birth to a daughter named Horatia, acknowledged by Nelson as his child. Hamilton himself had been recalled from his position at Naples in 1800, and died on April 6, 1803, in London.

JUDSON KNIGHT

Samuel Hearne
1745-1792
English Explorer

The first European to travel across the vast interior of northern Canada, Samuel Hearne also became the first to reach the Arctic Ocean from North America via an overland route. The findings of his journey, including the discovery of the Coppermine River, added to the growing knowledge that no viable Northwest Passage existed between the Atlantic and Arctic oceans.

Born in London in 1745, Hearne was the son of an engineer who died when Hearne was

Samuel Hearne. *(Library of Congress. Reproduced with permission.)*

three years old. By age 11, he was working as a servant on a Royal Navy ship, and would remain in the naval service until the age of 18. At that point he took a post as mate on a trading ship bound for Churchill, Manitoba.

Churchill was a fur-trading post, but at that time Hearne's employer, the Hudson's Bay Company, had begun looking for opportunities to diversify. Copper looked promising, and after proving himself by his service to the company at sea, Hearne was commissioned to undertake an expedition for copper deposits in the Canadian interior.

He made the first of three trips in November 1769, but his Native American guide deserted him, and Hearne was forced to turn back. Another attempt in the following February ended in disaster, and Hearne spent several months wandering alone. On the third try, however, he had a highly experienced guide, the Chipewyan chief Matonabbee.

Departing from Churchill in early December 1770, the party crossed a wide desert, living on buffalo and caribou, before arriving at the Coppermine River on July 14, 1771. Along the way, the Chipewyan met a band of Inuit, their enemies, and Hearne later recorded the slaughter of the Inuit that he witnessed. The party followed the Coppermine all the way to the Arctic Ocean, then turned upstream to look for copper.

Though in fact there are sizeable copper deposits in the area, Hearne ran out of time before he could locate them, and returned with nothing more than a lump of copper he had found. On his trip back south with Matonabbee and the others, he became the first white man to see and cross the Great Slave Lake. They reached Churchill at the end of June 1772.

Recognizing Hearne's abilities, the Hudson's Bay Company chose him to open its first inland trading post, and in the winter of 1774-75 he established Cumberland House, the first European settlement in what came to be the province of Saskatchewan. Later he became head of the post at Churchill, and in this capacity was charged with defending it during the American Revolution. The post was attacked by French allies of the Americans in August 1782, and Hearne was forced to surrender to the superior forces.

In the following year, Hearne rebuilt the fort at Churchill, and continued at his post until he retired to England in 1787. He later recounted his historic travels in *A Journey from Prince of Wales' Fort in Hudson's Bay to the Northern Ocean*. This book and his journals were published in 1795, three years after his death at the age of 47.

JUDSON KNIGHT

Charles-Marie de La Condamine
1701-1774
French Mathematician and Explorer

Charles-Marie de la Condamine led the first foreign scientific expedition into the Spanish territories of the New World. After an extended surveying expedition in the Andes mountains of what is now Ecuador, La Condamine became the first scientist to explore the Amazon River from its source in the mountains to the river's mouth at the Atlantic Ocean.

La Condamine was born into an aristocratic French family that made its fortune by speculating in mines in the French territory of Louisiana. During his childhood a series of wars broke out in Spain as various powers vied to install a king that would ally them with the huge Spanish Empire. The wars continued into La Condamine's adulthood, and he served as a soldier in the French armies that eventually succeeded in placing a French monarch on the throne of Spain. It was this same monarchy that would years later grant permission to La Condamine and his French scientists to conduct studies in Spain's South American territories.

Charles-Marie de la Condamine. *(The Granger Collection, Ltd. Reproduced with permission.)*

After his military service La Condamine continued his studies in mathematics and was elected into the French Academy of Science as a mathematician at the young age of 29. His mathematical skill also got him recognition by the eminent French intellectual of the time, Voltaire. When the French government instituted a lottery, La Condamine calculated that the government was not selling enough tickets for the prize, and that someone could, in theory, buy all the tickets and be assured of winning more than the price of the tickets. Voltaire put La Condamine's calculations to the test and gained half a million francs. With this money La Condamine gained a powerful ally in the French intellectual world.

La Condamine's South American odyssey began when the Academy of Sciences came to address two conflicting theories of the shape of Earth, one that Earth was elongated and narrow at the poles, the other that it was flat at the poles and bulging along the equator. The Academy appointed two teams of surveyors, one bound for the Scandinavian arctic and the other for the South American city of Quito in what is now Ecuador. Each team was to conduct precise measurements of Earth to settle the question for the Academy. After securing permission for the team of French scientists to travel within Spain's New World territories, the Academy took advantage of the unprecedented permission to conduct research in South America and appointed additional scientists to La Condamine's expedition.

The mission in Quito was plagued by delays that eventually made the data they were sent to collect outdated, as they learned when they received word that the Arctic expedition had returned to Paris in 1737 with data supporting Isaac Newton's (1642-1727) theory that Earth was flattened at the poles. Nevertheless, La Condamine's team continued their work and in 1743 took their last measurements, disbanding with only La Condamine returning to France to present their findings.

La Condamine decided to make the trip back to France by way of a mapping expedition down the Amazon River. During the course of his four-month trip down the river he made surprisingly accurate geographical measurements and made many corrections to the imperfect maps of the day. Most notably he dispelled some remaining myths of golden cities deep in the jungle and established the shape and size of Majoró Island at the mouth of the river. La Condamine also made important ethnographic and biological observations of the Amazon Basin.

When La Condamine finally returned to the Academy of Sciences after what had turned into a ten-year mission, he was honored, although the data he collected at the equator was contested by the scientists of the Academy. His observations of the Amazon remain one of the earliest and fullest accounts of the river and became the stimulus for a wave of scientific investigation in the region in the nineteenth century.

GEORGE SUAREZ

Jean-François de Galaup, comte de La Pérouse
1741-1788
French Navigator

Jean-François de Galaup, comte de La Pérouse, was a military hero and one of the greatest French explorers of the Pacific Ocean. While he made a number of significant discoveries, he is best known for his exploration of Sakhalin and the Kuril Islands and his discovery of the Tatar and La Pérouse straits, on either side of the island of Sakhalin.

La Pérouse was born in 1741 near Albi, in France. Little is known of his childhood until

Jean-François de Galaup, comte de La Pérouse. *(The Granger Collection, Ltd. Reproduced with permission.)*

age 15, when he joined the French Navy to fight the British in the Seven Years' War, later serving the French in North America, China, and India. He became famous in 1782 with his capture of two English forts along the Hudson Bay, and, shortly afterwards, married a woman he had met on the Ile de France (now known as Mauritius). In 1785, at the age of 44, he was placed in command of a two-ship expedition of discovery in the Pacific.

La Pérouse's ships, the *Astrolabe* and the *Boussole*, were supply ships, outfitted for exploration and classified as frigates for the expedition. At about 500 tons each, they were not large ships, but were considered adequate for the rigors of the upcoming expedition. A great admirer of English explorer James Cook (1728-1779), La Pérouse adopted many of Cook's practices, including several that were not common at the time: nearly 10% of his crew were trained as scientists, he treated his men well and was well-liked by them, and he made every effort to get along well with the Pacific Islanders with whom he came in contact. All of these traits helped to make him a successful commander, and helped make his expeditions unusually successful.

La Pérouse's voyage took him from Brest, around Cape Horn, and to Chile. From there, he sailed to Easter Island and Hawaii on his way to

Alaska, where he explored and mapped for some time before turning south again for Monterey Bay, where he investigated Spanish settlements and missions with some disapproval for the treatment of Native Americans.

From Monterey, he set forth across the Pacific, making his way to Macao and Manila. After a short stay, he left to explore the coasts of northeast Asia, visiting Korea and Sakhalin (a large island between the northern Sea of Japan and the Sea of Okhotsk). Here, he made some of his most important discoveries, attempting to sail through the Tatar Straight (which separates Sakhalin from the Asian mainland) and then through the Straits of La Pérouse, which passes between Sakhalin and the Japanese island of Hokkaido. From there, he continued north to the Kamchatka Peninsula, reaching the port of Petropavlovsk in September 1787.

In Petropavlovsk, where he received further orders via letter from Paris, La Pérouse and his crew rested for a short time before packing their notes, logs, and specimens to be returned to France and traveling through Siberia and Russia (a monumental year-long trip in itself). New orders in hand, La Pérouse then left for New South Wales (in what is now Australia) to investigate activities of the British. On his way, he stopped at the Navigator Islands (now called Samoa) where a dozen of his crew were killed during an attack. He reached Botany Bay in January 1788, just as the British commander was moving the colony to Port Jackson. Although the British were unable to help supply him with food (being short of supplies themselves) they did transport his journals and letters to France and provided him with wood and fresh water for the next leg of his trip. La Pérouse left in early 1788, bound for more exploration of the islands and coasts of the southern Pacific, and was never seen again.

In 1791, A. R. J. de Bruni, chevalier d'Entrecasteaux (1739-1793) was sent to find and, if possible, rescue La Pérouse and his crew. He found that both ships had broken up on the reefs of Vanikoro Island. He was able to determine that the crew tried to salvage what they could, unloading the ships and making a small boat from the wreckage of the *Astrolabe*. Although most of the crew were killed by local inhabitants, some left on the boat, vanishing without a trace.

P. ANDREW KARAM

Pierre Gaultier de Varennes et de La Vérendrye
1685-1749
French Explorer and Soldier

Pierre Gaultier de Varennes et de La Vérendrye was a very prolific, but largely unheralded, French explorer in the New World in the first half of the eighteenth century. He traveled widely through central Canada, visiting many of the places later seen by the Lewis and Clark expedition that was to take place 50 years later.

La Vérendrye was born in New France (now Canada) in 1685. Almost nothing is known of his early childhood except that it did not last long; at age 12 he joined the French army, fighting in raids against the British in the New World as well as fighting for France in the War of the Spanish Succession. Although taken prisoner in 1709, he was freed and returned to New France, where he married and started a family.

In 1726, La Vérendrye started work as a fur trader, starting a small outpost on Lake Nipigon, about 31 miles (50 km) north of Lake Superior. It was at these trading posts, looking at maps provided by Native Americans who came to trade, that La Vérendrye first heard of a huge river rumored to lead to a western sea.

Hearing this, La Vérendrye immediately thought it might be possible to follow this river to the Pacific Ocean, opening up an easy route to the Orient for French trade. Such a trade route would give the French a decided trade advantage against their British and Spanish rivals and could make him a wealthy man. With this in mind, he and his sons built a series of trading posts between 1731 and 1738 that reached from Fort Saint-Pierre in Ontario to what is now the city of Winnipeg in the province of Manitoba. Their goal was to help establish a French presence in the area and to learn more of the geography of western Canada. At these trading posts, he learned more about western waterways, including the promised great rivers.

In the fall of 1738, La Vérendrye and his sons pushed into what is now Montana and North Dakota, reaching the Mandan Indian villages that were to be so important to Lewis and Clark over 60 years later. From there, in 1743, two of his sons continued on through Nebraska, Montana, and perhaps Wyoming, possibly seeing the Rocky Mountains, although it is certain they did not cross them. In 1743, on their return trip, they placed a lead tablet near Pierre, South Dakota, claiming these lands for France. They then returned to New France, having explored further west than any white man to that time.

During his journeys, La Vérendrye lost one son, a nephew, and a Catholic priest at the hands of hostile Native Americans. In spite of his accomplishments, the French government criticized him for failing to find the western sea and for the losses his expeditions suffered. This is doubly ironic because, not only did he finance the expeditions himself, but he also sent at least 30,000 pelts annually for trade—pelts that would otherwise have gone to the British. Nonetheless, La Vérendrye made plans for yet another journey to the west, in spite of his age and failing health. Although he received permission from the government, he died before he was able to leave Montreal again.

During his travels, La Vérendrye made an impressive number of accomplishments. In addition to exploring many of the waterways in the North American interior (including parts of the Missouri River system), he helped establish trade relationships with a number of the Native American tribes, diverting trade from the British to the benefit of the French government. In addition, he established a string of trading posts that facilitated future trade, and his explorations helped to validate France's territorial claims to large parts of the North American interior. In spite of his accomplishments, he is among the least-heralded explorers of this period.

P. ANDREW KARAM

Alexander Mackenzie
1762-1820
Scottish Explorer

Alexander Mackenzie was a trader who explored northwestern Canada in an effort to open up new territories to the fur trade. In his explorations he traced the 1,100-mile (1,770-km) course of the river now named after him from the Great Slave Lake to the Arctic Ocean, near what is now the northern border between Canada and Alaska. He also became the first European explorer to cross the North American continent at its widest point north of Mexico.

Mackenzie was born in Stornoway on the Scottish island of Lewis to a prominent military family. At the age of 12 his family immigrated to New York as part of a large migration of Scots to North America. In New York during the American Revolution his father joined forces loyal to

Alexander Mackenzie. *(New York Public Library. Reproduced with permission.)*

the King of England and died during the revolution. As the war turned against England, the family moved to Montreal, Canada, where Mackenzie began working in the offices of the North West Company, one of the principal trading companies in Canada along with its main competitor, the Hudson's Bay Company.

In 1784 Mackenzie was offered a partnership in a fur trading form on the condition that he conduct an exploration of the Canadian interior the following year. The exploration was intended to increase the North West Company's knowledge of western Canada's geography and the Native American tribes that inhabited it. The fur trade at the time depended on the familiarity of traders with not only the land in which they trapped and transported animal skins but also the relationships that traders fostered with the tribes that they traded with.

As much a businessman as an explorer, Mackenzie reached Ile à la Crosse in 1785. There he was able to manage a large trading territory stretching from Lake Athabasca to the Great Slave Lake and the upper regions of the Churchill River. Control of the Churchill allowed the Native American trappers to do business with the North West Company rather than having to travel hundreds of miles further down the river to trade with the Hudson's Bay Company at Prince of Wales Fort.

Soon after, with a cousin, Mackenzie established Fort Chipewyan on the shores of Lake Athabasca. It was from Ft. Chipewyan that Mackenzie based his later expeditions down the Mackenzie River and to the Pacific coast.

On June 3, 1789, Mackenzie left Fort Chipewyan with a group of trappers and a member of the Chipewyan tribe named Nestabeck, whom Mackenzie called English Chief. Nestabeck had accompanied the Chipewyan leader Matonabbee from 1770 to 1772 when he guided Samuel Hearne (1745-1792) on his historic overland trip to the Arctic Ocean along the Coppermine River. Traveling down the river the expedition was able to cover about 75 miles (120.7 km) each day, and they reached the river's delta on the Arctic Ocean on July 14, 1789. The group realized they had reached the ocean only by accident when a rising tide swamped their baggage. The return upriver was slower, and they arrived back at Ft. Chipewyan on September 12 of the same year.

On his next expedition, conducted between 1792-93, Mackenzie set out from Fort Chipewyan for the Pacific coast of Canada in hopes of making contact with Russian traders there and fulfilling the goal of establishing a trading route across Canada to the far east, as Spanish and Portuguese explorers had hoped to centuries before. Rather than get rid of the Russian traders, Mackenzie thought they could become intermediaries in the lucrative Pacific trade with China. The expedition traveled west up the Peace River to the Continental Divide. After crossing the mountains the party descended along the Fraser River for 150 miles (241 km) and traveled overland to the Pacific coast of what is now British Columbia. This expedition and the voyage down the Mackenzie River combined to form the first crossing by a European of North America above Mexico.

Before his death in 1820, Mackenzie consulted John Franklin (1786-1847) in preparation for his disastrous expedition to map the northern coast of Canada. Mackenzie stressed the advantage of using members of the local tribes and Inuit as guides for their knowledge of the area. Almost 30 years later Franklin's disappearance in an Arctic naval expedition and the rescue attempts once and for all put to rest the search for a shipping route over the Canada's northern coast that would link the northern Atlantic trade routes with the Pacific.

GEORGE SUAREZ

Alejandro Malaspina
1754-1810
Italian Explorer and Naval Captain

Alejandro Malaspina sailed on voyages of exploration in the Pacific Ocean for the King of Spain on his way to circumnavigating the globe in the latter part of the eighteenth century. During his travels he was able to report that the long-sought Northwest Passage from the Atlantic to the Pacific did not exist. He also explored the American west coast, conducted hydrographic surveys of the Americas, and visited numerous ports in the Pacific.

Malaspina was born into a distinguished Italian family in the then-Spanish city of Parma. He entered the Spanish navy, rising through the ranks to become a captain while still in his 30s. He was then given command of a two-ship squadron, *Atrevida* (which meant bold) and *Descubierta* (or discovery). With these ships he was to complete a number of significant voyages during his explorations for the Spanish crown, including his explorations of the American Pacific Coast and his circumnavigation.

Leaving Spain in 1791 with orders from the King to search for the Northwest Passage, Malaspina sailed into the Pacific and made his way north to what is now the state of Washington, along Vancouver Island, and continued up to the Alaskan coast. Although he helped chart many hundreds of miles of coast, including much of the St. Elias Range, he eventually turned from this search, concluding that either a Northwest Passage did not exist or was not navigable. As we know today, there is, indeed, a Northwest Passage across the Arctic Ocean, but it is covered with ice for much of the year and is only marginally navigable with modern ships. In the days of wooden ships and sail, Malaspina was completely correct in his assessment.

During his visits to Spanish colonies, Malaspina grew concerned about ill-treatment of Native Americans at the hands of Spanish colonists. In a later report to the Spanish government he commented that, in the long run, Spain would be better served by forming a confederation of states, encouraging international trade with her colonies and the natives of other lands rather than simply conquering and plundering. He also suggested that Spain would be well-served by establishing a Pacific Rim trading bloc, with activities coordinated in the Spanish city of Acapulco. Unfortunately, his recommendations became caught up in political maneuvering and were completely ignored. In retrospect, especially in view of the success of former British colonies in which this approach was largely taken, Malaspina's comments seem quite prescient.

Leaving the American Northwest, Malaspina set sail for the South Seas, making a brief visit to Tonga in 1793. The first Spaniard to visit Tonga in 12 years, he claimed the islands for the King of Spain and departed. He continued on his journey, returning to Spain in 1795.

Upon his return, Malaspina presented his reports and recommendations to the Spanish government, but became embroiled in politics almost immediately. In spite of the accuracy of his observations and the perspicacity of his recommendations, his report was buried and his recommendations largely ignored. This culminated in his being imprisoned, where he died at the age of 55.

P. ANDREW KARAM

Michel-Gabriel Paccard
1757-1827
French Physician and Mountaineer

Michel-Gabriel Paccard and his porter Jacques Balmat (1762-1834) were the first to ascend Mont Blanc, the tallest peak in Europe. With that climb, they became known as the two men who initiated the sport of modern mountaineering. Paccard and Balmat made the climb in 1786, more than two decades after Genevan scientist Horace-Bénédict de Saussure (1740-1799) offered a monetary prize to the first climbers of the mountain. Mont Blanc had significance not only because of its height, but because it was the site of nearly 40 square miles of glaciers, which carried substantial interest for scientists. The glaciers were periodically active, having advanced to such a degree in the seventeenth century that they buried valley farmland and homes in the Chamonix area of France.

Paccard was a doctor in the town of Chamonix, which lies at the base of Mont Blanc. At the age of 29, he employed the services of the 24-year-old Balmat, to make an attempt on the nearly 15,800-ft-tall (4,807-m) peak, which was often scoured by extreme winds and sudden weather changes. Like many other educated people of the time, Paccard was intrigued by the works of nature and particularly enthralled by mountains that were considered some of nature's grandest achievements. The prize money offered by Saussure was an incentive that held little interest for Paccard. Saussure, who first saw the summit in 1760 on a trip to Chamonix, thought of climbing it himself, but offered the reward to ensure that, if not him, someone would make the ascent.

Paccard and Balmat made their attempt in August 1786, leaving Chamonix for the north face of Mont Blanc. Had the men tried the east or "Brenva" face, they would have been met by a steep ice- and snow-covered mountainside. The south face is even steeper and presents 1,500-foot-tall pillars of rock. Paccard and Balmat set out on August 7, lacking the tools of modern mountain climbers. Without even rope, they began the ascent up relatively gentle but strenuously long snowy slopes, traversing numerous dangerous crevasses along the way.

The excitement in Chamonix was great as residents set up telescopes and used binoculars to watch the men make their way to the top. When Paccard and Balmat reached the summit at 6:23 P.M. on August 8, the word spread through the small town, and the church bells heralded their success. Paccard set aside the thrill of the conquest long enough to take scientific measurements before he and Balmat began the trip down.

The two men were cheered as heroes when they returned to Chamonix. Balmat went to Geneva to claim the prize money from de Saussure, but Paccard instructed Balmat to keep all of the money. The relationship between the two men turned sour afterward. Balmat told acquaintances—reportedly after imbibing in a few alcoholic beverages—that he was the true hero of the ascent, and that he not only reached the top first, but had to climb partway down to haul an exhausted Paccard to the summit. That story gained widespread acceptance when the well-known author of *The Three Musketeers*, Alexandre Dumas, Sr., published Balmat's account in 1832, five years after Paccard's death. Historians have since found evidence, including documents from Balmat himself, that the unflattering comments were untrue and that Paccard was likely the first to the top of Mont Blanc.

Paccard's name was forever cleared when France honored him on the 200th anniversary of his and Balmat's initial climb. As part of the celebration, a monument was dedicated in his honor.

LESLIE A. MERTZ

Mungo Park
1771-1806
Scottish Explorer and Surgeon

Mungo Park was a Scottish surgeon best known for his explorations in Africa. In two expeditions he navigated large parts of the Niger River, writing a popular book between these journeys. Unfortunately, his second expedition ended disastrously, leading to the death of all members.

Mungo Park was born near the English-Scottish border in 1771. He studied medicine at the University of Edinburgh, graduating as a surgeon in 1791. He was then named an assistant surgeon on the *Worcester*, which sailed for Sumatra in 1792. During this journey he made many important observations about Sumatran plant and animal life, winning the respect of many British naturalists. This was sufficient to earn him command of a small expedition into Africa, which had the aim of following the Niger River as far as possible.

Park left Britain in May 1795 and, upon his arrival in Gambia, studied the local language for several months before setting out in December with an interpreter and a slave. By later standards, he had woefully little equipment for navigation, exploration, or scientific discovery. In fact, he left with a sextant, a compass, a thermometer, a few guns, and some pistols.

In spite of his relative lack of equipment and lack of knowledge of Africa, Park managed to make his way to the Niger and to ascend it for over 200 miles (322 km) into the African interior. During this part of his trip, he was forced to deal with thieves, obstinate village leaders, fearful Arab traders, and more. At times, only his medical training allowed him to pass, as he would treat villagers or village chiefs. Other times he was forced to use his supplies to bribe his way out of trouble. He finally realized that his original goal of reaching Timbuku was suicidal and, very ill and beset by unimaginable hardships (including being imprisoned for four months by an Arab chief), he halted and spent several months recovering from severe fevers. Much of this travel was on foot, his horse having died earlier. He finally made his way back to Britain in late 1797, where he wrote a book on his travels. This book, *Travels in the Interior Districts of Africa*, became an instant success and earned him no small degree of fame. Unfortunately, he was not very polished socially in ad-

Mungo Park. *(The Granger Collection, Ltd. Reproduced with permission.)*

dition to being a poor public speaker, which kept him from acceptance as either a public lecturer or in the society of the times, so he concentrated on trying to find a wife and passing his medical exams.

In the next few years, Park married and established a medical practice in Scotland. However, in 1805, he was again asked to lead an expedition along the Niger and, against the wishes of his wife, accepted. This time he was commissioned as a Captain in the Army and was given 40 men to lead on the expedition. Unfortunately, this trip was even more difficult than the first, and only 11 men even reached the Niger River, finding it near what is now the city of Bamako (in Mali). The survivors set off in canoes, after obtaining permission from a local ruler, reaching the village of Sansanding, slightly down-river from Segou. Here they rested, then set out again in November 1805. Although no expedition members were seen again, their final trip can be reconstructed somewhat.

According to later reports, mostly from an expedition made in 1812, they traveled nearly 1,000 miles (1,609 km) down-river before they were attacked by natives. During this attack, most expedition members were killed, and Park was drowned in the river.

P. ANDREW KARAM

David Thompson
1770-1857
English-Canadian Fur Trader, Surveyor and Explorer

David Thompson has been characterized as "one of the greatest practical land geographers the world has ever known," having done all his work in western North America, principally Canada and the northwestern United States.

Thompson was born in London, England, of Welsh immigrants. His father died when Thompson was only two years old. His formal education was limited to seven years of attendance at a London charity school for the poor, where he learned something of navigation and mathematics. In 1784 he went to Canada as an apprentice with the Hudson's Bay Company (HBC), landing at Fort Churchill on Hudson Bay. Though principally assigned to do clerical work, he soon became a competent woodsman, familiarizing himself with the countryside. He learned surveying, astronomy, and applied mathematics with the HBC surveyor Philip Turnor (1789-90). During this time, he also lost the sight in one eye. Over the years, he spent some time living and trading with the various western Native American tribes. In 1794 he became an HBC surveyor, having decided to make that his profession.

In 1797 he accepted employment with the North West Company (NWC), remaining with them for 15 years. His motives for leaving the HBC without the customary one-year notice were controversial. This probably happened due to friction with his supervisors, since he was in a good position to advance and earn bonuses within HBS. He was at first a surveyor with the NWC, then a clerk trader, and finally a partner for six years (1806-12). During his first year with the NWC, he surveyed the 49th parallel from Lake Superior west to Lake Winnipeg, as well as the headwaters of the Mississippi River and parts of what are now the northern plains states. Over a period of several decades, he surveyed and mapped many of the river systems in Western Canada, traveling some 50,000 miles (80,467 km) over sometimes difficult terrain through what are now the provinces of British Columbia, Alberta, Saskatchewan, Manitoba, and portions of the Northwest Territories. In June 1799 he contracted a "fur trade" (common law) marriage with 13-year-old Charlotte Small, daughter of a retired Scots NWC partner and a Cree woman, with whom he had 13 children. They remained very close, and their union was regularized in Montreal in 1812.

In 1807 he discovered the source of the Columbia River, and during 1811 he mapped it from its source to its mouth. This work was accomplished in part because of NWC concern about American explorations, surveys, and business enterprises (mainly fur trading) in what are now the states of Washington and Oregon. Following his retirement from the NWC in 1812, Thompson continued a private surveying practice. At various times between 1817 and 1827, he surveyed the border between Canada and the United States from the Lake of the Woods east to the St. Lawrence River. He compiled a large, extremely accurate and detailed manuscript map of the areas he had traversed and studied. Measuring more than 5 by 10 feet, it covered approximately 1,700,000 square miles (2,735,885 sq km) in both nations, more than two-thirds of them in Canada. It included information from his own surveys and those of other explorers, among them Meriwether Lewis (1774-1809) and William Clark (1770-1838). It has been carefully safeguarded at the National Archives of Canada in Ottawa. Thompson continued doing occasional smaller surveys during the 1830s. He purchased a farm in Ontario in 1815, and later was involved in several business enterprises, but afterward suffered a series of financial reverses. For a time in his mid-sixties, he was compelled to take up his old profession of surveying in order to stave off bankruptcy. Poverty-stricken in his last years, he began drafting a narrative of his experiences, which historians have found an invaluable view of exploration and fur-trading in the early nineteenth-century West. It is more interestingly written and refreshingly different from most of the official reports and histories put out over the years by the Hudson's Bay Company. Unfortunately, he was unable to complete it because he became completely blind in 1851. Fortunately, his wife was strongly supportive of him until the end. When he died in 1857, he was almost forgotten. His manuscript, dealing with events in his life and career down to 1812, was edited and published by J. B. Tyrell in 1916. In 1927, 70 years after his death, a monument was placed over his hitherto-unmarked gravesite in Montreal.

KEIR B. STERLING

Captain George Vancouver
1758-1798
English Navigator and Explorer

Captain George Vancouver commanded the British Royal Navy ship *Discovery* in 1791 on a four-year journey that surveyed and mapped the American Northwest coast. Vancouver also searched for the Northwest Passage, a navigable waterway rumored to flow from the West coast of America leading to the continent's Eastern seaboard. Through meticulous exploring and charting, Vancouver proved the passage did not exist.

Born in the English port village of King's Lynn, Vancouver spent his childhood near the sea. In 1772, at the age of fourteen, Vancouver left home to sail with famed English seaman Captain James Cook (1728-1779) on his voyage around the world. Vancouver later served as midshipman during Cook's explorations along the West coast of North America. These voyages honed Vancouver's navigational and surveying skills, and acquainted him with the many inlets, bays, and coves of the west coastline of America.

Vancouver was given command of the ship *Discovery* in 1791, then sent on a three-fold mission to the American Northwest coast. His orders were to map the coast, accept surrender documents from the Spanish post at Nootka in present day British Columbia, and to seek the existence of the navigable waterway across the North American continent. Such a waterway would help insure British naval superiority and open a trade route to the Orient. The British Admiralty also anticipated that accurate surveys of the Pacific Northwest would encourage lucrative commerce in the area. More importantly, accurate charts and maps would bolster British territorial claims.

Vancouver set sail aboard the *Discovery*, accompanied by a smaller ship, the *Chatham*, with crews that were well trained and armed with the best navigational equipment available. Land was sighted in April 1792, just north of present-day San Francisco. In an example of Vancouver's meticulous quest for accuracy, the crew took more than seventy-five sets of lunar observations before making landfall to determine an accurate point from which to begin the survey. Heading north, the crew mapped the Pacific coastline until reaching the Strait of Juan de Fuca (off present-day Washington state). Here Vancouver and his crew spent another three years using smaller boats to accurately map the numerous small inlets, islands, and intricacies of the area. He often named geography he mapped after family or crew members—Puget Sound, for example, was named after a lieutenant aboard the *Discovery*. Sailing further north, Vancouver discovered that Vancouver Island had no connection to the con-

George Vancouver. *(The Granger Collection, Ltd. Reproduced with permission.)*

tinent as was assumed, but was indeed an island. Finally, Vancouver and his crew, often facing adverse weather (the crew complained of continual mist with nowhere to dry themselves), and with weather-beaten boats, managed to survey the Pacific coast from Puget Sound north to Alaska with such accuracy that the charts would be used by mariners for the next 150 years. He returned to England, and published his account of the journey entitled *A Voyage of Discovery to the North Pacific Ocean and Round the World* in 1798.

Although Vancouver's survey correctly showed that a northwest passage did not exist, Vancouver did miss one important waterway farther south. In May 1792, American Captain Robert Gray named and claimed the Columbia River (bordering present-day Oregon and Washington) for the United States after telling Vancouver of its presence. Vancouver dismissed the river's inlet as unimportant and failed to investigate or find the Columbia. Vancouver returned later to map the river, and Gray's claim carried new significance fifty years later when Britain and the United States were embroiled in the Oregon Boundary Dispute.

Nevertheless, Vancouver's survey inspired further exploration of the American continent. At a time when little was known about North America from the colonial frontier to the west coast, Vancouver's charts were treasures to future explorers. American President Thomas Jefferson read them with interest and encouragement as he and others planned Lewis and Clark's westward crossing of the continent.

BRENDA WILMOTH LERNER

Pedro Vial

1746?-1814

French Explorer

Pedro Vial was one of the most important pathfinders in the American Southwest. His most important accomplishment was blazing the Santa Fe Trail, connecting New Orleans and Santa Fe, New Mexico. He was also involved in an abortive attempt by the Spanish to halt the Lewis and Clark expedition, traveling overland to try to persuade local Native Americans to stop the explorers before they could accomplish their mission.

Vial was born in Lyons, France, in the middle part of the eighteenth century, probably in 1746. Little is known of his childhood, but it is likely that he moved to North America while in his late teens or early twenties, based on comments he made about being familiar with pre-Revolutionary America. Unfortunately, there is virtually no certain knowledge of his life before 1779, when he first became known to the Spanish government in New Orleans and Natchitoches.

He appears to have remained on the move for the next several years, reaching San Antonio in 1784 and living among some of the Native American tribes at times. Because of his experiences with the Taovayas Indians, in the spring of 1784 the Spanish government asked him to learn what he could of the Comanche Indians, associates of the Taovayas. Vial spent the next several months with the Comanche, returning to San Antonio that autumn. The next year, the Spanish governor asked him for help again, this time to find the most direct route between San Antonio and Santa Fe.

Vial set out with a single companion, making notes on distances, directions traveled, Indians encountered and their numbers, and other pertinent information. At some point it is thought he became ill, straying from the most direct path towards the north. After spending some time with the Taovayas, he continued on along the Red and Canadian Rivers, reaching Santa Fe about six months after departing San Antonio and traveling about 1,000 miles (1,609 km). Staying in Santa Fe for nearly a year, Vial

returned to San Antonio via Natchitoches, then returned to Santa Fe again, for a round-trip distance of nearly 2,400 miles (3,862 km) in just over 14 months.

In 1792-93 Vial journeyed again to Santa Fe, this time ending up in St. Louis. From St. Louis to New Orleans was simply a boat ride down the Mississippi River, so Vial had effectively connected Santa Fe with not only San Antonio, but St. Louis and New Orleans, too. By so doing, he had forged a trail that linked the major political and economic centers of the Spanish and French territories in the New World.

The enormity of this accomplishment is difficult to imagine in today's world of quick, easy, and convenient transportation. Vial had covered over 2,400 miles, much of it on foot across desert, mountains, and plains. He successfully traversed lands claimed by several Native American tribes, many of which were not friendly to Europeans. All of this was accomplished with little or no help, while hunting for the food and water needed for survival.

For the next several years, Vial appears to have worked for the Spanish authorities in their New World territories, acting as interpreter and liaison with the Native American tribes. For a short time after returning from St. Louis, Vial lived with the Comanche, but he eventually moved back to the St. Louis area. He returned to Santa Fe in 1803, apparently living there until his death in 1814.

P. ANDREW KARAM

Samuel Wallis

1728-1795

English Explorer

Samuel Wallis is best known for his voyages as Captain of the HMS *Dolphin*, a British vessel that circumnavigated the world under his command between 1766 and 1768. During this voyage, the *Dolphin* discovered the island of Tahiti and many other South Pacific islands that were soon to become famous. His reports inspired the voyages of Captain James Cook (1728-1779) and those of several French explorers who followed.

Wallis was born in Lanteglos-by-Camelford in 1728. Not much is known of his childhood or early adulthood, although it is likely that he received some degree of formal education before joining the Navy. It is certain that he was appointed flag lieutenant for Admiral Boscawen due to his high levels of performance, and was given command of the HMS *Dolphin* in 1766, shortly after the *Dolphin's* return from a record-breaking circumnavigation under the command of John Byron (1723-1786).

Wallis was sent to the Pacific with instructions to search for a postulated great southern continent, thought to exist in order to counterbalance the weight of the great land masses in the Northern Hemisphere. Wallis set out with the *Dolphin* and a second ship, HMS *Swallow*, which was under the command of Lieutenant Philip Carteret (1733-1796) for the South Pacific. Unfortunately, the *Swallow* and *Dolphin* became separated in a storm while entering the Pacific and were not to rejoin each other for the duration of the voyage.

Wallis continued across the Pacific, trying to stay to low latitudes in search of the southern continent. Unfortunately, he did not go far enough south to find Antarctica, partly because the size the continent was thought to be large—on the same order of size as Asia and Europe—and he assumed it would extend to a lower latitude than is the case with Antarctica. For most of the next 20 months, Wallis searched for the phantom continent, discovering in the process Tahiti, the Society Islands, and some of the Tuamotu Islands.

Wallis finally gave up on his search in 1768 and returned to Britain to complete his circumnavigation. This voyage made *Dolphin* the first vessel to complete two such trips, accomplishing this feat in just over four years. Carteret returned to Britain 1769 after continuing his explorations at lower latitudes, where he discovered and mapped many islands in the vicinity of New Guinea and surrounding islands.

Sadly, although Wallis's achievements were important, they consisted primarily of negative accomplishments; that is, he was able to show where the southern continent was *not* located. Although it is valuable to discover that something does not exist, explorers were expected to find new lands, not to find more ocean. Because of this, Wallis has remained an under-appreciated explorer for more than two centuries. It is important to note, however, that Wallis's explorations and discoveries provided a great deal of impetus to launch the expeditions led by James Cook, considered by many to have been the greatest marine explorer of this time or, indeed, of any time.

Wallis's voyage also helped encourage the French to begin Pacific exploration in earnest. In fact, Louis-Antoine de Bougainville (1729-1811) had already departed for the Pacific when Wallis returned to England, but his confirmation of many of Wallis's findings combined with Wallis's reports of new lands helped persuade the French to invest in exploration, too, in order to keep up with her European rivals.

Following his return in 1768, Wallis continued serving Britain as a naval officer. He died in 1795, following a long and honorable career.

P. ANDREW KARAM

Biographical Mentions

George Anson
1697-1762

English explorer who circumnavigated the world in the early 1740s while leading a group of British warships during war against Spain. Anson's mission was primarily military in nature, designed to harass Spanish shipping in the Pacific. His squadron returned to England with nearly £1 million in treasure, the largest haul ever returned by an English privateer. Anson's voyage was also among the last of its kind.

Juan Bautista de Anza
1735?-1788

Spanish military leader who headed several exploratory expeditions through northern Mexico and California. In 1774 Anza established an overland route connecting Sonora in northern Mexico to Monterey in California. Two years later Anza headed another expedition with colonists and their cattle that led to the establishment of the city of San Francisco in 1776. Later as Governor of New Mexico, Anza brokered peace agreements between colonists and several tribes in the area.

Jacques Balmat
1762-1834

French gem cutter and fur trader who was part of a two-man team to first reach the top of the highest peak in western Europe. Jacques Balmat and Michel-Gabriel Paccard reached the summit of the 15,771-foot-tall (4,807-m) Mont Blanc on August 8, 1786. Their successful ascent spurred the ascent of the mountain a year later by aristocrat and well-known scientist Horace-Bénédict de Saussure, which in turn introduced mountaineering to the masses. Balmat, Paccard and Saussure are widely credited with instituting the modern sport of mountaineering.

Nicolas Baudin
1754-1803

French explorer who helped map the Australian coast during the late eighteenth and early nineteenth centuries. Baudin was among the first of many French sailors to contribute toward the mapping of the Australian coast and the islands of the South Pacific and Indian Ocean. Though their efforts provided the French with the potential to launch a huge colonial empire, this advantage was squandered during the French revolutions and preoccupations with continental European affairs.

Joseph de Beauchamp
1752-1801

French astronomer who conducted the first known excavation of a Middle Eastern site. At the instigation of his uncle Miroudat, the Bishop of Babylonia, he was appointed Vicar-General in Baghdad (1781). While exercising the functions of this office, Beauchamp examined numerous Mesopotamian ruins and initiated digs in Babylon (1786). His excavations showed the ease with which cuneiform tablets could be recovered. Reports of his discoveries created a sensation in France and England.

Jean-Pierre François Blanchard
1753-1809

French balloonist who made the first flight across the English Channel by balloon with physician John Jeffries on January 7, 1785. Blanchard made his first ascent by balloon on March 2, 1784 in Paris. He made the first balloon ascent in North America on January 9, 1793, which was observed by George Washington. Blanchard also experimented with parachuting from balloons, on the first occasion with a dog, and on later occasions by himself.

Juan Bodega y Quadra
1743-1794

Commander of the Spanish ship *Sonora* on its exploration of the northwest coast of North America. In 1775 Bodega y Quadra sailed north along the California coast as far as what is now Juneau, Alaska, claiming portions of the coastline for Spain. Distance from Spain's other territories made his claims north of Oregon hard to maintain. After years as head of the Spanish Navy in the area, he surrendered Spain's exclusive claims to the area in negotiations with England.

Napoleon Bonaparte
1769-1821

French general who had an active interest in science. He proposed and led a military expedition to Egypt in 1798 that included scientists and scholars. Much scientific and scholarly research was conducted in Egypt, including the discovery of the Rosetta Stone (1799), which later proved key to deciphering hieroglyphics. During his 14-year reign as emperor (1800-1814) Napoleon encouraged science-based industries and fundamental research through substantial prizes and commissions.

Daniel Boone
1734-1820

American explorer and frontiersman who helped open the lands beyond the Appalachian Mountains to settlement. A hunter, trapper, and surveyor, he is the best-known pioneer of this era of American westward expansion. He laid out the Wilderness Road through Cumberland Gap, which became the principal route to the West, and settled Boonesborough, the first permanent white settlement in Kentucky. He was an effective Indian fighter and also served as a militiaman, legislator, and sheriff.

Philip Carteret
1733-1810

English navigator who commanded the *Swallow,* which accompanied Samuel Wallis during his circumnavigation on the *Dolphin* from 1766-68. Carteret was uncertain about commanding the *Swallow,* questioning her seaworthiness, but completed the voyage in spite of becoming separated from Wallis and losing most of his crew to disease or desertion. He completed a great deal of exploration in the area of what is now Indonesia, and discovered Pitcairn Island, where the mutineers of the H.M.S *Bounty* later sought refuge.

Richard Chandler
1738-1810

British traveler and archaeologist who cataloged the ancient Greek ruins of Asia Minor. Chandler, a member of the London-based Society of Dilettanti, toured Asia Minor in 1756 on an expedition funded by the Society. His *Ionian Antiquities* (1769) and multi-volume *Antiquities of Ionia* (1769-97) presented detailed measurements and drawings of classical buildings, many of which have since been destroyed. His books helped to inspire the Greek Revival movement in eighteenth- and nineteenth-century Anglo-American architecture.

Pierre François Xavier de Charlevoix
1682-1761

French Jesuit historian and explorer best known for his travels and explorations in the American West. Charlevoix traveled from Quebec along the St. Lawrence River into central North America, then down the Mississippi River to New Orleans. From there, in spite of a shipwreck, he returned to France and published what was to become one of the first accounts of the American interior, and the best full description of that region until that time.

Alexsei Chirikov
1703-1748

Russian seafarer and explorer who, with Vitus Bering, explored the coasts of northern Siberia and southern Alaska. During the second of two Kamchatka expeditions, Chirikov became separated from the rest of the ships but continued on to the Aleutian Islands and, in 1741, was the first to reach Alaska from Siberia. In spite of an epidemic of scurvy on board, Chirikov returned to Petropavlovsk with his few surviving crew members after having been the first European to discover and name many of the Aleutian Islands.

Pierre de Clérambault
1651-1740

French archaeologist and Louis XIV's librarian and genealogist, who conducted the first known excavation of a medieval site. Captivated by classical antiquities and prehistoric Celtic culture, Europeans of the Enlightenment largely ignored the Middle Ages, which they viewed as a barbaric period. Attitudes gradually changed in the early eighteenth century, and articles on medieval antiquities began appearing. In this atmosphere of increasing medieval interest, Clérambault in 1727 excavated the thirteenth-century graveyard at Châtenay-Malabry in France.

Alexander Dalrymple
1737-1808

Scottish aristocrat and hydrographer, the first at the Royal Navy (1795), who made a name for himself by publishing several works about the unexplored Pacific in an effort to be named head of an expedition that ultimately was led by Captain James Cook (1728-1779). Before being posted to the Navy, Dalrymple was an administrator and staff hydrographer for the British East India Company in the Pacific. His best-known book was *An Historical Collection of the Several Voyages and Discoveries in the South Pacific Ocean* (1769).

James Dawkins

British traveler and archaeologist who, with Robert Wood, surveyed sites in the Eastern Mediterranean in 1750-51. Dawkins and Wood's expedition, which took them to Greece, Asia Minor, Syria, Palestine, and Egypt, was one of several in the mid-eighteenth century that revealed the depth and complexity of ancient history to the British public. These expeditions stimulated public interest in classical antiquities as objects of both historical and aesthetic interest.

Antoine Raymond Joseph de Bruni d'Entrecasteaux
1739-1793

French navigator and explorer who, while in search of the lost expedition of La Pérouse, mapped the coasts of southern Australia and Tasmania. Although La Pérouse was last seen leaving Botany Bay, Australia, in 1788, d'Entrecasteaux was not dispatched from France until 1791. In spite of many discoveries, the expedition never found any trace of La Pérouse. In addition, through poor luck, d'Entrecasteaux failed to discover important parts of Australia that might have resulted in French, rather than English, colonization of that continent. D'Entrecasteaux died of scurvy on the return voyage.

Silvestre Vélez de Escalante
1751-?

Franciscan priest sent to the Zuni pueblos in what is now western New Mexico. In the form of a diary Escalante provided the best record about the Escalante expedition that left Santa Fe in 1776 for Monterey, California. Instead of following its planned route, the expedition made an extensive 1,000-mile (1,609-km) exploration of what is now western Colorado and central Utah, including the discovery of Lake Utah and native descriptions of the Great Salt Lake.

Johann Friedrich Esper
1732-1781

German pastor and paleontologist who discovered the first human fossils. During Esper's lifetime the prevailing view was that fossils were created during the Biblical deluge by sedimentary processes. Esper discovered fossilized remains of extinct animals and human bones in the cave of Gaillenreuth in Barvaria. Though initially considering that man had lived side-by-side with these animals, Esper's *Detailed Account of Newly Discovered Zooliths* (1774) adopts the deluvian interpretation that the fossils were brought together during the Biblical flood.

Bryan Faussett
1720-1776

British clergyman and archaeologist who, in the last 20 years of his life, excavated hundreds of Anglo-Saxon burial mounds in his home county of Kent. Faussett was among the first archaeologists in Britain to mount a systematic program of fieldwork in pursuit of a specific goal: in this case a clear understanding of the mounds' origins and purpose. His methods, though hurried and clumsy by modern standards, laid the groundwork for later, more sophisticated excavations by others.

Matthew Flinders
1774-1814

British naval officer and explorer who was the first to circumnavigate Australia, publishing an illustrated account of his expedition, *A Voyage to Terra Australis* (1814) on the eve of his death. Flinders began his naval career in 1789, served from 1791 to 1793 under the famed Captain William Bligh (1754-1817), explored portions of Australia's coast from 1795-99, then commanded the expedition of the *Investigator* from 1801-03. The *Investigator* and its crew of sailors and scientists (who collected thousands of specimens during the expedition) completed the first circumnavigation of the continent, putting to rest several false notions about the geography of the island.

Johann Georg Adam Forster
1754-1794

German naturalist who, with his father Johann Reinhold Forster, accompanied Captain James Cook on his second major expedition (1772-75). Following their return to England after Cook's wildly successful voyages, the younger Forster published *A Voyage Around the World* (1777). This popular two-volume work preceded Cook's own account, but was charged with portraying Cook in an unfairly negative light. The younger Forster was also an important influence on German naturalist Alexander von Humboldt.

Johann Reinhold Forster
1729-1798

German naturalist who, accompanied by his son Johann Georg Adams Forster, served with Captain James Cook on his second major expedition (1772-75). Despite his ill-temper and acrimonious relationship with Cook, the elder Forster contributed much to knowledge of the flora and fauna of the South Pacific, as well as the native people. Though initially forbidden by Cook to publish anything about the voyage, Forster eventually published several significant accounts of his botanical, zoological, and anthropological observations.

John Frere
1740-1807

British landowner and archaeologist whose 1797 description of stone tools found near Hoxne, Suffolk, laid the conceptual foundations of prehistoric archaeology. Frere, a member of the London-based Society of Antiquaries, theorized in a letter to the society that the tools had been made by a people far older and far more primitive than the ancient Britons described in Roman chronicles. Frere's insistence on the tools' great antiquity, and his use of their geological context to roughly measure that antiquity, challenged established archaeological methods and theories and ultimately helped to create a new understanding of humankind's origins and early history.

Francisco Tomas Hermenegildo Garcés
1741-1781

Spanish-born Franciscan missionary educated in Mexico who was later sent to what is now Tucson, Arizona. In 1776 Garcés traveled from what is now Yuma, Arizona, up the Colorado River to the Grand Canyon and east across the Arizona deserts, where he eventually stopped at the edge of the New Mexico territory. The voyage established a route between older Spanish settlements in the Rio Grande Valley and new settlements in California.

Isabela Godin des Odonnais
1728-1792

Peruvian noblewoman and wife of Jean Godin des Odonnais, a member of the 1735 La Condamine expedition, which determined the circumference of the earth at the equator and was the first scientific expedition to explore the Amazon. Isabela Godin endured a 20-year separation from her husband after the expedition and returned to him via a journey of which she was the only survivor, traveling over the Andes mountains, through tropical jungles, and up the Amazon.

Robert Gray
1755-1806

American navigator who explored the northwest coast of North America and in 1792 discovered the Columbia River, which he named after his ship. Gray also circumnavigated the globe twice, becoming in 1790 the first American to lead a ship around the world. He was also influential in opening trade with China, to which he sailed from Oregon in 1789 and exchanged a load of furs for tea on his return voyage to Boston.

Mikhail Spiridonovich Gvozdev
169?-17??

Russian sea captain and explorer who in 1732 was the first European to discover Cape Prince of Wales, Alaska. Cape Prince of Wales is the nearest point on the North American continent to Asia, lying just across the Bering Strait from the eastern-most point of Siberia. Gvozdev's discovery ushered in a brief period of Russian exploration and colonization in Alaska, which ceased when the United States purchased the territory.

Juan Pérez Hernandez
1725?-1775

Spanish military officer who in 1774 captained the ship *Santiago* further north than any other Spanish explorer had before on its exploration of the Pacific coast to what is now the Canadian-Alaskan border. Unfortunately, shipboard illnesses and the icy waters of late winter prevented Pérez from landing in many areas and making accurate maps of the coastlines and islands he claimed as Spanish territories. Pérez died of scurvy the following year.

Bruno de Hezeta
c. 1750-?

Spanish military officer who commanded the Spanish ship *Santiago* on its exploration of the Northwest coast of North America. In 1775, accompanied by the *Sonora* and its commander, Bodega y Quadra, Hezeta and his ship sailed north along the California coast as far as what is now Juneau, Alaska, claiming portions of the coastline for Spain. Spain eventually surrendered its claims in the area.

Daniel Houghton
1740-1791

Irish explorer whose travels in Africa were both valuable and fatal. Houghton traveled inland along the Gambia River to the Niger River, one of Africa's great rivers. In the process he helped initiate efforts to identify and map some of the major rivers that traverse the African continent, offering routes of transportation and exploration. Unfortunately, Houghton was killed by one of his guides after being robbed.

Friedrich Heinrich Alexander, freiherr (baron) von Humboldt
1769-1859

German scientist-explorer considered by many the father of geophysics. His explorations in the Americas include navigating the length of the Orinoco, studying Pacific coastal currents, and setting a world record by climbing Chimorazo volcano at 19,279 feet (5,876 m). He was the first to link mountain sickness with oxygen deficiency, and introduced isobars and isotherms for weather maps. Humboldt also related geographical conditions to animal and vegetation distribution, studied global temperature and pressure variations, and discovered that Earth's magnetic field decreases from the poles to the equator.

John Jeffries
1744-1819

American scientist and physician who was the first to make scientific observations from a balloon. Jeffries made two ascensions by balloon with Pierre Blanchard, the first in 1784 and the second in 1785; they were the first to cross the English Channel in this manner. On the first ascent, after reaching an altitude of 9,309 feet (2,837 m), Jeffries took samples of air for analysis and made other observations with a thermometer, a barometer, and other instruments.

Estéban José Martinez
1742-1798

Spanish explorer who established the first Spanish base at what is now Vancouver Island on the Pacific coast of Canada. Juan Perez had claimed the island for Spain in 1774, but English and American traders began trapping animals in the area after the English explorer James Cook visited in 1778. In 1789 Martinez was sent to enforce Spain's trading rights to the island. Relations with foreign traders there quickly soured, and trade disputes eventually led to Spain abandoning its exclusive claims to territories north of what is now Oregon.

Jose Celestino Mutis
1732-1808

Spanish botanist, physician, astronomer, and priest who spent over 50 years studying and meticulously cataloging the biodiversity of New Granada (Colombia). In 1783 Mutis led a botanical expedition in which over 5,000 plants were illustrated and classified according to the newly developed Linnaeus system of modern scientific nomenclature. Mutis distinguished four species of chinchona, a valuable plant used for anti-malarial and heart medications. The Royal Botanical Gardens in Madrid house most of the Mutis-expedition illustrations.

Carsten Niebuhr
1733-1815

German explorer who was the only survivor of the 1761 Danish expedition to Arabia. While in Egypt, he was the first person to make an accurate copy of hieroglyphics, and in Persia his work led to the deciphering of cuneiform writing. His many maps and calculations were all remarkably precise. He returned to Denmark in 1767, then married and lived quietly as a council clerk in Germany for the remainder of his life.

William Pars
1732-1782

British artist who, with archaeologist James Chandler and architect Nicholas Revett, surveyed the antiquities of Ionia on a 1756 expedition funded by London's Society of Dilettanti. Pars's detailed illustrations of ancient Greek buildings for Chandler's multi-volume work, *Antiquities of Ionia* (1769-97), contributed to the Greek revival that shaped British and American architecture in the late eighteenth and early nineteenth centuries. They also preserved, for later archaeologists, a detailed record of structures long since demolished.

Giovanni Battista Piranesi
1720-1778

Italian draftsman, printmaker, architect, and art theorist also called Giambattista Piranesi, who was best known for his highly influential etchings of architectural views and his extraordinary imagination demonstrated in two series of engravings of imaginary prison scenes (the *Carceri d'Invenzione*, c. 1745). A prolific graphic artist, Piranesi produced more than 2,000 plates of intricate, textured, and highly contrasting imagery, which depicted a variety of subjects from ancient Roman monuments and ruins to imaginary reconstructions of ancient structures.

Richard Pococke
1704-1765

English bishop and explorer known for his *A Description of the East, and Some Other Countries* (1743, 1754). In this two-volume work Pococke produced systematic and complete descriptions of many archaeological sites he visited during his travels in Egypt and the Middle East, including the pyramids at Giza, Sakkara, and Dashur. His work grew in significance because many of the monuments he described were destroyed before the Napoleonic expedition of 1798.

Peter Pond

Canadian explorer who, in 1778, was the first white man to discover the area around what is now Athabasca, Canada. His explorations opened this part of the Canadian wilderness for the fur trade and helped extend British influence through central Canada. This began the process

of exploiting not only the area's natural resources, but natives living there, too.

Thomas Pownall
1722-1805

English colonial governor and antiquarian who sought to unite archaeology with history. Elected to membership in the Society of Antiquaries (1768), Pownall investigated prehistoric burial mounds and Roman ruins. He urged a reassessment of the objectives and methods most antiquarians employed for collecting artifacts, seeking to promote systematic historical writing informed by reliable archaeological research. Pownall is better known for advocating the union of Britain's North American territories in *Administration of the Colonies* (1764).

Francisco Requena
1743-1824

Spanish general and member of the commission formed to execute the 1777 Treaty of San Idelfonso, which set the borders between the Spanish and Portuguese empires in South America. Requena led an extensive expedition to map the tributaries of the Amazon River and settle the boundaries. He later became governor and commander of the Mainas territory in what is now Peru.

Nicholas Revett
1720-1804

English architect who, with the artist James "Athenian" Stuart, traveled to Athens to survey and draw the monuments and classical antiquities. Upon returning to England they were admitted to the Society of Dilettanti, whose members aided them in publishing *The Antiquities of Athens* (1761). Jealous that Stuart received the lion's share of the credit, Revett broke with him. Revett later surveyed the antiquities of Asia Minor.

Jacob Roggeveen
1659-1729

Dutch admiral who was the first European to discover Easter Island. On Easter Sunday in 1722, Jacob Roggeveen and his crew found the remote island off the coast of Chile. The island, which has a population of 3,000 people, became known mostly for the great stone heads that peer out over the ocean from its shoreline. The heads, constructed of volcanic ash, weighed up to 50 tons apiece and still remain on the island. Anthropologists and linguists have since also taken a keen interest in the written language of the island's inhabitants, who were believed to be of Polynesian descent.

Jean-François Pilatre de Rozier
1756-1785

French balloonist who, on November 21, 1783, became the first, along with François Laurent, marquis d'Arlandes, to fly a manned balloon over Paris. The balloon ascended to 500 feet (152.4 m), free of even a rope tied to the ground, and flew approximately 20 minutes, traveling 5.5 miles (8.85 km). Rozier was killed in a balloon of his own design, using hot air with a hydrogen balloon, as he attempted to cross the English Channel.

Junìpero Serra
1713-1784

Spanish missionary who founded a string of missions in California. Born Miguel José Serra, he entered the Franciscan order at Palma in 1730, and went with other missionaries to Mexico in 1749 to preach to the native people. In 1767 he went on to California (then known as Alta California) where he gained such monikers as "Apostle of California" and "walking friar" as he traveled on foot—despite an ulcerated leg—to establish nine missions from 1769 to 1782. They included missions in San Diego, San Luis Obispo, and San Juan Capistrano.

Daniel Charles Solander
1736-1782

Swedish geologist and botanist who assisted Joseph Banks, chief of scientists aboard the *HMS Endeavor* captained by James Cook, during their Pacific voyage from 1768 to 1771. Solander participated in the landing on Poverty Bay where he collected, described, and drew pictures of many plants. These pictures were used by Banks for the engravings of plants in his *Bank's Florilegium*. Two islands are named after Solander; one in the Mergui Archipelago and the other one south of New Zealand.

James Stuart
1713-1788

British architect and writer who helped establish the resurgence of interest in classic Greek art and architecture through his enormously influential *Antiquities of Athens*, a multi-volume work whose first volume was published in 1762. Stuart's popular volumes introduced Europe to the splendor and uniqueness of Greek art versus "derivative" Roman works, which had been the model of Classicism since the Renaissance. The books signaled the beginning of neoclassicism.

William Stukeley
1687-1765

English antiquarian and physician noted for his archeological work on Neolithic and Bronze Age sites. His studies of stone circles and earthworks, especially at Avebury and Stonehenge, led him to postulate elaborate theories connecting them to the ancient Celtic cult of the Druids, maintaining that the entire prehistoric landscape had been laid out in a sacred pattern. Stukeley's views were widely accepted in the late eighteenth century. His excellent field studies and surveys remain of interest.

Johann Heinrich Wilhelm Tischbein
1751-1829

German artist known as "Goethe" because of his celebrated portrait *Goethe in the Campagna* (1787). Tischbein began his career as a portrait painter but was equally well known for his engravings of classical antiquities. He studied art in Paris and then in Italy, where he was named Director of the Naples Academy of Art (1789). In 1799 Tischbein returned to his native Germany, where he developed an interest in German Romanticism before his death in 1829.

Charles Townley
1733-1805

British traveler and collector who stimulated interest in the archaeology of ancient Rome. Like many wealthy Englishmen of his era, Townley toured the Continent in order to absorb its history and culture. Visiting Naples in 1768, Townley met British diplomat Sir William Hamilton and, following his example, became an enthusiastic collector of ancient Roman sculpture. Willed to the British Museum upon his death, the Townley marbles became the core of the museum's Roman sculpture collection.

Johann Joachim Winckelmann
1717-1768

German scholar considered the father of art history and modern archaeology. Winckelmann was the first scholar to study works of art in context. His methodology and systematic classification of artistic styles explored geographical, social, and political circumstances to trace the history and explain the meaning of art objects. His scholarly methods profoundly impacted the development of classical archaeology. Winckelmann's idealized vision of Ancient Greece greatly influenced the development of European neoclassical fine arts and literature.

Robert Wood
1717-1771

British traveler and topographer who explored the ancient ruins of the Eastern Mediterranean in 1742 and 1750-51. He was the first person to systematically survey the supposed site of the lost city of Troy while working from the assumption that Homer's *Illiad* described real historical events. Wood's survey, though it did not locate the actual ruins of Troy, helped to pave the way for Heinrich Schliemann's discovery of them in the nineteenth century.

Bibliography of Primary Sources

Books

Bruce, James, and B. H. Latrobe. *Travels to Discover the Sources of the Nile, in the Years 1768, 1769, 1771, 1772, and 1773* (5 vols., 1790). This work describes Bruce's expedition to discover the source of the Nile River (he only discovered the Blue Nile). The book was marked by controversy, as Bruce refused to pay Latrobe for his efforts and refused to credit the Italian artist, Balugani, who had died in Ethiopia during the expedition, for his many superb drawings. In fact, Bruce took personal credit for the Italian's drawings. The book's reception was dismal.

Chandler, Richard. *Ionian Antiquities* (1769) and *Antiquities of Ionia* (1769-97). In these works Chandler presented detailed measurements and drawings of classical buildings, many of which have since been destroyed. His books helped to inspire the Greek Revival movement in eighteenth- and nineteenth-century Anglo-American architecture.

Gmelin, Johann. *Journey Across Siberia* (1751). This book consisted of Gmelin's diaries of his ten-year expedition to Siberia. Translated into numerous languages, the book became a best-seller.

Hearne, Samuel. *A Journey from Prince of Wales' Fort in Hudson's Bay to the Northern Ocean* (1795). In this work Hearne described his landmark expedition through northern Canada to the Arctic Ocean.

Niebuhr, Carsten. *Reisebeschreibung nach Arabien und andern umliegenden Ländern* (1774; English translation as "Travels through Arabia and Other Countries of the East" by R. Heron, 1799). This work described the first European scientific expedition to the Near East, dispatched in 1761, in which the entire crew save Niebuhr died. Niebuhr continued exploring and, upon his return, published several important reports. These included maps that were used for a century, cuneiform inscriptions copied from the ruins at Persepolis, and botanical data gathered by the expedition's lost naturalist.

Park, Mungo. *Travels into the Interior Districts of Africa* (1797). This book, written shortly after Park's first trip to explore the course of the Niger River, excited considerable public interest because it was one of the first accounts of Africa to be written based on actual personal experience, rather than based on the often-outlandish accounts of others. Park never finished tracing the course of the river. In fact, none of the discoveries made on his second trip made their way back to Britain because everyone in the party died.

Pococke, Richard. *A Description of the East, and Some Other Countries* (2 vols., 1743, 1754). In this work Pococke produced systematic and complete descriptions of many archaeological sites he visited during his travels in Egypt and the Middle East, including the pyramids at Giza, Sakkara, and Dashur. His work grew in significance because many of the monuments he described were destroyed before the Napoleonic expedition of 1798.

Stuart, James and Nicholas Revett. *Antiquities of Athens* (1762, 1789, 1795). These popular volumes, composed after Stuart and Revett traveled to Athens, introduced Europe to the splendor and uniqueness of Greek art versus "derivative" Roman works, which had been the model of Classicism since the Renaissance. The books signaled the beginning of neoclassicism. Revett later broke with Stuart because the latter received most of the credit for this work.

Vancouver, George. *Voyage of Discovery to the North Pacific Ocean and Round the World in the years 1790-1795* (3 vols., 1798). Complete with narrative, charts, maps, and drawings, this work describes Vancouver's expedition to chart the northwest coast of America and Canada. The accurate soundings and coastal information in Vancouver's book were invaluable. They included the location of dangerous rocks and offshore islands, sandbars, entrances to bays, and the location of viable, usable harbors. This ground-breaking work gave the English a significant claim to the northwest coast, at least enough to satisfy the king and his countrymen.

Wallis, Samuel. *Account of the Voyage Undertaken for Making Discoveries in the Southern Hemisphere* (1773). This book detailed Wallis's circumnavigation of the world in 1766-68. During this voyage, which was an attempt to find the famed and elusive southern continent, *terra australis,* Wallis encountered Tahiti.

Articles

Frere, John. "An Account of Flint Weapons Discovered at Hoxne, in Suffolk" (1800). This paper described Frere's discovery of a group of chipped-flint objects in a brick-earth quarry near Hoxne in 1790. What made the tools remarkable, first to Frere and later to others, was that they lay beneath 12 feet (3.66 m) of undisturbed soil and gravel, below (and thus older than) a sand layer containing shells that appeared to be marine and the bones of a large, apparently extinct mammal. Frere concluded that the tools and their unknown makers belonged to a time long before humans were thought to have existed. Frere's interpretation of the stone tools challenged current ideas about of the early history of the human race and set the stage for the evolution of a new understanding over the next 60 years.

NEIL SCHLAGER

Life Sciences and Medicine

Chronology

1713 The first dissecting room is established, in Berlin, Germany.

1725 *Histoire Physique de la Mer* by Luigi Ferdinando Marsili is the first scientific study of the oceans.

1730 Stephen Hales, an English botanist and chemist, discovers that the reflex movements of a frog's legs depend upon the spinal cord.

1735 Swedish botanist Carolus Linnaeus outlines his system for classifying living things, with a binomial nomenclature that includes generic and specific names.

1737 *Bijbel der Natuure* by Dutch naturalist Jan Swammerdam—later recognized as the father of entomology—is finally published, 100 years after his birth and long after his death.

1757 Albrecht von Haller of Switzerland begins publishing his eight-volume *Elementa Physiologiae Corporis Humani*, which marks the beginnings of modern physiology.

1759 Kaspar Friedrich Wolff, a German physiologist, launches modern embryology with his book *Theoria Generationis*.

1761 The publication of *De Sedibus et Causis Morborum per Anatomen Indagatis* by Giovanni Battista Morgagni leads to the establishment of pathology as a branch of science.

1763 Joseph Gottlieb Kolreuter, a German botanist, publishes his pioneering work on plant hybridization and gives the first account of his experiments on the artificial fertilization of plants.

1768 Lazzaro Spallanzani, who three years earlier had published data refuting the theory of spontaneous generation, concludes that boiling a sealed container prevents microorganisms from entering and spoiling its contents.

1771 English surgeon John Hunter founds scientific dentistry with his treatise *The Natural History of the Human Teeth*.

1779 Jan Ingenhousz, a Dutch plant physiologist, shows that light is necessary for plants to produce oxygen, and that sunlight is the energy source for photosynthesis.

1796 English physician Edward Jenner performs the first successful smallpox inoculation, on James Phipps.

Overview:
Life Sciences and Medicine 1700-1799

Previous Period

Though his research did not directly involve the life sciences, the seventeenth-century physicist Isaac Newton (1642-1727) greatly influenced how living things were studied in the eighteenth century because of his emphasis on finding simple laws underlying the complexities of movement. The earlier work of philosopher Francis Bacon (1561-1626) was also influential in the eighteenth century. Bacon stressed the importance of direct observation and of experimentation as keys to progress in science. In addition, the explorations around the globe in the sixteenth and seventeenth centuries, as well as the development of the microscope in the seventeenth century, opened up new worlds filled with fascinating life to be studied.

Basic Questions

As explorers sent back a treasure trove of new species to Europe, the problem of how to classify or organize them became urgent. While the British botanist John Ray (1628-1705) and others developed various classification schemes, it was the system of Swedish botanist Carolus Linnaeus (1707-1778) that became most widely accepted. He based his classification of plants on the structure of their flowers and gave each species two names, the first for the genus, or general category to which the organism belonged, and the second for the species itself. French zoologist Georges Buffon (1707-1788) published a multi-volume survey of the animal world, while other zoologists focused on particular groups of organisms, with René de Réaumur (1683-1757) publishing a massive work on insects.

One of the driving forces of seventeenth-century life science, particularly in Britain, had been natural theology, the attempt to learn about God by studying design in nature. There were, however, others who argued that there was no need to invoke divine intervention to explain life processes. Following the lead of the seventeenth-century French philosopher René Descartes (1596-1650), they developed a mechanistic concept of life, seeing it as governed by the same physical principles that Newton had identified as underlying movement in the nonliving world. Their ideas were opposed by vitalists, who saw life processes as involving principles unique to life and different from those in the nonliving world. These questions continued to be debated throughout the eighteenth century.

Another controversy in eighteenth-century life science involved development—how the form of an individual emerged. Some scholars, called preformationists, argued that in the case of human development, for example, the germ from which an individual arose contained a tiny replica of the human body, and all that happened during gestation was that this preformed figure increased in size. Others, who adopted epigenesis, contended that the human form was not preexisting, but instead only gradually developed and increased in complexity during that development.

Another debate at this time involved the question of spontaneous generation, of whether living organisms, particularly microscopic organisms, could arise from nonliving matter or whether they could only come from others of their kind. In 1767 Lazzaro Spallanzani (1729-1799) challenged the theory of spontaneous generation by demonstrating that microbes will not appear in flasks of broth that have been boiled and sealed. This controversy, however, would continue far into the nineteenth century.

Anatomy and Physiology

Francis Bacon's call for careful observation and experimentation influenced the development of anatomy and physiology. Marie-François Bichat (1771-1802) discovered that the body's organs are made up of many different kinds of tissues; his work opened the way for later studies on the cellular organization of tissues. There were also many efforts to discover the relationship between anatomy and disease, with Giovanni Morgagni (1682-1771) being particularly important in the growth of pathology, the study of abnormal anatomy.

Physiology, the study of function, also blossomed. In plant physiology, careful experimentation was done in a number of areas: Stephen Hales (1677-1761) studied the movement of water through plant tissues; Jan Ingenhousz (1730-1799) established that sunlight was necessary for the production of oxygen in leaves and thus for photosynthesis; Joseph Kölreuter (1733-1806) described how insects and birds carry pollen from one plant to another.

In animal physiology, digestion and movement were given particular attention and there was a great deal of study of the nervous system. Hales also made important contributions to this field with his discovery of the connection between the spinal cord and reflex movements in frogs. Albrecht von Haller (1708-1777) was the first to demonstrate that nerves stimulate muscle contraction, and Luigi Galvani (1737-1798) discovered the electrical basis of nerve impulses. The field of psychology, a term coined by David Hartley (1705-1757), began at this time and dealt with the study of the mind and behavior. A number of philosophers, including David Hume (1711-1776) and François Voltaire (1694-1778), originated theories of the mind and of the human sense of morality.

Medicine: Prevention of Disease

The eighteenth century saw a number of discoveries important to the prevention of disease, though in several cases the full impact of these discoveries was not felt for some time. For example, Percival Pott (1714-1788) found a high incidence of cancers of the scrotum and nasal cavity among chimney sweeps. This provided the first link between environmental factors and cancer, links that are still being investigated. James Lind (1716-1794) demonstrated experimentally that citrus fruits could prevent and cure scurvy, a nutritional deficiency common on long sea voyages and ultimately linked to a lack of vitamin C. At the end of the century, Edward Jenner (1749-1823) developed the first vaccine, an inoculation to prevent against the dreaded disease of smallpox. The vaccine contained cowpox, a virus that was less dangerous than the one causing smallpox, yet one that produced an immune response providing protection against the more serious infection.

It was also in the eighteenth century that the field of public health originated with the aim of improving living conditions, thus controlling and preventing disease. Johann Peter Frank (1745-1821) wrote a six-volume work on medical policy and identified poverty as the "mother of disease." John Howard (1726-1790) campaigned for improved cleanliness in poorhouses,

prisons, and hospitals, where sanitary conditions had been abominable. In general, there was a growth in the number and importance of hospitals, with a wave of construction occurring in Europe after 1730 and in America after 1750.

Medicine: Treatment Methods

There were changes in a number of branches of medicine during the eighteenth century. Franz Mesmer (1734-1815) developed a controversial technique of hypnotism to cure disease. Also, surgery became a more common and respected field as surgeons developed techniques for treating such conditions as ear infections and cataracts and for the removal of cancerous lymph glands. The British surgeon John Hunter (1728-1793) was a key figure in the transformation of surgery from a craft into an experimental science through his research on inflammation.

Along with the increased status of surgeons, there was a move in obstetrics from female to male domination of the field of midwifery, where new instruments such as the forceps were used in delivering babies. In dentistry, Pierre Fauchard (1678)-1761) made advances in techniques for filling cavities and Philipp Pfaff popularized casting methods for making false teeth.

The Future

By the close of the eighteenth century the life sciences were poised for the great advances that would occur in the next century. Jean Baptiste de Lamarck (1744-1829), building on the work of his teacher Buffon, would soon publish his theory of evolution, asserting that species change over time. Later, Charles Darwin (1809-1882) would offer an even more convincing evolutionary theory. Experimentation would continue to become more important in the study of organisms, with Claude Bernard (1813-1878) popularizing the experimental method in biology and doing significant research in physiology. Work toward improving sanitation would expand with the building of sewer systems and by providing cleaner water supplies in cities. Medicine would also see the development of antiseptic procedures in surgery and the establishment of the germ theory of infectious disease.

ROBERT HENDRICK

Natural Theology

Overview

Natural theology is a system of finding basic truths about the existence of God and human destiny by reason. "Natural" refers to the idea that reason is an essential faculty possessed by all thinking people. Thus, in this view, rational thinking may also provide a basis for revelation, so that reason and revelation go hand in hand. This approach supports the discovery of religious truths through rational argument, proofs, and reason, and is often concerned with two principal topics: 1) Can God's existence be logically and rationally proved?, and 2) Can the immortality of the soul be arrived at through logical, rational argument?

Several eighteenth-century philosophers and scientists, notably John Ray (1627-1705), Robert Boyle (1627-1691), William Derham (1657-1735), Nehemiah Grew (1641-1712), and Samuel Clarke (1675-1729), contributed to and developed these ideas. Natural theology had a great influence on the sciences, and biology in particular. Opponents to natural theology were led by Scottish philosopher David Hume (1711-1776), who pointed out that while the system had logic, it was also possible for things to fall together by chance.

Background

The idea of the existence of God in nature was present in the Greco-Roman world. Saint Paul, in his epistle to the Romans, wrote that since the creation of the world an eternal God is seen in things that are made. (Romans 1:20). Design in nature is a major supportive theme and argument. These organized designs, it was assumed, pointed toward a divine Creator who must have created the patterns. Anselm, an early church thinker, proposed that all humans possess the idea of God, which in itself implies the existence of a corresponding reality.

Thomas Aquinas (1225-1274) posited "five ways" of establishing the existence of God, so that from our experience as human beings we must conclude the existence of an infinite being on whom the world depends. According to his schema: 1) when looking at the world of motion, we must conclude there is a Prime Mover; 2) when looking at result or cause, we must conclude there is a First Cause; 3) when we look at the unsure happenings in life, we must conclude there is a Necessary Being; 4) when we look at the imperfections of life, we must conclude there is a Perfect Being; 5) when we look at order and design, we must conclude there is a Supreme Intelligence that created design and order.

The intellectual energy of the Renaissance sparked an adventuresome exploratory spirit in Europe. In 1509 Nicholaus Copernicus (1473-1543) proposed a revolutionary new view of the cosmos, positioning the sun—not the Earth—at the center of the solar system. This impact on traditional thought caused many to question the values of the Church and the existence of God.

Impact

The Enlightenment of the seventeenth and eighteenth centuries saw a period of great conflict between science and religion. Many of the Enlightenment thinkers were also Christians who did not want to undermine religious beliefs. The theologians of the time tried to adjust their beliefs to the scientific theories of Copernican astronomy and Newtonian physics. Many revived and revised St. Thomas's earlier five proofs. These scientists were involved in different areas of discovery.

John Ray, the son of a village blacksmith in Black Notley, Essex, England, came to Cambridge at a time when the organization of chemistry and anatomy was just beginning. A devout Puritan, he became respected and prospered as part of the regime of Oliver Cromwell. He received a bachelor's degree in 1648 and then spent the next 13 years as an academic scholar. With the Restoration of the monarchy, Ray hit unfortunate times since he was a dedicated Puritan. He had refused to sign the Act of Uniformity, leading to his dismissal from Cambridge in 1662. Some prosperous friends supported him for 43 years while he continued to work as a naturalist.

In 1660 Ray began to catalog plants growing around Cambridge and, after studying that small area, explored the rest of Great Britain. His dedication to taxonomy (the establishment of an orderly account of species) was fitting, as he was also interested in finding order in the world through natural theology. A turning point in his life occurred when he met Francis Willughby

(1635-1672), a fellow naturalist who convinced him to undertake a study of the complete natural history of living things. Ray would investigate all of the plants and Willughby all of the animals.

The two searched Europe for flora (plants) and fauna (animals). In 1670, just as Ray produced his "Catalog of English Plants," Willughby suddenly died, leaving Ray to finish both projects. He published *Ornithology* and *History of Fish,* giving all the credit and recognition to Willughby.

While he was continuing his major works in botany, Ray also published three volumes on religion. In an 1691 essay called "The Wisdom of God Manifested in the Works of Creation" he delineated the full range of his correlation of science with natural theology. The essay described how it was obvious that form and function in organic nature demonstrated there must be an all-knowing God. This argument of design supporting a static view was contrdicted by the development of evolution in the nineteenth century.

Also of importance was Robert Boyle, an Irish chemist and natural philosopher. Though known for his work on the properties of gases and the chemical elements, he was a great defender of the Christian faith. The fourteenth child of a family of wealth and influence, he went to Eton College and there began his experimental work while writing moral essays as well. At his estate in Ireland he was very interested in anatomical dissection. At the University of Oxford he came into contact with Robert Hooke (1635-1703), the scientist who discovered cells. Hooke was also an able inventor, and they worked to demonstrate the physical characteristics of air and the necessity of air in combustion, respiration, and the transmission of sound.

Boyle was dedicated to proving the Christian religion against what he called "notorious infidels." He was a devout Protestant and took a special interest in promoting the Christian religion abroad. In 1690 he wrote a book called *The Christian Virtuoso,* in which he sought to show the study of nature was a central religious duty. He believed that nature was a clock-like mechanism that had been made and set in motion by a divine Creator. This, it was believed, was done at the beginning of time and thereafter the world functioned according to secondary laws. In line with natural theology, Boyle asserted that science can reveal these secondary laws. Indeed, he believed the human soul was much more than just the circulation and moving blood cells in the body.

William Derham, a chaplain and ordained deacon of the Church of England, was very interested in the work of Boyle and delivered lectures on his ideas. He published many papers on meteorology, astronomy, and natural history, as well as an outstanding work on the sexes of wasps. His admiration of the works of Ray and Hooke led him to natural theology.

Though Derham published new editions of Ray's *Physico-Theological Discourses* and *Philosophical Letters,* it is for his own works that he is known. *The Artificial Clockmaker* and *Astro-Theology* were translated into French, Swedish, and German. "Astro-theology" concerns the attempt to argue from astronomy to God. Derham's works greatly influenced William Paley (1743-1805), the great natural theologian of the late eighteenth century.

Nehemiah Grew, an English physician and botanist, was one of the founders of the science of plant anatomy. As he pondered the growth and anatomy of plants and how they reproduce, he became convinced that the design must be part of the wisdom of God, the infinite designer. Grew was very prominent and was elected to the Royal Society in 1671. His most famous work was *The Anatomy of Plants* (1682), but he also wrote *Cosmologia Sacra,* or sacred cosmology, in 1701, in which he described the rational beliefs of natural theology.

Samuel Clarke, an English theologian, philosopher, and chaplain to Queen Anne, attempted to prove the existence of God by mathematical methods. He presented his ideas in *A Demonstration of the Being and Attributes of God* (1795). He also established a group of moral principles that aspired to the certainty of mathematical propositions, presented in the essay "A Discourse Concerning the Unchangeable Obligation of Natural Religion" (1706). Clarke was a friend and disciple of Isaac Newton (1642-1727) at the University of Cambridge, and was dedicated to spreading Newton's views. He defended Newton as well as Christian religion in four collected volumes of his works (1738-42). It was Clarke in particular who provoked Hume's criticism of religion.

Hume, one of the greatest critics of natural theology, is well known for his skepticism and empiricism. In his 1779 *Dialogues Concerning Natural Religion* he determined that the world was more or less a smoothly functioning system or it would not exist. He argued that the world may have come about by chance as particles falling into a temporary or permanent self-sustaining order. These chance developments have the look of design, but really are not. Hume also pointed out that the world is not perfect. There are many examples of human and animal suffering, and one cannot assume a good and powerful Creator would make such an imperfect world.

Less than a century later Charles Darwin (1809-1882) postulated that adaptations of life forms are the result of natural selection. Those species that adapted, survived; those that did not, died. And thus the "survival of the fittest" was ensured. Darwin's theory of evolution by natural selection contradicted the claims of natural theology. Evolution suggested that life originated and changed according to scientifically structured laws with timeframes and outcomes that directly refuted biblical claims.

Discussions about natural theology now are generally relegated to philosophers and theologians. While some groups invoke arguments of divine design, most modern scientists have reconciled and accepted a separation between science and religion.

EVELYN B. KELLY

Further Reading

Maddison, R.E.W. *The Life of the Honourable Robert Boyle.* London: Taylor & Francis, 1969.

Raven, Charles E. *John Ray, Naturalist: His Life and Works.* 2nd ed. Cambridge: Cambridge University Press, 1986.

Westfall, Richard S. *Science and Religion in the Seventeenth Century England.* Hamden, CT: Archon, 1970.

The Mechanical Philosophy: Mechanistic and Materialistic Conceptions of Life

Overview

Physiological studies from ancient times to the present have been guided implicitly or explicitly by a philosophical framework that has been either mechanistic or vitalistic. The mechanistic philosophy asserts that all life phenomena can be completely explained in terms of the physical-chemical laws that govern the inanimate world. Vitalist philosophy claims that the real entity of life is the soul, or vital force, and that the body exists for and through the soul, which is incomprehensible in strictly scientific terms. The triumph of Newtonian physics is reflected in the mechanistic materialism of the French *philosophes* of the Enlightenment and the mechanical philosophy adopted by many naturalists. Among the eighteenth-century naturalists who applied the concepts of mechanical philosophy to their scientific research and writings were Julien de la Mettrie (1709-51); Georges Louis Leclerc, comte de Buffon (1707-1788); and Erasmus Darwin (1731-1802).

Background

New ideas about chemistry, physiology, and medicine were promoted by teachers at Europe's leading medical schools, including Hermann Boerhaave (1668-1738), who taught chemistry, physics, botany, ophthalmology, and clinical medicine at the University of Leiden. Boerhaave's approach to anatomy, physics, and chemistry was widely disseminated by devoted students and disciples, such as Albrecht von Haller (1707-1772). Physiology owes a great debt to von Haller, who carried out an enormous number of experiments and summarized the state of physiology in his *Elements of Physiology.* For example, von Haller's experiments on the nervous system led him to conclude that the soul had nothing in common with the body, except for sensation and movement that seemed to have their source in the medulla; therefore, the medulla must be the seat of the soul. Eighteenth-century chemists also made major contributions to understanding the relationship between the circulation of the blood and the exchange of gases in respiration. Stephen Hales (1677-1761), who invented the pneumatic trough, investigated the hydraulics of the vascular system, the chemistry of gases, the respiration of plants and animals, and the measurement of blood pressure. Much of this work was reported in *Vegetable Staticks* (1727) and *Statistical Essays, containing Haemastaticks* (1733).

Erasmus Darwin. *(Archive Photos, Inc./Corbis. Reproduced with permission.)*

According to the mechanical philosophy, also known as materialism, all phenomena, even if not obviously mechanical in nature, can be reduced to primary qualities inherent in matter. In part, the mechanical philosophy represented the rejection of Aristotelian physics and the triumph of the Scientific Revolution, which led to new ways of thinking about the cosmos and the dynamic functions of the body. The mechanical philosophy systematized by the philosopher and mathematician René Descartes (1596-1650), who described animals and the human body as machines, is of special significance to the history of physiology.

The writings of Descartes provided the most influential philosophical framework for a mechanistic approach to physiology. The fundamental platform of Descartes's mechanical philosophy was that all natural phenomena could be explained in terms of matter and motion. Nevertheless, the mechanical philosophy was not necessarily atheistic or incompatible with religious beliefs in a Creator. Pious naturalists could assume that God was not directly involved in the ordinary motions of the universe or the normal activities of living beings.

Cartesian doctrine essentially treated animals as machines and explained their activities as the motions of material corpuscles and the heat generated by the heart. Even human be-

ings could be seen as earthly machines that differed from animals by virtue of the "rational soul," which governed thought, volition, conscious perception, memory, imagination, and reason. Descartes's disciples saw him as the first philosopher to dare to explain human beings in a purely mechanical manner. The Cartesian concept of animals and the human body as machines reached its ultimate expression in the writings of Julien de La Mettrie, author of *L'homme Machine* (1748). The mechanical philosophy allowed naturalists to investigate nature without relying on the vitalistic "soul" and "spirits" that had characterized ancient and Renaissance science. Only the rational soul of human beings remained as part of natural history. Some natural philosophers, such as La Mettrie, were willing to discard even this "soul" for a fully materialistic and atheistic system.

Impact

La Mettrie was quite willing to let his materialistic scientific theories conflict with or contradict Christian dogma. His materialistic interpretations of psychic and physiological phenomena were published in *A Natural History of the Soul* (1745), *L'Homme-machine, Discourse on Happiness* (1748), and *The Small Man in a Long Queue* (1751). La Mettrie argued that we cannot really understand either the soul or matter. Clarity of thought, however, was obviously affected by fevers and other illnesses. This proved that thought, allegedly a function of the rational soul, was actually a function of the brain and was dependent on physical conditions. Because of the problems caused by his unorthodox views, La Mettrie sought refuge in Holland. In *L'Homme-machine,* which he wrote in Holland, La Mettrie presented man as a machine whose actions were entirely the result of physical-chemical factors. Experiments on animals indicated that peristalsis continued even after death and that isolated muscles could be artificially stimulated to contract. La Mettrie reasoned that if these facts were true for animals, they must also be true for humans. In contrast to Descartes, however, La Mettrie denied the claim that humans were essentially different from animals. For La Mettrie, humans were essentially a variety of monkey, superior mainly by virtue of the power of language. Rejecting the mind-body dualism of Descartes, La Mettrie further argued that even the mind must depend directly on physiochemical processes. Substances such as opium, coffee, and alcohol affected both the body and the mind, affecting thoughts, mood, imagination, and volition. Moreover, diseases attacked the mind as well as the body. La Mettrie taught that atheism was the only road to happiness and that the purpose of human life was to experience the pleasures of the senses. Many aspects of his theories were incorporated into behaviorism and modern materialism. Eventually, La Mettrie found refuge at the court of Frederick II of Prussia, where he was able to write, lecture, and practice medicine until his untimely death.

Although Erasmus Darwin, English naturalist and physician, is now primarily remembered in connection with his grandson Charles Darwin (1809-1882), he was widely known in his own time. Indeed, his wide range of scientific interests established him as the leader of the Lunar Society, an association that included some of the most important British scientists and inventors of the eighteenth century. Erasmus Darwin's major treatise, *Zoonomia, or the Laws of Organic Life* (1794-1796), deals with medicine, pathology, and the mutability of species. Unlike Descartes, Erasmus Darwin attempted to prove that the same intellectual principle existed in humans and animals. His *Zoonomia* dealt with essentially all of the central questions of late eighteenth-century biology. He is generally given little credit for originality despite his sweeping speculations, particularly about evolution. According to Erasmus Darwin, all animals arose from a "living filament." In general, his approach to vital phenomena was mechanical, especially his ideas about epigenesis; he believed that all of nature could be explained in terms of matter and motion.

Georges Louis Leclerc, comte de Buffon, the great eighteenth-century French natural historian and mathematician, is primarily remembered for his encyclopedic *Natural History* (*Histoire naturelle, générale et particulière,* 44 volumes, 1749-1804), a systematic account of natural history, geology, and anthropology. Buffon was a disciple of Newtonian physics, but he was aware of certain difficulties involving a strict application of the mechanical philosophy to natural history. It seemed impossible to Buffon that simple attraction and repulsion between inert atoms could produce living beings. Instead, he suggested that living matter was made up of "organic molecules" that were directed by an "interior mold." He also postulated a special "penetrating force" that guided the organic molecules to their proper places in the interior mold. Although his writings were very popular, Buffon's ideas about geological history, the origin of the solar system, biological classification, the possibility of a common ancestor for humans and apes, and the concept of lost species were considered a challenge to orthodox religious doctrine. His *Epochs of Nature* and *Theory of the Earth* were especially controversial. Buffon was the first naturalist to construct geological history into a series of stages. He suggested that Earth might be much older than church doctrine allowed and speculated about major geological changes that were linked to the evolution of life on Earth. Buffon suggested that the seven days of Creation could be thought of as seven epochs of indeterminate length. These speculations led to an investigation by the theology faculty of the Sorbonne. Buffon avoided censure by publishing a recantation in which he asserted that he had not intended his account of the formation of the earth as a contradiction of scriptural truths.

Mechanical philosophy, however, did not work well when confronted with complex vital phenomena such as reproduction, development and differentiation, nutrition, and growth. Optimism about the explanatory power of mechanical philosophy began to decline by the middle of the eighteenth century. Many physiologists came to the conclusion that mechanical explanations for the activities of living beings were impossible and inappropriate. Naturalists argued that it was more useful to reduce complex vital (living) phenomena to simpler components that could be analyzed, or at least fully described and linked to each other. The new ideas and methods that emerged from the revolution in chemistry provided new ways of understanding some vital phenomena, such as digestion, respiration, and the role of the circulation of the blood. By the second half of the eighteenth century, new ideas from chemistry and experimental physics, especially the study of electricity, were exerting a profound influence on physiology and medicine.

LOIS N. MAGNER

Further Reading

Books

Fulton, John Farquhar, and L. G. Wilson, eds. *Selected Readings in the History of Physiology.* Chicago, IL: C. C. Thomas, 1966.

Hall, T. S. *A Sourcebook in Animal Biology.* Cambridge, MA: Harvard University Press, 1952.

Hall, T. S. *History of General Physiology 600 B.C. to A.D. 1900.* 2 vols. Chicago, IL: University of Chicago Press, 1975.

Hankins, Thomas L. *Science and the Enlightenment.* New York: Cambridge University Press, 1985.

McNeil, Maureen. *Under the Banner of Science: Erasmus Darwin and His Age.* Wolfeboro, NH: Manchester University Press, 1987.

Roger, Jacques. *Buffon: A Life in Natural History.* Ithaca, NY: Cornell University Press, 1997.

Rothschuh, Karl E. *History of Physiology.* Translated and edited by Guenter B. Risse. New York: Robert E. Krieger, 1973.

Thomson, Ann. *Materialism and Society in the Mid-eighteenth Century: La Mettrie's Discours préliminaire.* Genève: Droz, 1981.

Wellman, Kathleen Anne. *La Mettrie: Medicine, Philosophy, and Enlightenment.* Durham: Duke University Press, 1992.

Periodical Articles

Brown, Theodore M. "From Mechanism to Vitalism in Eighteenth-Century English Physiology." *Journal of the History of Biology* 7 (1974): 179-216.

The Search for New Systems of Classification

Overview

As naturalists began to expand their reach and study more and more of the Earth, they began to find such a bounty of plants and animals that the current and relatively informal practice of naming new discoveries quickly became outdated. Botanists and zoologists found themselves overwhelmed and unable to keep track of the different organisms. By the 1700s the problem had become severe. Swedish naturalist Carolus Linnaeus (1707-1778) recognized the need for a clear and logical method of classifying and naming living things. Thus, in his *Systema naturae* (1735) he introduced the practice of using two Latin names—one for genus, the other for species—to identify each different plant and animal known to man. This method, termed binomial nomenclature, is still used today. Baron Georges Cuvier (1769-1832) further refined the system of classification by providing a description for so-called "type" animals that are representatives of each species.

Carolus Linnaeus. *(Library of Congress. Reproduced with permission)*

Background

The 1700s saw a dramatic rise in the number of discoveries of living things. Botanists in particular introduced new species almost daily. At that time, however, the naming of plants was a very inexact science without universal guidelines. The result was a confusing tangle of plant names that made it difficult, if not impossible, both to determine how a new plant compared to other varieties and to simply ascertain whether a "new" plant had really already been discovered and given a different name by another botanist.

The same situation was faced by zoologists and animal species.

Linnaeus recognized the depth of the problem early in his career. With only four years of formal higher education behind him, he went to Lapland to conduct a five-month survey of its plants, animals, and minerals. During that five-month survey he wrote detailed descriptions of all of the organisms he encountered. Upon his return, as he summarized his findings and discoveries, he also drafted the first edition of *Systema naturae*.

The first edition of *Systema naturae* was a 15-page publication that presented a preliminary view of his thoughts on classification. Over the years, he published subsequent and much longer editions, including what would become the most influential—the 1,300-page 10th edition in 1758. In all, he published 12 editions. Under Linnaeus's system, each organism receives a generic (genus) and specific (species) name. The generic name helps botanists and zoologists to determine which organisms are closely related. Two species of cattails, for instance, share the genus *Typha*. The specific name is unique to that organism. The common cattail's specific name is *latifolia*, whereas the narrow-leaved cattail carries the specific name of *augustifolia*. Under the two-name system, or binomial nomenclature, the common cattail's scientific name, then, is *Typha latifolia,* while the narrow-leaved cattail is *Typha augustifolia.* (The genus name is often abbreviated to its first initial in subsequent references, such as *T. latifolia* and *T. augustifolia*.)

Linnaeus also developed higher levels of organization—with kingdom at the top, then class, order, and finally genus and species. He described three kingdoms: one for animals, one for plants and one for minerals. Each kingdom contained classes. His original classes under the animal kingdom, for example, included Amphibia (amphibians), Aves (birds), Insecta (insects), Mammalia (mammals), Pisces (fish), and Vermes (worms). While he arranged animals with an eye toward morphological similarities, he classified plants based on their flowers and other parts of their reproductive systems, such as the stamens and pistils.

Other eighteenth-century scientists also made important contributions to the classification system. French zoologist Cuvier introduced to Linnaeus's animal hierarchy the phylum level, which falls between kingdom and class. His four phyla included Vertebrata (animals with a backbone), Mollusca (including clams and squid), Articulata (including worms and insects), and Radiata (including sea anemones and jellyfish). Scientists Michel Adanson (1727-1806) and Georges Buffon (1707-1788), who was actually a critic of Linnaeus, furnished the classification system with the level of family between genus and order.

Cuvier also developed the concept of "types" to represent animal species. One particular species of grasshopper, then, would have one individual—the "type"—to represent and describe the entire species. That type would carry the species' different characteristics.

Impact

The classification system introduced by Linnaeus in the 1700s has shaped the way botanists and zoologists view the natural world. Linnaeus brought logic and organization to the naming of plants and animals, and also put in place a taxonomic system that encouraged scientists to look for similarities and differences among organisms.

In his *Systema naturae* Linnaeus described kingdom, class, order, genus, and species. All of these levels of hierarchy remain in use today, along with many of the titles he assigned. His original class names of Mammalia, Aves, Amphibia, and Insecta, along with his kingdoms of Plantae and Animalia, have survived the test of time. Scientists have since more than quadrupled the number of classes, and added many, many genera to Linnaeus's original list. His classification system, however, was so versatile that it allowed for this massive growth.

With his system—along with the additions in hierarchy levels, such as the family and phylum—scientists can now understand a great deal about an animal or plant just by knowing where it is placed taxonomically. For example, a new species can usually be immediately identified as belonging to only one of the multitude of orders. All frogs and toads, for example, fall under Order Anura. All members of the Order Anura share certain characteristics, so although a zoologist may not have done a complete study of that particular organism, he or she can initially assume that the specimen shares the order's characteristics, such as the presence of a shortened backbone or the use of external fertilization in reproduction. By noting other characteristics, which are often readily apparent, scientists can then quickly place the plant or animal in the correct family. For example, members of the Bufonidae (true toads) family of the Order Anura have thick, warty skin, and likely lay their eggs in long strings.

As this process of deduction continues, the identification follows down to the species level. Species are defined as groups of organisms that breed with one another and produce viable offspring. In all of the taxonomic levels, only the species is a natural unit. The others—from genus all the way to kingdom—are artificially created groupings formed to help scientists wend their way through the tangle of living things.

Zoologists and botanists officially accepted Linnaeus's classification system in the early twentieth century. Botanists regard Linnaeus's 1753 *Species plantarum* as the starting point for plant naming, and zoologists chose his 1758 tenth edition of *Systema naturae* as the starting point for animal names. For animals, the official recognition demanded that all names given to organisms become applicable only from 1758 hence, with the earlier names taking precedence over later names. In other words, the first name stands. Linnaeus provided a massive list of species in the 1758 publication. Some of the species he had collected himself, but many others came from the entourage of students who studied under him. Since Linnaeus named so many species, and his listing falls at the starting point of 1758, his genus and species names are still used today. Often the scientific literature will list the year the species was named along with the person who named it. An example is the northern pike, which may appear as *Esox lucius* Linnaeus 1758. If the genus has changed since the original naming, parentheses will surround the person's name in the listing. Linnaeus spent many years providing names for organisms, so it is common to find "Linnaeus" or simply "L." following scientific names.

The designation of a single, universal scientific name for each species was a giant step forward for zoologists and botanists. Scientists around the world used the same name to describe the same species. No matter how far apart they worked, scientists could finally exchange and share data with the confidence that all concerned were working with the same organism.

Both Linnaeus's organizational system and Cuvier's contribution of the "type" organism also proved to be great aids for zoologists and botanists trying to determine whether a newly found organism was actually a member of a new species. The taxonomical hierarchy provided a clear path to the correct group of comparison organisms, while the "type" provided one—and only one—accepted description for that species.

Linnaeus was also well known for his immense collection of plants and animals. Over his lifetime, he received specimens not only from his nearly 200 students but from botanists and zoologists worldwide. His work continued after his death most notably through his student Johan Christian Fabricius (1743-1808), a Danish zoologist. Fabricius began to classify the huge class of insects based on their mouthparts. His work encouraged early eighteenth-century entomologists to describe and arrange the invertebrates.

The Linnaean Society formed shortly after Linnaeus died, and became the caretaker of his library and specimen collection, which had become the largest in Europe. His manuscripts and collection are now housed in London.

Since Linnaeus developed his classification system, scientists have made new discoveries of organisms, and have proceeded with a wide range of morphological, genetic, and other studies. As new research presses on, the placement of plants and animals within the taxonomical hierarchy may change, but Linnaeus's classification system remains the foundation that directs the way scientists look at the natural world.

LESLIE A. MERTZ

Further Reading

Blunt, W. *The Compleat Naturalist: A Life of Linnaeus.* New York: Viking Press, 1971.

Byers, Paula K., ed. *Encyclopedia of World Biography.* 2nd ed. Detroit: Gale Research, 1998.

Coleman, W. *Georges Cuvier, Zoologist.* Cambridge, MA: Harvard University Press, 1964.

Simmons, John. *The Scientific 100: A Ranking of the Most Influential Scientists, Past and Present.* Secaucus, NJ: Citadel Press, 1996.

The Great Debate: Preformation versus Epigenesis

Overview

During the 1700s in Europe, embryology was the focus of a controversial and lively debate that involved many of the greatest and most celebrated scientists and philosophers of that time. Most scientists were convinced that all embryos existed since Creation as preformed miniatures, held within their parents, ready to simply grow larger and emerge when their time arrived. But a few scientists believed that each embryo was formed gradually, structure by structure, in a developmental series that started with the undifferentiated materials of the egg. Both sides of "The Great Debate" used the subjective philosophies of the Enlightenment period, such as rationalism and vitalism, to fill the gaps created by the limitations of their eyes and early microscopes, which left them unable to advance embryology any further. The Great Debate has come to symbolize the effects that cultural and nonscientific factors can have on scientists and their interpretation of scientific facts and theories.

Background

The biology of reproduction, or the ability to recreate new individual organisms, is possibly the single most remarkable phenomenon of life, and this process has remained a central issue of biology for several centuries. Man's quest to understand exactly how a new organism forms, referred to as "generation" in the past and currently termed "embryology," challenged the ability of many of history's greatest scientists and philosophers. The difficulty of studying such minute structures left anatomists and physiologists of the seventeenth and eighteenth centuries without the requisite facts needed to understand the events and processes of embryo formation. This period of scientific research and discovery was dominated by philosophers and scientists who were deeply affected by their own subjective interpretation of the world around them. The Great Debate that followed occurred among a group of men—women scientists were very rare during this period—who were passionate about science and the natural world, and many were devoted to God and what they saw as the illumination of his divine plan. Though concerned with the specific facts of embryo development, at its core the Great Debate questioned the role of God as divine Creator and ultimate source of all living organisms.

Most scientists and philosophers of the eighteenth century were preformationists, based on the works of researchers such as Joseph of Aromatari, who claimed to see miniature chick embryos in chicken eggs even before incubation (1626). Later, Jan Swammerdam (1637-1680) stated a distinct theory of preformation based on similar research with insect, chicken, and frog eggs (1669). Swammerdam's ideas on preformation were further expanded in 1674 by Nicolas de Malebranche, who postulated *emboîtement*, or

the encasement theory, in which completely preformed embryos were arranged in a box within a box system. Preformationists believed that all living organisms, plant and animal, were created in complete but miniature form, within the eggs that each parent contained. *Emboîtement* postulated that all future generations existed in a long series of miniature, complete embryos held in place and awaiting their appointed time to grow and emerge. During the eighteenth century Albrecht von Haller (1708-1777), an accomplished and distinguished scientist as well as a deeply religious man, became the preeminent advocate and supporter of preformation.

On the opposite side of this debate were the epigenesists, united in their complete rejection of preformation. They sought to explain the development of the embryo as a gradual formation of embryonic structures from the undifferentiated materials of the egg. One classic version of embryo formation was postulated by Aristotle (384-322 B.C.), who observed that the chicken embryo and its internal structures were not preformed but rather formed gradually. Two thousand years later in 1651, William Harvey (1578-1657), studying chicken and deer embryos, reached the same conclusion. Harvey stated that all life comes from an egg and that the embryo builds its organs and other parts individually, in succession and due order, in a process he termed "epigenesis." And like Aristotle, Harvey also postulated that a life force, which he called the "generative principle," initiated the embryo's growth after fertilization. Later, Caspar Friedrich Wolff (1738-1794) would take the lead in researching and advocating a theory of epigenesis that would counter the preformationists and their often implausible explanations.

Impact

Interestingly, the theory of preformation that developed in the eighteenth century was based on the apparently erroneous results reported by two researchers both considered by their contemporaries to be prime founders of preformation. First, Swammerdam reported dissecting complete butterflies from chrysalids as well as from within caterpillars. He went further in not only claiming that a complete butterfly would also be found in the egg, but that all animals are preformed as minute embryos awaiting an embryological development of simple enlargement. Perhaps more scientific than Swammerdam, but similarly in error, was Marcello Malpighi (1628-1694), who made detailed observations of chicken embryos at various developmental stages, including the presence of a beating heart at 38 hours of incubation. However, Malpighi also reported finding advanced organ development in fertilized but unincubated eggs. He believed (apparently incorrectly) that these eggs had not yet begun to undergo any development but still possessed the preformed rudimentary structures of the embryo. Other scientists and philosophers used these reports to bolster their support for and establishment of preformation.

After Harvey, further support for epigenesis came from Abraham Tremblay (1710-1784), who reported on the unusual ability of the hydra to reproduce through budding and by artificial means. Tremblay cut hydras into small pieces, which re-grew into whole new individuals. This discovery sent shock waves through the preformationists and their theories, including Haller, then a preformationist influenced by his university mentor. Haller now rethought his position and predicted the eventual abandonment of preformation theory due to several overwhelming observational facts. First, Tremblay's work revealed that the hydra possessed the ability to organize itself as necessary. It was believed that this power could also be seen in the embryo, when its original fluid state gradually changed into organs and internal structures, which continue to change over time. Finally, how could it be that offspring typically resemble both of their parents if they are the result of preformed embryos from one parent or the other? Still, for most scientists, the philosophy and religious implications of preformation held sway over both the ideas of epigenesis and the observed facts of embryo formation.

Previously, when Anton van Leeuwenhoek (1632-1723) confirmed the presence of "spermatic animalcules" in semen (1677), some scientists began to argue that the sperm held the preformed embryo, and some animalculists (spermatists) even imagined that a homunculus, or very minute human, was found in the head of the sperm. Animalculists, including Haller, saw the female as merely providing a supportive and nutritive environment for the enlarging sperm-fetus. Then in 1745 Charles Bonnet (1720-1793) reported that female aphids could produce miniature aphid offspring without fertilization, giving support for ovism, or the belief that the female egg housed the preformed embryo. In 1758 Haller, influenced by Bonnet and studying chicken embryos, returned to preformationism, supporting the ovist view, thus rejecting his view

of epigenesis. For a period of time, a major scientific conflict raged between animalculists and ovists, both groups doing vigorous battle despite their inability to accurately see most of the details of either germ cells or of embryo development. Instead, philosophical beliefs and rationalist ideas were substituted, even displacing the actual observed facts.

Epigenesis slowly began to gather renewed support in response to both the successes of the new mechanical laws of physics and astronomy and the apparent deficiencies and inconsistencies of the preformationist theories. With the advent of gravitational and mechanical laws, new approaches were being applied to living organisms, with the idea that matter and its "forces" could reasonably be applied to living organisms. Although epigenesists used more observational and empirical data to support their belief that embryos formed gradually from undifferentiated materials in the egg, they had great difficulty in providing explanations as to just how and why these changes actually happened. This was the great deficiency that discredited epigenesis in the minds of most scientists of the eighteenth century.

Haller was a well respected physician, anatomist, and physiologist as well as a botanist, poet, and noted writer when he concluded that the egg contained the preformed embryo. He would quickly become the foremost advocate and ardent supporter of preformation. Haller believed that science should aim to confirm the presence of God and his divine Creation, and must be carried on within the limits of religion, rejecting any theories that promoted materialism or atheism. Haller could not accept any theory in which matter itself possessed creative powers or formative abilities that were the domain of God alone. Thus Haller rejected epigenesis and embraced performation, despite the observed factual conflicts he once advocated. He argued that preformed embryos were simply too small to be seen before development, or that they were invisible until they gained their colored forms, and furthermore, there was a limit to what man could learn and understand about the world and nature, and that limit was set by God.

Pierre Louis Moreau de Maupertuis (1698-1759), already widely accomplished and celebrated for his research confirming Newtonian principles, published *Venus physique* (1745), in which he boldly supported epigenesis. He argued largely on the basis of the biparental human inheritance patterns he had studied in detail. Maupertuis believed that embryos resulted from the uniform mixing of particles present in the semen of the male and female, which represented all the body parts and structures. Embryos are thus composites of both parents, and the gradual formation of the embryo was the result of a cohesive natural force that directs this process. This position was a great challenge to Haller and the preformationists, who once again chose to ignore any conflicting observational data, while attacking and rejecting any idea that a force of nature not directly of the divine Creation could form or fashion a living organism, much less a human being.

Into this scientific and philosophical fray boldly stepped the youthful Caspar F. Wolff, who published his doctoral dissertation in 1758, advocating a "rational embryology" in which epigenesis could be the only process of true embryo formation. Wolff believed that only sound scientific methodology could lead to true knowledge of any biological question, and the empirical findings of detailed observation combined with the use of rational logic was the best method. Scientists, Wolff proposed, should aspire to gain philosophical knowledge based on their acquired historical knowledge. Wolff studied actual embryo formation in great detail, recognizing that embryos and their structures were not preformed but rather developed gradually from the undifferentiated egg. Wolff proposed that embryos form as a result of a solidification process of the liquid material in the egg, and this process is controlled by the *vis essentialis*, a life force conceived by Wolff and not very different from previous philosophers. Wolff continued to research embryology and epigenesis throughout his considerable career. But the influence of preformation would not be replaced by epigenesis until later in the nineteenth century, when it received support from another venerated German scientist, Johann Blumenbach (1752-1840), who confirmed the regeneration results of Tremblay and announced his acceptance of epigenesis.

Meanwhile, Haller responded to the challenge created by Wolff quickly and with rigor, if not the same arguments he employed previously. The preformed embryo is both too small and its components invisible to the human eye until it undergoes further growth, he countered once again. Haller and Wolff corresponded with each other by letter many times over the next decade, though neither side apparently moved the other to any degree. These two men could not have been more polar in their views: Haller, deeply convinced of divine Creation and the support that Newtonian principles gave to the existence of God's plan; and Wolff, entrenched in the German school of rationalism and in support of a modified type of vitalism. Haller, with the weight of his considerable stature, maintained general support for preformation beyond the end of the eighteenth century.

The Great Debate still serves to illustrate how arbitrary hypotheses lead to false systems of knowledge, and that scientific advancement and understanding requires sound scientific methodology, including proper experimental designs and observations, followed by empirical analysis. Both sides of this debate, striving to advance embryology, were unable to overcome the limitations of observation imposed by the size of the embryo, a situation that would not be resolved until the nineteenth century, when microscope efficiency improved dramatically. It is interesting to note further that preformation and epigenesis have been synthesized into a contemporary theory in which preformed molecular instructions, the DNA of the sperm and egg, are combined and used to create an embryo from the undifferentiated material in the egg.

KENNETH E. BARBER

Further Reading

Gardner, Eldon J. *History of Biology*. 3rd ed. Minneapolis, MN: Burgess Publishing Company, 1972.

Glass, Bentley, Owsei Temkin and William L. Strauss, Jr. *Forerunners of Darwin: 1745-1859*. Baltimore: Johns Hopkins University Press, 1968.

Needham, Joseph. *A History of Embryology*. Cambridge: Cambridge University Press, 1934.

Pinto-Correia, Clara. *The Ovary of Eve: Egg and Sperm and Preformation*. Chicago: University of Chicago Press, 1997.

Roe, Shirley A. *Matter, Life, and Generation*. Cambridge: Cambridge University Press, 1981.

The Spontaneous-Generation Debate

Overview

According to the ancient theory of spontaneous generation, living organisms could originate from nonliving matter. During the seventeenth and eighteenth centuries, however, naturalists began to conduct experiments that challenged the doctrine of spontaneous generation. After Francesco Redi published his *Experiments on the Generation of Insects,* the spontaneous-generation debate was essentially limited to microscopic forms of life. During the eighteenth century, the French naturalist Georges Buffon and the English microscopist John Turbeville Needham carried out a series of experiments that seemed to support the doctrine of spontaneous generation, but their conclusions and experimental methods were challenged by the Italian physiologist Lazzaro Spallanzani. After conducting a series of rigorous experiments on the growth of microorganisms, Spallanzani vigorously challenged the belief in spontaneous generation. Spallanzani claimed that Needham had not heated his tubes enough and that he had not sealed them properly.

Background

According to the theory of spontaneous generation, living organisms can originate from nonliving matter. Belief in the spontaneous generation of life was almost universal from the earliest times up to the seventeenth century. Small, lowly creatures and all sorts of vermin, which often appeared suddenly from no known parents, seemed to arise from lifeless materials. Insects, frogs, and even mice were thought to arise from slime, mud, and manure in conjunction with moisture and heat. In ancient times, this belief seemed to conform to common observations about the sudden appearance of insects, small animals, and parasites. Spontaneous generation also provided an answer to philosophical and religious questions about the origin of life.

During the seventeenth and eighteenth centuries, however, some naturalists began to conduct experiments that tested and challenged the doctrine of spontaneous generation. Italian physician and poet Francesco Redi (1626-1698) was one of the first to question the spontaneous origin of living things. In 1668 Redi initiated a now-famous experimental attack on the question of spontaneous generation. Redi discovered that when adult flies were excluded from rotting meat, maggots did not develop. If flies were not excluded, they laid eggs on the meat and the eggs developed into maggots. Thus, Redi demonstrated that maggots and flies were not spontaneously generated from rotting meats but instead developed from eggs that were deposited by adult flies. Although Redi's *Experiments on the Generation of Insects* (1668) did not totally discredit the doctrine of spontaneous generation, the eighteenth-century spontaneous-generation debate was essentially limited to microscopic forms of life.

Seventeenth-century microscopists were able to see a new world teeming with previously invisible entities, including protozoa, molds, yeasts, and bacteria. Many naturalists thought that the new world of microscopic "animalcules" discovered by the great microscopist Anton van Leeuwenhoek (1632-1723) provided proof that minute plants and animals were spontaneously generated in pond water or similar media. Some naturalists thought that these minute entities might even be the living molecules, or "monads," postulated by the mathematician and philosopher Gottfried Wilhelm Liebniz (1646-1716). While Leeuwenhoek was quite sure that he had discovered "little animals" that must have descended from parents like themselves, others took exception to this conclusion. Indeed, questions concerning the nature, origin, and activities of microorganisms were not clarified until the late nineteenth century.

Several interesting accounts of "infusoria" were, however, published in the eighteenth century. Louis Joblot (1645-1723), for example, confirmed the existence of some of Leeuwenhoek's animalcules. In 1718 Joblot published an illustrated treatise on the construction of microscopes that described his observations of the animalcules that could be found in various infusions. Joblot is now remembered primarily for his opposition to the doctrine of spontaneous generation. To prove that infusoria were not spontaneously generated, Joblot boiled his growth medium and divided it into two portions. A flask containing one portion was sealed off and the other sample was left uncovered. The open flask was soon teeming with microbial life, but the sealed vessel was free of infusoria. To prove that the medium was still susceptible to putrefaction, Joblot exposed it to the air and showed that infusoria were soon actively growing. Joblot concluded that something from the air had to enter the medium to produce microorganisms.

Impact

Joblot's experiments were repeated with many variations by other naturalists, but the results obtained were not consistent. Among the most notable eighteenth-century advocates of the doctrine of the spontaneous generation of microorganisms were the French naturalist Georges Buffon (1707-1788) and the English microscopist John Turbeville Needham (1713-1781). Together as well as separately, Needham and Buffon carried out a series of experiments to disprove the work of Joblot.

John Turberville Needham was a naturalist as well as a teacher and a clergyman. He was the first Roman Catholic to become a member of the Royal Society of London. In 1767 Needham retired to the English seminary in Paris. He devoted the rest of his life to his studies and experiments. Needham had decided to study natural history after reading accounts of "animalcules" and "infusoria" and philosophical speculations about microorganisms, spontaneous generation, and the origin of life. Having rejected mechanistic theories of physiology, Needham adopted vitalism (the idea that life processes cannot be explained by the laws of chemistry and physics) and the doctrine of spontaneous generation. In 1745 he published a book entitled *An Account of Some New Microscopical Discoveries,* in which he presented his experimental evidence for the theory of spontaneous generation.

According to Needham, many organisms developed in prepared infusions of various substances even if the infusions had been placed in sealed tubes and heated for 30 minutes. When Needham repeated Joblot's experiments, whether the flasks were open or closed and the medium boiled or not boiled, all vessels soon swarmed with microscopic life. Needham assumed that this heat treatment should have killed any living organisms that might have been in the original medium. According to Needham, a powerful vegetative force remained in every particle of matter that had previously been part of a living being. Therefore, when animals or plants died, they slowly decomposed and released the "common principle," which Needham thought of as a kind of universal semen from which new life arose. He concluded that the growth of microorganisms under his experimen-

tal conditions proved that spontaneous generation of microbial life had occurred. Published in the *Philosophical Transactions* of the Royal Society in 1748, Needham's views were well known. The claims of Needham and Buffon did not, however, stand unchallenged for very long.

When the Italian physiologist Lazzaro Spallanzani (1729-1799) repeated Needham's experiments, he obtained conflicting results. Spallanzani, who had studied philosophy, theology, law, and mathematics, was appointed professor of logic, metaphysics, and Greek at Reggio College in 1754. Six years later, he became professor of physics at the University of Modena. In 1769 he accepted a position at the University of Pavia and remained there until his death. (After attacking Needham and Buffon on the subject of spontaneous generation, Spallanzani investigated regeneration, transplantation, reproduction, generation, artificial insemination, the circulation of the blood, digestion, and the electric organ of the torpedo fish before returning to studies of microscopic plants and animals at the end of his career.) Like his friends Albrecht von Haller (1708-1777) and Charles Bonnet (1720-1793), Spallanzani supported an ovist preformationist view of generation and he attacked Buffon's mechanistic epigenetic theory.

After examining various forms of microscopic life, Spallanzani concluded that Leeuwenhoek had been correct in identifying these minute entities as living organisms. To prove that these entities were alive, he carried out a series of experiments in which he boiled rich growth media for fairly long periods of time. He found that, if he placed media that had been boiled for 30 minutes into phials and immediately sealed them by fusing the glass, no microorganisms were produced. He concluded, therefore, that the infusoria found in pond water and other preparations were actually living organisms.

In another series of experiments, Spallanzani exposed significant errors in the experiments conducted by Needham and Buffon. By heating a series of flasks for different lengths of time, Spallanzani determined that various sorts of microbes differed in their susceptibility to heat. Whereas some of the larger animalcules were destroyed by slight heating, other, very minute, entities seemed to survive in liquids that had been boiled for almost an hour. Further experiments convinced Spallanzani that all these little animals entered the media from the air. Convinced that a great variety of animalcular "eggs" must be disseminated through the air,

Spallanzani concluded that the air could either convey the germs to the infusions or assist in the multiplication of those germs already in them.

In 1767 Spallanzani published an account of his research on the growth of microorganisms and his criticism of the theory proposed by Buffon and Needham. According to Buffon and Needham, living things contained special "vital atoms" that were responsible for all physiological activities. They suggested that after the death of an individual these living atoms were released into the soil and water and taken up by plants. They thought that the "infusoria" that could be found in pond water or infusions of plant and animal material were actually evidence of these vital atoms.

Despite Spallanzani's criticisms, Needham and Buffon continued to champion the doctrine of spontaneous generation. Indeed, the debate was not resolved until the nineteenth century, when the great French chemist Louis Pasteur and the English physicist John Tyndall declared war on spontaneous generation. Although Spallanzani's experiments answered many of the questions raised by advocates of spontaneous generation as well as proved the importance of sterilization, his critics claimed that he had tortured the all important "vital force" out of the organic matter by his cruel treatment of his media. The vital force was, by definition, capricious and unstable, rendering it impossible to expect reproducibility in experiments involving organic matter.

During the nineteenth century, the design of experiments for and against spontaneous generation became increasingly sophisticated as proponents of the doctrine challenged the universality of negative experiments. Because any apparent exception could allow proponents of the theory to maintain that spontaneous generation only occurred under special conditions, however, opponents were always on the defensive. The work of Louis Pasteur (1822-1895) and John Tyndall (1820-1893) effectively proved that the existence of germs in the air was the critical issue in establishing the experimental basis of the debate. Pasteur was convinced that microbiology and medicine could only progress when the idea of spontaneous generation was totally vanquished. Although both knew that it is logically impossible to prove a universal negative, they demonstrated that under present conditions living beings arise from "parents" like themselves. Pasteur and Tyndall proved that the microbes that Needham and Buffon thought arose from the media actually came from microbes carried by particulate matter in the air. Pasteur proved that microorganisms come from the multiplication of parent microorganisms of their own kind. The experiments conducted by Pasteur and Tyndall did not deal with the question of the ultimate origin of life, but they did demonstrate that microbes do not arise *de novo* in properly sterilized media under the conditions prevailing today. Advocates of the doctrine of spontaneous generation have argued that some form of the doctrine is necessarily true in the sense that if life did not always exist on earth it must have been spontaneously generated at some point.

LOIS N. MAGNER

Further Reading

Brock, T. D., ed. *Milestones in Microbiology*. Englewood Cliffs, NY: Prentice-Hall, 1961.

Conant, J. B., ed. *Pasteur's and Tyndall's Study of Spontaneous Generation*. Cambridge, MA: Harvard University Press, 1953.

Doetsch, R. N., ed. *Microbiology: Historical Contributions from 1776-1908*. New Brunswick, NJ: Rutgers University Press, 1960.

Epstein, Sam. *Secret in a Sealed Bottle: Lazzaro Spallanzani's Work with Microbes*. New York: Coward, McCann & Geoghegan, 1979.

Farley, J. *The Spontaneous Generation Controversy from Descartes to Oparin*. Baltimore, MD: Johns Hopkins Press, 1977.

Lechevalier, H. A., and M. Solotorovsky. *Three Centuries of Microbiology*. New York: Dover, 1974.

Nigrelli, Ross F., ed. *Modern Ideas on Spontaneous Generation*. New York: New York Academy of Sciences, 1957.

Vandervliet, G. *Microbiology and the Spontaneous Generation Debate During the 1870s*. Lawrence, KS: Coronado University Press, 1971.

Abraham Trembley and the Hydra

Overview

The hydra is a small organism, often less than an inch (2.5 cm) long, that became the focus of much attention and debate during the eighteenth century. Abraham Trembley (1710-1784), along with a number of others, used the hydra to investigate basic issues concerning development, regeneration after damage, and the differences between plants and animals. Because of this and related work, by the end of the century the extents of both the plant and animal kingdoms were more clearly defined and how development occurs was more fully understood.

Background

Trembley, who was born in Geneva, Switzerland, was working as a tutor in Holland when he first encountered a green hydra called *Chlorohydra viridissima* in a sample of pond water. It was clinging to a plant, and at first he thought it was itself a plant because of its green color. But when he saw that its finger-like projections or tentacles actually moved, a characteristic of animal, not plant, life, he became uncertain as to how to classify this creature. Some few plants do move quickly when stimulated, as is the case with the Venus flytrap, which rapidly closes its leaf-shaped "trap" when a fly lands on it, but these quick movements are always triggered by some stimulus, they are never spontaneous as the movements of the hydra were.

To try to determine whether the hydra was a plant or an animal, Trembley decided to cut the organism in two. He reasoned that if it were an animal, this operation would kill it, but if it were a plant, each of the two parts was likely to survive. He split the hydra so that one part had all the tentacles. Over the next several days, Trembley continued to observe the hydra segments and found that each regenerated to the point where it looked like the original hydra. In another experiment, he removed the tentacles from several hydras, each of which he then put in a separate tank where they bore young by asexual budding with the buds separating from the parent, a separation that would not have occurred with plant buds. These experiments and the fact that the hydra had spontaneous movements made Trembley uneasy about classifying it as a plant.

To settle the question of whether the hydra was a plant or an animal, Trembley sent a letter describing his observations as well as some specimens to René Antoine Ferchault de Réaumur (1683-1757) at the Royal Academy of Sciences in Paris in 1741. Réaumur assured Trembley that the hydra was indeed an animal and that a related brown species, *Hydra vulgaris*, had been discovered by the great Dutch microscopist Anton van Leeuwenhoek (1632-1723) in 1702. After having Trembley's long letter read at two meetings of the Academy, Réaumur himself confirmed Trembley's results with his own experiments, as did two researchers in England. This work was exciting to those interested in zoology because this was the first case of an animal that could be multiplied by cutting it into pieces, though it was commonly known that many plant species could be propagated from cuttings. Here was a creature that seemed to be on the border between the plant and animal kingdoms.

Trembley continued to experiment with several species of hydra. He even managed to turn specimens of *Hydra vulgaris* inside out, and found that this reversed form could survive and feed without returning to its original form. He also performed the first permanent graft of animal tissues when he successfully fused two hydras by placing one inside the other: he pushed it in tail first through the mouth of the second individual.

His studies on the hydra were not Trembley's only contributions to the life sciences. He also was the first to identify the process of cell division by examining specimens of the one-celled diatom *Synedra*. This was cell division in the strict sense of the term, that is, a cell with one nucleus dividing to form two cells, each with a nucleus.

Impact

Trembley's research was important for a number of reasons. First, it focused attention on the hydra and other members of the phylum Cnidaria, formerly called Coelenterata, which includes jellyfish, corals, and sea anemones. Like the hydra, other cnidarian species fascinated students of nature because they seemed to have characteristics of both plants and animals. The branch-like structure of corals and the fact that most cnidarians spend part of their lifecycle attached to some solid surface made them seem plantlike. The sedentary form of the organism is called the polyp and the mobile form is the medusa, with jellyfish being the cnidarians with the most developed medusae. Extensive observations and experimentation, often following Trembley's lead, made it obvious that though they sometimes superficially resemble plants, these organisms were in fact animals since they could respond rapidly to stimuli, digest food, and reproduce in ways similar to other animals.

During the eighteenth century the idea of the great chain of being was still widely accepted. This held that all organisms could be organized into a chain from the simplest to the most complex—one-celled organisms would be at one end of the chain and humans at the other. It was further held that there were no breaks in the chain, that where there seemed to be a great difference between one link or organism type and the next, the assumption was that this was simply because missing links had yet to be discovered. This is where the term "missing link" arose; much later it was used in referring to an animal with a mixture of ape and human characteristics that was thought to have existed before the evolution of modern humans.

But the concept of a great chain of being was developed long before the idea of evolution became widely accepted, and in the seventeenth and eighteenth centuries missing links were sought not as support for evolution, but because it was thought that God had created a perfect chain of organisms, with all possible types. If this were the case, the apparent rather significant differences between plants and animals posed a problem, and some observers of nature saw the hydra and other cnidarians as links between the two. The debate over the proper way to classify cnidarians continued for some time until the evidence for their animal nature became overwhelming. Other research also made it clear that there were obviously major differences between plants and animals and no half-way organisms bridging the gap between the two kingdoms; so the idea of a great chain of being slowly lost support.

Trembley's studies were significant because they focused attention on the basic question of what makes plants different from animals. While this was usually an easy enough question to answer in the case of large plants and animals, at and near the microscopic level the distinctions were not so obvious. Also, at this time many botanists were seeking to understand plants in terms of animal anatomy and physiology. For example, they saw a similarity between the circulation of blood in animals and the movement of sap in plants, between the fertilization of eggs by sperm in animals and the fusion of eggs and pollen cells in plants. While the exploration of some of these similarities was useful, in many cases the assumption of similarity went too far. The careful observations of investigators like Trembley showed that surface similarities, such as green color, might hide deeper differences.

Another area that had long fascinated biologists is regeneration, the regrowth of structures that have been removed from an organism; it is interesting to see regrowth of a kind that is not possible in humans and mammals in general. While a hydra seems to be able to easily regrow a tentacle, we can't regrow our own limbs. The fact that a portion of the hydra could give rise to an entire organism was also appealing to researchers because it provided support for the idea of epigenesis, the concept that during development tissues are organized and changed radically in form as organs arise from shapeless masses of cells. Epigenesis was opposed by those who accepted preformation, the idea that in the plant seed or beginning animal embryo was a tiny copy of the adult of that species, so that all that happens during development is that the copy enlarges and its organs become functional.

Preformation seems a rather odd idea today, especially because it implies that within each tiny adult form is a still smaller copy representing the next generation. But preformation was widely accepted during the seventeenth and eighteenth centuries because many found it even more difficult to conceive of how an organism as complex as a human being could arise from a shapeless mass of cells. Experiments such as Trembley's, however, indicated that cells could indeed organize and reorganize themselves into different formations, that a whole hydra could develop from just a portion of one. If this were the case, then it became more likely that clumps of cells within an embryo could also organize themselves into more complex structures, as the theory of epigenesis suggested.

Finally, Trembley's work served as a model for the experimental study of organisms and for the use of the microscope in experimentation. The microscope had been developed in the seventeenth century and had opened up a whole new world of organisms to investigation. Before this time, no one had suspected that pond water, saliva, and milk were teaming with life. By the eighteenth century, looking through a microscope and being amazed by the great enlargement of everything from a hair to a snowflake became a popular form of entertainment, and the microscope became less a tool of science than of amusement. Investigators like Trembley showed the tremendous power of this instrument to also further understanding of living things. Some of the hydras he worked with were only a quarter of an inch long; it would have been impossible to manipulate these creatures as he did—cutting them into pieces, forcing one inside the other, and observing the results of these operations—without the aid of magnifying lenses. His research and that of his contemporaries foreshadowed the experiments on embryos that were done in the nineteenth century and that led to the flowering of the fields of experimental zoology and embryology.

MAURA C. FLANNERY

Further Reading

Baker, John. "Abraham Trembley." In *Dictionary of Scientific Biography*, vol. 13, edited by Charles Gillispie. New York: Scribner's, 1976: 457-458.

Gough, J.B. "René-Antoine Ferchault de Réaumur." In *Dictionary of Scientific Biography*, vol. 11, edited by Charles Gillispie. New York: Scribner's, 1975: 327-335.

Magner, Lois. *A History of the Life Sciences*. 2nd ed. New York: Marcel Dekker, 1994.

Ritterbush, Philip. *Overtures to Biology: The Speculations of Eighteenth-Century Naturalists*. New Haven, CT: Yale University Press, 1964.

Advances in Botany

Overview

The eighteenth century saw the development of a new approach to the study of plants: an experimental approach. Botanists were influenced by the great strides that had been made in physics as a result of Isaac Newton's (1642-1727) work on finding the basic principles underlying the complexities of motion, and they wished similarly to find unifying concepts governing the structure and activity of plants. One way to do this was to study plant physiology, that is, to devise experiments that would tease apart particular aspects of plant function. By the end of the century great progress had been made in determining how plants transport water, use sunlight to produce oxygen, and rely on insects and birds in pollination (the process by which pollen, containing the male sex cell, is transported from one plant to another of the same species).

Background

While at the beginning of the seventeenth century the English philosopher Francis Bacon (1561-1626) had encouraged experimentation as a way to discover information about the natural world, it took some time for those interested in nature to develop experimental techniques, to devise ways to test their ideas about the world under carefully controlled conditions. One of the eighteenth-century investigators who made impor-

Stephan Hales. *(Library of Congress. Reproduced with permission)*

tant contributions to this development and to botany was Stephen Hales (1677-1761). He is the founder of experimental plant physiology, although he also did significant research on animal blood pressure as well as important work on improving conditions in mines and on ships.

In 1727 Hales published *Vegetable Staticks* in which he reported on his studies of the movement of water in plants. At the time, some thought that water, in the form of sap, circulated through plants as blood circulates through animals. This was one of several such comparisons made during the eighteenth century between plant and animal physiology that were later found to be erroneous. Hales discovered that there was no pump comparable to the animal heart propelling water in plants, but that transpiration, the loss of water from the surface of leaves, caused more water to be drawn up into the leaves. There was also root pressure forcing water up into the stem from the roots.

Although the information he discovered about water movement in plants was important, what was even more significant about Hales's research is the way it was conducted. His work was not just descriptive, but quantitative. He took precise measurements to show not only that water evaporated from the leaves but the rate at which this occurred; to do this he covered tree branches with glass vessels in order to collect the water, and he worked with several different species to establish that this was a general phenomenon.

Hales also devised ways to measure plant growth. He drew equidistant marks on leaves, so the rate of expansion could be measured by the amount of displacement of the marks from each other over time. Though such techniques were simple, they had never been attempted before in botany. They led to many quantitative studies on plant growth, though it wasn't until the second half of the eighteenth century that significant progress beyond Hales's research took place.

Impact

Among the other areas of plant physiology he investigated, Hales studied the role of air in plant life, but he was hampered by the fact that the components of the air were not well understood at that time. He did make the important observation that air seemed involved in nourishing plants and that one component of the air was absorbed by the leaves. It was Jan Ingenhousz (1730-1799), a Dutch physician and botanist, who built on this work later in the century, publishing his *Experiments upon Vegetables* in 1779. A little earlier, the English chemist Joseph Priestley (1733-1804) had found that if he placed a mouse in a sealed container, it would quickly use up a component of air that was necessary for its life, though the rest of the air would remain. He also

Jan Ingenhousz. *(Library of Congress. Reproduced with permission)*

found that if a plant were then put into this container, it would sometimes restore to the air the component that the mouse had removed from it. Unfortunately, Priestley couldn't get consistent results; while this experiment sometimes worked, other times it didn't.

Ingenhousz used the techniques that Priestley developed to show that plants only restored the component of the air which the mouse used when they were exposed to light. He also showed that the generation of this component, which the French chemist Antoine-Laurent Lavoisier (1743-1794) named oxygen, didn't occur everywhere in the plant but only in the green parts, particularly the leaves. Finally, Ingenhousz demonstrated that plants absorbed the component of the air that the mouse had generated, called carbon dioxide, and that this gas was the source of carbon in plant material, not carbon in the soil. In other words, Ingenhousz worked out the basics of photosynthesis, the process by which plants use the Sun's energy to convert carbon dioxide and water into sugar and oxygen. At the beginning of the nineteenth century Nicholas de Saussure (1767-1845) built on Ingenhousz's work and showed that the amount of oxygen generated in photosynthesis was equal to the amount of carbon dioxide absorbed by a plant, indicating that the two gases were indeed involved in the same process.

Another area of botany that received a good deal of attention in the eighteenth century was plant reproduction. In 1694 Rudolf Camerarius (1665-1721) published a paper outlining his argument for the existence of sexual reproduction in plants. As was mentioned earlier, it was a popular idea at the time that there were similarities between plant and animal anatomy and physiology. While some attempts at showing likeness were misguided, Camerarius's work revealed that exploring similarities could sometimes be useful in understanding plants. He studied plant species such as mulberry that are called dioecious, meaning that there are two forms, one with stamens that produce pollen and one with pistils that produce seeds. He found that a mulberry plant with pistils won't produce any seeds unless there is a mulberry with stamens in the vicinity. He hypothesized that pollen grains were comparable to sperm in animals and were necessary for fertilization of eggs and their development into seeds within the pistils. Camerarius also studied the more common monoecious species, in which the stamen and the pistil are found on the same plant and often in the same flower structure, forming what is called a complete flower. By carefully removing the stamens from the flower of the castor bean, a monoecious species, he was able to prevent seed development, thus again showing the necessity of pollen for this process.

When Carolus Linnaeus (1707-1778) created his plant classification system in the mid-eighteenth century, he focused on the sexual organs found in flowers as the basis for his method. At the time of his research, significant advances were also being made on the work of Camerarius in terms of the mechanisms of fertilization. Joseph Gottlieb Kölreuter (1733-1806) dealt with a number of questions related to fertilization, using careful experimentation and observation. Camerarius had pointed out that sexuality in plants suggested that hybrids were likely, that is, that the pollen of one species could fertilize the egg of another, producing a plant with a mix of characteristics of both species. To investigate this question, Kölreuter developed a technique for artificial fertilization in plants. He removed the pollen-producing stamens from a *Nicotiana rustica* plant, and then brushed pollen from *Nicotiana paniculata* onto the pistil of the *N. rustica* plant. The hybrid offspring of this cross had distinct characteristics that were a mix of those of the two parent species. When the opposite cross was made—when pollen of *N. rustica* was applied to the pistil of *N. paniculata*—the offspring had the same characteristics as the offspring of the original cross, and these characteristics were stable, that is, always appearing when crosses were made between these two species. This suggested to Kölreuter that stable characteristics were at the basis of inheritance and that laws governing inheritance could therefore be discovered; in other words, a science of genetics was possible, though it would be a century before Gregor Mendel's (1822-1884) work on pea plants formed the basis for this science.

One important characteristic of the offspring of crosses between *Nicotiana* species was that they were sterile, that is, they could not reproduce. In later work on crosses between other species, Kölreuter found that most, though not all, of the hybrids were sterile, suggesting that sterility is what prevents such hybrids from becoming more common in nature. In the nineteenth century this finding became important to Charles Darwin's (1809-1882) theory of evolution because it helped to explain how the number of species increased over time: once new species came into existence, hybrid sterility helped to maintain their integrity.

Kölreuter also did microscopic studies on pollen, examining how the pollen adhered to the sticky stigma at the top of the egg-containing pistil. He disagreed with earlier botanists who had assumed that the pollen grains normally became swollen and burst. He correctly found that this only occurred when the grains absorbed abnormal amounts of water, and pollen normally didn't enlarge but instead sometimes grew an extension into the pistil. He didn't continue this research far enough to find, as later botanists did, that this extension, the pollen tube, carries the male sex cell down to the egg.

While Kölreuter made great strides using artificial fertilization techniques, he also did studies on natural fertilization. He found that few plant species were capable of self-fertilization in which pollen fertilizes an egg produced by the same plant; instead, pollen from a different plant of the same species was necessary for successful fertilization. This being the case, the question became how did pollen travel from one plant to another. Kölreuter's observations revealed that some pollen was wind-borne, but in other cases the pollen was carried from one plant to another by animals, most often by insects and birds. He also correctly hypothesized that the sugary liquid called nectar produced in some flowers was used to attract insect and bird pollinators. Kölreuter's long years of research contributed a great deal to botanists' understanding of fertilization, and this information was important in breeding studies that yielded many hybrids that became important food crops. Today, for example, most of the corn produced in the United States is hybrid corn and many of the most popular garden plants are also hybrids.

MAURA C. FLANNERY

Further Reading

Barth, Friedrich. *Insects and Flowers*. Princeton, NJ: Princeton University Press, 1985.

Gabriel, Mordecai, and Seymour Fogel, eds. *Great Experiments in Biology*. Englewood Cliffs, NJ: Prentice-Hall, 1955.

Guerlac, Henry. "Stephen Hales." In *Dictionary of Scientific Biography*, vol. 6, edited by Charles Gillispie. New York: Scribner's, 1972: 35-48.

Iseley, Duane. *One Hundred and One Botanists*. Ames, IA: Iowa State University Press, 1994.

Magner, Lois. *A History of the Life Sciences*. 2nd ed. New York: Marcel Dekker, 1994.

Morton, A.G. *History of Botanical Science*. New York: Academic Press, 1981.

Reed, Howard. *A Short History of the Plant Sciences*. New York: Ronald Press, 1942.

Ritterbush, Philip. *Overtures to Biology: The Speculations of Eighteenth-Century Naturalists*. New Haven, CT: Yale University Press, 1964.

Serafini, Anthony. *The Epic History of Biology*. New York: Plenum, 1993.

Van der Pas, P.W. "Jan Ingen-Housz." In *Dictionary of Scientific Biography*, vol. 11, edited by Charles Gillispie. New York: Scribner's, 1973: 11-16.

Toward the Science of Entomology

Overview

In the eighteenth century entomology, the study of insects, developed as a separate branch of the life sciences. At the beginning of the century, it was not clear precisely what an insect was, with many organisms including spiders and worms being referred to as insects. By the beginning of the next century it was clear that the class of insects was defined as those invertebrates (animals without backbones) having three pairs legs and segmented bodies. It was primarily through the work of two men, Jan Swammerdam (1637-1680) and René Antoine Ferchault de Réaumur (1683-1757), that the study of insects blossomed. Although Swammerdam's work was done in the previous century, most of it remained unpublished until 1737, so it was only then that it influenced other researchers' views on insects. Réaumur's six-volume study of insects appeared between 1734 and 1742.

Background

Swammerdam and Réaumur were interested in different aspects of insect life, with Swammerdam focusing on development and Réaumur on behavior. Swammerdam was Dutch and also did extensive work in animal physiology. He attempted to disprove the old assumption, originating with Aristotle, that insects were less perfect than other so-called "higher" animals such as those with backbones. This assumption was based on three ideas: insects lacked internal anatomy, they could form by spontaneous generation from nonliving matter, and they developed by metamorphosis, that is, their internal structure changed completely during development.

Swammerdam brought a wealth of observations to bear in his arguments against all three of these assumptions. He supported the views of Francesco Redi (1626-1698), who had demonstrated in 1668 that flies did not arise by spontaneous generation from rotten meat, but rather developed from eggs laid on the meat by adult flies. With the aid of the microscope, Swammerdam's careful dissections of insects revealed that they did indeed have a complex internal anatomy. His was the first major study of insect microanatomy. But it was in his investigations on insect development that he made his greatest contributions to entomology.

Swammerdam studied development in order to refute the idea of metamorphosis that was then current. Though we still use this word today to describe development in some insect species, namely in those that undergo an extreme transformation in body form such as that of a caterpillar metamorphosing into a butterfly, the word had a different meaning in Swammerdam's time. Then it meant a sudden and total change from one kind of organism into another, something like the transmutation of lead into gold that alchemists sought. Through dissections of caterpillars and of cocoons, which caterpillars create from fine fibers they spin and from which the adult butterflies emerge, Swammerdam showed that the change was not all that sudden, that the butterflies' wings and other body parts developed for some time within the cocoons.

Swammerdam studied development in so many different insect species that he was able to create a classification system based on the different types of development. He divided insects into four groups, with those in the first group hatching directly from the egg into the adult form. Development in the other three categories involves a pupal or immature stage. In the second group, the insect emerges from the egg without wings, but usually with its six legs, and then gradually develops adult traits; the mayfly, which Swammerdam studied in great detail, belongs to this group. For those species in the third category, the change during development is more extreme, and this group includes butterflies and moths. In the last group, the pupa is hidden within a case called the puparium, as with most flies where the maggot is found within a puparium. To summarize his findings, Swammerdam created a table showing the stages in insect development in the different groups. This approach to the organization of knowledge was less common at the time than it is today and made his results more noteworthy and convincing. For reasons that are unclear, Swammerdam didn't publish most of his results. Fifty years later, the Dutch physician Herman Boerhaave (1668-1738) bought Swammerdam's papers and published them in two volumes as *Bijbel der Natuure* (The Bible of Nature) (1737-38).

Impact

At the time that Swammerdam's books appeared, René de Réaumur was publishing his work on insects. Born in the French town of La Rochelle, Réaumur spent most of the rest of his life in Paris, where he eventually became an influential member of the prestigious Academy of Sciences. His interests were broad and included technological studies, particularly on the production of iron and steel.

Réaumur's greatest contribution to natural history was his six-volume study of insects. He was fascinated by bees, to which he devoted an entire volume. He discovered that in all hives there is only one queen that produces all the eggs. He kept track of individual bees by coloring them with various dyes, a technique that is still used by entomologists. Réaumur was one of the first to do comprehensive quantitative studies on insects. He found that by dunking a beehive in cold water, he could slow the activity of the bees to the point where he could separate them into several categories and count the number of each.

Caterpillars and butterflies were the focus of the first two volumes of Réaumur's work. He repeated Swammerdam's studies on the development of the butterfly within the cocoon. Like Swammerdam, he came to the conclusion that the fact that parts of the adult butterfly are present in a smaller size within the cocoon supported the idea of preformation. This was the concept that development simply involved the enlargement of a tiny, preformed adult found in the egg. This idea received a lot of support in the eighteenth century, in part because of the observations of Swammerdam and Réaumur on insect development.

Preformation was popular because it neatly explained how the complexities of organisms arose. With preformation, all these complexities were created in the original individuals of a species, which have been compared to those toy boxes which when opened reveal a smaller version of the same figure, which when opened reveals a still smaller version, and so on. The view opposite to preformation was epigenesis, meaning that a beginning embryo was merely a formless clump of cells that slowly became more complex and ordered as it grew, with cells becoming organized into tissues and organs as they developed. Epigenesis was difficult for many to accept; it seemed mechanistic and even atheistic since the tissue appeared to develop on its own without any help from a creator. On the other hand, preformation implied the existence of a creator who formed all the adults-within-adults at the time of the original creation.

Réaumur contributed to an important observation that seemed to support the idea of preformation. Studies on aphids, small insects that live on plants, suggested they could arise from eggs that had not been fertilized by sperm. This process is called parthenogenesis, and it implied that the form of the individual was already present in the egg and that the sperm did not add anything essential for development. Réaumur attempted to test this idea by raising aphids in isolation from birth, so that there would be no sexual contact. Though this work was unsuccessful because the insects died before they became sexually mature, his student Charles Bonnet (1720-

1793) was able to show that aphid eggs could indeed develop without fertilization. Réaumur did come up with another interesting finding on aphids. He calculated that, under ideal conditions, a single female aphid could produce 5 million offspring during her reproductive lifetime of about four to six weeks. Again, he used quantification to make a point about insect characteristics, in this case, about reproductive potential. It should be mentioned that aphids don't take over the world, despite being able to produce so many offspring, because not only would such numbers only be produced under ideal conditions, but most aphid offspring are eaten by predators or die from other dangers before they themselves become sexually mature.

Since Réaumur had done economically important work in technology and was interested in insect behavior, it's not surprising that he investigated the economic significance of insects. He argued that research on insects was important because of the destruction they can do attacking plants, the wood of houses and furniture, and even clothes. He himself did research on the lifecycle of the clothes moth in order to find the best way of eliminating it. But Réaumur also saw the positive benefits insects provide in the form of the products they produce—from the fibers of silkworms to the wax and honey of bees and the red dye produced from ground cochineal insects.

While Réaumur is the most noted entomologist of the eighteenth century, a number of others also made important contributions to the field. In 1760 Pierre Lyonnet (1707?-1789) published a comprehensive study of the goat-moth caterpillar, and this foreshadowed the increasing specialization and narrowing focus that was to be seen in the work of many nineteenth-century entomologists. C. de Geer (1720-1778) of Sweden continued Réaumur's work; he did further studies in insect classification that, in turn, in-

fluenced Carolus Linnaeus's (1707-1778) system of insect classification, and it was Linnaeus's work that put classification on a new, more organized footing. Réaumur's writings also influenced the studies of Georges Buffon (1707-1788), who produced a monumental 10-volume work on natural history during the second half of the eighteenth century. At the beginning of the nineteenth century two other French zoologists, Jean Baptiste de Lamarck (1744-1829) and Georges Cuvier (1769-1832), published works that built on Réaumur's entomological research, and these publications were followed by those of the English entomologist W. E. Leach. All three contributed to clarifying precisely which organisms qualified as insects and which belonged in other classes such as the Arachnida (spiders) and the Crustacea (lobsters and shrimp).

MAURA C. FLANNERY

Further Reading

Gough, J.B. "René-Antoine Ferchault de Réaumur." In *Dictionary of Scientific Biography,* vol. 11, edited by Charles Gillispie. New York: Scribner's, 1975: 327-335.

Hubbell, Sue. *Broadsides from the Other Orders: A Book of Bugs.* New York: Random House, 1993.

Lindeboom, G.A. "Hermann Boerhaave." In *Dictionary of Scientific Biography,* vol. 2, edited by Charles Gillispie. New York: Scribner's, 1970: 224-228.

Magner, Lois. *A History of the Life Sciences.* 2nd ed. New York: Marcel Dekker, 1994.

Ritterbush, Philip. *Overtures to Biology: The Speculations of Eighteenth-Century Naturalists.* New Haven, CT: Yale University Press, 1964.

Serafini, Anthony. *The Epic History of Biology.* New York: Plenum, 1993.

Wigglesworth, V.B. *The Life of Insects.* New York: World, 1964.

Winsor, Mary. "Jan Swammerdam." In *Dictionary of Scientific Biography,* vol. 13, edited by Charles Gillispie. New York: Scribner's, 1976: 168-175.

Experimental Physiology in the 1700s

Overview

Physiology is the study of the function of living organisms and their components, including the physical and chemical processes involved. The eighteenth century has been referred to as the Age of Enlightenment, or the Age of Reason, when knowledge in general advanced, and especially that relating to science and medicine. During this time, discoveries in physiology expanded due to a group of investigators known as experimental physiologists.

The study of human anatomy had been advanced by the daring work of Leonardo da Vinci (1452-1519) and Andreas Vesalius (1514-1564) in the previous two centuries, yet the processes by which the internal systems worked were still an enigma. Eighteenth-century scientists built upon this work. Unlike modern scientists who have distinct specialties, these early investigators would master several diverse fields. Scholars such as René de Réaumur (1683-1757) studied many disciplines in addition to physiology. Italian physiologist Lazzaro Spallanzani (1729-1799) contributed not only to human biology but made unusual discoveries concerning animal reproduction. In addition, Giovanni Borelli (1608-1679) studied movement, and Francis Glisson (1597-677) and Albrecht von Haller (1708-1777) studied irritability (muscle contraction).

The eighteenth century was rife with conflicts between science and religion. It saw the advent of mechanism, which envisioned living things as simple machines. This rivaled the ancient philosophy of vitalism, which proposed that all living things have an internal or "vital force" that is not accessible and cannot be measured. This force activates living things. Georg Ernst Stahl (1660-1734), German physician and chemist, supported vitalism, which held sway over biology until the early twentieth century. Stahl even lost his best friend, Friedrich Hoffmann (1660-1742), over differences of philosophical opinion. This quarrel between friends is just one example of how the optimism of their time was tempered by heated emotional debates over theory.

Background

The Age of Enlightenment had its roots in the brilliant breakthroughs of the previous centuries, including the work of Vesalius, a Flemish anatomist and physician who is often called the founder of anatomy. The Greek physician Galen A.D. 129-216?) had earlier established many of the basic principles of medicine accepted for centuries. Galen's beliefs about physiology were a composite of the ideas of Plato (427?-347 B.C.), Aristotle (384-322 B.C.), and Hippocrates (460?-377? B.C.). He believed that human health was the balance of four major humors, or bodily fluids. The humors were blood, yellow bile, black bile, and phlegm. Each humor contained two of the four elements (fire, earth, air, and water) and displayed two of the four of the primary qualities of hot, cold, wet, and dry. Galen proposed that humor imbalances could be in specific organs and the body as a whole. He prescribed

certain remedies to put the body back in balance. Galen's ideas held sway in all areas of medicine until the mid-1600s

In 1543 Vesalius dared to point out errors in the Galen's assumptions and proceeded to conduct dissections and experimental research. He showed that Galen had more information about animals than about humans, and also that translators throughout the ages had introduced many errors into Galen's texts. The ideas of physiology lasted until William Harvey (1578-1657) correctly explained the circulation of the blood.

Some scholars mark the beginning of the scientific revolution at 1543, the year that Versalius published *On the Fabric of the Human Body* and Nicholaus Copernicus (1473-1543) published *On the Revolutions of the Heavenly Spheres,* in which he challenged the old Ptolemaic belief that the Earth was the center of the universe. Both books did much to diminish the hold of classical Greek science, making possible the development of new ideas.

English physician William Harvey discovered how blood circulates in mammals, a finding that conflicted with the accepted ideas of Galen. Many of these achievements were impressive on paper, but did not impact the daily work of medicine. Major problems with disease existed in the beginning of the 1700s. War, plagues, and crowding in cities brought health hazards and high mortality rates.

Sir Isaac Newton (1642-1727) became an idol of the Enlightenment. His experiments and natural philosophy established a model. He led Herman Boerhaave (1668-1738), a Dutch professor, to apply physics to the expression of disease. According to Boerhaave, health and sickness were explained in terms of forces, weights, and hydrostatic pressure.

These views definitely encouraged experimentation. While the Church stymied new science and encouraged the acceptance of ancient truths, scientists began to observe, dissect, and challenge, leading to many new discoveries—as well as controversies. Physiology was only one of the fields advanced.

Impact

The impact of eighteenth-century science was not limited to changes in research methodology, but effected a fundamental new concept of reality with its foundation in observation and experiment. This basic change in perspective is called a paradigm shift, where one model is changed for another.

The eighteenth century as also an age of optimism. Even though war and disease took a toll, "enlightened" intellectuals in the Netherlands, Britain, France, and Germany proclaimed better times were on the way. Science and technology would give man control over nature, social progress, and disease. In the 1790s the marquis de Condorcet (1743-1794) declared that future medical advances, along with civil reason, would extend longevity even to the point of immortality.

Today, scientists typically focus their efforts in one field, or even a subspecialty of a particular field. For example, a molecular biologist may concentrate on one particular structure within the cells, such as the DNA of ribosomes. It is now difficult to imagine how diversely knowledgeable many of the early investigators were.

René de Réaumur was an eighteenth-century French scientist who researched and worked in a multitude of fields. In 1710 King Louis XIV put him in charge of compiling lists of all the industrial and natural resources in France. He devised a thermometer that bears his name, improved techniques for making iron and steel, and worked to uncover the secret of Chinese porcelain. He was one of the first to describe the regeneration of lost appendages by lobsters and crayfish. His work with animals led him to the study of digestion.

Scientists were beginning to ask key questions about how things worked. For example: How does digestion happen? By some internal vital force? By chemical action of gastric juice? By mechanical churning and pulverizing? Pondering these problems, Reaumur conceived of brilliant studies of digestion in birds. Since birds do not have teeth to grind food, he was convinced the structure called the gizzard acts to masticate and make food pieces smaller. He forced seed- or grain-eating birds to swallow tubes of glass and tin. The animals were dissected two days later. He was amazed to find the glass tubes shattered, but the action of the gizzard had smoothed and polished the pieces. The tin tubes were crushed and flattened.

Turning his attention to carnivorous or meat-eating birds, he trained a pet kite to swallow small food-filled tubes. He enclosed pieces of meat in a tube, closed the ends, then encouraged the bird to swallow. Later the bird would regurgitate the tube with the food in it. Analysis of the food showed that it had been partially digested and that digestion was more chemical than physical. He also attacked the prevalent idea of the day that putrefaction or rotting was the way food is digested. Réaumur had a great impact on the debate over digestion. He was one of the most esteemed members of the Academy of Sciences during the first part of the century, who wrote widely, and became known throughout the European community.

Lazzaro Spallanzani was an Italian physiologist who also contributed to many areas of science. Spallanzani received a sound classical education and became a professor of logic, metaphysics, Greek, and physics at the University of Modena in 1760. Although he worked during the day teaching, he devoted all of his leisure time to scientific investigation. Two scientists, Georges Buffon and John Turbeville Needham (1713-1781), had published a biological theory that all living things contain not only inanimate matter but "vital atoms" that go out into the soil and are recycled into plants. They claimed that the small moving objects found in pond water were the vital atoms from organic material. However, studying samples of pond water, Spallanzani was led to agree with Anton van Leeuwenhoek (1632-1722), the Dutch scientist who observed microscopic organisms. Spallanzani concluded that these moving objects were indeed small "beasties" or animals, and not just vital atoms.

Spallanzani was fascinated by the mechanism of how tails of salamanders, snails, and tadpoles regenerate. He devised several experiments to disprove the idea of spontaneous generation. This idea held that things like maggots came from nonliving decaying meat. Francesco Redi (1626-98) conducted a famous experiment with decaying meat that disproved this notion, but the idea of spontaneous generation persisted in many circles.

In 1773 Spallanzani did an important series of experiments on digestion. Using himself as the subject, he swallowed small linen bags that had different kinds of food. He would then regurgitate the bags and study the content. This enabled him to determine that digestive juice has special chemicals that target different kinds of foods. He also determined the solvent powers of saliva.

Spallanzani received much recognition for his study of reproduction. He also revealed that oxidation occurred in the blood. By 1800 it was accepted that oxygen combined with carbon in food to generate animal heat. Spallanzani took every opportunity to travel, exchange information with other scientists, and study new phenomena. This was also within the true spirit of the age of Enlightenment.

A tradition was developing among physiologists called "iatrochemistry," the chemistry of healing living things. This term might today be related to biochemistry. The iatrochemists explored links between respiration and combustion. Joseph Black (1728-1799), a Scottish professor, developed the idea of latent heat, which is given off in breathing. Later, Antoine Lavoisier (1743-1794) identified this as carbon dioxide. Black noted that this breathed-out air was not toxic, but could not be breathed to sustain life.

Giovanni Borelli was an Italian physiologist who first explained muscular movement. He studied muscular action, gland secretion, respiration, heart rhythm, and neural responses. He attempted to analyze these body movements in terms of physics. In his book *The Movement of Animals* he used the principles of mechanics to analyze movement. Looking at the contraction of muscles, he proposed that their operation was triggered by processes similar to chemical fermentation. He assumed respiration was a mechanical process, with the lungs driving air into the bloodstream. He described the bone structure as a series of levers, where muscles worked by pushing and pulling. He also believed air had something in it that sustained life. Air was a medium for elastic-like particles that went into the blood, enabling motion. According to Borelli, what could not be measured or weighed was mysticism and not scientific.

Borelli used the term iatrophysics, or iatrochemistry, to relate physics and chemistry to medicine. The word "iatro" is a Greek term that means to heal. One can see the word in "psychiatry," meaning to heal the mind, or in the word "pediatrics," meaning to heal children.

Francis Glisson was the first researcher to grapple with the idea of irritability. Glisson was a professor of physics at Cambridge and did a pioneering study on a "new" disease—rickets. As he studied the excretion of bile into the intestines, he noted that the reaction implied nerves were present. He became convinced that irritability (contraction) was a property of tissue and that nerves were independent structures. The idea of irritability was forgotten until Swiss biologist Albrecht von Haller (1708-1777) revived it in the next century. Glissons's observations were radical for his time, and prompted by Harvey's discoveries and mechanical philosophy.

Experimental physiology reached new heights through the work of Haller, a prolific writer who has been dubbed the father of experimental physiology. Born in Bern, Switzerland, he was a child prodigy who came under the influence of Hermann Boerhaave. One of his first labors of love was to revise an edition of Boerhaave's *Institutes of Medicine,* which showed that experimentation would produce new accounts of vitality and the relationship between the body and soul. Boerhaave's model of balance focused on the nervous system. He also became a strong advocate of the mechanistic philosophy.

Haller went to the University of Göttingen and served as professor of medicine, anatomy, surgery, and botany, He became very well known in the scientific community because of impressive experimental work at the newly formed university. He wrote an eight-volume encyclopedia called *Physiological Elements of the Human Body* (1757), regarded as a landmark in medical history. A devout Christian, Haller's writings evince the view that human beings consist of a physical body, which may be analyzed in terms of forces and matter, and an eternal soul.

Haller shocked the scientific community in 1753 when he suddenly resigned his prestigious position at Göttingen to return to Bern to continue his experiments, write, and develop a private practice. Experiments into breathing and circulation led Haller to be the first to recognize how respiration works and how the heart functions with it. He found that bile digests fats and identified many of the stages of embryonic development. He conducted anatomical studies of the brain and reproductive and circulatory systems.

However, his outstanding contribution was in the study of muscle contractions. Building on the results of 567 experiments—with 190 performed on himself—he showed that irritability is a special property of muscle. Irritability in medicine means capable of reacting to a stimulus or is sensitive to stimuli. He found a slight stimulus applied directly to a muscle causes a sharp contraction. He found that sensibility is the specific property of nerves. Sensibility is defined medically as the capacity to receive and respond to stimuli. In his book *On the Sensible and Irritable Parts of the Human Body,* he clearly established that irritability or contracting was a property of all muscular fibers, and that sensibility or carrying the message was the domain of nerves. His studies and writings became the foundation of neurology.

In his later years Haller spent much time cataloging scientific literature. His four volume *Bibliography of the Practice of Medicine* lists 52,000 publications on anatomy, botany, surgery, and medicine.

The idea of applying mathematics and measurements to science was also spurred by Boerhaave, who called the studies "iatromathematics." An Anglican clergyman, Stephen Hales (1677-1761), devised a unique experiment to measure the force of blood. He inserted a goose trachea, attached to a 11-ft (3.4 m) glass tube, into the jugular vein and carotid artery of a horse. This enabled him to measure how far the blood would extend into the tube.

Hales was fascinated with experiments with animals and used many different animals in inhumane ways. For example, interested in nerve action, he cut off the head of frogs and pricked the skin in different places to study nerve reflexes. His callous use of animals incited the strong disapproval of animal rights people of his day. These were called anti-vivisectionists, and they developed a strong voice in opposing physiology experimentation on animals. Led by Samuel Johnson, a great literary figure in Britain, the group proclaimed that such torture of animals had never healed anyone. This debate would continue into later centuries.

In the Age of Reason two philosophies of the biological nature of life were debated: vitalism and mechanism. The mechanistic theory viewed living organisms as machines. Thus, the whole of the body is the sum of its parts, programmed to run and operate on its own. This is called a reductionist position because it proposed that biology is simply reduced to physical and chemical laws. Vitalism, on the other hand, is a school of thought dating back to Aristotle. In this view the nature of life is seen as a result of a vital force peculiar to living organisms. This force is different from all other forces and controls the development and activities of the organism. One proponent of vitalism was Georg Ernst Stahl.

Stahl studied medicine at Jena and as a student became friends with Frederich Hoffman. Both went to the newly created medical school at Halle, where Stahl became a professor of practical medicine, anatomy, physics, and chemistry. When Hoffmann became an ardent convert to iatromechanics—a staunch mechanistic belief that living things were machines—he and Stahl parted ways after 20 years of friendship. Their philosophical differences had made the former friends bitter rivals.

Stahl was influenced by the German Pietists, a Protestant group that stressed devotion and the tolerant aspects of Christianity, similar to the Quakers or Friends. He insisted that mechanical and chemical laws were not enough to explain the mysteries of living beings. He determined that life must have direction by a special force and described this force using the Latin word *anima*. This gave rise to the name animism for the theory. Vitalism and anti-materialism are later words for this same view.

Stahl developed the constructs of the phlogiston theory. According to him, phlogiston was a substance that escaped from or was exchanged between materials involved in combustion. Phlogiston was from the Greek *phlogiston,* which means "to burn." Stahl built a practical theory that phlogiston was also a part of body physics and energy.

When Lavoisier discovered the role of oxygen in combustion, the phlogiston theory was proved wrong. However, modern researchers have praised Stahl for pioneering the use of molecular research in chemical analysis, and recognizing the difference between physical and chemical reactions.

Seeking to understand the workings of physiology, Enlightenment researchers made enormous contributions to the human capacity to master the environment. In searching for the answer to questions concerning the nature and internal processes of living things, they were able to advance knowledge of both the structure and the function of living systems.

Though historians note that the Age of Reason was marked by enthusiasm and hope, such great expectations were often tempered by disappointment, particularly when an overemphasis on theory and philosophical speculation actually inhibited scientific reason. However, the work of these eighteenth-century scientists laid the foundations for nineteenth-century advances in biology.

EVELYN B. KELLY

Further Reading

King, Lester F. *The Philosophy of Medicine: The Early 18th Century.* Cambridge, MA: Harvard University Press, 1978.

Koyre, Alexandre. *The Astronomical Revolution: Copernicus-Kepler-Borelli.* New York: Dover, 1992.

Lindeman, Mary. *Health and Healing in Eighteenth Century Germany.* Baltimore: Johns Hopkins University Press, 1996.

Porter, Roy. *The Greatest Benefit to Mankind A Medical History of Humanity.* New York: W. W. Norton, 1997.

Porter, Roy, ed. *Medicine in the Enlightenment.* Amsterdam: Rodolfe, 1995.

Spallane, J. *The Doctrine of Nerves.* London: Oxford Press, 1981.

Marie-François-Xavier Bichat and the Tissue Doctrine of General Anatomy

Overview

Long before the development of cell theory, philosophers and anatomists speculated about the nature of constituents of the human body that might exist below the level of ordinary vision. Even after the introduction of the microscope in the seventeenth century, however, investigators still argued about the level of resolution that might be applicable to studies of the human body. By the eighteenth century, many anatomists had abandoned humoral pathology and, in analyzing the structure and function of organs and organ systems, hoped to discover correlations between localized lesions and the process of disease. Tissue doctrine was elaborated by the great French anatomist Marie-François-Xavier Bichat (1771-1802) as an answer to the question about the constituents of the body. As a result of Bichat's ingenious approach to the study of the construction of the body, he is considered the founder of modern histology and tissue pathology. His pioneering work in anatomy and histology has been of lasting value to biomedical science. Bichat's approach involved studying the body in terms of organs, which were then dissected and analyzed into their fundamental structural and vital elements, called "tissues." This attempt to create a new system for understanding the structure of the body culminated in the tissue doctrine of general anatomy.

Background

As the son of a respected doctor, Bichat was expected to enter the same profession. After studying medicine at Montpellier, Bichat continued his surgical training at the Hôtel Dieu in Lyons. The turmoil caused by the French Revolution, however, forced him to leave the city for service in the army. In 1793 he was able to resume his studies in Paris and became a student of the eminent surgeon and anatomist Pierre-Joseph Desault (1744-95). In 1800 Bichat became physician at the Hôtel Dieu. One year later, he was appointed professor. Completely dedicated to anatomical and pathological research, Bichat essentially lived in the anatomical theater and dissection rooms of the Hôtel Dieu, where he performed at least 600 autopsies in one year. In 1802 he became ill with a fever and, barely 31 years of age, died before completing his last anatomical treatise.

Working in the autopsy rooms and wards of the hospitals of Paris, Bichat and his colleagues were committed to the goal of transforming the art of medicine into a true science. Bichat believed this goal could only occur when physicians adopted the method of philosophical analysis used in the other natural sciences. Research on the fundamental structure of the body would transform observations of complex phenomena into precise and distinct categories. This approach and the movement to link postmortem observations with clinical studies of disease were largely inspired by the work of the great French physician Philippe Pinel (1755-1826). Honored for his *Philosophical Nosography* (1798), Pinel argued that diseases must be understood not by references to humoral pathology but by tracing them back to the organic lesions that were their sources. Because organs were composed of different elements, research, in turn, must be directed towards revealing the fundamental constituents of organs.

Impact

Bichat reasoned that organs that manifested analogous traits in health or disease must share some common structural or functional components. Failing to find this analogy at the organ level, he conceived the idea that there might be some such analogy at a deeper level. His approach involved studying the body in terms of organs that could be broken down into their fundamental structural and vital elements, which he called "tissues." Organs had to be teased apart by dissection, maceration, cooking, drying, and exposure to chemical agents such as acids, alkalis, and alcohol. According to Bichat, the human body could be resolved into 21 different kinds of tissue, such as the nervous, vascular, connective, bone, fibrous, and cellular tissues. Organs, which were made up of assemblages of tissues, were, in turn, components of more complex entities known as organ systems.

Tissues were units of function as well as structure. The actions of tissues were explained in terms of irritability (the ability to react to stimuli), sensibility (the ability to perceive stimuli), and sympathy (the mutual effect parts of the body exert on each other in sickness and health). Obviously, Bichat's "simple" tissues were themselves complex; they were just simpler than or-

Marie-François-Xavier Bichat. *(Corbis Corporation. Reproduced with permission.)*

Bichat presented his objectives for a new science of anatomy and pathology. In essence, he believed that an accurate classification of the different tissues of the body was fundamental to the new science. An anatomist would have to know the distribution of tissues in the various organs and parts of the body and the susceptibilities of specific tissues to disease. These themes were developed further in his *General Anatomy, Applied to Physiology and Medicine,* a work which has been called one of the most important books in the history of medicine. Bichat's last great work, the five-volume *Treatise of Descriptive Anatomy,* was unfinished at the time of his death.

Clearly, Bichat's tissue theory of general anatomy is quite different from the cell theory that was elaborated in the nineteenth century by Matthias Jacob Schleiden (1804-1881) and Theodor Schwann (1810-1882). Cell theory is a fundamental aspect of modern biology and implicit in our concepts of the structure of the body, the mechanism of inheritance, development and differentiation, and evolutionary theory.

Many of Bichat's followers came to regard the tissue as the body's ultimate level of resolution. Among the more conservative French physicians—even after cell theory had been well established for both plants and animals—the tissue was still considered the natural unit of structure and function. Many agreed with Bichat's well-known skepticism concerning microscopic observations. The microscope was not a trustworthy tool for exploring the structure of the body, Bichat cautioned, because every person using it saw a different vision. Many microscopists had reported that biological materials were composed of various sorts of globules. While some of these entities might have been cells, in many cases they were probably just optical illusions or artifacts. Although Bichat's work is often regarded as the foundation for the science of histology, the word "histology" was actually coined about 20 years after his death.

LOIS N. MAGNER

gans, organ systems, or the body as a whole. Tissues, as Bichat himself acknowledged, consisted of combinations of interlaced vessels and fibers. Thus, Bichat's tissue theory of general anatomy provided no actual unit of basic structure that could not be further subdivided. Thus, Bichat's concept of the tissue is not like the concepts now associated with the cell or the atom. Nevertheless, Bichat hoped that his analysis of the structure of the human body would lead to a better understanding of the specific lesions of disease and improvements in therapeutic methods.

Embryology was essentially outside the boundaries of Bichat's own research program, and his account of the arrangement of animal tissues generally ignored the problem of tracing the origins of specific organs and tissues back through their embryological development. Bichat's goals and guiding principles were thus different from those that motivated the founders of cell theory. In formulating his doctrine, Bichat's objective was not merely to extend the knowledge of descriptive anatomy but to provide a scientific language with which to describe pathological changes. Through an understanding of the specific sites of disease, he expected better therapeutic methods eventually to emerge.

Bichat's dedicated disciples studied his writings and arranged to have them translated into other languages. In *Treatise on Tissues* (1800),

Further Reading

Books

Bichat, Xavier. *Physiological Researches on Life and Death.* Translated from the French by F. Gold. New York: Arno Press, 1977.

Haigh, Elizabeth. *Xavier Bichat and the Medical Theory of the Eighteenth Century.* London: Wellcome Institute for the History of Medicine, 1984.

Hall, Thomas S. *A Source Book in Animal Biology.* Cambridge, MA: Harvard University Press, 1951.

Hooke, R. *Micrographia* (1665). New York: Dover, reprint.

Hughes, Arthur. *A History of Cytology.* New York: Abelard-Schuman, 1959.

Periodical Articles

Haigh, Elizabeth. "The Roots of the Vitalism of Xavier Bichat." *Bulletin of the History of Medicine* 49 (1975): 72-86.

Neurology in the 1700s

Overview

Neurology is the study of the body's brain and nervous system. Although modern neurology was not established until the twentieth century when surgery, antibiotics, and imaging techniques were well known, the cornerstones for neurology were laid two hundred years earlier. During the eighteenth century, scientists in England and Europe began studies on the function of nerves and why they made muscles contract. Claims of "animal electricity" traveling through the nervous system sparked new discoveries in physiology and technology, and influenced the prevailing literature and culture. Eventually, eighteenth century scientists contemplated the spark of vitality itself, debating its natural, spiritual, or electrical origin in the perception of the human brain.

Background

Prior to the eighteenth century, the fledgling science of neurology, named by English anatomist Thomas Willis (1621-1675), consisted mostly of anatomical observations. Willis was the first to try to link the structure of the brain to its function. By removing the brain from the cranium, Willis was able to describe the brain more clearly, especially illustrating the arterial supply. Willis surmised that the disease of epilepsy and many paralytic conditions were linked with brain or nervous system dysfunction. With vague ideas of brain function, Willis attempted to map particular areas of the brain as responsible for certain mental functions such as vision, hearing, and touch. Willis also observed basic reflexes and attributed them to brain function.

The first scientific studies of nerve function were undertaken in the early 1700s by the English clergyman Stephen Hales (1677-1761). By 1730, Hales had discovered a connection between the spinal cord and reflex action. An avid animal experimenter, Hales was the first to measure blood pressure in an animal by attaching a goose's trachea to a living horse's carotid artery,

then observing how far up a glass tube the force of the heart pumped the horse's blood. The science of the day included the belief that muscle contraction was also caused by the force of the blood and its pressure in the muscles. Hales measured blood pressure in smaller and smaller blood vessels, and determined that blood pressure was so reduced in muscles it could not possibly be responsible for their movement. Instead, Hales suggested that muscle movement was electrical in origin. Turning to smaller animals for this experimentation, Hales studied muscle movement in decapitated frogs. By pricking the frog's skin and stimulating the frog's reflexes, Hales noted that the nerve action continued without the frog brain. Hales concluded that this action was related to the frog's extended nervous system which communicated with the spinal cord.

As to what force sparked the nervous system, it was thought at the time to be the domain of the soul. In his 1751 publication *On the Vital and other Involuntary Motions of Animals*, English physician Robert Whytt (1714-1766) explained the soul as the vital force from a rational and natural stance, rather than a religious one. This soul, Whytt reasoned, was equally distributed throughout the nervous system and was responsible for communicating sensory stimulus to muscles during a reflex action. Whytt demonstrated for the first time that the entire intact spinal cord is not necessary for a reflex action, but that only a small segment is necessary and sufficient to produce a reflex arc. Whytt also showed that certain reflexes could be destroyed (contraction of the pupils to light, known as Whytt's reflex) when cranial nerves were severed. In the clinical setting, Whytt observed and reported spinal shock and was the first to describe tuberculosis meningitis in children.

Swiss physiologist Albrecht von Haller (1708-1777), professor of medicine, anatomy, surgery, and botany at the University of Göttingen, made essential contributions to the developing study of neurology by delineating muscle

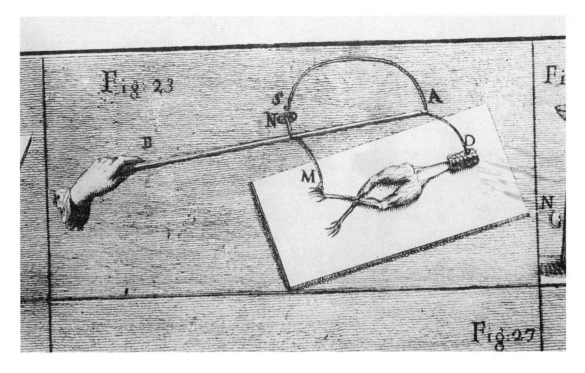

An illustration depicting Luigi Galvani's frog-leg experiments in 1771. *(Corbis Corporation. Reproduced with permission.)*

action from nerve action. Haller's animal experimentation on the irritability (contractility) of muscles and the sensitivity of nerves was extensive—Haller himself performed almost two hundred of over five hundred experiments. Based on the theory of general irritability proposed earlier by English scientist Francis Glisson (1597-1677), Haller showed in 1752 that contractility is a property inherent in muscle fibers, while sensibility is an exclusive property of nerves. By this experimentation, the fundamental division of fibers according to their reactive properties was established. Haller concluded that muscle fibers contracted independently of the nerve tissue around it. Haller thus explained why the heart beats: contraction of muscular tissue stimulated by the influx of blood. Even after the electrophysiology pertaining to the heart was understood more than one hundred years later, Haller's basic suppositions regarding tissue contractility remained essentially correct. Haller's experiments also showed that nerves were inherently unchanged when stimulated, but caused the muscle around it to contract, thus implying that nerves carry impulses which cause sensation. This distinction between nerve and muscle action illustrated by Haller provided the framework for the advent of modern neurology.

After an accidental discovery of the relationship between electricity and muscle movement while studying frogs, Italian physician Luigi Galvani (1737-1798) proposed a theory of "animal electricity." By suspending the legs of frogs with copper wire from an iron balustrade, Galvani noticed that the frog feet twitched when they came in contact with the iron balustrade. While dissecting frogs for study, Galvani also noticed that contact between certain metal instruments and the specimen's nerves provoked muscular contractions in the frog. Believing the contractions to be caused by electrostatic impulses, Galvani acquired a crude electrostatic machine and Leyden jar (used together to store static electricity), and began to experiment with muscular stimulation. Galvani also experimented with natural electrostatic-static occurrences. In 1786, Galvani observed muscular contractions in the legs of a specimen while touching a pair of scissors to the frog's lumbar nerve during a lightning storm. The detailed observations on Galvani's neurophysiological experiments on frogs, and his theories on muscular movement were not published until a decade after their coincidental discovery. In 1792, Galvani published *De Viribus Electricitatis in Motu Musculari* (On Electrical Powers in the Movement of Muscles). The work set forth Galvani's conclusions that nerves were detectors of minute differences in external electrical potential, and that animal tissues possessed electricity different from the "natural" electricity found in lightning or an electrostatic machine. Galvani later proved the existence of bioelectricity by stimulating muscular contractions with the use of only one metallic contact—a pool of mercury.

Impact

The accomplishments of eighteenth century scientists such as Hales, Whytt, Haller, and Galvani laid the foundation for the independent discipline of neurology. The fact that it was a new field, largely due to the recent knowledge of the underlying anatomy and physiology, made it an attractive choice for many outstanding clinicians of the early nineteenth century. From 1790 until the 1840s, hardly a year lapsed between major discoveries of the structure or physiology of the brain and nervous system. Aided by the eighteenth century invention of the microscope, neurons (nerve cells) were some of the first cells clearly described (in 1837 by J.E. Purkinje). Once the anatomy and elementary physiology were understood, scientists began to link various disease states solely to the brain and nervous system.

The fascination with "animal electricity" continued, although the minute electrical activity in the body was later proved to hold no special animal properties. Alessandro Volta (1745-1827), professor at Pavia, showed that a muscle could be made to contract continuously (as in tetany) by applying excessive electrical stimulation. Volta also developed the Voltic pile, the first electrical battery, in 1799, by following Galvani's example of alternating metallic conductors. In the late 1700s, some physicians used static electricity to treat a variety of ailments. The treatments were both fad and fake. The works of scientists such as Galvani and Volta inspired science-fiction fantasies. English author Mary Shelley (1797-1851) published an account of a monstrous life created with the aid of an electrical spark in *Frankenstein*, published in 1818.

The animal experiments of Hales, Galvani, and Haller were at times performed while the animals were still alive, provoking protest among some scientific and social contemporaries. Although a live animal was fundamental to some experiments, (Hale's measurement of blood pressure, for example) news of the protests against vivisection (live animal dissec-tion) reached the scientists. The anti-vivisectionists argued that the experimentation had not resulted in a cure for any maladies known to man. Even the straightforward and occasionally terse English writer Samuel Johnson (1709-1784) softened to the animal cause, claiming vivisection as cruel and denouncing the doctors who "extend the arts of torture." Haller, also a poet, later refrained from further animal research and turned his interests to compiling vast bibliographies on botany and physiology.

The prevailing philosophy of the Enlightenment encouraged most experimentation and reasoned intellectualism, both of which made acceptable the scientific study of the nervous system. Experimentation prompted new consideration for the nature of vitality. The prevailing philosophy of French physician René Descartes (1594-1650), held that the body functioned mechanically as a result of stimulus and reflex. Religion held that the soul supplied by God superseded and influenced the mechanical body. The new naturalism of the Enlightenment challenged both of these philosophies. Haller's work delineating the function of nerve and muscle tissue led to his assertion that man possesses a physical body whose function is explained by properties of matter and forces, and a soul which is spiritual rather than material. Whytt disagreed, fearing the new naturalism would encourage atheism. The philosophical arguments regarding man's brain and it's thought processes occupied much of the scientists' energy, perhaps explaining why, in an age of experimentation, few advances in clinical neurology were made.

BRENDA WILMOTH LERNER

Further Reading

Books

Gross, Charles G. *Brain, Vision, Memory: Tales in the History of Neurosciences.* Cambridge, Massachusetts: MIT Press, 1998.

King, Lester S. *The Medical World of the Eighteenth Century.* Chicago: University of Chicago Press, 1958.

Porter, Roy. *The Greatest Benefit to Mankind: A Medical History of Humanity.* New York: W.W. Norton, 1998.

Spillane, J. *The Doctrine of the Nerves.* London: Oxford University Press, 1981.

The Science of Human Nature

Overview

The revival of science during the European Renaissance included a renewed interest in the workings of the human mind. In the eighteenth century, there were two main schools of thought in the field that became known as "psychology": associationism, or empiricism, and the doctrine of mental faculties. Associationism was championed by such scientists as George Berkeley, David Hume, and David Hartley. This theory held that an individual developed mentally as simple sensations and ideas were associated by the mind into more complex concepts. The opposing theory was the doctrine of mental faculties. Immanuel Kant was among its chief advocates. The doctrine of mental faculties compartmentalized the mind into intellect, emotions, will and other attributes, each of which functioned more or less independently. Both concepts have contributed to our understanding of human nature.

Background

Psychology as an experimental science did not arise until the late nineteenth century. However, since earliest times humans have been speculating about the relationship between themselves and their environment.

Since so many things happened that they could not understand, prescientific humans ascribed them to hidden causes such as deities or spirits. These deities and spirits were thought to control the weather, the behavior of animals, and other natural forces. They could also be invoked to explain the puzzling actions of other people. This belief system is called *animism*.

Animism contributed to the idea that the ideas and sensations we perceive inside ourselves, which we think of as our *consciousness*, are also the product of a "spirit," called the *mind* or the *soul*. After many thousands of years, we have yet to completely understand the nature of consciousness.

The ancient Greek philosophers were interested in the relationship between the mind and the body. Plato (427-347 B.C.) recognized two types of ideas, those that were products of *perception* through the senses, and those that were innate and arose in the soul. Plato, by emphasizing rational deduction, encouraged the development of philosophy. Aristotle (384-322 B.C.), by contrast, stressed direct observation leading to the association of ideas formed by previous perception. His methods and observations of the natural world contributed to the development of empirical science.

During the Middle Ages, the promotion and enforcement of religious doctrine was deemed more important than scientific advancement, and little progress beyond Aristotle was made in most areas of science for about 1,800 years after his death. The intellectual boundaries were stretched during the Renaissance, and a great flowering of the arts, geographical exploration, and science began. This was the age of scientists such as Nicolas Copernicus (1473-1543) and Galileo Galilei (1564-1642) in astronomy, and William Harvey (1578-1657) in physiology. As these scientists relied on observation and experiment to prove their theories, philosophers of the mind began examining their own ideas as well.

René Descartes (1596-1650) dealt with the mind-body problem by compartmentalizing it. The body, he reasoned, works in a mechanistic fashion. The mind, on the other hand, consisting of everything that is not physical and mechanistic, he regarded as beyond scientific inquiry. Descartes envisioned the mind and body interacting at a single point. This *dualistic* or *Cartesian* concept of the mind-body relationship continues to influence our ideas today.

There were several other theories discussed by seventeenth century philosophers. One was *occasionalism*, in which God intervenes between mind and body as circumstances require. *Double aspect*, a theory of Baruch Spinoza (1632-1677) viewed mind and body as two attributes of the same substance. In *psychophysical parallelism*, the mind and body do not interact, but coordinate in their actions because they are influenced by the same external phenomena.

Impact

Among those who adopted the theory of psychophysical parallelism were a group of British philosophers known as *associationists*, or *empiricists*. The founders of this school of thought were Thomas Hobbes (1588-1679) and John Locke (1632-1704). They believed that the mind developed through experience, as simple ideas and simple sensations were synthesized into larger concepts. During the eighteenth century, Hobbes and Locke were followed by a number of philosopher-psychologists who advanced some form of *associationism*, without necessarily agreeing on the source of the ideas that the mind associated.

George Berkeley (1685-1753) was an Anglican bishop who tried to reconcile science and Christianity. He was a philosophical *idealist*, believing that physical things existed only as ideas or sensations in the mind. Considering how lukewarm water could feel cold to someone stepping out of a warm bath, and hot to someone coming in from the cold, Berkeley concluded that the water had no independent existence at all. If physical things existed only when being observed, and yet we believe them to continue to exist even when we are not observing them ourselves, some universally present mind, that is, God, must be observing them all the time.

Berkeley stressed the use of logical deduction in perception. For example, since the retina is two-dimensional, he reasoned that the human sense of sight could not visualize three-dimensional objects directly. He proposed that, rather than being present at birth, the ability to perceive depth was based on the association of certain visual images with the sense of touch. Modern psychologists agree that the ability to interpret some spatial cues does seem to be learned. However, other related abilities, such as detecting size differences, appear to be innate.

David Hartley (1705-1757) was the first to correlate physiological activity with associationism. Isaac Newton (1642-1727) had written that the vibration of light might cause analogous vibrations in the eye and brain, thereby producing the sensations of vision. Hartley advanced the theory that all human senses were caused by tiny vibrations in the nerves. He proposed further that learning was the result of the association of repetitive juxtapositions of these sensory vibrations with ideas formed in the mind. He believed that the mind was immortal, encoded with accumulated habits and patterns of thinking that persisted after death and, if malevolent, continued to torment the deceased.

Hartley, who was fairly orthodox in his religious views, resisted the idea that his theory had mechanistic implications for the development of the human mind. However, by proposing a specific physical mechanism for learning and adaptation, he provided a secular framework around which to discuss and understand human behavior.

David Hume (1711-1776) distinguished between impressions, which were felt, and ideas, which were thought. Ideas were simply "faint images," except when sleep, fever or insanity made ideas spring into the mind with such violence that they approached the force of impressions. Simple impressions, such as the color red, evoked correspondingly simple ideas. However, complex impressions and complex ideas did not necessarily correspond.

Impressions, Hume went on to propose, were of two kinds, sensation and reflection. He relegated further discussion of sensation to scientists, but was interested in reflection, which followed the formation of ideas about the sensations previously experienced, coupled with ideas about the pleasure or pain they may have caused. Impressions of "desire and aversion, hope and fear" were the result.

Hume's view of the way the mind processed ideas and impressions involved three principles: resemblance, proximity in space and time, and, most importantly, cause-and-effect. The mind tends to interpret an event that preceded another as its cause, and forms habits of expectation accordingly. However, Hume believed that it was impossible to actually prove causality. In fact Hume distrusted philosophical proofs in general, believing that knowledge was based strictly on the experience of the individual's consciousness. Thus, he argued, it was impossible to prove causality, the existence or nonexistence of God, or the existence of the physical world outside the self.

The school of thought opposed to associationism was the *doctrine of mental faculties*. This theory classified the mind into several properties, such as reasoning, memory, emotions and will, each of which operated independently. Among its proponents was the German philosopher Immanuel Kant (1724-1804).

Kant viewed the idealism of Berkeley as coarse and simplistic, and developed what he called *transcendental idealism*. He rejected the concept of the mind and physical objects as separate things. Instead, he proposed that the mind is involved in experiencing physical objects by organizing its experiences into patterns. Since everything may be arranged into the mind's patterns, we can have knowledge of that which we have not actually experienced.

The psychology of mental faculties led to the pseudo-science of *phrenology*, in which areas on the surface of the skull were identified with particular faculties. Bumps or hollows on the skull were interpreted as indicating the level of development of the associated faculties. A more useful effect of the doctrine of mental faculties was to spark the re-examination of textbooks and teaching methods used in schools. New materials were developed with the goal of "exercising the faculties."

SHERRI CHASIN CALVO

Further Reading

Allen, Richard C. *David Hartley on Human Nature*. Ithaca: State University of New York, 1999.

Buckingham, Hugh W. and Stanley Finger. "David Hartley's Psychobiological Associationism and the Legacy of Aristotle." *Journal of the History of the Neurosciences* vol. 6 no. 1 (1997): 21-37.

Stack, George J. *Berkeley's Analysis of Perception*. New York: Peter Lang Publishing, 1991.

Uncovering the Relationship Between Anatomy and Disease

Overview

The knowledge of basic human structure is known as the science of anatomy. The term anatomy comes from the Greek word *anatome*, which literally means dissection and often encompasses many other disciplines, such as physiology (the study of body function). While it is one of the oldest sciences and wrought with rich tradition, it is also one of the most disappointing from the standpoint of its development before the seventeenth century. As an example, it was known how the earth moved around the sun nearly one hundred years before it was known that blood moved through the body. One aspect of this slow growth in the knowledge base was the belief in the sanctity of the human body. Most dissections were performed on animals up until the sixteenth century because it was believed in most cultures that the human body must be preserved or protected after death, and that dissection was a defilement.

Background

The study of anatomy dates to antiquity. Herophilus of Chalcedon (330?-260? B.C.), who made detailed studies of dissections of the human body in the third century B.C., is often considered to be the founder of anatomy. His detailed descriptions of both the structure of the human body and how it related to function served as a template of study for his successors, the greatest of which was Galen of Pergamum (A.D. 129-216?).

Galen was the first person to tie disease and anatomy together. While Hippocrates (460?-377? B.C.) is often considered the most important early figure in the area of disease and medicine, he made no recognized contributions to anatomy. As the physician for Roman emperor Marcus Aurelius in the second century A.D., Galen was considered the foremost authority on anatomy and disease for the next 1,300 years. He made many important discoveries while exclusively studying animals. For example, he demonstrated that urine was formed in the kidneys and that damaging the spinal cord caused paralysis below the injury on the same side. He also made many mistakes, but so great was his influence that these went unquestioned for centuries. Such complacency and blind acceptance of received wisdom is often detrimental to the progression of science, and such was the case for the science of anatomy, which stagnated until the sixteenth century.

In 1543 Andreas Vesalius (1514-1564) published an account of human anatomy that proved to be the most significant publication in this area since Galen. This event is considered to be the point at which the science of anatomy enters the modern era. His work is significant because he published his own observations that were often is opposition to Galen. Thus, Vesalius opened the door for the progression of anatomy. Because he provided accurate descriptions of normal structures, it was then possible to begin to correlate disease with changes in structure. While very skillful and methodical, Vesalius was at a disadvantage because he performed his studies without the aid of any optical devices (gross anatomy), so he had to rely on his unaided observations, which could make out details no smaller than 0.019 in (0.48 mm). It was not until the microscope was invented that detailed study of cellular structures could be made.

William Harvey (1578-1657) made some of the most important advances in the history of medicine when in 1628 he published his idea that blood circulated through the body. This piece of information is so vital to our concept of modern medicine that it would not exist without it. As an example, it would be impossible to understand heart disease, which currently accounts for about one-third of all deaths in modern society. Harvey made other significant contributions, but he too was hampered by a lack of technology and could not prove the existence of capillaries, which was central to his idea of blood circulation. That missing link was provided by the Italian scholar Marcello Malpighi (1628-1694), who was the first to use a microscope in conjunction with the study of human anatomy.

DR. JOHN LETTSOM AND EIGHTEENTH-CENTURY MEDICAL PRACTICE

The goal of eighteenth-century medical intervention was to restore the balance of the four "humors" of the body. Physicians usually attempted to accomplish this task by ridding the body of corrupt humors through bleeding, purging, or sweating, as in this poem by Dr. John Lettsom: "I John Lettsom / Blister, purge and sweats 'em. / If after that they choose to die, / I, John, lets 'em."

LOIS N. MAGNER

Malpighi is considered the father of histology (the study of cells and tissues), and he provided basic information that allowed structure to be related to disease. In fact, modern diseases are often diagnosed through changes in cells and tissues that can only be evaluated histologically. With basic anatomy and physiology concepts now in hand, the stage was set for more advanced study of the relationship between structural changes in the body and disease.

Although the correlation between clinical symptoms and pathological changes was not made until the first half of the nineteenth century, the first textbook of morbid anatomy (pathology) was published in 1761 by an Italian scholar Giovanni Battista Morgagni (1682-1771). This book, *The Seats and Causes of Disease,* was the

first publication to link specific diseases with individual organs. He was the first to demonstrate the necessity of basing diagnosis, prognosis, and treatment on a comprehensive knowledge of anatomical conditions. Morgagni's investigations were directed chiefly toward the structure of diseased tissue, both during life and postmortem, in contrast to the normal anatomical structures. While this type of work was not originated by Morgagni, he did publish the largest, most accurate, and best illustrated work yet seen. From this time on, morbid anatomy became a recognized branch of the medical science.

Morgagni performed 640 dissections for his famous treatise. He was careful to link symptoms with structural changes induced by disease in the human body. For this reason, he is known as the father of pathology (the study of disease). He popularized the idea that observation was an important medical tool.

The Hunter family had a significant legacy on the development of thought regarding anatomy, disease, and treatment during the eighteenth century. William Hunter (1718-1783), his brother John Hunter (1728-1793), and their nephew Matthew Baillie (1761-1823) made important contributions during their lifetimes.

John Hunter has been referred to as the father of modern English anatomy. He was a renowned surgeon and researcher, especially in the areas of obstetrics, and was the first to discover that the mother and fetus have separate blood circulations. His brother William was responsible for making surgery a science. He incorporated scientific inquiry, physiology, and pathology into surgery, making it much more valuable than it had been previously. John Hunter was so influential that the history of surgical technique is often divided into two distinct categories, pre- and post-Hunter. Hunter was also influential because he had a famous pupil, Edward Jenner (1749-1823), who introduced the smallpox vaccination and is considered the founder of preventive medicine. Matthew Baillie was a Scottish pathologist whose *Morbid Anatomy of Some of the Most Important Parts of the Human Body* (1793) was the first publication in English on pathology as a separate subject and furthered the systematic study of pathology.

Impact

The clinical correlations between disease and anatomy paved the way for modern medicine to become much more effective in treating ailments. Ancient medical practices centered on relieving demonic possession, not investigating an organic cause for the disease. The medicine man and his modern forms (physicians) are one of the oldest professions on record, but lack of knowledge regarding the pathological basis for disease has plagued societies with poor and even unethical treatments. As an example, the oldest surgical procedure on record involves opening the skull with a sharp stone to let the evil spirit that is causing the disease out. If the patient died, it was blamed on the spirit, not the procedure. The evolution of modern treatment techniques have their roots firmly implanted in the initial research involving morbid anatomy.

Prior to Morgagni, disease was thought to be a general condition. That is, there was not much consideration regarding the cause. Morgagni's influence forever changed how doctors dealt with disease. It was Morgagni who first correlated the symptoms of the disease during life with the pathological changes in organs found at autopsy. He demonstrated that through a careful examination of the deceased body, one could begin to attribute outward changes that occurred with disease with specific organs in the body. With his careful and thoughtful work, doctors could think in specific terms such as heart or liver disease. Proper treatment for a specific disease would not

be available for most diseases unless the exact cause was known. Diseases could now be thought of in scientific terms and defined by the differences between normal and abnormal conditions of the body. With the technology available at that time, the only way of finding that out was through the careful examination and study of people who had passed away.

Much of the work that was performed in the eighteenth century provided the impetus for future study and helped to set up strict guidelines regarding the thoroughness of scientific inquiry in the area of pathology. It paved the road for subsequent researchers such as Rudolf Virchow (1821-1902), the father of modern pathology. In the nineteenth century Virchow showed that the structural changes in disease happen not in the organ as a whole, but rather in the cells for that organ.

JAMES J. HOFFMANN

Further Reading

Canguilhem, G. *The Normal and the Pathological.* Cambridge: MIT Press, 1991.

Cartwright, F. F., and M. Biddiss. *Disease and History.* New York: Sutton, 2000.

Major, R. *Classic Descriptions of Diseases.* New York: Charles C. Thomas, 1978.

Mesmerism: A Theory of the Soul

Overview

Mesmerism, named after its chief theoretician and practitioner, Franz Anton Mesmer (1734-1815), also known as Animal Magnetism, was one of the most popular medical theories of the late eighteenth and early nineteenth centuries. Today mesmerism is considered a form of medical quackery, but it is also seen as a precursor to the use of hypnosis in treating psychological disorders. In its time, however, mesmerism gained popularity because of the theories it put forth on the functioning of the body and the soul.

Background

In the eighteenth century medical practice was still attempting to differentiate itself from religious healing practices such as exorcism. In order to do so, medical practice needed to show that the object of its concern was different from that of religious practice by demonstrating, for example, that disease was the result of malfunctions within the body itself, and not the result of demonic possession. This meant that medical practice also needed to show that there were mechanical explanations for what happened within the body, instead of relying on religious explanations such as the exertion of spiritual forces. Some medical practitioners even attempted to explain in mechanical terms what the chief force was that animated the body—the force that religious practitioners refer to as a "soul." mesmerism was just such a theory of animating forces.

Mesmeric theory was based upon two ideas that had been in circulation for centuries. The first was that there were invisible, naturally occurring forces, such as the force of gravity that keeps planets in their orbits, that can also be held responsible for the animation of living bodies, such as animals and humans. The other idea was that the application of magnets could prove beneficial in the treatment of disease. In the case of mesmerism, what was new was the explanation put forth for the relationship between the forces of magnetism and the forces of life—and by extension, disease.

Though Mesmer would revise and expand on his theories throughout his life, as would his followers, the basic principles of "animal magnetism," as Mesmer referred to it, can be found in his medical degree dissertation of 1766, *De planetarum influxu* (Physical-Medical Treatise on the Influence of the Planets). Here he made reference to the work of Richard Mead (1673-1754), a British medical practitioner, as well as that of Sir Isaac Newton (1642-1727). Mead was a mechanist who sought mechanical explanations for the energy behind corporeal functions, and who suggested in his treatise of 1740, *Corpora Humana et Morbis inde Oriundis* (On the Influence of the Sun and Moon upon Human Bodies and the Diseases Arising Therefrom), that the same forces that caused tides and kept the Moon in rotation about Earth could also exert an influence over the functioning of human bodies. Mesmer, using both Mead's medical theories and Newton's theories of gravity as his examples, argued for the existence of a force he called "animal gravity," "a force which actually strains, relaxes and agitates the cohesion, elasticity, irritability, magnetism, and electricity in the smallest fluid and solid particles of our machine...." According to Mesmer, this force is generated by the actions of the particles of the body itself interacting with one another, just as gravity is the result of the interaction of the mass of two planets, so that the body becomes a kind of battery, capable of generating its own energy in order to animate itself. As Mesmer developed his theories further, he would suggest that magnets were useful in treating disease because disease was caused by a blockage in the flow of this energy, and magnets could exert an influence over this force.

Mesmeric treatment, then, concerned itself with attempting to restore balance in the flow of the "nervous fluid" of the body that was affected by the force of "animal gravity." At first, Mesmer treated patients by applying magnets to the affected parts of their bodies, but later he felt that, since all bodies generated these forces, and that the problem was an imbalance in the forces of the patient's body, a Mesmeric practitioner could simply lay hands on a patient and impart some of their own energy to the patient. Mesmer also felt that it was possible to "charge" certain objects, such as trees, metal poles, and jars of water, with energy, and that patients could treat themselves by drinking the water or holding onto the objects for a length of time.

Impact

The results of Mesmeric treatment were often spectacular, even if they did fail to cure physical disease. Since Mesmer believed that the cause of disease was a blockage in the flow of energy, a successful treatment would produce an "overcoming" of that blockage, with dramatic physical results. Patients under Mesmeric treatment would often report burning hot flashes and pains as the imagined blockage was overcome, and in his Paris treatment salon, Mesmer had special rooms set aside for those patients who swooned during the course of treatment. Today these results are interpreted as hysterical reactions brought about by the expectations of the patients, and as having little to do with the actual cure of physical disease.

The major breakthrough for mesmerism, and the one for which the term mesmerism is today most well known, came in 1780, when a disciple of Mesmer, the Marquis de Puységur, discovered that he could induce a state of "artificial somnambulism," what we would today refer to as a trance, in those undergoing mesmeric treatment. In this trance, patients displayed what was considered to be remarkable behavior; peasants could speak perfect aristocratic French or German, they could conduct everyday activities with their eyes closed, and the mesmerist could actually direct the flow of a patient's thoughts. In other words, the marquis had discovered hypnotism.

This discovery led Mesmer to speculate that the energy forces he had theorized earlier also bound all things together and enabled them, animate or inanimate, to have some communication with each other—how else could one explain, for example, the ability of a person with their eyes closed and seemingly asleep to pick up a broom and sweep out a room, then walk outside and perform other activities? While in the trance, it was as if the "somnambulist" was in communication with some higher plane of existence.

This idea caught fire in the imaginations of many philosophers of the time, such as the German Romantics, who felt that "artificial somnabulism" reflected the ability of all human beings to transcend to a higher "plane of unity" where all things were one. Throughout the early nineteenth century, then, one can find references to mesmerism in a vast array of literary works, from the writings of the German Romantic E.T.A. Hoffman to the works of the American writer Nathaniel Hawthorne. In all of these works, mesmerism is depicted as a strange, sometimes evil, force that has the power to change and manipulate the human soul.

Mesmerism was much more successful as a literary device than a medical practice, however. Because of the rather spectacular and indecorous reactions of patients, especially female patients, to mesmeric treatment, the French government undertook an investigation of mesmerism in 1784. The government concluded that not only was mesmerism a form of medical quackery but that it also constituted a genuine danger to patients, and in that same year the practice of mesmerism in France was banned. To this day, however, the use of magnets in medical treatment

persists, and hypnosis has become a standard tool in the treatment of psychological disorders, even though its functioning is little more understood today than it was in Mesmer's time. mesmerism may today be regarded as nothing more than a footnote in the history of medicine, but in its day it proved to be a powerful philosophical force, and it opened several doors in our thinking about the functioning of the human body and the energy that animates it.

PHILLIP A. GOCHENOUR

Further Reading

Darnton, Robert. *Mesmerism and the End of the Enlightenment in France.* Cambridge: Harvard University Press, 1968.

Mesmer, Franz Anton. *Mesmerism, a Translation of the Original Scientific and Medical Writings of F. A. Mesmer.* Translated by George J. Bloch. Los Altos, CA: W. Kaufmann, 1980.

Pattie, Frank A. *Mesmer and Animal Magnetism: A Chapter in the History of Medicine.* Hamilton, NY: Edmonston Publishing, 1994.

Scurvy and the Foundations of the Science of Nutrition

Overview

Scurvy, a serious nutritional disorder caused by a lack of vitamin C, became a significant problem aboard ships during the early modern era. The disease often killed sailors on lengthy voyages, and was so pervasive that ships frequently set sail with a double crew simply to have enough men to finish the voyage. In the mid-eighteenth century, a shipboard experiment by Scottish physician James Lind showed the value of citrus juice in preventing scurvy.

Background

Unlike most animals, humans cannot manufacture vitamin C and must eat a sufficient amount to stay healthy. Without enough vitamin C collagen begins to break down, causing a loss of bone, cartilage, and dentin—the calcerous material that makes up teeth. The early symptoms of scurvy, which begin to develop after about 12 weeks of poor diet, include lethargy and dry skin; more serious problems emerge after six to

seven months, including an inability to heal wounds, painful swollen joints, and the characteristic soft, bleeding gums and loose teeth.

Although descriptions of disease that resemble scurvy can be found in ancient writings, the first reliable accounts are medieval cases that were documented during the Crusades. Despite being recognized, the syndrome had no name until the early modern era, when long voyages of exploration caused extensive dietary insufficiencies. Vasco da Gama's landmark voyage of 1497-1499, in which he successfully sailed to India by first going south around the Cape of Good Hope in Africa, was plagued by scurvy, which killed 100 out of the original 170 crewmen during the voyage. On both the outgoing and return legs of this journey, many "suffered from their gums; their legs also swelled, and other parts of the body and these swellings spread until the sufferer died, without exhibiting symptoms of any other disease." In the years that followed, scurvy became a ubiquitous problem on long journeys for trade and exploration, often taking many lives.

Magellan's trans-Pacific voyage of 1519 set sail from Spain with five ships and 260 men. During the 3-year voyage their supplies ran out, causing rampant scurvy and horrendous hunger. A crewman's journal noted, "We ate only old biscuit reduced to powder, and full of grubs, and stinking from the dirt which the rats had made on it when eating the good biscuit, and we drank water that was yellow and stinking. The men were so hungry that if any of them caught a rat, he could sell it for a high price to someone who would eat it." Only 18 of the men returned to Spain in 1522.

Shipboard diets were notoriously poor, especially for the crew. Sailors had to rely on provisions that were easily stored in quantity and were as nonperishable as possible. In this era, such requirements meant a great deal of salted meat and hard biscuit. Fresh provisions were available only at landfalls, which often helped cure cases of scurvy and other health problems, but not always. Sailors were also prone to a number of other illnesses, including typhus and tuberculosis, so that they often suffered symptoms of scurvy along with other ailments.

Although scurvy was recognized as a illness, its causes were unknown. Some captains and ships surgeons, however, did make the connection between diet and health. Around the beginning of the seventeenth century two men, Sir Richard Hawkins and Captain Lancaster of the East India Company, recognized the value of lemon juice in preventing scurvy. Another East India Company physician published a book called *Surgeon's Mate* in which he encouraged its use as a cure for scurvy as well. But these treatments were by no means universally adopted. Because sailors endured filthy, poorly ventilated quarters, in addition to a limited diet, scurvy was often attributed to their appalling accommodations and simply accepted as a hazard of marine life. Since many sailors were convicts, criminals, or lowlifes plucked from the depths of society, their quarters were given minimal consideration.

James Lind is credited with proving the efficacy of citrus over other dietary treatments. In 1747 Lind, a British naval surgeon, conducted a shipboard experiment. Taking 12 sick sailors, each of whom showed similar symptoms of scurvy, Lind treated them with six different remedies common at that time. Keeping all other factors the same, Lind gave the men daily doses of either cider, vinegar, sea water, oranges and lemons, elixir vitriol (sulfuric acid), or an "electuary" (paste) made up of various herbs. He found that those men given the oranges and lemons most quickly improved, despite the fact that the supply of these fruits ran out after only six days. Second to begin recovery were those men given hard cider, which is known today to retain some vitamin C. Lind eventually followed up his experiment by publishing *A Treatise of the Scurvy* (1753), a definitive study of the disease and its remedy.

Impact

Lind's experiment is generally credited as the first clinical trial. Although he did not have an untreated "control" group by which to measure success, he nonetheless compared results achieved to the various remedies tried. Unfortunately, his findings did not result in the immediate adoption of his treatment of scurvy. More than 40 years would pass before the British Admiralty mandated a daily ration of lemon juice for each sailor.

The humoral theory of health, which had been developed by the ancient Greeks and was dominant in Europe through the eighteenth century, held that disease was caused by an imbalance in the body's four humors: blood, phlegm, black bile, and yellow bile. Diet had long been used by ancient physicians in treating illness, as it was believed that the humoral balance was directly affected by diet. (There was, however, no concept of disease caused by a *lack* of food.) Humors were also affected by environment, climate, and host of other individual variables. Medical theory of the time, therefore, attributed scurvy to both diet and the conditions in which sailors existed.

The most respected medical figure of the early eighteenth century, Herman Boerhaave (1668-1738), argued from theory (rather than experience) that sailors' salt-rich diets produced a corruption in their blood. This corruption could be counteracted by any of several acidic remedies. Lind himself believed that scurvy resulted from blocked pores in the skin, which prevented the body from eliminating dangerous vapors and humors through the skin. He thought that a cold, damp climate and the close "corrupted" air on board ships created this problem. Thus, the usefulness of citrus fruits, and particularly of the "rob" (a concentrated syrup that allowed lemon and other citrus juice to be preserved) was in their ability to counteract these environmental conditions by cleansing the body. Lind's treatise also recommended fresh air, exercise, and cleanliness as necessary for health. Although the treatment was effective, these theories on scurvy's causes pointed many away from the value of foods rich in vitamin C, and, in some cases, away from diet entirely.

A successful demonstration of the use of fresh provisions came with the explorations of Captain James Cook (1728-1779) from 1776-1780, during which the only death was a result of tuberculosis. Cook, who seems to have been well aware of Lind's treatise, is noted to have insisted on continually acquiring fresh provisions for his crew at each landfall. Again, however, the recognition of the importance of diet was tempered by an additional insistence on cleanliness, fresh air, and sufficient rest for the men, all of which were believed to be contributing factors to scurvy.

The use of citrus was not put into widespread effect until the end of the eighteenth century. Gilbert Blane (1749-1834), another Scottish physician trained in Edinburgh, was appointed as physician of the British fleet in 1781. Blane was also familiar with Lind's work and supported his findings. He immediately began to recommend rations of citrus for the fleet to prevent cases of scurvy. Other officials, however, held differing views on what was most useful for treating scurvy, and Blane's suggestions were disregarded. Finally, in 1796, after being appointed a commissioner to the Board of the Sick and Wounded Sailors, Blane convinced the Lord of the Admiralty to order that all ships be rationed sufficient quantities of lemon juice to give each sailor three-quarters of an ounce (20 ml) per day. This effort produced a dramatic drop in cases of scurvy and led to a new nickname for British sailors. As "lime" was a common term for both lemons and limes, they became known as "limeys."

While rations of lemon juice helped eliminate scurvy in the British naval fleet, it remained a constant problem for other nations' sailors. Scurvy has also been a problem anywhere that diet has been limited, including in the early New World colonies, in jails, and during sieges or other military campaigns. More recent scurvy outbreaks include the Potato Famine in Ireland (potatoes are rich in vitamin C) and the California Gold Rush of 1849. An epidemic of infantile scurvy occurred in Europe and North America in the late nineteenth century as well-to-do women increasingly switched from breast-feeding to using preserved milk, which did not contain any vitamin C. The current understanding of scurvy came only in the twentieth century. Not until after World War I was scurvy widely accepted as a deficiency disease and vitamin C isolated and identified as the causative factor.

KRISTY WILSON BOWERS

Further Reading

Carpenter, Kenneth J. *The History of Scurvy and Vitamin C.* Cambridge: Cambridge University Press, 1986.

Hughes, R. E. "James Lind and the Cure of Scurvy: An Experimental Approach." *Medical History* 19 (1975): 342-351.

Lind, James. *A Treatise of the Scurvy.* Edinburgh, 1753.

Roddis, Louis H. *James Lind: Founder of Nautical Medicine.* New York: Henry Schuman, 1950.

Percivall Pott and the Chimney Sweeps' Cancer

Overview

The English surgeon Percivall Pott (1714-1788) was the first to establish a causal link between cancer and exposure to a substance in the environment. In 1775 he described the occurrence of cancer of the scrotum in a number of his male patients, whose common history included employment as chimney sweeps when they were young. He related the malignancy to the occupation, and concluded that their prolonged exposure to soot was the cause. Pott's description of this disease, and his concern for the plight of these "chimney-boys," sparked a series of reports by other authors and brought to light a disgrace which took another hundred years to eliminate in England. Pott may legitimately be seen as a precursor to the modern investigators who seek to prevent occupational exposure to hazardous substances.

Background

Historians have variously referred to Pott's report on scrotal cancer as a milestone in the fields of chemical carcinogenesis (development of a cancer), preventive oncology, environmental health, and occupational medicine. From the time of the Greeks and Romans cancer was viewed as a systemic disease caused by an excess of black bile. Swellings caused by infection, injury, or cancer were indistinguishable.

In the eighteenth century there was an awakening of medical thought: a new physiology based on an understanding of the circulation of the blood, and a new surgery founded on accurate knowledge of human anatomy. With the rise of medical schools and the use of cadavers to teach human anatomy, observation and experience replaced traditional theories of cancer. Pott had read widely in ancient and medieval medical texts, but in his practice he chose to rely solely on personal experience and first-hand knowledge, and thus was able to discern a cause for cancer in the environment and to develop a cure with wide surgical excision.

Workers often have more intense and prolonged exposure to environmental chemicals and conditions than the general population. Therefore many environmental diseases are first noted in the workplace. The first recorded instance of an occupational disease may be the description of lead colic in metalworkers, attributed to the Greek physician Hippocrates (460-375 B.C.). The Swiss physician and chemist Paracelsus (1493-1541), who studied the health of miners, was responsible for the first book on the diseases of a specific occupational group. Bernardino Ramazzini (1633-1714), an Italian physician, authored the first systematic treatise on occupational diseases in 1700, in which he describes disorders associated with 54 different occupations.

The Industrial Revolution of the eighteenth century brought in its wake an alarming rise in occupational diseases. Crowded, unsanitary factories, mechanical accidents, and exposure to toxic materials were contributing factors. In eighteenth-century England it was the practice to construct long, narrow, and tortuous chimneys, often no more than 25-30 inches (63-76 cm) wide. This encouraged the employment of young boys, aged four to seven years, to clean the flues by hand, since they were thin and agile enough to maneuver inside the chimneys to loosen and remove the soot.

Called "climbing boys" or "chimney boys," some were abandoned children, others belonged to desperately poor families who apprenticed them to adults who exploited them. The climbers wore no clothing when working in the chimneys, subjecting them to abrasion and causing soot to become embedded in the skin. As bathing was infrequent for society in general, the climbers rarely washed; some reported an annual bath, and records at London's St Bartholomew's Hospital refer to children who were never washed for five or six years at a time. This was so common that the term "black as a sweep" became a national byword.

Pott's article, "Cancer Scroti," provides not only a clinical description of the cancer but also a compassionate picture of the daily lives of the child sweeps: "The fate of these people seems singularly hard; in their early infancy, they are most frequently treated with great brutality and almost starved with cold and hunger; they are thrust up narrow and sometimes hot chimnies where they are bruised, burned and almost suffocated and when they get to puberty, become particularly liable to a most noisome, painful and fatal disease." (*Chirurgical Observations*, 1775).

After a latent period of 20-25 years, a large number of these boys developed cancer of the skin of the scrotum, known at the time as "soot-wart." The disease also occurred in persons who were not chimney sweeps, but to Pott's credit he was able to recognize the special liability of that occupation, despite the long latency period.

Impact

The immediate impact of Pott's widely read article was to provoke an awareness among physicians. Following Pott's report, a frequently debated topic was why sweeps in other countries were not susceptible to this disease. Several reasons gradually became evident: coal was substituted for wood in England at an earlier date than elsewhere; in many European countries great care was taken to avoid contact with soot, including wearing protective clothing; German sweeps washed daily from head to foot and used leather trousers; in Scotland, chimneys were swept by lowering a weighted broom, whereas English sweeps worked with a brush from inside.

Besides the impact on diagnosis and treatment, the social effects of Pott's startling report were far-reaching. The depiction of the deplorable state of the child sweeps created a sympathetic response among physicians and clergy, exemplified by Jonas Hanway's book *A Sentimental History of Chimney Sweepers, in London and Westminster, Shewing the Necessity of Putting Them Under Regulation, To Prevent the Grossest Inhumanity to the Climbing Boys, With a Letter to a*

London Clergyman on Sunday Schools Calculated to the Preservation of the Children of the Poor (1785).

As a result of mounting public pressure, Parliament in 1788 passed *The Act for the Better Regulation of Chimney Sweeps and their Apprentices.* Although this was the first legislative action to address child labor, it was woefully inadequate, providing only that no boy under the age of eight should be apprenticed to a sweep. In 1803 a national organization was formed with many illustrious members from the nobility and legislature, called the *Society for Superseding the Necessity of Climbing Boys, by Encouraging a New Method of Sweeping Chimneys and for Improving the Condition of Children and Others Employed by Chimney Sweepers.* The Society advocated replacing chimney boys with mechanical sweeping devices, but faced strenuous opposition. Insurance companies as well as Master Sweeps opposed the idea, claiming that chimneys could only be properly cleaned and repaired by using small boys.

It was not until 1834 that further legislation was passed but it was scarcely an improvement, raising the apprenticeship age to 10, and introducing some standards for chimney construction. Bills presented to Parliament in 1817, 1818, and 1819 were thrown out by the House of Lords despite a successful passage through the House of Commons. Over the next 40 years further acts were passed raising the legal age of apprenticeship to 16, imposing regulations for protective clothing and hygiene along with prison sentences for offenders. None of these measures were effective and violations were pervasive. Finally in 1875, 100 years after Pott's report, the scandal was ended by successful legislation that established a system of licensing for sweeps, supported by police enforcement.

Occupational medicine was further advanced in England by the publication of Charles Turner Thackrah's *The Effects of the Principal Arts, Trades and Professions and of Civic States and Habits of Living on Health and Longevity* (1831), which played an important part in stimulating factory health and child labor legislation. In 1895 a statutory notification system was instituted in England that required the reporting of certain diseases. Other industrial nations followed, and legislative safeguards for worker health continued to be enacted throughout the nineteenth and twentieth centuries.

From Pott's time until the early twentieth century, knowledge of occupational cancer had developed due to the keen observations of individual clinicians and pathologists. After that time, the art of laboratory investigation progressively improved, increasing the chances of identifying causative agents and proving their effects, and even for predicting hazards before they caused disease. Along with this, sophisticated epidemiological studies, supported by modern technology, have greatly expanded the ability of physicians and scientists to detect disease patterns and their relationship to the environment.

A hundred years after Pott, the causative agent in soot was identified as coal tar. Forty years after that, another turning point was reached when in 1915 the Japanese scientist Katsusaburo Yamagiwa (1863-1930) produced the first experimental cancer in laboratory animals. Yamagiwa had been a student of Rudolf Virchow (1821-1902), who had emphasized the capacity of chronic chemical irritation to produce cancer, and called special attention to chimney sweeps' cancer. Virchow's theory gave rise to innumerable futile attempts to produce experimental cancer, leading to an atmosphere of failure and resignation. Yamagiwa's perseverance, along with the choice of an adequate chemical irritant, produced skin cancer in the rabbit ear by the repeated application of coal tar. The knowledge that it was possible to induce cancer in animals spurred worldwide research to isolate active carcinogens.

Pott's discovery became the model for many later investigations of workplace carcinogens, including observations of unusual cancers or a high incidence of common cancers, searches for responsible agents, experiments to demonstrate that agents cause cancer in laboratory animals, and finally implementation of preventive measures.

DIANE K. HAWKINS

Further Reading

Books

Gask, George E. "Percivall Pott." In *British Masters of Medicine.* Edited by D'Arcy Power. Freeport, NY: Books for Libraries Press, 1969.

Hunter, Donald. *The Diseases of Occupations.* 2nd ed. Boston: Little, Brown, 1955.

Pitot, Henry C. "Principles of Cancer Biology: Chemical Carcinogenesis." In *Cancer: Principles and Practice of Oncology.* Edited by Vincent T. DeVita, Jr. 2nd ed. Philadelphia: Lippincott, 1985.

Periodical Articles

Doll, Richard. "Part III: 7th Walter Hubert Lecture. Pott and the Prospects for Prevention." *British Journal of Cancer* 32, no. 2 (August 1975): 263-74.

Kipling, M.D. and H.A. Waldron. "Percivall Pott and Cancer Scroti." *British Journal of Industrial Medicine* 32, no. 3 (August 1975): 244-50.

Melicow, Meyer M. "Percivall Pott (1713-1788): 200th Anniversary of First Report of Occupation-Induced Cancer of Scrotum in Chimney Sweepers." *Urology* 6, no. 6 (December 1975): 745-49.

Miller, Elizabeth C. and James A. Miller. "Milestones in Chemical Carcinogenesis." *Oncology* 6, no. 4 (December 1979): 445-56.

The Rise and Practice of Inoculation in the 1700s

Overview

The 1700s saw the increased use of inoculation against disease as a medical practice. More importantly, the practice began to be used scientifically, with less chance of accidentally infecting those who were to be protected. By the end of the century, although some of the scientific principles were still not fully appreciated, inoculation and immunization had become more commonly used and were on their way to becoming the boons to public health they are today.

Background

Throughout human history infectious disease has claimed more lives than virtually any other cause. Even today, with the exception of the minority of people living in the developed world, infectious disease is the world's dominant killer. Smallpox, tuberculosis, rabies, polio, diphtheria, yellow fever, and many other illnesses have afflicted people for millennia, while emerging diseases such as Ebola, AIDS, and Legionnaires disease arise periodically.

It is also true that, for most of human history, we have had virtually no way to prevent or to combat infectious diseases. Thus, Europe was decimated by the Black Plague, Native Americans were destroyed by smallpox, and Africa is being torn apart by AIDS today. The discovery of germs, sterilization, and public health measures have helped immeasurably in the prevention of disease, while the discovery of antibiotics has helped cure many diseases that, earlier, were nearly always fatal. However, even these measures will only go so far because they leave individuals susceptible to disease organisms, and those organisms are always present, waiting to attack. Something more was needed.

At some time, unrecorded in any history, someone noticed that nobody became ill with smallpox more than one time. Those who had smallpox and survived were protected from the disease for the rest of their lives. It is not known how long ago this was first observed, but it is known that by the seventeenth century, Turkish medicine included inoculating healthy people with fluids from smallpox sufferers, hoping to transmit this immunity. We now consider this a dangerous practice and, indeed, some people did develop the full-blown disease. However, it was more typical for a person to develop a minor case of smallpox, recover, and never have to fear it again.

This practice of inoculation was brought to England by Lady Mary Montagu (1689-1762) in the early 1700s. She spent many years advocating its use, partly because she had survived smallpox years earlier and still bore its disfiguring scars herself. She enjoyed some limited degree of success, including seeing inoculation spread to the American colonies in the 1720s, when Zabdiel Boylston (1680-1766) introduced the practice in the wake of a smallpox epidemic in Boston. His efforts were encouraged by the preacher Cotton Mather (1663-1728), whose son nearly died of the disease.

Unfortunately, early efforts at inoculation were often met with responses ranging from skepticism to outright hostility. In Mather's case, at one point a bomb was thrown through the window to his house in protest against his support of inoculation. In England responses were somewhat more subdued, but no less skeptical. Some of the protest was religious in nature, especially from the more conservative believers who felt that disease and premature death were part of God's will, and it was sacrilege to attempt to forestall them (incidentally, some religions today continue to hold similar beliefs). In addition, because inoculations used live viruses from an infectious patient, some of those inoculated developed the disease and became infectious themselves. The knowledge that one could die from a measure that was supposed to protect helped stir fears and arouse anger in many of the general public. All of this helped to slow the spread of inoculation as a preventative medicine practice. However, by the latter part of the eighteenth century, the practice of inoculation was becoming increasingly accepted, even though it was not yet widespread.

In 1796 English physician Edward Jenner (1749-1823) discovered that cowpox sores, similar to those of smallpox, could be used to inoculate someone against smallpox. When finally accepted by physicians elsewhere, cowpox inoculations became increasingly common.

Impact

The rise of inoculation to help prevent smallpox was a qualified success in the short term. Its long-term impact, however, may be considered an almost unqualified success story. Nevertheless, in spite of the dramatic drop in death by infectious diseases since the advent of widespread immunizations, not all of the outcomes have been beneficial. Four major impacts of inoculation are:

1) The development of an ever-increasing number of vaccines

2) A dramatic reduction in death from infectious disease in most of the world, with a corresponding rise in death from heart disease and cancer

3) An increase in human lifespan in many nations and a decrease in child mortality rates

4) A corresponding increase in population in many nations

The most obvious impact stemming from inoculations is the subsequent development of vaccines to prevent other disease. Since that time, former scourges such as rabies, polio, measles, German measles, diphtheria, whooping cough, and yellow fever have been largely eliminated through the use of vaccines. We no longer expect children, at least in the developed world, to fall ill with an endless succession of potentially fatal diseases. Instead, we immunize them against the most common diseases and rely on medicine to cure those suffering from anything else.

It is primarily the developed world that enjoys this sort of protection, and it is primarily in the developed world where death from infec-

tious disease has fallen so dramatically. However, it must be noted that a large-scale immunization program was carried out against smallpox, resulting in it being declared eradicated by the United Nations in 1977.

As a result of this wider access to vaccines against a widening number of diseases, global death rates from infectious disease have been dropping for some time. There has recently been somewhat of a leveling off of these rates, primarily due to the AIDS epidemic, but the general trend continues to drop because of efforts to immunize children whenever possible.

A corollary of this drop in death by infectious disease is a corresponding rise in death from other causes, including cancer and heart disease in much of the developed world. Much of the reason for this is simple mathematics. Everyone is destined to die of something at some time. If any single cause of death is reduced greatly, then all others must increase. To look at it another way, suppose that there are only four causes of death, and they all are equally probable. This means that, say, cancer, heart disease, infectious disease, and accidents each cause 25% of all deaths. Now, suppose that there are no more deaths from infectious disease because medicine has developed perfect vaccines and antibiotics. This means that the death rate from the remaining three causes of death *must* increase to 33% each because they are still equally probable, and nobody lives forever. So, as a result of our progress against infectious disease, cancer and heart disease have become more prevalent in society, because people are now living long enough to develop them.

In addition, for perhaps the first time in human history, we can expect virtually every child to live long enough to reach a healthy adulthood. While this is still not the case in many parts of the developing world, it is true in the developed world, largely because of the ubiquity of childhood immunizations (increased nutrition and improved pre- and postnatal care are other important factors in this). On a related note, families are now smaller than even just a century ago, and population growth has slowed in many parts of the world.

However, population growth has not slowed in much of the developing world, and this is again partly due to improved vaccination programs (and, as above, also partly due to better nutrition and medical care). As a result, many developing nations find themselves with a booming population because families are still having a large number of children, most of whom are now living longer lives. As this increased population of children reach adulthood, they, too, have large families, and more of *their* children are surviving. So populations grow at an even faster rate, placing further stress on infrastructure, populations, and governments that can ill afford it.

Taken at its most simplistic, we can look at vaccines this way: they have helped to greatly reduce early death from infectious disease in favor of a later death from other causes. From this perspective, their primary contribution has been to lengthen human life in most parts of the world. In addition, vaccines against nonlethal diseases have saved people untold suffering from disease, and further suffering for those who would otherwise have spent much of their lives crippled or injured in some other manner by the disease. Inoculation and vaccination has helped millions of children to reach adulthood, and there is little doubt that some of them have gone on to live rich lives full of accomplishment. In this sense,

vaccines have helped make the world a better place, and have made lives longer and fuller than might otherwise be the case.

However, no blessing is unmixed. The irony is that, as vaccines have helped lengthen human life, they have also contributed to population growth, shortage of natural resources, the increase of heart disease and cancer as causes of death, and financial stress on third world governments.

P. ANDREW KARAM

Further Reading

Baxby, Derrick. *Jenner's Smallpox Vaccine: The Riddle of Vaccinia Virus and Its Origin.* London: Heinemann, 1981.

Bazin, Herve. *The Eradication of Smallpox: Edward Jenner and the First and Only Eradication of a Human Infectious Disease.* San Diego: Academic Press, 1999.

Stolley, Paul and Tamar Lasky. *Investigating Disease Patterns: The Science of Epidemiology.* New York: Scientific American Library, 1995.

Developments in Public Health

Overview

Public health became an important issue in the eighteenth century for two reasons. First, the Industrial Revolution was exerting a profound influence on the environment and the health of workers, and urbanization meant that more people were crowding into cities. With this growing population came increased incidences of disease. While reforms for industry and new sanitation practices became the focus of the public health movement early in the 1700s, the European and British public health movement did not gather its full strength until after 1750. In the U.S., serious efforts at improving the public's health did not get underway until the 1780s.

Background

The years 1700 to 1799 are also known as the period of the Enlightenment, an international intellectual movement that revolutionized political, philosophical, social, and scientific thinking. The great figures during the Enlightenment strove to improve the world and the circumstances of humankind. Betterment of the public

health was among the goals of many Enlightenment thinkers.

Both in Europe and the New World, science was increasingly employed to better human lives and the conditions under which people lived. The idea that unsanitary conditions could spawn and spread diseases, that germs were the cause of many illness, and that many illnesses were caused by contact with sick people—a concept called "contagion"—began to dominate medical thinking and change health practices.

This was also a period when hospital and prison conditions improved and when the combination of strong central yet democratic government—and an aroused public opinion—exerted their influence. Public health efforts in the New World were aimed at keeping diseases such as smallpox and yellow fever away from American shores while in Europe governments both local and national passed laws and acts aimed at sanitizing public water supplies, public streets, and public buildings.

Europe was the focus of the most rigorous advancements in public health. During the eighteenth century, the population of Prussia (as Germany was then called) increased dramatically. Berlin's population swelled five-fold during the years 1700 to 1797. Paris's population increased as well, but was a city with a reputation for unsanitary hospitals. People were concerned about the diseases of soldiers as well as the unhealthy conditions in institutions that housed the insane and criminals.

Impact

Germany

The most important European figure in the German public health movement was Johann Peter Frank (1745-1821), who established the concept of "medical police" with his publication of *The System of Medical Police* in 1779. The first known use of the term "medical police" was by German Wolfgang Thomas Rau (1668-1719) a few decades earlier, but Frank more clearly defined the concept in several volumes, outlining their responsibilities and making strong, enforceable recommendations for public health action.

Frank's idea of "medical police" meant that government medical policy should be implemented through enforced regulations. He presented medical policing as a responsibility of the state through a formal system of public and private hygiene laws with great reliance on control, regulation, and police enforcement. Many of his recommendations are out of place in a democracy, but fit the concept of "enlightened despotism," a concept alive and well at his time.

Population policy was a hallmark of Frank's medical policing recommendations. Marriage was promoted, "bachelor taxes" were levied, midwifery supported, and freedom from work and responsibility for new mothers was established. Frank thought that the state should support new mothers for the first six weeks after delivery. Frank was also concerned with the welfare of school children, from accident prevention to promoting cleanliness. Medical police were to monitor school buildings and the environment in school facilities, such as light, ventilation, and heat.

Frank's medical police dealt with problems of sanitation, such as sewage, garbage disposal, and maintaining clean water supplies. There was no more important task than keeping cities and towns clean, said Frank, who arranged for public restrooms in cities and public dumping grounds far outside of cities and towns. Frank published six volumes of his *System of Medical Police* in the 1700s. Three more volumes appeared early the next century.

France

In 1790, French physician Joseph Ignace Guillotine (1738-1814), for whom the guillotine was named, insisted that medical practice, medical education, "health police," sanitary services, and the control of diseases be controlled by a "health committee." The French Health Committee, led by Jean Gabriel Gallot, had jurisdiction over diseases and inoculation against diseases such as smallpox. He and others began planning a convention at which broader health issues would be addressed and the committee formalized.

Convention participants in 1793 passed a law that provided for the welfare and health of children and expectant mothers. The same convention decreed that every hospital patient should have his own bed and that the beds should be separated by 3 feet (7.6 cm). In the same year, French physician Philippe Pinel (1745-1826) similarly improved the treatment of the incarcerated insane, removing their chains and demonstrating that more humane treatment of the mentally ill had social value.

England

Like the population of Berlin, the population of London swelled quickly in the eighteenth century. Scottish philosopher and essayist Jonathan Swift was so concerned about the effect of population growth on the food supply that in 1789 he wrote *A Modest Proposal,* an essay satirically proposing that the answer to the increasing population was to eat babies.

Along with rising populations, treating disease was a primary concern in England during the 1700s. New hospitals and medical "dispensaries" opened in many parts of England, including York, Bristol, and Westminster; previously, most had been only in London. By the 1750s, preventing disease in the English populace was drawing the attention of Parliament. An early public health campaign waged by reformers was against the consumption of the alcoholic beverage gin, thought to be undermining the public's health. In turn, Parliament passed a series of acts which gave control of gin to magistrates while seeking to control its consumption. This control was associated with a decline in consumption and a concomitant fall in the death rates, especially for infants since alcohol has an effect on the developing fetus.

The British also attacked the problem of child mortality—as high as 80% in children younger than one year—through better care. In 1741, the Foundling Hospital of London was established to provide nursing and other care for children. In 1769, an Act of Parliament required sick children to be sent to the country for care.

Notable in the cause of British public health was James Lind (1716-1794), who helped reduce scurvy among British seamen by encouraging them to use lime juice (scurvy is a disease caused by a deficiency in vitamin C). Robert Willan described the various skin ailments of workers and connected them to work-related health risks.

Most notable among the British reformers was John Howard (1726-1790), who publicized and ultimately reformed the appallingly unsanitary conditions he found in prisons in England and Europe. In 1777, Howard published the *State of the Prisons*, which outlined his investigations of unsanitary British jails. Likewise, William Tuke, a tea and coffee merchant, reformed the British asylums by creating a "retreat" in York, England, that housed 30 mental patients and treated them more humanely than they were treated in the asylums, and with good results. His work paralleled that of French "lunacy" reformer Pinel.

Efforts in combating disease took a great leap in the 1790s when English country doctor Edward Jenner (1749-1823) noticed that milkmaids did not get smallpox, a disease ravaging the world. They did, however, get cow pox from the cows they milked, leading Jenner to think their exposure to cowpox protected them from smallpox. He began inoculating people with cowpox in attempt to save them from smallpox. His efforts lead to "vaccinations" which eventually eliminated many diseases, including smallpox.

United States of America

In the American colonies and later in the newly established United States of America, public health movements focused on preventing dreaded diseases, such as smallpox and yellow fever, from being brought ashore by visitors from the West Indies or Europe. Early public officials

sought control by policing arriving ships and quarantining passengers. In Boston, concerns about imported disease led to the establishment of a group of physicians called "The Selectmen," who inspected ships and crews for smallpox and other diseases. The Selectmen also were responsible for setting guards at houses of quarantine. The Selectmen also spearheaded efforts to drain stagnant water in streets and oversee street paving and provide cleaner water wells.

In New York, with population near 10,000 in the 1730s, the connection between dirt and disease occupied many educated and well-meaning citizens. Street cleaning laws were passed in 1731 and, in 1744, a formal Sanitation Act moved animal trades and slaughtering outside the city limits. Through the century, in New York, as in the other American colonies, smallpox and yellow fever were constant threats. In the middle decades, New Yorkers quarantined smallpox sufferers and refused docking to ships carrying small pox. In the 1740s and 1790s, yellow fever visited New York, as it had Philadelphia and Baltimore. New Yorker Cadwallader Colden recognized that yellow fever outbreaks were in proximity to swampy areas and made recommendations on draining these areas. It was not yet clear, however, that yellow fever was carried by mosquitoes that bred in swamps.

In the U.S., great public health projects came decades later than in Europe, but Philadelphia established a hospital in 1751, and New York in 1791. The U.S. Public Health Service and the office of the U.S. Surgeon General were established in 1798.

RANDOLPH FILLMORE

Further Reading

Books
Blake, John B. *Public Health in the Town of Boston, 1630-1822*. Cambridge: Harvard University Press, 1959.

Porter, Dorothy. *Health, Civilization and the State*. London: Rutledge, 1999.

Rosen, George. *A History of Public Health*. New York: MD Publications, 1958.

The Growth of Hospitals in the 1700s

Overview

Throughout the eighteenth century hospitals opened in the larger cities of Europe and America as industrialization developed and the middle class expanded in those countries. These hospitals were very different from the kinds of hospitals seen in Western and Arabic cultures since early in the Christian era. By 1800 the hospital in the West was becoming an integral part of medical care and education in urban areas. Within another century, the hospital would be at the center of medical practice.

Background

A hospital can be defined as a location for the acceptance, care, and treatment of the sick. In ancient Egypt, Greece, and Rome healing was often associated with the temples of particular deities. The staff of Aesculapius, a walking stick branched at the top with a snake wrapped around it, has long been a symbol of medicine, and today is the official insignia of the American Medical Association. Aesculapius was the god of medicine in Greek and Roman mythology, and the snake was his symbol. By the fifth century B.C. several temples to Aesculapius were active in Greece.

At about the same time, secular healing in the tradition of Hippocrates (c. 460-377 B.C.) developed in Greece, and public physicians appeared in Athens. As the Roman Empire grew in later centuries, provisions were made for the care of sick slaves and soldiers, since the culture was so dependent on both. Throughout Europe and the Middle East in the centuries before Christ's birth, travelers attending festivals or making pilgrimages to holy sites received medical care if they became ill.

Care of the sick became an important element of the faith from the earliest days of the Christian era. Jesus had included an obligation to visit the sick as among the essentials for salvation. By the third century A.D., as it spread rapidly throughout the Roman Empire, Christianity assimilated the secular healer, or physician, and the healing shrines from pagan cultures. Some churches, such as that in Antioch around 340 under Bishop Leonitios, operated houses for the poor that included medical care as needed. Monks living together in monasteries cared for their own sick members, and some-times constructed attached facilities for sick townspeople or travelers. Around 450, a monk named Theodosius built three structures near Jerusalem to house monks and poor and wealthy individuals who were ill. Thus by the middle of the first millennium A.D. buildings constructed especially for care of the sick had appeared in a number of locations in Christian Europe and the Middle East. These facilities seldom lasted more than a few years, often disappearing when a founder or sponsor died.

In the Arabic world, larger cities such as Cairo, Baghdad, and Damascus had hospitals by the tenth century. Over the next several centuries Arabic authors associated medical teaching with hospitals, while Christian writers did not. This practice appears to be one of several in which Islamic medicine eventually influenced its Western counterpart as empires and cultures rose and fell during the Middle Ages.

Throughout medieval Europe hospitals continued to appear as towns grew and trade increased. Most were associated with a church or monastery, and sponsored by a wealthy patron whose support demonstrated both social position and Christian charity. Patients in these small hospitals were either poor or elderly without local family support networks or travelers and pilgrims. Around 1400 several hundred such facilities operated in England alone; dozens of earlier such institutions had disappeared. These hospitals might have as few as two or three patients or as many as thirty.

A new model of a more complex hospital developed in such large trade centers as Florence, Italy, and Paris, France. By 1400 Florence had some thirty hospitals caring not only for the poor but also for growing numbers of tradesmen and merchants. Each hospital might be supported by a group of individuals or trade guilds instead of a single benefactor. The largest hospital in Florence at this time, S. Maria Nuova, had been founded in 1288 and by the fifteenth century had 230 beds served by six visiting physicians, a surgeon, and three junior staff members. Despite the size of this facility, medical care of this time was still extremely limited in its efficacy. Surgery would be performed for only a small number of acute injuries, since pain relief remained crude until the development of anesthesia in the 1840s.

Between 1106 and 1247 four hospitals—St. Bartholomew's, St. Thomas's, Christ's and Bethlehem—were founded in London in conjunction with monasteries. By the mid-1500s the monastery system was dissolving, and the hospitals had been sold to the city and reopened as more secular institutions. Support staff members were more likely to be recruited from former patients or local poor than from religious orders; this pattern would continue until well into the nineteenth century. St. Thomas had 203 beds by 1569, and several surgeons and physicians. Thus in the several centuries before 1700, hospitals in large European cities expanded in size and were transformed from largely religious to more secular institutions.

Impact

Another important aspect of hospital development in Europe appeared just after 1700 in the city of Leiden, in what is now the Netherlands. At the University of Leiden, medical professor Hermann Boerhaave (1668-1738) expanded and made systematic the practice of earlier teachers in both Leiden and Padua, Italy, who had taken their apprentices on rounds through patient wards at local hospitals. Boerhaave developed a ward of twelve patients at the Caecilia Gasthuis hospital for use in his medical teaching. His students followed him on rounds through the ward, observing his interactions with patients and listening to his comments on various diseases and conditions exhibited by the sick. This model of clinical rounds at university hospitals for the purpose of education continues today in medical schools around the world. Yet in Boerhaave's time interaction with patients was unusual in medical education.

Boerhaave's students soon spread his teaching methods to other European cities such as Vienna, Austria, and Edinburgh, Scotland. At the Royal Infirmary of Edinburgh in 1748, John Rutherford (1695-1779), a University of Edinburgh professor of medicine, persuaded the managers to let him organize formal medical teaching in the wards. Six years later, three more professors—Alexander Monro *secundus* (1733-1817), William Cullen (1710-1790), and Robert Whytt (1714-1766)—took over the separate ward organized by Rutherford and managed it in two-month rotations for several years.

In the late 1700s the French Revolution swept away many institutional structures in that country. Some historians have argued that hospital medicine changed radically as well. Other historians note that changes seen after the revolution, especially in the Parisian hospitals, had been developing in large European hospitals for decades or longer. Many of these changes did coalesce in the Paris hospitals by the early 1800s. Doctors in urban areas increasingly moved their practices and educational efforts into the hospitals. As physicians began to organize to protect their profession and increase their social status, hospitals served as a central location for these efforts. Parisian doctors such as Philippe Pinel (1745-1826), R.T.H. Laënnec (1781-1826), and Pierre Louis (1787-1872) began to correlate signs and symptoms in the sick with pathological conditions found during autopsies of deceased patients.

The number of hospitals in London and elsewhere in Britain continued to increase during the eighteenth century. Westminster opened in 1719 in London, followed by Guy's (1721), St. George's (1733), London Hospital (1740), and Middlesex (1745). In the provinces Winchester was established in 1736, and 20 more soon followed in major towns. Older towns tended to open hospitals first, followed by new industrial-era cities such as Manchester and Birmingham. In Scotland the Edinburgh Royal Infirmary opened in 1729. Medical staff were involved in these institutions from the beginning, and in Edinburgh provisions were also made for students.

In Austria, Emperor Joseph II initiated the Allgemeines Krankenhaus, a huge civic hospital complex near Vienna. Joseph planned to convert an almshouse just outside the city into a modern, 2000-bed hospital divided into sections for medical, surgical, venereal, and contagious patients. This hospital opened in August 1784, but immediately began to incur huge debts. Despite Joseph's death in 1790, the hospital remained open; by 1797 over 14,000 patients visited the hospital each year.

Hospital growth in the United States was much slower. By 1800 America had just five million in total population, with most people living in rural or frontier areas of the country. Only two significant hospitals had been established by that date. Pennsylvania Hospital in Philadelphia opened in 1752, and New York Hospital in 1771. In the first few decades after 1800 many more hospitals opened in the large cities of the northeast.

By the end of the eighteenth century the outlines of the modern Western hospital were in place in the urban areas of Europe and America. The religious impulse behind hospital creation in ancient and medieval times continued, but was supplemented by civic and private philanthropy. Doctors in urban areas moved some or most of their practices into the hospital setting. Many hospitals allowed local medical school faculty to use their facilities for clinical training to supplement classroom lectures or apprenticeships.

However, another century would pass before the modern hospital finally appeared. Throughout the nineteenth century hospitals remained a place primarily for the poor; wealthy individuals continued to receive care at home when possible. The development of anesthesia in the 1840s and the discovery of bacteria and the need for cleanliness several decades later meant that by 1900

more complex surgeries were possible in the hospital and survival rates were better. More complex anesthesia requiring bulky machines also developed by 1900, meaning that most surgeries had to be done in hospitals for the sake of both convenience and cleanliness.

A. J. WRIGHT

Further Reading

Granshaw, Lindsey. "The Hospital." In *Companion Encyclopedia of the History of Medicine*, ed. by W. F. Bynu and Roy Porter. London: Routledge, 1993: 1180-1203.

Risse, Guenter B. *Mending Bodies, Saving Souls: A History of Hospitals*. New York: Oxford University Press, 1999.

Obstetrics in the 1700s

Overview

Obstetrics, the medical specialty of caring for women and their babies during childbirth, arose in the mid-eighteenth century. Ordinarily, women in childbirth were attended by other women—mostly relatives, friends, or neighbors who offered support and practical aid. In addition, a midwife (meaning "with woman") was often employed to bring skilled assistance and the assurance of someone who had attended many births. The nearest surgeon was summoned only in the midst of dire complications. By 1750, physicians and surgeons sought opportunities to attend births and incorporate pregnancy and childbirth into the medical forum. Almost universally male, these man-midwives gained acceptance throughout the latter half of the eighteenth century. The increasing status of surgeons, advances in technology such as the invention of the obstetrical forceps, and the growing number of institutes and universities dedicated to medical knowledge and training of physicians enabled the birth of obstetrics as a medical specialty.

Background

Before the eighteenth century, childbirth was a women-only affair. Midwives received no formal training, and learned their skills mostly through informal apprenticeship or direct experience. Especially in English and European society, midwives were not viewed as part of the medical

community. Instead, the role of the midwife was seen as a benevolent and seemingly religious one. Superstitions regarding childbirth lingered from the Middle Ages. Attempting to prevent witchcraft associated with childbirth, bishops in the Church of England required midwives to obtain an Episcopal license which prohibited them from practicing magic, coercing fees, or concealing information about a birth. Midwives were obligated to treat the poor, and were able to perform baptism in emergencies. In the early 1700s, midwives in some American colonies were required to obtain a license similar in content to the Episcopal licenses of England. These rules of licensure kept the midwife's primary function a social one, and minimized her importance medically.

The first physician to place midwifery on sound scientific ground was William Smellie (1697-1763). British-born, Smellie practiced general medicine in Scotland and studied surgery in Paris before settling in London in 1739. Smellie, interested in obstetrics, conceived the idea of teaching the subject at his apothecary shop residence. He used a leather-covered mannequin of bones, and charged three guineas for the course. Smellie offered free care to London's poor women, thus providing him with clinical teaching material. As Smellie's patients allowed students to attend their deliveries, the trend toward medically trained persons attending childbirth was established. Smellie was also the first to teach obstetrics and midwifery on a scientific basis.

Smellie published his lectures, along with his *Treatise on the Theory and Practice of Midwifery* in 1752. Further advancing the practice of obstetrics, Smellie introduced obstetrical forceps that were properly designed and properly used. Made of wood or steel and padded with leather, the forceps were correctly curved, and had a simple lock design. Smellie's forceps were the length necessary to deliver the head of the baby only after it had descended into the pelvis, a standard for safety defined by Smellie and published in his works. Careful systematic measurements of pelvic capacity were taken by Smellie and his students, which led to standards of differentiating normal pelvic structure from the abnormal.

Smellie acquired a large practice, and among his pupils was English surgeon William Hunter (1718-1783). Hunter, per Smellie's example, also eventually gave private lecture courses on surgery, dissection, and obstetrics. Hunter trained many of London's man-midwives, and, due to his skill and courtly disposition, Hunter was also in demand as a man-midwife to London's social elite. Hunter's further contributions to obstetrics include a famous atlas of the pregnant uterus, praised in both scientific and artistic circles, and the discovery of a separate maternal and fetal circulation. Hunter also built the celebrated anatomical theater in London, which served as the training ground for the best surgeons of the day. Among these was John Hunter (1728-1793), brother to William Hunter. John Hunter, an enthusiastic experimentalist, became the leading surgeon-physiologist of his day, elevating surgery from a manual trade into a scientific discipline. As obstetricians trained as surgeons, the practice of obstetrics was elevated as well.

Impact

Like most of the medical advances of the eighteenth century, advances associated with childbirth and obstetrics were linked to the prevailing philosophy of the Enlightenment. A rational approach to the events surrounding childbirth was advocated by the new man-midwives. Previous superstitions and interventions labeled by physicians as unnecessary "midwife meddling" (binding breasts, for example) were abandoned in favor of allowing nature to accomplish much of the process. Ladies were encouraged to give birth in rooms with fresh air and sunlight. Newborn infants were no longer swaddled in restrictive linens as it was felt that allowing freedom of movement would promote muscle and bone de-

velopment. The importance of breastfeeding was championed as women were urged to dismiss their wet-nurses and bond with their infants. Women were admonished of the potential risks to their reproductivity brought about by wearing corsets, and one anatomist carried out a public campaign for women to abandon them. Not all Enlightenment-related opinions regarding obstetrics and child care were well received. In France, wet-nursing prevailed, and in Catholic Spain and Italy, male obstetricians made little progress as the Church demanded female modesty. Throughout the western world, most women kept their corsets.

By 1775, most women of the upper socioeconomic class in Europe and America chose the *accoucheur*, or male midwife, to attend them in childbirth. His confidence in allowing a natural approach to birthing was backed by a solid base in anatomy and the known physiology of the day. In addition, he had instrumentation to hasten emergency or difficult deliveries, and some advance knowledge of when he might need to use them. For instance, obstetricians were taught to identify pelvic deformities brought about by the fashionable low protein diet of wealthy women, as well as those caused by rickets in malnourished poorer women, and to anticipate a forceps delivery, or at worst, surgically dismember the fetus and deliver it in pieces to save the life of the mother. Maternal death from such complications was no longer a foregone conclusion. Fetal lives were also saved when birthing complications arose. Smellie was the first physician documented to successfully resuscitate an asphyxiated infant by inflating the lungs with a silver catheter.

Although the services of medically trained man-midwives were accepted and often preferred by childbearing women in the last half of the eighteenth century, the majority of babies were still delivered by traditional female midwives. Universities and institutions of medicine had simply not yet produced enough physicians and surgeons to meet the needs of the population. By the end of the century, university medical schools and private institutions of medicine that included obstetrics in the curriculum flourished across Europe. In colonial America, the Medical College of Philadelphia was founded in 1765, King's College (later Columbia) Medical School in 1767, and Harvard in 1782. No longer was it necessary for an aspiring American physician to travel to Europe to receive an excellent medical education. Obstetrics was the first medical specialty taught at these schools. Midwifery was assumed to be a cornerstone of American medical practice, since every physician would encounter childbearing patients after graduation. In America, physicians would essentially replace midwives in attending childbirth, although the transition would take more than another century, due to westward expansion and the enlarging frontier.

In other countries, notably Britain, there was sometimes considerable strife between midwives and their male physician counterparts. Besides competition for the financial reward (meager at times, and often settled in trade), midwives often felt male physicians should not be involved in such an intimate female process. One eighteenth-century midwife, Elizabeth Nihell, accused Smellie of insufficiently training midwives, so that physicians would have to be called in to birthing situations more often. Nevertheless, in Britain the practice of obstetrics flourished, but did not completely take the place of midwifery. The two professions often collaborated, as midwives received better training and the number of obstetricians increased.

Obstetrics in the eighteenth century made few advances profound enough to dramatically impact the health of the childbearing population as a whole. Childbirth, although now appreciated as a natural process, could still be a risky business. Puerperal fever (infection after childbirth) still afflicted women of all classes. In one epidemic year of 1772, mortality in Paris, Vienna, and other European centers rose as high as 20% of all new mothers. Few cesarean sections were attempted, and fewer were successful before anesthesia and Listerian antisepsis. Infections and other causes of maternal and infant mortality occurred at rates significant enough to lower the average life expectancy in Europe and America to well below 40 years. The most important accomplishment of obstetrics in the 1700s was that it began to transform perceptions of medicine's place in society. With Enlightenment thinking, the stage was set for future obstetrical and medical advances based on sound scientific reasoning and experimentation, rather than on religion or tradition.

BRENDA WILMOTH LERNER

Further Reading

Books

Odowd, Michael J. and Elliot E. Phillip. *History of Obstetrics and Gynaecology.* London: Parthenon Publishing, 1994.

Porter, Roy. *The Greatest Benefit To Mankind: A Medical History of Humanity.* New York: W.W. Norton, 1998.

Wilson, Adrian. *The Making of Man-Midwifery: Childbirth in England 1660-1770.* Lisse: Swets & Zeitlinger, 1995.

Surgery in the 1700s

Overview

Surgery has a long history in the healing arts. Its history dates back thousands of years to the great seats of early civilization in Athens, Rome, and Alexandria. However, amputations and invasive wound-healing procedures can even be traced to Upper Paleolithic peoples living tens of thousands of years ago, who, with apparently considerable knowledge about human anatomy, practiced bone-setting as well as minor surgical procedures.

The years between 1700 and 1799 stand out as important to medical history and surgical advancement because surgeons were willing—as must have been the patients—to attempt amputations as well as complicated and often heroic surgeries on major organs of the body, all without benefit of anesthesia, which did not come along until the nineteenth century. Eighteenth-century surgeons prided themselves on fast amputations, as well as procedures such as removing bladder stones, cancers, and even cataracts from the eye.

Not only were more hospitals established in England and Europe during the 1700s, their conditions vastly improved toward the end of the eighteenth century as reformers made strides in sanitary practices.

Background

By the eighteenth century, anatomists had long been practicing with cadavers to learn more about the body and its organs. In the developing hospitals and medical schools, anatomists instructed medical students in how the human body functioned, what could go wrong with the major systems, and how surgery might be used to correct bodily malfunctions or treat a variety of wounds, including those inflicted in war.

Amputations were by far the most frequent surgeries, practiced by surgeons who prided themselves on speedy procedures as patients, without benefit of anesthesia, were held down by the surgeons' strong-armed assistants. The great amputators controlled bleeding and infection as best as they could, and developed unique antiseptics.

The historical list of surgical advancements in the eighteenth century progresses organ by organ and system by system as surgeons in England and in Europe, particularly in France, developed new techniques to control bleeding, drain and close surgical wounds, or repair wounds that resulted from military combat—especially those from explosives and gunshot wounds. Surgeons also increased their abilities to repair malfunctions in the urinary system and the intestinal tract, perform Cesarean sections, amputate limbs, remove cataracts from the eyes, remove gall stones, do radical surgeries to cancerous tissue, remove glands and organs (such as the thyroid or spleen), and develop antiseptics.

During the eighteenth century, anatomists and surgeons not only perfected surgical techniques but published books that made a significant impact not only on the practice of surgery in their century, but in the next as well.

Impact

Advancements in the art and science of amputation top the list of surgical achievements in the eighteenth century. In 1718, French surgeon Jean Louis Petit (1674-1750) developed a tourniquet to stop bleeding during an amputation. Petit's tourniquet worked by twisting a screw compressor device that clamped down on the patient's leg and on the lower abdomen. It could stop the blood flow in the main artery and provide for higher leg amputations. Petit's tourniquet is considered by many sources to be one of the "most important surgical advancements before the advent of anesthesia."

Even with the Petit tourniquet to stop bleeding, quick amputations were desirable to spare unanesthetized patients prolonged agony. Speed became a source of pride for early surgeons. French surgeon Jacques Lisfranc became known for his ability to amputate at the thigh in

10 seconds. Scottish surgeon Benjamin Bell was able to cut through all but the bone in six seconds. It was said that James R. Wood, of New York, could amputate at the thigh in nine seconds. Other amputations, such as at the shoulder joint, may have taken longer. Many surgeons constantly practiced on cadavers to get the speediest amputations possible.

Quick and clean amputation was but one step in the art of surgery, however. What to do with the open wound once the amputation was finished and the tourniquet removed was yet another question facing eighteenth-century surgeons. If the patient were to live, wounds, whether caused by the surgeon's knife or the enemy's sword or gunshot, required cleaning and closing. How to clean and whether to close the wound were the subjects of great debate.

Although closing wounds by sewing them developed centuries earlier, closing wounds with sutures was not a standard practice in the eighteenth century. As far back as the sixteenth century, surgeons used percutaneous stitches to close amputations, but often with disastrous results from subsequent infection. Many surgeons preferred to let a wound, even an amputation, heal without drawing the edges of flesh together. Leaving the wound open to provide drainage, but keeping it somewhat covered, was often made possible by using an adhesive tape placed around the wound edges. Packing the wound with a porous lint-like material also allowed drainage.

Whether amputations actually saved lives was the subject of popular debate among surgeons. Many military surgeons preferred not to interfere by further amputation if a limb was amputated cleanly by gun or cannon shot. A large number of military surgeons advocated immediate amputation for gunshot-caused fracture, but even that procedure was debated. Some surgeons simply delayed amputation to see how well the fractures healed.

While between 45 and 65% mortality was average for hospital amputations, the Royal Infirmary at Edinburgh, Scotland, however, under the skills of the father-and-son duo Alexander Monro *primus* and *secundus*, boasted an 8% mortality rate during much of the eighteenth century.

It was widely known at this time that keeping wounds and amputations clean and draining freely helped lower surgical mortality rates. Military records show that in the next century, Napoleon's field surgeons introduced maggots, or fly larvae, into the soldiers' wounds. The maggots were allowed to eat the dead flesh and leave the healthier tissue. The maggots were then removed.

Surgeons, especially those who attended gunshot wounds, perfected several techniques for "debridement," the process of removing dead skin from a wound. This procedure was advocated by French surgeon Pierre Joseph Desault (1744-1795), who also coined the term. Debridement, taken from the French verb "brider," meaning to check or curb, removed dead skin in order to curb unhealthy wound closing and prevent infection. Desault debrided with a trimming blade.

Surgeons also practiced "preventive debridement" by dilating, or opening, gunshot and sword wounds wider to promote drainage. This technique was controversial in its time. The great British surgeon John Hunter (1728-1793) preferred to let a wound heal "under a scab." His French counterparts and some of his fellow Englishmen disagreed, preferring to open wounds wider.

A variety of antiseptic preparations were available during the 1700s. The word "antiseptic" is first thought to appear in a British pamphlet from 1721 in a discussion on how to prevent the purtrification of flesh. In 1752, British physician John Pringle (1707-1782) used the term when discussing the use of mineral acids in his experiments at stopping decomposition in freshly killed animals. So interested were French surgeons in antiseptics that the Academy of Sciences and Arts and Fine Letters of Dijon offered a prize for the best essay on the use of antiseptics. The winner discussed applications of turpentine, alcohol, benzoin, and aloe. With the success of antiseptics, many surgeons preferred to postpone amputations in favor of treatments with antiseptics.

The art of amputation was not restricted to limbs and war wounds. Surgeons in the 1700s grew increasingly adept at amputations in an attempt to prevent the spread of gangrene or cancer. The treatment of cancers by the removal of tumors and diseased organs occupied many eighteenth-century surgeons.

As early as 1756, British and French surgeons were performing "glossectomies," the removal of the tongue, to cure cancer of the tongue. Earlier surgical approaches to cancer of the tongue was tumor removal and the excitation of lesions. Surgeons also tried to "strangulate" tumors by ligations, tying off the tumor's

blood supply. Glossectomy became the treatment of choice and through ligation the surgery became bloodless and wound cauterization was accomplished by the use of hot iron.

Breast cancer was also treated surgically in the 1700s. Jean Louis Petit, the one who developed the Petit tourniquet, set the standards for mastectomy for breast cancer. He correctly believed that removing the breast and the nearby

THE SEPARATION OF ENGLISH SURGEONS AND BARBERS

In England in the mid-eighteenth century, surgeons and barbers—who performed minor surgeries—were members of the same union, or guild. In 1744 surgeons held a meeting at Court in London and expressed a desire to be separated from the barbers. The minutes of the Court showed that surgeons wanted that "each may be made a distinct and independent body free from each other..." In January 1745 the barbers, who had organized against the proposal, formally dissented. But the surgeons' proposal was submitted to Parliament, regardless. The committee hearing the proposal was headed by physician Charles Coates. British surgeon William Cheselden, Cotes' father-in-law, was one of the most active physicians campaigning for separation. Deciding that a separate company for physicians would advance the art and science of surgery, the committee under Coates sent a bill to Parliament that was approved in May 1745. With the bill's passage, barbers and surgeons separated. The "Company of Surgeons" held its first meeting in July 1745.

After separation, the surgeons wanted to rent a room in the Barber's Hall for dissecting. They offered a small sum, whereupon the barbers suggested a much larger amount, which the surgeons refused to pay. As a result, the surgeons did not get their own dissecting hall until 1753.

RANDOLPH FILLMORE

enlarged lymph glands prevented further spread of breast cancer.

The first hospitals devoted to treating cancer were opened in France in 1740, and in England in 1792. British surgeons treated breast cancer by removing the entire breast, pectoral muscles, and glands, as Benjamin Bell perpetuated Petit's views on mastectomy in his *System of Surgery,* published in 1784.

Ear surgeries led to early and usually unsuccessful attempts at brain surgery. The earliest ear surgeries were for an infected "mastoid," a boney area just behind the ear. In 1736, Petit drained an infected mastoid on a patient, who made a full recovery. This surgery was risky because of the mastoid's close proximity to the brain.

Treating the internal organ surgically was difficult, but early eighteenth-century surgeons made serious and successful attempts. In 1710, Parisian surgeon Alexis Littre experimented on a cadaver, making an artificial anus by resectioning the bowel out the abdominal wall. While intestinal resections were carried out as early as 1732 in London, in 1757 a Paris surgeon reconstructed part of the intestine. By the 1790s, surgeons were treating cancer of the colon by removing sections of the bowel and redirecting the anus to the abdomen (colostomy). The first recognized successful colostomy was performed in 1793 when French surgeon C. Duret attended a three-day-old infant with a perforated anus. Duret performed a colostomy and the patient lived to be 45 years old.

Renal surgery (surgery of the kidneys) progressed in the 1700s, thanks to British surgeon Charles Bernard—who was removing kidney stones as early as 1700—and the French surgeon Lafitte, who performed nephrecotomies in 1753. Other internal organs drew the surgeon's knife as well. The first successful removal of the spleen (splenectomy) was reported by English surgeon John Ferguson, who attended a man who was stabbed in the spleen. Ferguson reported that the spleen looked "quite cold, black and mortified." To save the man's life, Ferguson ligated the organ with a waxed thread and cut away the spleen.

Removing stones from the bladder (lithotomy) was a specialty practiced with good results in the first two decades of the eighteenth century when the surgeons cut open the bladder by making an incision in the abdomen above the pubic bone. Later eighteenth-century surgeons were able to cut open the patient's side to reach the bladder. Famous British surgeon William Chelseden (1688-1752) developed the lateral technique and was known for his low 17% mortality rate with the lateral incision. It was said that he could perform the whole operation in 45 seconds. Using Cheselden's technique, Paris surgeon S.F. Morand was said to be able to perform a lithotomy in 24 seconds. Attempts at removing growth from the bladder were made by the Englishman Warren in 1761 and by

French surgeon Desault. Desault twisted off cancerous tumors with forceps while performing lithotomies.

The first book on the surgery of the thyroid was published by Charles G. Lange in 1707. He suggested that a goiter, an enlargement of the thyroid, might be treated by ligation, by tying off its arteries. A German surgeon, Lorenz Heister (1683-1758), performed the first known successful thyroidectomy in 1752. Appendicitis also came under the surgeon's knife, first in London in 1736 and then in Paris 20 years later.

Attending to abscesses and abnormalities of organs without removing them also concerned eighteenth-century surgeons. The first operation to remove abscesses from the ovaries may have been performed in Scotland in 1701 when surgeon Robert Houston successfully drained an ovarian abscess.

Surgical specialization also began to come into its own in the 1700s. Plastic surgery was attempted in the eighteenth century as surgeons tried to heal and re-fashion tissue damaged either by burns or radical surgery by taking skin grafts and getting them to grow over wounds. In 1794, two English surgeons working in India took a flap of skin from a man's forehead to repair his nose amputation.

Surgeons had known for some time that cataracts on the eyes were due to a developing opaqueness of the eye's lens. Removing the opaque cover was a specialty of French surgeon Jacques Daviel (1696-1762), who extracted the covering over the lens rather than moving it aside (couching) as had prior eye surgeons. Daviel, who devised special instruments for his work, in 1747 extracted the lens from a patient's eye through a slit at the bottom of the cornea and cut through the cataract capsule with a sharp needle and squeezed it out. The "spoon" he made for this procedure was still in use in the twentieth century. Daviel, who traveled all over Europe removing cataracts, knew that cataracts could only be removed when they were mature and hard. In 1752 he described his technique to the surgical section of the Royal Academy in Paris.

Eighteenth-century surgery would not have progressed as quickly as it did without great anatomists, surgeons, teachers, hospitals, and academies of surgery. British leadership in surgical studies during the eighteenth century is recognized widely. Among the most famous British surgeon-teachers were Percival Pott (1714-1789), William Cheselden, and John

Hunter (1728-1793), who was perhaps the greatest surgeon and teacher of surgical techniques in his era.

At the age of 22, Londoner Percival Pott was admitted to the Barber-Surgeons Company, a guild that included both surgeons and barbers, who also performed minor surgeries. In 1745, Pott became Assistant Surgeon at London's St. Bartholomew Hospital. Pott researched, observed, and wrote many books, the most famous of which was *Injuries of the Head from External Violence*. Pott also wrote on fractures, ruptures, and palsy.

Cheselden published two important books—*The Anatomy of the Human Body* (1713) and *The Anatomy of Bones* (1733). Both books were used as texts for anatomy students for more than a century. Cheselden took an active role in the Barber-Surgeons Company and urged the separation of surgeons from barbers by petitioning the House of Commons. As a result, a new Corporation of Surgeons was formed in 1745, with Cheselden at its presidency. A skilled lithotomist (a specialist in removing stones from the bladder), it was said that he could perform a lithotomy in 54 seconds. Cheselden was a surgeon lecturer in anatomy at St. Thomas Hospital, where he trained a generation of surgeons, including John Hunter.

Hunter, of Glasgow, Scotland, has been credited with being the father of modern surgery. He went to London in 1748 to assist in the preparation of dissections for an anatomy course taught by his brother William, a famous obstetrician. He studied under William for 11 years and then studied under Cheselden, learning surgery at London's Chelsea Hospital. He became an army surgeon in 1760, returned to London in 1763 where he started his own surgical practice, and, in 1776, was appointed "physician extrordinare" to King George III. Hunter wrote three books—*The Natural History of Human Teeth* (1771), *A Treatise on Venereal Disease* (1786), and *A Treatise on the Blood, Inflammation and Gunshot Wounds* (1794). His last book, although published posthumously, made surgical history by differentiating between primary and secondary healing. Hunter also advocated enlarging gunshot wounds only if it was necessary to remove bone fragments. He attributed much of a person's infection to surgeons' unnecessary probing of wounds.

In France, Pierre Joseph Desault was Hunter's contemporary and nearly his equal, although they disagreed on key practices, such as debridement. Some sources suggest that it was really Desault, not Hunter, who was the founder of the modern treatment of war wounds. Desault, unlike Hunter, advocated wound debridement and taught it to his students. Desault believed in cleaning (preventive debridement) and "pruning" the edges of a wound with a blade, not simply opening a wound larger and causing a wound to bleed, and then closing wounds and leaving them undisturbed. He also preferred conservative treatments and is reputed to have said: "The sacrifice of a part for the preservation of the whole is the last recourse of the surgeon's art; before deciding on it the possibilities of restoring life and function to the sum total of the organs should be exhausted."

It seems that Hunter and Desault were in competition for the title of "the greatest surgeon in Europe." In 1785, within a few months of each other, both Hunter and Desault pioneered the ligation of the femoral artery to repair an aneurysm.

Desault was chief of surgery at the Hotel-Dieu in Paris from 1785 until his death in 1795. The Paris hospital was internationally famous for surgery, but the surgical preeminence of Paris began to wane with the French Revolution when extremists, aiming at destroying the privileged classes, threw many professions, including the medical professions, open to anyone, qualified or not.

Almost as important as the great teachers of surgery were the places in which they demonstrated their art and science, not only to medical students but often to any citizen who had an interest. Surgical amphitheaters made their debut in the seventeenth century but came into their own in the eighteenth century as most teaching hospitals adopted the amphitheater as a teaching tool.

Eighteenth-century surgery was also improved by hospital reformers who sought to improve the conditions in which patients were kept, thereby improving their chances for survival by providing clean beds and fresh air. Notable among British reformers was John Howard (1727-1790), who had spent most of his career reforming prison conditions before turning his attention to hospitals. Howard criticized conditions in Paris hospitals, including the famous Hotel-Dieu, where he reported seeing five or six patients to a bed.

Jacques Tenon (1724-1816) may have taken Howard's criticism to heart and embarked on an effort to reform and rebuild the hospital. Tenon reported seeing surgeries performed in the same rooms that held other patients waiting their surgeries. In a 1788 publication, Tenon urged that surgeons use separate rooms for surgeries.

The eighteenth century produced surgical giants who extended their expertise and imagination well into the next century, when their students started practices of their own. Their students, with the benefit of anesthesia, better antiseptics, and cleaner, more humane hospital practices—carried surgery to the edge of its modern applications.

RANDOLPH FILLMORE

Further Reading

Books

Cartwright, Frederick F. *The Development of Modern Surgery*. New York: Thomas Y. Crowell and Company, 1967.

Meade, Richard Hardaway. *An Introduction to the History of General Surgery*. Philadelphia: W.B. Saunders Company, 1968.

Wangenstein, O.H. and Sarah D. Wangenstein. *The Rise of Surgery*. Minneapolis: The University of Minnesota Press, 1978.

Zimmerman, Leo M. and Ilza Veith. *Great Ideas in the History of Surgery*. New York: Dover Publications, 1967.

Eighteenth-Century Advances in Dentistry

Overview

The eighteenth century, also known as the Age of Enlightenment or Age of Reason, saw dentistry advance from superstition and the showmanship of tooth-drawers to specialists who studied dentistry as a science. During this century France emerged as the leader in the dental field, a position it retained until about the middle of the nineteenth century when the United States became the leader.

The founder of modern dentistry was Frenchman Pierre Fauchard (1676-1761). He initiated a broad spectrum of advances and was the prime mover in establishing France as the front runner of the art. In Germany Philip Pfaff (1715-1767) described the process of tooth decay as originating from particles caught between the teeth. Meanwhile, English surgeon John Hunter (1728-1793) published several book on dentistry. The rise of American dentistry can be traced to the connection of colonial America to France.

John Hunter. *(Library of Congress. Reproduced with permission)*

Background

Dental problems have not changed much since the beginning of recorded history. Toothaches, decay, gum problems, periodontal disease, and tooth loss are documented in cave paintings and early recorded history. In about 5,000 B.C. ancient Sumerians recorded on clay tables their belief that toothaches were caused by the gnawing of tiny worms. The idea persisted for thousands of years.

The exact date at which dental art made its first appearance is not known. Some have speculated that dentistry was practiced before medicine. The first known dentist was Hesi-Re, the chief dentist to the Egyptian pharaohs around 3,000 B.C.. In addition to the Egyptians, the Etruscans of central Italy, Assyrians, and Chinese had some awareness of the problems of teeth. After the fall of Rome in A.D. 410 the scientific and technological advances of Western civilization largely vanished, and magic and superstition took over.

During the Middle Ages monks acted as physicians and dentists, with barbers as their assistants. The barbers had to take over completely when in 1163 the pope ruled that shedding of blood was not a priestly function. Guilds of barber-surgeons remained active until about 1745 in France, and later than that in America. The barber-surgeon would wrap the bloody rags after surgery around outside poles; this became the symbol of the barber-surgeon, which evolved into the modern red-and-white stripped barber pole.

Others were also pulling teeth. A band of vagabonds know as tooth-drawers roamed the towns of Europe pulling teeth. These tooth-drawers were entertainers who operated in the town squares and drew huge crowds. One famous showman was "le grand Thomas," who operated in Paris during the reign of Louis XV.

The Renaissance saw a rebirth in many areas of work, especially in the fields of anatomy. In the 1300s books about anatomy began to come out of major universities like Paris, Oxford, and Bologna. The famous French surgeon Guy de Chauliac (1300?-1368) wrote several chapters devoted to teeth and was the first to coin the term "dentator," which the English made into the word "dentist." Another mediaeval surgeon, Giovanni de Areoli (circa 1400), devised the pelican for extracting teeth and advised good cleaning habits. Leonardo da Vinci (1452-1519) presented the earliest accurate drawings of skull, teeth, and associated structures and maxillary sinus. His work inspired the dissector Andreas Vesalius (1514-1564), who accurately described the teeth and their chambers and laid the foundation for scientific investigation.

The first book describing work with teeth, titled *Artzney Buchlein,* appeared anonymously in Germany in 1530 and was used by the barber-surgeons. In 1563 Eustachio (1520-1574) wrote the *Book of the Teeth,* the first book on the anatomy of teeth.

French surgeons began to focus on dentistry, and in 1697 regulated the practice. It was in France that the term surgeon dentist was first used. The licensed dentists usually limited their practice to the wealthy. The common people still suffered with barbers, tooth-drawers, and fake doctors.

Impact

The eighteenth century saw the rise of France as the world leader in the field of dentistry. The outstanding dentist of the era was Pierre Fauchard, who was dubbed the father of scientific dentistry. Published in 1728, his two-volume book *The Surgeon Dentist* described dental anatomy, tooth decay, medicine, dental surgery, gum disease, and other aspects of dentistry.

LAUGHING GAS IN 1799

Nitrous oxide gas was first used for pain relief in dentistry in the 1840s and continues in wide use today for dental and other short surgical procedures. Isolated by Joseph Priestley in the 1770s, the first inhalation of the gas by humans took place in the Medical Pneumatic Institute near Bristol, England, in March 1799. A physician named Thomas Beddoes had founded the Institute six months earlier to continue the research he had begun years earlier on possible therapeutic uses of gas inhalation. By the spring of 1799 Beddoes's young research director, Humphry Davy, decided to try breathing nitrous oxide. At the time, medical experts considered nitrous oxide to be dangerous, but Davy and others at the Institute proved that humans could safely breath the gas. These researchers also discovered the impressive behavioral effects that breathing low doses of nitrous oxide can produce in man. Poet Robert Southey wrote after breathing the gas that it "made me laugh and tingle in every toe and finger tip. Davy has actually invented a new pleasure, for which language has no name." Others reported vivid hallucinations and other effects. In the 1790s Bristol was a center of literary activity in England, and many notables in addition to Southey visited the Institute to breath the gas or watch its effects in those who did. Southey's friend and fellow poet Samuel Taylor Coleridge participated, as did poets Anna Letietia Barbauld and George Burnett. Other participants included a young Peter Mark Roget, who decades later wrote his famous thesaurus; and inventor James Watt, his wife, and two of their sons. Among the observers of the experiments was novelist Maria Edgeworth. Nitrous oxide quickly became known as "laughing gas" because of its effects on these and other individuals breathing it at the Institute.

A. J. WRIGHT

Some of his treatments included the use of lead to fill cavities and oil of cloves and cinnamon for infection. He also made dentures with springs and real teeth.

Fauchard moved dentistry to a higher level of sophistication and advocated education for dentists. His amazing innovations were indeed unique for the time. For example, the earliest devices for removing decay were picks and enamel scissors. Instruments were then designed with two cutting edges that could be rotated between the fingers. Fauchard improved the drill in 1728, using catgut twisted around a cylinder to produce a rotating motion. He downplayed mere extraction to solve tooth problems.

Fauchard created techniques for filling, filing, and transplanting. With his interest in correcting cosmetic problems and straightening teeth, he became known as the originator of modern orthodontics. Another of his innovative ideas was the use of crowns and bridges to replace and cover missing parts of teeth. Fauchard worked and practiced until 1768 when he died at the age of 83.

Historcial records indicate that in 1789 a French apothecary named du Chateau had a ceramist make a set of teeth for him. Around this time a dentist named Dubois de Chemant made teeth from mineral paste. He took this specialized knowledge with him when he emigrated to England, and the English began to imitate the method. In 1791 de Chement was in America establishing the first dental clinic in New York City.

Inspired by French dentists, people in other parts of Europe began to investigate dentistry as a science. Philip Pfaff was the personal dentist to Frederick the Great of Prussia. Pfaff was the first to use gold foil to cap the pulp. He wrote a book called *Treatise on the Teeth* and proposed that dental decay was caused by the rotting or decay of food particles lodged in the teeth. Most people still believed that tooth decay was caused by tiny worms eating away at the teeth. Pfaff was the first to introduce plaster for pouring models of the teeth.

In 1766 Thomas Berdmore of London published the first English dental textbook in which he proposed that sugar was harmful to the teeth. As dentist to King George III, Berdmore trained many pupils who then emigrated to America.

John Hunter in Britain became the leading surgeon and dental physiologist in the late 1700s. He was a marvelous experimenter and worked to improve surgery in his country. His outstanding books *The Natural History of the Teeth* and *Practical Treatise on the Disease of the Teeth* (1771) described how teeth grow and develop. Dental development had not previously been explored. He also transplanted teeth and found he must remove the pulp before filling the teeth.

Writings about teeth began to grow in leaps and bounds. Joseph Fox, a pupil of Hunter, wrote *The Natural History and Diseases of the Human Teeth* (1797), in which he advocated that treatment of teeth be based on scientific knowledge and surgical experience. Also, the turn key, or English key, was developed in Europe. This surgical tool enabled surgeons to remove even the most difficult and deeply rooted teeth.

The American colonies were out of touch with major scientific dental developments during this time. According to records dated 1733-35, James Reading and James Mills were the first tooth-drawers in New York—and perhaps in America. In 1763 James Baker, M.D. and surgeon-dentist, became the earliest known qualified dentist to practice in Boston and America.

France was closely related to the American colonies and many dentists who emigrated to the colonies arrived with their trade. Two French dentists, Joseph Le Mayeur and James Gardette, were asked to teach their techniques to men in the revolutionary army. One person who came under the influence of the French was John Greenwood, who became the personal dentist to General George Washington. In 1790 Greenwood took his mother's foot-treadle spinning wheel, attached it to a drill, and used the modified device to remove decay. However, the idea did not catch on at time.

Greenwood began to use gold bases for dentures and made dentures for George Washington. It was known that Washington suffered terribly from tooth loss and complained about ill-fitting dentures.

At that time dental work was done without any pain killers or anesthesia. In 1772 Joseph Priestley (1733-1804) identified nitrous oxide, and a few years later Humphry Davy (1778-1829) noted that this gas not only made people laugh hilariously, but appeared to be capable of controlling pain. In his 1800 writings he suggested that this gas could possibly be used for surgery one day.

America, in its early years, was still the domain of tooth-drawers, barbers, and charlatans. However, by the middle of the nineteenth century it surpassed France as the world leader in dental treatment. With many of the basic concepts and approaches established during the eighteenth century, the field of dentistry witnessed tremendous advances during the nineteenth century.

EVELYN B. KELLY

Further Reading

Guerini, Vincenzo. *A History of Dentistry from the Most Ancient Times Until the End of the Eighteenth Century.* Boston: Longwood Press, 1977.

Ring, Malvin E. *Dentistry: An Illustrated History.* New York: Abrams, 1985.

Travis, B., ed. *World of Invention.* Detroit: Gale Research, 1994.

Biographical Sketches

Peter Artedi
1705-1735
Swedish Naturalist

Peter Artedi has been hailed by early historians as "the father of ichthyology" for his sole published work regarding the classification of fish. Artedi's untimely death (he drowned at the age of 30) cut short a very promising scientific career. Artedi's work was extremely important to Linnaeus (who is most famous for his system of classification of animals) and was used as a template for Carolus Linnaeus' (1707-1778) most famous work, *Systema naturae*. In fact, nearly every entry from this book dealing with fish has a reference to Artedi's work, which was extremely well thought out and detailed. Artedi and Linnaeus first met at the university and remained good friends throughout their lives.

Peter Artedi was originally born as Petrus Arctaedius in Anundsjö, Sweden, in 1705. His family had strong ties to the clergy, but ultimately, Petrus chose to study science. In 1716, his family moved to Nordmaling, Sweden so that his father could take over the parish duties from his grandfather whose health was failing. For a young, aspiring naturalist, this was an influential move because he was now situated close to the seashore with an abundance of interesting specimens to study.

In the same year, Petrus attended school at Hernösand where he had an unremarkable tenure. Most of his free time was spent in the dissection and study of fishes and the collecting of native plants, rather than in usual boyhood activities. He also learned Latin, which opened other avenues of interest for him. He was especially interested in the works of the early alchemists and even studied alchemy when he attended the University at Uppsala in 1724. Petrus depended largely on his own studies to advance his knowledge in the natural sciences, because there were few formal courses offered at the university in those areas.

Four years after Petrus had begun his university career, Linnaeus arrived at Uppsala with the intention of studying natural science. Upon his arrival he asked the names of other students engaged in similar studies and the name Petrus Arctaedius was foremost in everyone's mind. Petrus and Linnaeus did not meet at this time, however, because Petrus had been summoned homed to attend to his ill father. When they did finally meet, they became unfailing friends and helped each other extensively.

In 1734, after changing his family name to the more familiar Artedi, Peter had run out of scholarship money at the University of Uppsala and decided to travel abroad for research and educational purposes. First, Artedi sailed to London where he studied fish and began writing his famous manuscript. Later, he traveled to Holland and was in search of funding so that he would not have to return home. His friend Linnaeus introduced him to Albertus Seba, a wealthy Dutchman who hired Peter to study fish and other animals from his collection. Seba, a successful pharmacist and merchant, had a magnificent array of specimens to study. Artedi began working on this task. On September 27, 1734, Artedi spent the evening at Seba's house and started for his own home at one o'clock. Somewhere on this darkened route, he fell into one of the canals of Amsterdam and drowned.

Without doubt, Artedi's short life had been sufficient for him to produce work that influenced ichthyology as a science more than that of any other man, but his influence also extended far beyond this single aspect of zoology. He had significant influence on the life of Linnaeus, who had major influence on the classification system of animals. In addition, he excelled in languages, philosophy and medicine. It is hard to fathom the impact that he might have had on the world if he had not met such a premature death.

JAMES J. HOFFMANN

Matthew Baillie
1761-1823
Scottish Physician, Pathologist and Anatomist

Matthew Baillie wrote the first book on pathological anatomy as a separate science. His studies advanced the knowledge of diseases of the heart, lungs, liver, stomach, ovaries, and kidneys.

Baillie was born in Lanarkshire, Scotland. He was the son of Rev. James Baillie, later professor of divinity at the University of Glasgow, and Dorothy Hunter Baillie, the sister of prominent anatomists and surgeons John and William Hunter (1728-1793 and 1718-1783, respectively). He attended grammar school and Latin school in Hamilton, Scotland, then entered the University of Glasgow at the age of 13. In 1779 he was named a Snell Exhibitioner at Balliol College, Oxford University.

After Baillie's father died, his mother asked his uncle William to take charge of the boy's upbringing. In 1780 Baillie moved to London to live with his uncle. He studied anatomy and surgery with both Hunters, chemistry and medicine with George Fordyce, and obstetrics with Thomas Denman and William Osborn. He also attended lectures and rounds at St. George's Hospital in London. Even though he was living in London, he maintained his status as an Oxford student. From Oxford he received his B.A. in 1783, his M.A. and M.B. in 1786, and his M.D. in 1789. In 1787 he was appointed physician at St. George's. In 1791 he married Denman's daughter, Sophia.

When William Hunter died in 1783, Baillie inherited a great deal of wealth. William Hunter's will stipulated that Baillie and William Cruikshank together continue the anatomical lectures that Hunter had offered since 1768 at his Anatomical Theatre and Museum in Windmill Street, London.

In 1799 Baillie resigned from both St. George's and the Anatomical Theatre, gave up both writing and teaching, and thereafter devoted himself to private medical practice. He saw patients about 16 hours each day and soon became one of the most successful physicians in London. Toward the end of his career he was personal physician to King George III and several other members of the royal family.

Baillie is best known for writing *The Morbid Anatomy of Some of the Most Important Parts of the Human Body* (1793), the earliest scientific exposition of pathological anatomy, the anatomy of diseased tissue. It includes classic postmortem descriptions of heart valve disease, emphysema, tuberculosis, gastric ulcer, cirrhosis of the liver, ovarian cysts, and many other afflictions. Its illustrations, published separately as *A Series of Engravings, Accompanied with Explanations, which are Intended to Illustrate the Morbid Anatomy of Some of the Most Important Parts of the Human Body* (1799-1803), constituted the first significant pathological atlas.

Giovanni Battista Morgagni (1682-1771) founded pathological anatomy with his *De sedibus et causis morborum per anatomen indagatis ["The Seats and Causes of Diseases Investigated by Anatomy"]* (1761), but did not treat it as a separate science. Baillie's *Morbid Anatomy* established pathological anatomy as a discipline in itself, presented many new discoveries that soon had clinical importance, and set the stage for such eminent investigators as Jean Cruveilhier (1791-1874).

One traditional symbol of medicine is the gold-headed cane. In 1827 William Macmichael (1784-1839) published his account of a particular gold-headed cane that had belonged successively to six British physicians, John Radcliffe (1650-1714), Richard Mead (1673-1754), Anthony Askew (1722-1774), William Pitcairn (1712-1791), his nephew David Pitcairn (1749-1809), and Matthew Baillie. Baillie's widow donated the cane to the New College of Physicians, Pall Mall East, London, in 1823. Macmichael's book provides an insider's view of 140 years of British medicine, as well as biographical information about the owners of the cane.

ERIC V.D. LUFT

John Bartram
1699-1777
American Naturalist, Botanist and Explorer

John Bartram is best known as a naturalist and as the first American botanist. His explorations resulted in the identification, collection, and propagation of much of the flora of the American colonies.

Bartram was born into a Quaker family and grew up on a farm in Pennsylvania. He received no formal education but taught himself, learning by observing the natural world closely and reading everything he could obtain on biological subjects. In 1728 he established, on his farm, the first botanical garden in America. It was here that he performed the experiments that resulted

An illustration from the 1904 edition of *Bartram's Garden* by eighteenth-century botanist John Bartram.

in the first successful hybridization of flowering plants in America. He and Benjamin Franklin (1706-1790) became friends. He helped Franklin organize the American Philosophical Society and was one of its first members.

John Bartram traveled extensively in the wild areas away from the towns and settlements of the American colonies, exploring much of the region of the Allegheny, Blue Ridge, and Catskill mountains, the area around Lake Ontario, the forests and mountains as far west as the Ohio River, and portions of New Jersey, Virginia, North Carolina, South Carolina, and Florida. He made his exploratory trip in Florida in 1765-77 with his son, William (1739-1823), who became a noted botanist and naturalist in his own right. On these exploratory trips, he identified the native plants, collecting seeds and plants to add to his botanical garden in Pennsylvania. He and his son identified and cultivated more than 200 native American species. One species, *Franklinia altamaha*, that he discovered in Georgia and named after Benjamin Franklin, is now extinct in the wild; the only survivors are descendents of a specimen he brought back to Pennsylvania and planted in his botanical garden.

European botanists were anxious to receive descriptions and samples of the flora of the new world, most of which they had never seen before. Bartram corresponded with many of them

and sent them specimens and seed. Among his European correspondents and friends was Carolus Linnaeus (1707-1778), one of the best-known plant scientists of his day, who thought highly of his American colleague. Bartram's dissemination of information and specimens was improved in 1734 when his botanical expertise was recognized by Peter Collinson, a London merchant with numerous contacts. As a result of Collinson's encouragement, Bartram provided seed and specimens for the gardens of, among others, the Dukes of Richmond, Norfolk, Argyll, and Bedford. In 1765 King George III appointed him as the official royal botanist for the American colonies. He is credited with introduction of more than 100 new species into Europe.

In 1743 he served as a commissioner of the Crown and traveled to Canada to confer with the leaders of the American Indian League of Six Nations. On this trip he explored a section of Canada, identifying and collecting native plants of the region.

A true naturalist, he did not confine his interest to plants but became knowledgeable of the zoology and geology of the regions he explored as well. He carefully observed and collected samples and information on the shells, insects, birds, turtles, and snakes that he encountered in his travels. A moss genus, *Bartramia*, and the Bartramian sandpiper are named in his honor.

His interest in science and exposure to the European scientific thought of the day led him to become a deist (one who believes that God created the world but no longer has any direct interactions with it). As a result, the Quakers disowned him in 1757. He, however, never completely disengaged himself from them.

John Bartram has received well-deserved recognition as the father of American botany. His botanical garden remains in operation today and may be visited by the public.

J. WILLIAM MONCRIEF

George Berkeley
1685-1753
Irish Philosopher

George Berkeley is best remembered for his attempt to reconcile religion and science through his empiricist philosophy that separated a phenomenon's true existence from attempts to explain it through scientific laws and theories.

Berkeley received his formal education at Kilkenny College and at Trinity College in Dublin, receiving his B.A. degree in 1704. He was elected a fellow of Trinity in 1707 and remained associated with the College until 1724 when he became the Dean of Derry. In 1733 he was elevated to the position of Bishop of Cloyne.

Berkeley read widely and became thoroughly familiar with the scientific and philosophical developments that grew out of Newtonian physics. As a result of his deep religious convictions, he attempted to overcome the apparent conflict between the religious and scientific perceptions of the world. He became a spokesman for the Anglican church and a defender of the Christian faith in the effort to counteract attempts by some of science's more ardent supporters to use the ability of science to predict and explain natural phenomena, thereby dismissing the necessity of God. In the process, he criticized what he perceived as logical contradictions in Isaac Newton's (1642-1727) mechanics and attacked the deists. He also developed a unique philosophy of science and reality. His principal publications included *Treatise Concerning the Principles of Human Knowledge* in 1710, *Three Dialogues between Hylas and Philonous* in 1713, and *De motin* in 1721.

Philosophically Berkeley was an empiricist, believing that everything, except the spiritual, exists only as it is perceived by the senses. He was strongly opposed to the materialists who maintained that the world is made up only of material objects that act, and are acted on, mechanically and that obey laws of natural necessity that science can determine. He believed that objects and phenomena do not have a material existence but exist only as ideas in the mind of the perceiver. He asserted, however, that all such ideas are invariably in the mind of God and that humans receive the ideas of the phenomena that make up their existence through communion with God. Berkeley's philosophy, for this reason, is sometimes also labeled subjective idealism.

Berkeley's thought has been influential in the history of philosophy because it laid the foundation for later secular empiricists such as David Hume (1711-1776). It has had its greatest effect on the history and philosophy of science through his so-called instrumentalist views. He held that scientific theories are nothing more than tools for predicting the course of natural processes and are neither true nor false, merely useful. For him, Newton's equations are simply mathematical methods for the calculation of observed phenomena: they are not explanations of the phenomena. Scientific laws and theories are summaries of how objects behave under certain circumstances and predictions of how they will behave; they do not provide any real understanding of the phenomena they describe. He also pointed out that the observed data cannot uniquely determine a theory that explains them: more than one theory may be developed to explain the observations.

He believed that in order for humans to use knowledge, it must be linked to experiences that they can understand. In other words, our understanding of a new phenomenon is based on its relationship by analogy to experiences that we have already had; our understanding is dependent on the objects and events that we, as human beings, are capable of experiencing, and these are, of necessity, limited. Berkeley held that science is a "useful fiction" and that its explanations are mental constructs that are different from actual fact. His ideas have become more widely accepted as literal interpretations of scientific theory have became more and more difficult.

J. WILLIAM MONCRIEF

Marie-François-Xavier Bichat
1771-1802
French Anatomist, Physician and Pathologist

The great French anatomist and physician Marie-François-Xavier Bichat is remembered for his pioneering work in anatomy and histology. Bichat's attempt to create a new system for understanding the structure of the body culminated in the tissue doctrine of animal anatomy. Bichat's approach involved studying the body in terms of organs, which were then dissected and analyzed into their fundamental structural and vital elements, called "tissues." Bichat was the son of a respected doctor and was expected to enter the same profession. After studying medicine at Montpellier, Bichat continued his surgical training at the Hôtel Dieu in Lyons. The turmoil caused by the French Revolution, however, forced him to leave the city for service in the army. In 1793 he was able to resume his studies in Paris and became the disciple of the eminent surgeon and anatomist Pierre-Joseph Desault (1744-95). When his mentor died, Bichat spent several years editing the fourth volume of Desault's *Journal of Surgery*, to which he added a biographical memoir of the author.

Bichat began to give lectures in 1797 and became physician to the Hôtel Dieu in 1800. One year later, he was appointed professor. Completely dedicated to anatomical and pathological research, Bichat essentially lived in the anatomical theater and dissection rooms of the Hôtel Dieu, where he performed at least 600 autopsies in one year. In 1802 he became ill with a fever and died before completing his last anatomical treatise.

Tissue doctrine was an extremely influential theory on the construction of the body. Although many of his contemporaries still relied on Hippocratic humoral doctrines, Bichat and his Parisian colleagues believed that medicine could only become a true science if physicians adopted the methods used in the other natural sciences. Scientific investigations would then transform general observations of complex phenomena into precise and distinct categories. In other words, physicians should link clinical studies of disease to systematic postmortem observations. Bichat's approach was inspired by the work of another French physician, Philippe Pinel (1755-1826), who argued that diseases must be understood in terms of the organic lesions found at autopsy, rather than explained away in terms of humoral pathology.

Bichat reasoned that organs that manifested similar traits in health or disease must share some common structural or functional components. Because this common element did not exist at the organ level, Bichat was convinced that there must be some common element at a deeper level, that is, at a finer level of resolution.

Using various techniques to break up the organs, Bichat dissected out what he thought of as the fundamental elements of the body. Organs were teased apart by dissection, maceration, cooking, drying, and exposure to chemical agents such as acids, alkalis, and alcohol. Organs, therefore, were composed of a "web" of different elements, which Bichat referred to as "membranes" or "tissues."

According to Bichat, the human body was composed of some 21 different kinds of tissue, such as the nervous, vascular, connective, fibrous, and cellular tissues. Organs, which were aggregates of tissues, were also components of more complex systems, such as the respiratory, nervous, and digestive systems. The activities that characterized tissues were explained in terms of "irritability" (the ability to react to stimuli), "sensibility" (the ability to perceive stimuli), and "sympathy" (the mutual effect parts of the body exert on each other in sickness and health). Although Bichat could not resolve the tissues into more fundamental entities, he realized that they contained complex combinations of vessels and fibers. Thus, Bichat's tissue theory of general anatomy is quite different from the cell theory that was elaborated in the nineteenth century by Matthias Jacob Schleiden (1804-1881) and Theodor Schwann (1810-1882). Nevertheless, Bichat hoped that his analysis of the structure of the human body would lead to a better understanding of the specific lesions of disease and to improvements in therapeutic methods.

LOIS N. MAGNER

Johann Friedrich Blumenbach
1752-1840
German Anthropologist, Anatomist and Naturalist

Johann Friedrich Blumenbach had a primary role in founding the science of modern anthropology, was a pioneer in the field of comparative anatomy, and was a respected researcher and renowned teacher. Blumenbach advocated the unity and equality of the human race as a single species and directly attacked any use of anthropology as a means to promote social and political racial discrimination or abuse. He viewed the human species as a product of its natural history, and his publications provided a reliable survey of human geographical distributions and characteristics. Blumenbach wrote the first scientifically objective anthropology and comparative anatomy textbooks, both of which had tremendous influence on the course of development of these sciences for several generations.

Blumenbach's father was an assistant headmaster of a grammar school in Gotha, Germany, his mother's father was a city high official, and her grandfather a respected theologian. Thus Blumenbach began life in a well-to-do and cultured family, and was exposed to literature, the sciences, and a formal education from a young age. Later, when Blumenbach married in 1778, it was to the daughter of an influential administrator at the University of Göttingen, and his brother-in-law was Christian Gottlieb Heyne, a noted classics scholar.

During his education at the University of Göttingen, Blumenbach was greatly influenced by the natural historian Christian W. Büttner and his lectures on ethnography, exotic peoples, and cultures. This influence affected the direction and nature of Blumenbach's M.D. dissertation and the initiation of his ethnographic collection that became widely admired and celebrated, and included a large selection of literature by African writers. The first edition of Blumenbach's dissertation was published in 1776, and eventually included a third edition in 1795, becoming world famous as the most influential text on anthropology of its time. Blumenbach was appointed curator of the natural history collection at Göttingen in 1776, became a full professor of medicine in 1778, and in 1816 he became Professor primarius of the Faculty of Medicine. In the course of his career many of Blumenbach's students became successful in their own right, and Blumenbach was a member of over 70 academies and societies, as well as a scientist widely respected and consulted by his contemporaries.

Blumenbach was not the first to describe and classify the various "races" of mankind, and he was greatly impressed by the works of Carolus Linnaeus (1707-1778), Georges Buffon (1707-1788), and Gottfried Leibniz (1646-1716). However, he did base his work on objective physical characteristics, especially of the skull, and he used verified records of the geographical distributions of the known human groups. Blumenbach modified previous classifications to create a five-family distribution of the human species that has been tremendously influential and introduced the term "Caucasian" to define the group of Europeans. He hypothesized that the American and Mongolian families branched in one geographical direction and the Malaysians and Ethiopians branched in another. He stated that humans arose in the area of the Caucasus Mountains,

spread out over the world, and acquired variations in physical characteristics such as skin color, which are simply gradations of the same character. The scientific objectiveness and the quality of his work in his *De Generis Humani Varietate Nativa Liber* (1795) marks this text as the foundation of scientific anthropology.

Blumenbach was much more objective than previous scientists in his analysis and concluded without question that all human races belonged to one species and that all of humanity resided on the same level, regardless of cultural differences. Blumenbach regarded man as an object of natural history, though he did describe humans as "the most perfect of all domesticated animals." He argued against the prevailing opinion of European superiority and their divine selection, instead insisting that all races, especially African, were fully equal to Europeans. Blumenbach supported his assertion of equal intellect and morality of mankind with his famous collection of anthropological and ethnographical specimens, which included skulls, artwork, and literature of various peoples, and particularly of African writers and artists.

Blumenbach also published landmark comparative anatomy and physiology (1824) and natural history (1779) texts, which helped usher in new eras of modern zoology, natural history, and ecology. Blumenbach recognized that fossils represent extinct species, and he believed in a long geological history of nature and Earth, as well as the variability of nature, including the addition of new species over time. Blumenbach also worked on another of the great debates of the Enlightenment period, providing additional research evidence in favor of the theory of epigenesis and against the prevailing belief of preformation. Thus, Blumenbach's influence goes well beyond his scientific evidence and active stance in favor of racial equality, and includes his then-superior works of comparative anatomy and physiology, natural history, and human reproduction. His historical position in the period of the Enlightenment is as one of the most celebrated and honored scientists of his time, with reverberations and discussions that have continued into this century.

KENNETH E. BARBER

Hermann Boerhaave
1668-1738
Dutch Physician and Chemist

The Dutch physician and professor of medicine Hermann Boerhaave was widely known and revered for his exemplary role in teaching clinical, or "bedside," medicine. Although Boerhaave is not associated with any major discoveries in medicine or chemistry, his reputation as a teacher and writer was so great that his contemporaries considered him comparable to Sir Isaac Newton. He was considered the greatest clinical teacher of his time and the most eminent of European physicians.

Boerhaave, the son of a minister, was born in Verhout, Holland. He was expected to follow in his father's profession, but he became fascinated by chemistry and decided on a career in medicine. In 1684 he became a student at the University of Leiden and took courses in chemistry, botany, medicine, philosophy, languages, and philosophy. He graduated in natural philosophy in 1687. After qualifying in medicine at the University of Harderwijk in 1693, he returned to Leiden, where he began taking private pupils and teaching medicine and chemistry. He was offered a professorship at the University of Groningen in 1703 but declined it so that he could wait for an appointment to the University of Leiden. In 1709 he was appointed to the chair of medicine and botany at the University of Leiden. Boerhaave dedicated the rest of his professional life to the University of Leiden. In addition to serving as professor of chemistry, botany, medicine, and clinical medicine, he was also rector of the university.

Founded in 1575 by William of Orange, the University of Leiden was, by the seventeenth century, highly respected as a center of learning in theology, science, and medicine. During the eighteenth century, thanks largely to Boerhaave, the University of Leiden also achieved distinction in the study of medicine. Students came from all parts of Europe to the University of Leiden in order to hear Boerhaave's lectures. His students carried word of Boerhaave's brilliant lectures to their own medical communities. Many of his best pupils became teachers at prestigious medical schools, where they exerted a profound influence on the teaching of medicine. Boerhaave is generally credited with establishing the system of teaching medical students at the bedside of the patient. Boerhaave also published medical textbooks that were widely used long after his death.

Boerhaave's major works were *Medical Principles* (1708), *Aphorisms on the Recognition and Treatment of Diseases* (1709), *Index plantarum* (1710), and *Elements of Chemistry* (1734). Some of his lectures were published in unauthorized versions by his students, including *Historia plantarum* and one version of *Principles and Experiments in Chemistry*. Boerhaave explicitly repudiated the latter text. His own *Elements of Chemistry* was considered one of the most learned works on chemistry ever published. It was translated into many languages and remained in use for almost 100 years.

Physicians throughout the world were guided by Boerhaave's teachings and his published aphorisms. (Aphorisms are concise or pithy expressions of doctrines or generally accepted truths.) Since the time of Hippocrates, aphorisms dealing with the symptoms, diagnosis, and prognosis of diseases had been used as guides to students and physicians. Boerhaave's collection of medical aphorisms provides a valuable summary of essentially all the medical knowledge available at the time.

With his colleague Bernard Siegfied Albinus (1697-1770), Boerhaave also edited the writings of Andreas Vesalius (1514-1564) and William Harvey (1578-1657). Albinus was the first anatomist to demonstrate the connection between the vascular systems of the mother and the fetus. He had been appointed to the chair of anatomy, surgery, and medicine at Leiden in 1721.

As both a teacher and a writer, Boerhaave was respected for his ability to systematize and convey to his students and readers all the useful medical information that was available at the time. Boerhaave was so famous that a letter bearing only the address "To the Greatest Physician in the World" was delivered to him.

LOIS N. MAGNER

Charles Bonnet
1720-1793
Swiss Naturalist and Philosopher

Charles Bonnet, the eminent Swiss naturalist, is primarily remembered for his discovery of parthenogenesis, which is a form of reproduction without fertilization. His theories on embryological development and catastrophic evolution were also of great interest to eighteenth-century naturalists and philosophers.

Bonnet's wealthy family had immigrated to Switzerland to escape the persecution of Huguenots in their native France. Although Bonnet originally studied law and was elected to Geneva's town council, he was more interested in the natural sciences. Because of his family fortune, he was able to devote himself to the pursuit of his scientific and philosophical interests. Following the example of his teacher Réné Antoine Ferchault de Réaumur (1683-1757), Bonnet became interested in the life cycles of insects. This interest led to many significant contributions regarding insect metamorphosis and the publication of his *Treatise on Entomology* (1745). In addition to his studies of parthenogenesis in aphids, Bonnet discovered that caterpillars and butterflies breathe through pores, which he named stigmata. He also carried out investigations of the structures and functions of plant leaves. Unfortunately, a serious eye disease compelled him to give up research that required direct observation. Nevertheless, Bonnet made many other contributions to theoretical issues in natural science and to speculative issues in natural philosophy.

While studying the life cycle of aphids, Bonnet discovered that female aphids were able to reproduce without fertilization by the male. In his philosophical work *Considerations on Organized Bodies* (1762), Bonnet suggested that the eggs of each female organism contained the "germs" of all subsequent generations as an infinite series of preformed individuals.

One of the great debates in eighteenth-century natural science concerned the question of whether embryological development followed a preformationist or epigenetic model. According to preformationist theories, a miniature individual preexisted in either the egg or the sperm and began to grow as the result of the proper stimulus. Many preformationists believed that all the embryos that would eventually develop had been formed by God at the Creation. The theory of epigenesis, in contrast, was based on the belief that the embryo developed gradually from an undifferentiated mass by means of a series of steps during which new parts were added. Although Aristotle and the great seventeenth-century physiologist William Harvey had been advocates of epigenesis, many of the most distinguished eighteenth-century naturalists preferred preformationist theories. The increasing acceptance of the mechanical philosophy and microscopic observations during the late seventeenth century apparently favored preformationist theories. Preformationist theories, however, suggested that only one parent could serve as the source of the preformed individual. Some preformationists argued that God had implanted or encapsulated the "germ" of all human beings in the ovaries of Eve. Bonnet's studies of parthenogenesis supported ovist preformationism and allowed him to develop a sophisticated version of encapsulation theory. Thus, the pious Bonnet was encouraged to discover compelling evidence of harmony between natural philosophy and religious doctrine.

Bonnet found that female aphids that hatched during the summer gave birth to live offspring without fertilization. The new generation of males and females mated in the fall and the females laid eggs. By carefully segregating young females to avoid the possibility of contact with males, Bonnet was able to raise 30 generations of virgin aphids. According to Bonnet, the discovery of parthenogenesis proved that the ovaries of the first female of every species contained the miniature precursors for all future members of that species. In most species, the male seemed to provide the stimulus for growth of these preformed individuals. Once development was initiated, the unfolding and growth of the preformed individual could proceed as a purely mechanical process. Despite the difficulties posed by preformationist theory, Bonnet's work was still admired by many nineteenth-century naturalists.

Preformationism seemed to support the conclusion that God had created essentially immortal and immutable species. In response to reports that some species had become extinct in the distant past, Bonnet argued in his *Philosophical Revival* (1769) that periodically the Earth experienced universal catastrophes that destroyed almost all living beings. Following such catastrophes, the survivors advanced to a higher level on the evolutionary scale. Thus, combining his biological work and his philosophical speculations, Bonnet became the first natural philosopher to use the term "evolution" in a biological context. Further philosophical inquiries led to his *Essay on Psychology* (1754) and *Analytical Essay on the Powers of the Soul* (1760), which have been called pioneering works in physiological psychology.

LOIS N. MAGNER

Matthew Dobson
1731?-1784
English Physician

Matthew Dobson was a medical pioneer who discovered and described conditions such as hyperglycemia and sought further understanding of known diseases such as diabetes. Because medical knowledge of diabetes was limited in the 1700s, his research and observations aided other physicians in diagnosing patients' illnesses more accurately and in developing more effective treatments. Dobson's scientific writings provided a foundation for future generations of medical researchers.

Not much is known about Dobson's early life. Some biographical sources state that he was almost 40 years old when he died in 1784, suggesting that he was born around 1745, while others place his birth date a decade earlier, in the 1730s. The 1753 date of his first dissertation suggests that the latter estimate is correct. Biographical sketches of Dobson's wife, the notable translator and writer Susannah Dobson, say that her husband was from Liverpool and was buried at Bath, a town that was home to hot springs and health resorts where Dobson possibly sought treatment to relieve pain from a chronic condition. He might have been Scottish because he attended school in Glasgow and Edinburgh.

His 1753 dissertation at the University of Glasgow was entitled *Dissertatio philosophica inauguralis de natura hominis* and his 1756 dissertation at the University of Edinburgh was called *Dissertatio medica inauguralis, de menstruis,* suggesting that he may have been interested in gynecology and obstetrics. Medical histories locate Dobson at Yorkshire during his work on diabetes. He was named a fellow of the Royal Society.

Although he probably dealt daily with routine concerns such as fevers and infections, Dobson's specialty was what modern physicians label endocrine disorders, which concern the distribution of hormones in the bloodstream. The role of glands, which secrete hormones, including insulin (crucial for metabolizing sugar), was not understood when Dobson was conducting his investigations. Doctors did not comprehend the immediate and long-term health risks of diabetes, such as patients' being blinded, becoming comatose, and dying suddenly. Physicians such as Thomas Willis had commented that the urine of some diabetic patients tasted sweet. By 1776 Dobson had examined urine and tissue samples and demonstrated that diabetics had sugar in both their urine and blood serum, causing the sweetness.

Dobson is also cited as the discoverer of hyperglycemia, which is more commonly referred to as high blood sugar. He collected information on case studies of diabetes and wrote an article entitled "Experiments and Observations on the Urine in Diabetes." Medical historians refer to Dobson's work as being part of the diagnostic period of diabetes research, following the clinical era in which diabetes was merely described

and preceding the empirical treatment and experimental periods.

Dobson also contributed to early work in pneumatic medicine. He pioneered methods to treat lung abscesses with carbon dioxide and to relieve respiratory ailments with mineral waters. He also used his techniques to loosen hardened material in the urinary tract. Dobson's text discussing this work, *A Medical Commentary on Fixed Air,* was published in London in 1779. Like his patients, Dobson suffered from frail health and perhaps was even diabetic. He died either in his late 30s or early 50s, depending on his actual birth date.

ELIZABETH D. SCHAFER

Johann Christian Fabricius
1745-1808
Danish Entomologist and Economist

Johann Christian Fabricius was one of the most distinguished entomologists of the eighteenth century and an outstanding theoretical natural scientist. His research greatly increased and organized knowledge about insects as a class of animals. Fabricius established his scientific reputation in the field of descriptive taxonomy as the founder of a system of insect classification. Instead of basing his system on the characteristic of insect wings, he focused his attention on the structure of the organs of the mouth. His attempts to study the effect of the environment on the development of insect species led some scholars to call him the "Father of Lamarckism."

Johann Christian Fabricius was born in Tondern, Denmark. He was the son of a physician. Fabricius was educated in Altona and Copenhagen. After completing his studies, he spent two years with the great taxonomist Carl Linnaeus (1707-1778) in Uppsala. Although Fabricius considered himself a pupil of Linnaeus, the two formed a lifelong friendship based on mutual respect and admiration.

After returning to Denmark, Fabricius published several important works on entomology. Although he applied Linnaeus's methods to insects, Fabricius was more independent than most of Linnaeus's other disciples. Nevertheless, the titles of his works correspond to those of his mentor and the Linnaean system generally guided his own observations. Fabricius was highly respected in the scientific community. Although his most important work was in entomology, he

Johann Fabricius. *(Corbis Corporation. Reproduced with permission.)*

was professor of natural science and economics at the University of Copenhagen before accepting an appointment in 1775 as a professor at Kiel, Germany. Fabricius spent much of his time abroad and traveled over much of Europe.

In creating a system of insect classification, Fabricius was concerned with distinguishing between artificial and natural systems. He thought that the characteristics of the various structures of the mouth would provide the best approach to a natural system of taxonomy. Fabricius believed that the genus and the species were the major categories of entomological systematics. Genera could be regarded as natural combinations of related species, but he thought that "classes" and "orders" were artificial constructs. His objective was to create a system based on naturally defined insect genera. Even though he considered the task of creating a natural system more important and challenging than describing individual species, he carried out a prodigious amount of descriptive work. Fabricius is credited with naming and describing about 10,000 insects.

Despite his emphasis on insect systematics, Fabricius was interested in broader biological and philosophical problems, including evolution. Fabricius thought that humans might have originated from the great apes. In reflecting on the great diversity of existing species, he suggested that new species might come into existence through such means as hybridization, structural modification, and adaptation to the environment.

LOIS N. MAGNER

John Fothergill
1712-1780
English Physician, Botanist and Reformer

Fothergill added to the medical knowledge of diphtheria, scarlatina, trigeminal neuralgia, migraine, and other disorders. He created an important botanical garden in England and, through his philanthropy, helped found several educational and clinical institutions.

He was born into a yeoman Quaker family in Wensleydale, Yorkshire, England. His mother, Margaret, died when he was seven. His father, also named John, was a deeply religious man and spent much of his time on the circuit of Friends' meetings, even going to America three times.

Young John's education became the responsibility of his maternal uncle, Thomas Hough. After attending day school in Frodsham, Cheshire, and grammar school in Sedbergh, Yorkshire, he was apprenticed in 1728 to a Quaker apothecary, Benjamin Bartlett, in Bradford, Yorkshire. He served six years of his seven-year apprenticeship, then, with Bartlett's approval, decided to try the formal study of apothecary science. Because he was a Quaker, he was not allowed to attend institutions of higher learning in England. He therefore crossed the border into Scotland and enrolled at the University of Edinburgh in 1734.

At Edinburgh Fothergill immediately came under the kindly influence of Alexander Monro *primus* (1697-1767) who advised him to study medicine. (Monro, his son, and grandson, were all named Alexander, and were distinguished by the designation *primus, secondus,* and *tertius* [first, second, and third].) Fothergill received his Edinburgh M.D. in 1736 and spent the next four years at St. Thomas's Hospital, London. After traveling briefly in northern Europe, he established private medical practice in London in 1740.

His reputation was made with the publication of "An Account of the Sore Throat Attended with Ulcers" (1748), in which Fothergill presented careful observations of both diphtheria and scarlatina, although he failed to distinguish between the two. His work on these diseases was furthered by John Huxham (1692-1768) in "A Dissertation on the Malignant, Ulcerous Sore-Throat" (1757) and William Withering (1741-1799) in "An Account of the Scarlet Fever and Sore Throat, or Scarlatina Anginosa, Particularly as it Appeared at Birmingham in the Year 1778" (1779).

Fothergill published two important papers in a major journal of the time, *Medical Observations and Inquiries by a Society of Physicians in London*. In "Of a Painful Affection of the Face," (1776) the first description of trigeminal neuralgia, also known as Fothergill's neuralgia or tic douloureux, an excruciating pain along the path of the fifth cranial nerve. In "Remarks on that Complaint Commonly Known under the Name of the Sick Head-Ache" (1777) he offered one of the earliest accurate descriptions of migraine.

Fothergill was a close friend of Benjamin Franklin and a strong advocate of liberal policies toward the American colonies; until 1775 he strove for reconciliation between the two sides. Fothergill was an early supporter of the first hospital in America, the Pennsylvania Hospital, founded by Franklin in 1751. The books and anatomical teaching materials he donated to the Pennsylvania Hospital in the early 1760s helped William Shippen Jr. (1736-1808) become the first successful demonstrator of anatomy in America.

Encouraged by Edinburgh professor of botany Charles Alston, Fothergill made a serious hobby of botany. In 1762 he purchased a 30-acre country estate in Essex and created a botanical garden on the grounds. Carolus Linnaeus (1707-1778) gave the scientific name *Fothergilla* to a genus of witch hazel that Fothergill brought from America to Essex.

Concerned with the lack of educational opportunities for Quaker children, in 1779 Fothergill founded a coeducational Quaker boarding school, the Ackworth School, near Pontefract, Yorkshire. Twenty years later a group of Philadelphia Quakers, inspired by Fothergill's example, founded a sister school, the Westtown School, near West Chester, Pennsylvania.

Few physicians have been as revered for their kindheartedness as Fothergill. He had no harsh words for anyone and was devoted to Quakerism's gentle, peacemaking ways. When he died of prostate cancer he was mourned by all of London. Seventy carriages made up his funeral procession. The frontispiece of the first posthumous edition of his works (1781) depicts him as the Good Samaritan. Fothergill never

married. His maiden sister Ann lived with him the last 31 years of his life.

ERIC V.D. LUFT

Luigi Galvani
1737-1798
Italian Physician

Luigi Galvani was the pioneer of electrophysiology. A skilled anatomist, obstetrician, physician, and surgeon, Galvani conducted several experiments with animal nervous and muscular systems. After an accidental discovery of the relationship between electricity and muscle movement while dissecting a frog, Galvani proposed a theory of "animal electricity." Although his theory incorrectly identified the electro-conducting medium in animals as fluid, his research opened new lines of inquiry about the structure and function of the nervous system in both animals and humans.

Galvani was born on September 9, 1737 in Bologna, Italy. His original intention was to study theology and later enter a monastic order. However, Galvani was discouraged from pursuing monastic life in favor of continuing his studies of philosophy or medicine. He devoted his academic career at the University of Bologna to both interests and in 1759 received his degree in letters and medicine on the same day. In 1762, Galvani was named Professor of Anatomy at the university, and remained there for most of his career. His early works were primarily concerned with comparative anatomy.

In 1764, Galvani married Lucia Galleazzi, the daughter of a prominent member of the Bologna Academy of Science. Galvani's wife encouraged his independent research, and served as a counselor and guide for his experiments until her death. Drawing from his extensive training in anatomy and obstetrics, Galvani focused his research on the nature of muscular movement. While dissecting frogs for study, Galvani noticed that contact between certain metal instruments and the specimen's nerves provoked muscular contractions in the frog. Believing the contractions to be caused by electrostatic impulses, Galvani acquired a crude electrostatic machine and Leyden jar (used together to create and store static electricity), and began to experiment with muscular stimulation. He also experimented with natural electro-static occurrences. In 1786 Galvani observed muscular contractions in the legs of a specimen while touching a pair

Luigi Galvani. *(Corbis Corporation. Reproduced with permission.)*

of scissors to the frog's lumbar nerve during a lightening storm. He also noticed that a simple metallic arch which connected certain tissues could be substituted for the electrostatic machine in inducing muscular convulsions.

The detailed observations on Galvani's neurophysiological experiments on frogs, and his theories on muscular movement, were not published until a decade after their coincidental discovery. In 1792, Galvani published *De Viribus Electricitatis in Motu Musculari* (On Electrical Powers in the Movement of Muscles). The work set forth Galvani's theories on "animal electricity." He concluded that nerves were detectors of minute differences in external electrical potential, but that animal tissues and fluids must themselves possess electricity which is different from the "natural" electricity of lightening or an electrostatic machine. Galvani later proved the existence of bioelectricity by stimulating muscular contractions with the use of only one metallic contact—a pool of mercury.

Galvani's work altered the study of neurophysiology. Before his experiments, the nervous system was thought to be a system of ducts or water pipes, as proposed by Descartes. Galvani proved that there was a relationship between muscle movement and electricity and proposed that nerves were electric conductors. However, the nature of a different "animal electricity" which

Galvani proposed was disproved later by Italian physicist, Alessandro Volta. Galvani had experimented with using animal fluids and metallic conductors—hence the reference to his name in the term "galvanized"—to create an electric pile, or battery. However, in 1800, Volta created the first successful battery which could produce a sustained electrical current. Thus, Volta effectively rejected Galvani's theory of "animal electricity."

After the death of his wife, Galvani joined the Third Monastic Order of the Franciscans. He continued his academic research, but when Napoleon seized control of Bologna in 1796, Galvani refused to swear the oath of allegiance to the newly created Cisalpine Republic on the basis that it contradicted his political and religious beliefs. As a result, he was stripped of his professorship and pension. Galvani died on December 4, 1798.

BRENDA WILMOTH LERNER

Johann Wolfgang von Goethe
1749-1832
German Writer

Johann Wolfgang Goethe was the last of the great universal scholars. He was the greatest German poet and an outstanding figure in many fields of writing. In addition to criticism, poetry, novels, and plays, his scientific writings filled 14 volumes. In his 82 years he achieved masterful wisdom and accomplishments. He was a strong advocate of natural theology—the finding of God in nature.

Goethe was born August 28, 1749, in Frankfurt am Main, now Germany. At that time Frankfurt was a free city-state. His father, Johann Caspar Goethe, was an attorney who taught his son the intellectual and moral discipline characteristic of the cities of northern Germany. His mother, Katharine Textor Goethe, the daughter of the mayor of Frankfurt, opened up contacts for her son to have artistic and cultural opportunities. Eight children were born to the couple but only Wolfgang and his sister Cornelia survived.

At 16 he went to the University of Leipzig, and although the town was dubbed "Little Paris," he preferred lectures in the foundation of German and moral culture given by C. F. Gellert, poet and author of hymns. Serious illness forced him to leave Leipzig and return home. Here, he came in contact with Pietists circles. The Pietists were a religious group of German Lutherans who combined moral and ethical behavior with

Johann Wolfgang von Goethe. *(Archive Photos. Reproduced with permission.)*

reason. He also developed an interest in alchemy as a mystical endeavor and in the newly emerging field of chemistry.

Recovering from the illness, he went to Strasbourg to finish his law degree, but his illness had given a sentimental mood to his writing. He developed a sublime view of nature and art that became known as "Romantic." The combination of the mystical with the occult would appear later in the novel, *Faust*, a story of a young man who sells his soul to the devil.

In 1775 Goethe was betrothed to Anna Shonemann, a rich banker's daughter, but differences in age and temperament ended the engagement. Goethe had several stormy relationships, many of which appeared in his novels and as subjects of his poetry.

Goethe went to Weimar, home of the reigning duke Karl August, and quickly became his friend and adviser. The duke heaped a lot of responsibilities on his shoulders, including director of finance, supervisor of mines and military activities, and diplomatic duties. However, he did not neglect his writing and wrote poems and plays for numerous official occasions.

Advances in technology engaged Goethe in the natural sciences. His interest in art led him to the study of anatomy. In 1784 he discovered the intermaxillary bone, a segment of the upper

jaw. Art also led him to the study of optics, color, and light. His scientific studies took on a controversial bent when he tried to prove that Isaac Newton (1642-1727) was wrong and that light is one and indivisible and cannot be explained by the theory of particles.

In 1790 he wrote a classic text on botany and biology called *Versuch die Metamorphose der Pflanzen zu erklären* (Metamorphosis of Plants). The book was artistic and well written and showed how many parts of plants are modifications of a basic form.

While Goethe's scientific discoveries were not notable, he did have great insight into scientific method and the study of natural phenomena. He was a keen observer and knew how to construct theory. He insisted that knowledge of self should develop with knowledge of the world.

Goethe founded and named the branch of science called morphology, which is a systematic study of natural occurrences. He studied the formations and transformations of rocks, animals, plants, colors, and clouds. He was interested in the cultural happenings of human science. His interests were in connecting all nature.

His poem *Gott und Welt* (God and World) showed how Goethe saw god in the natural world. Actually, Goethe was not a traditional church person but was a believer in the roots of the Bible as his philosophical base. Some have tried to describe his beliefs as "panentheism," a belief in the divine and transcendent God; however, he disagreed, saying that his beliefs could not be put into a small category.

He died in Weimar, Saxe-Weimar, on March 22, 1832.

EVELYN B. KELLY

Stephen Hales
1677-1761
English Physiologist and Clergyman

Although trained as a clergyman with no formal education in physiology, chemistry, or medicine, Stephen Hales became one of the most important British scientists of the eighteenth century. He can be considered the father of plant physiology, and his studies of circulation in animals were the most important after William Harvey's (1578-1657) over a century earlier. Hales conducted experiments on gases with equipment of his own design that influenced the later work of Joseph Black (1728-

1799), Henry Cavendish (1731-1810), Joseph Priestley (1733-1804), and Antoine Lavoisier (1743-1794). He also did important research on respiration and in public health involving ventilation and fresh air.

Stephen Hales was born in Bekesbourne, county of Kent, England, on September 17, 1677, into an established, influential family. In 1696 he entered Benet College at Cambridge University, where he received a bachelor's degree and then a master's in 1703. In 1709 he was ordained as a priest in the Anglican Church and left Cambridge to become minister at Teddington, a small town on the Thames River between Twickenham and Hampton Court and about 15 miles (24 km) west of London. Hales kept this religious post until his death in January 1761, and did most of his research work at Teddington.

While at Cambridge Hales gained his first knowledge of scientific experimentation. In 1703 William Stukeley, later to become a well-known London physician, entered Benet College; and he and Hales developed a close friendship. In a room provided by Stukeley's tutor, the young men set up a laboratory where they conducted chemistry experiments and dissected small animals, including frogs. By 1707 Hales was attempting to measure blood flow in dogs. In these years before he left Cambridge, Hales also attended chemistry lectures by John Francis Vigani and his successor, John Waller.

After taking his religious post at Teddington, Hales spent several years attending to his parish duties before renewing his interest in science. By 1714 Hales began a study of the force of circulating blood—what we call blood pressure—in larger animals such as horses and deer. In 1719 he turned his attention to a study of the force of sap in plants. By this time Hales had been nominated to a fellowship in the Royal Society, a prestigious honor that validated his efforts. Membership in this body meant that oral accounts of Hales's research presented at the Society's meetings would be heard by the most important scientists and physicians in Britain at that time. Finally, Hales began to publish his work; in 1727 his book *Vegetable Staticks; Or, an Account of Some Statical Experiments on the Sap in Vegetables* was published in London. In 1733 *Haemastaticks; Or, an Account of Some Hydraulic and Hydrostatical Experiments Made on the Blood and Blood-Vessels of Animals* appeared.

These two works described in great detail Hales's meticulous experiments, which were characterized by his close observation and mea-

surement. His most famous experiment is described in the first few pages of *Haemastaticks*. Hales tied down a horse and cut open an artery about three inches from the stomach. Into this artery he inserted a brass pipe only 0.167 inch (0.424 cm) in diameter. Then he used another brass tube fitted to the first one to attach a glass tube 9 feet (2.74 m) in length. Hales could then watch the mare's blood rise with each heartbeat over 8 feet (2.43 m) into the glass pipe. After these measurements Hales removed the glass tube and measured the blood released until the horse died. Then he studied the heart and surrounding arteries to see how much blood remained in the carcass. Hales repeated these experiments in cows, deer, dogs, and sheep, and carefully recorded the results. He also conducted other experiments to detect the flow of blood through the lungs; in describing the tiny blood vessels in the lungs, Hales was apparently the first person to use the term "capillary." Hales demonstrated the same curiosity and attention to detail in his studies of sap and water flow in plants; he invented the U-tube manometer to aid him in this research.

Between 1725 and 1727 Hales conducted important research on gases. In his plant experiments Hales had noticed bubbles of air that appeared when a plant stem was cut. Hales developed several apparatuses that would allow him to measure the amount of air absorbed or released when various animal, plant, or chemical compounds were heated and cooled. His most famous device was the pneumatic trough, with which he could collect different gases. Although Hales did not differentiate one gas from another, this research and the pneumatic trough became very useful to later chemists who did manage to isolate the various gases.

In 1739 Hales received the Copley Prize from the Royal Society, which recognized the importance of his scientific work. Six years earlier he had been awarded the Doctor of Divinity degree from Cambridge. In the 1750s Hales helped develop a channel and drainage system that supplied Teddington with fresh water. He was also interested in the problem of fresh air in confined spaces such as ships, hospitals, and prisons, and he developed a ventilator and bellows system to supply fresh air in these areas. His ventilation systems were installed in British navy warships and in several prisons, including the notorious Newgate. Hales's ventilators did not remove bacteria, which were unknown at the time, but mortality among the sailors and prisoners was reduced. In addition to his clerical duties and his scientific researches, Hales also served as one of the original trustees of the American colony of Georgia. Although he never visited America, Hales left his entire library to the colonies in his will. Unfortunately, the books disappeared and have never been located.

A. J. WRIGHT

Albrecht von Haller
1708-1777
Swiss Physiologist and Poet

Albrecht von Haller. *(Library of Congress. Reproduced with permission)*

Albrecht von Haller was the father of modern experimental physiology. By uniting the fields of anatomy and "living anatomy," Haller created a functional approach to studying the physiology of the body. Haller's doctrine concerning the contractility or irritability of muscle tissue laid the foundation for the emerging field of neurology. An illustrious scholar and prolific writer, Haller influenced eighteenth century thought in poetry, botany, philosophy, and medicine.

Born in Bern, Switzerland, Haller was a physically frail but precocious student. He wrote scholarly articles by age eight, and finished a Greek dictionary by age ten. At seventeen, Haller studied medicine at The University of Leiden in the Netherlands under Hermann Boerhaave (1668-1728), the century's greatest medical teacher. After obtaining his medical degree in 1727, Haller continued his studies of anatomy, surgery, and mathematics in London and Paris. During this time, Haller began additional studies in botany after an inspiring trip collecting flora in the Swiss Alps. Haller eventually published a popular volume of poetry including "Die Alpen," a poem which illustrated the beauty of the Swiss Alps, inspired German pride, and influenced German literature.

Haller traveled to the University of Göttingen in 1736 to serve as professor of medicine, anatomy, surgery, and botany, a position he held for 17 years. At Göttingen, Haller performed extensive animal experimentation to improve the principle of general irritability, proposed earlier by English scientist Francis Glisson (1597-1677). Haller showed in 1752 that irritability (contractility) is a property inherent in muscle fibers, while sensibility is an exclusive quality of nerves. By this experimentation, the fundamental division of fibers according to their reactive properties was established. Haller concluded that muscular tissue contracted independently of the nerve tissue around it, and therefore explained why the heart

beats: contraction of muscular tissue stimulated by the influx of blood. Even after the electrophysiology pertaining to the heart was understood more than one hundred years later, Haller's basic suppositions regarding tissue contractility remained essentially correct. Haller's experiments also showed that nerves were inherently unchanged when stimulated, but caused the muscle around it to contract, thus implying that nerves carry impulses which cause sensation. This distinction between nerve and muscle action illustrated by Haller provided the framework for the advent of modern neurology.

Haller's additional contributions to physiology included works on respiration and embryology. Through experimentation, Haller refuted earlier teachings that the lungs contract independently. Haller was the first physician to use a watch to count the pulse. As scientific debate began between preformationists (those who believed that the human was totally preformed in miniature) and epigenesists (those who believed that not all organs were preformed, but appeared during the formation of the fetus), Haller studied developing chick embryos. Although Haller showed that the heart of a developing chick became apparent in the thirty-eighth hour, he sided with the preformationists, impeding the study of embryology for many years.

In 1753, Haller returned to Bern where he practiced medicine and continued his research, publishing extensive scientific works and sustaining a voluminous correspondence with scholars across Europe. Haller presented the knowledge of his time in anatomy and physiology grouped together in his eight-volume *Elementa Physiologiae Corporis Humani* (Physiological Elements of the Human Body). A milestone in medical history, the series of books is considered the first modern textbook of physiology. In his later years, Haller compiled twenty volumes of bibliographies on medicine, surgery, anatomy, and botany.

BRENDA WILMOTH LERNER

David Hartley
1705-1757
English Physician and Philosopher

David Hartley was an English physician who developed an influential philosophy of thinking known as associationism. Hartley strove to explain how the mind and consciousness functioned; and he speculated that the mind is able to work by the association of sensations and the cumulative experiences of life, or memories. Hartley was also influential in suggesting that the mind and body do not function independently of each other but work together in concert.

Hartley was the son of a poor Anglican clergyman; his mother died before he was a year old and his father died while he was a young boy. He entered Jesus College, Cambridge, in 1722 where he studied mathematics and divinity; he received his Bachelors in 1727 and his Masters in 1729. This course of education would have led to a career in the Anglican ministry, but Hartley was not willing to sign his commitment to the Thirty-nine Articles on the Church of England. Thus, having abandoned the ministry, he turned to medicine.

Hartley never obtained a degree in medicine, but he started his medical studies in Newark where he then began to practice. He later practiced medicine in Nottinghamshire, London, and finally Bath, where his second wife could take advantage of the nearby health spa for her illness. Hartley remained in Bath until his death.

Hartley's major work, *Observations on Man, His Frame, His Duty, and His Expectations*, was published in 1749 in two volumes. In this work, he further developed a concept from John Locke's *Essay Concerning Human Understanding*, namely that complex ideas are the result of simpler, earlier ideas tempered with memories and experience. Hartley then joined this idea with a concept of memory and sensation from Isaac Newton's *Opticks* in which Newton theorized that the vibrations that make up light entered the eye and traveled by way of vibration to the brain where they are recognized as light. Hartley forwarded the idea that all mental processing, especially that of memories, is based on vibrations in the brain and spinal column. The very existence of memories was based on "vibrantuncles," as Hartley called them, which resulted from the vibrations entering the brain from experiences. In this way, Hartley was able to formulate his idea that memories and experiences work together to make the brain operate.

The significance of Hartley's work *Observations* was realized later as other thinkers and scientists used his concept of associationism to explain psychological, social, and physiological concepts. Hartley wrote *Observations* intending it to be considered a work of philosophy, but it has been that and much more. Hartley's idea that things work together in order to function properly opened a floodgate of ideas that were cultivated by later scientists in areas such us biology, evolutionary theory, neurophysiology, psychiatry, social theory, and neurology. Hartley was the first to use the term "psychology" to describe the study of behavior and brain functioning.

Hartley's concept of association can be easily seen in use with the psychotherapeutic tool "free association" in which a patient reports the first image that comes to mind when the therapist says a random word. Brainstorming, an idea generating tool used in advertising and business, has its roots in associationism. Hartley's work *Observations*, based on his speculations regarding the functioning of the brain, is regarded as one of the pivotal pieces in the foundation of modern science, if not expressly for the principles of associationism.

MICHAEL T. YANCEY

John Howard
1726-1790
English Public Health Reformer

John Howard is remembered for his efforts and successes at reforming the appallingly unsanitary and inhumane conditions in hospitals and prisons both in England and in Europe during the eighteenth century. By visiting prisons first-hand, taking meticulous notes, and publishing books on the horrors he encountered, Howard raised public opinion. Howard was one of many mid-eighteenth century public health reformers, but recognized as one who obtained results and left a legacy that subsequent reformers could follow.

John Howard, the son of a wealthy man who inherited a fortune on his father's death in 1742, traveled widely in Europe. He suffered an unfortunate experience while traveling by ship to Spain in 1756 when his ship was seized by a French privateer. Howard, the ship's crew, and his fellow passengers were thrown into jail in France as prisoners of war. Although soon released, his suffering while imprisoned left a lasting impression. Twenty-five years later, after being elected high sheriff of Bedfordshire, England in 1773, he found conditions in his small Bedford county jail deplorable. Many prisoners, he found, were being kept in jail although they had been acquitted by the court but could not pay a "discharge fee."

In 1774, Howard persuaded the House of Commons to pass acts decreeing that discharged prisoners were to be set free, discharge fees abolished, and that justices were required to look after the health of prisoners. The acts were never fully enforced.

Embarking on a campaign, Howard toured jails in England, Wales, Scotland, and Ireland, and in 1777, published his findings in a book entitled *State of the Prisons* in which he outlined the horrors of prison life. Howard found that many prisons might have an inch or two of water on their floors, and that inmates often had no straw as bedding. Straw they did have was often wet. Very little water was afforded prisoners for drinking and washing. Even during the winter in London jails, Howard observed, there were no fires to keep inmates warm. The jailers were not paid by the local government but were maintained by fees paid by the prisoners. Howard found prisoners chained, often in darkness, for even minor offenses. Howard took methodical notes and even weighed the food allotted to prisoners. Howard's published work aroused public opinion and many reforms stemmed from his efforts.

His research in England led him to evaluate prisons all over Europe where he also investigated prison diseases, such as typhus and smallpox, known to run rampant in prisons. Through Howard's efforts, the relationship between unsanitary prison conditions and "jail fever," an illness that later killed him, became apparent.

Many of the reforms Howard sought and won prepared a path for other sanitary reformers in the nineteenth century by showing that once aroused, public opinion could change social conditions and improve the public's health.

Among the reforms Howard won in England in 1779 were statutes that authorized the building of two prisons where, through solitary confinement, supervised labor, and religious instruction, jailers attempted to rehabilitate prisoners.

Howard's public health reform interests also led him to investigate hospitals, particularly in France, where he reported that five or six patients, some of them dying, might share a bed. Howard recommended many hospital reforms, including fans to provide ventilation, and the separation of the sick based on their illnesses.

In 1786, Howard purposely traveled to Italy aboard a ship known to be carrying disease and was subsequently quarantined so that he could better understand the experience of quarantine. Howard was quarantined for forty-two days in quarters he reported were like the "sick wards of the worst hospitals."

His last investigation led him to Moscow and Saint Petersburg, Russia, and the Ukraine where, in 1790, while visiting Ukrainian military prisons he died of jail fever.

RANDOLPH FILLMORE

John Hunter
1728-1793
Scottish Surgeon

John Hunter was the first surgeon to dissect and examine cadavers to understand the function of the human body. Today he is considered the founder of pathological anatomy and remains among the world's greatest physiologists and surgeons.

John Hunter was born in rural Scotland in 1728. The brother of a suave London surgeon, the young Hunter was uneducated and considered crude by many. But his eclectic interest in the bits and pieces of the human body resulted in numerous contributions of great importance in surgery. By the time of his death in 1793, he had carried out many important studies and experiments in the comparative aspects of biology, anatomy, physiology, and pathology.

When Hunter was 20, his brother William gave him a job assisting in the preparation of dissections for an anatomy course he taught. By his second year in dissection, he had become so skillful that he was given charge of some of the classes in his brother's school. For many winters, he studied anatomy in the dissecting rooms, and later spent two summers learning surgical techniques from William Cheselden (1688-1752) at London's Chelsea Hospital.

He showed such talent there that his brother urged him off to Oxford. But Hunter was no scholar. He never completed a course of studies in any university and never really attempted to become a doctor. Yet his skills eventually launched him into medical prominence. In 1754 he became a surgeon's pupil in St. George's Hospital, and two years later was a house-surgeon.

One of Hunter's eye-opening, and certainly cruel, experiments occurred when he ruptured his Achilles tendon. To analyze the healing process, he cut the tendons of several dogs then killed each animal at different intervals. Hunter learned how bones and tendons mend and how scar tissue formed. His observations laid the foundation for the simple and effective operation for the cure of club feet and other tendon deformities.

The surgeon's most famous experiment was filled with incredible risk. Daring to do what no others would, Hunter infected himself with syphilis in an effort to demonstrate that syphilis and gonorrhea were manifestations of a single disease. His research far surpassed venereal disease; it also opened new doors into the body's ability to heal itself. Hunter was among the first to fully explain the role of inflammation in healing.

Hunter eventually settled into the country, where he kept a lively menagerie and a team of medical students. In the odd world of his own making, he began to expand medical understanding and became a persuasive advocate of investigation and experimentation. He performed a countless number of experiments, more, some say, than "any man engaged in professional practice has ever conducted."

In the early 1700s, Hunter began his own private lectures on the principles and practice of surgeries. He dreaded lecturing, but found himself forced to do so because he was constantly misquoted. He used infinite care in preparing each lecture, but because of his fear of public speaking, they were instructive rather than interesting.

Hunter's tireless pursuit of knowledge was rewarded with numerous honors and positions of responsibility. In 1776 he was named physician extraordinary to King George III; in 1783

he was elected a member of the Royal Society of Medicine and of the Royal Academy of Surgery at Paris; in 1786 he became deputy surgeon-general of the army; and in 1790 he was appointed surgeon-general and inspector-general of hospitals.

In his later years, Hunter suffered many ailments, possibly due to his self-infliction of syphilis. He also had bouts of angina, which flared often with his temper. In 1793 Hunter died from a heart attack after a heated debate with some colleagues.

KELLI MILLER

William Hunter
1718-1783
Scottish Anatomist, Obstetrician, Surgeon and Physician

Along with his teacher and fellow Lanarkshire Scotsman William Smellie (1697-1763), William Hunter was chiefly responsible for elevating obstetrics to the status of a rule-governed empirical science. Before his time it was generally the disrespected practice of ill-trained and superstitious midwives. He also made important contributions to surgical anatomy, forensic medicine, and cardiology.

Hunter was the seventh of 10 children of John Hunter, a farmer, and Agnes Paul Hunter. After attending grammar school in East Kilbride, Scotland, he entered the University of Glasgow in 1731. He made friends with William Cullen (1710-1790) and became his medical apprentice in 1736. On Cullen's advice, Hunter began anatomical studies under Alexander Monro *primus* (1697-1767) at the University of Edinburgh in 1739.

Hunter went to London to study obstetrics under Smellie in 1740. The following year he became the protégé of the distinguished obstetrician and anatomist James Douglas. He lived with Douglas, tutored his son William George, and became engaged to his daughter Jane Martha, but she died in 1744 before they could be married. Douglas enabled Hunter to study surgery at St. George's Hospital. On his deathbed in 1742, Douglas asked his wife to ensure that Hunter would have means to study anatomy on the Continent. Hunter spent much of 1743 and 1744 in Paris and Leiden.

In 1746 Hunter began offering the anatomical lectures that were the main source of his income for the rest of his life. They were an immediate success. His reputation as a surgeon and obstetrician also grew quickly and steadily. In 1750 the University of Glasgow awarded him an M.D. In 1762 he became the personal obstetrician and physician to Queen Charlotte, wife of King George III. In 1768 he founded the Anatomical Theatre and Museum, Great Windmill Street, London, where he and his assistants taught anatomy. Known as the Hunterian Medical School after his death, it remained open until 1839.

His articles in the leading medical periodical of its time, *Medical Observations and Inquiries by a Society of Physicians in London*, contained some of the earliest accurate descriptions of aneurysms, retroverted uterus, osteomalacia (softening of the bones), ovarian cysts, and several other conditions. His posthumously published article in the same journal, "On the Uncertainty of the Signs of Murder in the Case of Bastard Children" (1784), was the first important British contribution toward the medical jurisprudence of infanticide.

Hunter's masterpiece is *Anatomia uteri humani gravidi tabulis illustrata* (The Anatomy of the Human Gravid Uterus Exhibited in Figures) (1774). This book is among the best anatomical atlases ever produced, not only for its accuracy and detail, but also for its aesthetic qualities. It stands about 27 in (68 cm) high, and each of its 34 life-size copperplate engravings is magnificently executed. Hunter's text for this atlas was edited by Matthew Baillie (1761-1823) and published in 1794 as *An Anatomical Description of the Human Gravid Uterus and its Contents*.

In the 1750s, having become wealthy through his lectures and his practice, Hunter began collecting art, coins, anatomical and biological specimens, minerals, rare books, manuscripts, scientific instruments, and other valuable objects. His taste and connoisseurship were impeccable. He willed the entire collection and funds for its maintenance to the University of Glasgow, which in 1807 made the collection public with the opening of the Hunterian Museum, the first institution of its kind in Scotland.

Hunter was a nervous, paranoid, and eventually bitter man. He made enemies easily and picked many public quarrels. He was jealous of his younger brother John's greater success as a surgeon. He never married, did not socialize much, and generally spent his life at work. His dedication to the scientific study of anatomy bordered on monomania. Yet he was devoted to his nephew, Matthew Baillie, and the generous

terms of his will contributed much toward establishing Baillie's medical career.

ERIC V.D. LUFT

Jan Ingenhousz
1730-1799
Dutch Plant Physiologist and Physician

Jan Ingenhousz is best known for his discovery of photosynthesis, the process by which green plants absorb carbon dioxide in the presence of sunlight and release oxygen. Through an ingenious series of experiments, Ingenhousz proved that plant leaves need sunlight rather than heat in order to produce oxygen. He also discovered that plant leaves reverse this process in the dark and release carbon dioxide. Ingenhousz's remarkable observations on plant physiology and photosynthesis were published as *Experiments Upon Vegetables, Discovering Their Great Power of Purifying the Common Air in Sunshine, and of Injuring It in the Shade and at Night* (1779).

Ingenhousz was born in the Netherlands in 1730. He studied medicine, chemistry, and physics at the universities of Louvain and Leiden. In 1765 he visited London and established a successful medical practice there. He became well known as an early practitioner of inoculation against smallpox. In 1768 he was called to Vienna to inoculate the family of the Austrian empress Maria Theresa. His services were in great demand and he remained in Vienna as court physician and surgeon until 1779, when he returned to London. He was elected a Fellow of the Royal Society in 1769. Although best known for his experiments on plants, Ingenhousz was interested in many other areas of science. He invented a device that could be used to generate large amounts of static electricity and made the first quantitative measurements of heat conduction in metal rods. Ingenhousz carried out many experiments on electricity, magnetism, and the relationship between plants and animals. He died in Wiltshire, England, while visiting the Marquis of Lansdowne's estate.

The experiments on photosynthesis carried out by Ingenhousz were inspired by the work of Stephen Hales (1677-1761) and Joseph Priestly (1733-1804). Hales, the founder of experimental plant physiology, published his classic treatise *Vegetable Staticks* in 1727. While conducting quantitative experiments on the movement of water in plants, Hales discovered that some component in the air was essential to plant growth. Hales, however, did not fully understand that ordinary air is made up of a mixture of distinct gases. Having isolated and characterized several gases, including "fixed air" (carbon dioxide), Priestley demonstrated that plants have the ability to restore or "revivify" air that has been "damaged" by animal life or by combustion. Green plants, in other words, produce a substance that supports animal life and combustion. Priestley, however, did not always clarify the importance of sunlight in these experiments. Instead, it was Ingenhousz who proved that green plants must be exposed to light rather than heat in order to restore oxygen to the air. He also demonstrated that only the green parts of a plant are capable of performing photosynthesis. While green leaves exposed to sunlight produce large amounts of oxygen, all parts of a plant produce small amounts of carbon dioxide in the dark; plants, like animals, perform respiration and release carbon dioxide into the air. Ingenhousz thus helped to clarify the fundamental similarities and differences between plant and animal life.

LOIS N. MAGNER

Edward Jenner. *(Library of Congress. Reproduced with permission)*

Edward Jenner
1749-1823
English Physician and Scientist

Edward Jenner developed the medical procedure known as vaccination, or the inoculation of cowpox to prevent smallpox. He also conducted research on the cuckoo bird, promoted a new method of preparing emetic tartar, and observed the structural changes in the heart that are associated with angina pectoris.

Born May 17, 1749, Jenner was the third son of Reverend Stephen Jenner of Berkeley, Gloucestershire. After being educated locally, he entered into an apprenticeship with Daniel Ludlow, a surgeon-apothecary in Sodbury. In 1770 he traveled to London and became a house-pupil of surgeon and anatomist John Hunter (1728-1793) at St. George's Hospital. Jenner obtained a fellowship to the Royal Society in 1789, and received his medical degree in 1792 from St. Andrew's in Scotland. His personal and professional lives were centered in Cheltenham, where he lived with his wife, Catharine, and three children, and where he worked as a resident physician for nearly 20 years. He died in January 1823.

Jenner's most significant contribution to medicine was the development of smallpox vaccination. Smallpox is an infectious disease that may result in disfigurement, blindness, and death. Before the use of vaccination, men and women submitted to a method known as variolation, or the introduction of smallpox into the skin under controlled conditions. Chinese, Indian, and African practitioners were aware of and employed the procedure before it was widely used in Europe. Lady Mary Wortley Montagu (1689-1762) introduced variolation into England after observing and undergoing the procedure in Turkey in 1717. The goal of variolation was to produce a mild outbreak of the disease in order to confer acquired immunity on the patient. Variolation, however, posed the danger of a full-blown attack of the disease and had the potential of spreading the affliction among the general population.

Edward Jenner reported his findings on vaccination in *Inquiry into the Causes and Effects of Variolae Vaccinae* (1798). In his publication, Jenner claimed that cowpox provided protection from infection by smallpox. He came to this conclusion after observing 25 cases of cowpox that prevented 24 subjects from contracting smallpox. He also vaccinated a number of people, four of whom resisted smallpox after being variolated.

A keen observer, Jenner also noted changes in the heart brought on by angina pectoris, performing dissections that revealed the disease's artery hardening and blockage. In addition, he refined the composition of emetic tartar (a drug thought to rid the body of disease by inducing vomiting) in order to ensure a more exact dosage.

Finally, Jenner studied natural history. His interest in botany and the animal kingdom influenced John Hunter's decision to choose him as his pupil. Jenner paid particular attention to the cuckoo bird, a creature that lays its eggs in the nests of other birds. Jenner hoped to figure out how the newborns of the foster mother were evicted from the nest in order for the cuckoos to gain her full attention. He concluded that the body structure of the newborn cuckoo bird allows it to overwhelm the other birds and forcibly remove them from the nest.

KAROL KOVALOVICH WEAVER

Antoine-Laurent de Jussieu
1748-1836
French Physician and Botanist

Antoine-Laurent de Jussieu was an important influential botanist in France at the time of the French Revolution. He developed the principles that served as the basis for a system of classifying plants for over two centuries (in fact, it was only in 1999 that a major revision to this system was proposed), and he helped to completely reorganize the Natural History Museum during his tenure there.

Jussieu was born into a family of botanists. Three of his uncles, Antoine, Bernard, and Joseph, were respected botanists in pre-Revolutionary France, and his son, Adrien-Henri, followed the family tradition in the first half of the nineteenth century. At the age of 17, he traveled to Paris to live with his uncle Bernard while studying medicine. However, he was appointed a professor and demonstrator at the Jardins du Roi (the Royal Gardens) in 1770 and, thereafter, devoted himself to botany. He was named a botany professor at the University of Paris, remaining there until 1826, and he was elected to the French Academy of Sciences based on a paper he published when he was 25.

In fact, his Uncle Bernard had a significant impact on his own work, to the point that it is sometimes difficult to determine which man to credit for which discoveries. Part of this difficulty lies in the fact that some of his most important work was published before his uncle's death in 1777, and this is compounded by the fact that he

was virtually the only person to transcribe and publish his uncle's work. However, it is likely that Antoine-Laurent's most important contribution to botany, his classification scheme for plants, was entirely his work, although it followed from some of the work done by Bernard before his death.

SMALLPOX INOCULATION BEFORE JENNER

Edward Jenner did not invent inoculation. In the 1770s he originated the idea of using cowpox (vaccinia) to inoculate against smallpox. In 1796 he performed his first vaccination, and in 1798 published the results of his 23 case studies. But long before Jenner, doctors were using smallpox itself to prevent smallpox.

The technique of using dried matter from smallpox lesions to inoculate uninfected individuals against smallpox was known in ancient China and precolonial Africa. In 1714 Emanuel Timonius reported the same sort of procedure practiced among the Greeks and Turks. In 1715 Giacomo Pilarino reported inoculating three children in Constantinople, Turkey, in 1701. About the same time, the prominent American Puritan preacher Cotton Mather owned a slave named Onesimus. In his native Africa, Onesimus had learned traditional inoculation. He taught it to Mather, who in turn taught it to his friend, Boston physician Zabdiel Boylston. Influenced not only by Mather but also by Timonius and Pilarino and supported by Benjamin Colman (1673-1747), Boylston began transplanting smallpox lesions into healthy patients in 1721. Of the 240 people he inoculated during the 1721 Boston smallpox epidemic, just 6 died.

Despite Boylston's success, popular superstition prevailed and inoculation became a public scandal in Massachusetts. An anti-inoculation mob bombed Mather's house in 1723. Yet even in the face of opposition from both medical and nonmedical groups, smallpox inoculation became a fairly common preventive medical procedure in England and America by the mid-eighteenth century. Among its supporters were William Douglass, Charles Maitland, Richard Mead, Thomas Dimsdale, Benjamin Franklin, and Benjamin Rush.

ERIC V.D. LUFT

In a nutshell, Antoine-Laurent developed a system for classifying plants that proved much superior to the previous Linnaean system. The chief innovation lay in assigning unequal values to various traits or characteristics used to determine family affinities for plants. That is, when comparing leaf shape, stem branchings, flower petal arrangements, and other morphological characteristics of plants, not all of these were equally considered or weighted when determining which plants were more closely related. Using this scheme, he was able to group plants into three main categories, which were further divided into 15 classes and over 100 families. More importantly, his classification system made it possible to classify new plant species more quickly and accurately, an important advantage during this time of major exploration of the world with new plant specimens being returned from all corners of the globe.

Antoine-Laurent began his classification system by examining only a small number of plants, expanding it gradually during his tenure at the Jardins du Roi and the University of Paris. By the end of his life, it had been accepted by even British scientists (notoriously slow to accept new ideas from their political rivals, the French), and its use continues to this day. The system is not perfect, and mistakes have been made. However, it was not until 1999 that a serious challenge was posed, based on DNA evidence rather than on the physical characteristics of the plants. From this, we can infer that the system worked quite well, indeed, especially well, as it was developed during a time in which the mechanisms and process of heredity (i.e. genes and DNA) were not known.

Antoine-Laurent retired from his teaching duties in 1826, a decade before his death. He published widely during his working career and, even after retirement, continued working on a widely anticipated second edition of his most important book, *Genera Plantarum*. Unfortunately, he died after completing only a small part of this work, leaving his son, Adrien-Henri, to publish those portions he had completed after his death.

P. ANDREW KARAM

Josef Gottlieb Köhlreuter
1733-1806
German Botanist

Josef Gottlieb Köhlreuter was a German botanist who, in the mid-eighteenth century, made a number of extremely important discoveries about plant genetics and reproduction. His work in many ways foreshadowed that of Gregor Mendel (1822-1884) and helped to establish that plants, like animals, could be thought of as having two distinct sexes. He also made important progress in developing plant hybrids, and studied plant pollinization.

Köhlreuter was born in the city of Sulz in 1733. Educated at the universities of Berlin and Leipzig, he earned a degree in medicine from the latter. He was not, however, to make his reputation as a physician but, rather, as a botanist.

Beginning in 1761, Köhlreuter published a number of landmark papers on plant breeding that, unfortunately, were to go unrecognized until many years after his death. In his first papers, he reported the existence of sex in plants, noting that, like animals, plants have distinct sex organs that can be considered male and female. This was important because it helped show a link between reproduction in all forms of life, demonstrating that the gap between plants and animals was not too great after all.

In this context, it must be remembered that sexual reproduction simply means that two "parents" combine genetic information to form offspring. It is this combination of genes in a random fashion that gives such diversity to life. This is how, for example, a child can have some characteristics from her mother, some from her father, can resemble her aunt in other respects, and be completely unlike any relative in still other factors. This mixing of genes also helps species to evolve since it guarantees that every individual will be genetically unique.

Later in his career, Köhlreuter continued his work with plants by showing that plants could be artificially fertilized and by creating fertile hybrids between different species of plants. Both of these were important discoveries, and both are in common use in modern agriculture. In fact, some commercial crops are fertilized almost exclusively artificially, and hybridization is one of the more important tools used by horticulturists and plant breeders to develop plants with a combination of traits to make them more nutritious, hardier, more aesthetically pleasing, and so forth.

Köhlreuter was also the first to realize the importance of insects and the wind in helping plants pollinate. This realization followed almost directly his understanding of the existence of sexes in plants, and led him to the realization that plant pollen carried the genetic information from a male plant to the female. All of this helped to not only increase knowledge about genetics, but it also helped farmers and horticulturists develop new plants in a scientific manner, rather than by simple hit-or-miss. Also, by realizing how pollination normally takes place, Köhlreuter was able to use this knowledge to help make more consistent hybrids. For example, by placing a bag over a particular plant, he could keep it from being fertilized by insects or the wind. He could then collect pollen from a specific plant with desirable traits and spread this pollen on the plant he had isolated. The seeds, then, would have only the pollen from a known plant with known characteristics, and the plants they produced could be studied. By doing this with many plants over many years, he was able to better determine how certain traits were transferred from plant to plant, as well as selectively breeding plants with specific desired traits. This same method of plant breeding continues to be used today, with a high degree of success.

P. ANDREW KARAM

James Lind
1716-1794
Scottish Naval Surgeon

James Lind is associated with the elimination of scurvy, proving through an early clinical trial that citrus fruits were immediately effective in curing the symptoms of this disease, whereas other common remedies were not. Following this trial aboard the *Salisbury* in 1747, he published a definitive study of scurvy in 1753, *A Treatise of the Scurvy. Containing an inquiry into the Nature, Causes and Cure of the Disease. Together with a Critical and Chronological View of what has been published on the subject.*

James Lind was born in 1716 in Edinburgh, Scotland. His father, for whom he was named, was a well-to-do merchant. Lind received training in both Latin and Greek as a youth and was apprenticed to a local physician at age 15. Little else is known of Lind's youth or early training until he joined the British Navy as a surgeon's mate in 1739.

Lind began his work in the navy at a time when long-distance sea voyages were increasingly common, with scurvy a severe problem. Scurvy is a deficiency disease, caused by a lack of vitamin C in the diet. Shipboard diets were notoriously poor, generally consisting of salt pork or beef, hard biscuit, and beer supplemented with minimal amounts of cheese, butter, and dried fish. Fruits and vegetables, which provide most vitamin C, could not be kept on board and therefore were only sporadically available at ports of call.

Just after Lind joined the navy, Lord Anson headed an around-the-world voyage (1740-1744) that saw 977 men out of an original 1,955 die of scurvy. A few years after that voyage, Lind was witness to two outbreaks of the disease while serving aboard the *Salisbury*. The second outbreak, in 1747, prompted Lind to formulate an experiment to discover what remedies might cure the disease the quickest. Selecting 12 men under his charge, each of whom was exhibiting similar symptoms of scurvy, Lind divided the men into six groups. Maintaining each group in conditions as similar as possible, he changed only their diets, administering a different daily remedy to each group. Those men who showed the most immediate response and recovery had been given oranges and lemons, thus demonstrating their antiscorbutic properties. The remedy that showed the second-most promise was hard cider. The remaining remedies, which included salt water, vinegar, elixir vitriol (sulfuric acid), and a compound remedy made up of garlic, mustard seed, and other botanicals, were all shown to be ineffective.

Lind served in the navy for nearly 10 years, leaving in 1748 to return to Edinburgh. He received his doctorate in medicine from the University of Edinburgh the same year and became a fellow of the Royal College of Physicians. Practicing medicine in Edinburgh for the next 10 years, Lind also found time to write two treatises. The first was his volume on scurvy, *A Treatise of the Scurvy. Containing an inquiry into the Nature, Causes and Cure of the Disease. Together with a Critical and Chronological View of what has been published on the subject,* published in 1753 and dedicated to Lord Anson. The second was a related work, entitled *An Essay on Preserving the Health of Seamen in the Royal Navy,* published in 1757. Both of these treatises would become popular, undergoing several reprints in his lifetime. A final treatise, *Essay on Diseases Incidental to Europeans in Hot Climates,* was completed while Lind served as physician at the Royal Naval Hospital at Haslar, a position he received in 1758. Lind remained at Haslar until his resignation in 1783. He died in 1794.

KRISTY WILSON BOWERS

Carolus Linnaeus
1707-1778
Swedish Physician and Botanist

Carolus Linnaeus established the system of binomial nomenclature and a taxonomical hierarchy in the 1700s. Roundly acclaimed as the first successful attempt to classify and name living things, the system continues as the taxonomical system used today. Under this system, animals and plants have unique and universally accepted two-part names: the generic (or genus) name and the specific (or species) name. Linnaeus also designated higher levels of organization, such as classes and orders, to further organize the animals and plants into groupings that are useful for identification and study purposes.

Linnaeus was born Carl von Linné in Råshult, Sweden, on May 23, 1707. He became interested in botany at an early age. His father, Nils Linné (originally Nils Ingemarsson), had hoped his son would follow in his professional footsteps and become a pastor, but instead the young man showed more curiosity toward his father's hobby of gardening. Never an outstanding student, Linnaeus continued with his studies, eventually entering medical school at the University of Lund in 1727, but transferring a year later to the University of Uppsala, where he received encouragement to pursue his interests in botany.

By 1730, Linnaeus took an appointment as lecturer at Uppsala, and in 1732 went to Lapland to conduct a survey of its Arctic plants and animals. He wrote detailed accounts during his five-month survey and followed it with the publication of *Flora Lapponica* in 1737.

After his Lapland survey, Linnaeus accepted invitations to present lectures and to conduct other surveys. During this time, he also began work on a classification system for plants. At first he sorted and named the plants by the numbers of their floral parts, but soon began giving each two names: the first for the plant's genus and the second for its species. The genus name allowed him to indicate which plants were very closely related, and the species name gave each plant its own identity. Under this system, for example, the common butterwort has the name *Pinguicula vulgaris*. The genus name is *Pinguicula*, which derives from the Latin word for "somewhat fatty" and refers to the greasy feel of the leaves. The species name of *vulgaris* indicates that the plant is considered a common organism.

Linnaeus published a short work, called *Systema naturae*, on his binomial classification system in 1735, the same year he received a medical degree from the University of Harderwijk in Holland. At that time, scientists often sought the medical degree as a professional title and indication of societal status rather than as a means solely to practice medicine. To obtain the degree, Linnaeus spent just enough time at the university to pass examinations, and present and defend his thesis.

Within four years, he moved to Stockholm and married Sara Moraea, whom he had met while on a survey in 1734. In 1741, his life took another turn when he accepted a position as professor of botany at Uppsala. Nearly 190 of his students went on to defend dissertations. In addition to his duties as Uppsala, he accepted appointment as chief royal physician in 1747 and knighthood in 1758.

He also continued to refine his classification system for living things, and in 1758 published his most influential work, the 10th edition of *Systema naturae*. Although many zoologists had begun to employ his classification system, it was only in the twentieth century that they agreed to name it and the publication the official origination of scientific nomenclature for animals. Botanists did the same with Linnaeus' 1753 publication of *Species plantarum*. With these declarations, all namings of plants and animals became official beginning in 1753 and 1758, respectively. Linnaeus named many plants and animals in the two publications, and those names remain today.

Linnaeus, who reverted to his original name of Carl von Linné after the publication of *Systema naturae*, continued his work until his retirement in 1776. He survived a stroke in 1774, but a second in 1778 led to his death on January 10, 1778. He is buried in Uppsala. To honor his scientific contributions, the Linnaean Society in London was founded in 1788.

LESLIE A. MERTZ

Luigi Ferdinando Marsili
1658-1730
Italian Aristocrat, Scientist and Soldier

Marsili was an Italian aristocrat and soldier who was also a scientist with a wide range of interests from astronomy to geology. His contributions include writing a dissertation on marine biology as well as founding Academia delle Scienze dell'Instituto di Bologna or the Institute of Sciences and Arts.

Although Marsili was from a noble family, he did not complete his formal education. However, he did study informally with Marcello Malpighi (1628-1694), an Italian biologist, and other teachers to acquire basic knowledge of history, science, and mathematics.

As a young man in Rome, Marsili made powerful allies. Queen Christina of Sweden and Cardinal de Luca employed him with a diplomatic mission to Venice. It was the Queen to whom he would later dedicate his first work, and it was the cardinal who would later recommend him to Emperor Leopold I.

Marsili spent 22 years in the Emperor's army beginning in 1682. His role was that of an advisor on fortifications in Turkey as well as cartographer, engineer, and diplomat. He was second-in-command at his post when his service ended in dishonor due to his involvement in the surrender of the fortress of Briesach. After Marsili's discharge, upon his return to Bologna, the King of Spain restored his right to wear a sword. Marsili would then go on, in 1708, to head the papal army during the War of the Spanish Succession.

It was not until retirement that his dedication to science truly began. However, even during his time in the military, Marsili pursued his scientific interests. He made astronomical observations, studied rivers, and even collected specimens of fossils and indigenous fauna.

In 1712, he presented his collection to the Senate of Bologna, where Accademia delle Scienze dell'Instituto di Bologna was founded. Although the Senate supported Marsili's proposal for the new institute, it did not come up with the necessary capital to back his idea. Discouraged by the lack of support, he came very close to leaving Bologna with his collection. However, the Senate encouraged him to stay, which he did. Marsili found funding from the pope instead and the institute was born. This "Institute of Sciences and Arts" wasn't formally opened until 1715. Six professors were employed to head the different divisions.

Under Marsili's leadership, the institute became a center of scientific research, focusing mainly on natural history exploration around Bologna. The institute flourished and eventually absorbed Academia degli Inquieti. Marsili's dedication to his creation was so strong, he bequeathed his collection and even his home to the city.

In 1725, Marsili wrote *Histoire physique de la mer*, the world's first dissertation on oceanography. It was published by the Dutch East India Company. This influential work examined every imaginable aspect of the sea from geological formations to biological phenomena. Marsili explored the formation of basins as well as indigenous plants and fish. He dedicated the work to the Academie Royale.

Marsili also wrote *Danubius Pannonico-mysicus, observationibus* in 1726. This work explored one of Europe's greatest rivers and its surroundings. He made use of his military training as a cartographer here. This work includes 20 maps. Additional works are *Osservanzioni interne al Bosforo Tracio* (Rome, 1681) and *L'Etat militaire de l'empire ottoman* (Amsterdam, 1732).

Aside from his work as a scientist, Marsili was also a member of several scientific societies, including his induction into the Académie des Sciences in 1715. In 1722, upon being made a member of the Royal Society of London, Marsili was honored with a personal introduction by Sir Isaac Newton (1642-1727), who praised him as a famous scientist and founder of the Academy of Bologna.

AMY MARQUIS

Franz Anton Mesmer
1733-1815
Swiss Physician

The Swiss physician Franz Anton Mesmer was a well-known figure in the late eighteenth and early nineteenth centuries. The term mesmerized is derived from his name.

Franz Anton Mesmer was born in 1734 in the town of Iznang on the shore of Lake Constance in Switzerland. In 1766 Mesmer passed his medical examinations, completing a dissertation entitled *De planetarum influxu* (Physical-Medical Treatise On the Influence of the Planets), in which he first presented his theories on animal magnetism, a force which, according to him, existed between and within all bodies, and which Mesmer believed could be manipulated to cure disease.

Upon his graduation, Mesmer relocated to Vienna, where he was known as a patron of the arts, sponsoring a performance of an opera by the young Mozart in his garden. Following the lead of a local Jesuit, Father Maximillian Hell, Mesmer began to use magnets in the treatment of patients, a hallmark of his early practice and theory. He enjoyed a growing reputation among his wealthy patients, and was called in by the Elector of Bavaria to make a judgment in the case of Father Gassner, a priest who claimed he could cure disease by exorcism. In one of the few victories of his career, Mesmer stated that while the Father could cure disease, it was by the "scientific" manipulation of forces, not by casting out demons. In 1777, following this success, Mesmer was invited to consult in the case of Maria Theresa Paradis, a musical prodigy who had gone blind as a child. While the girl seemed to regain some of her sight while under Mesmer's care, various factors, including the doubts of his colleagues regarding Mesmer's medical ability, as well as the propriety of his relationship with Miss Paradis, eventually led to the young woman being withdrawn from Mesmer's care. Within months of beginning his treatment of her, Mesmer was ordered by a government commission investigating his methods to leave the city within 48 hours.

After this, Mesmer settled in Paris in 1778, where he enjoyed a tremendous popular reputation, but also suffered considerable scorn from the medical community. He taught courses, for a fee, on his theories and methods, and it was one of his students, the Marquis de Puysegur, who actually discovered what we refer to as "hypnotism" today. In 1794 a Royal Commission issued a ruling that the practice of mesmerism or animal magnetism was a danger to the public health, and these practices were banned in France. Much of this furor, however, resulted from the perceived lack of propriety between Mesmer and his female patients.

Following this ultimate condemnation of his theories and practices, Mesmer returned to his native Switzerland. In 1799 he published a final book, his *Memoire*, in which he presented his theories and hopes for an utopian world. This work enjoyed some popularity among German romantic writers, and Mesmer's theories lived on in popular culture even after his death in 1815. References to mesmerism can be found throughout American and European literature well into the nineteenth century, where it was often used as a device to explore the workings of the human mind and spirit. The use of magnets in medical practice still continues today, though much of the theory underlying their use has been discredited.

PHIL GOCHENOUR

Giovanni Battista Morgagni
1682-1771
Italian Physician and Anatomist

Giovanni Battista Morgagni holds a highly respected place in the history of medicine as the anatomist who established the science of morbid anatomy. Morgagni's anatomical and pathological research helped to replace the ancient humoral doctrine with a new approach to viewing disease.

Born in Forli, Italy, Morgagni studied medicine and philosophy at the University of Bologna. He graduated in 1701 and assisted Antonio Maria Valsalva with the publication of *Anatomy and Diseases of the Ear* (1704). When Valsalva took a position in Parma, Morgagni became demonstrator of anatomy at Bologna. Morgagni became professor of medicine at the University of Padua in 1710. Five years later, he was promoted to the chair of anatomy.

From 1706 to 1719, Morgagni published his anatomical observations under the title *Adversaria Anatomica.* These studies were highly regarded and widely praised by anatomists and physicians. Morgagni's remarkable pioneering study of morbid anatomy, *On the Seats and Causes of Diseases Investigated by Anatomy,* was published in 1761, when Morgagni was 80 years old. This massive five-volume landmark in the history of pathology is based on almost 700 cases. In his comments on each case, Morgagni attempted to correlate clinical observations of the signs and symptoms of disease with his postmortem investigations of specific lesions.

Although anatomy is an ancient and valued aspect of medical research, Morgagni's treatise is generally considered the first systematic textbook of morbid anatomy. His approach located diseases within individual organs rather than in disturbances of the humors. Meticulous observations on hundreds of patients before and after death allowed Morgagni to establish the hidden internal changes associated with the progression of diseases. Morgagni also carried out experiments and dissections on various animals in order to understand pathological processes in humans. He believed that systematic postmortem studies would determine the structural and functional changes associated with disease. Morgagni thought that such knowledge would eventually help physicians establish the causes of diseases and allow them to evaluate the efficacy of different remedies and therapeutic interventions.

Morgagni thought that normal human anatomy had been well established by his predecessors and contemporaries, but other anatomists had not yet explored the origin and seat of the diseases that caused the pathological changes observable in the cadaver at the postmortem autopsy. All parts of the body were described in some detail, but the longest section in Morgagni's treatise deals with disorders of the belly. After carefully describing each case history, Morgagni attempted to correlate observations of the symptoms of the illness with his findings at autopsy. Sometimes the autopsy revealed serious errors in diagnosis and inappropriate treatments that might have been directly related to the death of the patient. In one case, a physician had treated the patient for stomach problems, but the autopsy revealed that the patient had a normal stomach and diseased kidneys.

As a pioneer of morbid anatomy, Morgagni initiated a new epoch in medical science. Even though Morgagni did not directly challenge prevailing medical philosophy and essentially remained a humoralist in terms of clinical medicine, his work inevitably marked a departure from general humoral pathology towards the study of localized lesions and diseased organs. The new morbid anatomy encouraged physicians to think of disease in terms of localized pathological lesions rather than disorders of the humors. Thus, Morgagni's work encouraged a new attitude towards specific diagnostic categories and the efficacy of surgical interventions.

Morgagni's work helped to establish an anatomical orientation in pathology and a recognition that unseen anatomical changes in the living body were reflected in the clinical picture. He was the first to attempt a systematic examination of the connection between the symptoms of disease in the living body and postmortem results revealed only to the dedicated investigator. Confirmation of a diagnosis could only be found in the autopsy room, but recognition of the relationship between symptoms and internal lesions encouraged an interest in finding ways of "anatomizing" the living, that is, detecting hidden anatomical lesions in living patients.

Morgagni was honored and respected by students, colleagues, cardinals, and popes. When only 24 years of age, he was elected president of the Accademia Inquietorum. He was elected to many prestigious scientific societies, including the Accademia Naturae Curiosorum, the Royal Society, the Academy of Sciences of Paris, the Imperial Academy of St. Petersburg, and the Berlin Academy.

LOIS N. MAGNER

John Turberville Needham
1713-1781
English Naturalist

The English naturalist John Needham conducted a series of experiments that seemed to provide proof of spontaneous generation—the sudden appearance of organisms from nonliving ma-

terials. His work spurred that of Italian scientist Lazzaro Spallanzani (1729-1799), who conducted similar experiments, but had opposite results.

Needham was born in London in 1713. He left England in order to receive the education required for the Roman Catholic priesthood, which he completed in 1738. (Such schooling would have been difficult to obtain in England, which was under Protestant rule after a period of religious turmoil.) Rather than serve as a priest, however, Needham spent much of his life as a tutor to young English Catholics as they toured the European continent.

Needham had read about recent discoveries that had been made with microscopes, including the discovery of "animalcules" (which are now called microorganisms). He became fascinated with microscopy, the use of microscopes to make scientific observations. In 1745, he published *An Account of Some New Microscopical Discoveries*. This work included his observations of different types of pollen.

While studying microscopy in Paris, Needham became acquainted with Georges Buffon (1707-1788), a French naturalist who developed early ideas related to evolution. Buffon introduced Needham to some of the ideas of the German philosopher and mathematician Gottfried Wilhelm Leibniz (1646-1716). Leibniz had proposed the existence of living molecules, which he called monads. Needham not only accepted this idea, but he further believed that when organisms died and decayed, their individual molecules continued to live and could join together to form new living matter. He believed that a force, which he called the "vegetative force," brought these molecules together, much like oppositely charged atoms will be drawn together.

By Needham's time, it was generally accepted that animals could not form by spontaneous generation. However, the idea of spontaneous generation came into fashion with regard to "animalcules." Some naturalists did not believe that such tiny organisms would be able to produce even tinier offspring. They suggested that these organisms must therefore form spontaneously. Needham and Buffon were supporters of this view.

At Buffon's urging, Needham began a series of experiments in 1748 to test this hypothesis. He boiled broth containing meat and grain, presuming that the heat would kill any microorganisms that happened to be present. Then he poured the broth into glass containers and sealed them with corks. After several days, he opened the containers and looked at their contents under a microscope. He observed numerous microorganisms in the broth and concluded that they had arisen spontaneously from the nonliving material inside the containers. In 1750, he published a paper on the results of his findings.

Twenty years later, Lazzaro Spallanzani repeated Needham's experiments. He proposed that microorganisms had appeared in Needham's solutions because the containers were contaminated or because the broth had not been boiled long enough to be completely sterilized. In fact, Spallanzani found that some types of microorganisms could survive boiling for more than an hour. (Needham had boiled his broth for only a few minutes.) In addition, Spallanzani believed that Needham's containers were not necessarily airtight because they had been sealed with corks. To prevent this from happening, Spallanzani heated the ends of his containers and then pinched the softened glass together to form an airtight seal. No microorganisms appeared in Spallanzani's sealed vessels, and he concluded that spontaneous generation does not occur.

When Needham heard of these experiments, he repeated them as they were described in Spallanzani's published report. He quickly pointed out what he claimed were flaws in Spallanzani's procedures. Needham claimed that an hour of boiling would damage the vegetative force in the broth and prevent spontaneous generation from occurring. In addition, Needham heard air rushing into the sealed containers when he opened them. He claimed that this showed that the air inside the flasks had been damaged in some manner. (The rush of air was actually due to the heating of the glass as the seal was made.)

To counter Needham's arguments, Spallanzani conducted additional experiments. To show that the vegetative force (if it existed) was not damaged, he boiled broth for an hour and then left it exposed to the air. When he examined it a few days later, microorganisms had appeared. This showed that the broth was still capable of supporting life. To show that the air in the containers was not damaged, he designed a special flask with a long skinny neck that could be sealed very quickly. When he opened this container, no air rushed in. Therefore, the air could not have been damaged as Needham argued. Despite Spallanzani's experiments, Needham (and many other scientists) remained firmly convinced in the spontaneous generation of microorganisms. It was not until Louis Pasteur's (1822-1895) experiments in the nineteenth century that the question was finally settled.

In addition to his scientific work, Needham was also well-known for his writings on religion. In 1768, he was elected to the Royal Society, the oldest scientific society in Britain. In the same year, he moved to Brussels, Belgium, where he became the director of the Academy of Sciences. He died in Brussels in 1781.

STACEY R. MURRAY

Percivall Pott
1714-1788
English Physician and Surgeon

Percivall Pott was one of eighteenth-century's preeminent physicians and surgeons. In a time which predated surgical specialization, he made historic contributions to orthopedics, urology, neurosurgery, and oncology. No less than three disorders bear his name as a result of his precise and accurate clinical descriptions: Pott's disease, Pott's fracture, and Pott's puffy tumor.

Born in London in 1714 to a woman who was twice widowed, Pott nonetheless obtained a privileged education, developing a taste for classical knowledge and literature. Apprenticed at 16 to Edward Nourse, surgeon at London's St. Bartholomew's Hospital, he prepared Nourse's dissections, gaining invaluable training in anatomy. In 1736, at age 22, Pott was admitted to the Company of Barber Surgeons and established his own practice. At age 31 he was elected Assistant Surgeon at St. Bartholomew's, and four years later attained full surgeon, a post he held until 1787.

Pott discovered his propensity for writing as a result of an accident. Thrown from his horse in 1756, he suffered compound fractures of the tibia and fibula. He refused to be moved until a door had been purchased and two men procured to carry him home on an improvised stretcher. His colleagues advised immediate amputation, but as they were about to begin, Pott's old teacher Nourse arrived and saved the limb. During the long recovery Pott occupied himself by writing "A Treatise on Ruptures," an essay on hernia, considered by many to be his finest work.

In his 1760 and 1768 works on head injuries, Pott described the "puffy tumour," or the swelling of the scalp over the affected area, later named after him. He was one of the earliest to note the importance of neurological symptoms as markers of brain injury in the absence of external lacerations.

In 1768 Pott published his treatise on fractures and dislocations, providing the first definitive description of the various forms of ankle fractures and their attendant soft tissue injuries. Thus "Pott's fracture" is a generic term rather than the definition of a single fracture.

Because of Pott's observation in 1775 that cancer of the scrotum in chimney sweepers was caused by long-term exposure to soot, he is regarded as a pioneer in the fields of chemical carcinogenesis and occupational medicine.

In 1779 another of his significant works dealt with spinal curvature resulting in paralysis of the lower limbs. Although he did not recognize tuberculosis as the cause of the collapsed vertebrae, he was the first to give a complete clinical picture of what is now known as "Pott's disease."

Pott became a prosperous and sought-after surgeon, numbering among his patients David Garrick (1717-1779), Thomas Gainsborough (1727-1788), and Samuel Johnson (1709-1784). He was also a generous mentor and clinical teacher; one of his students was the brilliant surgeon John Hunter (1728-1793). Upon his retirement from St. Bartholomew's after 50 years, Pott continued to practice. Journeying in severe weather to see a patient, he caught a chill, developed pneumonia, and died in 1788.

In our highly specialized era it is hard to imagine how one surgeon could have excelled in so many areas. His books went through many editions and were translated into several European languages, underscoring his influence and position in eighteenth-century surgery.

His son-in-law, Sir James Earle, collected and published all of Pott's known works in *The Chirurgical Works of Percivall Pott, Including a Short Account of the Life of the Author* (1790). His writings are marked by precision and discipline carried over from his surgical training, and are models of clarity and elegance. His genuine concern for the welfare of patients and his insistence on skillful technique and good judgment, at a time when speed was the usual benchmark, advanced a more rational and humane approach to surgery.

DIANE K. HAWKINS

René Antoine Ferchault de Réaumur
1683-1757
French Naturalist, Physiologist and Physicist

French contemporaries referred to René de Réaumur as the "Pliny of the Eighteenth Century," and later authorities compared him to the English natural philosopher Francis Bacon (1561-1626). His obscurity today may be due to the fact that he considered everything scientific as having practical applications.

This French scientist was born at La Rochelle, France. His father died before René was two, and though little is known of his early education, it is thought that he received his early training from Oratorians or Jesuit priests at La Rochelle and Poitiers. An uncle then directed him to study law at Bourges, and he was later enrolled at the University of Paris. He early demonstrated an unusual aptitude for mathematics. In 1708, through the good offices of the mathematician Pierre Varignon, Réaumur was elected a "student geometer" to the French Academy of Sciences at the unusually early age of 25. By 1709, however, he had turned his attention to technology, natural history, and other subjects. In 1711 he was elected *pensionnaire mécanicien* of the academy. In 1713 he was selected to write a French industrial compendium, and he spent 10 years researching and writing memoirs concerning the fabrication of iron, steel, tin, and porcelain. His research led to the publication, among other works, of *L' art d'convertir le fer forgé en acier* (1722).

In 1731 Réaumur developed a new thermometer, attempting to resolve certain of the shortcomings of the ones previously invented by Gabriel Daniel Fahrenheit (1686-1736) and others. Unfortunately, many Réaumur thermometers were not correctly made according to his specifications, obliging modern scientists to use eighteenth-century temperature figures arrived at with his thermometers with care.

In the field of natural history he began with studies of the scales of fish and the manner of growth in the shells of bivalves. He also devoted considerable attention to the subject of regeneration in worms and crustaceans, and methods of locomotion in marine organisms, including starfish and mollusks. Between 1734 and 1742 he published six sections of his *Mémoires pour servir à l'histoire des insectes*, but this important work, particularly the section concerning ants, was still not completed when he died. A portion of what he had written about ants was finally published in an edition compiled by the American entomologist William Morton Wheeler (1865-1937) in 1926. His study of insects led him to develop a new system of classification based on an animal's behavior, rather than on morphological characteristics, as has been the practice since his day. It was he who discovered the role of the queen in bee colonies, and he did some of the earliest research into the ways in which bees communicate.

Réaumur was also interested in the artificial incubation of domestic birds' eggs, and he further carried out much research dealing with the digestive processes of birds, particularly domestic fowl and birds of prey. These efforts resulted in several published studies, culminating in *Sur la digestion des oiseaux*, which was printed in two parts the year before his death in 1757. Over a period of nearly half a century, Réaumur was elected a director of the French Academy on a dozen occasions and subdirector nine times, and he contributed 74 papers to its *Mémoires* (proceedings). He was also elected to at least five other European academies of science. He purchased several honorary posts in the Royal Military Order of St. Louis, which brought with them social standing that was the equivalent of a count. In 1755 he inherited the castle of La Bermondière and noble rank in the French province of Maine. His death at age 74 was the result of a fall from his horse.

KEIR B. STERLING

Benjamin Rush
1745-1813
American Physician, Politician and Medical Educator

Benjamin Rush was the most prominent American physician of his day. He was the first professor of medical chemistry in America, a signer of the Declaration of Independence, a hero of the yellow fever epidemics of the 1790s, and the founder of American psychiatry.

Rush was born on December 24, 1745 (Old Style) or January 4, 1746 (New Style) in Byberry, Pennsylvania. He prepared for college at his maternal uncle Rev. Samuel Finley's academy in West Nottingham, Pennsylvania, then graduated from the College of New Jersey, later called Princeton University, in 1760. For the next six years he was the medical apprentice of John Redman, a prominent Philadelphia physician. He then studied under Alexander Monro *secundus* (1733-1817), John Gregory, John Hope, Joseph Black (1728-1799), and William Cullen (1710-1790) at Edinburgh, where he received his M.D. in 1768.

Benjamin Rush. *(Corbis Corporation. Reproduced with permission.)*

In 1769 he began teaching chemistry as the fourth faculty member of the Medical Department of the College of Philadelphia, later called the School of Medicine of the University of Pennsylvania, the first medical school in America.

An ardent American patriot, Rush was Surgeon to the Pennsylvania Navy from 1775 to 1776, a member of the Continental Congress from 1776 to 1777, and Physician-General of the Military Hospitals of the Middle Department of the American Army from 1777 to 1778. He helped Thomas Paine publish *Common Sense*. In January 1776 he married Julia Stockton and six months later both he and his father-in-law, Richard Stockton, signed the Declaration of Independence.

From 1775 to 1777 Morgan, and from 1777 to 1781 Shippen, served successively as Director-General of Military Hospitals and Physician-in-Chief of the American Army. Throughout the war Rush plotted against both of them and alternately helped each to plot against the other.

In the wake of Carolus Linnaeus (1707-1778), many European and American intellectuals became obsessed with classification. Cullen's *Synopsis nosologiae methodicae* [Synopsis of Methodical Nosology] (1769) attempted to classify all diseases and regarded each disease as a separate entity to be treated with its own specific cure. Rush led other medical theorists in reaction against nosology. He criticized Cullen for treating the "name" of the disease, not its "proximate cause" or symptoms. He believed that there was only one disease, fever, existing in many manifestations and indicating only one set of cures: phlebotomy (bloodletting), emetics, purgatives, radical depletion, modifications of diet, lowering of room temperature, and drinking great quantities of water. These prescriptions would be adjusted for the qualities and severity of each manifestation.

A long debate arose. Confrontations between nosologists and "single-disease-single-cure" theorists were derisive, unrestrained, and occasionally even brutal. Alexander May, one of Rush's students at the University of Pennsylvania, wrote his medical dissertation in 1800 as a vicious personal attack on Cullen and his followers; but Rush himself, even though he strenuously opposed Cullen, never lost personal respect for him.

In 1791 Rush relinquished his professorship of chemistry to become professor of clinical medicine. As a clinician he was a follower of Thomas Sydenham (1624-1689), whose observations were keen but whose cures were often repugnant. Rush was one of twelve founding fellows of the College of Physicians of Philadelphia in 1787, but resigned in 1793 because the majority of fellows disapproved of bloodletting, mercurial purges, and other extreme methods Rush used to treat the yellow fever that was then devastating the city. Along with his letter of resignation to his old teacher Redman, President of the College, he sent a copy of Sydenham's works.

The epidemic resulted in his masterpiece, *An Account of the Bilious Remitting Yellow Fever as it Appeared in the City of Philadelphia in the Year 1793* (1794).

Toward the end of his life Rush turned his attention toward what is now called psychiatry. His *Medical Inquiries and Observations upon the Diseases of the Mind* (1812) was the first American book in the field.

Always politically active, he was a delegate to the Pennsylvania convention that adopted the federal Constitution in 1787 and from 1799 to 1813 he served as Treasurer of the United States Mint.

In many ways Rush remains an enigma. He was hot-tempered and sometimes bitter, yet he accepted Shippen's forgiveness and eulogized

Shippen when he died. He encouraged scientific research, yet he clung stubbornly to prescientific dogmas of bloodletting and violent purges. He often felt depressed, defeated, ready to give up,

ONE DISEASE OR MANY?

In *Sepulchretum, sive anatomia practica ex cadaveribus morbo denatis* (1679), Théophile Bonet included the first systematic attempt to classify all diseases. The usual term for disease classification, "nosology" (*nosologia*), was coined in the 1720s. Nosology grew in popularity throughout the eighteenth century. It dominated medical philosophy and medical research from the 1770s until the 1820s. The most influential nosologist was William Cullen, whose *Synopsis nosologiae methodicae* (1769) defined the field for 30 years. Other prominent nosologists were François Boissier de la Croix de Sauvages, Erasmus Darwin, Philippe Pinel, John Mason Good, David Hosack, and Thomas Young.

Because nosologists believed that every disease was a distinct "thing," they treated the relation between nosology and therapeutics as a metaphysical question; and because they believed that the accurate classification of diseases would suggest specific cures, they also treated it as a clinical question. They considered nosology primarily practical rather than theoretical. Against nosological medicine, the philosopher Immanuel Kant is supposed to have said, "Physicians think they do a lot for a patient when they give his disease a name." Clinicians no longer believe that the classification of diseases has anything to do with curing them.

Most physicians of the late eighteenth century accepted some form of nosological theory, but Benjamin Rush, one of Cullen's former students, inspired a large minority to assert against Cullen that there is only one disease, appearing as many different kinds of symptoms. Another of Cullen's former students who disagreed with him was the notorious John Brown, for whom all diseases were either "sthenic," i.e., marked by over-excitement, or "asthenic," i.e., marked by under-excitement. Whichever pole characterized the disease, the patient's body had to be moved in the other direction. Prescriptions generally included some combination of alcohol and opium (which were also Brown's personal drugs of choice).

ERIC V.D. LUFT

and worried that he would die young, yet he strived tirelessly in both politics and medicine and produced a tremendous amount of work.

ERIC V.D. LUFT

William Shippen Jr.
1736-1808

American Physician, Surgeon and Medical Educator

Shippen was the co-founder of the first medical school in America and the first professor of anatomy, surgery, and obstetrics in America.

The son of a prominent Philadelphia physician, William Shippen Jr. ("Billey") attended Rev. Samuel Finley's boarding school in West Nottingham, Pennsylvania, then the College of New Jersey, later called Princeton University, where he was valedictorian of the class of 1754. He was apprenticed to his father from 1754 until 1758, when he went to Great Britain to continue his medical training. In London until 1760, he studied anatomy under John (1728-1793) and William Hunter (1718-1783), obstetrics under Colin Mackenzie, and clinical medicine and surgery at St. Thomas's Hospital. Thereafter in Edinburgh he studied under William Cullen (1710-1790) and Alexander Monro *secundus* (1733-1817). He received an Edinburgh M.D. in September 1761. Before he returned to Philadelphia in 1762, he married Alice Lee, an aristocratic Virginian whom he had met in England.

In November 1762 the Shippens, father and son, began offering private anatomical lectures, but Billey was the prime mover. In 1765 he began offering private obstetrical lectures as well. During this period he was active in the development of the Pennsylvania Hospital, the first hospital in America, founded in 1751 by Benjamin Franklin and Thomas Bond. He helped to persuade Franklin's friend, London physician John Fothergill (1712-1780), to donate teaching materials to the hospital. Shippen used Fothergill's gifts in his anatomical lectures.

With John Morgan (1735-1789), a schoolmate from Finley's who had received his Edinburgh M.D. in 1763, Shippen co-founded in 1765 the first medical school in the Western Hemisphere, the Medical Department of the College of Philadelphia, later called the School of Medicine of the University of Pennsylvania. Morgan taught the theory and practice of physic while Shippen taught anatomy and surgery. The third professor was Adam Kuhn (1741-1817), who began teaching *materia medica* and medical botany in 1768. The fourth was Benjamin Rush (1745-1813), who taught chemistry starting in 1769.

The interaction of Shippen, Morgan, and Rush during the American Revolution was a tangled mess. The first Director-General of Military Hospitals and Physician-in-Chief of the American Army was Benjamin Church. This position later evolved into that of Surgeon-General. In November 1775 Church was court-martialed for treason and convicted. The Continental Congress appointed Morgan to replace him. Among his subordinates was Shippen. Among his supervisors was Rush, in his capacity as a member of the Continental Congress and chair of its medical committee. Partially because of Shippen's and Rush's complaints about his leadership, the Continental Congress summarily dismissed Morgan in January 1777. Morgan spent the next two years defending himself and received full exoneration in 1779.

In April 1777 the Continental Congress appointed Shippen as the third Director-General and Physician-in-Chief. Among his subordinates was Rush. Shippen lacked administrative ability, but his father had influence and was a member of the Continental Congress from 1778 to 1780. Rush accused Shippen of mismanaging hospitals, misappropriating supplies, neglecting patients, and running a slipshod department. Disgusted, Rush resigned his military commission in 1778 but continued his intrigues against Shippen. Rush and Morgan succeeded in having Shippen court-martialed in 1780. He was acquitted, but resigned in 1781 and returned to teaching.

After the war Shippen, Rush, and Morgan were all again members of the same medical faculty at the University of Pennsylvania. How they managed to co-exist harmoniously within a single educational department remains a mystery to historians, but the fact that they did is testimony to their professionalism. Shippen on his deathbed forgave Rush, and Rush presented a eulogy of Shippen to his medical students.

Along with Abraham Chovet, Gerardus Clarkson, Samuel Duffield, George Glentworth, James Hutchinson, John Jones, Kuhn, Morgan, Thomas Parke, John Redman, and Rush, Shippen was one of the 12 founding fellows of the College of Physicians of Philadelphia in January 1787.

ERIC V.D. LUFT

Lazzaro Spallanzani
1729-1799

Italian Physiologist

Lazzaro Spallanzani was an Italian physiologist who extensively studied animal biology and reproduction. He is probably most famous for his experiments that helped to disprove the theory of spontaneous generation, which helped to pave the way for future research by Louis Pasteur (1822-1895). Spallanzani was a creative and endlessly inquisitive researcher who studied subjects in biology as varied as sexual reproduction, blood pressure and echolocation in bats. He is also well known for his forays into other areas of the physical sciences. For instance, he studied lava flows inside an active volcano.

Spallanzani was born in 1729 in Scandiano. A son of a distinguished lawyer, Spallanzani was interested in science at an early age. He was given the nickname, "the astrologer" after he showed an early penchant for astronomy. At the age of fifteen, he attended a Jesuit seminary called Reggio Emilia. Spallanzani declined to join the order (but was eventually ordained) and went to the University of Bologna to study law. However, it turned out to be the natural world that most intrigued Spallanzani, so he began to exclusively pursue that area. He was granted his doctorate in 1754 and returned to the seminary to teach. In 1760 he became professor of physics at the University of Modena.

Although Spallanzani published an article critical of a new translation of the *Iliad* in 1760, he was a tireless scientific researcher. In 1766 he published a monograph on the mechanics of stones that bounce when thrown obliquely across water. His first published biological work was in 1767. It was a detailed description of hundreds of experiments that refuted the popular idea of spontaneous generation.

The theory of spontaneous generation asserted that living things could come into being without a living predecessor. Georges Buffon (1707-1788) and John Needham (1713-1781) largely championed these theories. They believed that all living things contain, in addition to inanimate matter, special "vital atoms" that are responsible for all physiological activities. After death, these "vital atoms" would escape into the soil and would be taken up by plants. The two men claimed that the small moving objects seen in pond water are not living organisms but merely "vital atoms" escaping from the organic material. Spallanzani designed elegant experiments that helped to support his theory that these were in fact small living microorganisms. In his most famous experiment, Spallanzani showed that a sealed container of boiled broth would not have any microorganisms present, while those that were left unsealed or at room temperature would have evidence of living creatures. He reasoned that if spontaneous generation really took place, then all flasks should have evidence of infestation. These experiments were also significant because they were the basic steps that Louis Pasteur initially followed in order to kill germs in milk without harming the liquid.

The range of Spallanzani's experimental interest expanded. He studied regeneration in a wide range of animals and concluded that lower animals have greater regenerative power than the higher, young individuals have a greater capacity for regeneration than the adults, and generally it is only superficial parts that can regenerate. He also successfully transplanted the head of one snail onto the body of another, investigated the circulation of the blood, did an important series of experiments on digestion, and studied the role of semen in reproduction. While he made the mistake of believing that sperm were actually parasites in the semen, he still made significant contributions in this area.

In 1799, after suffering from an enlarged prostate and a chronic bladder infection, Spallanzani lapsed into a coma and died within a week. His broad legacy laid the foundation for future scientific work and practical applications. As an example, his work lead directly to the practice of pasteurization of milk and the invention of food canning. There have been few scientists that have had an impact on such a wide range of scientific endeavors.

JAMES J. HOFFMANN

Abraham Trembley
1710-1784
Swiss Naturalist

Abraham Trembley was a pioneer in experimental morphology. He is primarily remembered for his experiments on the hydra, or freshwater polyp (*Chlorohydra viridissima*), tiny creatures about a quarter of an inch long. Trembley discovered that the polyp could produce separate and complete new animals after it had been cut in half. His experiments were described in *Memoir on the Natural History of a Species of Fresh Water, Horn-shaped Polyps* (1744). Trembley also made important observations of the reproduction of algae and protozoans.

Abraham Trembley was born into a prominent family in Geneva. His most famous discoveries, however, were made while he served as a private tutor to the children of wealthy families. He later accompanied the young Duke of Richmond on his grand tour of Europe (1752-1756). In recognition of Trembley's services, the duke granted him a pension. Trembley married in 1757 and settled near Geneva. The rest of his life was devoted to philosophical and religious studies.

Experiments on the hydra had a profound impact on eighteenth-century thinking about the nature of reproduction, development and differentiation, and regeneration. The tiny polyps had been observed by the great seventeenth-century microscopist Anton van Leeuwenhoek (1632-1723). Freshwater polyps are typically found among the roots of lily pads and other aquatic plants. Noting that the polyp reproduced by budding, Leeuwenhoek concluded that they were plants. After careful and patient observations, Trembley discovered that the hydra actually captured food in its tentacles. Moreover, the creatures had an interior stomach where the food was digested. Unlike typical plants, the hydra reacted to touch and used a foot-like part to move. Because the hydra could move and feed itself, Trembley and other eighteenth-century biologists thought that the polyp should be classified as an animal, although it seemed to represent the very bottom of the scale of animal forms. The polyp seemed to be at the border between the plant world and the animal world. According to modern taxonomists, *Hydra* is a genus of invertebrate freshwater animals of the class Hydrozoa (phylum Cnidaria). The genus includes about 30 different species, all of which feed on other small invertebrates, such as crustaceans.

Reasoning that plants and animals typically exhibited significant differences in their ability to regenerate parts, Trembley cut a polyp in half. He was amazed to see that each piece survived and produced a complete new specimen. To determine whether the new individuals needed specific parts of the original polyp, Trembley cut hydras lengthwise, crosswise, and even into several pieces. Much to his surprise, each piece of the hydra appeared to be capable of producing a new individual. This characteristic made the hydra more like a plant than an animal. The new hydras could also be cut up and the pieces would produce a third generation of hydras. In another series of experiments, Trembley succeeded in turning a hydra inside out, as if it were a glove, by inserting a bristle into the gut. These inside-out creatures survived and eventually appeared to become normal hydra. In other remarkable hydra experiments, Trembley made permanent grafts and observed cell division, long before the establishment of modern cell theory.

After Trembley published his experiments, other naturalists attempted to extend his observations to other species. René Antoine Ferchault de Réaumur (1683-1757) and Charles Bonnet (1720-1793) found that similar results could be obtained with freshwater worms. Naturalists were certain that freshwater worms were animals; therefore, the case for considering the hydra an animal was strengthened. If the polyp was indeed an animal, Trembley's discoveries led inexorably to a major philosophical dilemma. Other animals, such as salamanders and crabs, were able to regenerate missing parts. In such cases, the original individual survived and healed; the lost parts neither survived nor gave rise to new individuals. Thus, it could be said that the "soul" or "organizing principle" resided in the maimed individual. The ability of each piece of the polyp to produce a new individual challenged the belief in an organizing principle or soul. Of course, eighteenth-century naturalists could not solve this problem by invoking the modern idea that complex animals are composed of cells that contain the same genetic material in each of their nuclei. Some natural philosophers, such as Julien de La Mettrie (1709-1751) and Denis Diderot (1713-1784), concluded that Trembley's experiments supported a materialist and atheistic view of life. In other words, Trembley's polyp proved that there was no soul; "life" was distributed throughout the body.

The ability of the hydra to form new individuals when cut into pieces also had a profound impact on the debate between preformationists and epigenesists about the basis of generation. If pieces of an animal like the hydra could form two new individuals when cut in half, the existence of a preformed germ of the new individual in either the egg or the sperm seemed to be precluded.

LOIS N. MAGNER

William Withering
1741-1799
English Physician, Pharmacologist and Botanist

William Withering placed the medical use of digitalis on a firm scientific foundation. His 1785 book of digitalis cases, observations, and experiments is a classic in the history of therapeutics and pharmacology.

Withering was born in Wellington, Shropshire, England, the only son of Edmund Withering, an apothecary. He was tutored in Greek and

Latin at the home of Rev. Henry Wood in Ercall, then entered the University of Edinburgh in 1762. Upon receiving his M.D. at Edinburgh in 1766, he established a private medical practice in Stafford. In 1772 he married one of his patients, Helena Cooke. In 1775 he moved his practice to Birmingham, and soon was among the most prosperous physicians there.

In Birmingham Withering made friends with Joseph Priestley (1733-1804), Erasmus Darwin (1731-1802), and several other prominent scientists. He joined the Lunar Society (so called because it met monthly during the full moon), a scholarly fraternity of scientists, technologists, industrialists, and other professionals. Inspired by Priestley, he conducted investigations in chemistry, mineralogy, and meteorology. In 1782 he discovered that barium carbonate is distinct from other barium compounds. In 1796 the German mineralogist Abraham Gottlob Werner (1749-1817) honored him by naming the twin crystalline form of barium carbonate "witherite." In the 1790s Withering opposed the controversial phlogiston theory of combustion.

Although primarily a medical practitioner, Withering always had a strong interest in plants. He was a keen observer of physical phenomena and an astute classifier of his observations. In 1776 he published *A Botanical Arrangement of all the Vegetables Naturally Growing in Great Britain*, the first sustained attempt to catalog and describe British flora according to Linnaean principles. By 1830 it had gone into seven editions and was the standard work in its field.

The first of his two medical books dealt with epidemiology. In *An Account of the Scarlet Fever and Sore Throat, or Scarlatina Anginosa, Particularly as it Appeared at Birmingham in the Year 1778* (1779), he accurately described the symptoms, prognosis, and aftereffects of scarlet fever, and argued that containment of the contagion should be part of an effective treatment for this disease.

Edema or hydrops, the accumulation of water in various cavities of the human body, was then called "dropsy." In 1775 Withering learned a secret herbal treatment for dropsy from an old woman in Shropshire. He determined that among the twenty or so herbs she concocted, purple foxglove (*digitalis purpurea*) was the active ingredient. This ancient folk remedy proved effective in controlled clinical medicine. For ten years Withering carefully studied it, then presented his results in *An Account of the Foxglove and Some of its Medicinal Uses with Practical Remarks on Dropsy and Other Diseases* (1785). This

book is a model of sober, empirical medical writing. He reported the failures as well as the successes of digitalis therapy, developed guidelines to avoid overdosage and underdosage, and documented his findings with case histories.

Withering's pioneer work made possible such medical classics as Augustin Eugène Homolle's "Mémoire sur la digitale pourprée" [Memoir on the Purple Foxglove] (1845), Johann Ernst Oswald Schmiederberg's "Untersuchungen über die pharmakologisch wirksamen Bestandtheile der Digitalis Purpurea" [Investigations of the Pharmacologically Effective Components of the Purple Foxglove] (1875), Sir James McKenzie's "New Methods of Studying Affections of the Heart" (1905), Walter Straub's "Digitaliswirkung am isolierten Vorhof des Frosches" [Action of Digitalis on the Isolated Heart Auricle of the Frog] (1916), and Arthur Robertson Cushny's *The Action and Uses in Medicine of Digitalis and its Allies* (1925). The full significance of digitalis for heart disease was recognized only after McKenzie.

ERIC V.D. LUFT

Kaspar Friedrich Wolff
1733-1794
German Physiologist and Embryologist

Kaspar Friedrich Wolff, the author of *Theory of Generation* (1759), revived the theory of epigenesis during a period in which many of the most respected naturalists were advocates of preformationist theory. According to the theory of epigenesis, an embryo is gradually produced from an undifferentiated mass by means of a series of steps and stages during which new parts are added. Preformationist theories asserted that an embryo or miniature individual preexisted in either the egg or the sperm and began to grow when properly stimulated. Wolff, who was very much influenced by the mode of thought known as nature philosophy, engaged in a debate about epigenesis and preformationism with the great physiologist Albrecht von Haller (1708-1777).

Kaspar Friedrich Wolff was born in Berlin, where he studied at the *collegium medicochirurgicum*. He continued his studies at the University of Halle where he became fascinated by the ideas of Christian, freiherr von Wolff (1679-1754), professor of mathematics and philosophy. Wolff, who was known as the German spokesman of the Enlightenment, wrote numerous works in philosophy, theology, psychology, botany, and physics and a series of essays that all began with the title "Rational Ideas." Kaspar Wolff was impressed with his mentor's botanical work and his philosophical principle that everything that happens must have an adequate reason for doing so. According to the Wolffian system of philosophy, it was impossible for rational beings to believe that something might come out of nothing.

In 1759 Kaspar Wolff completed a dissertation entitled *Theory of Generation,* which became a landmark in the history of embryology. His work, however, brought him little recognition during his lifetime. After serving as an army surgeon, he gave private lectures on pathology and medicine in Berlin. Unable to obtain a professorial position at the Medical College of Berlin, in 1767 he accepted an invitation to go to St. Petersburg, Russia. He was appointed academician for anatomy and physiology and continued to conduct research for the remainder of his life.

Charles Bonnet (1720-1793), Lazzaro Spallanzani (1729-1799), and René Antoine Ferchault de Réaumur (1683-1757), as well as von Haller, were proponents of ovist preformationism. The debate between von Haller and Kaspar Wolff concerning the nature of embryological development provides important insights into the role of observation and metaphysical assumptions in scientific controversies. Von Haller's studies of the developing chick egg convinced him that the embryo could be found in the egg before the chicken laid the egg. He also accepted Bonnet's studies of parthenogenesis in aphids as compelling evidence for ovist preformationism.

Kaspar Wolff's theory of generation, in contrast, was founded on the philosophical assumption that development must occur by epigenesis. Unlike preformationists, Wolff believed that studies of generation could only be purely descriptive because it was impossible to determine the actual mechanism of development. When von Haller rejected Wolff's theory of generation on largely religious grounds, Wolff insisted that a scientist should only consider scientific evidence when investigating biological questions. Wolff, however, was also influenced by preconceptions that led him to seek out observations that supported epigenesis. Wolff and other advocates of nature philosophy believed that descriptive knowledge of the gradual development of different life forms was the basis for understanding vital (living) phenomena.

Wolff's essay on generation began with a series of definitions. Generation was defined as the formation of a body by the creation of its parts. The organs of the body did not exist at the beginning of gestation but were formed successively; each step led to the next step. Beginning with the assumption that the original undifferentiated materials were essentially the same in plants and animals, Wolff studied plant metamorphosis as well as the development of the chick egg.

Studies of the development of plants helped Wolff overcome the technological limitations of eighteenth-century microscopy since more details could be seen in plant materials than in unstained animal tissues. Wolff thought that liquid was drawn up to the growing parts of the plant where it became a kind of thin jelly. As liquid evaporated, small sacs, or vesicles, formed. Eventually, the ducts of the plant vascular system developed in the masses of vesicles. Wolff thought that similar processes occurred in the development of the chick embryo. It is not clear, though, whether the vesicles that Wolff reported seeing were actually cells. Wolff argued that development involved change as well as growth, but preformationism actually implied that development did not occur.

The growing influence of nature philosophy in the late eighteenth century made Wolff's philosophy more acceptable to naturalists than the mechanistic philosophy. The conflict between preformationism and epigenesis, however, was not resolved until the nineteenth century, when cell theory provided a new way to think about eggs, sperm, embryos, and organisms.

LOIS N. MAGNER

Biographical Mentions

John Abernethy
1764-1831

English surgeon who founded the school of surgery and anatomy at St. Bartholomew's Hospital in London. His lectures there on anatomy helped to train many other surgeons. Abernethy was one of the first surgeons to treat aneurysms (bulges in the wall of an artery) by tying off the artery with a ligature (a piece of thread or wire). In addition, Abernethy developed a new technique for the treatment of aneurysms in very large arteries. It involved using multiple ligatures spaced about 1 in (2.5 cm) apart.

George Adams Sr.
1704-1772

George Adams Jr.
1750-1795

English engineers who were famous for designing microscopes. One type of microscope they produced had interchangeable parts so that it could be easily converted from a simple microscope (containing one lens) to a compound microscope (containing more than one lens). They were commissioned to produce a collection of scientific instruments for England's King George III, including a microscope made from silver. In 1746, Adams Sr. published *Micrographia Illustrata*, a popular work about microscopes, and in 1787, his son published *Essays on the Microscope*.

John Aiken
1747-1822

English physician who translated Tacitus and who also wrote about history, poetry, and medicine. An important work, *Principles of Midwifery*, was published in 1785. He entered medical school in Edinburgh, Scotland, at the age of 18. After completing his residency in surgery in Manchester, England, Aiken then went to London in 1769 to study under Dr. William Hunter.

Bernard Siegfried Albinus
1697-1770

German anatomist and physician who first demonstrated that the vascular systems of mother and fetus are connected. Albinus became chair of the departments of anatomy, surgery, and medicine at the University of Leiden, Holland, in 1702. He is remembered for his drawings of the human muscular and skeletal systems and for co-editing the works of Vesalius and William Harvey.

Claudius Amyand
1681?-1740

French physician best known for performing the first documented appendectomy. Amyand, surgeon to the King of England, recorded removing the inflamed appendix of a 12-year-old boy in 1736, noting that the organ was perforated by a pin the boy had apparently swallowed. Today, the appendectomy is considered a routine surgical procedure and is performed quite frequently. Without this surgery, many, if not most, appendicitis patients would die because as the appendix becomes inflamed it can burst and infect the entire abdominal cavity.

Dominique Anel
1679-1730

French surgeon who was the first to perform the ligation operation for aneurysms (bulges in the wall of an artery) since the Greek physician Antyllus performed them in the second century A.D. In a ligation operation, the artery is tied off with thread or wire to keep it from bursting. Anel also invented several surgical instruments, including the fine point syringe in 1714.

John Arbuthnot
1667-1735

Scottish physician and writer best known as one of Queen Anne's physicians. In addition to his distinction in medicine, Arbuthnot was also well-known for his literary efforts, including a series of five pamphlets satirizing the Duke of Marlborough titled *The History of John Bull*. Arbuthnot wrote medical works in addition to his social satire.

Leopold Auenbrugger
1722-1809

Viennese physician who discovered and classified the various sounds that healthy and diseased chests make when struck. By striking different areas of the patient's chest gently but firmly with his fingers, he was able to determine specific sites and types of chest conditions. He called this diagnostic technique "percussion" and published his results in Latin in 1761, but his work was scorned until Jean Nicolas Corvisart des Marets translated it into French in 1808.

Henry Baker
1698-1774

English naturalist who received the Copley gold medal in 1744 for his microscopic observations of salt particles in a solution. Baker wrote *The Microscope Made Easy* in 1743 and *Employment for the Microscope* in 1753. He is also credited for bringing the alpine strawberry and rhubarb to England.

William Bartram
1739-1823

American naturalist, botanist, and artist who, continuing the work of his father, John, identified and cultivated the flora of the American colonies. He published his observations and drawings in *Travels through North & South Carolina, Georgia, East and West Florida* in 1791. His published work was poetic, showing delight in his experiences of nature. It significantly influenced the Hudson River School of American artists and the English romantic poets such as Wordsworth and Coleridge.

William Battie
1704-1776

English physician who published one of the first textbooks on psychiatry, the *Treatise on Madness*, in 1738. This marked the inauguration of psychiatry as a formal medical discipline. Psychiatry, which is now a recognized medical practice, has a number of tools to use in the fight against mental illness, considerably more than existed in the eighteenth century. Much of this progress began with the work of Battie.

Gaspard-Laurent Bayle
1774-1816

French physician who published a landmark treatise in 1810 describing the many clinical varieties of tuberculosis. In 1805 Bayle was appointed to the staff of the Hôpital de la Charité in Paris, where he recorded detailed histories of individual cases and conducted hundreds of autopsies. His *Research on Pulmonary Tuberculosis* (1810) described the tubercle, the characteristic starting point of the lesions of phthisis (pulmonary tuberculosis). Bayle suggested that tuberculosis was the result of some general defect, which he called the "tuberculous diathesis." He was the first to use the term "miliary" to describe a widely disseminated form of tuberculosis.

Giacomo Bartolomeo Beccari
1682-1766

Italian physician and naturalist who is known for his studies in sexual functioning, phosphorescence (the property of something giving off light without noticeable combustion or heat), the chemical composition of wheat and milk, and the effect of light on silver salts. He was a professor of physics at the Institute of Science and Art, a professor of medicine at the University of Bologna, and was the first professor of chemistry in Italy.

Thomas Beddoes
1760-1808

English physician who explored therapeutic uses of gas inhalation and befriended chemist Humphry Davy. After medical school in Edinburgh, Scotland, Beddoes returned to England to become one of the most popular professors in the history of Oxford University. However, his support of the American and French revolutions cost him a prestigious post at the university, so he resigned in 1792 and moved to Bristol to continue the research on gases he had begun at Ox-

ford. In 1798 he opened the Medical Pneumatic Institute and hired a young Davy as research director. The research on gases ended in early 1800 in failure and public ridicule, but Davy left for London and a career as a great chemist. Beddoes was highly critical of the medical practice of his time and suggested numerous remedies, including more published case studies by physicians and better statistics by hospitals.

Joseph Black
1728-1799

Scottish chemist who was one of the pioneers in the study of gases and quantitative chemistry. Black's M.D. thesis led to a series of experiments on alkalis and "fixed air" (carbon dioxide). Further investigations established the phenomenon of latent heat (the heat necessary to produce a change of state from solid to liquid or liquid to gas). Black also established the distinction between temperature and heat and demonstrated the principle of specific heat (the relative heat capacity of substances).

Sir Gilbert Blane
1749-1834

Scottish naval physician who introduced administering lemon juice to the British naval fleet, thus dramatically reducing the incidence of scurvy. A Scottish physician trained in Edinburgh, Blane became the physician to the fleet in 1781. In 1796 Blane's recommendation of rationing each sailor three-quarters of an ounce (20 ml) of lemon juice per day was adopted, and scurvy largely disappeared from the British fleet. As a result of these rations, often referred to as "lime juice," British seamen became known by the nickname "limeys." Blane was knighted in 1812.

François Boissier de la Croix de Sauvages
1706-1767

French physician and nosologist, that is, a classifier of diseases. Nosology, from two Greek words, *nosos* (disease) and *logos* (reason or explanation), was a major aspect of the philosophy of medicine in the eighteenth century. Boissier, a disciple of Georg Ernst Stahl, was among the earliest important contributors to it. His works include *Nouvelles classes de maladies* (1731), *Pathologia methodica* (1739), and *Nosologia methodica* (1763).

Théophile de Bordeu
1722-1776

French physician who theorized that organs may secrete substances into the blood to affect other areas of the body. This hypothesis is correct, and led eventually to scientific studies of internal secretion, hormonal balance, and endocrine function, but Bordeu himself performed no experiments. Instead, like many other physicians of his day, he just propounded a particular dogmatic system of medical philosophy. His was vitalism, which held that life arises from a special kind of force.

Zabdiel Boylston
1680-1766

American physician who performed experimental inoculations for smallpox in 1721, when New England was threatened by a smallpox epidemic. Boylston was the only doctor in the Boston area who agreed to work with the Reverend Cotton Mather (1663-1728) to test the safety and effectiveness of smallpox inoculations. In 1726 Boylston published *An Historical Account of the Small-pox Inoculated in New England.* His statistics demonstrated that inoculated smallpox was significantly safer than naturally acquired smallpox.

William Buchan
1729-1805

Scottish physician who wrote the most popular general medical guidebook of the late eighteenth and early nineteenth centuries. Intended for a lay audience, *Domestic Medicine, or the Family Physician* was first published in Edinburgh in 1769. In the next 80 years this home remedy manual appeared in over 100 British and American editions with over a dozen subtitles. It is the second best selling medical book of all time after Benjamin Spock's *Baby and Child Care.*

Georges Louis Leclerc, comte de Buffon
1707-1788

French natural historian and mathematician who is primarily remembered for his encyclopedic Natural History (*Histoire naturelle, générale et particulière,* 44 volumes, 1749-1804), a systematic account of natural history, geology, and anthropology. Buffon was the first naturalist to construct geological history into a series of stages. Although his writings were very popular, Buffon's ideas about geological history, the origin of the solar system, biological classification, the possibility of a common ancestor for humans and apes, and the concept of lost species were considered a challenge to orthodox religious doctrine.

Pieter Camper
1722-1789

Dutch naturalist who investigated how head shape might affect mental development. Camper won particular renown for "Camper's angle," the angle made by a line drawn from the chin to the forehead. According to Camper, the more vertical this line was, the more highly developed the person. Camper's facial angle would become part of the study of phrenology (the study of head shapes), now discredited as a scientific discipline.

Mark Catesby
1683-1749

English naturalist, botanist, and ornithologist who explored Virginia, the Blue Ridge Mountains, Bermuda, and Jamaica during several trips to the Western Hemisphere. He published his resulting observations of the flora and fauna of these regions, complete with excellent detailed illustrations, in *The Natural History of Carolina, Florida, and the Bahaman Islands.* The botanist Carolus Linnaeus used this book as a reference in his own work. Catesby was made a member of the Royal Society in 1732.

William Cheselden
1688-1752

English anatomist and surgeon who wrote two beautifully illustrated anatomical masterpieces, *The Anatomy of the Humane Body* (1713) and *Osteographia, or the Anatomy of the Bones* (1733). Cheselden was a student of William Cowper, and like Cowper, was often accused of plagiarism. As a surgeon at St. Thomas's Hospital, London, he improved procedures for lithotomy, restored eyesight by creating artificial pupils, and was renowned for his speed, which was one of the most important qualifications for a surgeon in era before anesthesia was available.

Vincenzo Chiarugi
1759-1820

Italian physician who performed landmark research into mental illnesses. One of Chiarugi's major accomplishments was establishing the "law of the insane," which specified proper treatment for the institutionalized insane. He also published a three-volume work on mental illness that was one of the first to try to link mental illness with physical damage to the structure of the brain. In many ways, his approach to these matters was very modern, anticipating several current practices.

Jane Farquhar Colden
1724-1766

American botanist and first woman in North America known to have a professional command of the Linnaean system of classification. Though her scientific contributions are minimal, much of her botanical enterprise probably contributed to the publications of her father, Cadwallader Colden. One of her surviving manuscripts, "Flora Nov-Eboracensis," capably describes and illustrates 341 plants growing near her New York State home. One of her plant descriptions was published by the naturalist Dr. Alexander Garden which led to the gardenia being named for him.

Peter Collinson
1694-1768

English naturalist and wealthy Quaker cloth merchant whose commercial contacts with the American colonies led to a beneficial two-way exchange of plants with co-religionists and others. Through his influence, John Bartram was appointed King's Botanist. Although best known for his patronage of John and William Bartram, Collinson also sent Benjamin Franklin news about German studies of electricity that inspired Franklin's kite experiment.

Domenico Cotugno
1736-1822

Italian physician best known for his investigations into the structure of the inner ear. In a paper published in 1761, Cotugno wrote the first detailed description of the inner ear, its different parts, and how they function. This was followed by a description of the role of certain nerves in causing sneezes, a description of some afflictions of the sciatic nerve (which helps control the legs), and more.

Adair Crawford
?-1795

British chemist who examined the question of heat generation in animals. His work helped to clarify some of the differences between warm-blooded (mostly mammals and birds) and cold-blooded animals, including the origin of body heat. Crawford also worked to develop methods for measuring the specific heat of a variety of substances, helping to set the stage for future developments in thermodynamics and some branches of engineering.

William Cullen
1710-1790

Scottish physician and educator, the preeminent medical teacher of the eighteenth century. He co-founded the Glasgow Medical School in 1744 and thereafter taught at both Glasgow and Edinburgh. In 1769 he published *Synopsis Nosologiae Methodicae*, an elaborate classification of all known diseases, modeled on the work of Carolus Linnaeus and François Boissier de la Croix de Sauvages. Cullen's *First Lines of the Practice of Physic* (1777) and *A Treatise of the Materia Medica* (1789) each went into many editions and remained standard textbooks long after his death.

John Dalton
1766-1844

English chemist who established the modern theory of the atom. Dalton formulated the theory in order to explain chemical reactions. Although both Dalton and the ancient Greeks considered the atom (taken from the Greek word for "indivisible") the smallest part of an element, Dalton's concept of the atom was based on his studies of mixtures of gases. Studying atoms of different weights allowed Dalton to suggest a physical explanation for the law of multiple proportions. His other interests included the study of color-blindness, a condition that affected him.

Erasmus Darwin
1731-1802

English naturalist and physician whose wide range of scientific interests established him as the leader of the Lunar Society, an association that included some of the most important British scientists and inventors of the eighteenth century. Although his grandson, Charles Darwin, denied being influenced by Erasmus Darwin's evolutionary speculations, Erasmus Darwin's work was widely known in its time. Erasmus Darwin's major treatise, *Zoonomia, or the Laws of Organic Life* (1794-96), deals with medicine, pathology, and the mutability of species.

Louis Jean Marie Daubenton
1716-1800

French comparative anatomist and physician who in 1742 was chosen by Georges-Louis Leclerc, comte de Buffon to help prepare anatomical descriptions of mammals for the latter's *Histoire naturelle* (published 1749-89). In 1744 Daubenton became Buffon's assistant in the Department of Natural History at the Jardin des Plantes (after 1793, the Museum of Natural History) in Paris. In 1753 Buffon and Daubenton set forth their Principle of the Unity of Composition, placing the skeletal structure of vertebrates within a comparative framework. A distinguished physiologist and descriptive paleontologist, Daubenton was named to the French Academy of Sciences in 1760, and in 1775 was appointed Lecturer in Natural History at the College of Medicine. In 1778 he became professor of zoology at the Collège de France. He introduced merino sheep into French agriculture. He was selected as the first director of the Museum of Natural History in 1793, and was named a member of the French Senate in December 1799, a month before his death.

Jacques Daviel
1696-1762

French ophthalmologist who performed the first known cataract surgery. In 1748 Daviel performed the first "modern" cataract surgery when he removed the entire cataract from the lens of a patient's eye while leaving portions of the lens capsule and attached muscles. This allowed the patient to retain some vision, though extremely limited. A cataract is the formation of opacities in the lens of the eye, making the world appear clouded or milky before blindness sets in.

Pierre Joseph Desault
1744-1795

French surgeon who improved techniques for treating fractures, dislocations, and aneurysms. Apprenticed to a barber-surgeon, he never earned a degree. His skill alone made his career. Together with François Chopart he originated urological surgery. They were the first to regard the entire genito-urinary system as a unified, separate entity. In 1791 Desault founded the world's first surgical periodical, *Journal de chirurgie*. He was a popular lecturer and teacher. Among his students was the anatomist Marie François Xavier Bichat.

William Douglass
1691-1752

American physician who initially opposed the inoculation against epidemic smallpox as practiced by Cotton Mather (1663-1728) and Zabdiel Boylston (1680-1766) on the grounds that inoculation was a new, untested, and potentially dangerous procedure. Based on evidence of the safety and effectiveness of inoculation, by 1730 Douglass had become a supporter of inoculation. Douglass published a pioneering account of the first cases of a scarlet fever epidemic that had appeared in Boston in 1735 and 1736.

Thomas Dover
1660-1742

English physician who is best known for introducing a popular remedy known as Dover's Powder. This potent medicine contained ipecac, opium, and potassium sulfate. It was prescribed as an analgesic, a sedative, and a sudorific or diaphoretic (an agent that increases perspiration). Ipecac (ipecacuanha) is a powerful emetic (a medicine that causes vomiting). Although Dover

practiced medicine in London, he was never certified by the College of Physicians. In *An Ancient Physician's Legacy to his Country,* Dover claimed that he had practiced medicine for 49 years.

Paul Dudley
1675-1751

American politician who suggested cross-fertilization techniques for corn. A Massachusetts lawyer, Dudley was a fellow of the Royal Society and published essays in *Philosophical Transactions* about New England natural history. He noted that Indian corn was yellow, white, red, and blue and that rows of corn positioned yards apart or separated by water canals often had stalks with varying kernel colors. Dudley hypothesized that wind transported reproductive material, which influenced corn color. Geneticists, however, ignored his theory until the twentieth century.

Gabriel David Fahrenheit
1686-1736

German physicist and instrument maker who invented the mercury thermometer (in 1714) and developed the Fahrenheit temperature scale. The thermometer, however, was not generally adopted as a clinical instrument until the 1860s. Because the zero point on Fahrenheit's original scale was the temperature of an ice-salt mixture, the freezing point of water on his scale is 32°, normal body temperature is 98.6°, and the boiling point of water is 212°. Although the Fahrenheit scale is still commonly used in the United States, scientists and most other nations have adopted the Celsius scale.

Pierre Fauchard
1678-1761

French physician, author, and dentist who is known as the "Father of Scientific Dentistry." During his career, Fauchard recognized the importance of good dental hygiene, wrote two influential books, and was among the first to remove all diseased parts of a tooth before filling the resulting cavity. He was also the first to use gold inlays to fill the tooth after performing a root canal. His invention of the dental drill was equally influential.

John Floyer
1649-1734

English physician who invented a "pulse-watch" to transform the ancient art of feeling the pulse into a quantitative measurement. Floyer invented a watch with a second hand in order to help physicians count the pulse. In 1707 he published his *Physician's Pulse Watch* as a way to help other physicians learn how to measure the pulse. Floyer believed that with accurate pulse measurements physicians could develop treatment regimens for their patients that would return an unhealthy pulse to a more normal state. Floyer also wrote about geriatrics, asthma, and emphysema.

Johann Peter Frank
1745-1821

German physician who is regarded as a pioneer in public hygiene. His classic treatise *A System of a Complete Medical Policy* (9 volumes, 1779-1827) is considered the first systematic treatise on public health and hygiene. Frank advocated strict and rigorous laws and "medical police" to protect the health of the people. His writings summarized all that was known about public health at the time and provided detailed models for hygienic laws. Frank believed that international codes of law were needed to control threats to public health.

Joseph Gaertner
1732-?

German biologist who made significant progress in the scientific classification of fruits and seeds. In fact, Gaertner's book, *Carpology: or treatise on the fruits and seeds of plants* (published in 1789), was among the first systematic and scientific treatments of this important matter. In this work, Gaertner discussed over 1,000 different species of plants and their fruits or seeds and illustrated nearly 200 of them.

Johann Gottlieb Gahn
1745-1818

Swedish chemist and mineralogist who, with Karl Scheele, discovered that phosphorus is essential to the structure of bone. In fact, bones and teeth are formed of a calcium phosphate mineral, apatite, secreted by bone-forming cells. This discovery was important to understanding how bone and teeth form and function, and in better understanding many fossils, which can consist of phosphatic minerals, too. Gahn also discovered the element manganese in 1774.

John Haygarth

English physician who was among the first to examine the role of infectious agents in the spread of disease. To contain the spread of disease, Haygarth practiced separating infectious patients from those with other ailments—at the time a novel idea. His insistence on cleanliness standards, well-ventilated rooms, and other innovations were viewed with skepticism, but his success rate led to their widespread adoption by other physicians. Haygarth was also active in debunking many questionable medical claims.

William Heberden
1710-1801

English physician who published a classic description of angina pectoris entitled "Some Account of the Disorder of the Breast" in 1772. Heberden was the first to use the term "angina pectoris" and to provide a complete and accurate description of the symptoms associated with this cardiovascular disease. The condition is now known to be caused by coronary artery occlusion. He was the founder and editor of *Medical Transactions* and one of the first physicians to distinguish carefully between chicken pox and smallpox. Heberden's nodes are named for him.

Lorenz Heister
1683-1758

German surgeon who improved the tourniquet and invented a spinal brace. His treatise on general surgery first appeared in German in 1718, was translated into many languages, and went through many editions as the most popular surgical textbook of the eighteenth century. In Heister's time the French, especially the Parisians, dominated surgery, but he is usually regarded as the founder of modern German surgery and the equal of any of his contemporary surgeons in France.

Claude Adrien Helvetius
1715-1771

French philosopher whose controversial book, *De l'esprit*, suggested that the source of all intellectual activity is the quest for sensation and that all human actions arise from self-interest. This book was burned in Paris, denounced by leading French intellectuals, and condemned by the English Parliament. Because of this, it was widely translated and read, forming part of the intellectual framework of the philosophical school of utilitarianism.

Friedrich Hoffmann
1660-1742

German physician who founded a famous eighteenth-century medical system. As the first professor of medicine at the new University of Halle, Hoffmann gave lectures on chemistry, physics, anatomy, surgery, and medical practice. He was widely respected as a teacher and a writer. Hoffman's medical system was based on the premise that health was the result of normal movement and disease was the result of disturbed movement. Therefore, the purpose of treatment was to restore normal movement with sedatives or tonics.

Christoph Wilhelm Hufeland
1762-1836

German physician who spent much of his career debunking medical myths, fallacies, and chicaneries. He attacked Franz Joseph Gall's phrenology, Franz Mesmer's animal magnetism, and John Brown's theory of tissue excitability, "Brunonianism." He supported and popularized serious science, defended Edward Jenner during the vaccination controversy, edited four medical journals, and encouraged cleanliness as a means to long life. He was the personal physician of Johann Wolfgang von Goethe at Weimar, and later taught at the universities of Jena and Berlin.

John Huxham
1692-1768

English physician, student of Herman Boerhaave. Huxham expanded medical knowledge of fevers and other infectious diseases such as diphtheria, colic, typhoid, typhus, and the common cold. He introduced the term "influenza" to the English-speaking medical world. Among his major works are *An Essay on Fevers* (1750) and *A Dissertation on the Malignant, Ulcerous Sore-Throat* (1757). He also recorded several decades of careful observations trying to prove a relationship between weather and health.

Bernard de Jussieu
1699-1777

French physician-botanist who was one of three brothers, all botanists, The brothers along with their descendants made major contributions to the field of plant science over at least four generations. Bernard accompanied his elder brother Antoine on a botanizing expedition to Spain in 1716. He received his M.D. at Montpellier University in 1720. In 1722 he was appointed Sub-Demonstrator of Plants at the Jardins du Roi in Paris, later becoming its director. In 1725 he was made a member of the French Academy of Sciences, the same year in which his *History of the Plants of the Environs of Paris* was published. In 1737 the Swedish botanist and taxonomist Carolus Linnaeus named the plant genus *Jussieua* in his honor. Jussieu was invited to become founding superintendent of the royal gardens at the Petit Trianon at Versailles in 1759. His arrangement of the gardens there was based on Linnaean principles. Jussieu published a natural classification of plants, based upon his study of plant embryos. It was later revised by Antoine de Jussieu, his nephew and successor at the Jardins du Roi, who went on to develop a new system.

Carl Friedrich Keilmeyer
1765-1844

German anatomist whose work in comparative biology helped advance that field considerably. Keilmeyer was one of the first to note that there exists among organisms some basic "body plans," around which organisms are organized. This observation led to the classification of organisms into family groupings and is one of the bases for classifying fossil organisms into their respective taxonomic groups. Keilmeyer was one of the more influential teachers of the great scientist Georges Cuvier.

Adam Kuhn
1741-1817

American physician and botanist who studied under Carolus Linnaeus in Sweden before receiving his M.D. at Edinburgh in 1767. He practiced medicine in Philadelphia, and in 1768 joined the faculty of the Medical Department of the College of Philadelphia, later called the School of Medicine of the University of Pennsylvania, as the first professor of medical botany in America. He was one of 12 founding fellows of the College of Physicians of Philadelphia in 1787.

Jean Baptiste Pierre Antoine de Monet, chevalier de Lamarck
1744-1829

French naturalist remembered for developing one of the first comprehensive theories of evolution—bodies developed in various ways to cope with environmental pressures, and such changes could be inherited. Regarded as naive, his theory was rejected. First to distinguish vertebrates from invertebrates by their bony spinal column, Lamarck established invertebrate zoology. In his seven-volume *Natural History of Invertebrates* (1815-22) he differentiated eight-legged arachnids from six-legged insects and categorized crustaceans and echinoderms (starfish, seaurchins, etc.).

Julien Offroy de La Mettrie
1709-1751

French physician and philosopher whose materialistic interpretations of psychic and physiological phenomena were published in *A Natural History of the Soul* (1745), *L'Homme-machine* (1747), *Discourse on Happiness* (1748), and *The Small Man in a Long Queue* (1751). La Mettrie taught that atheism was the only road to happiness and that the purpose of human life was to experience the pleasures of the senses. Many aspects of his theories were incorporated into behaviorism and modern materialism.

Giovanni Maria Lancisi
1654-1720

Italian physician and epidemiologist who served as personal physician to Pope Clement XI. His research added to the knowledge of syphilis, malaria, plague, contagious fevers, asthma, aneurysm, and heart disease. In *De noxiis paludum effluviis* (On Harmful Emissions from Marshes) (1717), he proposed draining swamps in order to prevent malaria and suggested that mosquitoes might be the transmitters of that disease. His *De subitaneis mortibus* (On Sudden Kinds of Death) (1707) is a classic of cardiology.

Henri François Le Dran
1685-1770

French surgeon who was among the first to recognize that cancer is a local, not systemic, affliction. He pioneered surgical treatments for cancer and understood that cancer surgery had to occur before the tumor could metastasize through the lymphatic system and affect other parts of the body. He improved methods of mastectomy, lithotomy, and military surgery for gunshot wounds. As senior surgeon at La Charité, Paris, he was influential throughout Europe.

Anton van Leeuwenhoek
1632-1723

Dutch microscopist and biologist who was among the first to study biological specimens with a microscope. He is regarded as one of the founders of modern biology. He was first to observe bacteria and protozoa; co-discovered spermatozoa; studied optic lens and red blood cells; observed the fine structure of muscles, nerves, insects, and plants; discovered parthenogenesis; and disproved spontaneous generation. Initially, his interest in microscopes was a hobby, but his observations made him world-renowned.

Abbé Charles-Michel de l'Epée

French priest and physician who worked extensively with the deaf. Embarrassed by his failure to recognize deafness in two girls he was trying to speak with, l'Epée devoted a great deal of attention to developing sign language and finger-spelling for the deaf. By so doing, he demonstrated that the deaf are capable of learning, something not recognized at the time. Since then, other advances such as the development of American Sign Language and other devices have become available to the deaf community.

John Coakley Lettsom
1744-1815

English physician and Quaker philanthropist who founded the Medical Society of London in 1773. Lettsom was one of Britain's most distinguished physicians and a friend of the American physicians Benjamin Rush and Benjamin Waterhouse. It was Lettsom who called Rush the "American Sydenham." In 1799 Lettsom sent a copy of Edward Jenner's paper on cowpox vaccination to Waterhouse, who published an account of Jennerian vaccination in an American newspaper. Lettsom's publications include *History of the Origin of Medicine* (1778) and papers on alcoholism, chlorosis (anemia) in boarding schools, and the value of drinking tea.

Benoit de Maillet
1656-1738

French natural philosopher whose controversial book of evolutionary speculations, entitled *Telliamed: Of Discourses Between an Indian Philosopher and a French Missionary on the Diminution of the Sea, the Formation of the Earth, the Origin of Men and Animals, etc.,* was published posthumously (1748). De Maillet, a diplomat and traveler, disguised his unorthodox ideas about evolution by ascribing them to an exotic sage named Telliamed ("de Maillet" spelled backwards). His book presents theories on cosmic and biological evolution in the form of a dialogue between Telliamed and a French missionary.

Andreas Marggraf
1709-1782

German biologist whose discovery of sugar in sugar beets made possible a revolution in food production. Prior to this discovery, sugar was a very rare commodity, imported from overseas. The ability to grow and produce sugar locally led to a huge increase in sugar production and consumption, leading in turn to an increase in tooth decay, obesity, and other problems related to excessive sugar and caloric intake. It also was crucial to the still-widespread popularity of industries devoted to satisfying the human sweet tooth.

Humphry Marshall
1722-1801

Nurseryman whose *Arbustum Americanum: The American Grove* (1785) was the earliest book of descriptive botany published by a native-born American author. Marshall's nursery, where he cultivated native and exotic plants obtained from others, became the main source of American plants for the European market after the death of his cousin John Bartram. Marshall also collected animal specimens for European naturalists and published his observations on sunspots.

George Martine

French physician and physicist who performed the first known tracheotomy to help treat serious respiratory problems. This procedure, which consists of making an external opening into the trachea, is frequently used today to create an airway for a person who otherwise cannot breathe, usually due to some obstruction. In Martine's case, the procedure was used to help diphtheria victims breathe. Martine also performed research in physics, showing that the heat contained by an object is not proportional to its volume.

Sebastian Menghini

Italian biologist who, in the mid-1700s, performed experiments on the effects of camphor on animals. Camphor has many uses, including as the active ingredient in mothballs, some explosives, liniments and other medicines, and for the production of many plastics. Menghini's work helped to elucidate some of these roles played by camphor, especially those dealing with its medical uses.

Alexander Monro primus
1697-1767

Scottish anatomist and educator, known as *primus*, because he was the first of the three eminent professors of anatomy and surgery in the Monro dynasty at the University of Edinburgh. Educated at leading European medical schools, Monro was an inspirational teacher and gifted technician, whose career spanned nearly 40 years. He introduced advances in surgery, instruments, and dressings, and was a tireless investigator whose work sparked the rise of Edinburgh as an international center of anatomical studies.

Alexander Monro secundus
1737-1817

Scottish physician and surgeon, called *secundus* to distinguish him from his father, Alexander Monro *primus* (1697-1767), and his son, Alexander Monro *tertius* (1773-1859). All three, in succession, were professors of anatomy at Edinburgh from 1720 to 1846. As teachers they contributed immeasurably toward making Edinburgh a major center of medical learning. *Secundus* is generally acknowledged as the keenest scientific mind of the three. In *Ob-servations on the Structure and Functions of the Nervous System* (1783) he reported his discovery of the foramen interventriculare, an important passage in the brain.

Mary Wortley Montagu
1689-1762

English writer who introduced smallpox inoculations to England. Born into nobility, Montagu grew up surrounded by influential people. She suffered smallpox, which scarred her face. Montagu later traveled with her husband Edward, an ambassador. While in Turkey, she became aware of local medical practices and arranged for her son to be inoculated against smallpox. Returning home, Montagu initiated smallpox inoculations in England and was praised for preventing thousands of deaths annually.

Sauveur François Morand
1697-1773

French ophthalmologist who was one of the first to describe the procedure by which cataracts could be treated. The operation performed by Morand—the removal of the lens of the eye—is still considered one of the treatment options for cataracts, although other techniques are now available. Left untreated, cataracts, which are opacities of the lens, will cause blindness in the patient. Cataract removal, while leaving the patient without full vision, is much preferred over blindness, and current treatments go far towards a full restoration of sight.

John Morgan
1735-1789

American physician and surgeon who received his M.D. at Edinburgh in 1763, then two years later co-founded with William Shippen the first medical school in the Western Hemisphere, the Medical Department of the College of Philadelphia, subsequently called the School of Medicine of the University of Pennsylvania. From 1775 to 1777 he served as Director-General of Military Hospitals and Physician-in-Chief of the American Army, a position equivalent to surgeon-general. In 1787 he became one of the 12 original fellows of the College of Physicians of Philadelphia.

Otto Müller
1730-1784

Danish biologist whose work on bacteria was essential in laying the foundations of the field later known as microbiology. Müller was the second person known to have seen bacteria through a microscope, the first having been Anton van

Leeuwenhoek a half-century earlier. However, Müller was the first to attempt to classify bacteria into various types, foreshadowing later work in this important field.

Jean Palfyn
1650-1730

French physician who was the first to use forceps to facilitate childbirth. At a time when physicians were often banned from childbirth, there were few implements or tools available to aid the mother in delivering difficult babies. Palfyn's introduction of forceps, which continue to be used to this day, was the first such device. By helping the physician or midwife ease the baby through the birth canal, the use of forceps helped make childbirth a little better for both mother and child.

William Dandridge Peck
1763-1822

Pioneer American naturalist. Although a Harvard graduate, Peck spent twenty years studying nature independently. After a European tour to acquaint himself with the current state of science, he accepted a Harvard professorship in natural science established for him in 1805. In 1818, he published a catalog of the plants in the botanic garden he had laid out in Cambridge, but the bulk of his publications were in economic entomology.

Thomas Percival
1740-1804

English physician who helped lay the foundations for modern medical ethics. Percival, and colleagues Benjamin Rush (an American) and the Scottish John Gregory, established the first clear set of ethics, codes of conduct for physicians, standards of practice, and expected physician etiquette and behavior. These were also the first standards to set the level of confidentiality expected between a patient and physician. Many of these ethical standards are still in use today.

Phillip Pfaff

German dentist who developed a method for casting models for making false teeth. This method, first described in 1756, helped those with rotting or missing teeth have dentures constructed that fit and worked better than the wood, metal, and ceramic false teeth of the day.

Marcus Antonius Plenciz

Physician who was among the first to advance the "germ theory" of disease causation and transmittal. Plenciz spoke of "seeds in the air" that spread disease from one patient to another, suggesting a different "seed" for each disease. Although he offered no experimental or observational proof of his theories, they became an important part of the germ theory that eventually was proven to be true.

Nöel-Antoine Pluche

French naturalist and philosopher who helped to popularize the philosophy of "natural theology," which became popular in France in the early and mid-eighteenth century. Following on the successes of Newtonian physics and other scientific advances of the Enlightenment, natural theology sought divinity in the laws and patterns of nature. Pluche's eight-volume work, *Spectacle of Nature*, was his most important work in this area.

Joseph Priestley
1733-1804

English clergyman, author, and chemist who first isolated such gases as oxygen, nitrous oxide, and sulfur dioxide. Priestley studied for the ministry and eventually became the best-known and most controversial Unitarian minister in Britain. In the 1760s he began scientific research into electricity and then optics, and also started work on the nature and characteristics of gases. Entirely self-taught in science and a very prolific author on theological, scientific, and political matters, Priestley read widely and corresponded with numerous scientists in Britain and abroad. Priestley remained the most famous chemist in the world until Frenchman Antoine Lavoisier's "new chemistry" supplanted his theories by the 1790s. An ardent supporter of the American and French revolutions, Priestley was forced to leave England in 1794 and settle in Pennsylvania.

John Pringle
1707-1782

Scottish physician who is considered the founder of modern military medicine. Based on the theory that putrefactive processes caused disease, Pringle believed that proper sanitary measures would maintain the health of the British army. He applied this concept to the administration of army camps and hospitals with considerable success. His ideas about hospital ventilation and camp sanitation are discussed in his *Observations on the Diseases of the Army* (1752). He also made important observations on dysentery, hospital and jail fevers, typhus fever, and influenza.

Giacomo Pylarini

Italian physician who became the first to attempt to prevent disease through inoculation by injecting children with smallpox in 1701. By so doing, nearly a century before Edward Jenner's experiments with cowpox, he hoped to save them from the more serious disease that strikes adults. For this, as well as some of his other scientific investigations, Pylarini is considered to be the world's first immunologist.

Johann Christian Reil
1759-1813

German physician, physiologist, and neurologist who in 1795 founded the first scientific journal of physiology, *Archive für die Physiologie*, and in 1805 the first journal of psychiatry, *Magazin für psychische Heilkunde*. His areas of research included tissue metabolism, tissue irritability, brain function, cerebral localization, mental diseases, the structure of nerves, and the so-called "life-force" behind biochemical processes. The island of Reil in the brain is named for him.

Jean Baptiste Robinet

French natural philosopher who was among the first to advocate concepts leading to our current theory of evolution in living organisms. Robinet suggested in his five-volume work, *On Nature*, that species gradually changed, one into another, forming a linear sequence of progress, with no gaps. Although this view of evolution is now felt to be overly simplistic, it was very important at the time, forming as it did the first published viewpoint of species as mutable over time.

Hilaire-Marin Roulle
?-1779

French biologist who investigated the chemical properties of blood. Roulle was the first to investigate the chemical components of vertebrate blood, discovering sodium chloride (salt), potassium chloride, and sodium carbonate to be present in the liquid portion of blood (the plasma). This led eventually to the discoveries that blood plasma resembles not only other bodily fluids, but also seawater in its chemical composition. Eventually, knowledge of blood chemistry became a powerful medical diagnostic tool.

Antonio Scarpa
1752-1832

Italian anatomist, surgeon, and artist, known for his careful research, many anatomical discoveries, and precise drawings. In his clinical practice he was especially skilled in ophthalmology and orthopedics. He developed new methods for treating hernia and clubfoot and was the first to recognize hardening of the interior of the arteries, or atherosclerosis. His greatest work is an illustrated anatomy of the nervous system, *Tabulae nevrologicae* (1794), which includes the first accurate description of the nerves of the heart.

Karl Wilhelm Scheele
1742-1786

Swedish chemist who made many contributions to chemistry, including the discovery of oxygen. He probably prepared oxygen, which he called "fire-air," as early as 1772, two years before Joseph Priestley. Scheele, however, published little and his work was not as well known. Scheele's book *Chemical Observations and Experiments on Air and Fire* (1777) presented evidence that the atmosphere is composed of two gases, one of which prevented combustion and another that supported combustion. Other substances discovered by Scheele include tartaric acid, arsenic acid, lactic acid, citric acid, oxalic acid, tungstic acid, and hydrocyanic acid.

Johann Scheuzer
?-1733

Swiss naturalist who was among the first to postulate that fossils were the remains of animals. His treatise, *Homo diluvii testis*, was an important contribution to early thinking that fossils were something other than rock and mineral formations. Although Sheuzer did not make the leap to considering fossils the remains of extinct animals, he did help to set the stage for future work in this area.

Armand Séguin
1765?-1835

French chemist who worked with Antoine Laurent Lavoisier (1743-1794) in pioneering studies on metabolism. Séguin and Lavoisier described experiments on human metabolism in their "Memoir on Animal Respiration," which was published in 1789. During these metabolic experiments, Séguin served as the experimental subject. They measured oxygen absorption when Séguin was fasting, eating, resting, or exercising. During the course of these experiments, Séguin and Lavoisier made several fundamental observations. They realized that oxidation in human beings is dependent on food, environmental temperature, and mechanical work. These experiments established the basic facts of the theory of energy transformation.

William Smellie
1697-1763

Scottish obstetrician who made fundamental advances toward understanding the anatomy and mechanics of safe birth. Despite the opposition of female midwives, he successfully taught obstetrics for over 20 years in his own home. He invented several varieties of obstetrical forceps and developed maneuvers for breech birth and version. His accurate descriptions and scrupulous judgments formed the basis of his two masterworks, *A Treatise on the Theory and Practice of Midwifery* (1752-1764) and its accompanying atlas, *A Sett* [sic] *of Anatomical Tables* (1754).

Christian Konrad Sprengel
1750-1816

German botanist who was among the first to study methods of plant fertilization. Sprengel found that plants could be fertilized by both wind and insects, helping to elucidate methods by which plants reproduced. Through this work, not only did he help to explain how plants and insects can depend on one another, but these discoveries also helped farmers and botanists to better control plant breeding experiments.

John Taylor
1703-1772

English physician who conducted early research into the nature and treatment of cataracts and other problems of the eye. He worked for the royal courts in England, Denmark, Poland, and Sweden, as well as for other European nobility before he was determined to be a charlatan.

Jacques René Tenon
1724-1816

French surgeon who helped describe the interior structure of the eye, several parts of which are named in his honor. This knowledge helped scientists to better understand how the eye and its parts function and made it possible to start conducting eye surgery. Although surgical techniques at that time were crude, it was still possible to conduct some surgery once the basic structure and function were known.

Samuel-Auguste-André-David Tissot
1728-1797

Swiss physician and medical writer who collaborated with a distinguished group of physicians interested in social and medical reforms. Tissot and his colleagues attempted to apply their medical knowledge to improve private and public health practices, professional training, and society in general. Tissot's work includes an important study of inoculation entitled *Inoculation Vindicated* (1754) as well as *Advice to the People on Their Health* (1761). His *Onanisme* (1760), a treatise on masturbation, however, has become the work by which he is most known today.

Gilbert White
1720-1793

English naturalist and clergyman who was the author of *The Natural History and Antiquities of Selborne* (1789). In 1752 White began keeping a journal of his observations of his garden in Selborne. He published his first book of observations, *A Calendar of Flora and the Garden,* in 1765 and followed this work with *Naturalist's Journal* (1768). Naturalists and nature lovers praised White for his keen observations and methodical approach and considered his book a true landmark in the progress of English natural history.

Robert Whytt
1714-1766

Scottish physiologist who was among the first to describe several mental afflictions and to suggest causes and cures for them. Whytt worked primarily with neurotic patients, categorizing them into hysteric, hypochondriacal, and those suffering from what he termed "nervous exhaustion." Although he attempted to develop a theory for these by discussing different aspects of the nerves (for example, he assumed women's nerves were more "motile" than men's), many of his theories were inaccurate.

Thomas Young
1773-1829

English physician and physicist who established the wave theory of light with his experiments on interference patterns. Young's interest in sense perception led to the discovery of the cause of astigmatism. He explained the colors of thin films and the phenomenon of polarization, calculated the approximate wavelengths of different colors, investigated color perception, and developed methods for measuring surface tension and the size of molecules. His ideas about color perception are acknowledged in the Young-Helmholtz three-color theory. "Young's modulus," a constant in the mathematical equation describing elasticity, was named for him. Young was also a distinguished Egyptologist and helped decipher the Rosetta Stone.

Bibliography of Primary Sources

Books

Baillie, Matthew. *The Morbid Anatomy of Some of the Most Important Parts of the Human Body* (1793). This was the earliest scientific exposition of pathological anatomy, the anatomy of diseased tissue. The book includes classic postmortem descriptions of heart valve disease, emphysema, tuberculosis, gastric ulcer, cirrhosis of the liver, ovarian cysts, and many other afflictions. Its illustrations, published separately as *A Series of Engravings, Accompanied with Explanations, which are Intended to Illustrate the Morbid Anatomy of Some of the Most Important Parts of the Human Body* (1799-1803), constituted the first significant pathological atlas.

Bartram, William. *Travels through North & South Carolina, Georgia, East and West Florida* (1791). Here Bartram, continuing the work of his father, John, identified and cultivated the flora of the American colonies. His published work was poetic, showing delight in his experiences of nature. It significantly influenced the Hudson River School of American artists and the English romantic poets such as Wordsworth and Coleridge.

Bayle, Gaspard-Laurent. *Research on Pulmonary Tuberculosis* (1810). In this book Bayle described the tubercle, the characteristic starting point of the lesions of phthisis (pulmonary tuberculosis). Bayle suggested that tuberculosis was the result of some general defect, which he called the "tuberculous diathesis."

Blumenbach, Johann. *De Generis Humani Varietate Nativa Liber* (1795). This text stands as the foundation of scientific anthropology. The work describes how Blumenbach modified previous classifications to create a five-family distribution of the human species that has been tremendously influential and introduced the term "Caucasian" to define the group of Europeans. He hypothesized that the American and Mongolian families branched in one geographical direction and the Malaysians and Ethiopians branched in another. He stated that humans arose in the area of the Caucacus Mountains, spread out over the world, and acquired variations in physical characteristics such as skin color, which are simply gradations of the same character.

Bonnet, Charles. *Considerations on Organized Bodies* (1762). While studying the life cycle of aphids, Bonnet discovered that female aphids were able to reproduce without fertilization by the male. In this 1762 work Bonnet suggested that the eggs of each female organism contained the "germs" of all subsequent generations as an infinite series of preformed individuals.

Bonnet, Charles. *Philosophical Revival* (1769). In response to reports that some species had become extinct in the distant past, Bonnet here argued that periodically Earth experienced universal catastrophes that destroyed almost all living beings. Following such catastrophes, the survivors advanced to a higher level on the evolutionary scale. Thus, combining his biological work and his philosophical speculations, Bonnet

became the first natural philosopher to use the term "evolution" in a biological context.

Boylston, Zabdiel. *An Historical Account of the Small-pox Inoculated in New England* (1726). Boylston's statistics in this book demonstrated that inoculated smallpox was significantly safer than naturally acquired smallpox.

Cullen, William. *Synopsis Nosologiae Methodicae* (1769). An elaborate classification of all known diseases, modeled on the work of Carolus Linnaeus and François Boissier de la Croix de Sauvages.

Darwin, Erasmus. *Zoonomia, or the Laws of Organic Life* (1794-96). This work, the author's most important, deals with medicine, pathology, and the mutability of species.

Dobson, Matthew. *A Medical Commentary on Fixed Air* (1779). Dobson here discussed his efforts to treat lung abscesses with carbon dioxide and to relieve respiratory ailments with mineral waters. He also used his techniques to loosen hardened material in the urinary tract.

Fauchard, Pierre. *The Surgeon Dentist* (2 vols., 1728). This book described dental anatomy, tooth decay, medicine, dental surgery, gum disease, and other aspects of dentistry. Some of Fauchard's treatments included the use of lead to fill cavities and oil of cloves and cinnamon for infection. He also made dentures with springs and real teeth.

Floyer, John. *Physician's Pulse Watch* (1707). Floyer published this work as a way to help other physicians learn how to measure the pulse. Floyer believed that with accurate pulse measurements physicians could develop treatment regimens for their patients that would return an unhealthy pulse to a more normal state.

Fothergill, John. *An Account of the Sore Throat Attended with Ulcers* (1748). In this book Fothergill presented careful observations of both diphtheria and scarlatina, although he failed to distinguish between the two.

Frank, Johann Peter. *A System of a Complete Medical Policy* (9 volumes, 1779-1827). This is considered the first systematic treatise on public health and hygiene. Frank advocated strict and rigorous laws and "medical police" to protect the health of the people. His writings summarized all that was known about public health at the time and provided detailed models for hygienic laws.

Goethe, Johann Wolfgang. *Versuch die Metamorphose der Pflanzen zu erklärchn* (Metamorphosis of Plants, 1790). A classic text on botany and biology, this book was artistic and well written and showed how parts of plants are modifications of a basic form.

Hales, Stephen. *Vegetable Staticks* (1727). This book reported Hales's groundbreaking work in plant physiology.

Hales, Stephen. *Statistical Essays, Containing Haemastaticks* (1733). This work was a highly important treatise about the physiology of blood circulation.

Haller, Albrecht von. *Physiological Elements of the Human Body* (8 vols., 1757-66). A landmark in medical history, this work summarized Haller's revolutionary experiments on the nervous system, which led him to conclude that the soul had nothing in common with the body, except for sensation and movement that seemed to have their source in the medulla; therefore, the medulla must be the seat of the soul.

Hunter, John. *The Natural History of the Teeth* and *Practical Treatise on the Disease of the Teeth* (1771). These books described how teeth grow and develop. Dental development had not previously been explored. Hunter also transplanted teeth and found he must remove the pulp before filling the teeth.

Hunter, William. *Anatomia uteri humani gravidi tabulis illustrata* (The Anatomy of the Human Gravid Uterus Exhibited in Figures) (1774). This book is among the best anatomical atlases ever produced, not only for its accuracy and detail, but also for its aesthetic qualities. It stands about 27 in (68 cm) high, and each of its 34 life-size copperplate engravings is magnificently executed. Hunter's text for this atlas was edited by Matthew Baillie and published in 1794 as *An Anatomical Description of the Human Gravid Uterus and its Contents*.

Ingenhousz, Jan. *Experiments Upon Vegetables, Discovering Their Great Power of Purifying the Common Air in Sunshine, and of Injuring It in the Shade and at Night* (1779). This work contains Ingenhousz's remarkable observations on plant physiology and photosynthesis, which he discovered.

Jenner, Edward. *Inquiry into the Causes and Effects of Variolae Vaccinae* (1798). In his publication, Jenner claimed that cowpox provided protection from infection by smallpox. He came to this conclusion after observing 25 cases of cowpox that prevented 24 subjects from contracting smallpox. He also vaccinated a number of people, four of whom resisted smallpox after being variolated.

La Mettrie, Julien de. *L'Homme-machine* (1748). Here La Mettrie presented man as a machine whose actions were entirely the result of physical-chemical factors.

Lind, James. *A Treatise of the Scurvy. Containing an inquiry into the Nature, Causes and Cure of the Disease. Together with a Critical and Chronological View of what has been published on the subject.* (1753). A definitive study of scurvy.

Linnaeus, Carolus. *Systema naturae* (1735). This was the first publication outlining Linnaeus's revolutionary binomial classification system. He continued to refine his classification system for living things, and in 1758 published his most influential work, the 10th edition of *Systema naturae*.

Linnaeus, Carolus. *Flora Lapponica* (1737). In 1732 Linnaeus went to Lapland to conduct a survey of its Arctic plants and animals. He wrote detailed accounts during his five-month survey, which were included in this publication.

Linnaeus, Carolus. *Species plantarum* (1753). As with his classification system for animals, Linnaeus's system for plants, described in this 1753 work, was a landmark achievement.

Maillet, Benoit de. *Telliamed: Of Discourses Between an Indian Philosopher and a French Missionary on the Diminution of the Sea, the Formation of the Earth, the Origin of Men and Animals, etc.* (1748). A controversial book of evolutionary speculations that presents theories on cosmic and biological evolution in the form of a dialogue between Telliamed and a French missionary.

Maupertuis, Pierre Louis Moreau de. *Venus physique* (1745). In this work the author boldly supported epigenesis. Maupertuis believed that embryos resulted from the uniform mixing of particles present in the semen of the male and female, which represented all the body parts and structures. Embryos are thus composites of both parents, and the gradual formation of the embryo was the result of a cohesive natural force that directs this process. This position was a great challenge to Albrecht von Haller and the preformationists.

Mead, Richard. *Corpora Humana et Morbis inde Oriundis* (On the Influence of the Sun and Moon upon Human Bodies and the Diseases Arising Therefrom) (1740). In this treatise Mead suggested that the same forces that caused tides and kept the Moon in rotation about Earth could also exert an influence over the functioning of human bodies.

Mesmer, Franz. *De planetarum influxu* (Physical-Medical Treatise on the Influence of the Planets) (1766). Though Mesmer would revise and expand on his theories throughout his life, as would his followers, the basic principles of "animal magnetism," as Mesmer referred to it, can be found in this work—his medical degree dissertation.

Monro *secundus*, Alexander. *Observations on the Structure and Functions of the Nervous System* (1783). Here Monro reported his discovery of the foramen interventriculare, an important passage in the brain.

Morgagni, Giovanni Battista. *On the Seats and Causes of Diseases Investigated by Anatomy* (5 vols., 1761). Morgagni's remarkable pioneering study of morbid anatomy and a landmark in the history of pathology is based on almost 700 cases. In his comments on each case, Morgagni attempted to correlate clinical observations of the signs and symptoms of disease with his postmortem investigations of specific lesions.

Needham, John Turberville. *An Account of Some New Microscopical Discoveries* (1745). Here Needham he presented his experimental evidence for the theory of spontaneous generation.

Pringle, John. *Observations on the Diseases of the Army*. His ideas about hospital ventilation and camp sanitation are discussed (1752).

Rush, Benjamin. *An Account of the Bilious Remitting Yellow Fever as it Appeared in the City of Philadelphia in the Year 1793* (1794). A masterpiece of medical reporting.

Rush, Benjamin. *Medical Inquiries and Observations upon the Diseases of the Mind* (1812). This was the first American book in what would become known as psychiatry.

Tremblay, Abraham. *Memoir on the Natural History of a Species of Fresh Water, Horn-shaped Polyps* (1744). This work contains Tremblay's pioneering experiments on the hydra, or freshwater polyp (*Chlorohydra viridissima*), tiny creatures about a quarter of an inch long. Trembley discovered that the polyp could produce separate and complete new animals after it had been cut in half.

White, Gilbert. *The Natural History and Antiquities of Selborne* (1789). In 1752 White began keeping a journal of his observations of his garden in Selborne. He published his first book of observations, *A Calendar of Flora and the Garden*, in 1765 and followed this work with *Naturalist's Journal* (1768). His 1789 book became a classic, as naturalists and nature lovers praised White for his keen observations and methodical approach and considered his book a true landmark in the progress of English natural history.

Withering, William. *A Botanical Arrangement of all the Vegetables Naturally Growing in Great Britain* (1776). This was the first sustained attempt to catalog and describe British flora according to Linnaean principles. By 1830 it had gone into seven editions and was the standard work in its field.

Withering, William. *An Account of the Scarlet Fever and Sore Throat, or Scarlatina Anginosa, Particularly as it Appeared at Birmingham in the Year 1778* (1779). Here Withering accurately described the symptoms, prognosis, and after-effects of scarlet fever, and argued that containment of the contagion should be part of an effective treatment for this disease.

Withering, William. *An Account of the Foxglove and Some of its Medicinal Uses with Practical Remarks on Dropsy and Other Diseases* (1785). This book is a model of sober, empirical medical writing, in which Withering described his experiments with foxglove to treat edema, then called dropsy.

Wolff, Kaspar Friedrich. *Theory of Generation* (1759). This work revived the theory of epigenesis during a period in which many of the most respected naturalists were advocates of preformationist theory. According to the theory of epigenesis, an embryo is gradually produced from an undifferentiated mass by means of a series of steps and stages during which new parts are added.

Articles

Fothergill, John. "Of a Painful Affection of the Face" (1776). In this article Fothergill was the first to describe trigeminal neuralgia, also known as Fothergill's neuralgia or tic douloureux, the excruciating pain along the path of the fifth cranial nerve.

Fothergill, John. "Remarks on that Complaint Commonly Known under the Name of the Sick Head-Ache" (1777). Here Fothergill offered one of the earliest accurate descriptions of migraine.

Heberden, William. "Some Account of the Disorder of the Breast" (1772). This article contains a classic description of angina pectoris. Heberden was the first to use the term "angina pectoris" and to provide a complete and accurate description of the symptoms associated with this cardiovascular disease. The condition is now known to be caused by coronary artery occlusion.

Hunter, William. "On the Uncertainty of the Signs of Murder in the Case of Bastard Children" (1784). This was the first important British contribution toward the medical jurisprudence of infanticide.

Pott, Percival. "Cancer Scroti" (1775). A landmark article that provides not only a clinical description of the cancer that Pott observed in chimney sweeps in London but also a compassionate picture of the daily lives of the child sweeps. After a latent period of 20-25 years, a large number of these boys developed cancer of the skin of the scrotum, known at the time as "soot-wart." The disease also occurred in persons who were not chimney sweeps, but to Pott's credit he was able to recognize the special liability of that occupation, despite the long latency period.

NEIL SCHLAGER

Mathematics

Chronology

1722 *Harmonia Mensurarum*, by the recently deceased Roger Cotes—of whom Sir Isaac Newton said, "If Cotes had lived, we might have known something"—is the first work containing the cycles of tangent and secant functions, and one of the first to recognize the periodicity of trigonometric functions.

1728 Daniel Bernoulli, studying the mathematics of oscillations, is the first to suggest the usefulness of resolving a compound motion into motions of translation and rotation.

1752 German geographer Anton Friedrich Busching lays the foundations for modern statistical geography with the initial volume of *Neue Beschreibung*, the first geographical study to stress statistics over description.

1755 Leonhard Euler, who 20 years earlier had established his fame by solving the famous "Königsberg bridge problem," publishes the first of two works (the second will appear in 1768) in which he establishes a paradigm for differential and integral calculus still in use two and a half centuries later.

1765 German mathematician Johann Heinrich Lambert gives the first systematic treatment of hyperbolic functions and their notation, and becomes the first to prove rigorously that the number for π is irrational.

1770 English mathematician Edward Waring states a number theory that becomes known as Waring's problem, and is the first to set forth a method for approximating the values of imaginary roots.

1777 Georges-Louis Leclerc, comte de Buffon devises his famous "needle problem" by which π may be approximated by probability methods.

1785 Mathematician and social thinker Marie-Jean-Antoine-Nicolas de Caritat, marquis de Condorcet devises a new theory of probability, and advances the use of mathematics in the social sciences.

1795 Gaspar Monge makes public his method for representing a solid in three-dimensional space on a two-dimensional plane— i.e., descriptive geometry, a military secret that had long been guarded by the French government.

1797 Norwegian mathematician Caspar Wessel produces the first useful and consistent graphical representation of complex numbers, though his work will go unrecognized for a century.

1797 Italian-French mathematician and astronomer Joseph-Louis Lagrange initiates the theory of functions of a real variable.

1797 Carl Friedrich Gauss writes his doctoral dissertation, in which he gives the first wholly satisfactory proof of the fundamental theorem of algebra—that every algebraic function must have at least one root, real or imaginary.

1799 French astronomer and mathematician Pierre-Simon de Laplace introduces the concept of potential in the first volume of his five-volume classic *Mécanique céleste.*

Overview:
Mathematics 1700-1799

Following on the resounding successes of the preceding century, eighteenth-century mathematics not only continued to break important new ground, but also paused to consolidate the gains made by Isaac Newton (1642-1727) and Gottfried Leibniz (1646-1716) through their invention of the calculus. Leonhard Euler (1707-1783), one of the greatest mathematicians of all time, helped to standardize and formalize mathematical notation, while also making important contributions to virtually every branch of the discipline. At the same time, mathematics education was revolutionized, resulting in a more standard approach to teaching mathematics to students at all levels and adding to the mathematical sophistication of the population who attended school. Add to these the development of fundamental tools, such as probability and statistical theory, imaginary numbers, and the continued development of algebraic tools, and it is apparent that the eighteenth century was a time of great mathematical progress, setting the stage for virtually all subsequent mathematical advances that were to follow in the next two centuries.

Eighteenth-Century Society

The eighteenth century was a century of war. As it opened Russia was at war with the Swedes, Saxony had invaded Livonia, European powers were fighting to determine who would become the King of Spain, and fighting in Europe, Asia, and the Americas continued through most of the century. The eighteenth century was also a century of the arts, with Bach, Handel, Haydn, Mozart, and other great composers writing some of the most famous and beautiful music ever heard. And the eighteenth century was a century of revolution. The philosophical principles of the Enlightenment inspired the American colonies to rebel against their mother country,

while Russian, English, French, and Dutch citizens also asserted their rights to self-government at one time or another.

In and around the political upheavals, the Industrial Revolution was taking place. As the growth of steam power began to replace humans and animals for many laborious tasks, the mechanization of industry and society began to take hold. People began to worry about machines taking the place of humans, concerned that they may be without work because of this. At the same time, engineers continued to develop ever-more sophisticated and elaborate machines, designed to do an increasing number of jobs. As these events took shape, the role of science in general, and mathematics in particular, became increasingly important to governments, the educated population, and society in general.

Eighteenth-Century Mathematics

At the dawn of the eighteenth century mathematics was in the process of taking its place as a formal discipline. This paralleled the emergence of science in general as a field in its own right, worthy of full-time pursuit at a professional level by highly trained practitioners. This recognition owed much to the successes of Newton, Leibniz, the Bernoulli family, and others who showed that science and mathematics were interwoven and that success in these fields could be a source of national pride. In fact, Newton is often considered the first scientist to win popular recognition during his lifetime, even though his discoveries were not widely understood by the great majority of the population. It is entirely possible that this very complexity also helped convince the general population of the legitimacy of science as a career, simply because so much study was necessary to understand the fundamental tools, such as the calculus.

One result of this recognition was the publication of mathematics textbooks. In previous centuries, many mathematicians viewed their craft as proprietary, hiding their discoveries from each other and from other scientists. With the growing importance of mathematics, however, it became apparent that this was not only unnecessary, but counterproductive. Some of the first mathematics textbooks were written in the eighteenth century, making it possible to teach advanced mathematics, in addition to basic arithmetic, in schools. This, in turn, helped turn out a more mathematically sophisticated population, at least among those who attended university classes.

Following these successes, it should be no surprise that scientists and mathematicians became increasingly well known throughout Europe. At times, in fact, mathematicians were almost placed on exhibit by royal courts. Euler was recruited by both Frederick the Great in Berlin and by Catherine the Great of Russia. Members of the Bernoulli family were also recruited by both of these leaders, and Jean d'Alembert (1717-1783) was influential in French mathematics as well as French politics.

At the same time, nations were becoming increasingly dependent on science and engineering, both of which required the development and use of increasingly sophisticated mathematical tools. Newton and Leibniz had taken the first steps to show that mathematics could reveal deep truths about the universe in which we live; Joseph Lagrange (1736-1813), Pierre Laplace (1749-1827), the Bernoulli family, Carl Gauss (1777-1855), and others followed this example in revealing the fundamental connections between mathematics and the physical world. And, as the Industrial Revolution progressed, it became obvious that the new machines upon which a nation's industry, economy, and military strength came to depend required engineers to be trained in the new mathematical tools. As the machines became more complex, they pushed the technology of the day to its limits, requiring mathematically trained engineers to carefully design improvements rather than simply building something that seemed like it should work. By century's end even the navy, often the bastion of traditional thinking, was training its captains to be "scientific sailors," using mathematics and physics to help plot courses, determine a ship's position, shoot cannons, and determine how to arrange the forces on a ship to maximize its sailing characteristics.

Leaving all social and technological impacts aside, the field of mathematics itself made giant strides during the eighteenth century. In one sense, much of eighteenth-century mathematics consisted of consolidating the advances in calculus made in the last years of the previous century. In particular, French mathematicians of the eighteenth century made impressive strides in understanding the calculus. Differential equations, infinite series, the wave equation, and the calculus of variations were all introduced during the this time—and all proved important to other fields as well as mathematics.

Outside of France, most mathematicians worked in the shadows cast by Euler and the members of the Bernoulli family. Both Euler and the Bernoullis contributed in so many areas that even a simple summary could fill an entire book. Indeed, Euler and the Bernoulli family contributed to, originated, or redefined virtually every major branch of mathematics at that time, and much of their work continues to be important today. Euler alone authored or co-authored over 500 books and scientific papers during his life, much of it after going blind in both eyes.

Finally, mathematicians began to recognize the importance of a number of mathematical concepts that are today recognized as vital in many fields. For example, they recognized that π is a number that will never terminate or repeat itself, forever ending hopes of "squaring the circle." Another step forward was the recognition that negative numbers can have square roots, and that they can be manipulated in much the same way as real numbers. This gave rise to the field of complex analysis, which today forms the mathematical foundation of much of the field of electronics. And, in the area of mathematical notation, Euler helped to standardize the symbols used to represent various concepts, making it easier for mathematicians to frame their thoughts and to explain these thoughts to each other in an unambiguous manner.

The eighteenth century is said to have come after the "Century of Genius" and before the "Golden Century" of mathematics. This is true to the extent that no discoveries were made to rival those of Newton and Leibniz, while the sheer volume of landmark mathematical work was far short of what was to come. However, eighteenth-century mathematics formed a crucial link between these two centuries. It was a time during which mathematicians began to develop the conceptual tools to take full advantage of the discoveries made during the previous century, and to construct a solid foundation upon which to build during the Golden Age of the nineteenth century.

P. ANDREW KARAM

France's Ecole Polytechnique Becomes the Most Influential Mathematics Institution of Its Time

Overview

The creation of France's Ecole Polytechnique in 1794 was one of the brighter consequences of the chaos of the early days of the French Revolution, and the effect of the Ecole Polytechnique on mathematics is brighter still. With its mathematical curriculum shaped at its inception by French mathematician and administrator Gaspard Monge (1746-1818), among others, the school was originally intended as a training-ground for civil engineers. In addition to engineering, though, the Ecole Polytechnique under the guidance of Monge and other distinguished mathematicians and teachers became a hotbed of mathematical (and, later in its history, revolutionary) activity, as students and professors formalized the teaching of mathematics. That process of formalization—along with the lectures, textbooks, exercises, and syllabi that accompanied it—reinvigorated mathematics as both an academic and theoretical discipline: an entire generation of Ecole Polytechnique graduates and instructors extended both the institution's influence on the teaching of mathematics and the range of pure mathematics itself. Descriptive and analytic geometry, for example, were first formally taught at the Ecole Polytechnique, rescuing geometry from a decline that some felt stretched back as far as Plato. Founded in revolution, the Ecole Polytechnique revolutionized the teaching of and the status of higher mathematics not only in France, but by example throughout the world.

Background

In the 1780s France poised precariously on the brink of chaos. Food shortages, an economic crisis, the heavy financial burden of the nobility and tens of thousands of pensioned military officers, additional financial demands by the Church and by debts incurred by three consecutive kings helped fuel a widespread social dissatisfaction. Intellectuals felt there was too much control of thought and expression; the middle class felt too heavy a tax burden, and the peasants felt themselves still imprisoned by feudal social structures. Public dissatisfaction reached its boiling point in May 1789, when King Louis XVI convened a legislative body, the Estates-General, to Versailles to approve a new tax code.

The convention quickly became something else as frustration and anger gathered force. By mid-June the Estates-General had proclaimed itself the National Assembly, and began re-shaping France. By July 14 the zeal for reform had become outright revolutionary fervor as the citizens of Paris stormed the Bastille. In August the National Assembly abolished the last vestiges of feudalism and serfdom, and shortly thereafter issued the Declaration of the Rights of Man. Within two years a new Constitution had been adopted; within four Louis XVI had been sentenced to the guillotine. Moderate reformers gave way to fanatic revolutionaries, and between fall 1793 and summer 1794 a Reign of Terror resulted in the deaths of between 20,000 and 40,000 people. As the tide of terror receded and calmer (un-guillotined!) heads prevailed, France had ceased to be a monarchy and become a republic.

It was not all terror. Among the revolutionary impulses was a call for educational reform on a grand scale. France's archaic university system was overhauled, with new institutions created to provide more specialized training than the universities offered. Already in pre-Revolutionary days some "Grandes Ecoles" had been created: in 1748 the Ecole du Génie Militaire (Army Corps of Engineers School) and the Ecoles de Constructeurs de Vasseaux Royaux (Royal Shipbuilding School) in1765.

The Revolution gave added impetus to the Ecole movement, and in 1794 the Conservatoire National des Arts et Métiers (National Conservatory for Arts and Crafts) and the Ecole Centrale des Travaux Publiques (Central School for Public Works) were established. The School for Public Works would include a curriculum stretched far beyond civil engineering and would soon be renamed the Ecole Polytechnique, under which name it would become France's, and the world's, leading center for mathematical thought and instruction.

The prospects for the new institution were greatly helped by the presence of Gaspard Monge on the committee setting out the charter for the school. A gifted mathematician and teacher, Monge also had a great deal of administrative and bureaucratic experience. He had been an important member of the French Académie des Sciences, a member of the

Académie Commission on Weights and Measures, and was well known for applying his knowledge of geometry to practical ends such as the design of fortifications.

As a mathematician Monge was startlingly original and devoted to extending the range of geometry and related mathematical disciplines. He published numerous papers on infinitesimal and descriptive geometry, as well as work in a variety of other scientific fields.

A supporter of the moderate forces of revolution, Monge was offered, in 1792, the post of Minister of the Navy, which he accepted. For once his administrative skills failed him, and his tenure as Minister was undistinguished and brief. He continued to be involved in the Académie, lobbying for educational reforms, but to little avail: the National Commission abolished the Académie in 1793. The next year the Ecole was founded, and Gaspard Monge named an instructor there, charged with teaching descriptive geometry and training those who would later become teachers at the Ecole. Also teaching geometry at the Ecole Normale (Normal School) for the training of teachers, Monge began assembling his lectures into a book, *Application de l'analyse a la géométrie*, on which he would work for much of his life, and which housed his insights into the use of analysis as a geometric tool.

Another book, *Géométrie Descriptive* (1795), adapted graphic drafting techniques into theoretical geometric reasoning. This book's popularity led to its adoption throughout Europe as the standard text on its subject. Monge was by all accounts a superb lecturer, and the Ecole quickly became a fount of exuberant students, many of whom went on, in turn, to write their own geometry texts. The Polytechnicians, as they came to be known, poured out books on analytical and other geometries, elevating geometry from its position in the shadow of the calculus to its preeminence in modern mathematics. The modern age of pure geometry had begun, and the Ecole Polytechnique was in many ways its starting point.

By 1797, after performing various administrative and semi-diplomatic services for Napoleon, whom he admired and knew well, Monge was appointed Director of the Ecole. He would not serve long: once Napoleon took control of the French government, Monge was appointed a senator for life, and he resigned his Ecole directorship.

But Monge was far from the only distinguished mathematician on the faculty of the

Joseph-Louis Lagrange. *(Library of Congress. Reproduced with permission.)*

Ecole Polytechnique (or, for that matter, the Ecole Normale; many professors taught at both institutions.) Among Monge's influential colleagues at the Ecole Polytechnique was Joseph-Louis Lagrange (1736-1813), who lectured on algebra and the calculus, particularly differential calculus. Named professor of analysis at the Ecole, Lagrange had a wide-ranging mathematical mind and had made many contributions, notably to the place of variables in the calculus, the nature of orbital positions in celestial mechanics, and the study of four-dimensional spaces, which would prove crucial to extending math to the study of the physical mechanics of solids and fluids. In 1793, as president of the weights and measures commission, Lagrange insisted that France adopt a 10-place, or decimal, rather than 12-place system; he is considered in many ways to be the father of modern metrics.

Lagrange's great contribution to education, to the mathematical discourse, had begun to take shape before the creation of the Ecole. In 1788 he published his book *Mécanique analytique*, in which he discussed the mathematics of physical mechanics and did so without recourse to a single drawing or diagram. Such an approach had never been taken: since the Greeks, all mathematical treatises had been leavened with illustrations. But the language of Lagrange's entire book was algebra, pure and unillustrated, and it would exert an enormous influence on future mathematical texts and treatises.

Also at the Ecole Polytechnique (and particularly at the Ecole Normale), Pierre-Simon Laplace (1749-1827) had built a substantial mathematical reputation on the strength of his work in probability studies and especially his work with celestial mechanics. While his studies of celestial mechanics are perhaps his greatest contribution to science (his theory that the planets coalesced out of clouds of gas is known as the nebular hypothesis), it was his awareness of the practical applications of probability to matters such as mortality rates, games theory, and even judicial decisions that exerted a strong and lasting effect on probability studies.

By the early nineteenth century the Ecole Polytechnique was well established as a mathematics center, with teachers, mathematical theorems, and books of both instruction and theory emerging from its halls at a steady pace. Geometry remained at the center of many of these endeavors, and the discipline's and the institution's debt to Monge was large.

Some things had changed. With the fall of Napoleon, Monge was disgraced, and in 1816 he was cast out of the French scientific community. Heartbroken, Monge fled France, but returned to die in Paris barely two years later. The French Government decreed that no notice be paid to honor the mathematician. The students at the Ecole, an institution born of revolution, paid no attention to the decree, and gathered to honor the passing of Gaspard Monge.

Impact

While it is an overstatement to attribute the explosion of nineteenth-century mathematics to an institution—Monge, Laplace, Lagrange, and others were doing serious work before the Ecole Polytechnique was founded, and would doubtless have made major contributions at other institutions—the Ecole's role as a central focus, an epicenter for the geometric and mathematical flowering, cannot be doubted. It was a place where ideas came together and were tested against each other by some of the finest mathematical minds of the time. That in itself helped generate excitement and refine emergent theories.

But because the Ecole's faculty also taught mathematicians who would go on to become professors themselves as well as writers of textbooks, its influence was enormous and ongoing. Primary among those influences was Monge's shepherding of geometry out of eclipse and into prominence as a mathematical discipline. Monge's textbook and the lectures it was based upon inspired even better geometry tests: between 1798 and 1802 no fewer than four major geometries appeared, inspired by Monge and the Ecole, and authored by leading mathematicians Lacroix, Puissant, Lefrançois, and Biot. Other texts followed, with analytic geometry blazing forth in ways it had not when presented earlier by Pierre de Fermat (1601-1665) and René Descartes (1596-1650).

Likewise, Lagrange's bold decision to write in purely mathematical terms without illustration was, although not a product of the Ecole Polytechnique per se, important to the teaching of mathematics. It placed the focus directly on the math, on pure mathematical analysis, on establishing properties and deducing them from the smallest number of principles.

The influence of the Ecole Polytechnique on mathematics would be felt throughout the nineteenth century. Its influence as a revolutionary body remained strong as well, with its students taking up swords and leaping to the barricades in the revolution of 1830 and the chaotic, near-revolutionary fervor of the June Days of 1848. In the best republican revolutionary tradition, these students of the Ecole pledged themselves to stand against oppression as workers of the mind, just as the "workers of the fist" stood.

Victor Hugo wrote: "Whoever wants to vanquish France must first destroy three institutions: the Institut, the Legion d'Honneur, and the Ecole de Polytechnique." Few institutions of learning have received higher praise, or done more for the cause of pure mathematics.

KEITH FERRELL

Further Reading

Carlyle, Thomas. *The French Revolution: A History*. New York: Wiley and Putnam, 1846.

Schama, Simon. *Citizens: A Chronicle of the French Revolution*. New York: Knopf, 1968.

Swetz, Frank J., ed. *From Five Fingers to Infinity: A Journey through the History of Mathematics*. Chicago and Lasalle: Open Court, 1994.

Eighteenth-Century Advances in Statistics and Probability Theory

Overview

Probability theory tells us how likely it is that an event will occur. Statistics tell us, among other things, how likely it is that a particular set of data accurately reflects reality. The two fields are closely linked because statistical results indicate the probability that our data is accurate. Both have had a profound impact on society, influencing such diverse ventures as quantum mechanics, the gambling industry, insurance, and the space shuttle.

Background

Games of chance date to early in human history, and include dice, cards, and other diversions that produce random results. If played fairly, all players have an equal chance of winning. These games inspired the first mathematical investigations into probability because mathematicians wondered if a deeper mathematical truth lay behind random events and, if so, if patterns in such events could be discerned and predicted.

Jerome Cardan (1501-1576), an Italian mathematician whose most significant contributions were in algebra, performed the first probability study in the sixteenth century. Unfortunately, his work was neglected and, a century later, reinvented by Blaise Pascal (1622-1662), the French mathematician and physicist, whose inspiration came from a dice game. Pascal's work was built upon over the years and, by the end of the eighteenth century, statistics had emerged as an independent field, albeit closely allied with probability. With the flowering of science during the Enlightenment, statistics became very much a field devoted to the analysis of data. All possible information was gleaned from the actual data, and much emphasis was placed on patterns that could be discerned from it.

Probability theory, on the other hand, became more closely allied with gaming until the advent of the Industrial Revolution. Then, as the industrialized nations became more mechanized, probability theory began to be used as a predictor, asking what the chance was that a particular piece of equipment, for example, would break down in a particular year. Still later, with the introduction of quantum mechanics, probability theory came into its own because, as it turned out, virtually all events at the level of individual atoms or particles are uncertain. Because of this, scientists could say, for example, that the probability of a particular atom undergoing radioactive decay in a particular period of time was 50%. In a sense, quantum mechanics helped rescue probability theory from the factories and gambling halls.

Also at about this time, the insurance industry was beginning to emerge. Insurance companies, in effect, bet that they will collect more money from their clients than they will spend on an unfortunate accident. It became very important for insurance companies to understand the odds (or the probability) that an event would (or would not) take place. This, in turn, produced actuaries, who are analysts that tell insurance companies how much to charge for their premiums. For example, an insurance company will charge an elderly person more for health insurance because statistics show that the elderly are more likely to become ill and that their illnesses generally cost more to treat. Without probability theory and good, solid statistics, such determinations would be very difficult, if not impossible to undertake.

Impact

Statistics underlies virtually all scientific research, and also influences most social science research, opinion polling, insurance rates, epidemiology, public health, and a number of other fields. Without statistics, our world would be very different, as would the way we see our world. Consider, for example, the following headlines:

> Violent crime rates drop for third
>   straight year
>
> Study shows drinking orange juice cuts
>   cancer risk
>
> Scientists discover new planet outside
>   the solar system
>
> New drug helps fight AIDS
>
> Physicists discover top quark
>
> Polls show public favors tax bill

All of these depend on statistical sampling or the statistical analysis of data. While statistics's relation to crime rates or public polls may be obvious, the others topics are equally dependent on statistics, because scientists must convince other scientists that their findings are correct and not merely some fluke.

For example, most planets found outside the solar system are detected because their gravitational pull causes their star to wobble very slightly. This wobble is barely noticeable, and can be detected only through a sensitive statistical analysis of data over the course of up to several years. Astronomers must consider random fluctuations due to turbulence in the Earth's atmosphere, the effects of our motion around our sun, effects due to changes in the telescope, and a host of other factors. When all of these have been considered, the astronomers must then show that, statistically speaking, there is more than a 95% chance that what they claim to have seen is an actual effect. This requires still more statistical treatment of the data, ending up with a claim that is likely to be accepted by the scientific community at large.

Similarly, every discovery of new subatomic particles must be shown to be statistically significant, as must discoveries of behavioral patterns, disease patterns, and so forth. Scientists in a wide variety of fields use very similar statistical methods to show that their data are valid and that their theories are likely to be correct interpretations of that data. In fact, a great deal of quantum mechanics, subatomic physics, and the fields of thermo- and gas dynamics depend completely on probability theory and the statistical treatment of physical phenomena. Although we cannot predict what an individual atom will do, we can predict quite nicely the behavior of tens of billions of atoms, using well know laws of statistics and probability.

Engineering uses a branch of statistics and probability theory called probabilistic risk assessment (PRA). In this field, a complex system, such as the space shuttle, is examined to determine the probability that a particular component will fail. Then, all the things that failure will affect are assessed and the probability that each of those will happen is analyzed. For each of *those* things, the assessors again try to figure out everything that can go wrong and, again, probabilities are assigned to each failure mode. This is propagated until either there are no more components that can fail, until all of the high-probability failure modes have been exhausted, or until some other stopping point is reached. Since many end results can have multiple causes (for example, a host of factors can cause an engine to shut down prematurely), the odds of a particular outcome are summed across *all* of the failure pathways to reach a final answer. This answer will tell the engineers the chance that, say, a main shuttle engine will stop for any reason.

In the case of PRA, many of the failure probabilities are determined by a statistical analysis of various components and systems that is usually compiled during design and testing. A thousand turbopump bearings, for example, may be subjected to the stresses of takeoff to see how many takeoffs they can survive before cracking or splitting. When all of the bearings have failed, a bell curve can be drawn, showing the mean lifespan of such a bearing under that level of stress. This curve, then, becomes input to a PRA analysis on the turbopump—one of the factor that can lead to pump failure during takeoff. The raw data come from statistics, the final answer from probability theory.

The other field in which probability and statistics are so closely linked is the insurance industry. As mentioned previously, the insurance companies are betting that customers will spend more on premiums than they will pay in claims. In order to attract customers, insurance rates must be low while, to stay in business, they cannot be so low that paying off policies costs more than the companies bring in.

This has led to the specialized field of actuarial science. Actuaries perform sophisticated statistical analyses of all sorts of factors, trying to arrive at a reasonable probability that certain events will take place. Once they know the likelihood that, for example, a certain type of car will be stolen, they can set an insurance rate. Say, for example, that 1% of all Corvettes are stolen and not recovered in a given year, and that the average Corvette costs $20,000. That means the insurance company has to charge 1% of the cost of the average Corvette, or about $200 each year simply to cover what it will cost to replace these Corvettes. If the company charges $150, it will lose money; if it charges $250 it will make a profit. Insurance companies perform similar calculations for everything—the incidence of various diseases at different stages of life, the risk of an airplane crash, or the chance that a satellite will not reach orbit. In these, as in so many other fields, a solid grasp of probability and statistics is essential.

P. ANDREW KARAM

Further Reading

Boyer, Carl, and Uta Merzbach. *A History of Mathematics.* New York: John Wiley & Sons, 1991.

Paulos, John Allen. *A Mathematician Reads the Newspaper.* New York: Anchor Books, 1995.

Key Mathematical Symbols
Begin to Find General Use

Overview

Mathematical symbols seem impenetrable to most nonmathematicians. A seemingly arbitrary collection of shapes and letters in several alphabets, each with arcane meanings (some with multiple meanings), these symbols often seem designed to obscure meaning rather than lead to a deeper understanding. However, to a mathematician, these symbols are as easy to read as this paragraph is to the average student. In addition, by representing complex or sophisticated concepts or mathematical operations in a short, easily-recognizable form, mathematical symbols make it easier to concentrate on the actual mathematical arguments taking place rather than the words describing what the symbols represent. Many of today's commonly used mathematical symbols came into being in the eighteenth century, and a disproportionate number of them were introduced by the great mathematician, Leonhard Euler. By formalizing the language of mathematics, Euler helped set the stage for the great flowering of mathematical thought that began in the latter part of the eighteenth century and carried forward to this day.

Background

A symbol represents something. Typically, a symbol is a way to quickly represent an object, an idea, or a concept. The first mathematical symbols were the numbers, which were a simple way to represent how many of any object there might be. The idea of using written symbols to represent numbers of objects dates back at least a few tens of thousands of years, and may be even more ancient. In fact, it is possible that numbers were the first symbols used by humans.

It wasn't until relatively recently that other mathematical symbols came into use. A few thousand years ago, the ancient Egyptians wrote some of the first arithmetic problems, using their equivalent of +, -, and fractions. Gradually, with time, other simple mathematical symbols came into being, all designed to make the communication of calculations more consistent and understandable.

As the field of mathematics grew in complexity, so did the need for ever more sophisticated symbols, describing these new mathematical concepts and operations. The greatest flower-

ing of mathematical symbolism followed the invention of the calculus by Isaac Newton (1642-1727) and Gottfried Leibniz (1646-1716). Each man developed calculus largely independently of the other, and each had his own preferred notation for writing problems and their solutions. It was, in a way, left to the great Swiss mathematician, Leonhard Euler (1707-1783) to standardize the notation and mathematical symbolism, giving rise to most of the symbols taught in college calculus classes to this day.

In a series of papers published during his long career in mathematics, Euler became the first to introduce use of the letter e to represent the base of the natural logarithms, the Greek letter π to represent the ratio of a circle's circumference to its diameter, and the first to use the symbol i to represent the square root of -1. In addition to these symbols, Euler was apparently the first to represent a mathematical function as f(x), which is instantly recognizable to anyone who has taken high school algebra. Other symbols and conventions pioneered by Euler were the abbreviations sin, cos, tan, sec, csc, cot, and so forth for the trigonometric functions and the Greek letter Σ to represent a summation of terms.

Each of these symbols carries as precise a meaning to those using it as the numeral 2 does to anyone writing it down. Each of these symbols, too, allows mathematicians to convey their thoughts to other mathematicians, physicists, engineers, and others in a completely unambiguous manner, so that their thoughts can be read, understood, and appreciated. By developing these symbols and giving them their current accepted meanings, Euler helped to advance the field of mathematics because, without clear communication, progress is much more difficult to achieve and convey to others.

Impact

It is difficult to understate the importance of mathematical symbolism on both the field of mathematics and on the other fields that use mathematics. The impact on mathematics will be discussed first, followed by the other effects of this development.

As mentioned above, symbols give us a way to quickly and accurately convey a complex idea to someone when we are communicating. If someone sees a Star of David, for example, they will think of the Jewish religion and everything that goes along with Judaism. Similarly, a cross symbolizes Christianity, an elephant symbolizes the American Republican Party, and a blue flag with a yellow cross symbolizes the nation of Sweden. All of these symbols, mathematical included, depend on recognition to succeed; a person unschooled in mathematics will see π simply as a Greek letter or, depending on their educational background, as simple a grouping of three lines. If a symbol is not recognized, then it is without meaning.

Similarly, a symbol must be unambiguous to succeed. If a symbol can have more than one meaning, then the reader has to be able to figure out the meaning from the context in which they see the symbol. For example, a symbol similar to the swastika was used as a symbol by Native Americans and in parts of Asia long before the rise of Nazi Germany. To most of the world, the swastika represents all of the evil that came from Adolf Hitler, in spite of the fact that, for several thousand years before Hitler was born, it was associated with peace. It is only through the consensus of those who experienced, read about, or in some other way know of Nazi Germany that the swastika has come to have the meaning it does for most of the world.

In the case of mathematical symbols, Euler's contribution was to take complex mathematical concepts and express them as simple, recognizable symbols. Because of his stature as one of the greatest mathematicians ever and because his symbols were easy to use, Euler's symbols became widely used.

To give a concrete example of the significance of this, consider the following equation:

$$f(x) = \sum_0^{10} x$$

For this equation to have meaning, we must be able to understand what each symbol means. We know that, for example, $f(x)$ means that this is a function where x is the variable. Thanks to Euler, we also know that the capital Greek letter sigma (Σ) means a summation and the zero and ten tell when to start and stop the summation. So, taken together, this short equation means this: "A variable, x is mathematically manipulated such that we add up (sum) all of the values of x from 0 to 10." This is a somewhat more cumbersome statement, however, than the elegant equation above.

Or, consider the statement "To find the surface area of a sphere, you multiply the number four by the ratio of a circle's circumference to its diameter, then by the square of a circle's radius, and divide this by three." Obviously, it's easier to write, instead, the equation:

$$SA = \frac{4}{3}\pi r^2$$

which is much simpler to express and to decipher.

The symbols introduced by Euler are among the most important and most commonly used symbols in mathematics, but they are also among the most important in many other fields that depend on mathematics. Physics, engineering, chemistry, and some aspects of geology, biology, social sciences, economics, and other disciplines that use mathematics all use Euler's notations. By making mathematics simpler to understand and communicate, these symbols make research in all of these fields easier to share with colleagues, even colleagues in other fields. This, in turn, has helped make possible many of the great advances in these fields that have taken place in recent centuries. Consider, for example, if only mathematicians used Euler's symbols and if every field of science had a unique way of expressing itself mathematically. This would make it nearly impossible, for example, for a geologist to learn from a physicist, and vice versa. The science of geologic dating uses concepts from both disciplines and is the foundation upon which our understanding of terrestrial and lunar evolution rests as well as providing the basis for our understanding of plate tectonics, evolution, and other important events. Would these concepts have gone undiscovered in the absence of consistent mathematical symbols? Of course not. But their discovery may well have been delayed by decades, waiting for the one person who could understand the mathematical symbolism of physics and the concepts of geology.

Many of the most fruitful human endeavors have relied on the ability of scientists in different specialties to understand the work of those in other fields. A team of geologists and physicists was the first to discover that an asteroid struck the earth, ending the dinosaur's reign. Geologists working with physicists, chemists, and biologists made a convincing set of arguments that the crust of the Earth was composed of plates that moved hither and yon through the ages. Scientists and engineers from several nations collaborated to build the international particle accelerator beneath the Alps that continues to generate important information about the structure of matter, itself. A team of chemists and physicists teased out the properties of radioactivity and radioactive elements, and mathematicians, chemists, and physicists developed the uranium, plutonium, and hydrogen bombs. While the benefits of some of these discoveries may be questioned, their impact on society is not. And the commonality of all is that scientists from disparate fields and, in many cases, from different nations were able to communicate complex, sophisticated ideas to one another using a single set of mathematical symbols and notation.

P. ANDREW KARAM

Further Reading

Boyer, Carl and Uta Merzbach. *A History of Mathematics.* New York: John Wiley and Sons, 1991.

Eighteenth-Century Advances in Understanding π

Overview

Mankind has been fascinated with π for millennia and attempts to calculate it exactly have taken place for nearly as long as mathematics has existed. In the eighteenth century, rapid advances in mathematics led to a deeper understanding of numbers in general and of π in specific. This enhanced understanding captured the public's attention, discouraged those trying to solve the ancient problem of "squaring the circle," and helped to advance mathematics enormously.

Background

Earlier than 2000 B.C. Babylonian and Egyptian mathematicians realized that the relationship between a circle's diameter and its circumference (which is what π is) was unchanging, regardless of the circle's size. They also quickly realized that this number fell somewhere between the values of 3 1/7 and 3 1/8, values that were to crop up repeatedly for the next three millennia.

One of the first formal mathematical attempts to calculate a value for π was conducted by Archimedes (287-212 B.C.) in the third century B.C. in his attempt to square the circle (described later in this essay). Archimedes realized that he could draw a polygon on the outside of a circle with the center of each side just touching the outer edge of the circle as it was drawn. He could repeat this process with a polygon on the circle's *interior*, in this case, with the "points" just touching the interior of the circle. The length of the sides of each polygon was easy to determine, and the length of the circle's circumference had to fall between the length of sides of the two polygons. Archimedes then went one step further and understood that, as the number of sides of these polygons increased, they came to resemble circles more and more closely until, with an infinite number of sides, they would be circles. Therefore, by adding sides repeatedly, Archimedes proposed to calculate an exact value for π, if only enough sides could be added. This approach is amazing because it anticipated the development of calculus by 2,000 years.

Unfortunately, Archimedes was killed by an invading soldier and was never able to finish his calculations, but he came up with the most accurate value for π that would be calculated for the next 2,000 years. In fact, Archimedes's method for calculating π was not improved upon for many centuries: a tribute to his genius. In fact, through the reign of the Roman Empire and the Dark Ages, very little original work was done on the problem of the nature of π or in calculating it more precisely than previous estimates.

Later efforts to determine the value of π precisely were no more successful than Archimedes and, in many cases, simply used his techniques or copied his numbers. One of the motivating factors behind these efforts to calculate π was the challenge of "squaring the circle," that is, using only an unmarked straight edge and a compass, to construct a square with exactly the same area as a circle. This remained one of the outstanding problems in mathematics from the time of the ancient Greeks until the late nineteenth century, when Ferdinand Lindemann (1852-1939) proved π to be a transcendental number. A transcendental number is one that cannot form the root for an algebraic equation. For example, the equation $x^2 - 2x + 1 = 0$ has solutions (or roots) that are equal to 1 and -1 because either of these numbers, in place of x, will make this equation correct. The solution to the equation $x^2 - 2 = 0$ is $\sqrt{2}$, an irrational number because it never terminates or repeats. However, this is an algebraic number, *not* a transcendental number because it forms the root of an algebraic equation. Since all transcendental numbers are also irrational (meaning they neither terminate nor repeat digits, unlike 1/2, 1/3, and so forth); this finding meant that π would never be calculated exactly. However, for the prospective circle-squarers, π's transcendence was of more fundamental importance; if π could never be calculated precisely and if it could not be forced to be the root of an algebraic equation, then it would be forever impossible to square the circle using only simple tools.

The next refinement in attempting to calculate π came in the late sixteenth century when François Viète (1540-1603), a French mathematician, became the first known person to attack the problem using trigonometric techniques. This resulted in an infinite series, that is, a never-ending series of terms, each related to the previous one algebraically and each smaller than the last. As each new term is calculated and added to the previous term, the series is said to "converge" on the correct answer to the problem, although an exact solution may never be reached. Using Viète's methods, and other infinite series, π was calculated to 140 places by the end of the eighteenth century.

The eighteenth century saw a number of significant advances in understanding π as a number. John Machin became the first person to calculate π to 100 decimal places in 1706. That same year saw the first use of the Greek letter π to describe the ratio of a circle's circumference to its diameter; prior to that time there was no formal name for this ratio. In 1719, the Frenchman Thomas de Lagny (1660-1734) calculated π to 127 decimal places, and in the middle of the century, the brilliant Swiss mathematician Leonhard Euler (1707-1783) developed an algorithm for calculating π that was superior to anything that preceded it.

At this same time, too, mathematicians were beginning to better appreciate some of the properties that made π so interesting and so intractable. In 1766, Johann Lambert (1728-1777) proved π to be an irrational number, that is, one that could not be expressed simply as the ratio of two integers. By proving this, Lambert also proved, as noted above, that any attempt to find an exact value for π was fruitless. A few years later, in 1775, Euler suggested that π might be transcendental, but at that time, transcendental numbers had not yet been proven to exist. This would not happen for another 65 years when Joseph Liouville (1809-1882), the great French mathematician, showed their existence. All of these events were important and all were destined to have an impact.

Impact

First, we should discuss the phenomenon of squaring the circle a little more. This problem was first posed by the Greeks, sometime before 400 B.C. It gained some degree of notoriety because it remained unsolved for so long a time. In fact, proposed solutions to this ancient problem had become so notorious that, in 1775, the Academy at Paris passed a resolution that solutions to this problem would no longer be officially considered by the Academy. However, in 1931 the American Carl Heisel self-published a book claiming to have successfully squared the circle, and claimants still come forward to this day. Precisely because of its longevity, this problem continues to fascinate, even in the face of clear mathematical proof of the impossibility of a solution.

The progress made in understanding π has helped to better understand irrational and transcendental numbers in general, numbers that are of great importance to mathematics, physics, engineering, and other fields of inquiry. As one example, the base for the Napieran logarithms, *e*, is a transcendental number that is also of exceptional importance in a wide variety of fields, including calculations as routine as determining the interest on a loan.

In the case of π, it turns out to be of fundamental importance to our understanding of many vital properties of physics and engineering. One reason for this is that so many objects are circular or spherical, and π is used to calculate the surface area and (if appropriate) volumes of such objects. As one example, in 1998 the radiation dose was measured from a gamma ray burst at a great distance from Earth. (As the burst expanded in space, one can think of it as filling the surface area of a sphere because the light from the burst expands equally in all directions.) Using the equation for calculating the surface area of a sphere, and knowing how much of the sphere is taken up by the instrument, one can then find out exactly how much energy was released by this gamma ray burst (as it turns out, about 10^{53} ergs of energy, making it the most energetic event ever witnessed). The solution to this calculation is utterly dependent on knowing the value of π to several decimal places.

Pi is also used to describe cyclic motion, such as the oscillations of a pendulum, a clock spring, a rotating CD, or the pistons in a car engine. Physicists in the eighteenth century realized that, for example, a swinging pendulum repeated the same positions at regular intervals, exactly as a spinning disk did. This meant that they could use the same equations to describe both rotating motion and regular oscillations. Because π is, by definition, the ratio of a circle's circumference to its diameter, any problems that deal with rotation or oscillations can only be solved by using accurate values of π.

Finally, in the present day, the fact of π's transcendence and irrationality make it an ideal candidate for proving the power of ever-faster computers. There seems to be some fascination on the part of the public to learn that the newest computer has just added another million places to the value of π, even though five or six places suffice for most serious computations. The fact that π never ends just makes it that much more appealing a target. This suggests, too, that in the last 2,000 years our fascination with π has not changed appreciably.

P. ANDREW KARAM

Further Reading

Books

Beckmann, Petr. *A History of Pi*. New York: St. Martin's Press, 1971.

Blatner, David. *The Joy of Pi*. Walker and Company, 1999.

Women in Eighteenth-Century Mathematics

Overview

Even before Aristotle (384-322 B.C.) declared women to be "imperfect men" and incapable of rational thought, women in the ancient world were denied education. There were exceptions, like Hypatia of Alexandria (A.D. 370?-415), a legendary mathematician and astronomer. During the Middle Ages women were educated in convents, giving them an opportunity, however limited, for intellectual expression. Hildegard von Bingen (1099-1179), was a twelfth-century mystic, writer, and composer—but her achievements, like Hypatia's, were well outside the norm.

During the Enlightenment, Parisian noblewomen presided over gatherings of scientists and thinkers in their *salons*, cultivating a social climate that became a driving force of progress in the Age of Reason. Throughout most of Europe, though, scholarly achievement for women remained unthinkable. A few remarkable women, however, managed to become prominent mathematicians, despite society's restrictions.

Background

During the 1700s the University of Bologna boasted several women professors, among them the famous mathematicians Laura Bassi (1711-78) and Maria Agnesi (1718-99), the anatomist Anna Manzolini (1716-1799), and Clotilda Tambroni (1758-1817), who taught Greek literature. Despite the university's liberal policies, Laura

Maria Agnesi. *(Corbis-Bettmann. Reproduced with permission.)*

Bassi had to fight the Bologonese senate to keep her position, and was forced to petition the administration for funds to conduct her experiments. She prevailed. Fortunately the climate in Italy was just tolerant enough for these exceptional women to succeed. Bassi's contemporary,

Maria Agnesi, in fact, was one of the most accomplished female mathematicians of all time.

Maria Gaetana Agenesi, born in Milan to a wealthy and educated family, was the eldest of 21 children. Both parents encouraged her extensive education in languages, philosophy, and mathematics. Agnesi was a brilliant student, especially in math. At age seventeen she wrote a paper about ballistics and planetary motion that was admired by contemporary scholars.

Agnesi authored *Analytical Institutions*, a classic mathematical text on differential and integral calculus in two volumes; the first volume was about mathematical process, the second about analysis. Her book was adopted as a textbook in France where the Academy of Sciences arranged for its translation into French. Still, the Academy could not accept Agnesi as a member because regulations prohibited females from joining.

At Cambridge University in England, mathematics professor John Colson was so impressed by Agnesi's text that he learned Italian in order to translate her book. It was Colson's translation error that caused Agnesi to be associated with the term "witch." Agnesi had used the Italian word *versiera* to refer to a versed sine curve. Colson mistakenly thought the Italian word was *versicra* which means "witch," and so the curve became known as "the witch of Agnesi." Agnesi held her position as honorary chair at the University of Bologna for two years. After her death, Italians named several streets, scholarships, and a school after Maria Agnesi.

Eighteenth-century France was somewhat different. Frenchwomen were denied access to universities, but noblewomen of the Enlightenment found another way to be part of the era's amazing intellectual developments. Wealthy women established and presided over *salons,* or parlors—glittering rooms in affluent homes that were frequent gathering places for the literate and educated. Despite the social protocol that limited them to playing hostess at these meetings, two Frenchwomen managed to became great mathematicians.

Emilie de Breteuil, marquise du Châtelet (1706-1749) was very tall and not considered good-looking as a child. As her parents were not optimistic about her marriage prospects, they agreed that she should have a tutor in order to have some life as a single woman. De Breteuil was clearly a genius; she quickly learned Latin, Greek, and English, and she excelled at and loved mathematics. At the age of 19 she married the marquis du Châtelet, who allowed her to continue to study mathematics. In order to meet

with other scientists, she would often disguise herself as a man because women were never allowed to join intellectual discussions. Her brilliance and wit were widely acknowledged and even won over Voltaire; the two were inseparable companions for years. Châtelet wrote several books, including the only French translation of Isaac Newton's (1642-1727) *Principia* and a text on introductory physics. She died at the early age of 43, days after the birth of her fourth child.

Sophie Germain (1776-1831), often referred to as "The Hypatia of the eighteenth century," was born in Paris. Germain decided to become a mathematician at age 13, and against her parents' wishes at first, she taught herself mathematics and calculus. She couldn't attend courses at the Polytechnique but did get the lecture notes. She submitted a research paper using a man's name, fearing that the professor would reject it if he knew that a woman had written it. The professor, however, was most impressed by her work and agreed to sponsor her. She still continued to write under a man's name and carried on a long correspondence with German mathematician Carl Friedrich Gauss (1777-1855), who acknowledged that Germain was a genius.

The French Academy of Sciences awarded Germain its grand prize in 1811 for her theory of elasticity. Her work in number theory led her to develop a theorem, known as Germain's theorem. Sophie Germain died from breast cancer at the age of 55, and despite her scientific accomplishments and prizes, she was referred to merely as a property holder on her death certificate. Today Germain is regarded as an important founder of mathematical physics and a pioneer in the area of elasticity.

German mathematician and astronomer Caroline Herschel (1750-1848) also faced parental disapproval as a girl when she wanted to study arithmetic. Born in Hanover, Hershel was told by her father that she would probably never marry because she was not beautiful. Her mother ignored her daughter's requests for a proper education and only wanted her to do the housework. Caroline's brother William (1738-1822), eleven years older, saved Caroline from this dreary existence. He had relocated in England, where he studied astronomy and wanted his sister to join him. Their mother acquiesced only when William agreed to provide a housemaid to replace Caroline.

Brother and sister left Germany for England in August 1772. Caroline quickly learned English, studied accounting, and enjoyed discussing astronomy with her brother, even helping him build a telescope. She diligently studied geometry and logarithmic tables, and was responsible for many calculations and reductions. She arranged calculations of 2,500 nebulae and the reorganization of a catalog that listed almost 3,000 stars.

After William's death, Caroline returned to Germany, where she remained until her death in 1848 at the age of 97. She received many awards, including an honorary memberships and a gold medal from the Royal Astronomical Society, election to the Royal Irish Academy, and a gold medal of science from the King of Prussia. Although her contributions to mathematics did not include any original work in pure mathematics, Herschel's work in applied mathematics is invaluable.

Impact

The accomplishments of eighteenth century mathematicians Maria Agnesi, the marquise du Châtelet, Sophie Germain, and Caroline Hershel cannot be overestimated. Prohibited a university education due to their gender, these extraordinary women nonetheless excelled in a field that was dominated by men. Mostly self-taught, their genius and resolve led them to produce exceptional and innovative research. Furthermore, their efforts eventually led universities to change their admittance policies and allow women to attend universities.

During the political revolutions of the 1700s, these women bravely waged an intellectual revolution. The struggle, however, was not an easy one. In the nineteenth century, a woman at Cambridge University who wanted to attend a mathematics lecture had to sit behind a screen at the rear of the classroom in order to be separated from the men. In Germany a woman student was warned not to enter the university lab when men were present. Nevertheless, several universities in the nineteenth century gradually started to admit women as students at the undergraduate and graduate levels. In 1876 the University of London began to grant degrees to women. In 1885 the University of Göttingen awarded the first official Ph.D. to a woman by a German university. Cambridge, however, did not allow women to receive degrees until 1948.

When women's colleges such as Girton at Cambridge and Bryn Mawr in the United States were established, these colleges accepted women as students and then as professors. This practice contrasted sharply to that of the major universities. In 1885 the only woman professor at a European university was Sofia Kovalevsky (1850-1891), who was teaching mathematics at the University of Stockholm. When the brilliant German mathematician Emmy Noether (1882-1935) was offered the chance to teach at Göttingen, it was an unpaid position. By the twentieth century, however, universities around the world had overcome their gender bigotry and accepted women not only as university students but also as professors. The bold crusade begun by the brave, bright women mathematicians of the eighteenth century had finally succeeded.

ELLEN ELGHOBASHI

Further Reading

Morrow, Charlene, and Teri Perl, eds. *Notable Women in Mathematics*. Westport, CN: Greenwood Press, 1998.

Osen, Lynn. M. *Women in Mathematics*. Cambridge, MA: MIT Press, 1974.

"4,000 Years of Women in Science." http://crux.astr.ua.edu/4000WS/4000WS.html

The Growing Use of Complex Numbers in Mathematics

Overview

Complex numbers are those numbers containing a term that is the square root of negative one. Initially viewed as impossible to solve, complex numbers were eventually shown to have deep significance and profound importance to our understanding of physics, particularly those parts of physics involving electricity and magnetism. Complex numbers were finally recognized as a legitimate branch of mathematics in the last years of the eighteenth century, and they have been regarded as indispensable to some fields since the early nineteenth century.

Background

In 1850 B.C. an unknown Egyptian mathematician wrote on a papyrus the formula for calculating the volume of a truncated cone, called a frustum [$V = \frac{1}{3} * H * (a^2 + ab + b^2)$]. Interestingly, the formula is identical to the one used today, although deriving and solving this formula is usually thought to require mathematics far more advanced than existed at that time. In later years, about A.D. 100, Hero of Alexandria (fl. A.D. 62) investigated this equation further. At some point in his calculations, he appears to have purposely overlooked the fact that, in an example he wrote down, he had to calculate a value for the square root of a negative number. In fact, his manuscript shows that the negative number was written as a positive number, making extraction of the square root possible by mathematics as it existed at that time. Hero made virtually no errors in his other work; it is nearly certain that he purposely chose not to solve this problem, possibly because negative numbers had no meaning to him, let alone the square root of such numbers. So Hero missed the historic opportunity to "discover" the square root of a negative number.

Negative numbers reared their heads repeatedly in subsequent centuries, continuing to bedevil mathematicians and students because their very existence seemed counter-intuitive. One could count stones, for example, but how could one count negative stones? It made no sense to have a number representing quantities that were less than nothing. Carrying this absurdity further, how could one then perform mathematical manipulations with such a number? Or, more precisely, how could one possibly propose to extract the square root of a negative number? These questions perplexed mathematicians for over 1,000 years after Hero's first-century manuscript. For the most part, mathematicians were simply not interested, so they tried to find some way to either avoid or ignore working with negative numbers.

One example of this is the work of the brilliant Italian mathematician Scipione del Ferro (1465-1526). Until del Ferro, it was widely thought that cubic equations (that is, equations in which one of the terms is a cube, or a variable raised to the third power) were impossible to solve. Del Ferro was able to show that a certain class of cubic equations could be solved, the "depressed cubics" in which one of the terms (the second degree, or squared term) was missing. Depressed cubics could be solved, but to do so threatened to generate both real and imaginary roots. The real roots were the ones corresponding to numbers that del Ferro understood and that, to him, had physical significance. However, in his solutions to problems of depressed cubics, del Ferro simply disregarded all except the real roots.

The next advances in solving cubic equations was made by Italian mathematician Girolamo Cardano (1501-1576), who developed a correct, but roundabout, method of extracting the roots of cubic equations. He didn't understand why his method worked; this was explained by Rafael Bombelli (1526-1572), but neither man yet understood the meaning or the reason for the numerous negative numbers and their square roots that showed up in these calculations. A century later, Gottfried Leibniz (1646-1716) arrived at some of the same conclusions as Bombelli and, not knowing of Bombelli's priority, thought he had come up with the thoughts first.

The first person to make a mathematically sophisticated effort to understand the significance of the imaginary roots was the Englishman John Wallis (1616-1703). Unfortunately, while Wallis earned a fine reputation in mathematics, his contributions towards a deeper understanding of what the square root of -1 actually meant were not terribly significant. However, Wallis did point out that negative numbers can have a physical meaning—they are simply the numbers that lie to the left of zero (in the negative direction) on a line of numbers (often called the number line). This explanation, while seemingly trivial today, was a great leap forward at the time because, previously, this explanation was simply not considered. This also suggested that, since negative numbers could now be visualized, perhaps their square roots could be, too.

During this time, too, some examples of problems requiring imaginary roots began to make themselves known. For example, if one person moving at a constant velocity is chasing a ball rolling downhill, it is not difficult to set up an equation to find out when the person will catch the ball. However, if the person is moving too slowly to do so, this equation will have an imaginary root; that is, a root expressed in terms of $\sqrt{(-1)}$. In a plot, such an equation might look like a parabola that doesn't quite reach the x-axis. Since the roots of an equation are those values that make the equation equal zero (i.e., where the plot intersects the x-axis), such an equation cannot have real roots and can only be solved using imaginary numbers.

The term "imaginary" to describe numbers that are the square roots of negative numbers comes from the great mathematician Leonhard Euler (1707-1783), who wrote "All such expressions as $\sqrt{(-1)}$...are consequently impossible or imaginary numbers, since they represent roots of negative quantities; and of such numbers we may truly assert that they are neither nothing, nor greater than nothing, nor less than nothing, which necessarily constitutes them imaginary or impossible."

Another term often heard in this branch of mathematics is complex numbers. A complex number is a number that contains both a real number and an imaginary one. The real part of the complex number might be zero, leaving just the imaginary part, or it can be any real number. Complex numbers are written to show both the "real" and "imaginary" parts. An example of this would be the complex number $2+3i$, in which 2 is the real part and $3i$ is the imaginary part of the complex number.

The person who finally helped make sense of this number was a Norwegian surveyor named Caspar Wessel (1745-1818). In a paper presented to the Royal Danish Academy of Sciences in 1797, Wessel proposed plotting complex numbers on a two-dimensional graph with the real part of the number along the x-axis and the complex part along the y-axis. This turned out to be a practical solution to the vexing question of trying to visualize something that was thought to have no physical meaning. By doing this, Wessel also led the way to developing techniques to actually use complex numbers, most typically in various branches of physics.

Impact

Wessel's paper generated little excitement outside of Denmark because Danish scientific journals received little attention. Nearly 100 years later, in 1895, Wessel's paper was rediscovered, and he received the credit he was due for this momentous interpretation of the meaning of imaginary and complex numbers and the manner in which they could be represented graphically. In the interim, others made the same discoveries, putting complex numbers to good use.

It turns out that complex numbers are a particularly useful means of mathematically handling operations such as multiplying vectors and angles. Because of this, they are ideally suited for describing the physics of alternating electrical current and AC circuitry, in which the voltage varies cyclically between positive and negative values, following a sine wave pattern. In fact, many physical phenomena behave in a similar manner. This means that, over time, the values for, say, electrical current voltage and direction cycle through the same values repeatedly, and, plotted in what is called polar coordinates, this forms a circle that repeats endlessly. At any instant in time, then, the AC current sine wave will have a specific voltage, whether positive or negative. These coordinate—voltage and time—can be described using complex numbers just as easily as by using conventional numbers. Drawing a line from the origin to a point representing the voltage at a particular time gives a vector, a line that represents a quantity with both magnitude (say, 25 volts) and a direction (say, 30 degrees up from the x-axis in the positive direction).

Of course, we can go through this exercise with real numbers, too, so the utility of complex numbers seems elusive. And, for simply describing quantities in this manner, there is not a tremendous amount of utility to complex numbers. However, when it comes time to add or multiply vectors, they become enormously useful, and complex numbers greatly simplify vector mathematics. This simplification is not just noticed in electrical circuits, either. Many physical phenomena are cyclic in nature, from the rotation of the Earth to the oscillation of pendulums to the orbits traced by satellites to the beating of our hearts. All of these phenomena are described easily and elegantly using complex analysis, giving greater insight to physicists and engineers into the workings of our world and helping them to design more useful machines.

P. ANDREW KARAM

Further Reading

Books

Devlin, Keith. *Mathematics.* New York: Columbia University Press, 1999.

Nahin, Paul. *An Imaginary Tale: The Story of* $\sqrt{(-1)}$. Princeton, NJ: Princeton University Press, 1998.

The Elaboration of the Calculus

Overview

Many of the most important and influential advances in mathematics during the eighteenth century involved the elaboration of the calculus, a branch of mathematical analysis that describes properties of functions (curves) associated with a limit process. Although the evolution of the techniques included in the calculus spanned the history of mathematics, calculus was formally developed during the last decades of the seventeenth century by English mathematician and physicist Sir Isaac Newton (1643-1727) and, independently, by German mathematician Gottfried Wilhelm von Leibniz (1646-1716). Although the logical underpinnings of calculus were hotly debated, the techniques of calculus were immediately applied to a variety of problems in physics, astronomy, and engineering. By the end of the eighteenth century, calculus had proved a powerful tool that allowed mathematicians and scientists to construct accurate mathematical models of physical phenomena ranging from orbital mechanics to particle dynamics.

Background

Although it is clear that Newton made his discoveries regarding calculus years before Leibniz, most historians of mathematics assert that Leibniz independently developed the techniques, symbolism, and nomenclature reflected in his preemptory publications of the calculus in 1684 and 1686. The controversy regarding credit for the origin of calculus quickly became more than a simple dispute between mathematicians. Supporters of Newton and Leibniz often argued along bitter and blatantly nationalistic lines and the feud itself had a profound influence on the subsequent development of calculus and other branches of mathematical analysis in England and in Continental Europe.

The first texts in calculus actually appeared in the last years of the seventeenth century. The publications and symbolism of Leibniz greatly influenced the mathematical work of two brothers, Swiss mathematicians Jakob Bernoulli (1654-1750) and Johann Bernoulli (1667-1748). Working separately, the Bernoulli brothers both improved and made wide application of calculus. Johann Bernoulli was the first to apply the term *integral* to a subset of calculus techniques allowing the determination of areas and volumes under a curve. During his travels Johann Bernoulli sparked intense interest in calculus among French mathematicians, and his influence was critical to the widespread use of Leibniz-based methodologies and nomenclature.

Impact

Physicists and mathematicians seized upon the new set of analytical techniques comprising the calculus. Advancements in methodologies usually found quick application and, correspondingly, fruitful results fueled further research and advancements.

Although the philosophical foundations of calculus remained in dispute, the arguments proved no hindrance to the application of calculus to problems of physics. The Bernoulli brothers, for example, quickly recognized the power of the calculus as a set of tools to be applied to a number of problems. Jakob Bernoulli's distribution theorem and theorems of probability and statistics, ultimately of great importance to the development of physics, incorporated calculus techniques. Johann Bernoulli's sons, Nikolaus Bernoulli (1695-1726), Daniel Bernoulli (1700-1782), and Johann Bernoulli II (1710-1790), all made contributions to the calculus. In particular, Johann Bernoulli II used calculus methodologies to develop important formulae regarding the properties of fluids and hydrodynamics.

The application of calculus to probability theory resulted in probability integrals. The refinement made immediate and significant contributions to the advancement of probability theory based on the late seventeenth-century work of French mathematician (and Huguenot) Abraham De Moivre (1667-1754).

English mathematician Brook Taylor (1685-1731) developed what became known as the Taylor expansion theorem and the Taylor series. Taylor's work was subsequently used by Swiss mathematician, Leonard Euler (1707-1783) in the extension of differential calculus and by French mathematician Joseph Louis Lagrange (1736-1813) in the development of his theory of functions.

Scottish mathematician Colin Maclaurin (1698-1746) advanced the expansion of a special case of a Taylor expansion (where x = 0). More importantly, in the face of developing criticism from Irish Bishop George Berkeley (1685-1753) regarding the logic of calculus, Maclaurin set out an important and influential defense of Newtonian fluxions and geometric analysis in his 1742 publication *Treatise on fluxions*.

The application of the calculus to many areas of math and science was most profoundly influenced by the work of Euler, a student of Johann Bernoulli. Euler was one of the most dedicated and productive mathematicians of the eighteenth century. Based on earlier work done by Newton and Jakob Bernoulli, in 1744 Euler developed an extension of calculus dealing with maxima and minima of definite integrals termed the calculus of variation (variational calculus). Among other applications, variational calculus techniques allow the determination of the shortest distance between two points on curved surfaces.

Euler also dramatically advanced the principle of least action formulated in 1746 by Pierre Louis Moreau de Maupertuis (1698-1759). In general, the principle asserts an economy in nature (i.e., an avoidance in natural systems of unnecessary expenditures of energy). Accordingly, Euler asserted that natural motions must always be such that they make the calculation of a minimum possible (i.e., nature always points the way to a minimum). The principle of least action quickly became an influential scientific and philosophical principle destined to find expression in later centuries in various laws and principles, including LeChatelier's principle regarding equilibrium reactions. The principle profoundly influenced nineteenth century studies of thermodynamics.

On the heels of an influential publication covering algebra, trigonometry and geometry (including the geometry of curved surfaces) Euler's 1755 publication, *Institutiones Calculi Differentialis*, influenced the teaching of calculus for more than two centuries. Euler followed with three volumes published from 1768 to 1770, titled *Institutiones Calculi Integralis*, which presented Euler's work on differential equations. Differential equations contain derivatives or differentials of a function. Partial differential equations (PDE) contain partial derivatives of a function of more than one variable. Ordinary differential equations (ODE) contain no partial derivatives. The wave equation, for example, is a second-order differential equation important in the description of many physical phenomena including pressure waves (e.g., water and sound waves). Euler and French mathematician Jean Le Rond d'Alembert (1717-1783) offered different

CREDIT FOR CALCULUS

During the later half of the seventeenth century, a set of analytical mathematical techniques eventually known as the calculus was developed by English mathematician and physicist Sir Isaac Newton. At the same time German mathematician Gottfried Wilhelm von Leibniz developed the calculus along different philosophical lines. Although the two approaches attacked similar problems they differed greatly in symbolism and nomenclature. In 1684 and 1686 Leibniz published his work, and a year later Newton's version appeared in *Philosophiae Naturalis Principia Mathematica* (Mathematical Principles of Natural Philosophy), a book destined to dominate the intellectual and scientific landscape for the next two centuries.

A feud over credit for calculus was on, and the dispute quickly became more than a scholarly tussle. Supporters of Newton or Leibniz often argued along bitter and sometimes nationalistic lines. Newton's scientific stature, especially within the influential British Royal Society, and the fact that Newton and Leibniz were in communication during the development of the calculus eventually resulted in a charge of plagiarism against Leibniz by members of the Royal Society. Supporters of Leibniz subsequently leveled similar charges against Newton. Leibniz petitioned the Royal Society for redress, but Newton hand-picked the investigating committee and prepared reports dealing with the controversy for committee members to sign. Leibniz's mathematical work became subsumed into the dispute over the invention of the calculus. Before the dispute was resolved, however, Leibniz died. Newton's anger at Leibniz remained unabated by the grave. In many of Newton's papers he continued to specifically set out mathematical and personal criticisms of Leibniz.

The feud over credit for calculus adversely affected communications regarding the development of calculus between English and continental European mathematicians. English mathematicians used Newton's "fluxion" notations exclusively when doing calculus. In contrast, European—especially Swiss and French—mathematicians used only Leibniz's dy/dx notation.

Although the notations and nomenclature used in modern calculus most directly trace back to the work of Leibniz, Newton has received the most credit for the development of the calculus in textbooks. Modern historians of science generally conclude that the feud between Newton and Leibniz was essentially groundless. A modern analysis of the notes of Newton and Leibniz clearly established that Newton secretly developed calculus some years before Leibniz published his version but that Leibniz independently developed the calculus so often credited exclusively to Newton.

K. LEE LERNER

Colin Maclaurin. *(Corbis-Bettmann. Reproduced with permission)*

perspectives regarding whether solutions to the wave equation should be, as argued by d'Alembert, continuous (i.e. derived from a single equation) or, as asserted by Euler, discontinuous (having functions formed from many curves). The refinement of the wave equation was of great value to nineteenth century scientists investigating the properties of electricity and magnetism that resulted in Scottish physicist James Clerk Maxwell's (1831-1879) development of equations that accurately described the electromagnetic wave. The disagreement between Euler and d'Alembert over the wave equation reflected the type of philosophical arguments and distinct views regarding the philosophical relationship of calculus to physical phenomena that developed during the eighteenth century.

Although both Newton and Leibniz developed techniques of differentiation and integration, the Newtonian tradition emphasized differentiation and the reduction to the infinitesimal. In contrast, the Leibniz tradition emphasized integration as a summation of infinitesimals. A third view of the calculus, mostly reflected in the work and writings of French mathematician Joseph Louis Lagrange (1736-1813) was more abstractly algebraic and depended upon the concept of the infinite series (i.e., a sum of an infinite sequence of terms). Converging series, for example, have sums that

tend to a limit as the number of terms increases. The differences regarding a grand design for calculus were not trivial. According to the Newtonian view, calculus derived from analysis of the dynamics of bodies (e.g., kinematics, velocities, and accelerations). Just as the properties of a velocity curve relate distance to time, in accord with the Newtonian view, the elaborations of calculus advanced applications where changing properties or states could accurately be related to one another (e.g., in defining planetary orbits, etc.). Calculus derived from the Newtonian tradition allowed the analysis of phenomena by artificially breaking properties associated with the phenomena into increasingly smaller parts. In the Leibniz tradition, calculus allowed accurate explanation of phenomena as the summed interaction of naturally very small components.

Lagrange's analytic treatment of mechanics in his 1788 publication, *Analytical Mechanics* (containing the Lagrange dynamics equations) placed important emphasis on the development of differential equations. Lagrange's work also profoundly influenced the work of another French mathematician, Pierre-Simon Laplace (1749-1827) who, near the end of the eighteenth century, began important and innovative work in celestial mechanics.

Despite the great success of eighteenth-century mathematicians in developing the techniques of calculus and in the application of those techniques to an increasingly wide variety of problems, a philosophical void remained with regard to the logical underpinnings of calculus. That calculus worked was apparent, but why it worked remained a question that eluded mathematicians. More importantly, the philosophical void in the logic of calculus open calculus to critical attacks. Most prominent among the critics of calculus in England was the influential Anglican Bishop, George Berkeley. The culmination of Berkeley's attacks on Newton's reasonings was formulated in a 1734 work titled, *The Analyst: or a discourse addressed to an infidel mathematician.* Berkeley, worried about the growing intellectual dominance and reliance upon mathematics and science to provide the most accurate depictions of the natural world, attempted to argue that apparent utilitarian accuracy of calculus was intellectually misleading. In particular, Berkeley argued that the theorems of calculus were derived from logical fallacies. Berkeley contended that the apparent accuracy of calculus resulted from the mutual cancellation of fundamental errors in reasoning.

Scholars took Berkeley's criticisms seriously and set out to vigorously support the logical foundations of calculus with well-reasoned rebuttals, including attempts to incorporate the rigorous mathematical arguments of the Greeks into the calculus. D'Alembert published two influential articles titled *Limite* and *Différentielle* published in the *Encyclopédie* that offered a strong rebuttal to Berkeley's arguments and defended the concept of differentiation and infinitesimals by discussing the notion of a limit. Regardless, the debates over the logic of calculus resulted in the introduction of new standards of rigor in mathematical analysis that laid the foundation for a subsequent rise in pure mathematics in the nineteenth century.

K. LEE LERNER

Further Reading

Boyer, Carl. *The History of the Calculus and Its Conceptual Development.* 2nd ed. New York: Dover, 1959.

Boyer, Carl. *A History of Mathematics,* New York: John Wiley and Sons, 1991.

Edwards, C. H. *The Historical Development of the Calculus.* Springer Press, 1979.

Hall, Rupert. *Philosophers at War: The Quarrel Between Newton and Leibniz.* Cambridge: Cambridge University Press, 1980.

Kuhn, Thomas S. *The Structure of Scientific Revolutions.* Chicago: University of Chicago Press, 1970.

Mathematics and the Eighteenth-Century Physical World

Overview

During the eighteenth century mathematicians and physicists embraced mathematics in general, and the calculus in particular, as an increasing powerful set of analytic techniques useful in the description of the physical world. Advancements in mathematical methods fueled increasingly detailed descriptions and investigations of the physical world. Caught between the rise of empirical science and the demands of the fledgling Industrial Revolution, the application of mathematics often outpaced its theoretical footing. Although there were sporadic attempts throughout the century to reconcile mathematical theory with its practical applications, it was a mostly French school of mathematicians in the later half of the century that undertook the task of producing a rigorous, self-consistent, mathematical system that was closely correlated to natural phenomena. The end result of these efforts to reintroduce rigorous analysis provided the foundation for the construction and rise of pure mathematics during the nineteenth century.

Background

A set of mathematical techniques known as the calculus was developed during the last decades of the seventeenth century by English mathematician and physicist Sir Isaac Newton (1642-1727) and, independently, German mathematician Gottfried Wilhelm von Leibniz (1646-1716). Although not fully developed in terms of logic, the calculus quickly found wide application, and its use provided a number of key new insights into the fields of physics, astronomy, and mathematics. The calculus became an essential tool to describe a clockwork universe that supported Western theological concepts of an unchanging, immutable God who ruled the universe through mechanistic laws.

In this regard the calculus fulfilled the intents of Newton, who developed his techniques and nomenclature in his important and influential *Philosophiae Naturalis Principia Mathematica* (Mathematical Principles of Natural Philosophy). In *Principia* Newton described the results of gravity (i.e. the effects of gravitational forces) in great detail. There was, however, no explanation of the underlying mechanisms of gravity. In a similar fashion, the calculus was shown to work in many situations, but the underlying mechanisms went unexplained. Although Leibniz's work was much more mathematically rigorous than Newton's, as the calculus developed throughout the eighteenth century it remained better used than it was understood. That calculus worked was often proved by its close correlation with natural phenomena (i.e. the predictions of calculus were in accord with experimental or observational evidence). Why calculus worked, however, remained a question that eluded mathematicians.

Despite great success reaped from the applications of the calculus, a philosophical void remained with regard to the logical underpinnings of calculus. Into this void stepped a number of critics, chief among them Anglican bishop George Berkeley (1685-1753), to cast doubt upon the logic of calculus. In his 1734 work titled *The Analyst; Or, a Discourse Addressed to an Infidel Mathematician*, Berkeley argued that the theorems of calculus were derived from logical fallacies and that its apparent harmony with the natural world resulted from the mutual cancellation of fundamental errors in reasoning. Through such arguments diminishing mathematics, critics hoped to reclaim some role for scripture as a basis of insight into natural phenomena. Mathematicians took these philosophical criticisms seriously and set out to support the logical foundations of calculus and other developing forms of analysis with well-reasoned proofs.

Impact

Eighteenth-century mathematics allowed mathematicians, scientists, and philosophers deep and precise insights into the workings of the natural world. Indeed, the Age of Enlightenment was chiefly characterized by an increasing reliance on scientific and mathematical descriptions of nature as opposed to purely philosophical or traditional theological descriptions. Among eighteenth-century scientists, mathematics triumphed over scripture as the preferred tool of reason.

Eighteenth-century mathematics emphasized a practical, engineering-like analysis of the material parts of physical systems. In Newtonian kinematics, for example, objects were often idealized as to shape, reduced to point masses, or treated only with regard to the motion of their center of mass. Instead of focusing on details, there was an emphasis on the elaboration of the behavior of systems as a whole.

Throughout the eighteenth century there was a continuing advancement of mathematical applications. Scottish mathematician John Napier's (1550-1617) seventeenth-century advancement of logarithms greatly simplified the mathematical descriptions of many phenomena, and during the eighteenth century Napier's work provided a mathematical basis for generations of important preelectronic calculating tools, including early versions of the slide rule. French mathematician Gaspard Monge (1746-1818) invented differen-

tial geometry. Swiss mathematicians Jakob Bernoulli (1654-1750) and Johann Bernoulli (1667-1748) both made substantial application of calculus to physical problems, and the Bernoulli brothers' work eventually allowed Swiss mathematician Leonhard Euler (1707-1783) to more fully develop variational calculus (calculus of variations) as an extension of calculus dealing with maxima and minima of definite integrals. Euler and French mathematician Jean Le Rond d'Alembert (1717-1783) applied theoretical, deductive thought to a variety of physical problems, in a process they termed "amixed mathematics," wherein pure mathematical analysis as embodied by such disciplines as algebra and geometry was separated from the "mixed" mathematics applied to astronomy, physics, mechanics, etc. In a very fundamental way, mathematics became increasingly linked with physical reality and, as such, separated from its logical and philosophical (latter to be specifically termed "logical") foundations.

Although the Bernoulli brothers, Euler, and other mathematicians used calculus to attack a variety of mathematical and physical problems ranging from thermodynamics to celestial mechanics, even Euler's texts devoted to the methods of calculus failed to offer formal mathematical proofs regarding the new techniques. This "looseness" with math was also reflected in its incorporation by the French philosopher Antoine Nicolas de Caritat, marquis de Condorcet (1743-1974) and by French economist Anne-Robert-Jacques, baron de L'auline Turgot (1727-1781) into sociological analysis. For the French mathematician and astronomer Pierre-Simon Laplace (1749-1827), mathematics was created as a tool to explain the universe.

For other mathematicians and scientists, however, the development of theory and proofs remained highly important. French mathematician Joseph-Louis Lagrange (1736-1813), in his 1788 work *Mécanique analytique* (Analytical Mechanics), applied differential equations to the field of mechanics. *Mécanique analytique* is notable for its attempt to provide a more rigorous logical framework for mathematical analysis. In fact, Lagrange's attempt to reintroduce analytical rigor in *Mécanique analytique* was not limited to calculus; Lagrange also made substantial effort toward the reintroduction of logical development in geometry. Lagrange subsequently attempted a theory of functions and the principles of differential calculus in his 1797 work titled *Théorie des fonctions analytiques*.

A modest disregard for formal mathematical proof was driven by the beginnings of the Industrial Revolution in the later half of the eighteenth century. The demands of navigation and especially the practical engineering embodied in Scottish inventor James Watt's (1736-1819) invention of the steam engine provided a rapid pace of innovation that placed emphasis on applications of mathematics as opposed to development of theory. The application of the calculus and other tools of mathematical analysis without a well-developed foundation of proof was, however, a departure from tradition and was indicative of a major philosophical shift. Driven by the practical need to measure and explore, eighteenth-century mathematicians and scientists essentially introduced a revolutionary epistemology (i.e. the study of human knowledge and the how truth is defined). Truth was increasingly defined by what worked—in other words, what results were in best accord with the natural world. More importantly, fully developed mathematical proofs were not required in order to apply tools of analysis to problems in order to obtain accurate, precise, and predictive data regarding the natural world. In essence, a working knowledge often supplanted philosophical reasoning.

With particular regard to the calculus, during the eighteenth century mathematicians and scientists failed to precisely define the derivative and the integral, nor were they able to provide proofs of the theorems underlying their use. Mathematicians were, however, able to determine and apply the correct derivatives of most functions. Although there was no satisfactory explanations as to why these derivatives were correct, calculations utilizing such derivatives proved useful and in accord with nature. Ironically, the tolerance for the working defense of calculus (i.e., it was mathematically and philosophically defensible simply because it worked) gained its ultimate expression in the early nineteenth-century work of French scientist Nicolas-Léonard-Sadi Carnot (1796-1832), who, despite his own emphasis on rigor and mathematical formalities, defended calculus as a "compensation of errors." Such defenses were, if not abhorrent, at least insufficient for many mathematicians.

During the eighteenth century explanations of the rational basis of mathematical analysis tended to be more descriptive than explanatory.

For example, the explanation of derivatives as a quotient of infinitesimals (an infinitely small but nonzero quantity) was not supportable by formal proof. The theoretical proof, involving epsilon-delta notations, that a function tends to a limit at a given point under certain conditions was not fully developed until the nineteenth-century work of French mathematician Augustin Louis Cauchy (1789-1857) and German mathematician Karl Theodor Weierstrass (1815-1897).

By the early 1820s Cauchy developed rigorous definitions of an integral, and by 1830 in his work *Leçons sur le Calcul Différential*, Cauchy set forth a coherent definition of a complex function of a complex variable. In addition, through his work on limits and continuity, Cauchy introduced rigor into calculus (and to other branches of mathematics as well) that enabled the rise of pure mathematics and modern analysis. Cauchy's work provided a philosophically and mathematically logical approach to calculus based upon the concepts of finite quantities and the limit. Interestingly, in the later half of the twentieth century German-born mathematician Abraham Robinson (1918-1974) developed nonstandard analysis, which provides a rationale for calculus based on a concept of infinitesimals similar to the reasonings of Newton and Leibniz.

By the end of the eighteenth century the renewed emphasis on mathematical rigor began to silence the critics of analysis, and in intellectual circles there was a growing acceptance of an understanding of nature based upon mathematics and empirical science. In turn the development of such Enlightenment thinking sent sweeping changes across the scientific, political, and social landscape.

K. LEE LERNER

Further Reading

Boyer, C. B. *A History of Mathematics*. Princeton, NJ: Princeton University Press, 1985.

Brooke, C., ed. *A History of the University of Cambridge*. Cambridge: University of Cambridge Press, 1988.

Dauben, J. A. W., ed. *The History of Mathematics from Antiquity to the Present*. New York: Garland, 1985.

Kiln, M. *Mathematical Thought from Ancient to Modern Times*. Oxford: Oxford University Press, 1972.

Enlightenment-Age Advances in Dynamics and Celestial Mechanics

Overview

Using equations based on Newton's laws, eighteenth century mathematicians were able to develop the symbolism and formulae needed to advance the study of dynamics (the study of motion). An important consequence of these advancements allowed astronomers and mathematicians to more accurately and precisely calculate and describe the real and apparent motions of astronomical bodies (celestial mechanics) as well as to propose the dynamics related to the formation of the solar system. The refined analysis of celestial mechanics carried profound theological and philosophical ramifications in the Age of Enlightenment. Mathematicians and scientists, particularly those associated with French schools of mathematics, argued that if the small perturbations and anomalies in celestial motions could be completely explained by an improved understanding of celestial mechanics, i.e., that the solar system was really stable within defined limits, such a finding mooted the concept of a God required adjust the celestial mechanism.

Background

Theories surrounding celestial mechanics grew and matured along with the Scientific Revolution and Age of Enlightenment. As seventeenth and eighteenth century scientists sought to explain the driving and controlling forces related to celestial motion, the various explanations found favor, including those that treated the planets as gigantic magnets that attracted and repelled each other in a cyclic dance. In his seminal work, *Philosophiae Naturalis Principia Mathematica* (Mathematical Principles of Natural Philosophy) English physicist Sir Isaac Newton's (1642-1727) formulation of the laws of gravitation, however, provided the first comprehensive and mathematically consistent explanation for the behavior of astronomical objects.

In the middle of the eighteenth century, French mathematician Jean Le Rond d'Alembert (1718-1783) wrote extensively for Denis Diderot's (1713-1784) *Encyclopédie* on various scientific subjects. In addition to the influence of Newton, during his education d'Alembert was heavily exposed to French mathematician René Descartes's (1596-1650) earlier vision of the physical world.

Although he later rejected much of Descartes's work, the concept of a mechanistic universe that could be described mathematically was an early and formative influence on d'Alembert.

D'Alembert's professional writings were important clarifications of problems in mathematical physics, especially problems related to the Newtonian concepts of kinetic energy. In 1753, d'Alembert published an influential work titled *Traité de dynamique* that set forth his principles of mechanics as derived primarily from mathematical analysis instead of observational data. What eventually became known as d'Alembert's principle was an insightful interpretation of Newton's third law of motion; and d'Alembert's contributions to the study of dynamics became widely known and influential with his elaboration of Newtonian concepts of force. D'Alembert's calculations regarding gravity extended the validity and acceptance of Newton's formulation of the inverse square law of force of gravity.

D'Alembert's philosophy of science tended toward the metaphysical and away from reliance on experimental data. In this regard, D'Alembert clearly ran counter to Enlightenment empiricism. Regardless, d'Alembert's range over treatments of dynamics was considerable, including work on the equilibrium state and fluid dynamics. D'Alembert's astronomical studies eventually offered solutions that accurately described the precession of the equinoxes in accord with Newtonian principles. In d'Alembert's writings for the *Encyclopédie*, he relied heavily on pure mathematical analysis and took little note of experimental data. As a result, d'Alembert's analysis of phenomena often proved faulty or fraught with exception. In addition, he often selected mathematical equations to describe phenomena that were the most elegant or pleasing to him, regardless of their conformity to the real world. Colleagues attacked d'Alembert for arguments based on eloquence rather than sound logic.

French mathematician Joseph-Louis Lagrange (1736-1813), born in Italy under the name Giuseppe Lodovico Lagrangia, also published important works on dynamics based on his principle of least action. Some of the most influential of Lagrange's work appeared between 1759 and 1766 in the journal, *Mélanges de Turin*.

Lagrange's work *Mécanique Analytique* (Analytical Mechanics), published in 1788, was notable among scholars for its clarity of notation and it was the first book to treat mechanics through purely mathematical analysis without resorting to the aid of diagrams.

In papers on the topic of fluid mechanics Lagrange advanced what would become known as the Lagrangian function and other methodologies to attack problems associated with observations of the orbital dynamics of Jupiter and Saturn. In 1763, Lagrange turned his attention to problems associated the lunar orbital dynamics that cause apparent oscillations in the observed positions of lunar features due to periodic movement of the lunar axis. Lagrange also proposed solutions to dynamics problems associated with the orbital motions of the moons of Jupiter and to problems related to perturbations in the orbits of comets caused by planets. In 1772, Lagrange predicted the existence and location (confirmed in the early part of the twentieth century) of two groups of asteroids at points of equilateral triangular stability (now termed Lagrangian points) formed by the Sun, Jupiter, and the asteroids (later termed the Trojan planets). Lagrange helped develop the use of differential equations to solve problems associated with mechanical analysis, and many of these techniques became useful in attacking a wide range of problems associated with celestial dynamics.

Another French mathematician, Pierre Simon, marquis de Laplace (1749-1827), worked to explain the small discrepancies between Newton's predicted and the observed orbits of the planets. Laplace understood that Newton's calculations had ignored the small yet significant gravitational influences of the other planets in the solar system. Newton largely discounted these perturbations in his mechanistic universe. In fact, Newton and natural theorists explained such aberrations as requiring the hand of God to constantly "wind the celestial watch, lest it run down" or to otherwise "reset" the mechanism of celestial mechanics. Laplace rejected this need for divine intervention, and strove to fully explain nature along mechanistic and deterministic lines.

In 1771, Laplace's work, *Recherches sur le calcul intégral aux différences infiniment petites, et aux différences finies*, contained formulations important to astronomers. In 1773, Laplace published a study titled *Traité de mécanique céleste* (Traetise on Celestial Mechanics) that set out exceedingly detailed mathematical calculations involving the eccentricities of planetary orbits. Sig-nificantly, Laplace's work accounted for the gravitational influences caused by multiple celestial bodies (planets) and involved an accounting of their mutual gravitational attraction.

In his 1779 book, *Exposition du système du monde* (The System of the World), Laplace set forth elegant calculations involving the orbital and rotational dynamics of bodies in a gravitational field. Laplace argued an early accretion (addition or gathering) hypothesis that allowed for the creation of the solar system from nebular gas constrained and contracted by gravity into the bodies observable today. Laplace specifically asserted that the planets in the solar system formed from the disruption and debris of a rotating, contracting and cooling solar nebula. In addition, Laplace was able to make very accurate predictions on the future positions of astronomical bodies. Many of Laplace's predictions were later confirmed by astronomers' identification of lunar positions and celestial objects in accord with Laplace's calculations.

In the later portion of the eighteenth century, Laplace began to develop and incorporate probability theory into his work on celestial mechanics. In 1786 Laplace demonstrated that observed eccentricities and irregularities in planetary orbits remained within predicted and defined limits. More importantly, Laplace showed these systems to be self-correcting.

Other physicists and mathematicians made notable contributions to the understanding of celestial dynamics. Swiss mathematician Leonhard Euler (1707-1783) studied lunar options and made detailed calculations regarding the interactive dynamics of the Sun, Earth, and Moon system. Euler also worked on problems associated with perturbations (small changes) in planetary orbits. Most importantly, Euler studied the dynamics of a three-body system in a gravitational field.

Impact

D'Alembert's metaphysical analysis culminated with his five-volume work *Mélanges de literature et de philosophie* that was published starting in 1753. Although d'Alembert's writing did not deny the existence of a God, his allowance for the existence of God was not based on belief in divine revelation, but rather on his opinion that man's intelligence could not be solely attributed to the natural interaction of matter. As he aged, however, d'Alembert gradually became a materialist and discounted deistic or natural philosophical arguments for a God ruling the mechanistic universe. This shift was to have profound influence on generations of French mathematicians mentored by d'Alembert. As a school, the French mathematicians took an increasingly skeptical or hostile attitude toward arguments in favor of a God needed to intervene in the workings of a mechanical universe.

D'Alembert's work, however, established that important physical laws, especially those associated with dynamics and mechanics, could in some circumstances still be deduced from pure mathematical analysis. In essence, during an age of empiricism, d'Alembert reasserted a role for purely mathematical analysis. Despite this departure from empiricism, d'Alembert's work helped extend both the range and power of Newtonian physics in eighteenth century European scientific circles.

Laplace's main contribution to the advancement of Newtonian physics was in his translation of Newton's geometrical analysis to a more widely understandable calculus- based analysis of mechanical dynamics. Although Laplace became the most important champion of a Newtonian-based understanding of celestial dynamics, French mathematicians argued that Laplace's work essentially removed the need for a god to tinker with—or reset—Newton's clockwork universe. Laplace made his assessments of the stability of the solar system by demonstrating the invariability of mean planetary motions (the motions of planets averaged over time).

Laplace's interpretation of celestial mechanics ran counter to philosophical and theological Enlightenment views of celestial mechanics as both proof of God as a "prime mover" and of the continued need for God's existence. Laplace argued for a completely deterministic universe, without a need for the intervention of God.

Laplace even asserted explanations for catastrophic events (e.g., flooding, comet impacts, extinctions, etc.) as the inevitable results of time and statistical probability.

The need for greater accuracy and precision in astronomical measurements spurred the development of improved telescopes and pendulum driven clocks. Consequently, the accuracy of mathematical predictions improved with each generation of instruments. More importantly, each generation of new data brought more general confirmation of Newtonian physics.

In general, despite important mathematical advances, observation outpaced prediction during the eighteenth century (e.g., English astronomer William Herschel's 1781 discovery of Uranus) and mathematicians were left to scramble for explanations consistent the emerging dominance of Newtonian physics. With very minor exceptions these explanations were always found. Accordingly, both observation and calculation accelerated the influence and rise of Newtonian physics as the basis for further advances in dynamics and celestial mechanics. The scientific world would have to await the development of relativity theory in the 20th century to fully explain away the minor discrepancies in Newtonian descriptions of the universe.

K. LEE LERNER

Further Reading

Bell, E. T. *Men of Mathematics: The Lives and Achievements of the Great Mathematicians from Zeno to Poincaré.* New York: Simon and Schuster, 1986.

Boyer, C. B. *A History of Mathematics,* 2nd ed. *New York: Wiley, 1968.*

Bronowski, J. *The Ascent of Man.* Boston: Little, Brown, 1973.

Hawking, S. *A Brief History of Time.* New York: Bantam Books, 1988.

Advances in the
Study of Curves and Surfaces

Overview

Eighteenth-century mathematicians enjoyed a vastly expanded set of techniques that could be applied to the study of curves and surfaces and a vastly expanded set of reasons to study them.

Problems of projectile and planetary motion required a renewed understanding of the conic sections. Problems from engineering and the need for accurate maps of Earth's curved surface drew special attention to the general problems of representing curves and surfaces by equations. The researches of Jacob Hermann, Leonhard Euler, Gaspard Monge, and others would lead to the new disciplines of descriptive and differential geometry.

Background

The ancient Greeks had a good understanding of those curves generated by the conic sections—the hyperbola, parabola, ellipse, and circle—but from a geometrical perspective only. With the invention of analytic geometry by the French mathematician and philosopher René Descartes (1596-1650) and French mathematician Pierre de Fermat (1601-1665), such curves were all understood to be described by algebraic equations. The study of curves and curved surfaces received new impetus in the eighteenth century from both science and technology. Given the laws of motion and gravitation of Sir Isaac Newton (1642-1727), it was now possible to calculate the trajectory of projectiles and planets with accuracy. Practical questions also arose about the shapes of freely hanging chains and beams under stress, the answers to which were important in the construction of buildings and bridges.

One significant innovation was the introduction of polar coordinates. In Cartesian coordinates the position of a point was specified by giving its distance from two, usually perpendicular, lines. In polar coordinates, the location of a point is specified by its distance from a single reference point and an angle of rotation measured with respect to a fixed line passing through the reference point. Polar coordinates provided a more natural description of curves like the circle and the ellipse, and were especially suited to describing the motion of particles around a center of gravitational attraction. The polar coordinate system had been introduced in 1671 by Newton, but as that work was not published in English until 1736, credit for the invention of polar coordinates is usually assigned to the Swiss mathematician Jakob Bernoulli (1654-1705), who published a paper on the subject in 1691. Bernoulli introduced a number of important curves, including the figure eight-shaped lemniscate and the logarithmic spiral. The general utility of polar coordinates was demonstrated by Swiss mathematician Jacob Hermann (1678-1729), who also gave the general relationship between Cartesian and polar coordinates. A second important innovation involved the parametric representation of curves. This method was introduced by the Swiss mathematician Leonhard Euler (1707-1783) in

1787 and allowed the specification of a curve given by a relation between two variables x and y, to be stated by separate relations between x and y and a third variable, say t. If x and y described the trajectory of a projectile, then t might represent the time.

While an equation in two variables would generally describe a curve in the plane, a single equation in three dimensions would define a surface in three-dimensional space. Three-dimensional Cartesian coordinates were essentially introduced in their modern form by Swiss mathematician Johann Bernoulli (1667-1748), brother of Jakob. With equations in three variables representing a surface, a question naturally arose as to whether two equations might represent the same surface, only with different coordinate axes. In 1848 Euler presented a general analysis of the second-degree algebraic equation in three variables:

$$Ax^2 + By^2 + Cz^2 + Dxy + Eyz + Fzx + Gx + Hy + Kz = L$$

in which he established a general procedure for rotating the axes of one three-dimensional coordinate system into those of another by a series of three rotations. The angles of the three rotations would come to be known as the Euler angles. By finding axes that simplified the equation through the elimination of one or more terms, Euler was able to classify all such surfaces as belonging to one of six special cases: cones, cylinders, ellipsoids, hyperboloids, hyperbolic paraboloids, and parabolic cylinders.

One of the most important figures in eighteenth-century geometry was the French engineer, scientist, and mathematician Gaspard Monge (1746-1818). The son of a merchant, Monge received some education at a local religious school, and at the age of 16 he was allowed to attend classes at Ecole Royale du Genie at Mezieres, a select military school that emphasized science and engineering. His admission to the school had come through the intervention of a military officer who was impressed by the quality of a map made by Monge using surveying instruments he had made himself. Within a few years Monge was appointed to the staff of the school, but because of his family's low social status, he was at first restricted to drafting and working as a technician. Assigned to design a fortification, Monge discovered a way to replace much of the tedious arithmetic work that had gone into calculating lines of fire and lines of sight with a much more direct geometrical method. He was appointed a professor in 1768 so that he could teach his methods to military engineers. In the process he developed the basic notions of descriptive geometry, the representation of three-dimensional objects through a combination of top and side projections that is now the basis of mechanical drawing. Monge's techniques would be treated as a military secret by France until 1794.

In 1771 Monge presented a number of memoirs to the Academy of Sciences in Paris. One of these concerned the geometric representation of the solutions to partial differential equations. While an algebraic equation in three variables would describe a single surface, a partial differential equation, involving three variables and their rates of change with respect to each other, would describe a family of related surfaces. For example, all of the spheres with a specified radius and centers in the xy plane would be allowed solutions to just one such equation, with the x and y coordinates of their centers serving to distinguish one sphere from another. Monge introduced additional geometrical concepts, such as that of the characteristic curve and characteristic cone, that allowed one in a sense to visualize all the solutions at once.

In 1775 Monge turned to the theory of developable surfaces. A developable surface is one that can be "flattened out" into a plane in such a way that distances between points are accurately represented. Euler had established the basic ideas of this field three years earlier. Interest in it was motivated by the fundamental problem of mapmakers, that of displaying the features of the Earth's surface on a flat piece of paper in such a way that the distance between points is accurately represented. The sphere is clearly not a developable surface, but there was considerable interest in finding developable surfaces that approximated portions of a sphere.

Like many of the thinkers of his time, Monge did not confine himself to mathematics but also was interested in physics and chemistry. He was the first scientist to produce water by burning hydrogen gas and he worked on the storage of hydrogen for use in balloons.

In 1794, following the French Revolution, Monge became a member of the Commission of Public Works, set up by the government to establish an institution of higher education for engineers. The school would become the Ecole Polytechnique, which would attract many of the best mathematical minds of the time to its faculty. There Monge himself taught in two different areas. The first was the descriptive geometry he had earlier invented. From his lecture notes it is

known that he addressed a number of fundamental mathematical problems, such as determining the curve defined by the intersection of two surfaces.

Monge also taught a course in what would come to be known as differential geometry, the application of the calculus to curves and surfaces in three dimensions. This was in part an outgrowth of his earlier work on characteristic curves of partial differential equations and on developable surfaces. His 1795 text on the applications of analysis to geometry was the first major work in the area.

Impact

Euler's investigation of the behavior of curves under rotation would have important implications for subsequent studies of the physics of rotating bodies. The Euler angles, in particular, form the basis of the theory of rotating bodies in physics, a theory that includes the rotational behavior of planets, spacecraft, gyroscopes, molecules, and atoms. The study of the behavior of curves and surfaces under coordinate transformations would, in the late nineteenth century, be turned into a study of "algebraic invariants," properties of an algebraic expression that would not change under certain types of coordinate transformation. This area of mathematics would combine with other approaches to the study of mathematical transformations to produce the modern mathematical theory of groups, which provides the most general mathematical framework for the description of symmetry and has turned into a powerful tool in elementary particle physics.

Both technical understanding of the behavior of curves and surfaces under rotations and the descriptive geometry of Monge are, of course, very important in the new field of computer graphics, and computer-aided design (CAD). In modern engineering practice the design of any high performance device, say a new jet engine, is in its initial stages almost entirely done using computer visualization software. Only once the components and their function are simulated and viewed from a great many vantage points are the first prototypes built.

DONALD R. FRANCESCHETTI

Further Reading

Boyer, Carl B. *A History of Mathematics*. New York: Wiley, 1968.

Kline, Morris. *Mathematics: The Loss of Certainty*. New York: Oxford University Press, 1980.

Taton Rene, ed. *History of Science: The Beginnings of Modern Science from 1450 to 1800*. New York: Basic Books, 1964.

Wolf, Abraham. *A History of Science, Technology and Philosophy in the Eighteenth Century*. Gloucester, MA: Peter Smith, 1968.

The Birth of Graph Theory: Leonhard Euler and the Königsberg Bridge Problem

Overview

The good people of Königsberg, Germany (now a part of Russia), had a puzzle that they liked to contemplate while on their Sunday afternoon walks through the village. The Preger River completely surrounded the central part of Königsberg, dividing it into two islands. These islands were connected to each other and to the mainland by seven bridges. The puzzle, which baffled the residents of Königsberg was this: Was it possible to pick a starting point in the town and find a walking route which would take them over each bridge exactly once? No one had ever found such a route; but did that mean that it did not exist? The problem caught the attention of the great Swiss mathematician, Leonhard Euler. Euler was able to prove that such a route did not exist, and in the process began the study of what was to be called graph theory.

Background

Leonhard Euler (1707-1783) is considered to be the most prolific mathematician in history. Originally educated for the ministry in order to follow in his father's footsteps, Euler discovered his talents in mathematics while attending the University of Basel. By 1726, the 19-year-old Euler had finished his work at Basel and published his first paper in mathematics. In 1727, Euler assumed a post in St. Petersburg, Russia, where he spent fourteen years working on his mathematics. Leaving St. Petersburg in 1741, Euler took up a post at the Berlin Academy of Science. Euler finally returned to St. Petersburg in 1766, where he would spend the rest of his long, productive life.

Euler worked, wrote, and published at a furious rate throughout his lifetime. Although blind for the last seventeen years of his life, he continued to work and write (with the help of various assistants and secretaries) at an astonishing rate. Euler worked in almost every branch of mathematics, both pure and applied, and his ability to perform complicated mathematical calculations in his head was legendary. Supported most of his life by the Berlin Academy and the St. Petersburg Academy, Euler worked with the famous Bernoulli family (Johannes, Daniel, and Nicolaus), as well as many other famous mathematicians. So it is no surprise that when Euler decided to analyze the problem of the Königsberg bridges, he not only found the answer, but also initiated the study of a brand new field in mathematics.

Here is how Euler went about solving the Königsberg Bridge Problem. The first step was to transform the actual diagram of the city and its bridges into a *graph*. The use of the word graph in this context may be different than what most people think of when they see the word graph. In this case, a graph must have *vertices* and *edges*. Furthermore, a graph must have a rule that tells how the edges join the various vertices. In the Königsberg Bridge Problem, the vertices represent the landmasses connected by the bridges, and the bridges themselves are represented by the edges of the graph. Finally, a *path* is a sequence of edges and vertices, just as the path taken by the people in Königsberg is a sequence of bridges and landmasses. Euler's problem was to prove that the graph contained no path that contained each edge (bridge) only once.

Actually, Euler had a larger problem in mind when he tackled the Königsberg Bridge Problem. He wanted to determine whether this walk would be possible for any number of bridges, not just the seven in Königsberg. To answer this question, Euler studied other graphs with various numbers of vertices and edges. Euler reached several conclusions. First, he found that if more than two of the land areas had an odd number of bridges leading to them, the journey was impossible. Secondly, Euler showed that if exactly two land areas had an odd number of bridges leading to them, the journey would be possible if it started in either of these two areas. Finally, Euler concluded that if there were no land areas with an odd number of bridges leading to them, the journey was possible starting from any area. At first glance, Euler's work seems to apply only to the trivial matter of walking through a city connected by bridges over a river. Of course, his conclusions were actually more general when the land areas were considered vertices of a graph and the bridges were the edges of the graph. Thus was born the modern mathematical study of graph theory. A path in which one may "walk" along each edge exactly once has come to be called a "Eulerian path," and an "Eulerian graph" is a graph in which such a path exists.

Impact

That such a seemingly trivial problem could lead to an entire branch of mathematics is not unusual. Although some areas of mathematics were developed to answer obviously important questions (for instance, calculus was developed by Isaac Newton (1642-1727) to help answer questions in physics and astronomy), others branches of mathematics had their origins in much less noble causes (the origination of probability is traced to letters exchanged by Pierre de Fermat (1601-1665) and Blaise Pascal (1623-1662) in which they discussed questions in gambling). Although the branch of mathematics known today as graph theory had its origins in a simple-minded puzzle that entertained the people of Königsberg, its eventual usefulness to mathematics has completely overshadowed its humble beginnings. For instance, chemists use graphical notation to represent chemical compounds; and physicists and engineers use graphical notation to represent electrical circuits. Graph theory is used in complex computer programs that control telephone switching systems.

Graph theory is a part of a larger field of mathematics called topology. Topology is the study of the properties of geometric figures that are invariant (do not change) when undergoing transformations such as stretching or compression. Imagine drawing a geometric figure, such as a square or a circle, on a sheet of flexible material like rubber, and then stretching or compressing the rubber sheet. The properties of the square and circle that do not change during this stretching or compression fall under the study of topology. For this reason, topology is sometimes referred to as "rubber-sheet geometry."

Euler called this type of mathematics "the geometry of position," because it did not address magnitude, as did traditional geometry. Topology and graph theory investigate position without the use of measurement (it does not matter how long the bridges are or how far the land masses are from each other.) In addition to his work on the Königsberg Bridge Problem, Euler proved that for any polyhedron, the number of vertices minus the number of edges plus the number of faces was always equal to two (v-e+f=2). For instance, a cube has eight vertices, twelve edges, and six faces (8-12+6=2). It is generally accepted that Euler's solution of the Königsberg Bridge Problem and his famous formula for a polyhedron form the foundation of the field of topology.

There are other problems similar to the Königsberg Bridge Problem that fall under the heading of graph theory. Euler worked on another of these famous problems called the "Knight's Tour Problem." This problem asks if a knight on a chessboard can be successively moved (the familiar two squares one direction and then one in the perpendicular direction) so that it visits each square on the board only once and then finishes on the same square as it began. Another of these problems is known as the "Traveling Salesman Problem." In this problem, a salesman is required to visit a certain number of cities, taking the path that will mean the shortest total distance, and ending up back in the city in which he started. With the cities being represented as vertices and the paths to each city represented as edges, this becomes a graph much like the Königsberg bridges.

Possibly the most famous application of graph theory is the Four-Color Problem. This problem was first proposed by Francis Guthrie through his brother, Frederick, to the London mathematician Augustus DeMorgan (1806-1871) nearly a century after Euler's death. The Four-Color Problem asks the seemingly simple question "How many colors must be used to color any map so that no two adjacent land areas (country, states, etc.) are the same color?" The problem may be posed as one in graph theory if the land areas are represented as vertices and the adjacent areas are connected by edges. Although the answer of four colors was suspected for over a century, it was not until 1976 that a proof was found that confirmed that four colors were required. The Four-Color Problem was the first major mathematical theorem whose proof depended in part on modern computers.

Each of these problems, although trivial on the surface, has led to incredible advances in mathematical theory and applications. From its innocent beginnings in the speculations of the people of Königsberg, to Euler's pioneering work, through modern applications, graph theory has had a profound effect on mathematics and its applications for over 250 years.

TODD TIMMONS

Further Reading

Books

Bell, E.T. *Men of Mathematics*. New York: Simon and Schuster, 1937.

Biggs, N.L., Loyd, E.K., and Wilson, R.J. *Graph Theory 1736-1936*. Oxford: Clarendon Press, 1976.

Euler, Leonhard. "From the Problem of the Seven Bridges of Königsberg." In *Classics of Mathematics*, Ronald Calinger, ed. Englewood Cliffs, NJ: Prentice Hall, 1995.

Periodical Articles

Sachs, H., Steibitz, M., and Wilson, R.J. "An Historical Note: Euler's Königsberg Letters." in *Journal of Graph Theory* 12 1 (1988): 133-39.

Mathematicians and Enlightenment Society

Overview

In the eighteenth century, mathematicians formed an integral part of society and culture. They exploited available avenues toward gaining patronage and prestige. Further, mathematicians such as Jean Le Rond d'Alembert (1717-1783) influenced the intellectual developments of the Enlightenment, which radiated out from France. The writers and thinkers of that time, in turn, relied upon mathematical language and logic in a unique manner.

Background

By 1700, the Scientific Revolution had culminated mathematically with Isaac Newton's (1642-1727) and Gottfried Wilhelm Leibniz's (1646-1716) invention of the calculus along with Newton's universal theory of gravitation and his study of optics. The next generation of mathematicians turned to finding physical confirmation of these mathematical theories and to applying new mathematical tools, such as differential equations. However, mathematicians who were not independently wealthy needed to somehow secure financial support for their work.

Professorships were not necessarily a viable option for employment at this time. European universities had declined in importance as centers of mathematical research, in part because jobs in universities were limited in number and often went to people who possessed political connections rather than intellectual ability. These professors, furthermore, often wasted time in pointless debates. Even Newton had left Cambridge University and went to work for the

Jean Le Rond d'Alembert. *(Library of Congress. Reproduced with permission)*

English Mint for thirty years, where he also gave positions to some of his scientific followers.

Mathematicians could alternatively look to scientific societies. The Royal Society of London and the Paris Academy of Sciences had been established in the seventeenth century as arbiters and promoters of scientific knowledge. Whereas the Royal Society offered little financial support beyond publishing some mathematical books for authors, mathematicians chosen for the Paris Academy were directly employed by the French state. Other European governments followed the French model of founding academies to create centralized communities of mathematicians and scientists who would devote some of their efforts to projects that would improve the state. Rulers often attached observatories to these academies as well. Installing the expensive telescopes and instruments needed to make observations further demonstrated the wealth and power of the state. Academies established by national governments provided a two-way relationship-this patronage gave legitimacy, credibility, and moral confirmation to mathematicians as well as to the state.

As these scientific societies judged and publicized new mathematical achievements, new knowledge became accessible on a popular level as never before. Eighteenth-century intellectuals studied and promoted mathematics and natural philosophy. For example, Voltaire (1694-1778) and Emilie du Châtelet (1706-1749) published a popularization of Newton's ideas in 1738 and a French translation of the *Principia* in 1759. In addition, authors adopted the rational reasoning behind the discoveries of the Scientific Revolution not only for the study of nature but also for philosophy. Their critical spirits embraced the values of toleration, freedom, reasonableness, and opposition to authoritarianism. In other words, appreciation for the powerful natural laws discovered in the seventeenth century was a driving force behind the tenets of the Enlightenment, which characterized the eighteenth century. To the philosophes, or the writers and thinkers of the Enlightenment in France, mathematics held the key to true philosophy.

While most European absolutist rulers did not adopt Enlightenment calls for governmental reform, they did covet the mathematical and scientific heroes of the Enlightenment. They became further convinced of the importance of patronizing scientific societies. Figures such as Frederick the Great of Prussia (1712-1786) and Catherine the Great of Russia (1729-1796) acquainted themselves with the mathematical members of their academies. They employed mathematicians at court, and they sponsored prizes for those who presented new results. Mathematicians, in turn, were generally willing to work on state projects or to tutor royal youths because these jobs gave them a route to wealth and prestige.

Impact

Thus, mathematicians and Enlightenment society were close partners in the eighteenth century.

Even a gargantuan talent such as Leonhard Euler (1707-1783) had to fit in with the literary and courtly culture to prosper financially. One of the most successful exploiters of the mutual relationship between mathematics and philosophy was d'Alembert, as he and Marie-Jean de Caritat, marquis de Condorcet (1743-1794) were the only intellectuals to function both as mathematicians and as philosophes.

First, d'Alembert established a reputation as a skilled mathematician by presenting his first mathematical paper, a criticism of Charles Reyneau's ideas, to the Paris Academy in 1739. He was elected to the Academy as an astronomy member in 1741, and he moved into the mathematics section in 1746. In 1743, d'Alembert published his first book, *Traité de dynamique*, which was an attempt to formalize dynamics. He also won a prize from the Berlin Academy of Sciences in 1746.

During the 1740s, d'Alembert had spent his evenings in the Paris salons of Madame Geoffrin and Madame du Deffand, which were the breeding grounds of Enlightenment culture. Topics of discussion in these informal gatherings included scientific discoveries and political or philosophical theories. Salon attendees were generally wealthy, well-educated young men who appreciated a clever turn of phrase as much as a profound idea. Thus, d'Alembert's abilities in witty conservation and at dramatically reading his correspondence won him favor with the other intellectuals.

D'Alembert and another philosophe, Denis Diderot (1713-1784), were also hired by a French publisher to oversee the development of an encyclopedia of all knowledge. D'Alembert wrote the scientific articles for the *Encyclopédie*, the first volume of which appeared in 1751. In addition, he prepared a "Preliminary discourse" which described how the history of the sciences might have been if they had been discovered in logical order. This paper helped to explain that the *Encyclopédie* was meant to lay down the principles of the Enlightenment; it also made d'Alembert well known as a literary talent. Still, d'Alembert left the project in 1758 due to the political difficulties caused by opponents of the *Encyclopédie*, while Diderot continued on and finished the final volume in 1772.

Furthermore, d'Alembert exerted his influence as a mathematician and philosophe in the academies. Although a scientific rivalry with Alexis Clairaut (1713-1765) hindered his personal advancement in the Paris Academy after his ally, Pierre-Louis Maupertuis (1698-1759), departed for the Berlin Academy in 1745, d'Alembert's enthusiastic campaign in favor of philosophy was one factor which flavored scientific societies across Europe with the cosmopolitan character of the French Enlightenment. More directly, d'Alembert was elected to the literary Académie Française in 1754 and became that society's perpetual secretary in 1772, a position that allowed him to see to the election of enough philosophes to form a majority of the members.

Prominence in the academies also provided d'Alembert with opportunities to make connections with and on behalf of other mathematicians. His friendship with Euler was rocky over the years, and after disagreeing over the definition of mathematical functions in 1749, the two did not reconcile until 1763, when d'Alembert was no longer an active mathematician. Still, d'Alembert's contacts did enable him to support younger mathematicians. He secured Pierre-Simon, marquis de de Laplace's (1749-1827) first teaching position at the Ecole Militaire. D'Alembert also mentored Joseph Lagrange (1736-1813) and Condorcet.

While d'Alembert's own mathematical research had no direct impact on Enlightenment philosophy, the discipline of mathematics was a major aspect of Enlightenment culture. The philosophes were inspired by the mathematization of natural philosophy to carry the rationalization of their subject even further in the later French Enlightenment. Condorcet and the Abbé de Condillac were two of the most notable figures in this regard. They argued that mathematical laws could be adapted to human reasoning and that algebra was an unambiguous language that should be the model for all communication. During the French Revolution, the former philosophes who gained control of the government instituted programs to decimalize all units of measurement and to rationalize education based on the mathematical approach to philosophy.

Laplace was one of the later eighteenth-century mathematicians who gained financially by adapting to shifting political situations during his lifetime. He had become one of the senior members of the Paris Academy by 1789, and he managed to remain friendly to each of the groups that gained political control over the course of the French Revolution. Around 1800, Napoleon Bonaparte (1769-1821) awarded Laplace the Grand Cross of the Imperial Order and put him in the Senate; but Laplace shifted his allegiances to Louis XVIII in 1814, just in time to benefit from the restoration of the French monarchy. One property Laplace purchased with his Senate salary was his house at Arcueil, where he and his protégés conducted research during the last two decades of his life.

In the nineteenth century, many mathematicians retreated from social, political, and philosophical involvements, partly because the status of academies and prizes became subordinate to that of research universities. Göttingen University, founded in the late eighteenth century, was one of the first to be oriented toward mathematical and scientific research. Simultaneously, internal developments in mathematics became increasingly specialized, and it was no longer possible for a person to know everything there was to know about mathematics. Thus, mathematicians turned to technical pursuits and away from the general intellectual culture.

In summary, eighteenth-century mathematicians were interconnected with Enlightenment society. The new mathematics impressed not only mathematicians, but also rulers who established academies to bring prestigious mathematicians to their courts, and the philosophes popularized mathematics and applied it to general human reasoning. Mathematicians benefited financially from the academies either directly or through contacts with new patrons; academy involvement also encouraged mathematicians to share their research openly. While d'Alembert was an archetype for a mathematician who was a proponent of Enlightenment philosophy, all of the most talented mathematicians on the European Continent during that time had to function within Enlightenment society.

AMY ACKERBERG-HASTINGS

Further Reading

Frängsmyr, Tore, J. L. Heilbron, and Robin E. Rider, ed. *The Quantifying Spirit in the 18th Century.* Berkeley: University of California Press, 1990.

Gillispie, Charles Coulston. *Science and Polity in France at the End of the Old Regime.* Princeton: Princeton University Press, 1980.

Hahn, Roger. *Anatomy of a Scientific Institution: The Paris Academy of Science, 1666-1803.* Berkeley: University of California Press, 1971.

Hankins, Thomas L. *Jean d'Alembert: Science and the Enlightenment.* Oxford: Clarendon Press, 1970.

Spielvogel, Jackson J. *Western Civilization.* 2d ed. Minneapolis/St. Paul: West Publishing Company, 1994.

The Algebraization of Analysis

Overview

While the principal concepts of the calculus were developed in the seventeenth century, a sound mathematical formulation of the subject would have to wait until the nineteenth. Critics of the early formulations pointed to the rather casual way in which infinitesimal and infinite quantities were defined. Lagrange attempted to replace the need for derivatives in the differential calculus with a requirement that the difference between the values of a function at two values of its argument be expressed as a power series. The theory of power series would eventually play an important role in the solution of differential equations and in the machine calculation of special functions.

Background

In the seventeenth century, mathematicians had arrived at the notion of a function as a well-defined combination of mathematical operations that allows for one quantity to be obtained from another quantity. Thus $f(x) = 2 \sin(x)$ would denote the function of replacing a number x, called the argument, with twice its trigonometric sine. Calculus is concerned with the changes that occur in the value of a function as the argument is changed. It was developed to provide a mathematical framework for the new physics emerging from the work of Galileo and Isaac Newton (1642-1727). The differential calculus permitted the calculation of instantaneous rates of change, making it possible to calculate the velocity of a particle at any point in time, given its position as a function of time. The integral calculus permitted calculation of the change in the value of a function over a time interval given its instantaneous rate of change as a function of time. The calculus was the nearly simultaneous discovery of two mathematicians, the English scientist Isaac Newton and the German scientist Gottfried Wilhelm Leibniz (1646-1716).

Both the integral and differential forms of the calculus involved considering the behavior of quantities as they became infinitesimally small. The instantaneous rate of change, or fluxion in Newton's terminology, was to be calculated by considering the ratio in the change in the value of the function to the change in its argument, as both changes became smaller and smaller. Many mathematicians objected to this procedure as it suggested a ratio of 0/0, which was understood to be indeterminate, that is, not equal to any definite number.

Criticism of the calculus continued into the eighteenth century. Possibly the most outspoken critic was not a mathematician, but the eminent George Berkeley (1685-1753), idealist philosopher and bishop in the Church of England. In a pamphlet published in 1734 he launched a major attack on the reasoning used in the calculus. Berkeley pointed out that no deductive basis had been provided for the calculus. He objected to the use of quantities that were arbitrarily close to zero and to the geometric interpretation being given to the fluxion as a tangent to a curve.

A number of mathematicians rushed to the defense of the calculus. James Jurin (1684-1750) quickly published a rebuttal relying on geometric intuition. Berkley then published a re-rebuttal pointing out that his objections had still not been met. In a treatise published in 1742, Scottish mathematician, Colin Maclaurin (1678-1746) attempted to establish the methods of calculus on a purely geometrical basis. In 1755, the influential Swiss mathematician, Leonhard Euler (1707-1783) published an algebraic treatment, introducing some formal rules to eliminate the necessity of considering the indeterminate form 0/0.

The Italian-French mathematician Joseph-Louis Lagrange took up the issue in a paper published in 1774 and in his book, *Theory of Functions*, published in 1797 revised in 1813. Lagrange argued that the difference in the value of a function at two points could be represented in a power series of the form:

$$f(x+h) - f(x) = a(x) h + b(x)h^2 + c(x) h^3 + ...,$$

where $a(x)$, $b(x)$, and $c(x)$, etc. are all functions of x. After making this assumption, which was certainly true for most of the functions to have received the attention of mathematicians, Lagrange noted that $a(x)$ was simply the fluxion of f and that twice $b(x)$ was the fluxion of $a(x)$, three times c was the fluxion of $b(x)$ and so on. In modern notation,

$$a(x) = f'(x), b(x) = f''(x) / 2 , c(x) = f'''(x) / 3!,$$
$$\text{and so on,}$$

where the exclamation point denotes the factorial, n!, the product of all positive integers less than or equal to n. Lagrange called the functions f'(x), f''(x), f'''(x), and so on derivative functions, from which we obtain the current term derivative.

To illustrate Lagrange's technique we will consider the simple function f(x) = x². By simple algebra, one can easily obtain the expression,

$$f(x+h) - f(x) = 2xh + h^2,$$

from which we can readily recognize the first derivative as 2x and the second derivative as 2.

Lagrange had begun teaching at the Ecole Polytechnique in 1794 and presumably taught his approach to his students and discussed it with his colleagues, who included many of the principal figures in French mathematics. The method was eventually abandoned, however, as it was realized that many functions of mathematical interest could not be expanded in the way assumed by Lagrange.

Impact

Lagrange is remembered today as a major mathematical thinker for his contributions to mathematical physics as well as numerous contributions to pure mathematics. His work on algebraicizing analysis is generally regarded as an unimportant byway. The American historian of mathematics, Eric Temple Bell (1883-1960) wrote:

"The contrast between what passed for valid reasoning then and what is demanded now is violent..... Some of Newton's successors who strove to make sense out of the calculus are among the greatest mathematicians of all time. Yet, as we follow their reasoning, we can only wonder whether our own will seem as puerile to our successors...."

In effect, one cannot anticipate the judgement of history. The calculus was put on a firm basis, beginning with the work of the French mathematician Augustin-Louis Cauchy (1789-1848) which provided a rigorous treatment of the limit concept. The notions of function, derivative, and integral have since been generalized and extended in a number of ways, sometimes introducing entire new fields for mathematical exploration.

The representation of functions by power series over small regions has come to play a role in modern practical mathematics. The solutions to most differential equations cannot be expressed in terms of simple functions like sines and cosines, and so an important approach to solving them is to assume a power series solution to be valid in some region and to work out

a general formula for the terms. One must then check that the series converges, or takes on a definite limiting value, at each point. Power series are also used in the machine generation of mathematical functions. When a key is pressed on a calculator to find the sine or the logarithm of a number, the microprocessor inside adds up enough terms in the series to produce an accurate estimate of the value. Power expansions are also of importance in that field of mathematics known as approximation theory, in which one seeks to understand how best to choose a function that is easy to evaluate to approximate a more complicated function or a function which is only known at certain points.

Modern readers may find it hard to imagine how developments in an area of mathematics could have been perceived as a threat to religion. It must be remembered, however, that the calculus was key to calculating the behavior of bodies in the new physics based on Newton's laws of motion. Before the publication of Newton's *Mathematical Principles of Natural Philosophy,* it was reasonable to view the motion of the planets as the result of divine will, or of angels appointed to push around different crystalline spheres bearing the planets. After Newton it was much more reasonable to consider the universe as a giant clockwork mechanism, which, once set into motion would continue according to the laws of motion without any further intervention. In short, there was not much left for God to do. Further, nothing that happened in the natural world was considered to be beyond human understanding. This view naturally clashed with the traditional teachings of the Christian churches about a personal God who had intervened in human history and answered prayers. It was not unreasonable for churchmen to perceive a threat, and to respond by seeking weaknesses in the arguments of their opponents.

The title of Berkeley's pamphlet is itself quite informative. It is "The Analyst, Or A Discourse Addressed to an Infidel Mathematician, Wherein It is examined whether the Object, Principles, and Inferences of the Modern Analysis are more distinctly conceived, or more evidently deduced, then Religious Mysteries and Points of Faith. 'First cast out the beam out of thine own Eye and then thou see clearly to cast out the mote of thy brothers Eye.'" The *infidel* in question was Edmond Halley (1656-1742), close friend and supporter of Isaac Newton, who had encouraged him to publish his researches into the laws of motion. Berkeley's attack was on the reasoning used by some of the new Newtonians who felt that their logic led to a surer form of truth than the teachings of the church.

An echo of this conflict might be seen in the reception accorded Charles Darwin's *Origin of Species* in 1859, which, by suggesting a mechanism for the appearance of new species, eliminated in some minds the need to invoke God as an explanation for the existence of living creatures. Again, the challenge to the new science was led by an Anglican bishop, in this case Samuel Wilberforce (1805-1873), who attacked the logic and evidence of the evolutionists. Remnants of this controversy can still be seen today in the debate over creation science in American education.

DONALD R. FRANCESCHETTI

Further Reading

Bell, Eric Temple. *Development of Mathematics*. New York: McGraw-Hill, 1945.

Boyer, Carl B. *A History of Mathematics*. New York: Wiley, 1968.

Kline, Morris. *Mathematical Thought from Ancient to Modern Times*. New York: Oxford University Press, 1972.

Mathematicians Reconsider Euclid's Parallel Postulate

Overview

Ever since the time of Euclid, mathematicians have felt that Euclid's fifth postulate, which lets only one straight line be drawn through a given point parallel to a given line, was a somewhat unnatural addition to the other, more intuitively appealing, postulates. Eighteenth-century mathematicians attempted to remove the problem either by deriving the postulate from the others, thus making it a theorem, or by replacing it with a simpler statement. Nineteenth-century mathematicians would change the postulate to generate logically consistent non-Euclidean geometries, which twentieth-century physicists would in turn propose as the true geometry of space and time.

Background

In his *Elements of Geometry,* the great Greek mathematician Euclid (335-270 B.C.) was forced to adopt a rather awkwardly worded fifth and final postulate:

> If a straight line falling on two straight lines makes the interior angles on the same side less than two right angles, the two straight lines, if extended indefinitely, meet on that side on which the angles are less than the two right angles.

It is clear that Euclid was not entirely comfortable with this postulate, as he postponed its use in proving theorems as long as he could. Over the centuries mathematicians speculated that the postulate could actually be proved as a theorem from the other axioms and postulates

or, failing that, that it could be reformulated in a far simpler way as befitted a self-evident truth about the nature of space.

The Greek historian Proclus (410-485) records that the first Ptolemy, ruler of Egypt at the time that Euclid was teaching in Alexandria, wrote a book on the fifth postulate including what he considered a proof from the other postulates. Proclus also identified the flaw in Ptolemy's argument. The Persian mathematician Nasir-Eddin (1201-1274), who translated *Elements* into Arabic also tried to prove the postulate, and his argument too had subtle flaws. The English mathematician John Wallis (1616-1703) offered a proof of his own in 1663, but this too was defective.

In 1733 Girolamo Saccheri (1667-1733), an Italian Jesuit priest and Professor of Mathematics at the University of Pavia, published a book entitled *Euclid Free of Every Flaw.* Saccheri, who had already published a book on logic, decided to prove the postulate using the logical technique of "reductio ad absurdum," in which one disproves a hypothesis by showing the preposterous results of carrying the proposition to its logical conclusion. Saccheri considered a quadrilateral with two adjacent right angles and two parallel sides each meeting the base at a right angle. If the fifth postulate were, in fact, a consequence of the other postulates, the remaining angles would have to be right angles. Saccheri began by assuming that one of the two other possibilities was true, that the angles were either obtuse or acute. There was no difficulty in proving that the assumption of obtuse angles led to a contradiction. The possibility of acute angles proved much more difficult to deal with. After much difficulty Saccheri was able to produce only a very subtle form of absurdity in which two parallel lines when extended to infinity would combine to form a single straight line with a common perpendicular. In fact, he had proved that the fifth postulate was independent of the others, and in the course of his investigation he had derived many of the important theorems of non-Euclidean geometry. Because he was not aware of the significance of his work, credit for the discovery of non-Euclidean geometry would go to others.

In a paper presented in 1763, Georg S. Klugel (1739-1812), a German professor of mathematics, suggested that the fifth postulate could not be proved. Klugel's suggestion was the stimulus for *The Theory of Parallel Lines* by the German mathematician Johann Heinrich Lambert (1728-1777), written in 1766 but not published during his lifetime. Lambert considered a quadrilateral with three right angles and investigated the consequences of assuming that the remaining angle was right, acute, or obtuse. Like Sacceri, Lambert obtained theorems that would be important in non-Euclidean geometry. He derived, for instance, formulas for the area of a triangle in terms of the difference of its angles and the sum of two right angles. Lambert also recognized that a geometry without a parallel postulate might apply to the surface of a sphere or to a saddle-like surface.

Joseph Fenn suggested a simple substitute for the fifth postulate in 1769. In the equivalent form proposed by John Playfair (1748-1819), it states:

> Through a given point P not on a line l there is only one line in the plane of P and l which does not meet l.

It is in this form that the axiom appears in modern texts.

In his attempt to derive as much of geometry as possible without using the fifth postulate, the great French mathematician Joseph Louis Lagrange (1736-1813) was able to prove that if the sum of the angles in any one triangle was equal to two right angles, then the sum of the angles in every triangle would be the same. Without the fifth postulate, however, the sum of the angles in any triangle could not be determined.

Impact

Work on the parallel postulate in the eighteenth century set the stage for the development of non-Euclidean geometries in the nineteenth century. Then it was discovered that totally consistent geometries could be obtained by assuming that through a given point either no lines or multiple lines could be drawn parallel to a given line. Perhaps the first significant mathematician to work in the area of non-Euclidean geometry was the great German Mathematician Carl Friedrich Gauss (1777-1855), who may also have been the first individual to suggest that the geometry of physical space might not be Euclidean. To investigate the latter possibility, Gauss measured the sum of the angles of a triangle defined by three mountain peaks. The uncertainties inherent in physical measurements prevented a definitive conclusion. Credit for the invention of non-Euclidean geometry is generally considered to be shared by Gauss, the Russian mathematician Nikolai Ivanovich Lobachevsky (1793-1856), and the Hungarian mathematician Janos Bolyai (1802-1860). With the groundwork established in the eighteenth century, non-Euclidean geometry was ripe for exploration.

With the exception of the problematic nature of the parallel postulate, the majority of mathematicians in 1800 regarded mathematics as a grand body of established truths. Over the course of the nineteenth century the emergence of several versions of non-Euclidean geometry, together with new problems concerning the concept of number and paradoxes of set theory and logic, would force a reappraisal of the very nature of mathematics. To the philosopher Plato (c. 427-347 B.C.), mathematics had described an abstract reality of which the world as perceived by humans was an imperfect reflection. To Galileo (1564-1642) and Isaac Newton (1642-1727) it was an accurate description of the nature of space. To the philosopher Immanuel Kant (1724-1804), it was one of the fundamental categories of human reason, an order imposed on experience by the human mind. Now it seemed that there was an element of arbitrariness in mathematics. It could no longer be regarded as a reliable means of deriving truths from self-evident propositions. Mathematicians and philosophers would now be divided into different camps by what they believed to be the substance of mathematics. Formalists considered it to be the manipulation of symbols according to specified rules— a sort of game. Intuitionists considered it to be the exploration of ideas that could be clearly conceived in the mind. Many mathematicians, of course, chose to continue proving theorems and solving problems without becoming alarmed about the uncertain foundations of their subject.

Even within the context of Euclidean geometry, there were still additional problems with Euclid's original text. Euclid had made many tacit (or unwritten) assumptions that needed to be made explicit. Logicians had come to realize the futility of trying to define every term that a mathematician might use. A more rigorous geometry text was published in 1882 by Moritz Pasch (1843-1930), a German mathematician. Pasch selected the concepts of point, line, plane, and congruence as undefined terms, and introduced axioms, which are statements about the undefined terms to be accepted without proof. Pasch's axioms included one stating that when a line intersects one side of a triangle it also intersects one of the other two sides. Euclid had assumed this to be the case, but not explicitly.

Pasch's geometry book was followed by others also attempting to provide a complete and logically sound exposition of geometric ideas. Noteworthy are *The Principles of Geometry,* published in 1889 by the Italian mathematician Giuseppe Peano (1858-1932), and the *Foundations of Geometry*, also first published in 1899 by the German Mathematician David Hilbert (1862-1943). Hilbert's book would run through seven editions.

In 1912, in his general theory of relativity, the German-born physicist Albert Einstein (1879-1955) proposed that the gravitational attraction between masses could be understood as resulting from a curvature of space and time. Straight lines were defined by the paths that a light beam would take, and the sum of the angles of a triangle would depend on the distribution of matter in that region of space. Einstein's equations gave a relationship between the the curvature of space and the matter and energy nearby. Applied to the universe as a whole, Einstein's equations allowed for three possibilities: Euclidean (flat) geometry, a space of positive curvature, like the surface of a sphere, or a space of negative curvature, like the surface of a saddle. Which geometry in fact applies is still to be resolved by observation.

DONALD R. FRANCESCHETTI

Further Reading

Boyer, Carl B. *A History of Mathematics*. New York: Wiley, 1968.

Kline, Morris. *Mathematical Thought from Ancient to Modern Times*. New York: Oxford University Press, 1972.

Kline, Morris. *Mathematics: The Loss of Certainty*. New York: Oxford University Press, 1980.

Wolf, Harold E. *Introduction to Non-Euclidean Geometry*. New York: Dreyden Press, 1945.

Symmetry and Solutions of Polynomial Equations

Overview

The search for the solutions of polynomial equations continued throughout the eighteenth century. Within a period of a couple of years around 1770, mathematicians in a number of countries made almost simultaneous advances that led to new ways of approaching the problem. There was disappointment in that none of the advances led to a general solution of the higher-degree equations, but some specific equations could be solved in the aftermath of this work. More importantly, the new approaches provided the means for answering the question about the solvability of polynomial equations in general early in the next century.

Background

An equation is simply an expression in which the two sides are asserted to be equal, which may not involve anything more than numbers. Early on in the development of algebra, the equations of greatest interest were those in which some symbol was used to stand for an unknown quantity whose value was to be determined on the basis of the information furnished by the equation. Linear equations (those involving only the variable to the first power) could be solved without much notation, and it was impressive that Babylonian mathematics was already up to the level of solving quadratic equations (those involving second powers of the variable as well). The description of the solution was usually in terms of words rather than symbols.

That same approach continued even into the sixteenth century with the solution of cubic equations (including third powers of the variable) and quartic equations (allowing fourth powers of the variable). These were solved by the efforts of Italian mathematicians, and the solutions were presented in words. The methods were not scrutinized carefully so long as the answers that were obtained worked in the original equation. Only with the work of François Viète (1540-1603) was there some attempt to apply the rigor of geometry to the techniques of algebra.

Despite Viète's efforts, however, no progress was made in the general question of solving quintic equations (those involving fifth powers of the variable). Towards the end of the eighteenth century Alexandre-Théophile Vandermonde (1735-1796) read a paper before the Académie des Sciences in Paris about a new approach to solving quintic equations and polynomial equations in general. He claimed that every equation in which a power of the variable is set equal to one could be solved by standard means. His work on the subject would be better known if he had been a member of the Academy at the time of delivery (1770), because it would have made it easier for him to publish his claims. As it was, his paper was not published until 1774, and in the meantime Joseph Louis Lagrange (1736-1813), one of the great mathematicians of the century, had published a couple of papers on the subject. Perhaps disappointed, knowing that being the first to publish in the scientific world is of crucial importance, Vandermonde did not pursue his researches into the solution of polynomial equations any further.

Another mathematician whose work anticipated some of Lagrange's was Edward Waring (1736-1798). In his case, the disadvantage that bedeviled his making further progress was the backward state of mathematics in England during his lifetime. As a result of a conscious decision by English mathematicians to follow the practices of Sir Isaac Newton (1642-1727) rather than the leading scholars in Europe, English mathematics drifted away from the advances made in the rest of the world. Waring's work received high praise from Lagrange and others, but he was not in an environment where progress was easy.

Lagrange, by contrast, spent time in the mathematical centers of Europe with the result that his work could have rapid dissemination. His approach to polynomial equations was partly historical, as he stepped back to examine how the Italian mathematicians had managed to come up with their solutions for cubic and quartic equations. On the basis of that analysis, he managed to formulate a research program for

Carl Friedrich Gauss. *(Corbis-Bettmann. Reproduced with permission.)*

how to solve quintic (and higher-degree) equations. As it turned out, even though he was unable to come up with anything by way of a general solution to the quintic equation, his reevaluation of his predecessors provided essential elements for the conclusion at which Evariste Galois (1811-1832) arrived, namely, that the quintic equation could not be solved by traditional means.

Impact

The idea that Lagrange brought to the solution of polynomial equations was that of symmetry. Instead of simply looking for solutions to the original equation, he looked for other relationships that the solutions must satisfy in the hope that they would be simpler than the original equation. So long as it was still possible to recover the original solutions from those of the new auxiliary equations, the method would work to supply those original solutions.

If one looks at a quadratic equation (with the coefficient of the squared term equal to one), then the sum of the solutions of the equation is equal to the negative of the coefficient of the first-order term. The product of the solutions is equal to the constant term. This approach would supply an alternative to the quadratic formula for solving a quadratic equation. Similar strategies give auxiliary equations for solving cubic and quartic equations, as the cubic equation can be reduced to a quadratic auxiliary and the quartic to a cubic auxiliary.

The problem with applying Lagrange's strategy to fifth-degree equations became evident in the work of Gian Francesco Malfatti (1731-1807). In a paper in 1770 he showed that if one tries to find an auxiliary equation for a quintic, the result will be of the sixth degree. Since the exponent of the variable in this auxiliary is higher than it was in the original, it has only made the problem worse. When sixth-degree equations showed up in solving the cubic equation, they could be treated as quadratics. The ones that arose from working with the quintic could not be reduced in that way.

Lagrange's work served as the basis for a new discipline of mathematics called group theory. This can be regarded as the study of symmetries, arising out of Lagrange's attempts to find symmetric functions of the solutions of an equation. A function is symmetric if some interchange of values of the variables representing the solutions does not change the value of the function. For example, $x + y$ retains the same value if x and y are interchanged.

Group theory proved to be the basis for the proof that there was no formula for solving all quintic equations as there had been for the lower powers of the variable. This did not mean that individual quintic equations could not be solved, as some of them had been handled centuries earlier. What the proof did guarantee was that any method of solution that claimed to tackle all quintic equations had to go beyond the standard algebraic operations to which attention had previously been limited. The entire branch of mathematics known as Galois theory is devoted to the application of group theory (based on symmetry) to the study of polynomial equations and their solutions.

At the end of the eighteenth century it might have seemed as though no progress had been made on the long-standing problem of solving polynomial equations of degree higher than four. In fact, however, the efforts of Lagrange and others had changed the problem into one that could be approached by the study of symmetry. Lagrange's clarification of what constituted a method of solution for a polynomial equation made it possible for different disciplines within mathematics to be applied to finding those solutions.

THOMAS DRUCKER

Further Reading

Katz, Victor J. *A History of Mathematics: An Introduction.* New York: HarperCollins, 1993.

Nov'y, Lubos˘. *Origins of Modern Algebra.* Prague: Academia Publishing House, 1983.

Van der Waerden, B.L. *A History of Algebra.* Berlin: Springer-Verlag, 1985.

Mathematical Textbooks and Teaching during the 1700s

Overview

The eighteenth century was a time of great progress in the field of mathematics. Included in this progress were advances in ways of teaching mathematics, including a number of textbooks at all levels of complexity by some of the world's greatest mathematicians. Because of these textbooks, mathematics education became more standardized and more formalized than had previously been the case, especially at introductory levels. This, in turn, helped in the education of still more mathematicians who were able to continue producing further mathematical research of very high quality.

Background

The great majority of students are not likely to consider mathematics textbooks a great boon and, in fact, are more likely inclined to curse the authors than praise them. However, any society that strives towards technical and scientific mastery needs to be able to teach its citizens the basic mathematical tools necessary for survival; at the very least, basic arithmetic and algebra skills and an appreciation of concepts in geometry and more advanced mathematical fields such as trigonometry should be realized by its citizens. These mathematical skills are necessary not just for future scientists and engineers, but for anyone hoping to work in a skilled job or profession. However, until the eighteenth century, mathematics education was not a highly developed field.

In fact, through most of the history of mathematics, math skills were taught to a society's elite: priests, rulers, and selected others. Much of this education most closely resembled an apprenticeship, in which a priest or private tutor taught a student either individually or in very small groups. Most students learned only basic arithmetic—addition, subtraction, multiplication, and division—because no greater knowledge would be required of them during their lives. A select few, destined for careers in finance, might learn the mysteries of interest calculations, and even fewer would learn sufficient math skills to calculate the answers to problems in physics or astronomy. However, the great majority of people learned virtually no math, and their lives were none the poorer for this because their world was a simple one.

Another factor to consider was that many mathematicians, instead of sharing their methods of solving problems, kept them secret. A very small number of mathematicians helped support themselves through mathematical "tricks," such as extracting square roots or performing rapid mental calculations. Teaching others their methods was akin to a magician giving away the secrets to his tricks; it only gave audiences and the "competition" the ability to do the same thing.

Finally, until relatively recently, there were very few students and even fewer books. The peasant class did not attend school, the aristocracy and royalty had private tutors, and religious orders trained their own students. It made no sense to write and publish textbooks unless students read and used them and, without enough students of mathematics, textbooks were simply not a necessary innovation.

This began to change in the seventeenth century, as university attendance began to increase (albeit still very strongly weighted towards the upper classes of society). In addition, scientific societies began to form, encouraging the teaching of science and the dissemination of knowledge, as opposed to its sequestration in the hands of a few select. Also, science began to be viewed as a field in its own right, capable of supporting full-time work by professors and scientists. And, as society took the first steps towards more financial and technological sophistication, the first steps were also taken in the process of training engineers, scientists, and accountants. In addition, the development of the calculus by Isaac Newton (1642-1727) and Gottfried Liebniz (1646-1716), their feud over priority, and the exploits of other famous mathematicians helped bring mathematics to the attention of the upper and merchant classes, raising its profile somewhat. Adding to this was Newton's stature as the world's first famous scientist, which further helped bring him and his mathematical discoveries to the forefront of public attention in England, while his Continental rivals did the same in Germany, France, and Switzerland. Finally, people began to realize that mathematics could be applied to solving everyday problems, again lending some credence towards teaching an increasing number of students at least the fundamentals of mathematics. All of these factors resulted in an effort to begin formalizing mathematics education, including publishing the first textbooks in this subject.

The first mathematics textbooks (as opposed to arithmetic primers) started coming out in the mid-1700s. Leonhard Euler (1707-1783), perhaps the greatest mathematician who ever lived, published a book, *Algebra*, in 1770-1772. This book went through six editions, and Euler's other textbook, on number theory, was popular as well. In Britain, Colin Maclaurin (1698-1746) wrote texts on both the elementary and advanced levels, including *Treatise of Algebra*, which also went through multiple editions in the latter half of the century. Other texts on algebra, geometry, and the mathematical treatments of physics were released during the eighteenth century, many by some of the era's greatest minds in mathematics.

Impact

The effects of this attention to mathematics education were significant.

1. First and foremost, the character of any subject is dictated in large part by the textbooks out of which students study because it is in the textbooks that students receive their first exposure to a topic.

2. In addition, these texts reflected the fact that a market for them existed. The growing number of students of mathematics encouraged the writing of texts, while the growing numbers of texts, many by famous mathematicians, encouraged students to study mathematics. As a result, European and American colleges and universities turned out increasing numbers of mathematically literate (or, at least, mathematically

aware) students, who went on to use these skills in their jobs.

3. Finally, much of today's mathematics texts and teaching methods are direct descendants of these first textbooks; we can trace the sequence in which we learn mathematics and many of the topics taught to these first texts. Each of these ideas will be developed in more detail in the following paragraphs.

To start with, textbooks have a vital role in shaping the manner in which any topic is taught and, in turn, with shaping the opinion of students. This fact is perhaps best appreciated by considering recent debates over teaching evolution in the classroom. Religious fundamentalists fully realize that the manner in which evolution, creationism, and creation science are presented in a textbook will affect a student's perception of these mutually exclusive systems. If evolution is taught without reference to any form of creationism, then students perceive that it is unchallenged as the correct theory surrounding the history of life. On the other hand, if evolution is taught as one of several possible theories, then competing theories gain credence and the case for evolution is weakened. Finally, if it is forbidden to teach evolution, regardless of the great weight of scientific evidence in its favor, it will then be perceived as lacking credibility, and a generation of students will consider it to be somewhat flawed.

Similarly, the content of mathematics texts helps to set the tone for how the educated public viewed mathematics in the eighteenth, and subsequent centuries. By portraying mathematics as a useful tool, these textbooks helped to encourage its acceptance by a wider array of merchants, engineers, and scientists than had previously been the case. When mathematics was taught by Mayan priests only to other Mayan priests, its utility was limited because it was a tool dedicated to calculating astronomical events by only a select few. When any literate person at a university could pick up a mathematics text and learn relatively sophisticated techniques, the subject was demystified and became much more useful for daily activities.

This, in turn, helped to advance the field of mathematics. Going back to the Mayans, mathematics barely progressed from a tool for astronomical calculations because that was its entire reason for being. However, as university students graduated and went on to careers in finance, engineering, mathematics, or science, they took with them the ability to adapt their mathematical tools to the uses at hand. This gave mathematics an infusion of new ideas, new applications, and new techniques, contributed by both mathematicians and nonmathematicians alike. Both mathematics and society benefited from this exchange of ideas.

Finally, the way that mathematics is taught today can, in many ways, be traced directly back to these first textbooks, suggesting that today's students continue to be influenced by the authors of the first such books, over two centuries ago. In spite of the occasional jokes (or furor) over new math, the fact is that students are still taught many topics in a manner their predecessors would recognize. Addition, subtraction, multiplication, and division continue to form the basis of elementary mathematics education, as do fractions and decimals. And with good reason; these are the skills that are easiest to learn and with the most applicability to a wide variety of circumstances and professions. After that, students start to learn basic algebra techniques, they may be taught the fundamentals of geometry, and they may study logarithms, trigonometry, and similar mathematical skills. This is a curriculum that Euler would have recognized, because these are among the most elementary tools in the mathematical repertoire. These tools will carry virtually anyone through most nontechnical careers, and constitute the foundation for further study of mathematics at a more advanced level in college. The primary difference between this education and that of the eighteenth century is that of timing; these topics are considered elementary topics to be taught to middle and high school students today, but were considered advanced topics for university students two centuries ago.

We can see, then, that the appearance of the first textbooks in mathematics was an important milestone in the history of mathematics. By writing these books, the authors helped to establish a standard framework that formed the basis for all mathematics education, not just at the time, but continuing until the present. In addition, by educating a larger number of students, these books also helped spread the utility of mathematics and, by so doing, to cross-fertilize mathematics with ideas from other fields. All of these had a significant impact on mathematics, education, and society.

P. ANDREW KARAM

Further Reading

Books

Boyer, Carl and Uta Merzbach. *A History of Mathematics.* New York: John Wiley & Sons, 1991.

Chinese and Japanese Mathematical Studies of the 1700s

Overview

The 1700s were a time of transition for China and Japan, in mathematics as in every other aspect of society and culture. Barriers to Western influence and information, in place for centuries, were beginning to crumble. Scholars and scientists in both countries, eager to learn more of Western science and technology, undertook ambitious programs of translation, aimed at making Western science—and the mathematics that underlay it—available to their countrymen. More importantly, the exposure to Western mathematics served as a spur to Chinese and Japanese mathematics, not in imitation or competition, but rather as a catalyst for rediscovering the roots and resources of their own mathematical traditions; both countries took long looks back at those roots, both extending traditional mathematics and melding those traditions with new skills acquired from the West. The West as well learned from Asian mathematical traditions, which had anticipated many Western discoveries. Mathematical education began to spread throughout both countries, becoming an important aspect of basic literacy.

Background

The Chinese mathematical tradition stretches back to the dawn of recorded history: oracle bones from the fourteenth century B.C. are etched with counting symbols. By the first century A.D. (to use Western notation) the Chinese were employing a decimal system, according to *Nine Chapters on the Mathematical Procedures*, the classic Chinese text of the time. There is a rich tradition of pictographic symbol-driven (rather than numerical) mathematics in China, with the unique addition of an auditory, or sound-based, mathematical component. Fundamentally algebraic—geometry seems not to have developed far there—Chinese math was at one time a powerful and sophisticated tool for the exploration of patterns, positions, and matrices of numbers. Indeed, Chinese mathematics anticipated some of Blaise Pascal's (1623-1662) findings by more than four centuries. Chinese mathematics also exerted a large influence over the development of Japanese mathematics.

By the 1700s, though, Chinese mathematics had largely abandoned theory to focus on practi-

cal matters, used for finance and agricultural and other records. The abacus had come into use in China during the 1600s, and found application in the counting of inventories and products. Also during the 1600s, Western approaches to mathematics, the calendar, and astronomy began to be imported and translated, along with the works of Euclid (c. 330-260 B.C.) and others.

This increasing contact with Western priests (particularly Jesuits) and traders helped re-ignite Chinese interest in pure mathematics, even as practical math was undergoing its own refinements and improvements. Scholars and mathematicians were delving into their country's mathematical past, much of the research and rediscovery prodded by K'ang-hsi (1654-1722), fourth emperor of the Manchu (Ch'ing) dynasty and a ruler devoted to scholarship and learning. At K'ang-hsi's request a vast encyclopedia, the *Ku-chin t'u-shu chi-ch'eng* (Synthesis of Books and Illustrations Past and Present or The Imperial Encyclopedia) was compiled, although not completed until after K'ang-hsi's death. With more than 10,000 entries, the first of the encyclopedia's main headings was "Calendar, Astronomy, Mathematics." Also at the emperor's behest, and with the assistance of Father Antoine Thomas, the basic Chinese unit of measurement, the li, was recalculated as a function of the meridian, nearly a century before such an approach was undertaken in the West.

Mathematical scholarship flowered during the 1700s, guided by the example of Mei Wenting (1632-1721), who assembled a large comparative study of Chinese and Western mathematics. The great Chinese scholar of historical mathematics during the period was Tai Chen (1723-1777), who applied himself to the study of the Chinese tradition of counting with rods. Tai Chen wrote books on the ability to use counting rods to solve equations with unknown quantities and variables, as well as a 1755 volume on the measurement of circles; he was also responsible for overseeing the republication of classical Chinese mathematical treatises.

The flow of mathematical knowledge traveled in both directions, and many European mathematicians and scholars were fascinated and influenced by Chinese mathematics. Notable among these was German philosopher

Gottfried Wilhelm Leibniz (1646-1716), who spent much of his final years applying himself to Chinese studies, particularly the ways in which Chinese mathematics either anticipated or independently duplicated Western findings. Much of Leibniz's energy was devoted to studies of early Chinese use of binary notation, although his scholarship was somewhat tainted by his determination to fit his findings into already existing philosophical conclusions. Leibniz also found superiorities in the Chinese symbology of mathematics over the Arabic numerical system. Some scholars feel that Leibniz's studies of Chinese mathematics stimulated the growth of mathematical logic in Europe.

While social disruption and cultural clashes would dominate the next three centuries of Chinese/Western relations, the 1700s were a period of sometimes wary exchange, with mathematics one of the "universal" languages of that exchange. China and the West learned from and were influenced by each other. In one area, astronomy, it should be noted that the Chinese (and the Japanese) did not suffer the social dislocation Europe had faced with the Copernican revolution; Chinese mathematics and astronomy, indeed Chinese philosophy overall, viewed the universe as a whole, with our world and its inhabitants merely a part of that whole. To the Chinese, Copernicus's removal of Earth from the center of the universe was a revelation, not a revolution: in Chinese science, math, and philosophy, humans had never been the absolute center.

The 1700s were a time of transition for Japanese mathematics as well, although the transition may have been more fitful, as Japan's barriers against Western influence were stronger and more rigid than those in China. Still, Jesuits and Dutch and Portuguese traders did establish a presence in Japan, and, slowly, the exchange of mathematical ideas began to flow, much as many Japanese mathematical traditions had flowed from China in the past. In fact, hewing to the prohibition against purely Western texts, many of the imported treatises were first translated into Chinese.

As in China under K'ang-hsi, Japan benefited from the scholarly interests of its leader, or shogun, Tokugawa Yoshimune (1684-1751.) Yoshimune was a great reformer who overruled—without overturning—longstanding edicts against the importation of foreign knowledge, particularly the publication of foreign texts. Yoshimune had a particular fascination with mathematics and its practical applications. Concerned over the accuracy of the Japanese calendar—whose social and religious importance was vast—and aware of new astronomical findings, Yoshimune encouraged the publication of European astronomical texts.

The importation of foreign mathematics resulted in a division in Japanese math studies: traditional Japanese methods of calculation came to be known as *wasan*, while mathematical tools and techniques imported from the West were referred to as *yosan*.

Perhaps the most influential, but in some ways least known, Japanese mathematician of the period was Seki Kowa (1642-1708), whose influence lasted throughout the century. A child of a samurai family, Seki Kowa made large contributions to Japanese calculus, as well as creating a theory of determinants that anticipated the work, shortly thereafter, of Leibniz.

Practical mathematics outweighed the theoretical in Japan at the time. Application of math to navigation and cartography took precedence over pure math.

Much of the mathematical focus of the period was on education, with mathematical training being extended throughout Japanese (male) society. The *terakoya*, or one-room school, common to Japanese villages included instruction in the use of the abacus, as well as reading and writing. Japanese primers, or *orai*, included volumes devoted to calculation, particularly useful for the *tedai*, or clerk class. As commerce became more and more central to Japanese daily life, so did mathematics.

Scholars of the time noted the increasing importance of math for all citizens, at least male citizens. Confucian scholar Ito Jinsai (1627-1705) had written at the beginning of the century that "It goes without saying that those of low status should also learn writing and arithmetic." That insight continued to guide the spread of mathematics education throughout eighteenth-century Japan.

Impact

As the eighteenth century closed, both China and Japan stood on the brink of full discourse—sometimes violent—with the West. Barriers would be raised and lowered more than once in the decades to come, but the process of cultural and intellectual exchange had begun, and could not be stopped.

The increasing Westernization of Japanese and Chinese mathematics was a cause of some concern for scholars and philosophers in both countries. There was a tendency among scientists, bedazzled by the scientific and technical achievements of the West, to belittle the equally vast achievements of China and Japan. Gradually, Western approaches to mathematics, particularly the introduction of geometry, began to exert more sway than traditional methods.

In practical matters as well, Western mathematics moved to the forefront as China and, particularly, Japan, began the race to technological and military modernity.

Yet those local traditions possessed great inherent as well as mathematical value, and the challenge for scholars, historians, and mathematicians was to strike a balance between the two, to find, as it were, the best of both mathematical worlds.

KEITH FERRELL

Further Reading

Gernet, Jacques. *A History of Chinese Civilization.* Cambridge: Cambridge University Press, 1996.

Hall, John Whitney, ed. *The Cambridge History of Japan.* Volume 4: *Early Modern Japan.* Cambridge: Cambridge University Press, 1991.

Japan: An Illustrated Encyclopedia. Tokyo: Kodansha, 1993.

Needham, Joseph. *Science and Civilization in China.* Volume 3: *Mathematics and the Sciences of the Heavens and the Earth.* Cambridge: Cambridge University Press, 1959.

Sansom, G. B. *The Western World and Japan: A Study in the Interaction of European and Asiatic Cultures.* New York: Alfred A. Knopf, 1970.

Biographical Sketches

Maria Gaëtana Agnesi
1718-1799
Italian Mathematician

The daughter of a mathematics professor at the University of Bologna, Maria Gaëtana Agnesi grew up possessing an enormous command of mathematics, languages, and the sciences. She later dedicated herself to the education of her many younger siblings, a project that resulted in her most important work, *Instituzione analitiche.* This work includes discussion of the curve referred to as the "Witch of Agnesi."

Agnesi spent her entire life in the northern Italian city of Milan. Her father taught mathematics at the University of Bologna, and appreciation for learning permeated the family's home life. Recognizing his daughter's promise as a scholar, the father hired a tutor for her; meanwhile, she excelled in languages all on her own. The many visitors to her parents' home carried on their conversations in Latin, still the language of learned discourse. Not only did young Maria master that tongue, but for guests who did not speak it, she could also converse and translate into a variety of other languages, including Greek and Hebrew.

At the age of 17, Agnesi wrote a commentary on a study of conic sections by an earlier mathematician, the marquis de l'Hospital. Three years later, in 1738, she published a book of essays entitled *Propositiones philosophicae,* in which she discussed a number of subjects, including the theory of universal gravitation put forth in the previous century by Sir Isaac Newton (1642-1727). Yet, when her mother died, Agnesi expressed a desire to enter a convent and put an end to her intellectual pursuits. Her father, however, requested that she devote herself to the education of her younger siblings, of which she had many (Agnesi was the eldest), and this became her life's work.

As part of her effort to teach mathematics to her younger brothers and sisters, Agnesi wrote and in 1748 published *Instituzione analitiche ad uso della gioventu' italiana.* This mathematics textbook for youngsters provided an introduction to algebra, analysis, and both differential and integral calculus. Among the topics it discusses is the curve incorrectly identified as the "Witch of Agnesi," so named due to a mistranslation; the Italian word for curve, *versiera,* looked like *versicra,* or "witch." The "Witch of Agnesi," studied earlier by Pierre de Fermat (1601-1665) and Guido Grandi (1671-1742), was represented by the equation $y(x^2 + a^2) = a^3$, where a is a constant. As y tends toward 0, x will tend toward ∞. In 1801, John Colson of Cambridge, who learned Italian specifically for the purpose, translated Agnesi's book as *Analytical Institutions.*

Recognizing the contributions to learning made by the 32-year-old Agnesi, Pope Benedict XIV named her professor of mathematics and natural philosophy at Bologna—her father's old school. Her father, however, had become seriously ill, and Agnesi devoted herself to caring for him. Two years later, he died, and rather than accept the appointment at Bologna, Agnesi threw herself into religious work. In 1771 she opened a home for the poor and infirm, and devoted herself to this project for the remaining 18 years of her life.

JUDSON KNIGHT

Thomas Bayes
1702?-1761
English Mathematician

Thomas Bayes spent most of his career as a Presbyterian minister overseeing his flock at Tunbridge Wells, Kent, yet in his spare time, he produced a number of intriguing and influential mathematical papers. Of particular interest were two published posthumously by the Royal Society of London. In one of these, Bayes offered the first discussion of asymptotic behavior by series expansions; and in the other—a treatise that continues to exert an influence in a variety of statistical applications—he laid the groundwork for what became known as Bayesian statistical estimation.

The circumstances of Bayes's life, including the exact year of his birth, are a mystery. He was born in 1701 or 1702, either in London or in Hertfordshire. At least the identity of his parents, Joshua and Ann Carpenter Bayes, is known, as is the fact that Thomas was the first of six children. His father was one of the first Nonconformist ministers ordained in England. Just as the Anglicans had broken with Rome in the sixteenth century, in the seventeenth century the Nonconformists had begun splitting with the Anglicans, and this movement would ultimately spawn the Baptist, Methodist, and Congregationalist churches.

The nature of Bayes's education is also open to question, but it appears he attended the University of Edinburgh. In 1727, at about the age of 25, he was ordained, and went to work assisting his father at Leather Lane. Later he took charge of the Presbyterian meeting house called Mount Sion in Tunbridge Wells, where he would remain for the rest of his career.

In 1742 Bayes was elected a Fellow of the Royal Society of London, which might seem an unusual honor for a rural minister, but Bayes was no ordinary country pastor. He is widely credited with the authorship of *An Introduction to the Doctrine of Fluxions and Defence of the Mathematicians Against the Objections of the Author of the Analyst,* a tract written in response to *The Analyst* by George Berkeley (1685-1753) Berkeley's 1734 publication attacked Isaac Newton (1642-1727) and his writings on differentials or fluxions. It was in honor of this work, wherein Bayes defined the "business of the mathematician" as one of making deductions rather than propounding a specific theory, that Bayes received his election to the Royal Society.

Bayes lived his life quietly, but in secret he recorded a number of fascinating observations, as demonstrated by a notebook of his that offers a model for an electrifying machine and other intriguing ideas. He retired from the ministry in 1750, and died on April 17, 1761, after which a Unitarian minister and friend of the family, Richard Price (1723-1791), went through his papers.

Thanks to Price's efforts, in 1764 the *Philosophical Transactions of the Royal Society* included a paper by Bayes on the subject of series expansions. According to Price, Bayes wrote the paper in an effort to underscore what philosophers typically call the "argument from design": that is, the defense of God's existence on the basis of his creation and its intricacy.

In the same issue of the *Philosophical Transactions* was a second Bayes piece, "An Essay Towards Solving a Problem in the Doctrine of Chances." In this tract, Price explained, Bayes sought "to find out a method by which we might judge concerning the probability that an event has to happen, in given circumstances, upon supposition that we know nothing concerning it but that, under the same circumstances, it has happened a certain number of times, and failed a certain other number of times." This helped lay the foundations for statistics in general, and for Bayesian statistical estimation—still used today for a variety of applications—in particular. Price further developed Bayes's ideas in "A Demonstration of the Second Rule in the Essay . . .", published in 1765 by the Royal Society.

JUDSON KNIGHT

Daniel Bernoulli
1700-1782
Swiss Mathematician and Physicist

Daniel Bernoulli belongs to that rarefied upper echelon of mathematicians and scientists for whom distinctions between disciplines are largely academic. Trained as a mathematician, he is best known for the application of

Daniel Bernoulli. *(Corbis Corporation. Reproduced with permission.)*

his ideas to physics as expressed in Bernoulli's principle, which concerns the inverse relationship of pressure and velocity in fluids. Numerous and groundbreaking as his contributions in the field of hydrodynamics were, however, these were far from the full range of the discoveries made by a genius whose work explored realms as diverse as probability, medicine, and music.

Throughout his life, Bernoulli would be locked in competition with his father Johann (1667-1748), professor of mathematics at the University of Groningen in the Netherlands when Bernoulli was born. As he became aware of his second son's great talents, Johann would become increasingly jealous; yet the Bernoulli family, whose native land was Switzerland, was filled with men of distinction. Johann's brother Jakob (1654-1705) held the chair of mathematics at the University of Basel, a position Johann took over when Daniel was five.

As he progressed with his education, Daniel demonstrated abilities that made him a standout even within the Bernoulli family. At age 16, he had already earned a master's degree, which so inflamed Johann's jealousy that he forbade his son from a career in the sciences. The closest Daniel could come to a scientific career was to study medicine, and there too he excelled. His first love was mathematics, however, and he continued to study the subject in Italy during the early 1720s.

He gained his first fame in 1724, with the publication of *Exercitationes quaedam mathematicae,* which contained his first discussions in the areas of probability and fluid motion—areas in which he was destined to make his most lasting impact.

On the strength of this widely recognized publication, Bernoulli received an appointment to the St. Petersburg Academy in Russia, where he was joined by his brother Nikolaus (1695-1726). More honors were to follow, as in 1725 he won the first of nine prizes from the French Académie Royale des Sciences. While in St. Petersburg, Bernoulli produced important studies in the field of probability, and began a correspondence with his distinguished compatriot Leonhard Euler (1707-1783). Eventually he grew homesick, and in 1737 accepted a professorship in botany—a subject of limited interest to him—simply because it would give him an opportunity to return to Basel.

Back in Switzerland, Bernoulli devoted himself to a wide array of undertakings, the most notable of which was *Hydrodynamica,* published in 1738. (Ever jealous, Johann later published his own *Hydraulica,* which he retroactively dated to 1732 and in which he appropriated many of his son's discoveries as his own.) *Hydrodynamica* contains Bernoulli's principle, which states that the pressure of a fluid decreases as its velocity increases. Also in *Hydrodynamica,* Bernoulli confirmed Boyle's law on the inverse relationship of pressure and volume; and by relating the ideas of pressure, motion, and temperature, he laid the foundations for the kinetic theory of gases that would emerge in the following century.

Bernoulli became a professor of physiology in 1743, and seven years later was appointed to the chair of natural philosophy, or physics, in Basel. He remained in that position until his retirement in 1776, and during those years continued to develop his seminal ideas on kinetic energy—then called *vis viva,* or "living force." He also conducted a number of musical experiments, calculating frequencies and demonstrating the relationship between mathematics and music. Bernoulli died in Basel on March 17, 1782.

JUDSON KNIGHT

Nikolaus Bernoulli
1695-1726
Swiss Mathematician

The first and favorite son of Johann Bernoulli (1667-1748), Nikolaus Bernoulli was destined to be overshadowed by his younger brother Daniel (1700-1782). However, working in collaboration his brother Daniel, Nikolaus contributed to the formulation of the famous St. Petersburg Paradox, a question of probability.

The Bernoullis were a distinguished family of Swiss mathematicians and scientists dating back to Nikolaus's grandfather, Nikolas (1623-1708). Grandfather Nikolas had three sons—Jakob (1654-1705), Nikolaus (1662-1716), and Johann (1667-1748). Jakob, a professor of mathematics in Basel, Switzerland, died without leaving any notable offspring of his own. Nikolaus, often referred to as "Nikolaus I" to distinguish him from the two other men of that name in the family, had a son named Nikolaus (1687-1759), typically designated as "Nikolaus II". Adding to the confusion is the fact that this Nikolaus, a professor of mathematics at Padua and later of law and logic at Basel, contributed to the study of probability and infinite series—as did "Nikolaus III," eldest son of Johann.

"Nikolaus III" was born on February 6, 1695, in Basel, but later the family moved to the Netherlands, where his father Johann took a position as professor of mathematics at the University of Gröningen. Nikolaus, who always enjoyed a close relationship with his father, was five years old when Daniel—against whom the jealous father would place himself in competition throughout his later career—was born. Five years later, the family moved back to Switzerland, where Johann took over the mathematics chair vacated by his recently deceased brother Jakob.

Like his younger brother, Nikolaus proved himself an exceptional student, but unlike Daniel, Nikolaus's brilliance did not excite his father's envy. At the age of 13 Nikolaus entered the University of Basel, where he studied both mathematics and law. Seven years later, at age 20, he became a licentiate in jurisprudence.

Nikolaus also worked with his father, acting as his assistant in correspondence. He was particularly involved in writing letters to address a dispute between Isaac Newton (1642-1727) and Gottfried Leibniz (1646-1716), and exchanged letters with the English mathematician Brook Taylor (1685-1731). During this time, Nikolaus also worked on problems involving curves, differential equations, and probabilities.

In 1725, at about the time Daniel—by then emerging as the most outstanding member of the talented family—received an appointment to the St. Petersburg Academy in Russia, Nikolaus was likewise assigned to a post there. In Russia, the two brothers worked on the "St. Petersburg Paradox," which involves a coin toss game in which a player bets on the number of tosses needed before a coin turns up heads. Given that he will receive 2^n dollars if the coin comes up heads on the nth toss, he is theoretically incapable of losing money.

On July 31, 1726, just eight months after arriving in St. Petersburg, Nikolaus died of what was then called a "hectic fever." Thus a promising career was cut short at the age of 31.

JUDSON KNIGHT

Etienne Bézout
1739-1783
French Mathematician

Though he conducted important research in algebraic equations, Etienne Bézout is remembered primarily for his contributions to mathematics education. Most notable was *Cours de mathematiques,* a six-volume textbook that arose from his work as teacher for the French military, and which in turn influenced the teaching of mathematics on both sides of the Atlantic.

The second son of Pierre and Helene Filz Bézout, Etienne Bézout was born on March 31, 1739, in Nemours, France. His family was well-connected politically, and both his father and grandfather had served as district magistrates. Rather than follow in their footsteps, however, Bézout—who at an early age was deeply influenced by Leonhard Euler (1707-1783)—chose a career in mathematics.

At the age of 19, Bézout became an adjunct member of the Académie Royale des Sciences, and 10 years later was elected to full membership. Much of his work was distinctly practical in nature, a fact that arose partly from his marriage at the age of 24: with a family to support, Bézout availed himself of the steady income offered by a position as teacher and examiner to naval cadets in the Gardes du Pavillon et de la Marine. Five years later, in 1768, he began working with artillery officer candidates as well.

These teaching responsibilities required that Bézout eschew a theoretical approach in favor of a practical one, and as a result he developed a highly understandable course of study. This became the basis for the six-volume *Cours de mathematiques: A l'usage des Gardes du Pavillon est de la Marine,* which became an enormously popular textbook. Over the years that followed, it would reach the United States, where it became a favored text at both Harvard University and West Point Military Academy.

Many observers admired Bézout for his simplification of difficult subjects, and for preserving the important discoveries of such recent mathematicians as Euler and Jean le Rond d'Alembert (1717-1783). Other commentators, however, faulted Bézout for what they regarded as a simplistic approach to the discipline. In fact Bézout himself can hardly be dismissed as a mathematical lightweight: during the 1760s, he devoted himself to challenging questions involving determinants in algebraic equations, and particularly to elimination theory.

Bézout died in 1783 in Basses-Loges, France, but his influence extended far into the next century. *Cours de mathematiques,* the English edition of which saw numerous reprints, remained an important text in America; meanwhile, Bézout's work had an impact on the mathematician Gaspard Monge (1746-1818), the military leader Lazare Carnot (1753-1823), and others.

JUDSON KNIGHT

Gabrielle-Emilie Le Tonnelier de Breteuil, marquise du Châtelet
1706-1749
French Mathematician, Chemist and Physicist

Despite her many achievements as a scientist and mathematician, Gabrielle-Emilie, marquise du Châtelet is most famous for being the lover of François Marie Arouet, who is best known as Voltaire (1694-1778). This fact is unfortunate because Châtelet was a highly talented thinker in her own right and would probably have become much more famous if she had not been a woman in eighteenth-century France.

She was born Gabrielle-Emilie le Tonnelier de Breteuil in Paris on December 17, 1706. Her aristocratic upbringing allowed her an opportunity to enjoy an education beyond the reach of most girls, and she studied science, music, and literature with a series of tutors. At 19, Gabrielle-Emilie married the marquis du Châtelet and was thenceforth known as the marquise du Châtelet.

The couple had three children but her marriage to the marquis did not stop the marquise from seeking romance elsewhere—in the arms of Voltaire, who became her lover in 1733. Voltaire encouraged her to study mathematics, which she

did under the tutelage of the scientist Pierre-Louis Moreau de Maupertuis (1698-1759). In 1734 Voltaire, facing trouble for his outspoken criticism of the French royal house, moved in with the marquise and, as a result, their chateau became a magnet for many of the era's great minds. During this time, Châtelet assisted Voltaire in a study on Newton (1642-1727) called *Elements of Newton's Philosophy* (1738).

Châtelet became intrigued with a number of scientific questions, among them the nature of combustion. Combustion, a much-debated topic at the time, was at the subject of a 1737 contest sponsored by the Académie Royale des Sciences, for which Châtelet submitted a paper. She rightly rejected the prevailing phlogiston theory, arguing correctly that heat was not a substance and, with even more prescience, suggested that light and heat are related. Châtelet did not win, but the fact that Leonhard Euler (1707-1783) walked off with the prize suggests the kind of competition she faced.

Châtelet also translated Newton's *Principia mathematica* into French and might have accomplished many more things. She became pregnant again at the age of 42, however, and died of a puerperal fever shortly after giving birth to a daughter on September 4, 1749.

JUDSON KNIGHT

Alexis-Claude Clairaut
1713-1765
French Mathematician

Alexis-Claude Clairaut is best remembered for his book *Théorie de la figure de la terre,* an outgrowth of studies he made during a 1736 journey to Lapland. In it, he discussed the curvature of the Earth, and thus proved Sir Isaac Newton's (1642-1727) theory that the shape of the Earth is that of an oblate ellipsoid. This in turn led to widespread acceptance of Newton's gravitational theory.

Born in Paris on May 7, 1713, Clairaut was the son of Jean-Baptiste, a mathematics teacher, and Catherine Petit Clairaut. The couple had some 20 children, of which few survived. Their son made an early display of his talent, studying calculus at age 10 and writing erudite mathematical treatises when he was barely a teenager. He went on to conduct a lively correspondence with some of the leading mathematical figures of the day, including Leonhard Euler (1707-1783) and Johann Bernoulli (1667-1748).

When he was 23 years old, Clairaut traveled to Lapland in the far north as part of an expedition directed by Pierre Louis Moreau de Maupertuis (1698-1759). The group's mission was, in part, to measure the curvature of the Earth inside the Arctic Circle, and later Clairaut published the results of their activities in *Théorie de la figure de la terre* (1743). In addition to ensuring the acceptance of Newton's ideas concerning gravitation and the shape of the Earth, the book presented a means for determining the shape of the Earth by timing the swings of a pendulum. Clairaut also analyzed the effects created by gravity and centrifugal force on a rotating body, putting forth a proposition later known as Clairaut's theorem.

The unmarried Clairaut maintained an active social life, and became heavily involved in societies to promote the advancement of young mathematicians. Most prominent was his involvement with the distinguished Académie Royale des Sciences, to which he was elected a member at the age of 18. Later studies on tides won him an award from the Académie in 1740, and three years later he was appointed associate director of the organization. He also became a Fellow of the Royal Society of London.

Among Clairaut's other ventures into applied mathematics was his measurement of Venus, the first accurate reckoning of that planet's size. He also conducted important studies on the gravitational relationship between the Earth and the Moon, as well as predictions regarding Halley's comet, and published a number of works. Clairaut died on May 17, 1765, after a brief illness, at the age of 52.

JUDSON KNIGHT

Marie-Jean-Antoine-Nicolas de Caritat, marquis de Condorcet
1743-1794
French Mathematician and Social Philosopher

Marie-Jean-Antoine-Nicolas de Caritat, better known by his title, marquis de Condorcet, was among the last of the social commentators known as *philosophes,* a group that had included Jean-Jacques Rousseau (1712-1778) and Voltaire (1694-1778). A nobleman trained as a mathematician, Condorcet sought to apply mathematical study to social problems, and is considered among the founding fathers of the social sciences.

Born at Ribemont in Picardy on September 17, 1743, de Caritat would later receive the title marquis de Condorcet, which referred to the town of Condorcet in Dauphiné. He studied at the Jesuit college in Reims, and later at the Collège de Navarre in Paris before going on to the Collège Mazarin, also in Paris.

Condorcet published his first significant work, *Essai sur le calcul intégral,* in 1765, when he was just 22 years old. Four years later, he was elected to the Académie des Sciences. During this time, he produced a number of important works, including a 1772 piece on the integral calculus that Joseph-Louis Lagrange (1736-1813) later described as "filled with sublime and fruitful ideas."

In the early 1770s Condorcet became friends with Anne-Robert-Jacques Turgot (1727-1781), an economist who accepted a series of important administrative positions under the French monarchy. After Louis XVI appointed Turgot comptroller general of finance in 1774, Turgot arranged for his friend to become inspector general of the mint. Two years later, Turgot was dismissed and Condorcet attempted to resign, but the king refused his resignation, and Condorcet retained the post until 1791.

Appointed secretary of the Académie des Sciences in 1777, Condorcet continued to study probability and the philosophy of mathematics, concepts he applied to social problems in *Essay on the Application of Analysis to the Probability of Majority Decisions* (1785). Among his breakthrough observations was the Condorcet Paradox, which demonstrated that if the majority prefers option A over option B and option B over option C, it is still possible for a majority to prefer option C over option A—in other words, the statement "majority prefers" is not transitive.

In 1786 Condorcet produced a biography of Turgot, and in 1789 one for Voltaire. From this point forward, however, his attention would increasingly be consumed with politics, and he took an active role among the moderate Girondist faction in the French Revolution. As a Paris representative in the Legislative Assembly, he called for universal male suffrage, proportional representation, and local self-government—liberal ideas for which the radical Jacobin faction had no use.

By 1793 the Reign of Terror was in full swing and Condorcet was on the run. While in hiding from the Jacobins, who then controlled the government, he produced a philosophical

Marie-Jean-Antoine-Nicolas de Caritat, marquis de Condorcet. *(Corbis Corporation. Reproduced with permission.)*

work, *Esquisse d'un tableau historiques des progrès de l'esprit humain* (1795). By the time of its publication, however, Condorcet was dead. Fleeing from a house where he had been hiding in Paris, he was captured and imprisoned on March 27, 1794. Two days later, he was found dead in his prison cell, though it is unclear whether he died of natural causes, suicide, or murder.

JUDSON KNIGHT

Roger Cotes
1682-1716
English Mathematician and Astronomer

An English mathematician who worked closely with Sir Isaac Newton (1642-1727) on the second edition of Newton's *Principia,* Roger Cotes is also remembered for advancing understanding of trigonometric functions. In his entire career, he produced just one full mathematical paper, which served as ample evidence of his genius.

Cotes was born in Burbage, England, in 1682, son of the Reverend Robert and Grace Cotes. Cotes's uncle John Smith, also a minister, took an interest in the precocious boy, and encouraged him to obtain a higher education. First at Leicester School, and later at St. Paul's School in London, Cotes carried on a correspondence

with his uncle, discussing the latest advances in science and mathematics.

In 1699, Cotes entered Trinity College at Cambridge, where he earned his undergraduate degree in 1702. Three years later, he was named a fellow of Trinity College, and in 1706 earned his master's degree. At that point, 24-year-old Cotes was appointed as Cambridge's first Plumian professor of astronomy and natural philosophy. He then solicited funds for the construction of an observatory over King's Gate, and lived in the uncompleted building with his cousin, Robert Smith (1689-1768).

Cotes became friends with Newton, and the two men corresponded on subjects such as telescopes, clocks, and the movement of heavenly bodies. Beginning in 1709, Cotes worked with Newton for three years on the second edition of *Philosophae naturalis principia mathematica,* in which the great physicist refined his theory of universal gravitation. Cotes, who received no financial remuneration for his years of labor on the project, wrote a preface in which he defended Newton's ideas on empirical grounds.

Cotes established a school of physical sciences at Trinity, where he and fellow scientist William Whiston conducted a number of experiments. The experiments were published posthumously as *Hydrostatical and Pneumatical Lectures by Roger Cotes* (1738). Robert Smith also published a collection of Cotes's papers as *Harmonia mensurarum* (1722).

In 1714, Cotes published his sole mathematical paper, "Logometria," which presented innovative methods for calculating logarithms and for evaluating the surface area of an ellipsoid of revolution. In 1715, he made detailed notes on a complete eclipse of the Sun. He died of a fever during the following year, before having reached his 34th birthday. "Had Cotes lived," wrote Newton, a man 40 years his senior who nonetheless outlived him by nearly a decade, "we might have known something."

JUDSON KNIGHT

Gabriel Cramer
1704-1752
Swiss Mathematician

The name Gabriel Cramer is associated with Cramer's rule and Cramer's paradox, as well as with the introduction of the concept of utility to mathematics. Yet perhaps Cramer's greatest contributions to learning emerged from his sup-

Gabriel Cramer. *(The Granger Collection, Ltd. Reproduced with permission.)*

port of other talented contemporaries, specifically in his work as editor of their writings.

Born in Geneva on July 31, 1704, Cramer was the son of Jean, a physician, and Anne Mallet Cramer. He came from a family of three brothers, one of whom became a doctor and the other a professor of law. By the age of 18, Cramer had earned his doctorate with a dissertation on the qualities of sound, and two years later was appointed co-chair of mathematics at the Académie de la Rive. He shared both the position and the salary with Giovanni Ludovico Calandrini, and the two men broke new ground by allowing students who knew no Latin to recite in French instead.

Encouraged by the Académie to travel as a means of expanding his knowledge, in 1727 Cramer spent five months in Basel, where he became acquainted with Johann Bernoulli (1667-1748). Over the two years that followed, he visited London, Leiden, and Paris. Five years after his return to Geneva in 1729, Cramer was appointed full chair of the department after Calandrini received an appointment to a philosophy professorship.

Along with his brothers, Cramer had long taken part in local politics, and he likewise displayed a sense of community spirit where the world of mathematics was concerned. Among the mathematicians and thinkers whose work he edited were Johann and his brother Jakob Bernoulli (1654-1705), as well as the German philosopher and mathematician Christian Wolff (1679-1754). Thus he helped spread these men's ideas, and contributed greatly to their success.

In 1750, Cramer was appointed professor of philosophy at the Académie (Calandrini had left his post to serve the Swiss government), and published the four-volume *Introduction à l'analyse des lignes courbes algébriques*. The work contained Cramer's rule, which governed the solutions of linear equations, and Cramer's paradox, which clarified a proposition first put forth by Colin Maclaurin (1698-1746) regarding points and cubic curves. In addition, Cramer introduced the concept of utility, a key principle today linking probability theory and mathematical economics.

A year after publishing his most important mathematical work, Cramer suffered a fall from a carriage. The scholar had long been overworked and suffering from fatigue, so his doctor recommended that he rest in the south of France. On January 4, 1952, however, Cramer died on his way to the town of Bagnoles.

JUDSON KNIGHT

Jean Le Rond d'Alembert
1717-1783
French Mathematician and Physicist

The name of Jean Le Rond d'Alembert belongs among the most honored of the *philosophes,* French thinkers whose ideas exemplified the Enlightenment. D'Alembert is most readily associated with Denis Diderot (1713-1784), to whose *Encyclopédie* he was a significant contributor. As a mathematician and physicist, his contributions include d'Alembert's principle, an extension of Newton's third law of motion.

D'Alembert's early years were not happy ones. His mother, Claudine-Alexandrine Guérin, marquise de Tencin, was a former nun who lived with a number of men—including, for a short time at least, Louis-Camus, chevalier Destouches-Canon, a military officer. From this union, a son was born in Paris on November 17, 1717, but the mother regarded her pregnancy as an unpleasant interruption in her affairs, and abandoned the infant on the steps of the church at Saint-Jean-le-Rond.

Thus the boy was baptized as Jean Le Rond, and afterward was sent to live in a foster home at Picardy. (Later, in college, he began calling himself Jean-Baptiste Daremberg, and this was eventually shortened to d'Alembert.) Unlike d'Alembert's mother, his father continued to care for him, and later arranged for him to be raised by a Madame Rousseau, a working-class woman who d'Alembert came to regard as his true mother. He lived in her home until he was nearly 50 years old. The father died when d'Alembert was just nine, leaving him with an income of 1,200 livres a year. These funds permitted him the independence he needed to engage in his later scholarly pursuits.

D'Alembert attended Mazarin College, or the Collège des Quatre-Nations, from which he received his bachelor's degree in 1735. Three years later, he earned his license to practice law, then went on to study medicine before rejecting both careers in favor of mathematics. He submitted his first paper to the Académie Royale des Sciences in 1739, and over the next two years bombarded the Académie with papers until in 1741 he was admitted to membership. In the years that followed, he became heavily involved in the world of the salons, social gatherings in which a number of philosophes came to prominence. He became particularly close to Julie de Lespinasse, a popular hostess, and though they never married, they were intimate for many years.

During the early 1740s, d'Alembert studied questions of dynamics, or the effects of force on moving bodies. In *Traité de dynamique* (1743), he introduced d'Alembert's principle, which maintains that for an object in motion to resist acceleration, the force of this resistance must be equal and opposite to the force producing the acceleration. This extended to moving bodies, the application of Newton's third law of motion, which holds that for every action, there is an equal and opposite reaction. In 1747, *Réflexions sur la cause générales des vents* marked the first use of partial differential equations in mathematical physics, and won d'Alembert a prize from the Prussian Academy. Also in 1747, an article on the motion of vibrating strings contained the first use of a wave equation in physics.

D'Alembert followed these writings with works on astronomy, but his attention was turning from mathematics and science to other areas. During the 1750s, he devoted himself to work on Diderot's *Encyclopédie*, producing first the *Discours préliminaire* (1751), an introduction that showed the links between various disciplines. D'Alembert later contributed some 1,500 articles to the *Encyclopédie*, but later resigned in the face of a scandal surrounding a 1757 article

on Geneva, Switzerland, that criticized both Catholics and Calvinists.

During the two decades from 1761 to 1780, d'Alembert produced scientific and mathematical works on a wide array of subjects, but his work suffered due to physical and personal problems. Deathly ill in 1765, he moved in with Julie de Lespinasse, who nursed him back to health. The two lived together until her death in 1776, after which he discovered that she had long maintained affairs with other men. Lonely and bitter, he lived his remaining eight years in a Paris apartment provided by the French Académie (he had been accepted for membership in 1754), and died on October 29, 1783.

JUDSON KNIGHT

Leonhard Euler
1707-1783
Swiss Mathematician

The mathematician Leonhard Euler (pronounced "oiler") was one of the great geniuses of his age, a thinker who explored virtually every known area of his chosen field. Nearly 900 books and papers are attributed to him, many of them written in the last 17 years of his life, when he was completely blind. (Euler had been blind in one eye for the preceding 31 years.) His interests ranged from analytic geometry to differential and integral calculus to the calculus of variations. Like many mathematicians of his day, his work also involved questions of applied mathematics: on the day he died, Euler had just finished calculating the orbit of the recently discovered planet Uranus.

Born in Basel, Switzerland, on April 17, 1707, the future mathematician was the son of Paul, a Calvinist pastor, and Marguerite Brucker Euler. Soon after his birth, the family moved to the town of Reichen, where his father had a parish. Paul wanted his son to follow him in the ministry; but the gifted boy's tutors, the brothers Jakob (1654-1705) and Johann (1667-1748) Bernoulli, convinced him that God had called Leonhard to a different path, as evidenced by his demonstrated abilities.

As promising as his talents seemed, Euler's first attempts to make a name for himself met with little success. Because he came from landlocked Switzerland and had never seen a ship, he failed to win a contest regarding the optimum placement of masts on sailing ships, sponsored by the Académie Royale des Sciences in France.

In time he would win a dozen awards from the distinguished institution; meanwhile, he had difficulty just getting a job. Rejected by the University of Basel, where he hoped to obtain a professorship, he jumped at an offer from Johann's son Daniel Bernoulli (1700-1782) to help him obtain a position teaching medicine at the St. Petersburg Academy in Russia.

By 1733, six years after Euler arrived in St. Petersburg, Bernoulli had returned to Switzerland, and Euler took his place as director of the mathematics department at St. Petersburg. Also in 1733, Euler married Catharina Gsell, the daughter of a Swiss painter who served in the court of Peter the Great. Eventually the couple had 13 children, of whom three sons and two daughters survived. In 1735, Euler lost the sight in his right eye, probably while studying the Sun to solve a problem put forth by the French Académie.

In the 1730s, Euler distinguished himself by solving a number of practical problems on matters such as weights and measures for the Russian government. Less practical, but perhaps more noteworthy, was his solution to the famous Königsberg Bridge Problem. The puzzle involved an attempt to cross seven bridges without backtracking, and Euler's solution helped spawn the field of graph theory. Also during this period, Euler wrote one of his most important works, *Mechanica* (1736-37), in which he became the first to apply mathematical analyses to the laws of dynamics discovered by Sir Isaac Newton (1642-1727).

Dissatisfied with the ever-present state of political oppression in Russia, Euler in 1740 accepted an appointment by Frederick the Great of Prussia to the Berlin Academy. He would remain in the service of the German king for nearly a quarter-century, again assisting with practical matters such as pension plans and the water supply system. It was during this time that Euler produced his famous proof of Pierre de Fermat's (1601-1665) Last Theorem. Though in fact his proof contained an error, Euler's work on the problem helped make possible the development of a successful proof some 250 years later. Also while in Berlin, Euler wrote a number of significant works, of which perhaps the greatest was *Methodus inveniendi lineas curvas maximi minimive proprietate gaudentes* (1744). This book on the calculus of variations led to his election as a Fellow of the Royal Society of London.

Euler had maintained good relations with his former Russian hosts throughout his time in Germany, making it easy to return to Russia in

Leonhard Euler. *(Corbis Corporation. Reproduced with permission.)*

1766, after a falling-out with Frederick. Catherine the Great, now the czarina, presented Euler with an estate and servants; and during this last phase of his life, the mathematician produced some of his greatest work, including *Institutiones calculi integralis* (1768-70), a classic on integral calculus. This was particularly impressive in light of the fact that he had lost the vision in his left eye to a cataract, and was thus completely blind.

More tragedy followed in 1776, when his wife Catharina died; but in the following year Euler married her aunt and half-sister, Salome Gsell. He continued to work on a number of problems related to astronomy, including questions involving moon phases, right up to the time of his death. Immensely fond of children, Euler died of a stroke while playing with his grandson on September 18, 1783.

JUDSON KNIGHT

Carl Friedrich Gauss
1777-1855
German Mathematician and Astronomer

Carl Friedrich Gauss holds a place among history's greatest geniuses, such as Sir Isaac Newton (1642-1727) or Albert Einstein (1879-1955). He advanced number theory with his formulation of the complex-number system; laid the groundwork for modern probability theory, topology, and vector analysis; and, like a few other mathematicians of his day, contributed extensively to the knowledge of astronomy. Gauss is also credited with the invention of a trigonometric measuring device called the heliotrope as well as the bifilar magnetometer and a prototype for the electric telegraph. This interrelation between mathematics and science and the applications of both were central to the worldview of Gauss, who saw mathematics itself as a science.

Gauss's story is particularly intriguing given the fact that he was born into a humble family; his father, Gebhard, was a laborer and merchant and his mother, Dorothea, a functionally illiterate servant woman. His parents, who lived in Brunswick, capital of the German duchy of Braunschweig, did not even record the exact date of his birth. Gauss, their only son, however, was able to calculate the date later because his mother remembered that it was eight days after the Catholic Feast of the Ascension in 1777—April 30.

The boy soon proved himself a prodigy, first by catching his father in an addition mistake when he was just three and, later, by a feat at school. A teacher, hoping to keep the students busy for hours, ordered them to add all the integers from 1 to 100. Gauss, however, independently derived a formula dating back to the time of Pythagoras (c.580-500 B.C.), $S = n (n + 1) / 2$. He added the smallest and largest integers together in succession, coming up with 50 sets of 101s (1 + 100, 2 + 99, 3 + 98, etc.).

Though his father initially discouraged him from a career in mathematics, Gauss managed to enter secondary school in 1788 and went on to Caroline College in Brunswick. He then entered the University of Göttingen in 1795. Again he showed himself a highly promising mathematician, in 1799 proving that every polynomial equation has a root in the form of the complex number $a + bi$. Also in 1799, he earned his doctorate from the University of Helmstedt, and, two years later, he published his thesis, *Disquisitiones arithmeticae*. In 1801, the year of the publication of his thesis, Gauss discovered Ceres, the largest of the asteroids orbiting around the Sun.

Gauss married Johanna Osthoff in 1805 and, though at first he had trouble finding work, in 1807 secured a dual position as director of the university and professor of mathematics at Göttingen. Tragedy struck in 1810, when the couple's third child died soon after birth, with Johanna following shortly thereafter. Gauss was only a widower for a short time before asking Friederica

Waldeck, the daughter of a faculty colleague, to marry him. They also had three children, and Gauss would later be widowed a second time when Friederica died of tuberculosis.

Despite misfortunes at home, Gauss continued to conduct highly fruitful research in the realms of mathematics and astronomy. In 1809 he published his most significant work on applied mathematics, *Theoria motus corporum celestium*. He also performed a number of jobs for various principalities and duchies in Central Europe, traveling and making geodetic surveys. To aid him in his work, he invented a measuring device, the heliotrope, in 1821.

In 1827 Gauss published *Disquisitiones generales circa superficies curvas,* in which he anticipated Einstein's work by speculating that space was curved. He also continued to teach, but he is not remembered as an especially generous mentor; though he corresponded with Sophie Germain (1776- 1831), he did little to cultivate the young minds under his tutelage at Göttingen. Furthermore, he discouraged his own highly talented son, Eugene, from a career in mathematics.

Gauss died on February 23, 1855, of a heart attack. In his lifetime, he received some 75 official honors, including membership in the Royal Society of England. The gauss, a magnetic unit of measurement, is named after him.

JUDSON KNIGHT

Sophie Germain
1776-1831
French Mathematician

Among the achievements of Sophie Germain's career was her attempt at a proof of Fermat's Last Theorem. Her successes were all the more remarkable in light of the fact that she was largely self-taught and practiced at the fringes of a mathematical establishment largely closed to women at that time.

She was born Marie-Sophie Germain, the middle child of Ambroise-François and Marie-Madeleine Germain, in Paris on April 1, 1776. Later, she would drop the "Marie" because both her mother and her older sister also bore that name in hyphenated form.

In time, both of her sisters would marry successful men, but Sophie's own future—which held neither marriage nor children—was influenced by events that took place when she was 13 years old. That year was when the French

Sophie Germain. *(The Granger Collection, Ltd. Reproduced with permission.)*

Her correspondence with Adrien-Marie Legendre (1752-1833) led to a collaboration of sorts, and Legendre in 1830 published Germain's proof of Fermat's Last Theorem in *Théorie des nombres.* The proof, like all attempts through the early 1990s, turned out to be incorrect, but Germain nonetheless advanced the understanding of the problem.

Between 1811 and 1815, Germain was the sole contender in three contests sponsored by the Institut de France, which challenged contestants to "formulate a mathematical theory of elastic surfaces and indicate just how it agrees with empirical evidence." Germain's first two solutions were rejected but, although the third had some questionable aspects, it nonetheless won the prize.

Germain became the first woman not married to a member of the Académie des Sciences to attend its meetings, a measure of the respect that her efforts had gained for her. She died of breast cancer in Paris on June 27, 1831.

JUDSON KNIGHT

Comte Joseph-Louis Lagrange
1736-1813
French-Italian Mathematician and Astronomer

Comte Joseph-Louis Lagrange ranks among the most important eighteenth-century Continental mathematicians. His best-known feat is the establishment of the metric standard, but this accomplishment came only after a number of other achievements, including contributions to algebra and number theory. Lagrange paved the way for modern mechanics by helping to establish a link between solid and fluid forms and he invented the calculus of variations when he was just 19 years old.

Born in Turin, Italy, on January 25, 1736, Lagrange was the son of aristocrats. His father served as treasurer of war in Sardinia and his mother came from a wealthy Italian family. Although Joseph-Louis was their eleventh and last child, he was the only one to survive childhood.

Initially, Lagrange showed little interest in mathematics, but, after reading an essay on calculus by Sir Edmond Halley (1656-1742), the course of his life was set. At age 16, the boy genius was sufficiently knowledgeable on the subject of calculus to serve as professor of mathematics at the artillery school in Turin. Three years later, while working on a set of isoperimetric problems then very much under discussion

Revolution began, and, although her father served in the revolutionary legislative assembly, the tumultuous events of 1789 took their toll on Sophie. She quite literally hid in the family library, where, for the next six years, she undertook a campaign of self-education that would have been impressive even under conditions of peace. Not only did she teach herself Latin and Greek but she also managed to learn algebra, geometry, and calculus.

Germain's formal education was slight but she submitted papers—under the pseudonym Monsieur Le Blanc (Mr. White)—for a course taught by Joseph-Louis Lagrange (1736-1813) at the Ecole Polytechnique. So great was Lagrange's respect for the work of "le Blanc" that he sought out the mystery man and, when he discovered that le Blanc was a woman, he extolled her abilities.

Lagrange was but one of many distinguished mathematicians who, to one degree or another, took Germain under their wings. She carried on a lengthy correspondence with Carl Friedrich Gauss (1777-1855) and in 1806 used her family's political connections to help protect the German mathematician from Napoleonic soldiers invading his homeland. Germain was also acquainted with Augustin-Louis Cauchy (1789-1857) and Jean-Baptiste-Joseph Fourier (1769-1830).

by the leading mathematicians of Europe, Lagrange developed the calculus of variations. After he sent a letter describing his work to Leonhard Euler (1707-1783) in Berlin, the two struck up a correspondence. Euler, nearly 30 years his senior, helped arrange Lagrange's election to the Berlin Academy.

Lagrange continued to make great strides in mathematics throughout the 1750s and 1760s. In 1759 he contributed three papers to a book published by the Turin Academy of Sciences and in the last of these papers used a partial differential equation as a means of providing a mathematical description of string vibrations. With this paper, he helped settle a controversy of long standing between Euler and Jean Le Rond d'Alembert (1717-1783). In 1764 the French Académie Royale des Sciences sponsored a contest to determine the gravitational forces that cause the Moon to present a relatively consistent face to Earth. Lagrange won the grand prize and another one two years later in a contest regarding the planet Jupiter.

The second académie prize was but one of many significant events for Lagrange in 1766, the year in which he married and also took the position, recently vacated by Euler, of director of mathematics at the Berlin Academy. A year later, Lagrange published *On the Solution of Numerical Equations,* in which he discussed universal methods for reducing equations from higher to lower degrees, thus preparing the way for modern developments in algebra. He proved several of Fermat's theorems, thereby contributing to number theory in its nascent stages, and in 1772 won his third académie grand prize for a study of the gravitational forces involving Earth, the Sun, and the Moon. Two more grand prizes followed, in 1774 and 1778.

During the 1780s, even as his *Mécanique Analytique* (1788) revolutionized the understanding of mechanics, Lagrange was mired in a period of depression. He had earlier lost his wife to an illness and remained widowed for many years. In 1792, at the age of 56, he again married, this time to the teenaged daughter of a friend, the astronomer Pierre-Charles Lemonnier. Neither of his marriages produced children.

The French Revolution, which began in 1789, seemed to shake Lagrange out of the period of malaise that had begun nearly a decade before. By then, he had long since moved to Paris. During the revolution, Lagrange was appointed to a committee for establishing standards of weights and measures; at this period, he developed the metric system.

A second fruitful period of Lagrange's life extended from the time of his second marriage to his death 21 years later. During this time, he wrote extensively and maintained professorships first at the Ecole Normale and later the Ecole Polytechnique. Lagrange died on April 10, 1813.

JUDSON KNIGHT

Johann Heinrich Lambert
1728-1777
Swiss-German Mathematician

Overcoming enormous social barriers, Johann Heinrich Lambert entered the ranks of Europe's foremost mathematicians. He did so without becoming part of the Continental academic establishment and, partly for that reason, much of his most advanced work was lost for many years, including his studies of what came to be known as non-Euclidean geometry. Nonetheless, he contributed to the knowledge of pi and the theory of errors.

Born on August 25, 1728, in Mulhouse, an independent town allied with Switzerland, Lambert was the son of Lukas (a tailor) and Elisabeth Lambert. Because of the family's poverty, he had to leave school at age 12 to help his father, but, during the years that followed, he progressed to more professional positions. Thus, by 1745, when he was 17, Lambert was working as secretary to the editor of a newspaper in Basel, Switzerland. Two years later, when his father died, he obtained employment with a wealthy family as tutor to their children.

During the period from 1756 to 1758, Lambert and his young charges traveled around the continent, giving him an opportunity to advance his knowledge of mathematics. He was elected a corresponding member of the Learned Society at Göttingen, and, although he was offered a position at St. Petersburg, where he would have joined some of the leading intellectual figures of the day, he chose to remain in western Europe. By 1761, he had been proposed for membership in the Prussian Academy and, four years later, began receiving a full salary from the academy.

Lambert's pursuits with the Prussian Academy were varied. Influenced by the work of Gottfried Leibniz (1646-1716), he took part in an ultimately fruitless attempt to realize the philosopher-mathematician's dream of a universal language for reasoning. More successful were Lambert's efforts on a much older problem, that of "squaring the circle," that is, creating a square of the exact same size as a given circle. To do so requires an understanding of the number pi, and Lambert's achievement was to prove that π is not a rational number; in other words, π cannot be expressed as the ratio of two whole numbers. Also during this period of Lambert's career, when he produced some 150 papers under the aegis of the Prussian Academy, he worked on an early version of non-Euclidean geometry.

Lambert's work on non-Euclidean geometry was well ahead of its time, as was his attempt at a general theory of errors. While measuring light, he developed a method for calculating the probability distribution of errors, a method that would be used by statisticians a century later. Anticipating various aspects of chaos mathematics, he analyzed the weather, treating it as the outcome of an apparently limitless number of unknown causes. The mathematician died in Berlin on September 25, 1777.

JUDSON KNIGHT

Adrien-Marie Legendre
1752-1833
French Mathematician

Though he was overshadowed by his older contemporaries Joseph-Louis Lagrange (1736-1813) and Pierre-Simon de Laplace (1749-1827), Adrien-Marie Legendre played a significant role in the world of eighteenth-century mathematics. Not only was he the author of the highly popular *Eléments de géométrie* (1794), but in the field of research he founded the theory of elliptic functions and contributed to the understanding of number theory and celestial mechanics.

Born in Paris on September 18, 1752, Legendre was the son of wealthy parents. He studied at the Collège Mazarin in Paris, and in 1770, when he was 18 years old, defended his theses in mathematics and physics. Four years later he published a work on mechanics, the first of many publications. In 1775, Legendre took a position at the Ecole Militaire in Paris, where he would remain for five years.

Legendre first distinguished himself internationally when in 1782 his essay on projectile paths and air resistance won a prize from the Berlin Academy. This attracted the attention of both Lagrange, who took an interest in the much younger man, and Laplace, who would become a jealous rival.

By 1783, Legendre was ready to present his first paper before the Académie Royale des Sciences in Paris. The topic was the gravitational attractions of planetary spheroids— i.e., sphere-like celestial objects that are not perfectly round--and the paper helped win him election to the Académie later that year. A year later, another paper on celestial mechanics introduced what came to be known as "Legendre polynomials," solutions to a specific type of differential equation that would prove highly useful in applied mathematics.

From the 1780s onward, Legendre concerned himself primarily with two areas of interest, number theory and elliptic functions. He also participated in a 1787 geodesic survey conducted jointly by observatories in Paris and Greenwich, England, that resulted in a theorem stating the properties of a triangle when moved from the surface of a sphere to that of a plane.

The outbreak of the French Revolution in 1789 led to the suppression of the Académie Royale, and this had a negative impact on Legendre's career. He had recently married Marguerite Couhin (the couple had no children), and his financial situation became threatened as his modest inheritance began to diminish. He took a number of jobs in public service during the years that followed, first by serving on a commission that in 1791 converted the measurement of angles to the decimal system. By 1799, he had taken the place of Laplace as examiner in mathematics for artillery students.

Following his publication of *Eléments de géométrie,* which popularized aspects of the *Elements* by Euclid (c. 325-c. 250 B.C.), Legendre issued several appendices to his highly successful book. Among these was a proof of the irrationality of π and its square. His *Essai sur la théorie des nombres,* in which he discussed number theory, was first published in 1789, but again, Legendre continued to add appendices over the course of later years.

In 1811, Legendre published the first volume of *Exercises de calcul intégral,* and during the period from 1825 to 1828 brought out the three volumes of his most significant work, *Traité des fonctions elliptiques.* The latter expanded on earlier work in elliptic function research, but with the later efforts of Karl Gustav Jacob Jacobi (1804-1851) and Niels Henrik Abel (1802-1829), Legendre realized he had approached the problem incorrectly, failing to recognize the potential of elliptic function research to be considered as a distinct branch of analysis. In his last years, he published supplements to the seminal work, discussing corrections by the younger mathematicians and humbly paying homage to their improvements on his work. Legendre died in Paris on January 9, 1833.

JUDSON KNIGHT

Colin Maclaurin
1698-1746
Scottish Mathematician and Physicist

First mathematician to provide systematic proof of Sir Isaac Newton's (1642-1727) theorems, Colin Maclaurin was also noted for his advances in geometry and applied physics. On the one hand, Maclaurin belonged solidly to the Age of Reason, with talents as diverse as those of more famous figures from the period, such as Thomas Jefferson; on the other hand, he remained a man of faith to the end of his life.

Maclaurin, who was born in Kilmodan, Scotland, in 1698, lost both of his parents early in life. His father, a highly learned minister named John Maclaurin, died when the boy was just six years old, and the mother followed by three years. Maclaurin was left in the care of his uncle Daniel.

As a college student, initially Maclaurin studied divinity at the University of Glasgow, but a professor encouraged his interest in mathematics. In 1715, Maclaurin defended his thesis, "On the Power of Gravity," which contained his first discussions of Newtonian principles. After earning his master of arts degree, in 1716 he was appointed professor of mathematics at Marischal College in Aberdeen.

In 1719, Maclaurin had the opportunity to meet his hero Newton, and in the following year he published *Geometrica Organica, sive descriptio linearum curvarum universalis,* a discussion of higher planes and conics that proved a number of Newton's theories. He took a job as tutor to the son of a British diplomat, Lord Polwarth, in 1722, and traveled to France, where he produced his *On the Percussion of Bodies.* The latter won him a prize from the French Académie Royale des Sciences in 1724.

Polwarth's son died shortly thereafter, and Maclaurin tried to return to his old position at Marischal, only to discover that it had been filled. Instead he took a chair in mathematics at the University of Edinburgh in 1725. Eight years later, at the age of 35, Maclaurin married Anne

Stewart, daughter of Scotland's solicitor general, with whom he had seven children.

With a 1740 essay called "On the Tides," in which he described the tides in Newtonian terms as an ellipsoid revolving around an inner point, Maclaurin entered a competition sponsored by the Académie Royale des Sciences. He shared the Académie's prize with two distinguished contemporaries, Leonhard Euler (1707-1783) and Daniel Bernoulli (1700-1782), and as a result attracted international attention.

In 1742, Maclaurin published his *Treatise of Fluxions,* which—like another work published that same year by Thomas Bayes (1702?-1761)—was a response to George Berkeley's *The Analyst, A Letter Addressed to an Infidel Mathematician* (1734). With his defense of Newton, Maclaurin persuaded British scientists to adopt their distinguished compatriot's geometrical methods. This proved a rather dubious success in the long run, because as a result the British remained in the dark regarding advances in analytical calculus taking place on the continent during the late 1700s.

Political events soon intervened in Maclaurin's life, as in 1745 a revolt broke out among the Jacobites, a faction who claimed that the Catholic line descended from James II, second son of Charles I, were the rightful heirs to the British throne rather than the descendants of James's elder brother Charles II. Defending the city of Edinburgh against a Jacobite attack, Maclaurin succumbed to fatigue, and died a year later at the age of 48. He was still writing practically on his deathbed, and his last writings give firm indication of his belief in life after death.

JUDSON KNIGHT

Abraham de Moivre
1667-1754
French-English Mathematician

Because of problems related to nationality and religion, Abraham de Moivre never had an opportunity to teach mathematics at a university. Nonetheless, he enjoyed fruitful interactions with Sir Isaac Newton (1642-1727) and others and later published texts in which he advanced the understanding of probability theory and other areas of mathematics.

Born in Vitry-le-François, France, on May 26, 1667, de Moivre (he apparently adopted the aristocratic "de" after moving to England) was the son of a surgeon. His family belonged to the

Huguenot sect, a Protestant group protected since 1598 by the Edict of Nantes, which ensured their limited freedom within the Catholic nation. Thus, de Moivre was educated in both Catholic and Protestant schools before going to Paris, where he studied under the renowned teacher Jacques Ozanam.

When de Moivre was in his 20s, however, the Crown revoked the Edict of Nantes, changing the course of his career. After a short imprisonment, he fled to England, where he would spend the remainder of his life.

Though his foreign ancestry made it difficult for him to obtain employment through most recognized channels, de Moivre made a good living by providing mathematical advice to gamblers and underwriters, who visited him at Slaughter's Coffee House on Fleet Street in London. Newton's support further ensured a steady stream of tutorial students. In 1695 another influential friend, Edmond Halley (1656-1742), presented a paper by de Moivre before the Royal Society, which accepted him for membership two years later. During the next half-century, de Moivre would have some 15 papers published in the *Philosophical Transactions of the Royal Society*.

As a mathematician, de Moivre concerned himself with areas ranging from probability theory to calculus. Among his contributions to mathematics was de Moivre's theorem, which helped establish the close connection between the algebra and geometry of complex numbers. His writings on probability, translated into English as *Doctrine of Chances* (1718), would prove highly popular (de Moivre wrote in an easy, understandable style geared toward nonmathematicians) and highly influential. The book, in which he improved on ideas presented earlier by Jakob Bernoulli (1654-1705) and Christiann Huygens (1629-1695), is often referred to as the first modern probability textbook.

The purpose of de Moivre's work was one of pressing concern in the Enlightenment, as he and other men of faith attempted to establish a rational basis for a belief in God. Thus, he hoped to use mathematics to prove the so-called "Argument from Design," which maintains that the evidence of order in the universe proves the existence of God.

In 1735 de Moivre was honored with membership in the Berlin Academy and, almost two decades later, by something much more surprising: membership in the Paris Académie in the country that had once expelled him. By then, de

Moivre was in his last days and he died in London on November 27, 1754.

JUDSON KNIGHT

Gaspard Monge
1746-1818
French Mathematician, Physicist and Chemist

The most widely recognized of the many achievements attributed to Gaspard Monge, sometimes known as the comte de Péluse, was his development of descriptive geometry as a means of representing three-dimensional objects in two dimensions. So valued was this technique for its military applications that the French government pledged him to secrecy. Monge also contributed to the adoption of the metric system, and his other work as a mathematician, physicist, and chemist took him into a variety of arenas, including caloric theory, acoustics, and optics.

Born on May 9, 1746, in Beaune, France, Monge was the son of Jacques and Jeanne Rousseaux Monge. The father, a knife grinder and peddler, believed in his son's ability to rise above his humble beginnings through education, and at an early age the talented youth was enrolled in the College of the Orations at Beaune. From Beaune in 1762 he went on to Lyons for two more years of college, at which point the 18-year-old Monge was declared a physics instructor. On vacation in his hometown, he created a detailed map of Beaune using projection methods and surveying equipment he developed himself, and this so impressed an army officer that he was recommended for the military school at Mézières.

As a young military engineer, Monge soon developed a revolutionary plan for gun emplacements in a proposed fortress, substituting a geometric process for the cumbersome arithmetic calculations then in use. This was the birth of descriptive geometry, and though Monge's commanding officer initially ignored it as a form of trick, its value for representing three- dimensional objects in two dimensions was soon recognized. For many years thereafter, Monge was ordered to keep this valuable military secret to himself.

Beginning in 1768, Monge held a dual professorship in physics and mathematics at Mézières, a position he would hold for the next 15 years. He married Catherine Huart in 1777, and together they had three daughters. Elected to the French Académie Royale des Sciences in 1780, his center of activity began to shift from the

as a pariah under the restored monarchy. So great was Monge's reputation, however, that upon his death in 1818, many defied a government ban and placed wreaths on his grave.

JUDSON KNIGHT

Gaspard Monge. *(Library of Congress. Reproduced with permission)*

provinces to the capital, and when in 1783 he was appointed examiner of naval cadets, he was forced to relocate to Paris.

Monge was a supporter of the French Revolution, which broke out in 1789, but his liberal and republican sympathies put him at odds with the radical Jacobin faction that took control in 1793. Thus he was removed from his position as minister of the navy, to which he had been appointed by the revolutionary leadership, after just eight months of service. He did help establish the metric system, as well as the Ecole Polytechnique, which opened in 1795. Later some of his lectures at the school were published in what came to be known as *Application de l'analyse a la géométrie*. In the latter, Monge explained the algebraic principles of three-dimensional geometry, ideas that would have an enormous impact on engineering design.

He served in a mission to Italy in 1796, and from there followed Napoleon Bonaparte on the latter's campaign in Egypt, where Monge helped establish the Institut d'Egypt in Cairo. Over the years that followed, as Napoleon held sway over France and Europe, Monge enjoyed great success and honors, including the title of comte or count, bestowed on him in 1808. But when Napoleon's star fell, so did Monge's. In 1816 he was stripped of all the honors accorded him by Napoleon, and spent the last two years of his life

Paolo Ruffini
1765-1822
Italian Mathematician and Physician

The Abel-Ruffini theorem states that quintic equations, or algebraic equations greater than the fourth degree, cannot be solved using only radicals—i.e., square roots, cube roots, and other roots. It is named after Niels Henrik Abel (1802-1829) and Paolo Ruffini, whose findings Abel confirmed. Ruffini also contributed to the development of group theory, and was known for his work as a philosopher and physician.

When Ruffini was born on September 22, 1765, his hometown of Valentano was a part of the Papal States, a geopolitical entity on the Italian peninsula that had existed for a thousand years. His own lifetime, however, would witness political upheaval that brought the existence of the states to a temporary end under the invading forces of Napoleon Bonaparte. While he was a teenager, his parents—Basilio, a physician, and Maria Ippoliti Ruffini—moved the family to Modena, and Ruffini later studied mathematics, medicine, and other subjects at the city's university. Among his mathematics instructors were the geometer Luigi Fantini, and Paolo Cassiani, who taught infinitesimal calculus. Ruffini himself was such a talented student that he took over instruction of Cassiani's foundations of analysis course when the latter had to take a leave of absence in 1787.

In 1788, Ruffini earned his degrees in philosophy and medicine, and followed this with a degree in mathematics soon afterward. He became a professor teaching the foundations of analysis, but in 1791 replaced Fantini as professor for the elements of mathematics. He also began practicing medicine that year.

Napoleon's troops occupied Modena in 1796, when Ruffini was 31 years old, and Ruffini was forced to become a local representative in the junior council of the newly declared "Cisalpine Republic." Two years later, he was allowed to leave that post, but his troubles continued due to the fact that, for religious reasons, he refused to swear an oath of allegiance to the republic.

Despite the troubles of this period, intellectually it was a fruitful time for Ruffini. In 1799 he published *Teoria generale delle equazioni,* which contained the first statement of what became known as the Abel-Ruffini theorem. Several mathematicians initially rejected Ruffini's ideas, but Augustin-Louis Cauchy (1789-1857) was an outspoken supporter of his findings. Abel confirmed the theorem in 1824.

Ruffini is also credited with developing the theory of substitutions, vital to the theory of equations, which in turn helped lay the foundations for group theory. Additionally, he served as president of the Societa Italiana dei Quaranta, which advanced mathematical learning, but in his later years he increasingly turned his attention to the two other fields of primary interest to him, philosophy and medicine.

By 1815, Napoleon was defeated, but other challenges appeared. During the 1817-18 typhus epidemic in Italy, Ruffini contracted that disease while treating patients in Modena, and after his recovery wrote about his experiences as both a patient and a doctor. A few years later, he contracted chronic pericarditis, along with a fever, and died on May 10, 1822.

JUDSON KNIGHT

Giovanni Girolamo Saccheri
1667-1733
Italian Mathematician

In *Euclides ab omni naevo vindicatus* (1733), Girolamo Saccheri was the first mathematician to suggest the possibilities of non-Euclidean geometry. He did not follow through with his speculations, however, and thus it would be more than a century before other mathematicians took Saccheri's ideas further.

Saccheri was born on September 5, 1667, in San Remo, Italy, which was then part of lands controlled by Genoa. In 1685, when he was 18 years old, he entered the Jesuit order of priests, and five years later went to Milan, where he studied philosophy and theology at Brera, the Jesuit college. Tomasso Ceva (1648-1737), brother of the more famous Giovanni Ceva (1647?-1734), happened to be a professor of mathematics at Brera, and encouraged the young Saccheri to take up the discipline.

Ordained at Como in 1694, Saccheri went on to teach at a number of Jesuit-sponsored colleges throughout Italy. At Turin from 1694 to 1697, he taught philosophy before moving on to Pavia, where he taught philosophy and theology from 1697. The latter town, where he held the chair in mathematics from 1699, would remain his home for the rest of his life.

Saccheri's first mathematical publication came in 1693, with *Quaesita geometrica.* The latter shows the influence of Tomasso Ceva, who continued as a friend and mentor for many years. (In fact the hardy Ceva outlived his student.) Tomasso introduced him to his brother Giovanni, as well as to Vincenzo Viviani (1622-1703), a mathematician who had worked with Galileo and Torricelli. Saccheri corresponded with all three men.

Viviani had published an Italian version of the *Elements* by Euclid (c. 325-250 B.C.), a work that had stood as Europe's foremost geometry text for 2,000 years. With *Logica demonstrativa* (1697), which discussed mathematical logic by use of definitions, postulates, and demonstrations, Saccheri himself emulated the style of the great Greek mathematician.

In 1708, Saccheri published *Neo-statica,* a work on the subject of statics; but his most important writing did not appear until shortly after his death on October 25, 1733, in Milan. This was *Euclides ab omni naevo vindicatus,* a discussion of Euclid's geometry. In it, Saccheri became the first mathematician to discuss the consequences of defying Euclid's fifth postulate, concerning parallel lines. More significant was his suggestion that a non-Euclidean geometry, independent of the fifth postulate, might be possible.

Saccheri did not use the term "non-Euclidean geometry," nor did he even see his idea as such, and was either unwilling or unable to pursue it further. Nonetheless, his work paved the way for groundbreaking achievements many years later.

JUDSON KNIGHT

Brook Taylor
1685-1731
English Mathematician

Two works published by Brook Taylor in 1715 indicate the scope of his interests and accomplishments. The first was *Methodus incrementorum directa et inversa,* in which he became the first mathematician to discuss the calculus of finite differences. The book also introduced the "Taylor series," concerning the expansion of functions of a single variable in a finite series, for which Taylor became famous. Also in 1715, Tay-

Brook Taylor. *(Archive Photos. Reproduced with permission.)*

lor published *Linear Perspective,* which contained the first general treatment of the principle of vanishing points.

Born in Edmonton, Middlesex, England, on August 18, 1685, Taylor was the oldest child of John and Olivia Tempest Taylor. The family were members of the minor nobility, and John encouraged his son to take an interest in music and the visual arts. Taylor entered St. John's College, Cambridge, in 1709, and later earned his bachelor of laws degree, which he followed with a doctorate of laws in 1714.

In the meantime, Taylor had been elected to membership in the Royal Society in 1712. He later served as the Society's secretary for four years, and during the period from 1712 to 1724 published 13 articles on a variety of subjects in the *Philosophical Transactions of the Royal Society.* His first significant paper, a solution to the problem of the center of oscillation, appeared in 1714. Not everyone was impressed with Taylor's work, however: Swiss mathematician Johann Bernoulli (1667-1748), who would remain a lifetime rival, attacked the piece.

In Taylor's watershed year of 1715, his *Methodus* (its full title means "Direct and Indirect Methods of Incrementation") became the first work to discuss the calculus of finite differences. The book presented the Taylor series, a theorem that provided the first general expression for the

expansions of functions of a single variable in finite series. Once again, Taylor was challenged by Bernoulli, who claimed he had discovered the idea first—a claim that turned out to be untrue. (Bernoulli attempted to do much the same thing to his own son, Daniel [1700-1782].) Only in 1772, when Joseph-Louis Lagrange (1736-1813) declared the Taylor series the basic principle of differential calculus, were the implications of Taylor's work fully appreciated.

The *Methodus* also contained groundbreaking discussions regarding the vibration of strings, and of light rays, concepts that touched—at least obliquely—on the other interests of Taylor, who was said to be a gifted visual artist and musician. His fascination with the visual arts led to *Linear Perspective,* which he updated in 1719 with *New Linear Perspective.* He is also reputed to have authored an unpublished manuscript entitled *On Musick.*

Family matters took much of Taylor's attention during the 1720s, which would turn out to be the last decade of his short life. In 1721, he married a woman of whom his father disapproved, and was estranged from the family. Taylor returned to the fold two years later, when his wife died in childbirth, and in 1725 he married again, this time with the family's approval. His father died in 1729, and Taylor inherited the family estate in Kent. A year later, his second wife died, also in childbirth, and Taylor followed her by another year.

The daughter from the second marriage, Elizabeth, lived. Years after Taylor's death, her son wrote a biography of his grandfather and thus helped ensure that Taylor's achievements gained much wider recognition than they had enjoyed in his lifetime.

JUDSON KNIGHT

Biographical Mentions

Gottfried Achenwall
1719-1772

German mathematician and political scientist who coined the term "statistics." Achenwall's use of the term, however, referred strictly to the comprehensive summation of a nation's political, economic, and social aspects. Since then, the term and field of statistics have expanded substantially to include virtually all fields of inquiry, attaining great significance in science, politics, economics, actuarial science, and many other disciplines.

Louis François Antoine Arbogast
1759-1803

French mathematician who achieved renown as a historian of mathematics and for his work on the philosophy of mathematics. Arbogast collected the papers of Martin Mersenne, Pierre de Fermat, René Descartes, Guillaume-François-Antoine de l'Hospital, and other influential mathematicians of his day, which now form the backbone of important collections in both Paris and Florence. In addition, Arbogast was among the first to investigate the calculus of operators (symbols that represent entire sets of mathematical operations), and helped make communication between mathematicians more effective.

John Arbuthnot
1667-1735

Scottish mathematician best remembered for his witty and satirical writings which were aimed at poor quality literature. He earned a medical degree at the University of St. Andrews and was one of Queen Anne's personal physicians. Along with Jonathan Swift, Alexander Pope, and John Gay, he founded the famous (or infamous) Scriblerus Club, which ridiculed contemporary writers in the form of *The History of John Bull*, which eventually became a cartoon and literary symbol of England.

George Berkeley
1685-1753

Irish philosopher and Bishop of Cloyne known for his philosophico-scientific treatises. Berkeley argued that science, no less than religion, relies on metaphysical speculation and lack of rigor as demonstrated by the development of calculus. In *The Analyst* (1734) he noted that both Newton's fluxions and Leibniz's infinitesimals rely on demonstrations requiring increments with finite magnitudes which nevertheless vanish. Berkeley's attack on these "differential entities" was only intended to reveal calculus' faulty foundation, not to impugn its practical utility.

Erland Samuel Bring
1736-1798

Swedish mathematician best known for his discovery of an important transformation that simplified solutions of quintic equations. Bring was born in Ausas, Sweden. After studying at Lundh for seven years, he taught history there, first as a reader, then as professor beginning in 1779. The

library at Lund houses eight volumes—in his own handwriting—posing questions and possible answers to problems in geometry, algebra, analysis, and astronomy.

Lazare Nicolas Marguerit Carnot
1753-1823

French military engineer who is known in France as the military genius of the French Revolution (1787-99). Carnot also made significant contributions to mathematics, especially geometry. After Napoleon's defeat at Waterloo, Carnot was exiled and eventually settled in Prussia. His study of the steam engine most certainly influenced his son Sadi Carnot, who later wrote the classic work on thermodynamics.

Giovanni Ceva
1647-1734

Italian mathematician who spent most of his life teaching mathematics at the University of Mantua—a post he obtained through his employer, the Duke of Mantua. He was the author of four works: *Concerning Straight Lines* (1678), *A Short Mathematical Work* (1682), *The Geometry of Motion* (1692), and *Concerning Money Matters* (1711). The latter is considered one of the earliest works ever published on mathematical economics. There is also a geometry theorem bearing his name that concerns lines intersecting at a common point when they are drawn through triangle vertices.

Giulio Carlo Fagnano dei Toschi
1682-1766

Italian mathematician who made important contributions in finding solutions for second, third, and fourth order equations. These equations, in which a variable is raised to the second, third, or fourth power, respectively, are often exceedingly difficult to solve because of the wide range of possible answers. Fagnano dei Toschi helped develop more reliable ways to approach and find solutions for such equations. He also investigated complex numbers and elliptic functions, two other issues of great mathematical importance.

Bernard Le Bouvier de Fontenelle
1657-1757

French mathematician and scientist who was educated by the Jesuits. He produced writings on the philosophy of both mathematics and science. Because he was able to assess the value of work being done by his contemporaries, he provided future scholars with valuable information about the scientists at work during his lifetime. He received numerous honors including a membership in the Royal Society, election to the Académie Française, and the post of permanent secretary to the Académie des Sciences.

Christian Goldbach
1690-1764

Russian mathematician best known for the Goldbach conjecture, which states that "every number greater than 2 is an aggregate of three prime numbers." (At that time, 1 was considered a prime number.) He was named professor of mathematics at the Imperial Academy at St. Petersburg but left this position to serve as tutor to Tsar Peter II in Moscow. He was also well known for his work on the theory of curves and the integration of differential equations.

Isaac Greenwood
1701-1745

American mathematician who wrote *Arithmetic vulgar and decimal,* the first mathematics text by an American-born author, published in Boston in 1729. Greenwood was also the first mathematics professor to work at Harvard University and seems to have devoted his life to teaching mathematics, making several significant contributions in this field.

William Jones
1675-1749

Welsh mathematician and historian. Although his own contributions to the science of mathematics was minimal, his correspondence with most of the famous mathematicians of his era survived and provided valuable information to future students. He served England by teaching mathematics aboard ships for seven years, then teaching in London, where he was elected a Fellow of the Royal Society in 1711. When he died in London in 1749, his estate included an imposing collection of letters and manuscripts that were apparently intended for future publication.

Sylvestre François Lacroix
1765-1843

French mathematician who was born to poor parents who could not afford to educate him properly. Fortunately, he came to the attention of famed mathematician Gaspard Monge, an influential educator who secured a professorship for him in mathematics at the Ecole Gardes de Marine at Rochefort. After teaching at several other schools in France, Lacroix held the chair of analysis at the Ecole Polytechnique and later at the Sorbonne and the Collège de France. In addition to a 3-volume text on calculus, Lacroix published a 10-volume edition titled *Cours de*

Mathématique (1797-99).

Thomas Fantet de Lagny
1660-1718

French mathematician who was tutored by an uncle before entering a Jesuit college in Lyon. While tutoring in mathematics for a private family over a period of 10 years, he collaborated with l'Hospital in Paris. After he had been active in the Académie des Sciences, he was offered a professorship at Rochefort, where he worked on trigonometry and made use of binary arithmetic. Streets in Paris have been named for him—passage de Lagny and rue de Lagny—and he was awarded the honor of a fellowship of the Royal Society in 1718.

John Landen
1719-1790

British mathematician whose research and subsequent writing added greatly to the general knowledge of the time on elliptic integrals, calculus, and rotary motion. His studies also included astronomy and physics. He wrote and published a two-volume set called *Mathematical Memoirs* (1780-89) in which the still notable Landen's theorem appeared. He also tried to simplify the study of calculus by applying it to the accepted principles used in geometry and algebra in his era. Landen was elected to the Royal Society of London in 1766 and lived another 24 years until he died in Northamptonshire in 1790.

Pierre-Simon Laplace
1749-1827

French astronomer and mathematician whose manifold contributions to the exact sciences make him one of the greatest scientists of all time. Laplace's *Celestial Mechanics* (1799-1825) generalized and applied the laws of Newtonian mechanics to planetary motions and established the solar system's stability. Uninterested in pure mathematics, Laplace nevertheless made many fundamental contributions to the field including the Laplace transform method for solving differential equations (1777), research on solutions for the potential function, and establishing probability theory on a rigorous basis.

Gottfried Wilhelm Leibniz
1646-1716

German philosopher and mathematician famous for formulating differential and integral calculus (1684). Though Newton discovered the calculus around 1665, there is little doubt Leibniz arrived at his results independently. His notation is still preferred to Newton's. Leibniz established symbolic logic, probability theory, and combinatorial analysis; and he produced a practical calculating machine that could add, subtract, multiple, divide, and extract roots (1672). He also developed the concept of kinetic energy and the principle of conservation of momentum.

Simon Antoine Jean Lhuilier
1750-1840

Swiss mathematician. As a result of winning an academic competition in mathematics, he wrote a math textbook intended for use in Polish schools. This effort resulted in his appointment to tutor members of the Polish royal family. This honor was followed by an offer to chair mathematics departments in both Germany and Switzerland. While teaching, he conducted his own research and made discoveries that were important in the development of topology.

John Machin
1680-1751

English mathematician who made significant progress towards understanding the number pi (π), which he calculated to over 100 decimal places. Machin developed one of the first efficient trigonometric identities that could be used to calculate the value of pi to a large number of places in a reasonable amount of time. This formula was one of the primary mathematical tools used by investigators in this area for two centuries.

Lorenzo Mascheroni
1750-1800

Italian mathematician best known for calculating Euler's constant to 32 decimal places. Although only 19 places were correct, his approach helped advance understanding of Euler's calculus. Mascheroni taught algebra and geometry in Pavia, where he was ordained as a priest. He also proved that all Euclidean shapes and objects can be constructed using only a compass, a discovery that was first made in 1672 by the relatively unknown Dutch mathematician Georg Mohr.

Pierre Rémond de Montmort
1678-1719

French mathematician who helped start the field of combinatorial mathematics. He published an influential book in which he examined the mathematics behind games of chance and showed that this was a respectable field of inquiry for mathematicians. By so doing, he also helped advance the field of game theory. Montmort was almost unique in maintaining friendly relations with followers and friends of both Isaac Newton and Gottfried Leibniz, even at the height of the controversy surrounding their respective claims for having invented calculus.

Jean Etienne Montucla
1725-1799

French mathematician who wrote several influential books on the history of mathematics and on interesting mathematical problems of his day. Montucla was also friends with many of the finest French mathematicians of his time, reviewing papers and collaborating with Jean Baptiste Fourier, Denis Diderot, Jean d'Alembert, and others. Unfortunately, he had a position with the French government, so much of his work ground to a halt during the French Revolution.

Sir Isaac Newton
1642-1727

English mathematician and physicist who, independent of Gottfried Leibniz, invented the calculus. Newton's contribution to physics was perhaps more influential than that of any other person in history. His shared, though bitterly contested, discovery of the calculus ranks as one of the greatest human intellectual achievements of all time. Using calculus, Newton developed his theories of motion, which are the underpinnings of Newtonian physics, still used to determine many important facts about the solar system, the universe, and the behavior of objects virtually anywhere.

Nicolas Pike
1743-1819

American mathematician who published the first American textbook on mathematics. Pike's book, *An American Arithmetic* (1788), was just the third book in the United States to receive an American copyright. It was used in many schools as a preparation for college algebra, and was even adopted for use at both Harvard and Yale. The book was, for the day, remarkably thorough in its scope of coverage, ranging from basic arithmetic through conic sections.

Vincenzo Riccati
1707-1775

Italian mathematician and engineer best known for his work on hyperbolic functions. Riccati, the son of a mathematician, used hyperbolic functions to find the roots of cubic equations, among other uses. In fact, Riccati was the first to use hyperbolic functions, and led the way toward finding ways to add them and determine their derivatives. He was also a talented hydraulic engineer who contributed to a project to help control flooding in the regions near Venice and Bologna.

Benjamin Robins
1707-1751

English engineer and mathematician esteemed for his work on the laws of motion and military engineering. Robins's primary mathematical significance, however, involves his support of Newton's differential calculus against Newton's detractors. In addition, using his knowledge of mechanics and calculus, Robins helped develop methods for determining projectile velocities and ranges, in part through his development of the ballistic pendulum. Robins died in India while helping to make plans for the defense of Madras against invading armies.

Abraham Sharp
1651-1742

English astronomer and mathematician who in 1705 calculated the value of π to an unprecedented 72 decimal places. Sharp was the first to use the new tools of calculus, recently introduced by Isaac Newton, to perform such calculations. His tremendous achievement surpassed the previous record by 27 decimal places. Sharp's record, however, was broken the following year John Machin, who calculated 100 digits of pi.

Thomas Simpson
1710-1761

English mathematician best known for his work on the numerical integration of equations and mathematical interpolation. Simpson also made significant contributions to probability theory, including work that showed the arithmetic mean of a number of observations is more appropriate than the results of a single observation. Simpson wrote several mathematics textbooks, including *The Doctrine and Application of Fluxions* ("fluxions" was Newton's term for derivatives of mathematical equations).

James Stirling
1692-1770

Scottish mathematician best known for explaining and extending the scope of Newton's calculus. Stirling also published a treatise on convergent series that helped to develop methods of calculating their sums more accurately and efficiently. Stirling's work on mathematical interpolation was also influential, as was his major work, *Methodus Differentialis*. In addition to these accomplishments, Stirling produced important work showing the shape of the Earth to be an oblate spheroid rather than a perfect sphere.

Alexandre Théophile Vandermonde
1735-1796

French musician and mathematician whose four memoirs, published in 1771-72, were influential to subsequent mathematics research. Vandermonde turned to mathematics at the relatively late age of 35, following a career in music. He began by working on the theory of equations and determinants, but later branched out and discovered a mathematical solution to a problem called "The Knight's Tour." This work, and more, was contained in his memoirs.

Edward Waring
1734-1798

English physician and mathematician who contributed significantly to the fields of number theory and algebraic curves. Waring's theorem, stated in 1777, held that every integer is equal to the sum of no more than nine cubes and not more than 19 fourth powers. Although he gave no mathematical proof of this, it has since been shown to be true many times. David Hilbert's proof of this theorem in 1909 led to profound insights in number theory.

Caspar Wessel
1745-1818

Norwegian surveyor and mathematician whose experience with cartography led him to interesting insights concerning relationships that exist among algebra, trigonometry, and geometry. Wessel's sole published paper, on the geometrical interpretation of complex numbers, was to prove exceptionally influential. His knowledge of surveying and mathematics also led him to construct one of the first accurate maps of Denmark, which, at the time, was united with Norway.

Bibliography of Primary Sources

Books

Agnesi, Maria. *Propositiones philosophicae* (1738). A volume of essays that address a number of subjects, including the theory of universal gravitation put forth in the previous century by Isaac Newton.

Agnesi, Maria. *Instituzione analitiche ad uso della gioventu' italiana* (Analytical Institutions, 1748). A mathematics textbook for youngsters that includes discussion of the curve known (due to a mistranslation) as the "Witch of Agnesi." The book also provided an introduction to algebra, analysis, and both differential and integral calculus.

Bayes, Thomas. *An Introduction to the Doctrine of Fluxions and Defence of the Mathematicians Against the Objections of the Author of the Analyst* (1736). This tract was written in response to *The Analyst* by George Berkeley, which attacked Isaac Newton's writings on differentials or fluxions.

Berkeley, George. *The Analyst* (1734). Here Berkeley noted that both Newton's fluxions and Leibniz's infinitesimals rely on demonstrations requiring increments with finite magnitudes which nevertheless vanish. Berkeley's attack on these "differential entities" was only intended to reveal calculus' faulty foundation, not to impugn its practical utility.

Bernoulli, David. *Exercitationes quaedam mathematicae* (1724). This work contained Bernoulli's first discussions in the areas of probability and fluid motion—areas in which he was destined to make his most lasting impact.

Bézout, Etienne. *Cours de mathematiques a l'usage des Gardes du Pavillon est de la Marine* (4 vols., 1764-67). A mathematics textbook that influenced the teaching of mathematics on both sides of the Atlantic. The work became a favored text at both Harvard University and West Point Military Academy. The English edition saw numerous reprints and remained an important text in America.

Ceva, Giovanni. *Concerning Money Matters* (1711). Considered one of the earliest works ever published on mathematical economics.

Clairaut, Alexis-Claude. *Théorie de la figure de la terre* (1743). An outgrowth of studies he made during a 1736 journey to Lapland, in the book Clairaut discussed the curvature of Earth and proved Sir Isaac Newton's theory that the shape of Earth is that of an oblate ellipsoid. This in turn led to widespread acceptance of Newton's gravitational theory. In addition to ensuring the acceptance of Newton's ideas concerning gravitation and the shape of Earth, the book presented a means for determining the shape of Earth by timing the swings of a pendulum. Clairaut also analyzed the effects created by gravity and centrifugal force on a rotating body, putting forth a proposition later known as Clairaut's theorem.

Cramer, Gabriel. *Introduction à l'analyse des lignes courbes algebriques.* (4 vols., 1750). This work contained Cramer's rule, which governed the solutions of linear equations, and Cramer's paradox, which clarified a proposition first put forth by Colin Maclaurin regarding points and cubic curves. In addition, Cramer introduced the concept of utility, a key principle today linking probability theory and mathematical economics.

d'Alembert, Jean Le Rond. *Traité de dynamique* (1743). This book represented an attempt to formalize dynamics.

d'Alembert, Jean Le Rond. *Réflexions sur la cause générales des vents* (1747). This book marked the first use of partial differential equations in mathematical physics, and won d'Alembert a prize from the Prussian Academy.

Euler, Leonhard. *Mechanica* (1736-37). Here Euler became the first to apply mathematical analyses to the laws of dynamics discovered by Sir Isaac Newton.

Euler, Leonhard. *Methodus inveniendi lineas curvas maximi minimive proprietate gaudentes* (1744). This book on the calculus of variations led to Euler's election as a Fellow of the Royal Society of London.

Euler, Leonhard. *Institutiones calculi integralis* (1768-70). A classic work on integral calculus.

Euler, Leonhard. *Algebra* (1770-72). One of the earliest math textbooks (as opposed to primers), this book went through six editions.

Lagrange, Joseph-Louis. *On the Solution of Numerical Equations* (1767). Here Lagrange discussed universal methods for reducing equations from higher to lower degrees, thus preparing the way for modern developments in algebra. He proved several of Fermat's theorems, thereby contributing to number theory in its nascent stages.

Lagrange, Joseph-Louis. *Théorie des fonctions analytiques* (1797). In this work Lagrange presented his ideas about a theory of functions and the principles of differential calculus.

Legendre, Adrien-Marie. *Eléments de géométrie* (1794). This book popularized aspects of the *Elements* by Euclid. Legendre later issued several appendices to his highly successful book. Among these was a proof of the irrationality of π and its square.

Maclaurin, Colin. *Geometrica Organica, sive descriptio linearum curvarum universalis* (1720). This book included a discussion of higher planes and conics that proved a number of Isaac Newton's theories.

Maclaurin, Colin. *Treatise of Fluxions* (1742). Like a 1736 book by Thomas Bayes, this work was a response to George Berkeley's *The Analyst, A Letter Addressed to an Infidel Mathematician* (1734). With his defense of Newton, Maclaurin persuaded British scientists to adopt their distinguished compatriot's geometrical methods. This proved a rather dubious success in the long run, because as a result the British remained in the dark regarding advances in analytical calculus taking place on the continent during the late 1700s.

Moivre, Abraham de. *Doctrine of Chances* (1718). Moivre's writings on probability would prove highly popular (de Moivre wrote in an easy, understandable style geared toward nonmathematicians) and highly influential. The book, in which he improved on ideas presented earlier by Jakob Bernoulli and Christiann Huygens, is often referred to as the first modern probability textbook.

Ruffini, Paolo. *Teoria generale delle equazioni* (1799). This work contained the first statement of what became known as the Abel-Ruffini theorem. Several mathematicians initially rejected Ruffini's ideas, but Augustin-Louis Cauchy was an outspoken supporter of his findings. Niels Henrik Abel confirmed the theorem in 1824.

Saccheri, Giovanni. *Euclides ab omni naevo vindicatus* (1733). This book was a discussion of Euclid's geometry. In it, Saccheri became the first mathematician to discuss the consequences of defying Euclid's fifth postulate, concerning parallel lines. More significant was his suggestion that a non-Euclidean geometry, independent of the fifth postulate, might be possible.

Taylor, Brook. *Methodus incrementorum directa et inversa,* (1715). Here, Taylor he became the first mathematician to discuss the calculus of finite differences. The book also introduced the "Taylor series," concerning the expansion of functions of a single variable in a finite series, for which Taylor became famous.

Taylor, Brook. *Linear Perspective* (1715). This book contained the first general treatment of the principle of vanishing points.

Articles

Thomas Bayes. "An Essay Towards Solving a Problem in the Doctrine of Chances" (1763). This tract helped lay the foundations for statistics in general, and for Bayesian statistical estimation—still used today for a variety of applications—in particular.

Cotes, Roger. "Logometria" (1714). This paper, Cotes's sole one, presented innovative methods for calculating logarithms and for evaluating the surface area of an ellipsoid of revolution.

Maclaurin, Colin. "On the Tides" (1740). In this essay Maclaurin described the tides in Newtonian terms as an ellipsoid revolving around an inner point.

NEIL SCHLAGER

Physical Sciences

Chronology

1705 Edmond Halley predicts that the comet he observed in 1682 will return again in roughly 75 years—as it did, in 1758.

1714 Gabriel Daniel Fahrenheit invents the first accurate thermometer, along with the scale that bears his name; in 1730 Réné Antoine Ferchault de Réaumur develops his own thermometer and scale, and in 1742 Anders Celsius introduces the centigrade scale later adopted by scientists internationally.

1738 Establishing theoretical hydrodynamics as a discipline, Swiss mathematician Daniel Bernoulli presents what comes to be called Bernoulli's principle—as the velocity of fluid increases, its pressure decreases.

1754 Scottish chemist Joseph Black proves by quantitative experiments the existence of carbon dioxide.

1774 Joseph Priestley, an English chemist, isolates oxygen. Swedish chemist Karl Scheele independently isolated the gas in 1772 but failed to publish his results until after Priestley, who thus is given sole credit for the discovery.

1781 German-English astronomer William Herschel discovers Uranus, the first planet discovered in historic times.

1783 Antoine-Laurent Lavoisier identifies hydrogen.

1785 Charles Augustin Coulomb publishes the first of seven papers in which he establishes the basic laws of electrostatics and magnetism.

1785 James Hutton reads his first paper, "Concerning the System of the Earth," which marks the beginnings of geology as a real science, before the Royal Society of Edinburgh.

1787 German geologist Abraham Gottlob Werner publishes *Kurze Klassifikation,* the first systematic treatise on geologic formations.

1791 Pierre Prévost of Switzerland presents his theory of exchanges, in which he correctly states that all bodies of all temperatures radiate heat.

1791 The French National Assembly adopts the metric system.

1796 Pierre-Simon Laplace publishes his *Exposition du Système du Monde,* in which he states his "nebular hypothesis"—that the Sun and later the planets originated from a giant rotating cloud of gas.

1798 English chemist and physicist Henry Cavendish becomes the first to calculate Earth's mass, and in so doing also becomes the first to find a value for Newton's gravitational constant.

Overview: Physical Sciences 1700-1799

Overview

During the 1500s, Polish astronomer Nicolas Copernicus (1473-1543) proposed that Earth orbits around the sun rather than vice versa. Copernicus's view was seen as a challenge to widely held beliefs about the universe. These beliefs, however, were based on traditional interpretations of the Bible rather than on scientific observations. The Italian astronomer Galileo Galilei (1564-1642) defended Copernicus's ideas in 1632. In the process, he helped to promote the idea that science should be based on experiment and observation as opposed to faith or reasoning alone. This way of approaching science came to be known as the scientific method and had a great impact on the physical scientists of the 1700s.

The eighteenth century is often referred to as the Enlightenment or the Age of Reason. Scientists stopped relying on untested ideas in ancient texts and began conducting their own experiments and making their own observations. In 1671 Isaac Newton (1642-1727) published his three laws of motion and his law of gravitation. His work showed that the forces of nature are not random, but operated by certain rules. Afterward, scientists came to believe that similar rules would be discovered to explain every phenomenon in the universe.

Physics

During the 1700s, scientists made discoveries about the basic nature of phenomena such as gravity, electricity, and heat. For instance, Benjamin Franklin (1706-1790) suggested the idea of positive and negative electrical charges, and he performed his famous kite experiment in which he showed that lightning is electricity. He also (incorrectly) proposed that electricity involved the exchange of an invisible electrical fluid. (It is now known that electricity involves the exchange or movement of electrons.)

Charles Coulomb (1736-1806) showed that the electrical force between two oppositely charged objects depends on both the size of the charges and on the distance between the objects. These observations mirrored Newton's law of gravitation, which states that the force of gravity between two objects depends on their mass and the distance between them. Coulomb went on to show that a similar law applied to magnetic force.

Another phenomenon physicists investigated during the 1700s was heat. The first step toward determining the nature of heat was the creation of instruments to measure temperature. In 1714 Gabriel Fahrenheit (1686-1736) created a mercury-filled thermometer that was based on the boiling and freezing points of water. In 1742 Anders Celsius (1701-1744) created a similar thermometer, but unlike Fahrenheit, he separated the interval between the boiling and freezing points of water into 100 equal divisions.

Once temperature could be accurately measured, scientists could then perform more useful experiments with heat. Pierre Prévost (1751-1839), for example, determined that cold is the absence of heat and that all objects radiate heat regardless of temperature. (He also believed that heat is an invisible fluid, an idea that persisted into the 1800s.) Joseph Black (1728-1799) found that it takes about five times as much heat to turn boiling water into steam as it does to bring water to the boiling point. Using Black's work, James Watt (1736-1819) developed a practical steam engine. The steam engine eventually became a cheap source of power for both transportation and industry.

New fields of physics developed during the 1700s as well. Daniel Bernoulli (1700-1782) established hydrodynamics—the study of the motion of fluids—and Ernst Chladni (1756-1827) became known as the father of acoustics—the study of sound waves.

Chemistry

In the 1600s Georg Stahl (1660-1734) proposed that objects release an invisible fluid called phlogiston when they burn. This belief held throughout much of the 1700s. Then, in 1791, Antoine Lavoisier (1743-1794) published the results of experiments that disproved the existence of phlogiston. Lavoisier showed that burning results in the gain of oxygen from the air rather than a loss of invisible fluid. (The fluids used to describe electricity, heat, and burning were called imponderable fluids; none of these fluids actually exists.) Lavoisier's experiments also provided evidence for the law of conservation of mass—the idea that the mass of the products of a chemical reaction is equal to the mass of the reactants.

As chemists began to adopt the scientific method during the eighteenth century, they discovered many new compounds. For instance, Joseph Priestly (1733-1804) was the first to discover numerous gases, including ammonia, carbon monoxide, and hydrogen sulfide (the gas responsible for the smell of rotten eggs). Most importantly, however, he is generally credited with the discovery of oxygen.

In addition to gases, many new elements were also isolated during this time period, many of them by Swedish scientists. Georg Brandt (1694-1768) discovered cobalt, Johann Gahn (1745-1818) discovered manganese, Peter Hjelm (1746-1813) discovered molybdenum, and Baron Axel Cronstedt (1722-1765) discovered nickel. Another Swedish scientist who worked with many of these men and encouraged them in their work was Karl Scheele (1742-1786). Although he is not generally credited with the discovery of any element, he made contributions to the discovery of several. He was also the first to isolate many types of acids.

Astronomy

Newton's laws of motion and gravitation could be used to describe the motion of planets and moons. For instance, in 1789 Benjamin Banneker (1731-1806) accurately predicted a solar eclipse. Edmond Halley (1656-1742) showed that Newton's laws applied to comets as well. He analyzed records of past comet appearances and found what seemed to be a pattern. Based on his observations, he correctly predicted that one of these comets (now known as Halley's comet) would appear again in 1758.

Some astronomers in the eighteenth century rather naively assumed that Newton's laws left little for them to discover. In 1781 William Herschel (1738-1822) proved them wrong. Using a large telescope of his own design, he discovered the planet Uranus, a finding that greatly renewed interest in the field of astronomy.

Some astronomers began to consider the origin of the solar system from a purely scientific point of view rather than from a religious or spiritual perspective. In 1796 Pierre Laplace (1749-1827) suggested that the Sun had formed from a spinning cloud of gas that had condensed as it increased in speed. A modified version of Laplace's idea came to be accepted in the twentieth century.

Earth Sciences

In the 1700s geologists began to consider the age of Earth itself. In 1654 James Ussher (1581-1656) concluded that Earth was created in 4004 B.C., based on the timeline of the Bible. One of the first scientists to challenge Ussher's work was Georges Buffon (1707-1788). In 1749 Buffon proposed that Earth was between 75,000 and 100,000 years old. He believed that Earth originated as the result of a comet collision with the Sun, and he based his estimate of Earth's age on the time it would take the planet to cool after its formation from such a collision. (Earth is now estimated to be about 4.6 billion years old.)

Geologists also began to investigate the origin of Earth's rocks and landforms. Abraham Werner (1750-1817) proposed that the main force that shaped Earth was erosion, which wore away the land and carried sediments to the ocean. These sediments then settled out of the water, and pressure changed them to sedimentary rock. Werner recognized that sedimentary rock formed in layers and that the deepest layers were the oldest.

Like Werner, James Hutton (1726-1797) believed that erosion was a major force in the formation of Earth's surface. However, he also believed that volcanoes played an equally important role. He stated that these two forces (erosion and volcanic activity) were at work in the past just as they are in the present. Over long periods, he proposed, these forces could account for the appearance of Earth today. Like Buffon, Hutton came to the conclusion that Earth must be much older than scientists had previously believed. Both men faced opposition from those who saw their work as a challenge to the story of Earth's creation in the Bible. Nevertheless, many of Hutton's ideas came to form the basis of the new science of geology.

Conclusion

By the end of the 1700s physical scientists had used the scientific method to make significant strides toward understanding some of the large questions of science. Astronomers had shown that the solar system was larger and more complex than previously thought. Geologists had started to form theories about Earth's origin and age, despite the religious opposition that they sometimes faced. Physicists had begun to resolve some of the mysteries surrounding heat, electricity, and magnetism, and chemists had carefully catalogued many different types of compounds and elements.

Science began to expand in many directions during the Age of Reason, and the body of knowledge increased dramatically. As a result, scientists would become more specialized in the following century. During the 1800s, further discoveries would be made regarding the nature of heat, magnetism, and especially electricity. Chemists would begin to investigate the nature of the atom, and geologists would develop a timeline of Earth's history. And with the invention of better instruments, astronomers would make discoveries about the composition of and distances between stars. All areas of physical science, however, would continue to rely upon the scientific method.

STACEY R. MURRAY

The Rise of Experiment

Overview

The publication in 1704 of Isaac Newton's (1642-1727) second major treatise, the *Opticks*, came at a significant moment in the development of experimental science. Enthusiasm for empirical study of nature, and skill in manipulating experiments, had been building throughout the previous century, and Newton's account of his own experimental studies of light and of his experimental methods fell on fertile ground. The eighteenth century saw a blossoming of experimental efforts in the study of the physical world, especially in exciting new fields such as electricity.

Background

Galileo's (1564-1642) telescopic investigations and his experiments with inclined planes and falling objects were some of the most remarkable discoveries in the history of seventeenth century science, but they were far from isolated events. From planets to plankton, natural philosophers were studying the world around them by interacting with it in new and important ways. The older tradition of contemplative science, which relied on logic and speculation about causes, was replaced by a new tradition aimed at confronting the world directly through the senses and experience. New instruments, such as telescopes, microscopes, and air pumps were invented and used to investigate realms never seen or created before. New practices such as repeated testing of configurations of equipment and systematic improvements to experiments and devices led to sharing of ideas, opinions, and "hypotheses" about experimental outcomes and their meaning. Natural philosophers observed animals, plants and other things in nature with new care. They also manipulated nature through tests and experiments to learn yet more about the physical world.

By the start of the eighteenth century, *empiricism*—the philosophical outlook that interacting with the physical world is the most effective way to learn about it—and experimental practices of many kinds were well established in the scientific societies of Europe. Curricula changed more slowly but gradually new instruments and an emphasis on experiment made their way into the universities as well. Air pumps (instruments that evacuated the air out of a space to allow for experiments with vacuums), telescopes, thermometers, globes, and orreries (moveable models of the Newtonian universe) were common tools in eighteenth-century university classrooms in British and American universities. Learning through demonstration extended beyond the university: artisans, tradesmen, and others could learn about the Newtonian science from specialized itinerant lecturers at special classes designed to bring together Newton's science with the practical problems of the new industries.

These demonstration experiments were often designed to illustrate to audiences not familiar with advanced mathematics the mathematical laws governing motion presented in Newton's major treatise, *The Mathematical Principles of Natural Philosophy*. But Newton's work extended to other subjects as well. The *Opticks* is an account of Newton's experiments with light and colors, his experimental techniques, and his open-ended queries about subjects for future investigation. Newton was not able to determine mathematical relationships to explain his observations of optical phenomena, but he expected that careful experimentation would permit such laws to be determined in the future. The importance of pursuing this goal was on his mind when Newton assumed the presidency of London's Royal Society in 1703. One of his first acts there was to re-establish the practice of weekly demonstration experiments that the Society had let lapse. Some of these demonstrations over the next decades involved the long-popular air pump and the effects of vacuum on various creatures and experimental arrangements. But many others drew more explicitly on the material in Newton's *Opticks*, and included various investigations into the nature of light and of the relationship between light and electricity. Newton was one among several important figures to emphasize the value of careful experimental practice and the replicability of experiments. By insisting on a high level of precision in performance and reporting, the knowledge produced by experiments became more reliable and could more easily be shared among different investigators. Increasing standards of precision became a hallmark of progress in physical science.

Electricity provides one of the most interesting examples of experimental science in the eighteenth century. The phenomenon of electricity was known but not at all understood in 1700. The simplest experiments with static electricity showing the ability of glass, when rubbed, to attract various light objects and to produce a glow were performed for the Royal Society soon after the start of Newton's presidency. Starting from these modest performances, electrical experiments became more dramatic as investigators learned how to store charges in devices such as Leyden jars. Curiosity about electricity was widespread, and anyone with the means to produce a modest experimental setup could try to advance knowledge of the field. A small number of electrical enthusiasts were killed or seriously injured by experiments that involved natural lightning or very large stored charges, but such martyrdom was rare. One of the most important electrical investigators was the multi-talented Benjamin Franklin (1706-1790). Franklin, working with a group of friends, performed a large number of electrical experiments, helped popularize the use of lightning rods on buildings in Philadelphia, and wrote an influential treatise on the nature of electricity.

Electricity did not remain the private entertainment of sober scientific groups like the Royal Society or dedicated individuals like Franklin. The grand sparks, shocks, and artificial lightning produced pleased all kinds of audiences. Traveling demonstrators took electrical shows on the road, and dramatic electrical games in which shocks were passed through a crowd or various tricks were performed became extremely popular in fashionable circles throughout Europe. The public had previously flocked to see displays of curiosities such as automata and unusual animals, but electricity was a new and fascinating kind of interactive entertainment. It gave audiences a tantalizing sense of the power and control an experimenter could wield over even the most mysterious elements of nature. By getting shocked or wearing a crown of sparks, they could share in the experimenter's mastery as well as his curiosity.

Impact

Interest in experiment and empiricism more generally was a vital quality of seventeenth century natural philosophy. Scientific societies and journals were established in several countries to facilitate the performance of experiments and the exchange of ideas about them. As science found a wider audience after 1700, so did the belief that the natural world could be understood by careful observation and manipulation. Interest in the study of nature and the practice of experiment spread from an elite circle of enthusiasts to a broad cross-section of the general population. Attendance at public lectures and demonstrations stimulated curiosity among people not privileged enough to have attended university, including tradesmen and entrepreneurs who applied the ideas demonstrated by clever experimenters to the practical challenges of their working lives. Women were also part of the growing audience for scientific ideas. In France and Italy as well as Britain, women were both consumers and producers of popular eighteenth century treatises about Newtonian science.

Between the time of Newton and the end of the eighteenth century, the scientific discipline that came to be known as physics was born. For natural philosophers prior to Newton's time, the study of the physical world was not restricted to the subjects later associated with physics but was considered to include topics such as anatomy and botany. With the dissemination of Newton's ideas came a separation. The study of mechanics, statics, dynamics, light, and electricity came together under a new umbrella called "physics." This newly rigorous field was not entirely uniform, however. Some areas of physics, such as mechanics, were highly mathematical by the eighteenth century. While their principles could be illustrated by demonstration and experiment, real understanding and innovation required advanced mathematical skills. Other subjects, such as electricity, still had no mathematical laws, and were approached through experiment and qualitative theory. What brought these subjects together as physics, and excluded others, was their accessibility to experimental investigation and a shared optimism that such investigations would lead eventually to mathematical relationships and laws.

The study of the principles and experiments of Newton's physics and their applications to the construction of machines large and small helped transform intellectual, commercial, and daily life in myriad ways. The embrace of Newtonian ideas and the principle of empirical investigation brought about generations of improvements to the tools, machines, and techniques of mechanics and craftsmen, leading to the development of the practice of engineering and the evolution of precision equipment of all kinds. These in turn brought diverse new experiences to daily life, as inventions such as the steam engine transformed transportation, production, and work itself.

By the late nineteenth century, scientists and engineers had come together to create several industries wholly based in science. Experimental knowledge and prowess from chemistry helped bring about the pharmaceutical, dye, and munitions industries. Knowledge and skill gained from experimental investigations in physics were essential to the development of the telegraph, telephone, electrical and optical industries, and to the establishment of precision measurement and standards that facilitated international trade in technology. These advances, which were multiplied many times over in the next century, had deep roots in the enthusiasm for experimental knowledge and practice that flourished in the decades after Newton.

LOREN BUTLER FEFFER

Further Reading

Books

Cantor, Geoffrey. *Optics after Newton.* Manchester: Manchester University Press, 1983.

Clark, William, et al., eds. *The Sciences in Enlightened Europe.* Chicago: University of Chicago Press, 1999.

Dobbs, Betty Jo Teeter and Margaret Jacob. *Newton and the Culture of Newtonianism.* Atlantic Highlands, NJ: Humanities Press, 1995.

Gooding, David, Trevor Pinch, and Simon Schaffer, eds. *The Uses of Experiment: Studies in the Natural Sciences.* Cambridge: Cambridge University Press, 1989.

Heilbron, J. L. *Electricity in the 17th & 18th Centuries.* Los Angeles: University of California Press, 1979.

Kuhn, Thomas. *The Essential Tension: Selected Studies in Scientific Tradition and Change.* Chicago: University of Chicago Press, 1977.

Shapin, Steven. *A Social History of Truth: Civility and Science in Seventeenth Century England.* Chicago: University of Chicago Press, 1994.

The Cultural Context of Newtonianism

Overview

The enthusiastic reception by natural philosophers of Isaac Newton's (1642-1727) difficult *Philosophiae Naturalis Principia Mathematica (Mathematical Principles of Natural Philosophy)*, published in 1687, eventually spread Newton's name and influence far beyond the small community of scholars of physics and mathematics. Throughout the eighteenth century, "Newtonian" ideas were put forward in many social and cultural arenas, in England and its colonies in America. Although Newton's name was used to describe many different, even conflicting, viewpoints, most proponents of "Newtonianism" thought of Newton as a paradigm of rational belief and of his work as a model for understanding all aspects of the universe in terms of laws.

Background

It would be impossible to exaggerate Newton's influence on the history of physical science, but his indirect influence on developments in religious, cultural, intellectual, and political life was also profound. How did this happen? Newton's scientific works were notoriously difficult, accessible only to mathematically literate investigators and those who had, like Newton, made forays into careful and intense experimentation. Newton himself was a difficult and at times troubled man, uncomfortable with public interactions and controversies. Although late in his life Newton did take up a public position as Master of the Mint and participate authoritatively in the scientific life of London, throughout the earlier decades of his career Newton led a mostly solitary life at Cambridge University, punctuated primarily by scientific exchanges with a small circle of devoted peers. He struggled mightily for years with private demons entangled with his brilliantly creative ideas. Much of Newton's "scientific" work was in fact deeply concerned with theology and with alchemy, including the more mystical aspects of that ancient practice.

Newton's broad cultural influence, then, did not stem from Newton himself but rather from the inspiration that his work gave to others searching for new ways to approach their own spheres of thought. One of the most direct channels of influence of Newton's work was upon the philosophical architects of the French Enlightenment movement, a group of writers who were known as the "philosophers" and who included the author François-Marie Arouet Voltaire (1694-1778). Voltaire was an outspoken critic of Catholicism, a "deist" who professed general belief in a supreme being but rejected the principles and institutions of organized religion. Exiled to England for three years in the 1720s, Voltaire marveled at the relative freedom of cultural, social, and intellectual exchange he found in London. Voltaire associated these freedoms with Newton's scientific ideas; he embraced Newton's name, and endorsed Newtonian rationality as an antidote to what he labeled religious intolerance and ignorance in his native France. Voltaire was largely responsible for popularizing Newtonian ideas in Europe, and those who saw Newtonian ideas as a threat to Christianity often drew their interpretation from Voltaire. In addition to Voltaire's own work, his mistress, Madame du Châtelet (1706-1749), made the only translation of Newton's *Principia* into French and helped popularize Newtonian ideas to a female audience.

Newton's influence on religion was extraordinarily complex. At the same time Voltaire and others were claiming that Newtonianism supported their deism or even atheism, many others were using Newtonian language to defend Christianity. Although he rejected Anglican orthodoxy, Newton himself very much wanted his science to show the power of the Creator, and a number of Newtonians, especially in Britain, did make use of Newton's scientific legacy to support religious belief. Protestant theologians such as Samuel Clarke and Richard Bentley argued passionately that Newton's work had revealed laws of the physical universe that decisively demonstrated its design and order—design and order that confirmed God's existence and revealed his handiwork. Newtonian philosophy was also used by these theologians and others to

Isaac Newton. *(Archive Photos. Reproduced with permission.)*

support a kind of political agenda—the stable, balanced, law-governed authority of Church and state that had emerged in the aftermath of the Revolution of 1688-89. This kind of "natural theology" that looked to science for evidence of design in the universe to support Christianity became quite entrenched in England and the American colonies. Its impact was embraced by members of the Anglican majority and was multifaceted, including such diverse developments as the Unitarian Church and Freemasonry.

Impact

The cultural force of Newton's name and his ideas remained irresistible throughout the eighteenth century and well into the nineteenth. The impact of this singular exemplar on the progress of scientific work was mixed. In Newton's own country, Newtonian fervor settled into a kind of orthodoxy that eventually had a stultifying effect on the advance of scientific investigation. This was felt most strongly in the field of mathematics, where nationalistic support of Newton's claim of priority in the discovery of calculus caused English schools to teach Newton's awkward fluxions notation rather than Gottfried Leibniz's (1646-1716) more workable differentials. The presence of Newton himself as the President of the Royal Society for many years had the predictably intimidating effect on the members of that organization. England was, however, noteworthy for the rapidity and breadth of the dissemination of Newtonian ideas to engineers and artisans who applied his principles to important industrial problems in mechanics, hydrostatics, and pneumatics. Widespread enthusiasm for and understanding of the value of careful experiment and the search for mathematical relationships to solve practical physical problems probably helped England achieve its early start in industrialization. This work in applied mechanics was free of theological implications that might have arisen from more theoretical or speculative investigations, and coexisted quietly with the widespread intertwining of Newtonian and Christian ideas.

France, perhaps more than any other country, had a blossoming of mathematical and conceptual talent inspired by Newton and his work. In France, an ideological struggle between Catholic-tinged Cartesian science and the Newtonian science supported by deists such as Voltaire waged for sometime. French scientific institutions such as the important Academie des Sciences did not officially embrace Newtonian science until about 1740. Long before then, however, skilled mathematicians such as Pierre-Louis de Maupertuis (1698-1759) were solving problems posed by the *Principia* and testing the experiments described in Newton's *Opticks* (1704), all the while simply ignoring theological disputes about Newton's ideas. (Maupertuis also helped bring Newtonianism to German-speaking countries when he was appointed to head the Berlin Academy of Sciences, where Voltaire was also a visiting member.) Newton's physics of the earth and planets was greatly advanced by French mathematicians such as Maupertuis and Pierre-Simon Laplace (1749-1827), among many others, who worked even through the upheaval of the French Revolution. By century's end they had transformed Newton's physical laws into a sophisticated system of celestial mechanics—a wholly mathematical system that was completely free of the religious ideas that had animated Newton's own work, and that had been the source of squabbles in French institutions earlier in the century. While these French mathematicians may have ignored religious questions initially to avoid confrontation with established opinion, the result of their work made theology superfluous to physics. In effect, they redefined the field of inquiry about the physical universe to leave questions about a Creator entirely aside.

Newton's ideas surely generated Newtonianism, but that term came to mean many things far removed from Newton's own beliefs. Newton's principle ideas of instrumentalism, of motion determined mathematically by accelerative forces among microscopic elements, and of the continuous mathematics of calculus had far-reaching implications for understanding the natural world. Newton's ideas and methods inspired by their success, and others saw them as useful analogies and examples for their own purposes. Newton's personal theological beliefs were for him an underpinning of his scientific endeavors, but they proved to be highly malleable, and eventually superfluous. Those looking for a law-bound but godless universe found it easily in Newton's science; those looking to science to support religion against its challengers found support from Newton as well in the argument from design. Although the view always had well-spoken critics, when England's next scientific giant—Charles Darwin (1809-1882)—emerged in the mid-nineteenth century, the design-based vision of natural theology received a blow far more damaging than that of the godless physics of Laplace. After Darwin, it was no longer straightforward to use science to support religion; where natural theologians had seemingly deliberate designs, Darwin recognized the handiwork only of immense time, variation, and the forces of selection. Those who wished to balance science and belief could still do so, but the easy identification of scientific understanding with God's works that Newton had inspired lost its power to convince, and religion in England once again stood without the support of science.

LOREN BUTLER FEFFER

Further Reading

Books

Brooke, John Hedley. *Science and Religion: Some Historical Perspectives.* Cambridge: Cambridge University Press, 1991.

Cohen, I.B. *The Newtonian Revolution.* Cambridge: Cambridge University Press, 1980.

Dobbs, Betty Jo Teeter and Margaret Jacob. *Newton and the Culture of Newtonianism.* Atlantic Highlands, NJ: Humanities Press, 1995.

Guerlac, Henry. *Newton on the Continent.* Ithaca: Cornell University Press, 1976.

Hankins, Thomas. *Science in the Enlightenment.* Cambridge: Cambridge University Press, 1985.

Heilbron, John L. *Physics at the Royal Society.* Los Angeles: University of California Press, 1983.

Wallis, Peter and Ruth Wallis. *Newton and Newtoniana, 1672-1975: A Bibliography.* Folkstone: Dawson, 1977.

Westfall, Richard. *Never at Rest.* Cambridge: Cambridge University Press, 1984.

Astronomers Argue for the Existence of God

Overview

At the dawn of the eighteenth century, scientific and Western theology was based on the concept of an unchanging, immutable God ruling a static universe. For theologians, Newtonian physics and the rise of mechanistic explanations of the natural world held forth the promise of a deeper understanding of the inner workings of the Cosmos and, accordingly, of the nature of God. During the course of the eighteenth century, however, there was a major conceptual rift between science and theology that was reflected in a growing scientific disregard for understanding based upon divine revelation and growing acceptance of an understanding of nature based upon natural theology. By the end of the century, experimentation had replaced scripture as the determinant authority in science. Enlightenment thinking, spurred by advances in the physical sciences, sent sweeping changes across the political and social landscape.

Background

Throughout the eighteenth century, English physicist Sir Isaac Newton's (1642-1727) *Philosophiae Naturalis Principia Mathematica* ("Mathematical Principles of Natural Philosophy"), first published in 1687, dominated the intellectual landscape. Moreover, Newton actively wrote and modified his observations during the first quarter of the eighteenth century. In addition to the elaboration of physics and calculus, however, Newton also concerned himself with the relationship between science and theology. Without question, Newton was the culminating figure in the Scientific Revolution of the sixteenth and seventeenth centuries and the leading articulator of the mechanistic vision of the physical world initially put forth by French mathematician René Descartes (1596-1650). Within his own lifetime Newton saw the rise and triumph of Newtonian physics and the widespread acceptance of a mechanistic concept regarding the workings of the universe among philosophers and scientists.

Newtonian laws—and a well-functioning clockwork universe—depended upon the deterministic effects of gravity, electricity, and magnetism. In such a universe matter was passive, moved about and controlled by "active principles." For Newton, who rejected the mainstream Trinitarian concepts of Christianity, the order and beauty found in the universe was God. Newton argued that God set the Cosmos in motion, and to account for small differences between predicted and observed results, God actively intervened from time to time to reset or "restore" the mechanism.

Theologians and scientists were deeply concerned about the moral implications of a scientific theory that explained everything as the inevitable consequence of mechanical principles. Accordingly, much effort was expended to reconcile Newtonian physics—and a clockwork universe— with conventional theology to provide an on-going and active role for God. Objective evidence regarding the universe was often sifted through theological filters that evaluated whether a set of facts or theories tended to prove or disprove the existence of God. Ironically, it was this interplay between religion and science that led many to insist subsequently on a strong scientific objectivity that largely discounted religious subjectivity.

Although the eighteenth century is often cited as the Golden Age for classical science, the full impact of Newtonian physics and reductionist philosophy was assured with Newton's seventeenth-century correspondence with and influence upon English philosopher John Locke (1632-1704). Locke, particularly through his work *An Essay Concerning Human Understanding*, broadened Newtonian concepts into a range of a mechanistic explanations regarding human knowledge and action that profoundly influenced the social and political thoughts of such Enlightenment thinkers as Scottish philosopher and economist Adam Smith (1723-1790) and American political philosophers and revolutionaries Thomas Jefferson (1743-1826), Thomas Paine (1737-1809), and Benjamin Franklin (1737-1809).

The Age of Enlightenment proved turbulent for theology, as the rise of natural theology clashed with traditional Christian concepts. The conceptual shifts that occurred during the eighteenth century diverged substantially from the subtle reasonings of Locke to the more radical arguments put forth by French scientists, astronomers, and philosophers. These arguments essentially eliminated God and all divine revelation from scientific cosmology—in other words, all theories regarding the nature and origin of the universe. There were, however, notable attempts to heal the growing schism and to reestablish religious truths in accord with and based upon scientific fact and reasoning.

Within the theological community, as an alternative to fundamentalist rejection of science as holding truth counter to divine relation, theologians and astronomers argued that scientific evidence proved the existence of God. It is interesting to note that English Anglican bishop George Berkeley (1685-1753) asserted a more radical defense against mechanistic reductionism. Berkeley asserted that there was no proof that matter truly exists and that God acted by shaping our perceptions of matter. Within the scientific communities the rise of Deism was supported with increasingly detailed evidence regarding the scope and scale of the universe. In both camps, excepting Berkeley's assertions, the detailed workings of the celestial machinery were put forth as arguments for the existence of God.

Although arguments for the existence of God based on the grandeur and expanse of the natural world reached back into antiquity, the premise that the design of the universe also revealed the mindset and intent of the Creator achieved a new formality in the writings and arguments of English astronomer and scientist William Derham (1657-1735) and reached powerful cohesion in the work of the Scottish philosopher David Hume (1711-1776), especially in his 1779 treatise, *Dialogues Concerning Natural Religion.* As early as 1701 microscopist Nehemiah Grew (1641-1712) set forth arguments for the existence of God from a set of natural proofs in his work *Cosmologia sacra* ("Sacred Cosmology"). John Ray's (1627-1705) publication of *The Wisdom of God* stirred a revival of interest in natural history. George Cheyne (1671-1743) took up the argument with his 1705 publication of *Philosophical Principles of Natural Religion*—a work designed by Cheyne to reveal the wonders of God's creation through natural science. In 1713 Derham published *Physico-theology*, designed to demonstrate and ascertain the attributes of God by careful study and observation of the natural world. Derham also published lists of nebulous objects in the *Philosophical Transactions of the Royal Society* (many were later found to be nonexistent) designed to advance the scale and grandeur of the universe as supporting argument for a God of infinite scope and majesty.

In 1705 Edmond Halley (1656-1742) noticed that three previous comets were seemingly the same body. Accordingly, Halley, noting the periodicity of the observations, predicted the return of what became known as Halley's comet in 1758. Halley also measured the motion of stars and put forth an argument against an infinite universe similar to the modern version of Olber's paradox (i.e., if the universe were infinite, every line of sight would end on a star and hence the night sky would be brightly illuminated). Most importantly, Halley's prediction weakened the interpretation of comets as a "sign" or "miracle" and placed the movements of comets within the predictable mechanistic universe.

Throughout the eighteenth century empiricism (the testing of theory by experiment or observation) and rationalism grew in practice and influence. Astronomers of the time devoted considerable effort toward solving the major problems of the era. Astronomical observations and data were of paramount importance for safe navigation on the seas and the concurrent growth of trade needed for economic development. There was a high public regard for astronomy both for its practical value, as evidenced by the intense interest in observations of the transit of Venus in 1769. All of this lent credibility both to astronomical observation and to the interpretations of the nature of the Cosmos put forth by astronomers.

If Newton was the dominant intellectual force of the century, following Newton's death English astronomer Sir William Herschel (1738-1822) assumed the mantle of the era's preeminent astronomer. The German-born Herschel, an accomplished musician, was arguably the greatest astronomer and telescope builder of the eighteenth century. His telescopes were the largest and most powerful, and they enabled Herschel to discover previously unseen stars and nebulae. Accordingly, estimates of the size of the universe grew—and became, for some, an important argument for the existence of an infinite, all-powerful God.

In 1781 Herschel's discovery of Uranus profoundly affected philosophical perceptions of an immutable and known Cosmos. Herschel's advancement of concepts originating with German philosopher Immanuel Kant (1724-1804) resulted in measurements of the Milky Way galaxy and of the nature of "island universes" consisting of other galaxies.

In the last quarter of the century, the idea of a divine intervention to correct the anomalies associated with the predicted orbits of the planets was unacceptable. French mathematician and scientist Pierre Simon de Laplace (1749-1827) asserted that celestial motions could be fully explained without any reliance on the divine. By performing exact mathematical calculations regarding the eccentricities of planetary orbits (e.g., taking into account their mutual gravitational attraction as well as accounting for the gravitational influence of the Sun), Laplace's work left little for God to do in a mechanistic universe. In his *Exposition du système du monde* ("The System of the World") published in 1796, Laplace hypothesized that the Cosmos had begun as nebular gas, concentrated and contracted by gravity.

Impact

If Galileo, Descartes, and Newton sowed, even if inadvertently, the seeds of dispute between modern science and theology, eighteenth-century advancements provided the soil in which those arguments bloomed. Not only were the facts of science challenging to conventional theology, especially Western Christianity, the very nature of what was accepted as reason, evidence, and truth (epistemology) was anathema to classical theology. Deterministic interpretations of Newtonian physics stripped God of personality and sovereign action. God was regulated to the force associated with first movement—the original creator of a mechanistic universe. Accordingly, whether God intervened in the mechanism of the universe through miracles or signs (such as comets) became a topic of lively philosophical and theological debate. Moreover, without day-to-day responsibilities for the mechanistic universe, God became increasingly identified with the eternity or infinity of the universe.

The immutability of a static universe—recently radically revised by Nicolaus Copernicus (1473-1543), Galileo, and Johannes Kepler (1571-1630)—once again was advanced as proof of the existence of God. Accordingly, astronomers arguing for a static universe simply applied the formulations found in Newton's philosophy to the established interpretations of divine revelation. For the first three quarters of the century, the mechanistic clockwork of the heavens, regulated in accord with knowable physical laws, was reconciled with conventional theology and offered as confirmation of the existence of a law-giver or God of infinite power. According to popular interpretations of Newtonian physics, astronomers argued that God was the "prime mover" of the Cosmos.

As the eighteenth century proceeded, scientists and philosophers increasingly sought to explain "miracles" in terms of natural events. In accord with the development of natural theology, scientists and philosophers argued that only as a part of a greater clockwork universe could such celestial phenomena as comets be interpreted as acts of—or signs from—God. Corresponding to this reliance on material and rational explanations of the Cosmos, the very nature of God could only be understood within the laws of science. For other scientists, however, the revelations of a mechanistic universe left no place for God, and they discarded their religious views.

K. LEE LERNER

Further Reading

Bronowski, J. *The Ascent of Man*. Boston: Little, Brown, 1973.

Cragg, G. R. *Reason and Authority in the Eighteenth Century*. London: Cambridge University Press, 1964.

Deason, G. B. "Reformation Theology and the Mechanistic Conception of Nature." In *God and Nature*, ed. by D. C. Lindberg. and R. L. Numbers. Berkeley: University of California Press, 1986.

Hawking, Stephen. *A Brief History of Time*. New York: Bantam Books, 1988.

Hoyle, Fred. *Astronomy*. New York: Crescent Books, 1962.

Sagan, Carl. *Cosmos*. New York: Random House, 1980.

Edmond Halley Successfully Predicts the Return of the Great Comet of 1682

Overview

On Christmas night, 1758, a comet appeared in the skies over Europe. First seen by German amateur astronomer Johann Georg Palitzsch, its appearance was a landmark event in the history of both astronomy and physics. For the first time, using Isaac Newton's (1642-1727) laws of gravitation and motion, a comet's appearance had been successfully predicted. By so doing, Edmond Halley (1656-1742) had not only launched a new era in predictive astronomy, but had also proven the accuracy and value of Newton's revolutionary new physics.

Background

For millennia, comets were viewed as harbingers of change, both good and bad. A comet appeared when Julius Caesar was murdered; another comet was said to have marked the death of Alexander the Great. Throughout history, comets were seen as messengers, portents, or omens.

During these same millennia, arguments raged over where comets were located. Some, including the Greek philosopher Aristotle (384-322 B.C.), felt that comets were part of the Earth's atmosphere while others thought they existed in the realm of the planets. This debate was finally settled in the sixteenth century, when Ptolemy (c. 100-170 A.D.) proved that comets were at least as distant as the moon, proving that they must lie beyond the atmosphere.

However, even in the early seventeenth century, little more was known about comets than in previous times. As late as 1619, Johannes Kepler (1571-1630) thought that comets moved in straight lines across the heavens, and it was not until Newton's 1687 publication of the *Principia* that this was realized to be impossible. In fact, Newton found that his laws of gravity and motion required all bodies to move in paths that belonged to a family of shapes called "conic sections," the shapes one gets from cutting a cone at different angles. Conic sections are circles, ellipses, parabolas, and hyperbolas, but do not include straight lines. For Newton to be right, either Kepler was wrong or comets obeyed a different set of rules.

Shortly after publishing the *Principia*, Newton used the observed positions of the Great Comet of 1680 to calculate its orbit as a parabola. A few years later, Newton's friend, Edmond Halley, took Newton's calculations to the next level, calculating the orbits of no fewer than 24 comets, assuming in advance that they were parabolic. In so doing, he noticed that one comet in particular, the Great Comet of 1682, seemed to have an orbit identical to that of two previous comets, the comets of 1531 and 1607. Intrigued, Halley also noticed that the periods of these three comets were almost identical. However, Halley also realized that it was impossible for a comet following a parabolic orbit to return to the inner solar system; parabolic orbits are not "closed," they continue on forever without returning.

Halley eventually realized that, contrary to expectation, many comets follow elliptical orbits that are so elongated that they appear parabolic at their closest approach to the Sun, which is all that was visible from the Earth. Casting aside the assumption that his orbits were parabolas, Halley recalculated the orbital parameters of these three comets, finding them to be a nearly exact match. Based on this, in 1705 he predicted that another great comet would appear in the skies in 1758. Unfortunately, Halley died in 1742 without seeing the return of "his" comet.

Halley was proven correct on Christmas night in 1758 when a large new comet was sighted. Coming close to the time predicted and in the correct location in the sky, there was little doubt that this was the same comet for which Halley had calculated an orbit. Its reappearance marked the start of astronomy as a predictive science, able to use universal laws to predict the positions of planets, stars, moons, and comets across the sky and through the solar system.

Impact

Halley's prediction resonated on several levels, both scientifically and within society. First, and perhaps most important, he had shown the predictive and descriptive power of Newton's physics and mathematics in a dramatic fashion. This would not be equaled for nearly a century, when Neptune was discovered from mathematical calculations based on Newton's laws of gravity and motion. Indeed, ironically, the hunt for Neptune was launched following the 1835 appearance of Halley's comet, when astronomers noticed that it failed to follow the predicted orbit during that return.

In addition to confirming the accuracy of Newtonian mechanics, Halley's successful prediction made it possible to finally resolve the place of comets in the solar system. And, from the standpoint of society, this prediction was extremely important because it helped demonstrate the power of the human intellect. Using nothing more than a pencil, Newton and Halley had described how gravity held the solar system together, forcing planets and comets alike to follow the paths laid out for them since the solar system first formed. This was heady knowledge that gave people pause. The next time that such a fundamental reassessment of our ability to describe the universe would occur was in the early 1900s, when Albert Einstein's (1879-1955) theory of relativity was confirmed by noting changes in apparent star positions during a solar eclipse.

Edmond Halley's correct prediction that the Great Comet of 1682 would return in 1758 was a high point in eighteenth-century astronomy. The comet was renamed for Halley. *(NASA. Reproduced with permission.)*

In both cases, the consensus seemed to be that such predictions were a triumph of science.

The scientific implications of Halley's successful prediction were immense. Although the accuracy of Newton's laws were not doubted with respect to objects on earth, and they appeared to govern the movements of the observed planets, this was the first time they had been used to predict the motions of an object that was no longer visible. In a sense, the comet had vanished from human senses and was "visible" only mathematically. Its dramatic reappearance proved, beyond any doubt, that the solar system and, indeed, the whole of the universe, operated according to specific laws that had been elucidated by Newton. The success of Newtonian mathematics and physics helped encourage scientists in these and other disciplines that the universe was, fundamentally, logical and understandable. While this is hardly a surprising statement today, it was almost heretical during the seventeenth and previous centuries.

It is also difficult to comprehend today that scientists have not always been able to make predictions like the one made by Halley. And many point to the Egyptians, Maya, the builders of Stonehenge, and their intellectual kin, noting that they made astronomy into a predictive science with elaborate and highly accurate calendars. However, these early astronomers, for all of their successes, lacked a real understanding of the objects of their study. They could, for example, predict the location of planets in the sky with uncanny accuracy, but they didn't know what the planets were, where they were located in space, or what rules led them to appear in their designated places year after year. In a sense, they simply developed rules to explain when a particular point of light would appear in a particular place in the sky, based on watching the sky for centuries.

Before Newton, Galileo, Copernicus, and others came to the realization that the planets were other worlds, like the Earth, in orbit around the Sun. Newton explained how they remained in orbit, made more accurate calculations of their predicted locations over the years, and helped place planetary motions into a solid, consistent physical framework. He also showed that gravity's influence extended from the Earth's surface throughout the solar system, and that it worked the same everywhere. And, most importantly, unlike ancient astronomers, he was able to erect a theoretical structure that allowed the tracking of the motions of invisible objects, as well as calculating future motions of asteroids, comets, and other bodies, based only on a few observations rather than on years or centuries of studying the heavens. Once confirmed by Halley's successful prediction, this firmly established astronomy and physics as legitimate, predictive, theoretically based fields of science rather than collections of observations made over the years.

In addition, Halley's successful prediction helped give comets their rightful place in science and in the solar system. No longer were they seen as extraordinary portents of terrestrial calamity or as celestial bodies obeying their own set of laws. Instead, they were now known to be simply a different kind of object in orbit around the Sun, and obeying the same laws of nature that ruled the rest of the solar system. This is not to say that superstitious thinking about comets ceased. However, at least scientists and the scientifically aware public began looking at comets as ordinary objects following well-understood laws of nature. In the case of Halley's comet, this culminated in the 1986 flotilla of scientific

spacecraft rendezvousing with it during its return to the inner solar system in that year.

Finally, this was a philosophical breakthrough. Coming as it did during the Enlightenment, when human thought was taking center stage in man's intellectual universe, this scientific triumph served to reinforce the emerging vision of the universe as being something that, eventually, could be fully understood by man. This was one stage in a revolution in human thought.

P. ANDREW KARAM

Further Reading

Books

Kippenhahn, Rudolph. *Bound to the Sun.* W.H. Freeman and Company, 1990.

Schaff, Fred and Guy Ottewell. *Comet of the Century: From Halley to Hale-Bop.* Copernicus Books, 1996.

William Herschel and the Discovery of the Planet Uranus

Overview

William Herschel (1738-1822) discovered the planet Uranus in 1781. It was the first planet discovered since the beginning of recorded history. The discovery of Uranus brought Herschel much fame, which enabled him to carry out his unconventional astronomical research. His discovery of Uranus, to a small degree, even consoled England for its loss of the 13 colonies in the American Revolutionary War. Perhaps most importantly, the discovery of Uranus opened up a new phase in the discovery of the planets of our solar system.

Background

The discovery of the planets in our solar system can be said to have two distinct phases. The discovery of the planet Uranus marks the boundary between these two phases. The first phase began before recorded history. In this phase the only known planets were the five that were visible to the naked eye. Nearly every culture had knowledge of the planets we call Mercury, Venus, Mars, Jupiter, and Saturn. These five objects were first called *planets* by the Greeks; *planet* was the Greek word for *wanderer*. Planets were called wanderers because, unlike the stars (which do

not appear to change their positions relative to one another), the planets were seen to move (or change position) relative to the fixed stars. We call this motion the *orbit* of planets.

Prior to the death of Nicholas Copernicus in 1543 our Earth was not considered to be one of the planets. Instead, it was thought to be the center of the universe with the Moon, Sun, planets, and stars revolving around it. No one suspected that there might be other planets besides Mercury, Venus, Mars, Jupiter, and Saturn. But there are, as we now know—Uranus, Neptune, and Pluto. And the discovery of the first planet beyond Saturn, the planet Uranus, marks the beginning of the second phase of the discovery of the planets in our solar system. Uranus was discovered in 1781 by a German astronomer living in England. His name was William Herschel.

Herschel was born in Hanover, Germany, to a family of musicians. In 1757 he moved to England. In 1766 he obtained a permanent position as an organist in the English city called Bath. But William Herschel was a man of wide interests outside of music, and astronomy was one of them. He read and mastered a number of books on astronomy. In 1767 he began to observe the night sky with small telescopes. But they were poor instru-

Uranus was discovered in 1781 by William Herschel. *(NASA. Reproduced with permission.)*

ments and soon frustrated him. Unfortunately for Herschel, larger telescopes were not available. So, in 1773, he began to build his own. Working at night when his musical duties were over, he built his own telescopes and observed the heavens. Before long, with the help of his brother Alexander and sister Caroline, he was building the best telescopes in the world and seeing farther and farther into space. He was soon much more interested in astronomy than in music.

Herschel observed many different objects in the heavens. He searched for double stars, and also for the fuzzy-looking objects called nebulae (which we now know to be either star clusters, clouds of gas, or distant galaxies). He was very much like a naturalist, someone who collects and categorizes specimens. He "collected" double stars and nebulae in the same way that naturalists collect insects or plants. In this way he was rather unlike other astronomers of his day. They were interested in charting the positions of stars or calculating the orbits of planets according to the laws of gravity.

William Herschel also observed the Moon, Sun, and planets of our solar system. He believed, as did many people during his day, that the planets, Moon, and even the Sun were inhabited. It was widely believed at the time that God would have created intelligent life on other worlds. Herschel believed very strongly in the

possibility of extra-terrestrial intelligent life, and he spent many nights with his telescope watching the Moon, Sun, and the planets for signs of it.

Impact

Throughout the 1770s William Herschel spent most nights observing nebulae, looking for double stars, and searching for signs of life on other worlds. Then, on the night of 13 March 1781, Herschel made an interesting discovery. While observing the area in the constellation of Gemini, he noticed a large object that he thought was a nebula or a comet. Four nights later he noticed that this object had moved, and so he concluded it was in fact a comet. He wrote of this new "comet" to astronomers in London. But the comet was difficult for those astronomers to detect because their telescopes were inferior to Herschel's. They simply couldn't see it at first. But eventually the supposed comet was observed by both English and French astronomers. They were astounded that Herschel had even seen the object, as it was so faint in their own telescopes.

About a month later the Astronomer Royal in London, Nevil Maskelyne (1732-1811), suggested that the object might be a planet. But it was hard to tell. In order to be sure, the object had to be observed over time to chart its orbit. And then calculations of that orbit had to be made. This took a few more months. Finally, in the summer of 1781, Anders Johan Lexell (1740-1784), a Russian astronomer who was visiting London, calculated an orbit for the comet. Lexell's calculations convinced most astronomers that Herschel's "comet" was in fact a planet—the first planet discovered since the beginning of recorded history!

Herschel's discovery of the planet made him instantly famous. He named the planet *Georgium sidus*, or "George's Star," for King George III of England. George III granted Herschel an annual salary, which meant that he could give up music for a living and turn to astronomy full time. Herschel moved from Bath to Windsor, near London, where the king lived. Herschel was also made a member of the Royal Society of London and given its highest award, the Copley Medal. His telescopes were suddenly in demand, and he made a small fortune selling telescopes throughout Europe. His customers included the King of Spain as well as Lucien Bonaparte, the brother of Napoleon Bonaparte.

The naming of the new planet is an interesting story. Initially, as has been mentioned, the planet was called "George's Star." In time, other names for the planet were suggested, because many astronomers outside of England thought the name "George's Star" was simply too English, an unfair extension of English terrestrial imperialism into the heavens. Some astronomers suggested calling the planet "Herschel," and others liked the name "Neptune." But in the end the name "Uranus" was chosen. This kept with the theme of naming the planets after the Roman gods: Mercury, Venus, Mars, Jupiter, and Saturn. Using the name Uranus continued the family lineage of those gods. In Roman mythology Uranus was the father of Saturn and the grandfather of Jupiter. Thus Uranus was a perfect name, because the planet Uranus was the next planet after Saturn and the second planet after Jupiter—grandfather, son, and grandson.

Even though William Herschel had become famous for discovering Uranus, he was still seen as a very different sort of astronomer from other astronomers of his day. The goal of most astronomers was to chart the positions of planets and stars in the sky using mathematics. Herschel, however, became interested in the structure of heavenly objects—how objects in the heavens came to be formed. In other words, he was interested in the evolution of stars, and not their positions. Other astronomers thought he was an oddball. Because he had discovered Uranus, however, he had money and re-

spectability, and was able to continue with his unique program of astronomical observation. For Herschel, the discovery of Uranus was not the high point of his career. It was instead a fortunate event that brought him enough wealth to study independently

It is interesting to note how much time William Herschel spent looking for extra-terrestrial life. He was a firm believer that the Moon, Sun and planets of our solar system were inhabited. This belief was shared by many others of his day, and it became a political issue as well. In 1781, when Herschel discovered Uranus, England was at the end of the war with its colonies in North America—the war known as the American Revolution. When the war finally ended, the American colonists were victorious and England had lost a large portion of its empire. But some Englishmen thought that William Herschel had in a way opened up new territory to English control—the territory of the heavens. William Herschel's discovery of Uranus was a consolation for the loss of the American colonies. As one English physician said: "It is true that we have lost the *terra firma* of the Thirteen Colonies in America, but we ought to be satisfied with having gained in return by the generalship of Dr. Herschel a *terra incognita* of much greater extent...."

William Herschel's discovery of Uranus freed him to do astronomy his own way. And his discovery also consoled the English after their loss of the American colonies. Perhaps most important for the science of astronomy is that the discovery of Uranus began the search for other undiscovered planets in our solar system. It was the beginning of a second phase in the discovery of planets, during which Neptune and Pluto were discovered. Neptune was discovered in 1846 when astronomers were trying solve a problem with their calculations of Uranus's orbit. And Pluto, the last planet in our solar system, was discovered in 1930 with the aid of a telescope and photographic equipment. But that was not the end of the discovery of new planets. It might be said that a third phase has begun—the search recently and successfully undertaken to discover planets orbiting other stars.

STEVE RUSKIN

Further Reading

Books

Armitage, Angus. *William Herschel*. New York: Doubleday and Company, 1963.

Clerke, Agnes M. *The Herschels and Modern Astronomy.* New York: MacMillan and Company, 1895.

Crowe, Michael J. *The Extraterrestrial Life Debate, 1750-1900: The Idea of a Plurality of Worlds from Kant to Lowell.* New York: Cambridge University Press, 1988.

Hoskin, Michael, ed. *The Cambridge Illustrated History of Astronomy.* Cambridge, England: Cambridge University Press, 1997.

North, John. *The Norton History of Astronomy and Cosmology.* New York: W. W. Norton and Company, 1995.

Laplace Theorizes That the Solar System Originated from a Cloud of Gas

Overview

In *Exposition du système du monde* (Exposition of the System of the World) (1796), the French astronomer Marquis Pierre Simon de Laplace (1749-1827) briefly stated his "nebular hypothesis" that the Sun, planets, and their moons began as a whirling cloud of gas. This hypothesis sparked controversy among theologians and politicians as well as astronomers and physicists.

Background

After the pioneer astronomers Nicolaus Copernicus (1473-1543), Galileo Galilei (1564-1642), and Johannes Kepler (1571-1630) put forth their respective heliocentric (sun-centered) theories of the relationships among celestial bodies, Christianity was hard pressed to defend its traditional geocentric (earth-centered) cosmology. In the sixteenth and seventeenth centuries, the church suppressed heliocentric astronomy. Many scientists became disenchanted with religion. By the late seventeenth and early eighteenth centuries, astronomers were arguing publicly among themselves for and against the immanence of God, i.e., the presence and activity of God in the world.

Accepting heliocentrism did not entail disbelief in either God or Christianity. Nevertheless, some adjustments in the content of Christian doctrine became necessary because of heliocentrism. It prompted a Christian theological debate that lasted about two centuries. In the light of the new cosmology, some thinkers even came to question the reality of God.

By the beginning of the eighteenth century, atheism had become such a viable alternative that even some astronomers were engaged in arguing for the reality of God. Among them were Isaac Newton (1642-1727) and William Derham (1657-1735), an astronomer, physicist, entomologist, and Anglican priest who published several passionate defenses of God's immanence. Throughout history, relatively few great physicists and astronomers have been atheists. Albert Einstein (1879-1955) insisted that God is real and does not "play dice with the universe."

The eighteenth-century debate more clearly defined three possible stances: theism, the belief that God created the world and remains active in it; deism, the belief that God created the world but then left it alone; and atheism, which asserts that there is no God. Theists, concerned with God's immanence, sometimes accused deists of atheism. Atheists, convinced that there is no place for God at all, sometimes accused deists of theism. By the late eighteenth century, deism was dominant among intellectuals, so that the argument was no longer primarily between theism and deism, but between deism and atheism. It was in this theologically skeptical environment that Laplace arose.

The two greatest astronomers of Revolutionary and Napoleonic France were Laplace and his rival, Joseph Jérôme Le Français de Lalande (1732-1807), director of the Paris Observatory. Both were atheists. Lalande is reported to have said that because he had searched throughout the heavens with his telescope and found no God, therefore there is no God. He popularized his science and took his case for atheism to the public. Laplace thought Lalande was foolish to involve the common people in matters of astronomy. Laplace preferred the company of other scientists, intellectuals, and political leaders. He is supposed to have told Napoleon privately that he did not mention God in his *Traité de mécanique céleste* (Treatise on Celestial Mechanics) (1799-1825) because there was no need of that hypothesis.

A cogent, scientific, non-theological explanation of the origin of the solar system was not possible until scientists had absorbed the full import of Newton's law of universal gravitation and three laws of motion. Newton published these four laws in 1687 in *Philosophiae naturalis principia mathematica* (Mathematical Principles of Natural Philosophy), one of the most important books of all time. Newtonian physics was also a key part of Laplace's environment.

Laplace attacked the traditional "design argument" for the reality of God. The *locus classicus* of this argument is the "Fifth Way" of Thomas Aquinas. It claims that God can be known by contemplating the natural phenomena of the world and reasoning that the world must have been designed by an intelligent being in order for these phenomena to interact the way they do. Both theism and deism are compatible with the design argument. According to the former, God continues actively in time as creator and sustainer after the primordial design and creation; but according to the latter, God designs and creates the world once, before time, then, like an architect whose commission is complete, abandons it to its fate.

Not considering any role for God at all, either as architect or as creator, Laplace suggested a possible origin of the solar system. He argued that as a hot nebula cooled, it would require a proportionately faster rate of rotation in order to conserve its angular momentum, and the increased centrifugal force resulting from this increased speed would throw out material which would eventually condense into planets. This suggestion is known as the "nebular hypothesis." Laplace did not speculate to the ultimate origin of the materials in the nebula.

Laplace was not the first thinker to propose a nebular hypothesis. In 1755 the German philosopher Immanuel Kant (1724-1804) published *Allgemeine Naturgeschichte und Theorie des Himmels* (General Natural History and Theory of the Heavens) in which he suggested that the gravitational forces in a slowly rotating nebula would gradually flatten it and create within it several denser clouds of gas which would separately compact themselves into distinct spheres, the Sun, and planets.

Kant's astronomical speculations may have been influenced by the Swedish theologian Emanuel Swedenborg (1688-1772), who in 1734 asserted that the solar system originated as a rapidly rotating nebula condensing and coalescing into the Sun and planets. Swedenborg claimed that this revelation came to him in a seance, not as the result of scientific inquiry. Neither Swedenborg nor Kant could say what

started the gas cloud rotating in the first place. Gravity alone is not sufficient to do that.

Laplace derived his nebular hypothesis in part from the work of William Herschel (1738-1822) on the origin of stars. He was probably unaware of Kant's cosmology when he theorized about the solar system in the 1790s. Even though Kant and Laplace never collaborated, and even though there are significant differences between their respective versions of the nebular hypothesis, it is sometimes called the "Kant-Laplace hypothesis."

Impact

The Swedenborgian and Kantian versions of the nebular hypothesis were completely disproved in the twentieth century. The Laplacean version was still workable in 2000, and despite frequent refutations, it has just as frequently been revived and refined. The Hubble telescope has found evidence in the Orion Nebula, the Eagle Nebula, and other nebulae that supports it. When it is in favor, science progresses by deliberating it; when it is out of favor, its discredit provides fodder for creationists.

James Clerk Maxwell (1831-1879) and Sir James Hopwood Jeans (1877-1946) criticized Laplace's hypothesis on the grounds that the material thrown out by centrifugal force would not have enough mass to generate enough gravity to pull the material together into planets. In support of this criticism, Moulton noted that the Sun has 99.9% of the mass of the solar system while the planets have 99% of its angular momentum. Angular momentum measures the intensity of the motion of a rotating body. Moulton calculated that the Sun would have to rotate over 100 times faster than it actually does in order to conserve the angular momentum demanded by the nebular hypothesis. Therefore, he reasoned, the hypothesis must be false.

The nebular hypothesis was the first coherent nontheological explanation proposed for the origin of the solar system. Several others, some ephemeral, some with lasting importance, arose in the eighteenth century. The most important of these was the "collision hypothesis," proposed in 1778 by Count Georges Louis Leclerc de Buffon (1707-1788) and revived in various forms in the early twentieth century by Thomas Chrowder Chamberlin (1843-1928), Forest Ray Moulton (1872-1952), Jeans, Sir Harold Jeffreys (1891-1989), and others. It stated that the solar system could have arisen from the collision or near miss of two stars or a star and a comet, with the smaller debris cooling into planets and the larger debris becoming the Sun.

In 1900 Chamberlin and Moulton theorized that small meteors or particles of solid matter, not gas, pulled off from the Sun by a passing star or comet would eventually accumulate into planets by their mutual gravitational attraction. This was the "planetesimal hypothesis." In 1917 Jeans and Jeffreys proposed their "tidal hypothesis" that the gravity of a massive celestial body passing close to the Sun would pull out from the Sun a long stream of gaseous beads, which would subsequently cool into planets. Both the planetesimal and tidal hypotheses were attacked on the grounds that the material pulled off would be so hot that it would more likely disperse into space than coalesce into planets.

In 1944 Carl Friedrich von Weizsäcker (b. 1912), disappointed in both the planetesimal and the tidal hypotheses, resurrected Laplace's nebular hypothesis. Weizsäcker argued that the core of a nebula would collapse to form a sun while the outlying gases, being more turbulent because of the collapse of the core, would form eddies that would eventually condense into planets. He also tried to explain the origins of the nebulae themselves. Drawing upon the "big bang" theory of the origin of the universe that Georges Lemaître (1894-1966) introduced in the early 1930s, Weizsäcker claimed that expanding gases from the primeval explosion would randomly clump together and that the gravity within these clumps would condense them into nebulae.

Many subsequent theories of the origin of the solar system, such as the "cloud hypothesis" proposed in 1948 by Fred Whipple (1906-) and the "protoplanet hypothesis" proposed in 1951 by Gerald P. Kuiper (1905-1973), depend to significant extents upon Weizsäcker's revival of Laplace's theory.

ERIC V.D. LUFT

Further Reading

Books

Bennett, J.H. Hobart. *Genesis of Worlds.* Springfield, IL: H.W. Rokker, 1900.

Brush, Stephen G. *Fruitful Encounters: The Origin of the Solar System and of the Moon from Chamberlin to Apollo.* New York: Cambridge University Press, 1996.

Brush, Stephen G. *Nebulous Earth: The Origin of the Solar System and the Core of the Earth from Laplace to Jeffreys.* New York: Cambridge University Press, 1996.

Chamberlin, Thomas Chrowder. *An Attempt to Test the Nebular Hypothesis by the Relations of Masses and Momenta.* Chicago: University of Chicago Press, 1900.

Elliott, James. *The Nebular Hypothesis: Untenable.* Carlisle, PA: Sentinel Printery, 1908.

Gillispie, Charles Coulston. *Pierre-Simon Laplace, 1749-1827: A Life in Exact Science.* Princeton: Princeton University Press, 1997.

Hahn, Roger. "Laplace and the Vanishing Role of God in the Physical Universe." In *The Analytic Spirit: Essays in the History of Science in Honor of Henry Guerlac.* Woolf, Harry, ed. Ithaca: Cornell, 1981: 85-95.

Hastie, W., ed. *Kant's Cosmogony.* Bristol, England: Thoemmes, 1993.

Jeans, James Hopwood. *The Nebular Hypothesis and Modern Cosmogony, Being the Halley Lecture Delivered on 23 May, 1922.* Oxford: Clarendon; London: Milford, 1923.

Jeffreys, Harold. *The Earth: Its Origin, History, and Physical Constitution.* Cambridge: Cambridge University Press, 1976.

Kant, Immanuel. *Universal Natural History and Theory of the Heavens.* Ann Arbor: University of Michigan Press, 1969.

Melosh, H.J., ed. *Origins of Planets and Life.* Palo Alto, CA: Annual Reviews, 1997.

Möhlmann, Diedrich, and Heinz Stiller, eds. *Origin and Evolution of Planetary and Satellite Systems.* Berlin: Akademie-Verlag, 1989.

Numbers, Ronald L. *Creation by Natural Law: Laplace's Nebular Hypothesis in American Thought.* Seattle: University of Washington Press, 1977.

Pickering, James S. *Captives of the Sun: The Story of the Planets.* Garden City, NY: Natural History Press, 1964.

Whipple, Fred Lawrence. *Orbiting the Sun: Planets and Satellites of the Solar System.* Cambridge, MA: Harvard University Press, 1981.

The Work and Impact of Benjamin Banneker

Overview

Benjamin Banneker (1731-1806) demonstrated that African Americans were capable of scientific and technological achievements. During the time that Banneker lived, the fledgling United States was attempting to create order from late eighteenth-century chaos. Although the American Revolution had secured political independence, the former colonies, merged into a confederation of state governments, experienced strife between and within the states. Entrenched social patterns, particularly that of slavery, prevented many individuals from aspiring to attain personal goals and contribute to society's improvement. Many white Americans, to maintain power, perpetuated untruths about blacks, especially concerning their intelligence and ingenuity. Banneker proved the falsehood of cultural myths about African Americans held during the early republic. Although he did not directly contribute to scientific theory, Banneker advanced American science through his example.

Background

Curious colonists pursued scientific investigations regarding natural phenomena in their nearby environments. Most early American scientific activities consisted of amateur observations about wildlife, plants, and weather. Individuals wrote essays for local newspapers and British scientific journals and published their own pamphlets, commenting on what they had seen. Many articles presented new theories about unexplained events.

Most early American scientists were white males. Native Americans relied on spiritual and magical explanations for natural occurrences. Slaves and indentured whites lacked the time, literacy, and freedom to pursue scientific inquiries such as those their masters were able to initiate. The eighteenth-century Enlightenment encouraged scientific activity among the privileged. Benjamin Franklin (1706-1790) is probably the best known colonial scientist. His experiments with lightning attracted a great deal of public attention. Franklin's *Poor Richard's Almanac* began publication in 1732, seven years after Boston physician Nathaniel Ames issued the first *Astronomical Diary and Almanac*. John Winthrop (1714-1779), a Harvard College mathematician, was the second most eminent

colonial scientist, after Franklin. He experimented with electricity and magnetism in his laboratory and pioneered research in seismology. Colonial scientific endeavors resulted in the creation of the American Philosophical Society in 1745, and libraries and colleges collected scientific writings and equipment.

Many Americans resisted scientific developments, considering intellectual pursuits unnecessary and a waste of time. People doubted scientists' ability to discover new information that would influence daily life. Religious groups regarded scientific inquiry with suspicion and encouraged their congregations to be cautious of accepting such work. Individuals instead valued practical experience and believed in myths, such as how the moon's phases affected crops. Only by the mid-1800s did Americans begin to realize how scientific knowledge could be transformed into monetary profits. People became more familiar with science because fairs exhibited scientific displays and technological innovations; contemporary periodicals such as *Scientific American* printed explanatory illustrations; and museums featured scientific specimens. Industrialization created demand for scientific engineering as well.

Because landowners were often uninvolved with the daily functioning of their plantations, slaves, especially those who had been trained as artisans, creatively solved scientific and technological problems and were skilled in practical engineering. Slaves helped build railroads, bridges, and waterways. Prior to (or concurrent with) Banneker's work, a few African Americans had contributed to mainstream medicine, including Cesar, Lucas Santomee, Onesimus, and James Durham. They devised cures for diseases or poisoning, such as snake bites, or inoculations for smallpox. Early African-American inventors such as James Forten Sr., Banneker's contemporary, invested their profits in abolitionist causes. The first recorded black female inventors were Sarah E. Goode, Ellen F. Eglin, and Miriam E. Benjamin, all who developed devices after Banneker's death.

Impact

When Benjamin Banneker was born, his family consisted of freed slaves, who did not fulfill the criteria early Americans expected of scientists. Banneker ultimately acquired many scientific titles: inventor, mathematician, surveyor, and astronomer. His work inspired both black and white scientists. Born free near Baltimore, Maryland, Banneker's childhood was unlike most African Americans in the late eighteenth century. Although his father and grandfather had been enslaved, they were emancipated before his birth, and Banneker refused to comply with whites' racist dictates. Also, the Banneker family's prosperity assured that Banneker was be treated with a certain degree of respect by Maryland's economy-savvy population. Isolated on his family farm, Banneker did not experience the overt racism that other blacks suffered.

Educated in an integrated community school during winters, Banneker also studied books loaned to him by neighboring Quakers, who encouraged him to develop his academic talents. These prosperous members of the community valued Banneker's abilities and work ethic. He yearned to improve his intellect and continued to seek self-education, specifically in science and mathematics. Banneker especially enjoyed solving mathematical puzzles and composed his own problems. He worked on his family's tobacco farm, where he applied scientific concepts to solving practical problems, such as diverting natural springs for irrigation during droughts.

Banneker especially was intrigued by mechanical objects. Few Americans at the time owned watches or clocks because they were scarce and expensive. Traditional accounts say that in 1753 Banneker borrowed a wealthy neighbor's pocketwatch. He disassembled the watch and drew each part. Banneker used his sketches to carve a wooden clock with a knife. He calculated ratios to make the clock larger than the watch. Banneker carefully determined how the gears should be fitted together and how many teeth were required on each gear to replicate the timing mechanism in the watch. The clock accurately counted time, striking each hour. Banneker's clock was the first striking time device made in the United States, and his cleverness attracted public attention.

Interested in nature, Banneker watched bees and locusts and estimated a 17-year life cycle for the latter. Learning to use a telescope at his neighbor George Ellicott's house in 1788, Banneker also diligently recorded his observations and measurements of celestial objects and their movements. His astronomical calculations resulted in the successful prediction of a 1789 eclipse. Such achievements resulted in more whites becoming aware of Banneker's work.

A portrait of Benjamin Banneker on an American postal stamp. *(Corbis Corporation. Reproduced with permission.)*

George Ellicott's cousin Major Andrew Ellicott admired Banneker's mathematical prowess and insisted that he assist him in surveying the 10-square-mile area procured from Maryland and Virginia that formed the site of the nation's new capital. In 1791 Banneker and Ellicott joined Pierre L'Enfant in assessing the land. Banneker monitored an astronomical clock and collected data about the times different stars crossed the meridian in order to establish latitudes. His use of sophisticated scientific instruments impressed area residents and the *Georgetown Weekly Ledger* praised him. When L'Enfant departed with crucial sketches after a conflict, Banneker reproduced the drawings from memory. Historians sometimes refer to Banneker as having saved Washington because, without his maps, the survey would have taken longer to recreate.

Banneker returned to his farm to resume his astronomical activities, consulting books and using scientific instruments loaned by his Quaker friends to document solar and lunar cycles. He slept during the day in order to work all night, a habit that resulted in his being falsely accused of laziness. Although he was not the first American astronomer (David Rittenhouse, 1732-1796, preceded him), Banneker was the first recognized African-American astronomer.

In 1792 Banneker distributed *Benjamin Banneker's Pennsylvania, Delaware, Maryland and Vir-*

ginia *Almanack and Ephemeris,* which was updated annually through 1797. This work was the first scientific book published by an African American as well as the first almanac compiled by an African American. Senator James McHenry penned a biographical sketch of Banneker, stating that Banneker proved why slavery should be abolished. The 1795 almanac included an engraving of Banneker and the editors praised him, commenting "If Africa's sons to genius are unknown, / For Banneker has prov'd they may acquire a name / As bright, as lasting, as your own."

Readers did not seem concerned that the almanac's author was black. Recommended by abolitionist groups, the almanacs sold well throughout the United States, territories, and Europe and Banneker acquired international acclaim. His astronomical information, tide calculations, and weather predictions were especially useful for farmers and sailors. By reprinting anti-slavery material, he emphasized the injustices that African Americans encountered.

He became an outspoken abolitionist, denouncing slavery and striving to improve conditions for African Americans. Anti-slavery advocates presented Banneker's almanacs as examples of blacks' capabilities; when public support of abolitionism waned and the Maryland Society for the Abolition of Slavery closed, however, Banneker was unable to find a publisher for his work. He also was occasionally the target of local thieves and harassed by threatening gunfire.

Popularly known as the "Sable Astronomer" and referred to as a "wizard" because of his ingenuity, Banneker was often cited as an example showing that African Americans were intellectually competent. He used his fame to press for opportunities for African Americans. On August 19, 1791, Banneker boldly wrote Secretary of State Thomas Jefferson, who had stated that he believed blacks were mentally inferior to whites and incapable of scientific comprehension. Eloquently expressing his outrage at racism, Banneker asked Jefferson to use his political and social power to sway popular opinion regarding African Americans. Quoting the *Declaration of Independence*'s assurances of liberty and humanity, he pleaded with Jefferson to recognize his hypocrisy and "embrace every opportunity, to eradicate that train of absurd and false ideas and opinions, which so generally prevails with respect to us." Banneker expressed abhorrence of "that state of tyrannical thraldom, and inhuman captivity, to which too many of my brethren are doomed." He included a draft copy of his almanac, "the production of my arduous study," which fulfilled his "unbounded desires to become acquainted with the secrets of nature." He told Jefferson that his work helped "gratify my curiosity" despite "the many difficulties and disadvantages, which I have had to encounter." Jefferson responded favorably in his reply: "No body wishes more than I do, to see such proofs as you exhibit, that nature has given to our black brethren talents equal to those of the other colors of men." To advance Banneker's career, he forwarded the almanac to the French Academy of Sciences, "because," he continued, "I considered it as a document, to which your whole color had a right for their justification, against the doubts which have been entertained of them."

When Banneker died, his clock was still functioning accurately, demonstrating the quality of work he had performed. Although some of his peers recognized his intellectual merits, Banneker was mostly overlooked or discredited. The reaction to his achievements reveals the rigid cultural patterns and racial attitudes of the Federalist and Jeffersonian eras. Although the Banneker Institute was opened in 1853, Banneker did not receive the recognition he deserved until the twentieth century. During the Civil Rights Movement, landmarks related to Banneker were located and identified in Maryland, and history books, especially those focusing on African-American pioneers, began including his noteworthy achievements. The United States Postal Service designed a stamp featuring Banneker, the first American astronomer so honored. The Maryland Historical Society sponsors the Banneker-Douglass Museum and educational facilities are named in his honor. Organizations have appropriated his name, including the Benjamin Banneker Association, Inc., a nonprofit organization that promotes mathematics education for African-American children and scientific opportunities for blacks.

Almost a century after Banneker's scientific achievements, Edward A. Bouchet was the first African American to earn a science doctorate and, another century later, David H. Blackwell became the first African-American member of the National Academy of Sciences. "The most sensible of those who make scientific researches, is he who believes himself the farthest from the goal, &...studies as if he knew nothing and marches as if he were only yet beginning to make his first advance," wrote Banneker in 1795, foreshadowing his scientific legacy.

ELIZABETH D. SCHAFER

Further Reading

Bedini, Silvio A. *The Life of Benjamin Banneker: The First African-American Man of Science*. Baltimore: Maryland Historical Society, 1999.

Haber, Louis. *Black Pioneers of Science and Invention*. San Diego: Harcourt Brace Jovanovich, 1991.

James, Portia P. *The Real McCoy: African American Invention and Innovation, 1619-1930*. Washington, D.C.: Smithsonian Institute Press, 1989.

Jordan, Winthrop D. *The White Man's Burden: Historical Origins of Racism in the United States*. New York: Oxford University Press, 1974.

Kaplan, Sidney, and Emma Nogrady Kaplan. *The Black Presence in the Era of the American Revolution*. Amherst: University of Massachusetts Press, 1989.

Russell, Dick. *Black Genius and the American Experience*. Foreword by Alvin F. Poussaint. New York: Carroll & Graf Publishers, 1998.

Stearns, Raymond Phineas. *Science in the British Colonies of America*. Urbana: University of Illinois Press, 1970.

Webster, Raymond B. *African American Firsts in Science and Technology*. Foreword by Wesley L. Harris. Detroit, MI: Gale Group, 1999.

The Emergence of Swedish Chemists during the Eighteenth Century

Overview

Despite Sweden being one of the largest countries in Europe, extending northward into the Arctic Circle and for a time including most of Finland, its achievements in science somehow escape attention. Yet from the beginning of the eighteenth century, the belief that scientific methods could help to improve mining, metallurgy, and agriculture was a spur to the development of Swedish chemistry. A substantial innovation came in the form of the blowpipe, which was introduced in 1738 and made it possible to increase the heat of a flame. Applying the heated flame to a mineral revealed information about the mineral's nature and composition. Swedish chemists also concentrated their efforts on pharmacy and the applications of chemistry to medicine.

Background

Georg Brandt (1694-1768) inherited his interest in chemistry and metallurgy from his father, a mine owner and former pharmacist. Brandt began his studies at Uppsala University. He went abroad to study chemistry and medicine, and on his way back to Sweden trained in the Hartz mountains in mining and smelting. On his arrival home he was made director of the chemical laboratory of the Council of Mines. Brandt's professional life was multifaceted: in addition to being an able administrator and teacher, he was also an expert chemical experimenter. He was the first to establish the metallic nature of arsenic, but his discovery of cobalt is the work for which he is best known. In 1751 he proved that

the brittleness of iron when hot is due to its sulfur content.

Johan Gottschalk Wallerius (1709-1785) was instrumental in opening new avenues for the practical application of chemistry. Born into a family of clergymen, Wallerius studied astronomy and mathematics before turning to medicine. He was employed on the medical faculty first at the University of Lund (where he had only three students) and then at the University of Uppsala. His lectures at Uppsala covered medicine, physiology, and materia medica and were attended by students of both medicine and mining science. In addition to his stated duties, Wallerius set up a private chemical laboratory so he could do experiments and lecture in chemistry and mineralogy. In 1750 he was appointed to a professorship in chemistry, metallurgy, and pharmacy at the University of Uppsala, the first of its kind in Sweden. With this new respectability, he succeeded in getting the university to fund a chemical laboratory, which was completed in 1754. A book he published in 1761 on the chemical foundations of agriculture was the most widely used textbook on the subject. Failing health forced his retirement, in 1767, to his farm south of Uppsala, where he continued his scientific writing.

The leading eighteenth-century Swedish chemist was Torbern Bergman (1735-1784), who achieved an international reputation for his skill as a chemical analyst and his studies of attraction between chemicals. When Bergman's health began to fail in 1770, he was advised to drink foreign mineral waters. But these waters were expensive and often arrived in Sweden in poor condition. Bergman decided to study the composition of the waters with the aim of preparing similar mixtures locally. Although his results were not very accurate, his guiding principles were sound and led to the law of equivalent proportions formulated by German chemist J. B. Richter (b. 1762) in the 1790s. Another problem that interested Bergman was how one element in a compound could be displaced by another. He reasoned that such changes must be due to a difference in the attractive forces between the elements (he called these attractive forces "affinities"). His work in this area was an important step in understanding how chemicals combine.

In 1770, Bergman met Karl Wilhelm Scheele (1742-1786), an apothecary's assistant who impressed Bergman with his knowledge of chemistry. Bergman suggested that the younger man study the properties of a recently discovered ore called pyrolusite. Scheele's investigations led to the discovery of chlorine, which was announced in 1773 and revolutionized the textile bleaching process in early industrial England. On heating pyrolusite, Scheele found that it gave off a green choking gas that attacked virtually any matter. Scheele also made observations regarding the formation of soda that suggested new ways of making soda for glass or soap manufacture. Between 1770 and 1773, Scheele discovered what he called fire air, which we know as oxygen, and which he described as colorless, odorless, and tasteless. Although his discovery was significant, his adherence to the phlogiston theory prevented him from understanding the role of oxygen in combustion, later clarified by Antoine Laurent Lavoisier (1743-1794) in France. Scheele questioned the theory, but he did not abandon it. Oxygen theory was not published in Sweden until 1778.

Johann Gottlieb Gahn (1745-1818) studied physics and chemistry at Uppsala and became Torbern Bergman's laboratory assistant there in 1767. Gahn later took up a position at the College of Mining, where he worked on improving the processes used in copper smelting at a local mine. Gahn carried out his experiments in a laboratory that he installed in his own garden and paid for himself. Gahn published almost nothing, but knowledge of his expertise traveled far, and he was much in demand. Gahn collaborated with both Scheele and Bergman, and he was responsible for introducing Scheele to Bergman.

Bergman was generous in crediting Gahn with extracting manganese from pyrolusite in 1774. Bergman had doubted that the mineral contained any metal. Although Scheele reigned supreme as a chemical experimenter, no one surpassed Gahn in blowpipe analysis. Gahn's blowpipe experiments with inorganic substances in animal bones led later to Scheele's method of obtaining phosphorus from animal bones.

Peter Jacob Hjelm (1746-1813) was educated at the University of Uppsala and in 1794 was appointed chemical director of the Swedish bureau of mines. Following work laid earlier by Scheele, in 1782, Hjelm heated a paste prepared from molybdic oxide and linseed oil at high temperatures in a crucible and became the first person to produce pure metallic molybdenum. The first practical application of the metal, however, would await World War I, when a shortage of tungsten provoked its use in the manufacture of arms.

Impact

Science in Sweden both represents some of the most important advances in the field and serves as an example of how ideas develop under unusual conditions. Although a Catholic country during the Middle Ages, Sweden emerged from the Protestant Reformation firmly committed to Lutheranism. This unity in religion was a major influence on the development of science in Sweden.

The backdrop for scientific exploration in Sweden was the University of Uppsala, one of the oldest universities in Scandinavia. By 1620 it had become a major seat of learning, with professors imported from Germany. Progress in science, however, was inhibited by a stubborn adherence to the Aristotelian view of the universe, which centered on the idea of an immobile Earth. As more and more Aristotelians came to fill chairs at the university, new thinking was stifled. Moreover, an alliance formed between Aristotelianism and the Lutheran faith, which further cemented the influence of established views.

The teachings of the French philosopher René Descartes (1596-1650), though a late arrival to Sweden, opened the way for new scientific ideas from the south, and along with them the introduction of scientific instruments such as the air pump, the thermometer, and the barometer. The end of absolute monarchy in 1718 likewise signaled an age of democracy that fostered the rise of science. Because the policies of the

University of Uppsala were subject to government interference, the university was required to provide training in physics and chemistry for civil service degrees. As a result, the first university appointments in science were established.

Further evidence of a changing climate in Sweden was the formation of other centers of science, for example, the Royal Society of Science at Uppsala, first convened in 1710 as a "College of the Curious" in response to the plague. Among the luminaries who served as secretaries were Anders Celsius (1701-1744) and Carolus Linnaeus (1707-1778). In 1739, a national academy of science was established to promote the "useful sciences" that came to include all the well-known names in Swedish science, for example, Emanuel Swedenborg (1866-1772), Cronstedt, Bergman, and Scheele. Today the Royal Swedish Academy of Sciences awards and administers the Nobel Prizes in chemistry and physics.

The importance of Swedish chemistry in the eighteenth century is undisputed, and remarkable considering that modern science was not accepted in the country until almost the end of the 1700s. Forty percent of the chemical elements found since the Middle Ages were discovered in Sweden, among them cobalt, nickel, oxygen, chlorine, lithium, and manganese. Discoveries were also made of nonelementary substances such as formic acid, a substance found in the bodies of red ants and used in dyeing textiles, and the chemical nature of many minerals was elucidated.

Swedish investigations into chemistry were as hazardous to their practitioners as they were impressive: the process of determining the properties of unknown substances required chemists to touch, breathe in, and taste a wealth of toxic materials. Bergman and Scheele both died young. Central to Sweden's success in chemistry were its vast mineral resources—a virtually unlimited store of materials to investigate—an emphasis on careful measurement, and the innovation of the blowpipe for use in experiments.

GISELLE WEISS

Further Reading

Books

Goodman, David C. and Colin A. Russell. *The Rise of Scientific Europe, 1500-1800*. Sevenoaks, Kent, UK: Open University Press, 1991.

Lindroth, Sten. *Swedish Men of Science, 1650-1950*. Stockholm: The Swedish Institute, 1952.

The Rise and Fall of the Phlogiston Theory of Fire

Overview

At the beginning of the eighteenth century the Phlogiston theory of fire dominated. By the end of the eighteenth century, however, the Phlogiston theory had been overturned by the new concept of the combustion of oxygen. The overthrow of the Phlogiston theory of fire is often presented as a shining example of the triumph of good science over bad, yet the saga is one of many false starts, false experiments, and false assumptions. Personalities, social and cultural influences, and the new emphasis on experimental analysis and natural causes combined to challenge and replace the Phlogiston theory.

Background

The Greek philosophers considered fire to be one of the basic elements of nature, offering a number of different interpretations. Heraclitus of Ephesus (about 535-475 B.C.) made fire the universal force of creation. Aristotle (384-322 B.C.) called fire one of the great principles of all things. Plato (427-347 B.C.), Aristotle's teacher, suggested that burnable objects contained within them some inflammable principle, a substance that made them burn, but it was Aristotle's ideas that dominated medieval European thought.

Aristotle's fire was part of a four-element system consisting of air, earth, fire, and water. A substance such as wood was made up of a combination of the four elements. When it burned the flame was the element of fire escaping, any vapor was air, any moisture water, and the ash that remained was earth.

The sixteenth century Renaissance rediscovered the works of Plato, as part of a wider intellectual movement of rediscovering the classical past. Plato's notion of a burnable principle within in a substance fit well with the alchemical ideas of the period. Plato's concept was modified and alchemists came to regard sulphur, or "some vague spirit of sulphur," as the inflammable principle. Sulphur burned almost completely, therefore sulphur was seen as fire itself, or something closely related to fire. A new system of elements was constructed, with substances explained by a combination of sulphur, mercury and salt. So wood burned because it contained sulphur, gave off flame because it contained mercury, and left ash because it contained salt.

In the mid-seventeenth century the observations, experiments, and philosophy of Johann Joachim Becher (1635-1682) and his pupil Georg Ernst Stahl (1660-1734) led them to suggest a new interpretation of sulphur. They proposed that sulphur was actually made from a combination of sulphuric acid plus a new substance they called phlogiston. Phlogiston (pronounced FLO-jis-ton) was actually the principle of fire, not sulphur, and Stahl suggested that phlogiston was released by all substances when they burned. Hence, as wood burns it releases phlogiston into the air, leaving ash behind. Ash was therefore wood minus phlogiston. Sulphur and materials like charcoal and fat burned well because they contained a great deal of phlogiston.

Impact

The phlogiston theory quickly became popular, and was very robust, explaining a wide variety of phenomena. It explained the rusting of metals. As the metal rusted, it gave off phlogiston into the air, so a metal was a combination of its rust and phlogiston. The breathing of animals could also be explained. As food was 'burned' inside the body, phlogiston was released and expelled out of the body by the lungs. Phlogiston was the "motive power of fire," the foundation of color, the principle of inflammability, indestructible, and an "extremely subtle matter." It could easily be used to explain observed results in experiments. For example, experiments showed that if you burned a stick of wood in a confined space, such as a jar, after a short time the combustion would stop. This was explained by suggesting that air could only contain a certain amount of phlogiston, and once it reached its limit then no more combustion could take place.

The theory of phlogiston was very successful, and was so broad in its scope and acceptance it became one of the first unifying hypotheses of the chemical sciences. However, sci-

entists began to have problems explaining some new experimental results. One reason was that the theory tried to explain too many things. The more the theory was modified by its supporters to explain one particular observed behavior, the more difficulty they had explaining others.

The whole method of inquiry into nature was changing. The reliance on the past was shattered by new discoveries and inventions. Challenges to ancient science occurred at the same time as challenges were presented to traditional religion, economics, social structures, and governments. The eighteenth century was a period of revolutions, including the American Revolution, the French Revolution, and between these a revolution in the chemical sciences.

As the theory of phlogiston developed, the nature and properties of the mysterious substance began to be described in different ways. Whereas Stahl had considered phlogiston as a vague principle, the followers of his theory began to assign physical properties such as weight to phlogiston. At first, this seemed only to strengthen the logic of the theory. When wood burns it leaves a lighter substance, ash, behind. Therefore, the missing weight is the escaped phlogiston. When a metal such as iron rusts, the rust appears lighter, so once again the missing weight was the escaped phlogiston.

However, careful experimenters noticed that while the rust of metals appeared lighter, or at least less dense, than the metal it had come from, in fact the rust weighed more. This resulted in more tinkering with the theory. Some supporters suggested phlogiston had a negative weight, and so when it left a substance it made the result heavier. The phlogiston theory began to become unwieldy and overly complicated. Explanations of its properties began to be contradictory. To explain certain properties, sometimes it had to have no weight, sometimes positive weight, and sometimes negative.

Further problems for the phlogiston theory resulted from new experiments and research conducted into gases. An international group of experimenters began working on gases, exchanging research, and publishing and translating experimental results, each bringing their own perspective and assumptions to the results they observed.

In England during the 1770s Joseph Preistley (1733-1804) was a dedicated supporter of phlogiston, but he was also a careful experimenter. He isolated a new gas by heating the rust of mercury.

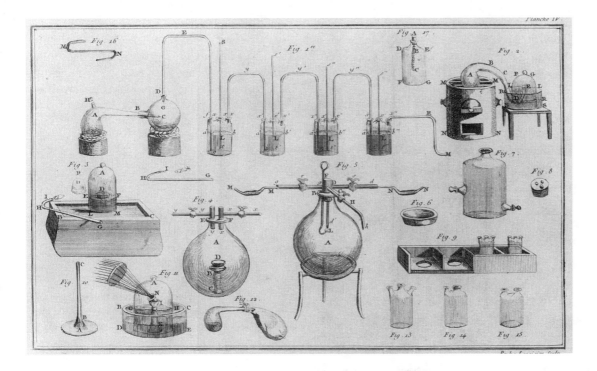

Instruments from Antoine Lavoisier's laboratory.*(Library of Congress. Reproduced with permission.)*

When heated the rust gave off the new gas, and left behind the metal mercury. This new gas made things burn brighter and longer than normal air. Mice sealed in jars of this new gas could breathe for longer than in normal air. Preistley sought an explanation that would remain consistent with the phlogiston theory, so he speculated that this new gas was particularly good at absorbing phlogiston. Ordinary air, he suggested, already contained some phlogiston, and so could quickly be filled up with more phlogiston, making combustion, rusting, and breathing impossible. This new air, which Priestley called dephlogisticated air, was completely free of phlogiston, so it took much longer to fill up.

In France Antoine Lavoisier (1743-1794) performed similar experiments with the same substances. He got the same results as Priestley, but he was seeking a new explanation of combustion, so he saw his results from a different perspective. Lavoisier suggested that rather than phlogiston being given off when a metal rusted, or a substance burned, a more simple explanation was that Priestley's new gas, which he called oxygen, was being absorbed from the air.

While both theories explained the observed results well Lavoisier's explanation had one major advantage over Priestley's, it gave a mechanism for the gain in weight of rusts. The rust of a metal was the metal combined with oxygen, producing a heavier substance called an oxide. This was a revolutionary approach to the problem, breaking with the previous traditions that stretched back to Plato. While common sense suggested burning or rusting an object results in something escaping, Lavoisier's careful experimental analysis showed that in fact oxygen was being absorbed.

However, Lavoisier could not explain the nature of heat and fire, and was forced to invent a strange new substance, which he called caloric. Caloric had a number of similarities to phlogiston in that it was a principle of fire, just as sulphur and phlogiston had been previously considered.

Further experimental work with other metals, their rusts, and other new gases slowly began to develop a more coherent picture of what occurred during rusting and burning. Another breakthrough came with the realization that water was the combination of the gases hydrogen and oxygen. If you burn hydrogen, it produces water. Lavoisier's theory gained support as more and more experiments gave favorable results.

Lavoisier's main opponent, Priestley, outlived him, but was not able to overturn the trend to the 'new chemistry' of Lavoisier. Priestley's last book, published in 1796, still strongly supported the Phlogiston Theory, but did contain a note of surrender to the prevailing opinions of others. He wrote, "There have been few, if any, revolutions in science so great, so sudden, and so general, as the prevalence of what is now usually termed the new system of chemistry, or that of the Antiphlogistons, over the doctrine of Stahl, which was at one time thought to have been the greatest discovery that had ever been made in the science."

While many historians have characterized Priestly as a stubborn, foolish defender of an outdated theory, the acceptance of Lavoisier's ideas in such a short time is more surprising. Critics rightly pointed out that Lavoisier's theory was incomplete, and could not explain all observed results. However, over time the theory grew stronger and more complete, without losing its simplicity. Some accused him of merely substituting Stahl's phlogiston with his own caloric, a substance at least as mysterious. But caloric was not central to Lavoisier's ideas.

The new theory of combustion had several key points in its favor. It was simple, consistent, did not invoke negative weights or other seemingly arcane concepts, and was based firmly on experimental analysis. There remained a few supporters of phlogiston here and there, but the evidence for Lavoisier's theory kept mounting. However, it was not until the twentieth century that the last legacy of phlogiston, Lavoisier's caloric, was explained away. Heat was revealed to be a form of energy, and the mysterious and mythical ideas of caloric and phlogiston were no longer necessary.

DAVID TULLOCH

Further Reading

Books

Conant, James Bryant. *The Overthrow of the Phlogiston Theory-The Chemical Revolution of 1775-1789.* Cambridge, MA: Harvard University Press, 1956.

Lavoisier, Antoine. *Essays Physical and Chemical.* Thomas Henry, trans. 2nd edition. London: Cass, 1970.

White, John Henry. *The History of The Phlogiston Theory.* London: E. Arnold, 1932, reprinted by AMS Press (New York), 1973.

Internet Sites

Selected Classic Papers from the History of Chemistry. http://maple.lemoyne.edu/~giunta/papers.html. Includes several papers by Lavoisier.

Geology and Chemistry
Emerge as Distinct Disciplines

Overview

By the beginning of the eighteenth century, the pursuit of scientific knowledge was becoming a full-time occupation for some. As knowledge in many areas increased, early scientists began to specialize, and natural philosophy began to fragment into various scientific disciplines. Among the first sciences to which this happened were geology and chemistry, two studies with a surprising amount of common ground. This increasing specialization allowed individuals to devote themselves to a particular area of study, thus making more significant progress by focusing their efforts, and has resulted in today's plethora of scientific disciplines.

Background

Science, as such, is a relatively recent phenomenon in human history. In spite of the advances of the ancient Greeks, Sumerians, Maya, and others, the hallmarks of science were largely absent until the Renaissance. Even then, though, science was descriptive, consisting of observations about the world and trying to fit them into some sort of pattern. However, one of the triumphs of what we call science is its ability to explain phenomena based on a set of consistent rules and, also based on these rules, to predict future events. An example of this can be found in predicting lunar eclipses. The Maya are thought to have been able to predict eclipses, and astronomers from the sixteenth century certainly could. However, these predictions were based on observations and not on any understanding of the causes; it was not until Isaac Newton published his laws of motion and gravity that the reason was known. After that point, science could describe both the why and the when of eclipses, giving a much deeper understanding of the events.

A good argument can be made that Isaac Newton (1642-1727) was the world's first true scientist in the modern sense. In physics, mathematics, and chemistry he set about to try to develop simple, consistent rules that explained why the world around him worked in the manner observed. He met with notable success in physics and mathematics, but his explorations in chemistry were less impressive. Nonetheless, his work in all three fields helped to pave the way for other scientists who were to follow.

During the eighteenth century, mining became increasingly important to many nations, including Britain and France. Because of this, there came a breed of men (for early scientists were almost exclusively men) who began to specialize in geology, the study of the Earth. Initially, their chief interests lay in trying to predict where mineral ores, coal, and other important rock and mineral deposits could be found, and this led to attempts to understand the way in which rocks lay beneath the ground and even under water. In addition, as the early geologists studied these rocks, they began to wonder how the rocks they saw could be made to conform to their biblical cosmology.

At the same time, miners were using evermore sophisticated techniques to determine the relative value of different ore deposits. By treating rocks and minerals with a number of different solutions and observing the effects, they began to describe rocks in terms of a few distinct groups. Further, they could begin to determine rough metal content of various ores, the beginnings of analytical chemistry. These techniques were borrowed by less practical chemists and refined, serving both miners and academics well over time. By the latter part of the century, Antoine Lavoisier (1743-1794) had developed a theory of chemistry that required accounting for the masses of all reactants, helping to place chemistry in a more mathematically formal setting and making it more of a predictive science.

As the eighteenth century drew to a close, both geology and chemistry had emerged as two legitimate and separate sciences, each worthy of full-time pursuit by an increasing number of practitioners. In following decades and centuries, these were but the first of what is now a bewildering array of specialties and sub-specialties in the ever-growing structure of modern science.

Impact

The most immediate impact of this development was, of course, on the scientists themselves and on the science they practiced. In addition, miners benefited from this increase in knowledge specific to their needs, as did the markets served by their efforts. Society itself was impacted, too; as people began to consider science a legitimate profession, their respect its practitioners grew even as their ability to understand their results diminished. Finally, and perhaps most important, the recognition of these two sciences as distinct disciplines was to set the stage for the continuing subdivision of science into a bevy of increasingly focused fields.

As noted above, scientists were among the first to notice the effects of specialization. Until the time of Newton, and even somewhat beyond, a single person could make significant contributions to knowledge in a number of fields. Our knowledge of the world was sufficiently small that a single person could understand almost everything that was known in the world of the seventeenth century. As our level and depth of understanding grew, however, this changed, and it became more and more difficult for a single person to grasp the subtleties across the entirety of science. Of necessity, people began to specialize, concentrating their efforts on one topic or another.

With this concentration came the realization that any one of a number of areas held enough unanswered questions to keep many scientists busy studying them for a lifetime. Understanding this, and making the conscious decision to devote one's professional life to the study of geology, made it possible for a single person to learn an incredible amount about the Earth, albeit at the expense of a broad appreciation of the whole of science. However, those who did specialize quickly found they could make several lifetimes worth of progress by devoting all their efforts in a single direction. In addition, through specialization, there was now room for a larger number of scientists, and the rate of learning increased dramatically. This spread to other specialties with time, resulting in the creation of a large number of scientific fields.

In addition to the general gains made by scientists, there were very specific impacts on miners as well. The more rapid increase in geologic knowledge helped provide greater understanding of the phenomena controlling mineral deposits. This, in turn, made it somewhat easier to predict in advance where minerals might be found, rather than digging fruitless holes or missing rich deposits that did not manifest themselves at the Earth's surface. In conjunction with developments in economic geology, the rapid growth of knowledge in chemistry also began to provide more accurate tests for ore materials. While mining is still far from an exact science, it is now far more exact than it had been, and this increase in precision is allowing the recovery of ores that would previously have been overlooked.

From a societal standpoint, the average "man in the street" was not terribly aware of most scientific advances and, indeed, even today the degree of scientific illiteracy is alarming. However, as the numbers of scientists increased and their discoveries became more common, the number of people aware of their activities increased, too. The growth of the middle and leisure classes in Britain and France also led an increasing number of people to become aware of and interested in the pursuit of science. This led to a relatively large number of "gentlemen scientists" and scientifically literate clergymen, especially in nineteenth-century Britain, many of whom made significant contributions to their chosen science.

Finally, as mentioned briefly above, geology and chemistry were but the first of what is now a bewildering number of scientific specialties. In just the field of geology, for example, there are now specialties in petrology (the study of rocks), economic geology, sedimentology, stratigraphy, hydrogeology, geophysics, geochemistry, isotope geology, paleontology, glacial geology, tectonics, geomorphology, mineralogy, environmental geology, geomagnetism, petroleum geology, marine geology, and more. People spend their entire working lives delving into the intricacies of a single point in space or time, so a person might work for 30 years or more to fully describe and understand the micro-fossils found in a single rock formation. This same degree of specialization can be found in many other fields of study because, as we discover more and more about any single area, we also find that there are an increasing number of questions that remain to be answered. From this perspective, it is apparent that the emergence of geology and chemistry as independent disciplines was just the first step in this process of increasing fragmentation and specialization that continues in science today.

There is a down side to this. Although researchers are making discoveries in their specialties at an ever-increasing pace, science has become fragmented, and it is not uncommon for two scientists in separate academic disciplines to be working on almost identical problems, publishing their results in different journals, and to remain completely unaware of the other's work. It is also common to see a scientist struggling to develop the intellectual or experimental tools to attack a problem, unaware that these techniques already exist in another, related field. For these reasons, there has recently been an increasing interest in promoting interdisciplinary research in which specialists from a number of fields all collaborate to solve interesting or important problems. In this way, by aggregating specialists, we are in a sense on the road back to reassembling the sciences from their current fragmented state.

P. ANDREW KARAM

Further Reading
Books

Gohau, Gabriel. *A History of Geology*. Rutgers University Press, 1990.

Oldroyd, David. *Thinking about the Earth: A History of Ideas in Geology*. Cambridge: Harvard University Press, 1996.

Rudwick, Martin. *The Great Devonian Controversy*. University of Chicago Press, 1985.

Johann Gottlob Lehmann Advances the Understanding of Rock Formations

Overview

Just as a book can be called the fundamental unit of a library, so can rock formations be thought of as the fundamental units of historical geology. However, until the eighteenth century, few people really noticed that rocks come in discrete "packages," and nobody thought to try to use these formations to reconstruct the geologic history of a region until Johann Lehmann (1719-1776). Although Lehmann's interpretation of the rocks has since fallen out of favor, his insights have proven fruitful and, over two centuries later, the concept of rock formations continues to be used in much the same manner and for much the same purpose as originally proposed.

Background

The first geologists are lost in the depths of time. They were our most distant ancestors who noticed that flint and obsidian made better knives and scrapers than sandstone or shale. Combing their surroundings, they found that some locations were better suited for finding these rocks, and they returned there repeatedly for their raw materials. Lacking any idea as to why one place was better than another, they could not predict in advance which locations would prove rich or poor; they could simply look, find, and remember. Possessing only descriptive power, however, geology was not yet a science. It was simply a collection of observations about rocks.

One thing they undoubtedly noticed was that like stones tend to congregate, often in beds that form layers stacked one atop the other. Sometimes flat, sometimes folded, these rocks formed the foundation of the world, underlying mountains and plains, hills and valleys. As time passed and people began digging into the earth to bring up valuable rocks, they also must have noticed that similar rocks tend to congregate even underground. The same patterns and associations a person sees above ground continues in the sub-surface. However, this remained an observation only (albeit a valuable one) for many more years, still lacking predictive or explanatory power.

One of the hallmarks of any science is its ability to explain what we see in the world around us in such a way that a scientist can accurately predict what he or she will find elsewhere, in a yet-unseen part of the world. The early "geologists" at this point still only observed. Using their observations to explain and predict properties of the Earth still lay far in the future.

One of the first steps towards making geology a science came with Nicolaus Steno's (1638-1686) recognition of fossils as the remains of once-living organisms and, more importantly, his use of fossils to help understand the relative positions of various beds of rock. These breakthroughs took place in the late seventeenth century. Less than a century later came another breakthrough: Lehmann's recognition that certain groups of rocks tended to be associated with each other, often over large distances. For example, a certain red sandstone might lie atop a distinctive limestone and beneath a particular green shale, and these rocks might also appear in another location a great distance away. Such rock formations are chapters in the history of the Earth, and reading them provides great insights into the geological history of particular regions or of the entire Earth.

In the middle of the eighteenth century, Lehmann published some interesting work. First, he identified three separate types of mountains—primary, secondary, and tertiary—each with a distinct set of geological characteristics and characteristic rock types. He then extended this by using these characteristics to try to determine the conditions under which the mountains formed, and he developed a set of drawings to try to show how these formations continued underground. Finally, he suggested that these rock formations and the rules he had developed to describe their characteristics could be extended throughout the world instead of being a purely local phenomenon. Although these observations and theories were designed as a way to help the mining and mineral industries, it is also significant that Lehmann regarded himself as a historian of the Earth, and was apparently the first person to try to use the rock record in this fashion. In spite of his attempts to place all that he saw in a biblical framework, his work helped introduce concepts that continue to be useful to this day.

Impact

Lehmann's findings resonated on a number of levels and were to influence geology and society for many years.

1. By introducing the concept of a rock formation, he helped give future geologists a fairly powerful tool with which to develop their science and their understanding of the Earth and its structures.

2. In addition, this was one of the first steps taken towards making geology a legitimate science because he used his rock formations to predict not only what the rocks looked like beneath the Earth's surface, but also to suggest how they might be arranged in other parts of the world. Even though many of his interpretations of the rock record were later shown to be wrong, many of the tools he developed are still in use.

3. Finally, although Lehmann felt strongly that the biblical account of the Earth's history was correct, his methods were later used by others to help show that the Earth is far more ancient.

As mentioned in the overview, the rock formation can be considered a fundamental unit of geological interpretation. By looking at a formation, and at the relationship between a series of rock formations, geologists can interpret the geological history of a region. Piecing together the histories of many regions, they can extend this understanding across a continent, between continents, and throughout the world. This process of understanding the Earth's history is still taking place. Geologists have roughed out the basics, but they are still mapping, interpreting, and puzzling out these histories. In this quest, they have developed any number of concepts ranging from mundane to revolutionary. As one example, the theory of plate tectonics was bolstered by noting the existence of nearly identical rock formations on both sides of the Atlantic Ocean and other identical formations shared by Africa and South America, Australia and Antarctica, and so forth. Without the basic concept of what a geological formation was, this process would have been much more difficult to follow. On a smaller scale, economic geologists use the relationships between various rock formations to help predict where to best drill for oil, dig for gold, or search for other economically valuable minerals that are necessary for our industrial society. In fact, the Industrial Revolution was powered in part by coal found in this manner, by tracing coal-bearing formations across England and Continental Europe.

In doing this, the early geologists were beginning to turn geology from a purely descriptive pursuit into a predictive and explanatory science. By predicting where coal could be found, for example, they helped save time, money, and lives; just as modern economic geologists help mining and petroleum companies better determine where to look for minerals and oil. This helps to save the companies money, but also helps reduce costs to manufacturers and consumers because raw materials can be sold for less. All of society benefits.

In addition, Lehmann consciously strove to use his rock formations to determine the history of the Earth and to describe why it looks the way it does. Using rock formations in this manner, he used geology to explain, in a relatively self-consistent manner, the observed physical features of the Earth. As mentioned above, his interpretations are now known to have been incorrect, but it was a decent first step. In fact, geology was to become one of the first recognized sciences in its own right towards the end of the eighteenth century, based partly on these newly developed abilities.

Lehmann's observations were later extended impressively to help unravel the geologic history of England during the nineteenth century. Most of the currently recognized geologic eras, based initially on the existence of similar rock formations with similar assemblages of fossils, was developed at this time, with radiological dating later used to determine the ages of the rocks. It is likely that Lehmann would have been pleased to discover that, although individual rock formations are usually not as wide-spread as he had envisioned, there is enough overlap of these formations to allow the tracing of geologic periods across the planet.

Finally, as knowledge of geology grew, it became obvious that the Earth had to be ancient. Finding the same rock formations on either side of the Atlantic Ocean, for example, implied that either the Atlantic Ocean had opened up, rifting Europe and the Americas apart, or that there was once a continent there that had eroded away, leaving identical rocks on either side of the ocean. Both of these scenarios required the Earth to be quite old because of the vast amounts of time required for either event. Similarly, finding formations of marine rocks high in the Alps suggested that, somehow, those rocks had either been covered with water that subsequently drained off the land or that they had been lifted into place at some time in the past. Again, either scenario required a lot of time. Ironically, a tool first used to help set rocks into a biblical context was used to strike some of the first blows in favor of an old, nonbiblical Earth forced into its current shape by huge, but decidedly, natural forces. This conception of the Earth's antiquity made its way back into the public consciousness, and most of the educated public gradually came to accept that the Old Testament could be interpreted as allegory as well as fact. This, in turn, sparked a debate that continues to this day in schools and courts.

P. ANDREW KARAM

Further Reading
Books

Gohau, Gabriel. *A History of Geology.* Rutgers University Press, 1990.

Oldroyd, David. *Thinking about the Earth: A History of Ideas in Geology.* Cambridge: Harvard University Press, 1996.

Rudwick, Martin. *The Great Devonian Controversy.* Chicago: University of Chicago Press, 1985.

Abraham Gottlob Werner's Neptunist Stratigraphy: An Incorrect Theory Advances the Geological Sciences

Overview

Abraham Gottlob Werner (1749-1817) is often remembered as the mistaken champion of a false theory about the structure of Earth's crust. However, his water-based hypothesis of the formation of rock strata was more than a wrong idea. Werner's theory was the first well-ordered geological description of the strata of Earth based on physical evidence that accounted for Earth's history. While his ideas were eventually overturned, often by his own students, he established a new way of thinking about the formation of Earth based on observation, experiment, and an attempt to understand historical geological processes.

Background

Earth's strata are the layers of various rock and mineral deposits that exist in a cross-section of Earth's crust. Strata look like the layers of a cut onion, or the side of a hamburger when you take a bite out of it. These layers are sometimes revealed by erosion, in the case of the Grand Canyon, or by landslides and man-made excavations.

The idea that water was the main force in the creation of Earth's surface dates back to at least the tenth-century Arab philosophers. In medieval Europe water-based theories became popular, as they explained the biblical flood of Noah. Such theories often received the blessing of the church, whereas other ideas could be condemned. The eighteenth century saw a new wave of water-based, or "neptunist," theories (after Neptune the ancient god of water). Many such theories took the Bible as their starting point, while others were based on abstract philosophical ideas or random notions. However, Werner's neptunist theory was an interpretation of the physical evidence that described the formation of Earth's crust as a long historical process.

Werner's family had a long association with iron-working, and his father was the inspector of the Duke of Solm's ironworks. Werner seemed destined to follow his father's career. However, while gaining the required legal education at the University of Leipzig, he became sidetracked by mineralogy. Werner abandoned his law degree in 1774, but by then he had published his first book, *Von den äusserlichen Kennzeichen der Fossilien,* which was a simple and orderly mineral identification manual. On the strength of this book Werner secured a teaching position at the Mining Academy at Freidburg in Saxony. He remained there for the rest of his life, developing and teaching his theory of the origin of the strata of Earth.

Werner's theory was not completely original. He combined several popular ideas, in particular the work of Johann Gottlob Lehmann (1719-1767) and Georg Christian Füchsel (1722-1773). Lehmann had stressed the importance of the order of rock strata and used the practical knowledge of quarrymen and miners to formulate his theories. Füchsel suggested that strata were formed successively over time.

Werner's theory added the new dimension of a historical explanation for the observable strata. Firmly basing his theory on the geological knowledge of his day, he proposed a global scheme that accounted for the origin and distribution of the entire Earth's surface. Werner proposed that the rock layers of Earth had been laid down at different times by an all-encompassing, universal ocean. The rocks that made up the crust of Earth had been formed from particles that settled out of the murky universal ocean, as precipitates or sediments. The differences in rock type and layering visible in the world were explained by rises and falls in the level of the universal ocean, as well as the turbulence or calmness of the waters.

Werner's first version of his theory broke down history into four periods: the primitive, floetz, volcanic, and alluvial, each with specific conditions to explain the rock types formed. In the primitive period, for example, the universal ocean was deep and calm, producing a solidly packed, smoothly distributed granite layer. Later the waters became more turbulent, so later rocks were less smoothly distributed. No life was supposed to have existed in the primitive period, so granite rocks must be free of fossils.

Impact

Werner's ideas had immediate and wide appeal. The theory was a simple yet elegant explanation of the evidence, unlike many rival theories that seemed to disregard observations. The theory appeared complete, and a number of very successful predictions were made from it. Unlike theories that merely described a localized region, Werner's concept explained the strata of the whole Earth. At the same time, the theory was flexible enough to account for local variations. With additional after-effects, such as cave-ins and erosion, the theory could explain virtually all the geological phenomena observable in Werner's time.

The theory was also flexible in the sense that Werner was prepared to modify the exact details to accommodate new evidence. Without abandoning the fundamentals, Werner and his pupils explained away seemingly contrary data with new twists to the basic concept. For example, Werner added a new age, the transitional period, to explain the presence of fossils in some granite deposits. Instead of threatening the theory, contrary observations were used to help refine the overall concept.

Unlike many earlier theorists on the formation of Earth, Werner did not feel compelled to make his theory fit the biblical creation. His ideas implied that Earth was over a million years old, to allow time for the formation of rocks by the slow processes of sedimentation and precipitation from the universal ocean. This time span was far greater than the calculations of biblical scholars, who suggested about 6,000 years. Yet Werner was not attacked by church authorities, as his concept of a universal ocean was taken by many as supporting evidence of the biblical flood. Indeed, Werner's theory was championed by many religious figures long after it had ceased to be used by geologists.

Werner wrote down few of his ideas and published even fewer. His most influential work, *Kurze Klassifikation und Beschreibung der verschiedenen Gebirgsarten,* was finished in 1777 but not published until 1786, and it was only 28 pages long. He never published a complete version of his theory, and it was only through unpublished documents and the lecture notes of his students that a full theory was reconstructed. Werner's lecturing was enthusiastic and engaging. He had a personal magnetism that attracted students from all over Europe and beyond. His teaching produced large numbers of eager disciples who spread the gospel of Wernerian neptunism.

Despite the popularity of Werner and his ideas, there was no shortage of critics. The perceptive Italian Scipione Breislak (1750-1826)

asked where all the water had gone. Werner suggested outer space but gave no reason. The theory was also criticized for not explaining the formation of the entire Earth, only the crust.

There was opposition to neptunism in general. Many Italian and French scientists were convinced that volcanic action, not water, was responsible for rock formation. A division developed between vulcanists—named after Vulcan, the ancient god of volcanoes—and neptunists that was partly to do with the geography native to each scientist. Particularly active volcanic regions, such as the Italian peninsula, tended to produce vulcanists, whereas in England and the German states local geology produced supporters of neptunism.

In his *Kurze Klassifikation* booklet Werner declared all basalt to be of watery origin, sparking the great basalt controversy. Basalt is a common and widely distributed rock type, so it was keenly debated whether fire or water was the key creative agent. Werner regarded volcanoes as recent events, giving them no role in the historical formation of Earth's deeper strata. The local strata of Saxony seemed to support his ideas, or at least not contradict them. Werner's poor health meant he could not travel, and so the vast majority of his supporting evidence came from his immediate surroundings. However, this did not stop Werner from proposing that his stratigraphy applied to the entire surface of Earth.

A number of Werner's students were inspired to go further afield and gather evidence from other regions, which they hoped would support their teacher's theories. However, for many the effect was to turn them away from Werner's ideas. After visiting volcanic areas in Europe, Jean François d'Aubuisson de Voisins (1769-1819) became convinced that neptunism was incorrect. "The facts which I saw spoke too plainly to be mistaken," he wrote, "the truth revealed itself before my eyes." For Leopold von Buch (1774-1853) the volcanic rocks around Rome sowed the first seeds of doubt, and later trips to Scandinavia converted him completely. Alexander von Humboldt (1769-1859) traveled all over South America trying to find supporting evidence for Werner, but again his observations persuaded him otherwise. It is to Werner's credit as a teacher that his own students, who had originally gone out to prove him right, allowed the evidence to shape their conclusions.

Die-hard supporters of Werner's ideas tried to accommodate the new observations into the theory by adding more and more fluctuations in the universal ocean. However, the theory began to lose its simplicity and overall structure. Other theories, such as James Hutton's plutonism (after Pluto the god of the underworld), which stressed the role of underground heat in rock formation, began to gain in popularity as Werner's faded. Neptunism continued to be taught in Germany after Werner's death, and also remained popular in England. However, by the mid-1820s neptunism was all but dead.

Werner's lasting legacy to geology was his approach to the field, not the form his theory took. His stress on explaining physical evidence, and the sense of history he injected into geological theories, gave the science a new direction. He passed his passion for the field to his students, many of whom, ironically, used the skills they had been taught to overturn their master's theories.

DAVID TULLOCH

Further Reading

Faul, Henry and Carol Faul. *It Began with a Stone: A History of Geology from the Stone Age to the Age of Plate Tectonics.* New York: John Wiley & Sons, 1983.

Gohau, Gabriel. *A History of Geology.* Albert V. Carozzi and Marguerite Carozzi (trans.). New Brunswick, NJ: Rutgers University Press, 1990.

Hallam, A. *Great Geological Controversies.* Oxford: Oxford University Press, 1983.

Genesis vs. Geology

Overview

In the eighteenth century the dominance of biblical geology as stated in the Book of Genesis was challenged by new discoveries, the undermining of ancient authority, and a general spirit of revolution. Natural causes for the formation and shaping of the geological features were stressed by some, the power of unnatural catastrophes by others. Uneasy battle lines were drawn between those who took the bible literally, those who sought compromise, and those who rejected biblical creation. Often debates and theories were long on words, and short on evidence. Creationists won the short-term battles, and geologists retreated to less controversial areas, until the argument resurfaced with the evolutionary theories of Charles Darwin (1809-1882).

Background

The ancient Greek philosophers speculated widely on the creation and formation of the Earth and its geological features. The most enduring ideas where those of Plato (427-347 B.C.), as passed down and modified by his pupil Aristotle (384-322 B.C.). They held that the natural order of the world was eternal and unchanging.

The rise of Christianity modified the Platonic model to a one directional history, with a fixed end in sight, the time when God judges and destroys the world. This implied there was no need to understand the 'fallen' world, as it was almost at an end. However, as the centuries passed, intellectuals in the church began to seriously consider the origin of the earth. The most influential was St. Augustine (354-430), whose ideas dominated Christian theology for centuries.

St. Augustine's starting point was searching for the ultimate truth of the Bible and the questions he considered were: Were the beasts of prey and venomous animals created before, or after, the fall of Adam? If before how can their creation be reconciled with God's goodness; if afterwards, how can their creation be reconciled to the letter of God's Word? Why did the Creator not say "Be fruitful and multiply," to plants as well as to animals? This style of questioning formed the framework for such debates for centuries to follow.

One persistent challenge to biblical accounts of creation was that posed by fossils. Shapes that resembled animals and plants were found inside rocks. Some closely resembled living organisms, others were unlike any creature known. Even more puzzling were the fossils of sea creatures found high in mountains. Some suggested that rocks naturally generated reproductions of living organisms. Others thought the striking similarities to living beings implied they must have once been plants and animals. But if so how did they bore into solid rock, and how did sea creatures climb mountains?

Before the sixteenth century the general view was that the older a text the more reliable it was. The discovery of the New World by Christopher Columbus (1451-1506) was just one of many revolutions that threatened the authority of ancient texts. The Americas contained a vast number of new plants and animals, which made the biblical description of Noah's ark seem less believable. How had all these new species fit into the ark? There were different flora and fauna in different parts of the world, yet they had supposedly all come from the one source.

The era of exploration and discovery was also a time of social, religious and economic upheavals. Revolts and rebellions, and a new sense of radical change combined to make the period seem a total break with the past. This provoked a number of thinkers to reconsider the history of the Earth. Often the theories were very speculative, based on little physical evidence, and reflective of the upheavals of the time.

In the seventeenth century René Descartes (1596-1650) suggested that the Earth had been a fiery ball, which formed a crust as it cooled over deep waters, which were released when the crust collapsed. This theory became quite influential, as it supported the biblical flood, and formed the basis of many later theories.

Nicolaus Steno (1638-1686) studied a group of fossils known as tongue stones, and realized that they resembled shark teeth. This led him to propose that fossils were plants and animals trapped in river sediments, which over time hardened into rock.

However, the majority of new theories were attempts to fit the new evidence within the framework of the biblical creation. Thomas Burnet's (1635-1715) *Sacred Theory of the Earth* (1681) added to Descartes' concept of the crusted earth. He proposed that the Earth had once been a smooth sphere, with no seas, valleys or mountains. No rain more severe than gentle dew fell, until the Great Flood. This explained why only Noah's family had survived the deluge, as the ark was the first boat. In 1696 William Whiston (1667-1752) used Newtonian ideas to modify Burnet's theory, suggesting a comet had struck the earth to release the underground waters.

Impact

From the middle of the eighteenth century, a further wave of creation theories began to emerge. The expansion of mining, quarrying, and the building of canals had unearthed more and more evidence of fossils and sedimentary strata. There was a new trend to mechanical ex-

planations of the processes of the Earth, with experimental and mathematical analysis to support the theories.

In 1774, Georges Louis de Buffon (1707-88) produced a massive multi-volume work on natural history. He used the Newtonian ideas of natural forces and empirical causes to explain the observed geology of the world. His experiments on the cooling of iron spheres suggested that the earth was 75,000-years-old, much older than the Bible suggested. His manuscript notes however, suggest he actually thought the earth was billions of years old, but did not think such a figure would be understood.

As it was his ideas caused enough trouble. The faculty of the Sorbonne, where Buffon held a post, declared that he was opposing "the sacred deposit of truth committed to the Church." He was dismissed from his position and forced to print a complete retraction which read: "I declare that I had no intention to contradict the text of Scripture; that I believe most firmly all therein related about the creation, both as to order of time and matter of fact. I abandon everything in my book respecting the formation of the earth, and generally all which may be contrary to the narrative of Moses."

In England James Hutton (1726-1797) printed his *Theory of the Earth* (1795), which suggested that the weathering effects of water produced the sedimentary layers. This sediment was raised through volcanic action, and the erosion of wind, rain and rivers sculpted valleys and plains, starting the cycle again. However, based on observation and experimentation of river flow and mud content, Hutton realized this process would require much longer than 6,000 years. Hutton's theory attempted to explain the geological structure of the earth without resorting to catastrophic events such as worldwide flooding, comets, or massive earthquakes. This emphasis of slow, gradual change over time came to be known as Uniformitarianism.

Hutton's views were opposed not only by the literal defenders of the bible, but also by another group of geologists who thought that violent catastrophes were responsible for the shaping of the earth. The social and political revolutions that were occurring in the period provided a backdrop for the geological catastrophists. The old Platonic idea of an unchanging world had seemed less valid in a period of dramatic change. If the once secure institutions of religion and government could undergo revolutions, then why not the Earth? The divisions between

Georges Buffon.

catastrophists and those believing in a slow process were further sub-divided into those who stressed the role of water (neptunists) and those who favored volcanic action (vulcanists).

The divided nature of the new challengers to biblical creation made them easier targets for their religious opponents. There was a strong backlash against those who sought "nothing less than to depose the Almighty Creator of the universe from his office." The poet William Cowper neatly summarized the traditionalist viewpoint. "Some drill and bore /The solid earth, and from the strata there / Extract a register, by which we learn / That He who made it, and revealed its date / To Moses, was mistaken in its age!"

Geologists began to avoid questions of the historical earth, and focused more on activity they could observe directly. In doing so they accumulated much more solid evidence of the natural forces that alter the earth. However, there were some unexpected consequences of the continuing attacks on the new geology. The repeated condemnation of Hutton's views actually served to popularize his once obscure ideas. It was through their exposure to criticism that the stronger theories developed and gained supporters.

In 1830 the debate flared up again with Charles Lyell's (1797-1875) *Principles of Geology.* Lyell revived and popularized Hutton's Uniformitarianism. However, the most impassioned battles between science and religion came when Charles Darwin (1809-1882) applied the principles of Uniformitarianism to living creatures, suggesting not only that the earth was very old, but also that the animals of the present had evolved from earlier forms.

Final evidence for the age of the Earth was not uncovered until the early twentieth century when the discovery of radioactivity, and the principle of radioactive half-life, allowed for the age of the Earth to be calculated by atomic physics.

The late eighteenth century debates between religion and the new science of geology resulted in an uneasy retreat by many geologists. Some early theories were poorly constructed, based on sketchy evidence, and the scientific community presented a fractured and uncoordinated case for a creation other than the Judeo-Christian Book of Genesis. However, the early work of Buffon, Hutton and others was expanded upon and refined in the following century, with new evidence and experimentation. This opened up a new way of viewing time and the history of the earth. Just as microscopes showed the world of the very small, geology revealed the world of the very old.

DAVID TULLOCH

Further Reading

Books

Barnes, Barry and Shapin, Steven, eds. *Natural Order-Historical Studies of Scientific Culture.* London: Sage, 1979.

Hallam, A. *Great Geological Controversies* Oxford: Oxford University Press, 1983.

Moore, John A. *Science as a Way of Knowing: The Foundations of Modern Biology.* Cambride, MA: Harvard University Press, 1993.

Porter, Roy. *The Making of Geology: Earth Science in Britain 1660-1815.* Cambridge: Cambridge University Press, 1977.

The French Revolution
and the Crisis of Science

Overview

The eighteenth century belonged to the period known as the Enlightenment. Thinkers of the time, such as Jeremy Bentham (1748-1832) in England and Jean-Jacques Rousseau (1712-1778) in France, were influenced by the experimental science of Sir Isaac Newton (1642-1727) and the mathematical rigor of René Descartes, among others. According to Enlightenment thinkers, reason—as opposed to spiritual revelation—enables humankind to make sense of the world around them and to better their condition. The aim of a rational society is knowledge, freedom, and happiness. These convictions led to criticism of the old order and visions of a better future. In England, these ideas stimulated reform. In America and France, they led to revolution.

Background

Begun in 1789 with the Declaration of the Rights of Man and Citizens, the French Revolution was the polar opposite of the peaceful beginnings of the Industrial Revolution in England. Although eventually the French Revolution would signal a crisis for science, at first the relationship between government and science was cooperative. In 1790, the National Assembly established by the revolutionaries asked the venerable Académie Royale des Sciences, a product of the seventeenth century, to reform the chaotic system of weights and measures. The result was the metric system of basic units that we know today as the meter, gram, and liter.

Following the overthrow of the monarchy in August 1792, a governing assembly of businessmen, tradesmen, and professional men called the Convention was elected to replace the National Assembly and to provide a new constitution for the country. In September the Convention formally abolished the monarchy and declared France a republic. Because at the time France was at war with England, Austria, and Prussia, a Committee of Public Safety was organized to mobilize scientists to defend the new republic. The official position of relating science to politics resulted in concentrated efforts to get things done. For example, French scientists demonstrated unusual resourcefulness in finding and extracting saltpeter for use in making gunpowder. Similar energy was applied to research into steel making, munitions, copper, and sodium carbonate (a compound used in manufacturing glass and soap).

In 1793, on the premise that science was undemocratic in principle, the Convention closed the Académie, which had always been considered a seat of aristocratic privilege, as well as other learned academies of France. With the closure of the few schools that taught science and technology, the era of so-called aristocratic science was over, and for a time research in France was in a state of disarray.

In September 1793 the Convention effectively revoked the rights of individuals, and a period of extreme violence known as the Terror began. At its conclusion 10 months later, a new government reconstituted the Académie in a new guise, that of the Institut de France. Acceptance to the institute was based on merit, not inherited wealth, and because other centers in France were also doing research and acting as consultants, its members no longer formed an isolated group. The old idea that science was sufficient to itself was replaced by the expectation that science be useful. Increasingly, science became less and less like art. This differentiation was the beginning of the professionalization of science. It became possible to conceive of it as something a person could make a career of.

A national system of secondary schools was established, with an emphasis on mathematics. In 1794 the Ecole Polytechnique was formed to train engineers to defend the republic. Mathematics and chemistry were taught by renowned teachers such as Pierre Simon de Laplace (1749-1827), Joseph Louis Lagrange (1736-1813), and Claude Louis Berthollet (1748-1822). Systematic laboratory instruction, virtually unknown elsewhere in Europe, was a feature of the polytechnic. Its students were without peer, and some became first-rate scientists.

In the decade before the revolution remade French society, the chemist Antoine Laurent Lavoisier (1743-1794) took the first dramatic steps toward what would turn out to be an equally momentous development in chemistry. Up to Lavoisier's time, chemistry was an ambiguous discipline owing to its tradition of

Antoine Lavoisier. *(Library of Congress. Reproduced with permission.)*

1804). Before 1700, the idea of gases as chemical entities was hardly known. At the age of 37, Priestley undertook to make a special study of different kinds of gases, which he called airs, by heating or mixing substances, and submitting them to simple tests to better describe them. At a meeting with Lavoisier in Paris in 1774, Priestley announced that he had obtained a new kind of air in which a candle burned more brightly than in normal air. But Priestley did not try to explain his discovery in terms other than phlogiston theory.

Building on Priestley's experiments, Lavoisier was able to show that air was not itself an element but that it consisted of the element nitrogen and of Priestley's "dephlogisticated air," which Lavoisier called oxygen. His findings meant that rather than being a dumping ground for phlogiston, air combined with the substance being burned to form an oxide. For example, burning zinc causes it to combine with oxygen from the air to form zinc oxide. Understanding the role of oxygen in combustion allowed Lavoisier to determine the composition of many substances.

Because Lavoisier's "antiphlogistic" theories went against the mainstream, he needed to build support for them. One way to do that was to try to organize the disordered nomenclature of chemistry into a single system of naming things. Based on the new discoveries and theories, and published in 1787, the work listed 55 elements—that is, bodies that could not be decomposed—among them oxygen, nitrogen, hydrogen, carbon, and sulfur. Because so much importance was attached to how substances are named and classified, in practical terms, adopting Lavoisier's nomenclature amounted to the same thing as adopting his theories.

Impact

Owing to the immense social and political upheaval caused by the French Revolution, it is tempting to interpret everything connected with it in terms of a break with the past. But that was not the case. For example, the Académie des Sciences had always been considered elitist, even among some of its members such as Lavoisier, and the government's antipathy toward it and other learned academies of France did not represent a complete rejection of science. The central administration of science and the use of scientists as consulting experts was kept intact, and although schools closed, the curriculum did not change. Despite its abuse by various revolution-

alchemy, useful art, and pharmaceuticals. The legacy of alchemy—a philosophy of the Middle Ages that asserted that base metals could be changed into gold—was a persistent belief in the transmutation of the elements. For example, whether water could be changed into earth was a topic of discussion at a public meeting of the Académie in 1767. Alchemical notions also endured in the theory of phlogiston, which dominated the science of chemistry from the mid-1600s until Lavoisier proved it wrong.

The theory of phlogiston stated that when a substance is burned, something is given off into the atmosphere. By definition, all combustible materials contained this "something," called phlogiston. What was left when a material was "dephlogisticated" was the true material. So, for example, ashes were wood minus phlogiston. Likewise, zinc, when burned, released phlogiston into the air and left a solid residue called calx. Phlogiston as a theory became a unifying principle in chemistry, and very difficult to dislodge. It seemed to provide a satisfactory explanation of the nature of burnt matter. The problems with it—why, for instance, materials gained weight on burning—were illuminated by Lavoisier's careful experimentation.

Lavoisier's pioneering work would not have been possible had the groundwork not been set by the Englishman Joseph Priestley (1733-

ary factions, the Declaration of the Rights of Man and of Citizens guaranteed freedom of the press, which in the early days of the revolution stimulated the publication of new scientific journals to promote science. And freedom of association enabled the formation of societies to propagate scientific knowledge.

Napoleon Bonaparte (1769-1821) modified many of the scientific institutions created in the wake of the Terror and brought them under more centralized control. Although his administration emphasized military training and cut budgets for research and for laboratories, the system he inaugurated in France persists to this day. In contrast, in the United States after the War for Independence (1775-1778) the balance of powers and the rights of individual states led to decentralized universities and institutions. A National Academy of Sciences did not appear in America until the time of the Civil War (1861-1865).

In debunking the theory of phlogiston and in establishing the value of quantitative measurements, Lavoisier elevated chemistry to science. His nomenclature brought order where chaos had reigned, and a treatise he wrote on chemistry in 1789 became a model for teaching the subject for many years. His discoveries changed perceptions of chemistry as surely as the French Revo-

lution changed perceptions of monarchy and individual rights. The creator of his own scientific revolution, the supporter of his country's, Lavoisier nonetheless was arrested toward the end of the Terror and executed on the guillotine at the age of 51. At the news of his death, Lagrange is said to have commented, "It took them only an instant to cut off that head, and a hundred years may not produce another like it."

Priestley's support of the early stages of the French Revolution and his outspoken criticism of the established Church of England earned him his own share of enemies. Following several years of harassment by a hostile mob, Priestley emigrated to the United States in 1794. His chemical apparatus resides in the Smithsonian Museum in Washington, D.C.

GISELLE WEISS

Further Reading

Books

Hoffmann, Roald, and Vivian Torrence. *Chemistry Imagined: Reflections on Science*. Washington, D.C.: Smithsonian Institution Press, 1993.

Marks, John. *Science and the Making of the Modern World*. London: Heinemann, 1983.

Russell, Colin. *Science and Social Change, 1700-1900*. London: Macmillan, 1983.

Joseph Priestley Isolates Many New Gases and Begins a European Craze for Soda Water

Overview

Until the late eighteenth century, the accepted theory of chemical reaction was the "phlogiston theory." This theory, whose name was coined by German chemist Georg Stahl (1660-1734) in the early 1700s, stated that a substance "phlogiston," which is Greek for "burned," was liberated when any material underwent combustion or when a metal was oxidized. By the late 1700s chemists, especially Antoine Lavoisier (1743-1794), had applied quantitative measurements and demonstrated that oxidized metals, or metals that had rusted, weighed more than nonoxidized ones, thus proving that phlogiston did not exist. Proponents of the phlogiston theory then countered that phlogiston was a negative quantity. Finally by 1800, after several new gases had been discovered and their role in combustion and oxida-

tion identified, the phlogiston theory was overthrown and modern chemistry was born.

The groundbreaking experiments of one Englishman spurred this scientific revolution. Working alone, this minister and teacher isolated ten new gases and noted their properties. Scientific curiosity led him to dissolve his new gases, or "airs" as he called them, in water. The result with one gas was a carbonated drink, which he called "soda water." He was also among the first to observe and write about the process of photosynthesis and about a plant's respiratory cycle. In addition to his discovery of several gases and their properties, he was also the first to confirm that graphite could conduct electricity. When others could find no use for a New World substance called "India gum," he did, calling it "rubber" because it could rub out or erase pencil marks. In addition to his passion for scientific discovery, he had liberal views of religion and politics, openly criticizing the Church of England and the British government. He was steadfast in his support of both the American and French revolutions. A pioneer in science, religion, education, and politics, Joseph Priestley (1733-1804) was indeed a genius.

Background

Born in Leeds, England, Joseph Priestley was the eldest of six children. His father was a clothier and a member of the Calvinist religion. Young Priestley loved to read and taught himself several languages. His family applauded his choice of the ministry as a career and encouraged him to be a member of the Dissenting congregation, whose members were opposed to the Church of England. In 1752 Priestley entered the Nonconformist Academy and studied literature and natural philosophy.

Priestley entered the ministry by the age of 22; later, as an assistant minister, he established a school that became well known for its high educational standards. By 1761 he had accepted a teaching post at an academy, where a colleague who asked Priestley to assist him in his chemistry classes sparked the young minister's interest in chemistry.

As chair of languages and literature, Priestley's reputation grew. In 1765 he wrote about his philosophy of education in "An Essay on a Course of Liberal Education for Civil and Active Life," and the University of Edinburgh gave him an honorary Doctor of Laws degree. During this same year Priestley met Benjamin Franklin (1706-1790) in London, and the two men became lifelong friends. After Franklin introduced Priestley to electricity, the Englishman's fascination led him to conduct many electrical experiments. At Franklin's behest, Priestley agreed to write a history of electricity. In 1767 he published *The History and Present State of Electricity with Original Experiments.* Because of his work with electricity, Priestley was elected a fellow of the Royal Society.

Priestley returned to the ministry in 1767, accepting a congregation in Leeds. There he continued to express his liberal religious and political views. He published several articles, one of which condemned the British government for its treatment of the American colonies and another of which criticized the Church of England.

In Leeds the Priestley home was situated near a brewery. Whenever he walked by the brewery, Priestley observed an unusual phenomenon. He noticed that "fixed air" (carbon dioxide) was released in the process of fermentation and that this new "air" would extinguish burning pieces of wood and then drift to the ground. At home Priestley was able to prepare the new gas he observed at the brewery. When he tried to dissolve it in water, the result was a drink with a pleasant, rather tangy taste. Priestley had made carbonated water, or "artificial Pyrmont water" as he called it. Although he was not the first one to make this bubbly water, his methods were the most successful.

The first Lord of the Admiralty, Lord Sandwich, and his colleagues believed that this new water was more healthful than ordinary drinking water and might even prevent scurvy at sea. Officials were impressed by Priestley's experiments before the College of Physicians, and two Royal Naval battleships were subsequently equipped with the necessary machinery to make Priestley's soda water. He was asked to accompany Captain James Cook (1728-1779) to the South Seas, but Priestley's extreme religious views made this unacceptable to the Royal Navy. In 1772 Joseph Priestley published his famous booklet that described how to make his soda water. Entitled "Directions for Impregnating Water with Fixed Air," the booklet cost one schilling.

Priestley's scientific work began to earn him international recognition. He was elected to the French Academy of Sciences in 1772, and he received the Copley Medal in 1773 from the Royal Society of London for his articles about fixed air and water solutions and for his experiments on gases.

Priestley then became the tutor to the son of the Earl of Shelburne, a position that allowed him more free time for his experiments and writing. He started a 14-year project, writing a 6-volume work, *Experiments and Observations of Different Kinds of Air,* and also continued his experiments. On August 1, 1774, Priestley conducted one experiment that led to a scientific breakthrough. Using a magnifying glass to heat mercuric oxide, he collected a colorless gas above the mercury and noticed that this gas would vigorously support a burning candle. He did further tests that showed that this new gas, compared to ordinary air, could double the lifespan of mice.

Priestley had, of course, discovered oxygen. (Swedish chemist Karl Scheele [1742-1786] had discovered oxygen a year earlier, but Priestley published his findings first and is therefore credited with the discovery.) As his new "air" sustained combustion, Priestley postulated that it was low in phlogiston; he therefore called it "de-phlogisticated air."

When he traveled to the European Continent in October 1774, Priestley met with Antoine Lavoisier, the brilliant French chemist who immediately realized the importance and significance of Priestley's discovery. Lavoisier conducted his own tests and identified the new gas as an element vital to combustion and respiration. In 1789 he gave it the name "oxygine," later "oxygen," which is Greek for "acid former" because he mistakenly thought that all acids contained this gas.

Joseph Priestley maintained his belief in phlogiston, however, and continued his experiments. He discovered several new "airs": ammonia, sulfur dioxide, silicon, nitrogen, and carbon monoxide. He observed photosynthesis when he experimented with plants. He saw that if his containers were left in sunlight, a green matter grew inside the container, and that this green matter emitted the same "air" that had been produced in his mercuric oxide experiments. In another experiment, he made water by combining oxygen and hydrogen gases with a spark of electricity.

In 1779 Priestley returned to the ministry with a parish in Birmingham. By then, his vigorous support of the American and French independence movements and his treatise, *History of the Corruptions of Christianity,* antagonized both political and religious factions and made him many enemies. On July 14, 1791, the second anniversary of the storming of the Bastille, an angry mob in Birmingham attacked Priestley's home, which, along with his laboratory and library, was burned to the ground.

Priestley and his family left Birmingham to live in London. Three years later, he and his wife emigrated from England to join their three sons in America. They chose to live in Northumberland, Pennsylvania, where he continued to publish religious and political articles and work on his scientific experiments. Among his friends and acquaintances in America were George Washington, John Adams, and Thomas Jefferson, the latter two sharing Priestley's liberal religious and political views. Joseph Priestley died on February 6, 1804, in Pennsylvania, still clinging to the phlogiston theory.

Impact

After Priestley met with Antoine Lavoisier in 1774, the French chemist continued laboratory testing of the gas he had named "oxygen." Lavoisier's experiments not only conclusively disproved the phlogiston theory but also showed that oxygen formed 20% of air and that the element was a component of combustion. Lavoisier then defined the nature of an element as a substance that cannot be broken down further by any chemical means; he proposed naming elements with the nomenclature still used today.

Lavoisier, whose contributions to chemistry are legion, was inspired by the findings of Joseph Priestley. The English theologian has rightly been called one of the most important men in the history of science. Although Priestley's scientific background was minimal, his inquisitive mind and innovative experiments launched the field of modern chemistry. Scientists have forgiven Joseph Priestley for his stubborn belief in the existence of phlogiston and acknowledge his contribution to the foundation of chemistry. The American Chemical Society honored this remarkable man and his outstanding accomplishments by naming its highest award the Priestley Medal.

ELLEN ELGHOBASHI

Further Reading
Books
Gillam, John Graham. *The Crucible: The Story of Joseph Priestley LL.D., F.R.S.* London: Robert Haled Ltd, 1954.

Holt, Anne. *A Life of Joseph Priestley.* Westport, CT: Greenwood Press, 1931.

Roberts, Royston M. *Serendipity: Accidental Discoveries in Science.* New York: John Wiley & Sons, 1989.

Zumdahl, Steven S. *Chemistry.* D.C. Heath and Co., 1989.

Internet Sites
"Biography of Joseph Priestley." http://home.ptd.net/ ~sjrubin/uucsv/carter.htm

"Joseph Priestley." http://home.earthlink.net/~afriedman/ priestly.html

"Joseph Priestley: The King of Serendipity." http://home. nycap.rr.com/useless/priestly/priestly.html

Daniel Bernoulli Establishes the Field of Hydrodynamics

Overview

Swiss mathematician and physicist Daniel Bernoulli (1700-1782) brought a high level of mathematical rigor to the study of natural phenomena, particularly phenomena associated with liquids and gases. He was especially fascinated with the behavior of fluids. Having been trained as a physician, Bernoulli combined his medical knowledge with his mathematical and analytical skills to study the speed with which a liquid flows through an enclosed space such as a blood vessel. Bernoulli's observations resulted in the first effective method of measuring blood pressure but, more importantly, put Bernoulli on the path that would lead to his derivation of the fluid equation. That equation showed that a fluid's pressure is determined by its velocity, the pressure decreasing as the velocity increases. Bernoulli's formulation related pressure to the law of conservation of energy, placing the behavior of fluids (and by extension gases) squarely within the realm of mathematical and mechanical physics. Bernoulli's insight would affect all studies of fluids and gases for centuries, guiding studies of the behavior of water under pressure, the hull designs of ships, problems of turbulence in turbines, and explaining, for example, the behavior of air traveling over and beneath an airplane's wing.

Background

Daniel Bernoulli was born into a distinguished family of mathematicians and scientists. His father, Jean (Johann) Bernoulli (1667-1748), was a mathematician, chemist, and professor; his uncle, Jakob Bernoulli (1654-1705), was a mathematician and natural philosopher. Both of the elder Bernoullis made substantial contributions to calculus and to probability theory.

Despite this family history, Daniel Bernoulli faced obstacles to becoming a mathematician. His father, in particular, aware that mathematics offered little in the way of financial reward, was opposed to Daniel's interest, and sought to steer his son into a career as a merchant. Daniel raised strenuous objections and his father finally relented, permitting the boy to study medicine but continuing to prohibit mathematics. Daniel entered medical school and obtained his degree in 1721. But he never gave up his fascination with

mathematics, and ultimately even his father could not hold the young man back. Jean Bernoulli himself began at last to tutor Daniel in advanced mathematics. Daniel began applying the mathematical lessons to his medical observations and studies, specifically the flow of blood through the body's circulatory system.

Unable to find the teaching post he desired, Daniel Bernoulli began to travel around Europe. By 1723 Bernoulli was in Italy, where he was taken ill. As he recuperated he turned his attention to the design of an hourglass whose flow of sand would remain constant even aboard a ship in storm-tossed waters. The French Academy awarded Bernoulli's hourglass first prize in a design competition. Bernoulli also published his first book, *Some Mathematical Exercises*.

Upon turning 25, Bernoulli at last received a teaching commission—from no less an employer than Empress Catherine I of Russia, who offered Bernoulli the position of mathematics professor at the Imperial Academy of St. Petersburg. Bernoulli accepted the position, and his brother Nikolas was given the Academy's second mathematics chair, although Nikolas died within a year of their arrival in Russia.

Working with Leonhard Euler (1707-1783), Bernoulli tackled questions of fluid behavior. Studies of the properties of water were nearly as old as science—around 250 B.C. the great Greek scientist Archimedes (c. 287-212 B.C.) supposedly noticed the displacement of water by his body when he entered his bath. Whether or not Archimedes actually exclaimed "Eureka!" upon observing the displacement, his observation did found the science of studying water (or other fluids) at rest, or hydrostatics.

Over the ensuing centuries hydrostatics remained a relatively active field of study and inquiry. Displacement of fluids by solids, and particularly different displacement properties of different materials—density—received much scientific attention. Properties of surface tension also attracted scientific study. But the careful, systematic study of fluids in motion—hydrodynamics—awaited the arrival of Daniel Bernoulli. Applying himself, along with Euler, first to the behavior of blood in veins, Bernoulli essentially created the field of hydrodynamics in a very few years.

Using a thin-walled pipe to simulate a blood vessel and running water through it at various speeds, Bernoulli discovered that a hollow straw, inserted through the wall of the pipe, drew water to different levels depending upon the pressure of the fluid flowing through the pipe. Blood pressure could be measured in similar fashion, using a hollow glass needle inserted directly into a vein. (Not until 1896 was a less painful method of measuring blood pressure discovered!)

With the lessons learned from his blood studies, Bernoulli turned to the larger question of the relationship of fluid pressure to the laws of conservation of energy. Understanding that kinetic, or moving, energy becomes potential energy when a body is raised in height, Bernoulli achieved his great insight into the behavior of fluids: when a fluid moves, its kinetic energy is transferred to pressure. If the fluid's velocity increases, the pressure decreases; if the velocity decreases, the pressure rises.

Bernoulli captured his insight in an equation:

$$1 \div pu^2 + p = \text{constant}$$

where p is pressure and u is its velocity. This equation still bears the name Bernoulli's principle.

Quickly seeing the potential for his insights, Bernoulli set to work cataloging and extending his observations, applying his mathematical skills to his studies. Equally important, he extended those studies to the properties of gases.

In doing so, Bernoulli assumed that gases are composed of tiny particles. By viewing gases in this way, Bernoulli was able to speculate, mathematically, about their properties, much as he had about liquids. Bernoulli's research and studies of gas properties laid the foundation for the kinetic theory of gases that would be developed in the next century. On a practical level, Bernoulli's exploration of gas properties enabled him to create a mathematical means for measuring atmospheric pressure at different altitudes, showing that the pressure changed as altitude changed.

Bernoulli's ability to combine careful scientific observation and measurement with a high level of mathematical analysis enabled him to create, by 1738, his treatise *Hydrodynamica*, which is considered to be one of the great achievements of eighteenth-century science. In this book Bernoulli distilled his insights and mathematical analyses of the properties of fluids and gases and in one stroke, as it were, created a major branch of physics. Ironically and sadly,

Bernoulli's father grew obsessively jealous of his son's accomplishments and published his own book, *Hydraulics*, which many felt plagiarized Daniel's work. The two remained estranged, and that estrangement seems to have drained Daniel of interest in further pursuing hydrodynamics.

Bernoulli devoted the rest of his life to anatomy, botany, and pure mathematics, making contributions to each field. But it is his establishment of hydrodynamics as a field of mathematical inquiry that remains the greatest of Daniel Bernoulli's many accomplishments, a field that remains vigorous and important to this day.

Impact

Bernoulli's creation of hydrodynamics (or, as it is more commonly referred to today, *fluid dynamics*) is a scientific contribution of great importance. First, he applied himself to the mechanical properties of fluids in motion, a relatively unexplored but increasingly important field. Next, he quantified the study of fluids by bringing mathematical attention to bear on their behavior in various circumstances. That in itself was a major contribution to establishing the field as a pure science rather than an area of natural philosophy. Equally important, Bernoulli extended the study of fluid behavior to include the behavior of gases, thus establishing principles on which further inquiry and study could be built.

As a result of the foundation Bernoulli established, hydrodynamics was able to investigate and explain not only the behavior of fluids and gases in motion, but also the effects of variables on that motion. Turbulence, which occurs when pressure and velocity undergo irregular fluctuations, is perhaps the most important of the areas of inquiry that derived from Bernoulli's starting points. Most natural flows—rivers, air currents—are, obviously, irregular or turbulent. By creating the mathematical and scientific framework for studying turbulence, scientists were able to create a more accurate picture of the ways in which physical laws are revealed in the natural world.

Bernoulli's contributions also extended to technology. In medicine, of course, he contributed to the understanding of blood flow and blood pressure. But as the Industrial Revolution gathered force and advances in metallurgy made possible the creation of high-pressure steam (and water) vessels, the ability to predict the behavior of liquids and gases flowing under pressure became critical. In turn, studies of turbulence became crucial to the design of turbine blades turned by steam. *Hydrodynamica* itself contained Bernoulli's insights into the effect of fluid dynamics on ship propulsion and hull, lessons that were put to work by the naval industries.

In the twentieth century Bernoulli's contribution found applications that he would have perhaps considered pure fantasy, as aerodynamics was created as a sub-discipline of hydrodynamics. Aerodynamics uses mathematical tools to study the effects of air flow over wings—to study the physics of flight, and to use those studies to create aircraft that take advantage of fluid dynamic principles in order to lift themselves from the ground and into the skies. Aerodynamic studies also affect the design of automobile bodies.

KEITH FERRELL

Further Reading

Bernoulli, Daniel. *Hydrodynamica. Argentorati, sumptibus. J. R. Dulseckeri,* 1738.

Quinney, D. A. "Daniel Bernoulli and the Making of the Fluid Equation." *PLUS* 1 (January 1997).

Wilson, Derek Henry. *Hydrodynamics.* London: E. Arnold, 1959.

The Cavendish Experiment and the Quest to Determine the Gravitational Constant

Overview

The determination of a precise value for the gravitational constant (G) has proved a frustrating, but fruitful, exercise for scientists since the constant was first described by English physicist Sir Isaac Newton (1642-1727) in his influential 1687 work, *Philosophiae Naturalis Principia Mathematica* ("Mathematical Principles of Natural Philosophy"). In many ways as enigmatic as mathematicians' search for a proof to Fermat's last theorem (proved only in the last decade of the twentieth century), the determination of an exact value of the gravitational constant has eluded physicists for more than 300 years. The quest for "G" provides a continuing challenge to the experimental ingenuity of physicists and often spurs new generations of physicists to recapture the inventiveness and delicacy of measurement first embodied in the elegant experiments conducted by English physicist Henry Cavendish (1731-1810).

Background

In *Principia* Newton put forth a grand synthesis of theory regarding the physical universe. According to Newtonian theory, the universe was bound together by the mutual gravitational attraction of its constituent particles. With regard to gravity, Newton formulated that the gravitational attraction between two bodies was directly proportional to the masses, and inversely proportional to the square of the distance between the masses. Accordingly, if one doubled a mass one would double the gravitational attraction; if one doubled the distance between masses one would reduce the gravitational attraction to one-fourth of its former value. What was missing from Newton's formulation, however, was a value for a gravitational constant that would accurately translate these fundamental qualitative relationships into experimentally verifiable numbers.

The gravitational constant essentially measures the strength of gravity. Newton's law of gravitation is mathematically expressed as $F = (G)(m_1 m_2)/r^2$ where F is the gravitational force between two bodies, m_1 and m_2 are the two masses, and r is the radius or distance between the two masses. The gravitational constant has been experimentally calculated by modern methods to approximately 6.6726×10^{-11} m^3 kg^{-1} s^{-2}. Prior to the determination of the Cavendish value for the gravitational constant (usually referred to as the "Cavendish constant"), however, scientists could only derive a value for the product of the gravitational constant and the mass of the Earth—there was no practical way to separate them. Essentially, although it was possible to calculate the product of the gravitational constant and the mass (GM) of the Earth, there seemed no easy way to determine the value of G and M separately.

The units of the gravitational constant (m^3 kg^{-1} s^{-2}) dictate that, by itself, the gravitational constant is not a tangible physical entity. Accordingly, the determination of G requires experimental ingenuity. Prior to Cavendish, English astronomer Nevil Maskelyne (1732-1811) attempted to determine the gravitational constant using a plumb bob in an experiment inspired by Newton's *Principia*, an estimated mass of Mount Shiehallion. Maskelyne's experiments failed, however, to yield a practical value for the gravitational constant.

Although Cavendish was wealthy, he devoted himself to scientific works and study. Considered eccentric by his colleagues, he worked quietly and apart from the academic mainstream. Cavendish attended Cambridge University but did not complete his studies. He was so reclusive that he kept many of his scientific discoveries to himself. Much of Cavendish's work with electricity, for example, remained unpublished for almost another century, when his notes were compiled and published by Scottish physicist James Clerk Maxwell (1831-1879).

In 1798 Cavendish performed an ingenious experiment that led to the determination of the gravitational constant. Cavendish used a carefully constructed experiment that utilized a torsion balance to measure the very small gravitational attraction between two masses suspended by a thin fiber support. (Cavendish actually measured the restoring torque of the fiber support.) Cavendish's experimental methodology and device design was not novel. A similar apparatus had been designed by French physicist Charles Coulomb (1736-1806) and others for electrical measurements and calibrations. Cavendish's use of the torsion balance to measure the gravitational constant of the Earth, however, was a triumph of empirical skill.

Cavendish balanced his apparatus by placing balls of identical mass at both ends of a crossbar suspended by a thin wire. By using lead balls of known mass, Cavendish was able to account for both the masses in the Newtonian calculation and thereby allow a determination of the gravitational constant. The Cavendish experiment worked because not much force was required to twist the wire suspending the balance. In addition, Cavendish brought relatively large masses close to the smaller weights—actually on symmetrically opposite sides of the weights—so as to double the actual force and make the small effects more readily observable. Over time, due to the mutual gravitational attraction of the weights, the smaller balls moved toward the larger masses. The smaller balls moved because of their smaller mass and inertia (resistance to movement). Cavendish was able to measure the force of the gravitational attraction as a function of the

Henry Cavendish. *(Library of Congress. Reproduced with permission.)*

time it took to produce and given the twist in the suspending wire. The value of the gravitational constant determined by this method was not precise by modern standards but was an exceptional value for the eighteenth century given the small forces being measured. Because all objects exert a gravitational "pull," precision in Cavendish-type experiments is often hampered by a number of factors, including underlying geology or factors as subtle as movements of furniture or objects near the experiment.

Because of his age (Cavendish was nearly 70 years old) the tedious and intricate experiment was taxing on Cavendish, and the determination of the gravitational constant proved to be his last great experiment. Decades after his death, a portion of the substantial Cavendish estate was used to fund the endowment of the great Cavendish Laboratory at Cambridge University, where so much of the fundamental work on the detailed structure of the atom and of the small scale forces underlying modern physics was to take place.

Impact

The effort to determine a precise and accurate value for the gravitational constant was a natural consequence to the fact that, by the end of the eighteenth century, Newtonian physics dominated Western scientific and intellectual thought. Cavendish's experiment essentially allowed scientists to "weigh" Earth (more properly to determine its mass and density). Once the value of the gravitational constant was determined, the mass of Earth could be calculated from the experimentally determined gravitational acceleration of 9.8 m/s^2.

An accurate value for the gravitational constant was essential to determine the mass of the Sun, Moon, planets, and other astronomical bodies. As a result of the Cavendish determination, revised applications of Kepler's third law allowed for subsequent refined estimates of the mass of the Sun and planets.

Eighteenth-century astronomical estimations of the radius and of the distances between the planets were accurate enough to make reasonably good estimates of the size and, hence, the volume of the planets. One unexpected outcome of the eighteenth-century quest for G was the ability to more accurately measure the density of Earth. Maskelyne's data established a greater than expected density of Earth—and inferred that Earth must have a core much denser than its crust. Cavendish's values for G indicated an even greater density for Earth (approximately 5500 kg / m^3) than Maskelyne had proposed.

The gravitational constant is a fundamental quantity of the universe. It was the first great universal constant of physics (the others subsequently being the speed of light and Planck's constant), and modern physicists still argue its importance and relationship to cosmology. Regardless, almost all the major theoretical frameworks dictate that the value for the gravitational constant is in some regard related to the large-scale structure of the cosmos. Ironically, despite centuries of research, the gravitational constant is—by a substantial margin—the least understood, most difficult to determine, and least precisely known fundamental constant value.

Although profoundly influential and powerful on the cosmic scale, the force of gravity is weak in terms of human dimensions. Accordingly, the masses must be very large before gravitational effects can be easily measured. Even using modern methods, different laboratories often report significantly different values for G.

The Cavendish experiment was, therefore, a milestone in the advancement of scientific empiricism. In fact, accuracy of the Cavendish determination remained unimproved for almost another century until Charles Vernon Boys (1855-1944) used the Cavendish balance to make a more accurate determination of the gravitational constant. More importantly, the Cavendish experiment proved that scientists could construct experiments that were able to measure very small forces. Cavendish's work spurred analysis of the fundamental force of electromagnetism (a fundamental force far stronger than gravity) and gave confidence to the scientific community that Newton's laws were not only valid but testable on exceedingly small scales.

The ability to measure the mass and movements of the planets more critically became increasingly important throughout the eighteenth century in philosophical and theological circles. The precise delineation of planetary movements under the influence of gravity was an issue often raised by all sides in the increasingly heated debate about whether there was any role for God in a mechanical, clockwork cosmos.

In modern physics, the speed of light, Planck's constant, and the gravitational constant are among the most important of fundamental constants. According to relativity theory, G is related to the amount of space-time curvature caused by a given mass. Modern concepts of gravity and of the ramifications of the value of the gravitational constant are subject to seemingly constant revision as scientists aim to extend the linkage between the gravitational constant and other fundamental constants.

K. LEE LERNER

Further Reading

Bronowski, J. *The Ascent of Man*. Boston: Little, Brown, 1973.

Hoyle, F. *Astronomy*. Crescent Books, 1962.

Jungnickel, C. and R. McCormmach. "Cavendish." *Memoirs of the American Philosophical Society Series*. American Philosophical Society, 1996.

Sparks and Lightning:
Electrical Theories from the "Electrician"
Dufay to the Scientist Coulomb

Overview

To relate the history of electricity and magnetism during the eighteenth century is to give a good general outlook of how early modern physics matured from being a mostly qualitative topic of study to a demanding quantitative and mathematical science. Scientific instruments played a great deal in accomplishing this conceptual transformation, which is part of what the French call *l'esprit géométrique des Lumières*, or the quantifying spirit of the Enlightenment. Woven into this story is the emergence of a popular scientific culture and awareness that foregrounds the educational practices conducted today by our modern science centers.

Background

The most ancient reference pertaining to electricity that we know of comes from Antiquity. Thales of Miletus (c. 624-547 B.C.), one of the earliest Greek philosophers, noted in circa 600 B.C. that rubbing amber and a few other substances attracted feathers or bits of straw or leaves to them; the first written statement on this experiment, however, was recorded 200 years later in Plato's dialog entitled *Timaeus*. In Rome, Pliny the Elder (A.D. 23-79) wrote about similar experiments in his famous *Natural History*, where he also mentioned shocks given by torpedo fishes.

We cannot say much about the European Middle Ages, except for the appearance of the magnetic compass late in the twelfth century, originating from China. This device, along with the log-line and the hourglass, helped sailors find their way by making use of the navigational technique called "dead reckoning." A discrepancy had been observed between the geographic and magnetic north poles, as indicated by the compass, the difference in degrees being named "declination." Christopher Columbus (1451?-1506) was the first, however, to recognize during his 1492 epoch-making voyage of discovery that somewhere in the Atlantic the declination came to zero and then, by going further westward, increased again but in the opposite direction. (Edmond Halley [1656-1742] later tried to use this phenomenon in order to find a solution

to the long-sought problem of the determination of longitude at sea, but to no avail.)

Even though some applications were found for electricity and magnetism, a theoretical explanation was still needed. The first one came in 1600 with the publication of a book entitled *De Magnete* ("On the Magnet"). The author, the English scientist William Gilbert (1544-1603), was in fact the first to distinguish what he called "the amber effect" (electricity) from magnetism. He worked out a complete theory on the latter, associating, for example, the Earth to a big lodestone and thus explaining the gravitational attraction between the planets. Gilbert discovered also many other "electrics"—substances like common glass, resin, sulphur, precious stones, and sealing wax that attract light objects after being rubbed—and "nonelectrics"—that is, substances that could not be electrified by friction. Today, these nonelectrics are named conductors.

During the seventeenth century electrical experiments were carried out on the whole by the Jesuits and by the members of the Italian Accademia del Cimento. While René Descartes (1596-1650) tried to understand electrical attraction while working on his scheme of ethereal vortices, Robert Boyle (1627-91) investigated with the help of vacuum pumps whether air was mechanically involved in electrical attraction or whether the action resulted from an independent electrical effluvium. It was Francis Hauksbee (1666?-1713), however, who inaugurated the Enlightenment's successful train of electrical researches with his study of the "barometric light" effect—an occasional flashing observed in the vacuum tube of a mercury barometer. He discovered later that neither the barometer nor the mercury was needed to produce the flashes; it is basically an electrical discharge through a gas under low pressure, like our contemporary neon signs. Although he could associate this effect to electricity, he was unable to explain it.

More than 20 years later, in 1729, Stephen Gray (1666-1736) became the first experimenter to ascertain that electricity could be communicated by contact. Helped by Granville Wheler (d. 1770), Gray used his orchard as an experimental test-bench—in other words, by suspending a wire from silk cords mounted on poles, he managed to carry electricity over more than 650 feet (198.12 m). The conduction of electricity gave rise to spectacular and entertaining demonstrations. Gray and Wheler, for instance, suspended a young boy from the ceiling and as soon as he became electrified by conduction, he started attracting objects with all parts of his body. Gray's experiments proved to be revolutionary. All that was needed now was a sound theoretical explanation for all of these phenomena.

Impact

In 1733 Gray and Wheler's reports caught the interest of Charles-François de Cisternay Dufay (1698-1739). In a more methodical and organized way than Gray, Dufay began a systematic survey of all the substances that could be electrified. Except for metals and materials too soft or fluid to be rubbed, he found that electricity was an almost universal property of matter. Furthermore, playing with electrified gold leaves led Dufay to the "bold hypothesis" of *two* kinds of electricity: a "vitreous" one, produced by vitreous substances like glass; and a "resinous" one, produced by resinous substances like amber and copal. He discovered that electricity of the same kind repelled while it attracted the opposite kind. This became the basis of a two-fluid electrical theory worked out by the most prominent French electrician during the Enlightenment, abbé Jean-Antoine Nollet (1700-1770).

In 1745 Nollet published in the *Mémoires* of the Paris Academy of Science (and more thoroughly a year later in a book) an electrical theory based on two electricities, not *qualitatively* different as Dufay discovered but distinct in the strength and direction of current flow. Attraction and repulsion were respectively explained by "affluent" and "effluent" fluxes, two opposing currents of electrical fluids emerging in jets from an electrified body. Until 1752 Nollet's system enjoyed the widest consensus any electrical theory had yet received, even though it could not account for one of the foremost discoveries of the century: the Leyden jar.

Devised by Ewald Georg von Kleist (1700?-48) but best described by Pieter van Musschenbroek (1692-1761), the Leyden jar — a glass flask filled with water — was the first condenser. A wire extends from the water inside through a stopper that seals the flask. The wire conducts electricity from another source into the jar, where it is held. When a conductor (like someone's hand) touches the wire, the electricity

WHEN LIGHTNING STRIKES—ALMOST

In the eighteenth century getting shocked by static electricity was nothing short of a fad. However, even those who experimented with electricity—known as "electricator physicists"—having taken sizeable jolts were not aware how dangerous electric charge could be. Among the latter was American statesman/scientist Benjamin Franklin, admired by the French—the most avid experimenters—for his more realistic and practical experiments. Studying the point discharge phenomena, Franklin invented the lightning rod in 1747—some French even had lightning rod hats. Franklin was not the first, but he believed lightning was the same as the electrostatic spark. He wrote a paper describing an experiment to "draw down the electric fire" from a cloud. This entailed a rod or wire about 40 feet (12 m) long connected to a Leyden jar (an early capacitor) inside a supporting insulated enclosure, like a sentry box, to protect the observer.

Others took on the experiment before Franklin tried it himself. In May 1752 Thomas Francois D'alibard used a 50-foot (15 m) vertical rod and successfully drew sparks from the rod in Paris. A week later a M. Delor repeated the success also in Paris. Franklin's most ardent electrical theory supporter, John Canton, was able to duplicate the feat the same year in England. From the descriptions it appears these did not involve an actual lightning stroke—fortunately. Franklin's turn came in June, and he did not stand in the pouring rain with kite in hand as romantic paintings show. He stood under a shed roof with a portion of his silk kite's string kept dry. A key was tied to the string and from there connected to the Leyden jar. Sparks jumped from the key to Franklin's knuckles and did charge the jar. Franklin was lucky that is all that happened.

Then in mid-1753 a Swedish experimenter, Georg Wilhelm Richmann, tried the experiment in his room in St. Petersburg, Russia. Franklin had cautioned to ground the Leyden jar. Lightning actually traveled down an erected wire with Richmann and a servant waiting by the Leyden jar. But Richmann, with discharge test rod in hand, had not grounded the jar, and unfortunately became the target. A foot-long spark leapt from the jar to Richmann's head, killing him instantly and bowling over his dazed servant. In the same city a few days later, Russia's eminent chemist Mikhail V. Lomonosov performed the experiment successfully, but now it was known that playing with lightning could be fatal.

WILLIAM J. MCPEAK

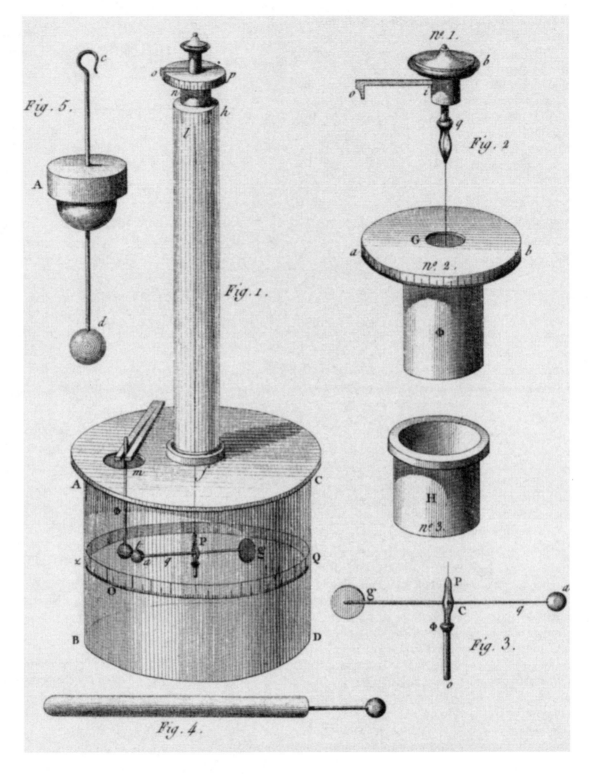

The invention of the torsion balance by Charles Coulomb was an important advance in eighteenth-century understanding of electricity and magnetism. *(Corbis Corporation. Reproduced with permission.)*

flows out. Depending on the charge stored inside the jar, this can create quite a force, as Musschenbroek discovered: "My whole body quivered just like someone hit by lightning. . . . The arm and the entire body are affected so terribly I can't describe it. I thought I was done for." Putting several Leyden jars in parallel increased the effect, which permitted Nollet to use the device in spectacular demonstrations, for instance by simultaneously jolting 180 soldiers for the entertainment of the king and court. These kinds of experiments anticipated the electrical shows given today in many science centers by Van de Graaf electrostatic generators.

The first person to give an explanation of the Leyden jar was Benjamin Franklin (1706-1790), one of the earliest American scientists. In 1747 he proposed a one-fluid theory of electricity based on accounting, meaning that if someone transfers electricity to a friend, the latter will have an excess of the fluid (be charged positively) while the former will have a lack of it (be charged negatively), the total sum being zero (or conserved). Therefore, this single electrical fluid could neither be created nor destroyed, only transferred from one body to the other. For Franklin, the effect of the Leyden jar was readily explained by the interior of the bottle being charged positively with electrical fluid while the exterior was charged negatively by a similar amount.

Franklin is also credited with the first useful technology associated to electricity, the lightning rod, used to slowly draw electricity from the sky before it could strike as lightning. Later, with his famous kite experiment (done by French experimenters before Franklin did it), the American proved (as did Nollet) that electricity naturally produced in the sky was identical to the one artificially created by friction with an apparatus. From 1752 on, Franklin's system made numerous proselytes in Europe, gradually pushing out of the way Nollet's system. Improved later by numerous "electricians," Franklin's system gave birth to our modern general principles of electricity.

Electricity was then recognized as a "subtle" or "imponderable" (weightless) fluid, which means it is a substance that possesses physical properties but is not like ordinary matter—heat, light, gravitation, and magnetism were also regarded as subtle fluids. Numerous electrical effects had been found, but how could they be quantified? Was there a mathematical law of electricity? Charles-Augustin Coulomb (1736-1806) attacked the problem in 1785. Using a torsion balance of his own conception, he made precise measurements of angles, enabling him to affirm that particles of electrical and magnetic fluids interacted in a manner identical to Isaac Newton's (1642-1727) law of gravitation. Called Coulomb's law, the principle he worked out states that the force between two electric charges is inversely proportional to the square of the distance between them. According to the historian of science John Heilbron, "It was this move [the transfer from celestial mechanics to terrestrial physics], and a similar simultaneous one in magnetic theory, that established a model, and presaged a future, for classical mathematical physics."

Besides Franklin's lightning rod, few applications for electricity were devised in the eighteenth century. These had to wait until the turn of the century and Alessandro Volta's (1745-1827) discovery of current electricity, first produced by his silver and zinc battery. The glorious years that saw Dufay, Nollet, and Franklin's success were soon replaced by a new era characterized by a scientific and instrumental ethos in the manner of Coulomb, whose rigor tried to master matter itself. The quantifying spirit tidal wave that broke against the shore of enlightened Europe required a level of mathematical and technical skills only within the reach of professional scientists. The artistic and gentlemanly work that had previously characterized European science gave way to a nascent modern science.

JEAN-FRANÇOIS GAUVIN

Further Reading

Books

Frängsmyr, Tore, John L. Heilbron, and Robin E. Rider, eds. *The Quantifying Spirit in the 18th Century.* Berkeley: University of California Press, 1990.

Hackmann, Willem D. *Electricity from Glass: The History of the Frictional Electrical Machine, 1600-1850.* Amsterdam: Kluwer Academic Publishers, 1978.

Heilbron, John L. *Electricity in the 17th and 18th Centuries: A Study in Early Modern Physics.* Berkeley: University of California Press, 1979.

Internet Sites
"The Franklin Institute Online." http://www.fi.edu/qa99/spotlight3/index.html

Eighteenth-Century Development
of Temperature Scales

Overview

Development during the early eighteenth century of practical thermometers with stable temperature scales by Daniel Fahrenheit (1686-1736), Anders Celsius (1701-1744), and others made possible reproducible, intercomparable temperature measurements. This had immediate practical applications in many branches of science as well as catalyzing work on the nature of heat and temperature and leading to the formulation of the concepts of specific and latent heat. Further efforts to refine these scales eventually led to the development of an absolute temperature scale.

Background

Devices that merely indicate changes of temperature—known as thermoscopes—have existed since antiquity. Such devices are not thermometers since they provide no means of identifying specific temperatures or quantifying temperature changes. This requires a temperature scale. Galileo Galilei (1564-1642) is generally credited with constructing the first thermometer (1592), although Cornelius Drebbel (1572-1633) may have preceded him. Regardless, Santorio Santorre (1531-1636) was the first to publish a description of a thermometer (1612), and by the middle of the seventeenth century it was a well-known and widely used device. However, early temperature scales were completely arbitrary. They were often attached to thermometer stems capriciously or calibrated against a single temperature, such as the hottest day of the summer, with the degree being some arbitrarily selected distance. Consequently, even measurements made with instruments produced by the same maker were not guaranteed to be comparable.

Two strategies were pursued to overcome this state of affairs: calibrating instruments against some standard reference instrument and constructing fixed-point-scales from first principles. The former approach was undertaken by the Academia del Cimento around 1654 and the Royal Society of London about 1665. Distribution of their instruments made possible the first intelligible meteorological temperature records, but intercomparison with instruments not calibrated to their standards was difficult if not impossible. The latter approach was pursued with scales having either one or two fixed points. Robert Hooke (1635-1702) constructed a scale using one fixed-point and having degrees correspond to equal increments of the thermometric-substance volume. Scales with two fixed points had their degree markings defined as some fraction of the distance between the fixed points. Carlo Rinaldini (1615-1698) first suggested the melting point of ice and boiling point of water as appropriate fixed points. He divided his scale into twelve degrees (1694).

Though much progress was made, confusion was still the order of the day. Because volume is that property of substances which is most readily perceived to change with temperature, it was the preferred means of measuring temperature in early thermometers. However, instrument makers employed different thermometric substances. Failure to understand the difference between heat and temperature made it difficult to select appropriate fixed points and standardize calibration procedures. Furthermore, lack of appreciation for how scales depend on the properties of thermometric fluids made intercomparison between different types of thermometers, even when well calibrated, difficult. Thus, by the beginning of the eighteenth century we find air, alcohol, mercury, and even linseed oil employed as thermometric substances with scales of anywhere from 50 to 300 degrees and calibrated with respect to anything from the freezing point of water to the melting point of butter.

Daniel Fahrenheit was the first to manufacture uniform thermometers with stable, reproducible scales. His scale evolved from Ole Röer's (1644-1710), who took the freezing and boiling points of water as fixed. Fahrenheit mistakenly took Röer's upper fixed point to be normal body temperature. Having done so he fixed the freezing point at 32° and upper point at 96°. The latter point was calibrated by placing a thermometer in a healthy person's mouth. Water's boiling point on this scale was approximately 212°. Though Röer clearly never took water's boiling point as fixed, the Royal Society of London officially established it as such in 1777, fixing the value at 212°. (Normal body temperature on this scale is 98.6°.) The revised scale was quickly adopted in England and Holland and became the standard throughout the English-speaking world.

Anders Celsius also made reliable thermometers with two fixed points. His lower fixed point was determined by immersing the instrument in ice, the upper point by placing it in boiling water. He set the upper point at 0° and the lower at 100°, thus producing the first centigrade thermometer (1741). Carolus Linnaeus (1707-1778) inverted the scale shortly after Celsius' death and is sometimes given credit for the present centigrade scale. But a centigrade thermometer with the freezing point at 0° had been built sometime before 1743 by Jean Pierre Christian (1683-1755).

Impact

The existence of stable, reproducible scales such as those of Fahrenheit and Celsius made it possible to compare readings from thermometers with different scales. George Martine (1702-1741) took advantage of this to prepare conversion tables, which he included in his *Essays Medical and Philosophical* (1740). Subsequently, by mid-century thermometers were widely used in chemistry, medicine, meteorology, oceanography, and many other disciplines as well as finding industrial applications in London breweries and elsewhere. The development of reliable temperature scales also allowed heat to be appropriately distinguished from temperature and led to the formulation of the concepts of specific and latent heat.

Eighteenth-century opinions about heat were largely derived from the medical writings of Galen (c. 130-200), who conceived of bodies as possessing differing amounts of heat, which was thought to be a material substance much like a fluid. Thermometers were thought to measure the quantity of heat in bodies, with temperature being proportional to the amount of fluid present. The amount of fluid in turn was believed to be proportional to volume. Joseph Black (1728-1799) challenged these views armed with the experimental results of Fahrenheit and others.

Fahrenheit had shown that a mixture of three parts water and two parts mercury reached an equilibrium temperature midway between the two initial temperatures—a result only possible on the fluid view if initial volumes were equal. Additionally, George Martine had experimentally demonstrated that when equal volumes of water and mercury are heated the mercury temperature rises faster (1739)—a result not possible if volume alone determines the amount of heat an object can absorb. Black concluded that the quantity of heat in an object was not proportional to its volume or mass (1760). He argued that different substances have different affinities or capacities for heat and that thermometers actually measure the density of heat in objects. Thus, mercury's temperature rises faster than water's because its affinity for heat is greater, meaning mercury accumulates heat faster than water. Black established that the amount of heat in any body is proportional to its temperature, mass, and heat capacity, defining "heat capacity" as amount of heat required to raise the temperature of a unit mass of a given substance one degree. Johann Carl Wilcke (1732-1796) independently arrived at the same conclusion in 1781, though referring to a substance's heat affinity as "specific heat."

The ability to clearly conceive of heat as a measurable physical quantity distinct from, though related to, temperature also made it possible to formulate the concept of latent heat, which ultimately helped undermine the fluid theory of heat. Employing Fahrenheit's mercury thermometers, Black showed that when substances go through phase changes (e.g. from solid to liquid, as when ice turns into water) they absorb heat without changing temperature. According to the fluid theory this heat in some way combines with the substance so as to remain concealed from the thermometer. Because this hidden heat could still potentially affect a thermometer Black referred to it as "latent heat."

If heat were a material substance as maintained by fluid theorists, then it would be expected to have weight. It follows that the weight of substances acquiring latent heat should increase, e.g., a fixed volume of water should weigh more as ice than liquid. George Fordyce (1736-1802) claimed to have measured such increases. Benjamin Thompson (1753-1814), better known as Count Rumford, repeated Fordyce's experiments with greater care and detected no such difference (1787). He concluded that heat must be a mode of motion, not a substance. Rumford provided a more convincing proof when he observed that as long as the mechanical action required to bore a cannon continued, heat was generated (1798). This suggested that heat was a form of motion. It also undermined the fluid theory, which assumed conservation of heat.

As thermometers saw increasing application and were required to measure temperatures over a wider range, it became evident that the values assigned to temperatures by the scales of Fahrenheit and others had no absolute physical significance. In 1739 René-Antoine de Réaumur (1683-1757) realized that the scales of alcohol and mercury thermometers do not coincide because each substance has a different thermal expansion. It was also recognized that the working range for any thermometric substance was restricted by thermal resistance. For example, mercury-in-glass thermometers have a lower range defined by the freezing point of mercury and upper range circumscribed by the glass resistance. To cover the range of temperatures that scientists wished to measure, instruments with different working substances were required. This, however, introduced insuperable practical problems. The ideal solution would be a scale valid over the entire temperature range and independent of any particular thermometric substance.

William Thomson (1824-1907), also called Lord Kelvin, provided such a scale with his thermodynamic definition of temperature based on the efficiency of a reversible Carnot cycle (1848). Known as the Kelvin scale, any thermal state can be assigned a thermodynamic temperature using it. It thus provides a standard and meanings by which to unify thermometry. The problem is that the Kelvin scale is impossible to realize in practice. Fortunately, the Kelvin scale

is equivalent to the ideal-gas scale, which, though impossible to realize in practice as well, is nicely approximated by hydrogen thermometers except at very low temperatures.

STEPHEN D. NORTON

Further Reading
Books

Middleton, W. E. Knowles. *The History of the Thermometer and Its Use in Meteorology*. Baltimore: The Johns Hopkins Press, 1966.

Quinn, T. J. *Temperature*. London: Academic Press, 1990.

Roller, Duane. "The Early Development of the Concepts of Temperature and Heat." Case 3 in James B. Conant, ed., *Harvard Case Histories in Experimental Science*. Cambridge: Harvard University Press, 1950: 9-106.

Van der Star, Pieter. "Life and Works." In P. van der Star, ed., *Fahrenheit's Letters to Leibniz and Boerhaave*. Amsterdam: CIP-Gegevens, 1983.

Periodical Articles
Landsberg, H. E. "A Note on the History of Thermometer Scales." *Weather* 19 (1964): 2-7.

Middleton, W. E. Knowles. "More On 'Celsius or Linnaeus.'" *Bulletin of the American Meteorological Society* 59 (1978): 190.

Taylor, F. Sherwood. "The Origin of the Thermometer." *Annals of Science* 5 (1942): 129-56.

Joseph Black's Pioneering Discoveries about Heat

Overview

In 1761, Joseph Black (1728-1799), an English chemist, discovered that ice, while it was in the process of melting, did not warm up until it was completely melted. He later made the same discovery about boiling water; it stayed stubbornly at 212 degrees while boiling, regardless of the amount of heat applied to the pot. These discoveries, simple though they seem, led directly to major discoveries in the science of heat transfer, called thermodynamics, and to greater efficiencies in the steam engines that powered the Industrial Revolution. These same principles continue to be taught to physics students and engineers today, and are frequently used in operating and designing air conditioning, refrigeration, jet engines, nuclear power plants, and many others.

Background

The Industrial Revolution was as much a revolution in power supplies as in manufacturing techniques, although both were vitally important. Power was originally supplied by wind, water, or muscle because no other sources existed. Unfortunately, wind power was unreliable, water power was limited to very specific areas near rivers or streams, and muscles wore out and needed to sleep (rather, the animals supplying the muscles needed sleep). So, at the turn of the eighteenth century, all the world's inventiveness was dependent on manufacturing technologies powered by sources of energy virtually unchanged in over 1,000 years.

This began to change with Thomas Savery's (c. 1650-1715) Miner's Friend, the first useful invention that used steam as the motive force. In theory, early steam engines were simple—boiling water produced steam, which takes up a great deal more volume than the water did—1,000 times more volume. As the steam formed and expanded, it could be used to push a piston, which was attached to a rod of some sort. On the other end of the rod was another piston that was designed to let water in on top of it. When the steam pushed the piston, it forced the water into a pipe that took it to the ground surface. Then the steam escaped, the piston returned to its original position, and the cycle started again.

Over time, engineers learned that keeping the water under pressure would produce higher pressure steam, which could do more work. They also learned that, at elevated pressures, higher temperatures were needed to make the steam. By trial and error, engineers and physicists developed a number of rules of thumb for making steam engines increasingly efficient and powerful, but they really didn't know how or why water boiled. Without that knowledge, they could not design an efficient steam engine except by trial and error.

In 1761, English chemist Joseph Black gathered the first bits of information that would shed light on this problem. In his laboratory, Black noticed that the temperature of melting ice stayed exactly at the same temperature from the time the ice started to melt until it was completely gone. Later, he noticed that boiling water also maintained the same temperature from the time it started boiling until the container was dry. Both of these were completely independent of the amount of heat being added to the containers of ice or water. In other words, "turning up the flame" had absolutely no effect on the temperature of melting ice or boiling water. Later work would show the converse to be true, too; freezing water and condensing steam also maintained the same temperature.

This discovery led Black to a fundamental understanding of the nature of heat and its applications to making steam. He understood that, up to a point, heat would make water (or ice) warmer, but that after a certain point was reached, heat instead went into melting (or boiling) rather than warming. This meant that two phenomena were occurring, and that each used heat in a different way. Black referred to two different types of heat that he called "sensible heat" and "latent heat."

Sensible heat is the heat that raises water temperature. Latent heat is the heat needed to

Joseph Black. *(Library of Congress. Reproduced with permission.)*

change water (or any other substance) from one phase to another (that is, from solid to liquid or from liquid to vapor). So, heating a tub of water from freezing to boiling requires the addition of sensible heat. Actually boiling the water takes latent heat—in this case, the latent heat of vaporization. Later scientists realized that sensible heat goes into making the water molecules move more quickly while latent heat actually makes them move so much more quickly that they break away from the solid (in melting) or from the liquid (in boiling). They also realized that the only difference between sensible and latent heats are in their effects on the liquid.

Black made another important discovery: he found out how much heat it took to boil (or melt) a given quantity of water. This was the crucial step in learning how to make heat engines more efficient—by knowing exactly how much heat had to be added to a given quantity of water to turn it into steam, engineers could start to design engines along scientific principles rather than by hit-or-miss techniques. By so doing, they could design machines that had a specific power output, or they could begin to control the output of a given engine. In short, Black's discoveries did not cause steam engines to be constructed, but it made it possible to construct them intelligently, for specific purposes.

Impact

Black's discoveries had impacts on science, society, and European politics. Many of these impacts reverberate to the present day.

In the scientific arena, Black's discoveries formed one of the first foundation blocks of what was to become the science of thermodynamics. This field of study, elevated to an independent discipline by William Thomson, Lord Kelvin (1824-1907), has proven enormously fruitful over the intervening two centuries. Simply put, thermodynamics is as its name suggests, the study of heat in motion. However, it has expanded incredibly since its first appearance, and thermodynamic principles now encompass molecular and chemical reactions, refrigeration plants, and engines of all sorts. For example, for a chemical reaction to take place, some chemical bonds must be formed while others are broken. In some cases, heat must be added to the chemicals for this to occur, while in other cases, heat is produced by these reactions. By understanding the thermodynamics of the chemicals involved in these reactions, a chemical engineer can determine what sort of reaction vessel to use—one that will heat the chemicals or one that is refrigerated to keep them from boiling. Some of the finest minds in science for the next century devoted themselves in whole or in part to further elucidating the science of thermodynamics, including James Watt (1736-1819), Rudolf Clausius (1822-1888), Ludwig Boltzmann (1844-1906), Lord Kelvin, James Joule (1818-1889), and Sadi Carnot (1796-1832).

Similarly, electricity is produced in most parts of the world by steam turbines. Other thermodynamic principles can be used to design these power plants to be as efficient as possible, wringing every last kilowatt out of the steam used. This helps to save fuel, reduce emissions, and cut energy bills.

It is in this last area that Black's contribution was most immediately important. He was able to help engineers determine how to make a better steam engine because he helped them to learn more about what happens when water boils. Specifically, he was able to show them that neither water nor steam in contact with the water can be heated to a temperature higher than boiling until all of the water is gone. Since the amount of work that can be extracted from steam depends on its temperature, this meant that a limited amount of work could be done by any steam engine, unless a way could be found to heat the steam even further. To do this, water is boiled and the steam is drawn off into a second chamber with no water in it. In this chamber, the steam is "superheated," since it's no longer in contact with the water, raising its temperature to several hundred degrees. At this point, steam engines become much more efficient, letting the same size engine do much more work with only a minor increase in fuel used.

This translated directly into benefit to society. More efficient steam engines made railroads possible. Factories could become more efficient, mills could grind grain without water or horse power, mine shafts could remain de-watered more efficiently, ships could be propelled without dependence on the wind, and much more. The Industrial Revolution was powered by steam, and steam engines were more efficient because of Black's discoveries.

Finally, steam power had a significant impact on European politics of the late eighteenth and early nineteenth centuries. As the first nation to develop steam power, Britain enjoyed an industrial and military advantage over most of the rest of Europe for several decades. These translated into economic and political advantages, much to the dismay of Napoleon. In fact, Sadi Carnot, a great French scientist who fought under Napoleon, came to the conclusion that only Britain's supremacy in steam power had led to France's defeat in their wars because steam power helped Britain mine coal, fabricate iron, saw lumber, and weave cloth. Because of this, Britain made more arms, built more ships, fed its people and soldiers better, made more sails, and, with a stronger economy, could better afford a war. Steam power became military, economic, and political power. Britain's supremacy in these areas led her to become a great world power for over a century, and she remains a world center for finance and trade to this day.

P. ANDREW KARAM

Further Reading

Books

Atkins, P.W. *The Second Law: Energy, Chaos, and Form.* Scientific American Library, 1994.

Fitt, William C. *Steam and Stirling : Engines You Can Build.* 1980.

Sawford, E.H. *Steaming On: Engines & Wagons from the Golden Age of Steam Power.* Sutton Alan Publishing Inc, 1988.

The Flow of Heat

Overview

In 1791, the Swiss physicist Pierre Prévost (1751-1839) published a theory of heat exchanges, which described how heat is transferred from one object to another. This theory, which is still accepted today, formed a basis for other scientists who studied heat transfer. Prévost also supported the caloric theory of heat—the idea that heat is a liquid. His work helped to convince other scientists of caloric theory, and it was not until late in the 1800s that this idea was finally disproved.

Background

In the eighteenth century, certain phenomena, such as heat, light, electricity, and magnetism, were considered to be imponderable fluids. Scientists used the term *imponderable* to mean "weightless." Imponderable fluids were supposedly composed of weightless, invisible particles that could flow from one object to another. (It is now known that such fluids do not exist.) Scientists thought that the imponderable fluid was responsible for heat *caloric*, and they believed that the greater the temperature of an object, the more particles of caloric it contained. This idea is known as the caloric theory of heat.

In 1789, Antoine Lavoisier (1743-1794) published the first chemistry textbook, called *Elementary Treatise on Chemistry*. In this book, he included a list of all of the chemical elements known at that time. Two items Lavoisier incorrectly listed as elements were light and heat. He included them because he believed them to be imponderable fluids.

Ironically, Lavoisier had disproved the existence of one imponderable fluid, that of phlogiston. Phlogiston was supposedly responsible for combustion, or burning. Prior to his work, scientists believed that a burning object released the fluid phlogiston. Lavoisier, however, showed that combustion involved a gain of oxygen from the air rather than a loss of phlogiston. Partly because of this work, Lavoisier was highly respected by other chemists. For this reason, his belief that heat was a fluid was very influential.

One scientist who accepted the caloric theory of heat was a Swiss physicist named Pierre Prévost. Prévost was a professor of philosophy and physics at the University of Geneva, where he conducted experiments involving heat. At the time, most scientists believed that cold, like heat, was an imponderable fluid. A cold object was said to contain an abundance of "cold" particles. Prévost, however, noted that all observations concerning heat and cold could be explained by a single fluid rather than a pair of fluids. Therefore, he argued that cold was simply a loss of heat.

Suppose, for example, that a person puts his or her hand in snow. Before Prévost's work, many scientists believed that the person would feel the sensation of coldness because cold was entering his or her hand. Prévost, however, claimed that the person's hand would feel cold because of a loss of heat, not because of a gain of cold. In other words, he believed that caloric flowed from the hand to the snow, rather than that cold flowed from the snow to the hand.

Prévost also showed that all objects give off heat, regardless of their temperature. He observed that the hotter an object is, the more heat it gives off, or radiates. For this reason, he argued that caloric would always flow from a hot object to a cold one. In addition, he conducted experiments into the nature of heat radiation. For instance, he heated groups of nearly identical objects to the same initial temperature. He found that dark objects radiated more heat than light-colored objects and that rough objects radiated more heat than smooth objects.

In 1791, Prévost used the results of his experiments to develop a theory of heat exchanges. He proposed that when objects at different temperatures are placed in contact with each other, they will exchange heat by radiation until they are all at the same temperature. The objects would then remain at this temperature as long as the amount of heat they radiated equaled the amount of heat they received from their surroundings.

For instance, suppose that a spoon at room temperature is placed in a hot cup of coffee. Because the coffee is at a higher temperature, heat will flow from the coffee to the spoon. The temperature of the coffee will decrease (because it is losing heat) and the temperature of the spoon will increase (because it is gaining heat) until both the coffee and the spoon are at the same temperature. Eventually, both will fall to room temperature as they radiate heat to their cooler surroundings.

Impact

Many of Prévost's ideas are still accepted today, including his theory of heat exchanges and his observation that coldness is due to a loss of heat. However, the caloric theory of heat (the idea that heat is a fluid) has long been abandoned. Nevertheless, Prévost's observations agreed with the caloric theory, and his results were offered as proof of it at the time. Within a few years, however, a different theory of heat began to emerge. (Prévost's observations also agree with this second theory of heat, which is the one accepted today.)

In 1798, the British physicist Count Benjamin Thomas Rumford (1753-1814) proposed that heat is not a fluid, but a type of motion. He based this conclusion on his observations of cannon making in Munich, Germany. Workers used tools to bore tubes down the center of brass cannons. Rumford noted that an enormous amount of heat was given off during this process. The cannons had to be cooled with water as the boring tool pounded against the brass. According to caloric theory, this heat resulted from the release of caloric as the brass was broken down into shavings.

Rumford, however, believed differently. He developed an apparatus that could be used to measure the temperature of a cannon as it was being bored. From his measurements, he concluded that the total amount of heat given off during the boring process was more than enough to melt the cannon itself if it were not cooled with water. In other words, more caloric seemed to be released from the brass than it could have possibly contained. In fact, there seemed to be almost no limit to the amount of heat that could be produced by this process.

Rumford repeated his experiment with a boring tool that was so dull it did not produce metal shavings. According to caloric theory, the temperature of the cannon should not have changed because caloric would only have been released if the metal were being broken into smaller pieces. However, Rumford found that the cannon continued to heat up as before. (In fact, even more heat was produced). From these observations, Rumford concluded that the motion of the boring tool was being converted to heat. Therefore, heat was a form of motion.

In 1797, the English chemist Humphrey Davy (1778-1829) read Lavoisier's textbook. After learning about caloric theory, he began his own experiments with heat the following year. In one of these, a block of ice was rubbed while its surroundings were kept just below the freezing point of water. According to caloric theory, there should not have been enough heat present to melt the ice. However, Davy observed that the ice did melt. He concluded that the rubbing motion was converted to heat and that this heat was sufficient to melt the ice. (Some scientists now doubt that Davy's experiment could have worked as he described it. However, Davy was convinced and published his results.) Neither Rumford nor Davy's work persuaded most scientists to abandon the caloric theory. Instead, the conclusions drawn by Prévost and the opinions of the respected physicists who supported him seemed more likely to be true.

Then, in 1843, the English physicist James Joule (1818-1889) measured the amount of heat that was produced by a given quantity of mechanical energy (motion). At first, his work was little read, and it was not until the second half of the nineteenth century that the Scottish physicist James Clerk Maxwell (1831-1879) helped to convince the scientific community of the kinetic theory of heat. The word *kinetic* refers to motion; the faster an object moves, the more kinetic energy it has.

Maxwell showed that the average velocity of gas molecules increases with an increase in temperature. In other words, the greater the temperature of a gas is, the faster its molecules move. Maxwell's findings demonstrated that temperature and heat could be described by the movement of molecules, rather than an invisible, weightless fluid.

Even though the majority of scientists in the eighteenth and nineteenth centuries did not know the exact nature of heat, advances were made on Prévost's work regarding the exchange of heat between objects. For example, the French mathematician Jean Fourier (1786-1830) began to investigate the conduction of heat within an object.

He found that this type of heat transfer depended on many variables, including the difference in temperature between the warmest and coolest parts of the object and the object's shape and conductivity. (Conductivity is the ability of a material to transfer heat. For example, metal has a greater conductivity than wood.) Fourier was able to describe this process mathematically, and he published his findings in a book titled *Analytic Theory of Heat* in 1822. The mathematics

Fourier developed for studying heat transfer were later used in studies of sound and light.

STACEY R. MURRAY

Further Reading

Books

Asimov, Isaac. *Asimov's Biographical Encyclopedia of Science and Technology.* 2nd Rev. Ed. New York: Doubleday & Company, Inc., 1982.

Fox, Robert. *The Caloric Theory of Gases: From Lavoisier to Regnault.* Oxford: Clarendon Press, 1971.

Spangenburg, Ray and Diane K. Moser. *The History of Science in the Eighteenth Century.* New York: Facts on File, Inc., 1993.

Other

Lynds, Beverly T. "About Temperature." http://www.unidata.ucar.edu/staff/blynds/tmp.html The University Corporation for Atmospheric Research, 1995.

Wilson, Fred L. "History of Science: Mechanical Theory of Heat." http://www.rit.edu/~flwstv/heat.html Rochester Institute of Technology, 1996.

Ernst Chladni's Researches in Acoustics

Overview

Ernst Chladni, an amateur musician and inventor of musical instruments, studied the production of sound by vibrating solid objects, particularly solid plates. Using sand sprinkled on the plates, he was able to discover the nodal lines associated with the different modes of vibration. The intricate patterns of curved lines, called Chladni figures, generated much popular interest and stimulated mathematical research that would have important implications for the physical sciences and engineering.

Background

The ancient Greeks understood that sounds originated in the vibrations of solid bodies. The Greek mathematician Pythagoras (c. 580-500 B.C.) and his disciples knew that strings (under equal tension) with lengths in the ratio of small whole numbers produced combinations of sounds pleasing to the ear. Around the year 1700 the French mathematician Joseph Sauveur (1653-1716) demonstrated that in its simplest vibrations the string would exhibit nodes (points of no movement) and antinodes (points of maximum movement) at equally spaced locations. For the lowest frequency motion, nodes only occurred at the ends, where the string was fastened to its supports. The overtone, an octave higher in pitch, had an additional node at the midpoint; the second overtone, two nodes, and so on.

A mathematical theory of the vibrating string was developed by the Swiss mathematician Johann Bernoulli (1667-1748) as one of the first applications on the new mechanics of Isaac Newton (1642-1727). A full understanding of the vibrating string would come with the work of the French mathematician Joseph Fourier (1768-1827), who provided the mathematical framework needed to understand the general case of vibration as a combination, or superposition, of the individual simple modes.

But strings are not the only source of musical sound. Relatively pure tones can be obtained from a membrane stretched tightly across a cylindrical frame, from the air column in a flute or an organ pipe, and from bells or flat plates such as are used in the modern xylophone. The vibrations of a drumhead were explained by the Swiss mathematician Leonhard Euler (1707-1783), and later independently by the French mathematical physicist Simeon Denis Poisson (1781-1840). The study of sound generation by wind instruments was taken up by the Swiss mathematician Daniel Bernoulli (1700-1782), Euler, and the Italian-French mathematician Joseph Louis Lagrange (1736-1813). Euler also began the study of the vibrations of bells and solid rods but did not progress far, finding the fourth order differential equations required beyond the capacity of the mathematical techniques of the time.

Ernst Florenz Friedrich Chladni (1756-1827) undertook the study of vibrating plates by experiment. Chladni was born in Germany, the son of a lawyer who forced him to study law against his wishes. On his father's death, he abandoned the law for the study of acoustics and the invention of musical instruments. By 1787 he had published his first experimental study. Much as Sauveur had demonstrated the existence of nodes or points of no motion in the different modes of vibration of a string, Chladni discovered nodal lines or curves on the two-dimensional vibrating surface. To excite the different modes of vibration he would draw a violin bow across the edge of the plate at different locations. The plate would be clamped at one location. To visualize the nodal lines and curves he sprinkled sand on the surface of the vibrating plate. The sand would only aggregate at locations where there was no motion. The resulting figures soon became known as Chladni figures. By changing the location of the clamp and the placement of the bow, many different figures could be obtained. Chladni also conducted investigations of the velocity of sound in different solid media.

Chladni spent much of his time touring through Europe, giving demonstrations of his figures and also of two new keyboard instruments he had invented that were based on the glass harmonica. When he visited the French Academy of Sciences in 1708, the Emperor Napoleon himself was present and was so intrigued by Chladni's intricate sand figures that he persuaded the Academy to issue a cash prize to the scientist or mathematician who could provide an explanation for what was observed.

The challenge was accepted by Sophie Germain (1776-1831), a remarkable French woman mathematician who had taught herself mathematics from lecture notes borrowed from male students attending the Ecole Polytechnique. In 1811 she submitted a memoir on the subject under a pseudonym to the academy but it was rejected. In 1913 a second entry of hers received honorable mention. In 1816 a paper submitted in her own name in which Germain solved the problem by applying the calculus of variations to a fourth order differential equation finally won the grand prize. Germain immediately gained world renown for this work, but died in 1831 before she could accept the honorary doctoral degree that would have been bestowed on her by the University of Göttingen.

Chladni's figures also stimulated further experimental work. The French physicist Felix Savart (1791-1841) extended Chladni's work using powders such as lycopodium, with much smaller particles than Chladni's sand. A much more extensive study was reported by the great English physicist Michael Faraday (1791-1867) in 1831.

Impact

The histories of music, the visual arts, science, and mathematics have been interconnected at many points. Early Greek notions of proportion entered into both music and geometry as well as into sculpture and architecture. The notion of a harmonious relationship based on numbers played an important role in early astronomical speculation. Innovations in musical instruments often followed discoveries in acoustics. Chladni's interest in music certainly played a role in motivating his discoveries and in popularizing them.

Chladni's researches presented a challenge to the mathematicians and physicists interested in the mathematical description of vibrating bodies. Compared to the description of the vibrating string, which involves a differential equation of second degree in one spatial dimension, the case of the vibrating plate requires a fourth-degree equation in two spatial dimensions. Fully understanding the mathematics of the vibrating plate would require not only the work of Germain, but further major contributions by the French civil engineer Claude Louis Navier (1785-1836), Poisson, and French Mathematician Augustin Louis Cauchy (1787-1857).

The understanding of the vibrations of solid plates that grew out of the researches of Chladni, Germain, and others has had important implications for science and technology. Many large structures include elements that are, to a first approximation, rigid flat plates. A very important design consideration in building a bridge, a skyscraper, or an airplane wing or fuselage is avoiding frequencies of vibration that might become excited by the wind or other disturbances. If the amplitude of such vibrations became large enough, the results could be unpleasant or dangerous for people near these structures. Special considerations apply to the design of auditoriums and speaker systems. The mathematics of Germain and Poisson has also been extended to Earth's crust and is used by seismologists studying wave motion.

The use of a visual characterization of a phenomenon in the absence of an adequate and usable mathematical theory is a recurrent theme in the history of physics. In Chladni's case the mathematical techniques were not yet available. Faraday was certainly acquainted with Chladni's results when he introduced the notion of lines of force to represent the electric and magnetic fields. As in Chladni's case, the required mathematics—the calculus of vector fields—had not been fully developed. In a much more recent example, the American theoretical physicist Richard Feynman (1918-1988) introduced a set of diagrams to represent the interactions between elementary problems. While the mathematical formalism underlying the use of Feynman diagrams has been established, the thinking and communication among particle physicists is mainly in terms of the diagrams rather than the equations.

By the mid-eighteenth century public demonstrations of scientific principles like those held by Chladni were becoming increasingly common. The demonstration of electrostatic principles and static electricity machines were becoming popular both in Europe and in colonial America. They would become increasingly more important in the nineteenth century, with the founding of the Royal Institution in London and similar activities in other countries.

Chladni's demonstrations also attracted some attention from the new writers of what would come to be called science fiction. In a short story, "The Atoms of Chladni," by J. D. Whelpley, published in *Harper's New Monthly Magazine* in 1859, a mentally unbalanced scientist builds a device based on Chladni's experiments that can record conversations. The magazine was almost certainly one that would have been sold by the young American inventor,

Thomas Edison (1847-1931), as a 12-year-old newsboy on the Grand Trunk Railroad. While there is no solid evidence that Edison had read the story, it is a tantalizing speculation that Chladni's experiments might have, perhaps unconsciously, played a role in Edison's invention of the phonograph.

DONALD R. FRANCESCHETTI

Further Reading

Franklin, H. Bruce. *Future Perfect: American Science Fiction of the Nineteenth Century*. New York: Oxford University Press, 1966.

Kline, Morris. *Mathematical Thought from Ancient to Modern Times*. New York: Oxford University Press, 1972.

Levinson, Thomas. *Measure for Measure: A Musical History of Science*. New York: Simon and Schuster, 1994.

Lindsay, R. Bruce, ed. *Acoustics: Historical and Philosophical Development*. Stroudsburg, PN: Dowden, Hutchinson and Ross, 1972.

Taton Rene, ed. *History of Science: The Beginnings of Modern Science, From 1450 to 1800*. New York: Basic Books, 1964.

Wolf, Abraham. *A History of Science, Technology and Philosophy in the Eighteenth Century*. Gloucester, MA: Peter Smith, 1968.

Eighteenth-Century Meteorological Theory and Experiment

Overview

Eighteenth-century meteorological study in Europe marked interest in methodical observations on a larger scale, providing a more systematic and cumulative database to formulate early theories of atmospheric movement and phenomena. With innovative instrumental designs, laboratory experiment and resulting new fundamental laws of physics helped the progress of meteorology into field application. Expanding theory included further delving into the general circulation of the atmosphere, the nature of lightning, and some delineation of the basic physical extension of atmospheric parameters or indicators (pressure, temperature, and humidity). Perhaps the most significant and fundamental advance marking eighteenth-century meteorology was early attempts at mathematical correlation through the study of fluid flow.

Background

While seventeenth-century interest in the atmosphere had mainly dealt with basic conception of atmospheric parameters and devising experimental instrumentation to record these parameters, methodical data gathering had not spread widely enough. An almost faddish interest in meteorological measurement pervaded the eighteenth century, and now well-established scientific academies from the previous century turned to more concerted daily observations and made attempts at cooperative networks of atmospheric data recording from city to city. Meteorological data collecting at sea proceeded in much expanded fashion from the seventeenth century, with wind data of particular interest for nautical charts. Comprehensive enough meteorological data prompted some attempts at dealing with theory of the general circulation of the atmosphere, initially by the otherwise obscure Ralph Bohun (d.

1716) and later astronomer Edmond Halley (1656-1742). Halley had the right idea about the trade winds of the tropics arising from upward flow of equatorial air with north and south air flow induced to replace it, but he could not explain the easterly orientation of this air flow.

The seventeenth century had initialized the empirical and instrumental inspection of the atmosphere. Galilei Galileo's (1564-1642) profound experimental acumen had included questions dealing with atmospheric pressure and credit for the first glass-bulb thermometer. England's Francis Bacon (1561-1626) with his dedication to inductive philosophy used his own fascination with the wind as the one completed subject for his method. Rene Descartes (1596-1650), who introduced the corpuscular theory of matter and motion, applied it to investigating the phenomena of the atmosphere, particularly the rainbow.

Groundwork in researching the nature of gases was one hallmark of the seventeenth century. And important correlation in methodology between laboratory experimentation and instrumentation was followed by meteorological applications probing the parameters of the atmosphere. Galileo's student Evangelista Torricelli (1608-1647) invented the glass tube mercury barometer (there were also water barometers) for measuring atmospheric pressure. Robert Boyle (1627-1691) would coin the term "barometer" and use the instrument in the laboratory and in weather observation along with various hygrometers (indicating atmospheric humidity) in observations of changing weather. He and other familiar names—fellow Englishman Robert Hooke (1635-1695) and Dutch Christian Huygens (1629-1695)—all studied relevant temperature scales centering on freezing and boiling points of water. Hooke and architect/scientist Christopher Wren (1632-1723) invented rain gauges and the first attempted multiple functioning weather instruments, or weather clocks, using a clock as a drive to register remote readings—still 200 years in the future. Gottlieb Wilhelm Leibnitz (1646-1716) conceived the aneroid or vacuum-chamber barometer, and Pierre Daniel Huet (c. 1721) developed the idea of the pressure-tube anemometer for measuring wind speed.

Impact

Why did the tropical winds move with a easterly slant toward the equator, not north and south as Halley had said? In 1735 a London lawyer who dabbled in natural philosophy and made regular weather observations, George Hadley (1685-1768), was the first thinker to understand that the rotation of the Earth was the cause, and set it down in a paper titled "Concerning the Cause of the General Trade Winds." Hadley thought in terms of equatorial atmospheric air cycling by warming and rising then flowing out and sinking. The flow became known as the Hadley Cell. And his effort was the first limited but correct explanation of tropical circulation of the atmosphere.

Wind observations with respect to storm movement prompted ideas about the scale of weather patterns. Again, seafarers were well acquainted with changing weather and the bad winds that accompanied the worst. They had logs of observations on these and the rarer great whirling storms in tropical oceans: hurricanes, cyclones, typhoons. American political philosopher/scientist Benjamin Franklin (1706-1790), who included weather studies among his polymathic interests, was an acute observer. On observing the passage of a storm in October of 1743 at Philadelphia not reaching Boston to the northeast until the next day, he studied other observations of the storm and other similar storms and reasoned correctly that the prevailing trajectory of east coast weather was from the southwest to the northeast. Investigating storm movement and differentiating between local surface winds and wider scale storm winds would become a meteorological focus of the next century.

Atmospheric phenomena had always prompted curiosity and certainly superstitious fears. Nothing in the atmosphere put more fear into people and conjured more allusions to myth and an avenging God as lightning and thunder. The true nature of lightning became known before the middle of the eighteenth century, prompted by the century's mania for demonstrating electrostatic phenomena, whether as a study in the laboratory or as a diversion in the parlor. One of the most dedicated researchers was the celebrated Abbé Jean Antoine Nollet (1700-1770), whose electrical research, starting about 1730, had included collaboration with fellow Frenchmen chemist Charles Francois Dufay (1698-1739) and technologist René Antoine de Réaumur (1683-1757). Dufay had discovered the attractive-repulsive nature of electricity and believed that this indicated the existence of two separate electrical fluids moving in opposite directions and inducing charge on bodies. Nollet agreed.

Various simple electrical apparati were used in studying the nature of static electricity from conductive materials to containers to hold electric charge, the Leyden Jar, and measuring gauges. Charging these jars and discharging them in ever more spectacular ways became a favorite demonstration. Nollet, one of the first, and others constructed "electroscopes" or electrometers—crude gauges which would indicate the presence of electric charge by the separation of suspended pith balls or strips of metal foil. It was not long before the spark of discharging static electricity prompted thinkers to conjecture that lightning might also be electricity.

Franklin, best known for his electrical experiments, began research in 1746, deciding (1751) that there was one fluid of electricity leaving surfaces positively or negatively charged. Franklin also decided that lightning was but a huge electric spark. With the basic equipment being used, he proposed (1752) a method for carrying out an experiment with lightning—never realizing sufficiently how dangerous that was. During a thunderstorm Franklin flew a kite—even more dangerous—high enough that electric charge traveled down the string and charged a Leyden Jar. Lightning was an electric discharge. The experiment was actually carried out by several researchers in differing circumstances before Franklin.

Probing the atmosphere in the field was yet another interesting aspect of eighteenth-century meteorological investigation, starting with much weather observing at the systematic level. The serious eighteenth-century observer realized the importance of some sort of systemization to accurate recording of weather observations. Coordinated weather observing was carried out by Britain's Royal Society, France's Academie des Sciences, in central Europe, and Russia. In 1780 under the auspices of the Societas Meteorologica Palatina of Mannheim, Germany, a network of 39 weather-observing stations began operations with the first standardizations of mounting and reading instruments, which was later followed and imitated by other groups.

There were many observer/instrument enthusiasts taking extended observations but largely known for other accomplishments. There was John Dalton (1766-1844), famous for his atomic theory and interest in the properties of gases (partial pressure), who was an avid weather observer. He acquired the interest from his Quaker schoolmaster Elihu Robinson, who observed and also built weather instruments. Dalton kept daily weather records from his childhood until his death and from this developed an acute interest in forecasting weather "with tolerable precision" as important "advantages" to humanity. He published his ideas about observing and making instruments in *Meteorological Observations and Essays* (1793). Benjamin Franklin and fellow Americans George Washington and Thomas Jefferson were weather observers. More central figures in meteorology were Jean-André Deluc (1727-1817) and Horace-Benedict Saussure (1740-1799). Saussure, the inventor of the hair hygrometer (1783), disputed theories of evaporative cooling with De Luc, who invented a gut hygrometer and shared with Joseph Black the discovery of the latent heat property.

Other observing went further afield. And this was important to the early conceptions of the spacial delineation of atmospheric parameters: pressure, temperature, and humidity. Climbing mountains was one way to ascend higher to appraise characteristics of the atmosphere. The only problem was that with few exceptions, prior to the mid-seventeenth century, no one wanted to climb mountains, which were considered anomalies of nature. But mountaineering interest and meteorological field research came together about the same time late in the eighteenth century. Mont Blanc was climbed in 1786. Saussure followed suit with his own ascent of the highest mountain in the Alps the next year with scientific equipment and a large contingent. Taking barometers, hygrometers, and thermometers on mountain excursions became a popular pursuit. In the process, the fact that atmospheric pressure, humidity, and temperature decreased with height prompted the urge to venture into the atmosphere itself.

Humanity was already heading into the free atmosphere in 1783 with the first hot air balloon flights. Among other intrepid balloonists was physicist Jacques Charles (1746-1823), who would add the factor of temperature (Charles Law, warming/expanding and cooling/contracting of a gas) to Boyle's inverse law of pressure and volume. Experimenting with hydrogen-filled balloons, he finally made a flight at the end of 1783 in his own designed balloon, taking a barometer to use as an altimeter. Two years later American Dr. John Jeffries (1744-1819) took a barometer, hygrometer, and thermometer on a balloon trip up to 9,000 feet (2,743 m), recording those parameters. On subsequent balloon flights by others, more would lie in store than was bargained: freezing temperatures and violent storms. The atmosphere was being observed—sometimes painfully—but at the source and with accurate instruments.

Interpreting meteorological data mathematically was a slow process, but a string of brilliant mathematicians put the process underway in the eighteenth century through the development of the science and mathematics of hydrodynamics. The first of these was the most celebrated of the Swiss Bernoulli family, Daniel (1700-1782), whose greatest published work was his *Hydrodynamica* (1738). He explained both hydrostatics and dynamics via the mechanics of Newton with many examples of forces in fluid flow. Following him was the one Frenchman of the group, Jean le Rond D'Alembert (1717-1783), whose ideas centered on analyzing problems of dynamics by static means (bodies at rest and acted on by forces). He applied this method to coming up with rather involved equations for showing the planar motion of fluids, further applying this to the ocean and the atmosphere. The latter was an important confirmation of what many had suspected since the seventeenth century: the air behaved like a fluid. The most advanced ideas were those of Swiss Leonhard Euler (1707-1783). He was the first to elucidate on the force of pressure in fluid flow. In three papers between 1753 and 1755, he formulated the basic equations and concepts of fluid mechanics.

The study of weather and basic meteorological parameters stimulated advances in understanding the workings of the atmosphere—and also some fundamental scientific method. In the latter case, Swiss physicist/mathematician Johann Heinrich Lambert (1728-1777) included among his varied research the first ordering of recorded atmospheric measurements with the use of graphical plots to interpret the data, a fundamental method of interpreting the relationships of parameters in modern science.

WILLIAM J. MCPEAK

Further Reading

Books

Frisinger, H. Howard. *The History of Meteorology to 1800.* New York: Science History Publications, 1977.

Good, Gregory A., ed. *Sciences of the Earth: An Encyclopedia of Events, People, and Phenomena.* 2 vols. New York: Garland Publishing, Inc., 1998.

Middleton, W.E. Knowles. *Invention of Meteorological Instruments.* 3rd ed. Baltimore: John Hopkins University Press, 1969.

Wolf, A. *A History of Science, Technology and Philosophy in the 18th Century.* 2 vols. New York: Peter Smith, 1963.

Biographical Sketches

Benjamin Banneker
1731-1806

African American Inventor, Astronomer and Mathematician

Benjamin Banneker is known as the first African-American scientist in the United States. Until the 1980s his life and work were virtually ignored by historians. He made important contributions the fields of inventing, surveying, agriculture, and the sciences. He corresponded with some of the leading political figures during the revolutionary war period, and his work began the slow process of gaining respect for African-American scientific contributions.

Banneker was born in Ellicott, Maryland, on November 9, 1731, and was the first of three children. The only son of two freed slaves, he grew up on the family farm where he cultivated early in his childhood a fascination with the mechanical nature of how things work. Despite having to work hard to help support the family, he received eight years of formal schooling from a nearby Quaker school. During his childhood his grandmother read and discussed the Bible with him, and he read many books by William Shakespeare, John Milton, Alexander Pope, and other popular authors of the day. In addition, he enjoyed studying the movement of the stars and planets and creating and solving math puzzles. It is said that he owned no books of his own until he was nearly 32 years old.

Because he grew up on a tobacco farm, Banneker was very familiar with the problems of farmers. He designed a system of irrigation that countered the effects of dry periods that had previously been devastating for local farmers. During the Revolutionary War, wheat grown on a farm designed by Banneker is credited with saving the revolutionary troops from near starvation.

Banneker was also known as a mechanical genius and a knowledgeable mathematician. At age 21 he took apart a friend's watch in hopes of understanding how it worked. With the watch as a model, he worked for two years, building a wooden clock that gained him a measure of fame. The clock kept time and struck on the hour for over 20 years.

Banneker also taught himself astronomy through books he borrowed from friends. He built a work cabin with a skylight and soon was able to predict solar and lunar eclipses. In 1791 he compiled his information into a book titled *Pennsylvania, Delaware, Maryland, and Virginia Almanac and Ephemeris,* an almanac that contained information about tides, eclipses, history, literature, astrology, and medicine. This popular volume ran for six years and remained in publication for nearly ten years. Banneker sent a copy to Thomas Jefferson as proof that black people, if given better living conditions and an education, were capable of intellectual accomplishments. The book impressed Jefferson, and he passed it along to the French Academy of Sciences and to the President of the United States, George Washington.

Later in 1791, as a direct result of Banneker's correspondence with Thomas Jefferson, President Washington appointed Banneker to a six-person planning team surveying the Territory of Columbia in preparation for the future American capital to be built there. When Pierre-Charles L'Enfant (1754-1825), the project architect, was terminated, he took the only set of plans with him. Within two days Banneker was able to recreate the layout of the streets, parks, and major buildings of L'Enfant's plans from memory. This amazing effort saved the fledgling United States government incalculable time and effort, as it was able to use Banneker's recreated plans to build Washington, D.C.

Banneker also was involved in the anti-slavery movement. He wrote pamphlets and essays, the most significant being *A Plan of Peace-Office for the United States.* His opinion was well respected by many people opposed to slavery. Later in his life he sold off parcels of his farm, maintaining only enough funds to finance his scientific experiments. He died in poverty at his farm in October 1806. As his body was being interred, his house caught fire, destroying his books and notes and his prized wooden clock.

LESLIE HUTCHINSON

Joseph Black
1728-1799
Scottish Chemist, Physicist and Physician

Joseph Black is best known for his work with gases, alkaline substances, and heat, including the development of the concepts of latent heat and specific heat. His groundbreaking work in the quantitative study of chemical reactions places him among the founders of modern chemistry.

Joseph Black was born in France. His father, who was of Scottish descent, was a wine merchant. He received his early education in Belfast, Ireland, and completed advanced studies in chemistry and medicine at the universities of Glasgow and Edinburgh in Scotland. He was appointed professor of chemistry and anatomy at the University of Glasgow in 1756 and subsequently was given the chair of medicine. He became professor of chemistry at Edinburgh University in 1766 and remained in that position until his death. He was a dedicated and thorough researcher and a very popular lecturer, attracting nonscientists as well as scientists. He also practiced medicine during his early career.

Carbon dioxide, known as "fixed air," had been discovered previously by Stephen Hales (1677-1761), but had not been studied until it was rediscovered by Black. He performed extensive quantitative studies on this gas and on the alkaline materials, known as carbonates, that give off carbon dioxide when heated. He showed that carbon dioxide acts as an acid when dissolved in water and discovered that mild alkaline carbonates become more alkaline when they lose carbon dioxide, while strong alkaline carbonates become less alkaline when they absorb carbon dioxide. He proved that this gas is produced in fermentation, respiration, and the burning of charcoal. His clear demonstration that "fixed air" differs chemically from ordinary air resulted in a clarification of the concept of gases as a state of matter and led to the subsequent discovery of a variety of other gases.

Black's development of the concepts of latent heat and specific heat were the first important discoveries in the study of heat in modern times. Observing that when ice melts or when water evaporates, heat is taken up but the temperature is not changed, he concluded that the absorbed heat becomes "latent": it is contained in the substance but is not active (i.e. does not change the temperature), yet it is available to be given off when water freezes or condenses. The amount of heat needed to melt one gram of a solid substance is its latent heat of fusion. Similarly, the heat needed to evaporate one gram of a liquid substance is its latent heat of evaporation. He also noted that when the temperature of the identical quantities of different substances is raised by the same number of degrees, different amounts of heat are required. He defined the amount of heat required to raise one gram of a substance one degree as the specific heat of that substance. To facilitate his experiments, he developed the methods of calorimetry, the quantitative measurement of the heat given off or absorbed in chemical and physical processes. His friend James Watt (1736-1819) was influenced by these ideas on heat in the development of the steam engine.

Black's careful experimental studies played a major role in the development of chemistry as a modern quantitative science, especially in the development of physical chemistry. In addition to his careful experiments, he made significant contributions to the basic understanding of important concepts in chemistry and physics. Perhaps his most notable contribution was the clarification of the concept of chemical reaction. Black applied to chemistry the Newtonian physics idea that if one agent acts upon another, the second agent will exhibit a reaction to this action. Using this terminology, he explained the chemical change that occurs when substances are mixed to be the result of the "reaction" of these substances as they act upon each other.

J. WILLIAM MONCRIEF

Johann Elert Bode
1749-1826
German Astronomer and Mathematician

German mathematician and astronomer Johann Elert Bode was seemingly born to teach the world about the wonders of astronomy. The child of well-educated parents, Bode early on developed a passion for astronomy and higher mathematics, and only slightly later began publishing accounts of what he had learned and observed. A prodigy, Bode published his first astronomical treatises while he was a teenager.

In 1772, still in his early twenties, Bode accepted a position as a mathematician at the prestigious Berlin Academy. The key benefit for Bode was his supervision of the Academy's yearbook, an annual publication aimed at distilling astronomical information and charts, as well as recapitulating leading scientific discoveries of the time. Before Bode, the yearbook had lost some of its luster, but he applied himself to restoring its reputation for accuracy and thoroughness, with the result that sales of the yearbook increased dramatically.

In that same year, 1772, Bode published the mathematical formula that bears his name. A

Johann Bode. *(Library of Congress. Reproduced with permission.)*

scrupulously honest scholar, Bode never claimed to have invented the formula. That accomplishment took place in 1766 by Prussian astronomer Johann Daniel Titius (1729-1796), who discovered a fascinating relationship among certain numbers, a relationship that seemed to have astronomical significance. According to Titius , adding 4 to each of the numbers in the sequence 0, 3, 6, 12, 24,48, 96, 192, and so on, and then dividing each number by ten, would roughly derive the distances of the planets from the Sun (in astronomical units).

Bode's contribution was to present the formula to a wider public; honoring its originator, Bode named the formula the Titius-Bode law, although it came to be known almost universally as Bode's law. Under either name the formula provided astronomers with a tool for deducing the likeliest locations of planets not yet discovered. The formula, for example, helped astronomers locate the asteroid belt, which occupied an orbit that, according to Titius-Bode, should have held a planet. In 1846 the discovery of Neptune in an orbit that violated the formula caused the law to fade from scientific currency, although it has lost little of its popular appeal. The relationship between the numbers and the planets remained a fascinating coincidence—the law was never found to have a provable theoretical basis.

By 1776 Bode had transformed the *Astronomische Jahrbuch* ("Astronomic Yearbook") into one of the world's most respected catalogs of astronomical data. He published other works frequently, including massive star charts and concordances, one of which, the 1801 *Uranographia*, was a major contribution, cataloging more than 17,000 stars and nebulae. (In 1781 it had been Bode who assigned the name Uranus to the planet newly discovered by William Herschel [1738-1822].)

Bode's reputation continued to grow, and in 1784 he was made Royal Astronomer. In 1786 Bode was named Director of Berlin's Astronomical Observatory and labored to make the institution a world-class facility, although, lacking the best equipment, the facility continued to lag behind other observatories. His own observational achievement rests on his catalogs of astronomical objects visible to the naked eye. But it was not as an observational astronomer that Bode would be remembered: rather, posterity salutes his tireless efforts to centralize and systematize astronomical knowledge.

Bode remained Director of the observatory until his retirement in 1825. He died a year later, still working on the astronomy yearbooks that had occupied him for more than half a century, an all but unparalleled commitment to the ongoing, systematic synthesis of the world's astronomical findings, a crucial contribution during a period when observational astronomy was undergoing a great flowering. Much of the public's information about astronomy and related sciences came from publications overseen by the field's great popularizer, Johann Elert Bode.

KEITH FERRELL

Georges Louis Leclerc, comte de Buffon
1707-1788
French Geologist and Naturalist

Georges Louis Leclerc, comte de Buffon, is referred to as Buffon. He is primarily remembered for his encyclopedic *Natural History* (*Histoire naturelle, générale et particulière,* 44 volumes, 1749-1804). His father, Benjamin François Leclerc, was a state official in Burgundy. Buffon liked to say that he inherited his intelligence as well as his fortune from his mother. While a student at the Jesuit's College of Godrans in Dijon, Buffon demonstrated an aptitude for mathematics. Although his father wanted him to prepare for a career as a lawyer, Buffon was drawn to the study of mathematics and as-

tronomy. A youthful love affair and a duel interrupted his studies and forced him to leave France for tours of Italy and England with the Duke of Kingston. While in England, Buffon was elected a member of the Royal Society.

When his mother died, Buffon returned to France and used his inheritance to pursue his interests in natural history and mathematics. Buffon explored the binomial theorem and the calculus of probability. His first publication on games of chance introduced differential and integral calculus into probability theory. He conducted a famous probability experiment by throwing loaves of French bread over his shoulder onto a tiled floor. The number of times the loaves fell across the lines between the tiles were then counted and compared with the value reached by theoretical calculations. This experiment, which could also be conducted with needles and a chessboard, created great interest among mathematicians and led to the proposition now known as "Buffon's needle problem," which asks the probability that a needle of a given length will fall on a line when a piece of paper is ruled with parallel lines a specific distance apart.

In 1735 Buffon published a translation of Stephen Hales's *Vegetable Staticks*. In the preface to this study of plant physiology, Buffon discussed his own ideas about the scientific method. Five years later, he published a translation of Sir Isaac Newton's *Fluxions*. In 1739 Buffon was appointed keeper of the Royal Botanical Garden and Natural History Museum. Assembling a catalog of the royal collections in natural history provided the impetus for Buffon's attempt to create the first modern systematic account of natural history, geology, and anthropology. Buffon began working on his great *Natural History* in 1749. (The last eight volumes of the beautifully illustrated 44-volume encyclopedia were published after Buffon's death by the Count de Lacépède [1756-1825].) In 1753 Buffon became a member of the prestigious French Academy and presented his famous "Discourse on Style." He was made a count in 1773. Buffon's only son was claimed by the guillotine during the French Revolution (in 1794).

Despite the great popularity of his writings, Buffon had many critics among scientists, scholars, and theologians. Buffon's ideas about geological history, the origin of the solar system, biological classification, the possibility of a common ancestor for humans and apes, and the concept of lost species were considered a challenge to religious orthodoxy. His *Epochs of Nature* and *Theory of the Earth* were especially controversial. Buffon was the first naturalist to construct geological history into a series of stages. He suggested that Earth might be much older than church doctrine allowed and speculated about major geological changes that were linked to the evolution of life on Earth. Buffon suggested that the seven days of Creation could be thought of as seven epochs of indeterminate length. He proposed that Earth and the other planets were formed during the first epoch as the result of a collision between the Sun and a comet. During the second epoch, Earth cooled and became a solid body. During the third epoch, Earth was covered by a universal ocean. In subsequent epochs, the waters subsided, volcanoes erupted, dry land was exposed, and plants and animals appeared. The original landmass broke up and the continents separated from each other. Finally, human beings appeared. These speculations led to an investigation by the theology faculty of the Sorbonne. Buffon avoided censure by publishing a recantation in which he asserted that he had not intended his account of the formation of the earth as a contradiction of Scriptural truths.

LOIS N. MAGNER

Henry Cavendish
1731-1810
English Chemist and Physicist

Henry Cavendish made many significant contributions to a wide range of scientific endeavors and is regarded as one of the greatest scientist of his day. He is best known for his work with the chemistry of gases, the discovery of hydrogen, the determination of the composition of water, the synthesis of water, and his contributions to electrical theory.

Henry Cavendish was born into one of England's most prominent families; two of his grandfathers were dukes. When he received his inheritance, he became one of the wealthiest people of his time. He was educated at exclusive Hackney School and attended Peterhouse College at Cambridge University. He never received a degree, refusing to declare his acceptance of the Church of England, a requirement of all graduates. Instead he returned home to assist his father in the laboratory that the elder Cavendish, a well-known amateur scientist, maintained in

their home. He subsequently followed in his father's footsteps, working for the rest of his life in his own private laboratory in his home.

Cavendish was an extremely shy and eccentric person, an introvert and recluse who seldom left his house. It is not surprising that he also was reluctant to publish the results of his scientific work, most of which remained unpublished until the Scottish physicist James Clerk Maxwell (1831-1879) edited and published it after Cavendish's death.

He was among the first scientists to use quantitative methods in chemistry. He performed extensive experiments with gases, especially hydrogen ("inflammable air") and carbon dioxide ("fixed air"). He determined that hydrogen is a separate substance and that water is not an element but is made up of hydrogen and oxygen. He succeeded in synthesizing water from its constituent elements. He also discovered nitric acid.

Cavendish made significant groundbreaking contributions to the study of electricity. He suggested that an electric atmosphere surrounds a charged substance, providing the basis for the development of electric field theory. He proposed the concept of electrical potential and discovered that the potential across a conductor is proportional to the current through it. He was one of several scientists who independently worked out the inverse square law of electrical attraction and repulsion: the attraction between opposite charges or the repulsion between two charges having the same sign is inversely proportional to the distance between the charges. He was first to determine and study the electrical conductivity of salt solutions. It is interesting to note that Cavendish had no instrument to measure electric current. Instead, he used the reaction of his own body to the electrical shock to measure the intensity of the current, developing a quantitative scale based on the intensity of his reaction to the shock.

As a part of a research effort he undertook late in life, he invented an extremely sensitive torsion balance and used it to determine the constant in Isaac Newton's (1642-1727) universal law of gravitation. He was also able to measure the density of Earth using data obtained with this balance.

Cavendish also performed extensive experiments with heat. His work with the calibration of thermometers, the measurement of vapor pressures, latent heats, and specific heats is particularly noteworthy. He rejected the theory that heat is a substance that flows from one object to another. Instead he championed the explanation of heat as the motion of particles within substances.

Cavendish died at age 78, a recluse with neither close friends nor any direct descendants to inherit his vast wealth. The various relatives who became his heirs honored him by endowing the Cavendish Laboratory at Cambridge University. This laboratory quickly became a major center for developments in experimental and theoretical physics and remains so today.

J. WILLIAM MONCRIEF

Anders Celsius
1701-1744
Swedish Astronomer and Mathematician

Anders Celsius opened Sweden to modern European science and initiated reforms in his country's astronomy curriculum. Known as the founder of Swedish astronomy, he is today remembered for establishing the centigrade scale, which bears his name.

Celsius was born in Uppsala on November 27, 1701. He studied astronomy and mathematics, and in 1725 became secretary of Uppsala's Scientific Society. After teaching mathematics for a few years he succeeded his father as professor of astronomy at Uppsala University (1730).

An ecclesiastical ban on teaching of Copernican theory and lack of astronomical instruments in Sweden obliged Celsius to travel abroad to complete his studies and practical training. His first stop was the new observatory in Berlin where he assisted Christfried Kirch (1694-1740) in taking observations (1732-33). He then traveled to Nuremberg where he initiated an international astronomical review and published his aurora boerealis observations. He then visited Venice, Padua, and Bologne before proceeding to Rome and then Paris.

His arrival in Paris coincided with an ongoing debate over Earth's shape. Isaac Newton (1642-1727) had argued Earth's axial rotation would cause bulging at the equator and flattening at the poles where as René Descartes' (1596-1650) vortex theory implied Earth would be flattened about the equator and elongated along the polar axis. Most Academy of Science members supported Descartes' theory; but, Pierre Louis Maupertuis (1698-1759) spoke out in support of Newton (1732). More accurate measurements were required to settle the matter. In

1733 Charles-Marie La Condamine (1701-1774) proposed an expedition to measure Earth's curvature where it was expected to be greatest—the equator. His expedition departed in May 1735.

Maupertuis proposed a similar expedition to measure Earth's curvature in the Arctic. Celsius, having already made Maupertuis' acquaintance, was consulted about a suitable location. Torneå in Swedish-Finnish Lapland was selected, and Celsius was invited to join the expedition.

Preliminary preparations made, Celsius left for London in July 1735 with the dual purpose of continuing with his studies and purchasing instruments for the expedition. He departed England in April 1736 and soon after sailed from Dunkerque with Maupertuis' expedition. Their work was completed in less than a year, with their measurements supporting Newton's theory. But their work was challenged, and Celsius participated in the ensuing debate. The matter was only decisively settled in Newton's favor with the return of La Condamine's expedition (1744).

Upon resuming his academic post at Uppsala in 1737, Celsius undertook the establishment of Sweden's first modern observatory, which opened in December 1742. While engaged in this project, Celsius also attempted to determine the magnitude of stars in the constellation Aries by purely photometric means (1740) and worked at reforming academic instruction of astronomy.

From early on Celsius began taking daily meteorological measurements. His efforts were however hampered by inaccurate thermometers, and he sought to make a more reliable instrument by employing a fixed scale based on two invariable, naturally occurring points. His lower fixed point was determined by immersing the instrument in melting ice, the upper point by placing it in boiling water. He set the upper point at 0° and the lower at 100° thus producing the first centigrade thermometer. His instrument was ready for use on December 25, 1741.

Confusion still exists over priority for the centigrade thermometer. Carolus Linnaeus (1707-1778) inverted the scale shortly after Celsius' death in 1744 and is sometimes given credit. But a centigrade thermometer with the freezing point at 0° had been built sometime before 1743 by Jean Pierre Christian (1683-1755). If one considers the scale with a freezing point of 100° as the one used today then Celsius was first. If not, the strongest priority claim is Christian's.

STEPHEN D. NORTON

Ernst Florens Friedrich Chladni
1756-1827
German Physicist

Ernst Chladni worked in the fields of acoustics (the study of sound waves) and meteoritics (the study of meteorites). At the time, these areas of science were still in their in-

---

THE TRUTH IS OUT THERE

Before Chladni's research on meteorites, members of the general public (usually peasants) had reported seeing stones falling from the sky. Scientists tended to disregard these reports as being tall tales told by the ignorant. However, after Chladni's book was published and especially after Biot's dramatic report of the meteorites at l'Aigle, these falling stones became a new source of fascination for scientists and nonscientists alike.

For instance, when a shower of stones fell on Siena, Italy, local peasants sold many of them to eager English tourists. When demand for the meteorites became known, many ordinary stones were sold as "meteorites" to the unsuspecting. Meanwhile, Major Edward Topham, who owned the property on which the Wold Cottage meteor fell, was exhibiting the stone in a coffee shop. Visitors paid a small fee to look at the stone and to read the sworn testimony of those who had witnessed its fall. Articles began appearing in magazines about people who claimed to have seen falling stones, but had been afraid of coming forward previously for fear of public ridicule.

Some scientists were dismayed at these new meteorite enthusiasts, believing that the public was being led to accept nonsense. Speaking on this topic, the French scientist Eugène Patrin claimed that the "love of the marvelous is the most dangerous enemy of natural science." Within a decade, however, scientists generally accepted Chladni's ideas.

STACEY R. MURRAY

---

fancy, and Chladni is sometimes known as the "father" of both of these disciplines.

Ernst Chladni was born in Wittenberg, Germany, in 1756. His father was a lawyer, and Chladni was expected to become a lawyer as well. He studied law at the Universities of Leipzig and Wittenberg and received his degree in 1782. His father died soon after Chladni

graduated, and as a result, he was free to devote himself to his true love—science. He gave up the practice of law and stayed on at the University of Wittenberg for several more years in order to study mathematics and science.

Chladni had an interest in music and soon began to investigate acoustics, the study of sound waves. Through his experiments, Chladni discovered an intriguing phenomenon. First, he covered a circular or rectangular metal plate with a thin layer of sand. Then he caused the plate to vibrate by rubbing it with a violin bow. The plate vibrated in a complex pattern, with certain lines on the plate remaining motionless. These lines are called nodal lines. The sand from the vibrating areas of the plates moved onto these lines, producing a pattern of sand that matched the pattern of vibration. He exhibited these patterns, now called Chladni figures, to a group of scientists in Paris in 1809. Today, Chladni figures are still used in demonstrations in high school and college physics classes.

Chladni also conducted experiments involving the speed of sound. In the seventeenth century, a French scientist named Pierre Gassendi (1592-1655) had measured the speed of sound in air. Chladni extended his work by developing a method of measuring the speed of sound in other gases. He did so by filling an organ pipe with a particular gas. Then he measured the pitch (the highness or lowness) of the sound produced by the pipe when it was played. By knowing the pitch, he could then calculate the speed of the sound in the gas.

Besides his contributions to the field of acoustics, Chladni also helped to establish the science of meteoritics—the study of meteorites. He was one of the first to propose that meteorites have an extraterrestrial origin. In Chladni's time, people had observed what appeared to be fireballs streaking across the sky. Various explanations were offered for these objects. For instance, some believed that they were an unusual form of lightning or somehow connected to the northern lights.

Scientists had also discovered unusual masses of stone or metal that seemed very different from the other rocks in the areas where they were found. (We now know that these stones are meteorites.) Again, various explanations were offered for the presence of these stones. Some believed they were formed from lightning strikes; others believed they came from volcanic explosions. However, scientists had not yet made the connection between fireballs and the unusual stones.

Chladni became interested in a 1,600-pound (726 kg) iron meteorite that had been found in Siberia. He was able to obtain samples of it, which he compared to other iron meteorites from around the world. Because of the similarities among them, Chladni concluded that they must all have the same source. Since they were all found far from iron deposits, he argued that these stones were not from Earth, but had fallen from space. He presented his findings in a book titled *Concerning the Origin of the Mass of Iron Discovered by Pallas and Others similar to it, and Concerning a few Natural Phenomena Connected therewith* in 1794.

Chladni also suggested that fireballs were caused when these stones entered Earth's atmosphere. The meteorites Chladni studied showed evidence of intense heating. He proposed that when a meteorite entered Earth's atmosphere at high speed, friction caused it to glow, producing a fireball that could be seen from the ground.

When his book was first published, other scientists rejected his ideas, in large part because they went against traditional scientific thinking. The year after Chladni's book was published, however, a large 56-pound (25 kg) meteorite fell into the English village of Wold Cottage. Several villagers observed its fall, and when scientists analyzed its composition, they found that it had similarities to other types of meteorites.

Just a few years later, the French town of L'Aigle was hit by about 3,000 meteorites. Jean Baptiste Biot (1774-1862) investigated this occurrence in 1803. He observed impact craters from the stones and tree branches that had been broken by the falls. He also interviewed villagers who had witnessed the event. When his findings were published, scientists began to accept Chladni's explanation of the extraterrestrial origin of meteorites.

The areas of science in which Chladni did his research—acoustics and meteoritics—were very new fields. Therefore, he had great difficulty finding a position as a science professor. As a result, he made a living by giving scientific lectures and performing on the euphonium, a musical instrument he had invented. It consisted of glass rods and steel bars that produced sounds when vibrated. (Chladni's euphonium was completely different from the modern euphonium, which is related to the tuba.) He died in 1827.

STACEY R. MURRAY

Charles Augustin Coulomb
1736-1806
French Physicist and Engineer

Charles Augustin Coulomb was one of eighteenth-century Europe's greatest engineers. However, he is better known for inventing the torsion balance and discovering the inverse-square laws of attraction and repulsion for both electric and magnetic phenomena.

Coulomb was born in Angoulême, France on June 14, 1736. His family moved to Paris where he attended lectures at the Collège Mazarin and Collège de France. Following his father to Montpellier, he joined the Société des Sciences de Montpellier (1757). He entered the Ecole du Génie at Mézières (1761) to train as a military engineer and graduated a First Lieutenant in the Corps du Génie. Coulomb served at Brest (1761-1764) before being posted to Martinique (1764-1772). French postings at Bouchain, Cherbourg, and Rochefort followed before his election to the Académie des Sciences (1781) assured him permanent residence in Paris. Henceforth, he devoted himself to fundamental researches in physics.

Coulomb presented numerous memoirs before and participated in hundreds of civil engineering, instrumentation, and machinery committee reports to the Academy. With the coming of the Revolution Coulomb was obliged to resign his commission (1791). He later withdrew from Paris (1793) but returned as a member of the new Institut de France (1795). In 1802 Napoleon appointed him inspector-general of Public Instruction, a position he held until his death.

Coulomb's early work was largely in mechanics, producing treatises dealing with structural supports, beam flexure and rupture, friction, soil mechanics, and ergonomics. He introduced the concept of the "thrust line" and provided a general analysis of its application in controlling for oblique forces in the construction of buildings. In considering friction and cohesion he applied Guillaume Amontons' (1663-1705) law that friction is proportional to load, noting the law does not strictly hold and that coefficients of friction vary from material to material (1773).

Coulomb's further researches on friction appeared in his prize-winning essay "Théorie des machines simples" (1781). His investigations of both static and dynamic frictional forces under various conditions allowed him to confirm Amonton's earlier work that friction is approximately proportional to the normal pressure. He extended this work by considering effects due to variations in load, material, lubrication, velocity, and more. Coulomb's work remained the standard until superseded by early twentieth-century molecular studies of friction.

In another Academy prize-winning essay, Coulomb first presented his work on torsion (1777). He demonstrated how torsional forces in thin threads are proportional to the angle of twist and that torsion suspension could be used to measure extremely small forces, which he succeeded in doing.

Coulomb adapted his torsion balance to study electric and magnetic phenomena. His instrument consisted of a silken thread supporting a carefully balanced, wax-covered straw to which a charged (or magnetized) sphere could be attached. Isolated within a glass tube whose circumference was appropriately marked, another charged (or magnetized) sphere would be brought near the balance and the straw's rotation observed. Forces equivalent to 10-5 grams could be thus measured.

Coulomb presented the results of his researches in a series of memoirs read before the Academy (1785-1791). In the first, he demonstrated that electrical forces of repulsion are inversely proportional to the square of the distance separating the charges (1785). In the next memoir he extended these results by showing that the inverse-square laws of attraction and repulsion hold for both electric and magnetic forces (1787). Although he established the inverse-square laws, Coulomb never experimentally demonstrated electric or magnetic forces are proportional to the products of either charges or pole strengths. In the remaining memoirs Coulomb showed that charge leakage is proportional to charge, and that static charge is distributed on conductor surfaces (not their interiors); and he produced a fully-developed magnetic theory.

STEPHEN D. NORTON

Charles-François de Cisternay Du Fay
1698-1739
French Physicist

Charles-François de Cisternay Du Fay is best known for his discovery of positive and negative electrical charges and some of their properties. He was one of the first to use an electroscope in his research, and his work inspired Benjamin Franklin's (1706-1790) famous kite experiment.

Du Fay was born in France in 1698, and virtually nothing is known about him until he began his experiments with electricity in the 1730s. Prior to that time, although electricity had been known for over 2,000 years (the Greeks first noticed it at about 400 B.C.), virtually nothing had been learned. The best theories of the day were that electricity, like "heaviness," was simply a property of all solid materials. However, Du Fay quickly discovered that there were at least two distinct types of electricity, obtained from rubbing amber and sulfur. The electricity from rubbing amber, when transferred to a cloth, would repel electricity from another piece of amber, but attracted that from sulfur. Du Fay named these "vitreous" and "resinous" electricity, from the types of materials they came from (sulfur being vitreous and amber resinous).

Working in collaboration with Jean Antoine Nollet (1700-1770), a priest, Du Fay tried to better characterize the "electrical fluids" he was investigating. And, at this point, he seriously thought he might be looking at two separate fluids since they behaved differently. To test his theory, he had to find some way to capture enough "fluid" to perform experiments. At the time, there was a storage device named the Leyden jar. This was a glass jar, partially filled with water, and covered inside and out with tin foil. A wire ran through a cork stopper and, as electricity was generated, it passed through the wire into the jar, where it was stored.

Since the Leyden jar could store a fairly large amount of electricity, Du Fay was able to perform some interesting (and potentially deadly) demonstrations. In one, he passed an electrical discharge through 180 soldiers who had joined hands in a circle (which also gave rise to the term "circuit") by having the first soldier hold the jab and the last soldier touch the center wire. He (and others) apparently enjoyed themselves greatly, charging Leyden jars and shocking their friends and relatives. Unfortunately, they did not know that such shocks could be deadly.

Some of Du Fay's experiments were performed with an electroscope, a device he did not invent, but was one of the first to use in research. An electroscope makes use of the fact that like-charges repel each other and that the strength of the repulsive force is proportional to the electrical charge. Therefore, two small metal leaves will be held at an angle against the pull of gravity if they both have the same electrical charge, and the angle separating them will be greater if the charge is greater. While simple in theory and in operation, electroscopes still find extensive use today in fields ranging from physics to radiation safety.

Du Fay's work was influential in Benjamin Franklin's later work with electricity, and he also influenced many other prominent scientists of the day. Today, we understand the importance of electricity and have found innumerable uses for it. This makes it difficult to realize that, until just a few hundred years ago, it was thought of as a curiosity and nothing more. It was not until Du Fay and a few others began investigating its properties in a systematic manner that its nature and potential began to be fully appreciated.

P. ANDREW KARAM

Daniel Gabriel Fahrenheit
1686-1736
German-Dutch Instrument Maker

Daniel Fahrenheit is famous for the temperature scale that bears his name and for developing the first mercury thermometer. He also established quantitatively that boiling-point temperatures vary with pressure, and he discovered supercooling of water.

Daniel Gabriel Fahrenheit was born in Danzig (now Gdansk) on May 24, 1686 to a wealthy merchant family. Demonstrating a gift for learning, he was to have attended the Danziger Gymnasium. However, his parents died before he matriculated. Much against his will, his guardians sent him to Amsterdam to complete a business apprenticeship. He failed to complete it, and a warrant for his arrest was issued with the intent of sending him to the East Indies (1707).

Thus pursued, Fahrenheit traveled throughout Europe, visiting scientists and instrument makers. He later settled in Amsterdam (1717) and established himself as a manufacturer of thermometers, barometers, and areometers. Throughout his life, Fahrenheit devoted all his resources to inventing instruments, including a mercury clock for determining longitude at sea and a pumping device intended for draining Dutch polders. Though highly esteemed, he died impoverished on September 16, 1736.

In 1708 Fahrenheit visited Ole Römer (1644-1710). Since at least 1702 Römer had been making alcohol thermometers with two fixed points and a scale divided into equal increments. He impressed upon Fahrenheit the scientific importance of reproducible, intercomparable measurements by different thermometers. Inspired, Fahrenheit eventually devised a method for commercially producing such thermometers.

Fahrenheit conducted experiments to determine the best thermometric substance. Settling on mercury, he produced the first mercury thermometer (1713). Further investigations led him to discover that glass expansion varies depending on its composition. He eventually devised a method for determining expansion coefficients for different glasses (1729), but had by then developed a method for calibrating thermometers regardless. Fahrenheit accurately determined boiling points for many liquids and quantitatively established their variation with barometric pressure (1723). Based on his discoveries he constructed the first hypso-barometer—an instrument for measuring barometric pressure from changes in water's boiling point (1724). Fahrenheit also discovered the supercooling of water—a phenomenon whereby water is chilled a few degrees below its freezing point without solidifying.

The present Fahrenheit temperature scale evolved from Römer's. Römer calibrated his standard thermometer in melting ice and boiling water, marking the ice point $7\frac{1}{2}$ and boiling point 60. Römer took precautions to avoid negative measurements. However, Römer noticed temperature measurements never went above 20°. Since most of the thermometer's length was thus unnecessary, he calibrated shorter thermometers. Fixing the ice point as before, he placed shorter instruments in $22\frac{1}{2}$° water-baths, as determined by his standard thermometer, to set the upper point.

Observing this procedure, and unaware Römer's second fixed point was actually that of boiling water, Fahrenheit took the upper fixed point to be $22\frac{1}{2}$°, which was approximately normal body temperature. Fahrenheit applied Römer's $22\frac{1}{2}$°-scale to his own thermometers but subdivided it into quarters to yield a 90°-scale, which he later altered to 96°. The freezing point was fixed at 32° and upper at 96°, which was calibrated by placing a thermometer in the mouth of a healthy person. Depending on the barometer reading, Fahrenheit measured water's boiling point from $205\frac{1}{2}$° to $212\frac{1}{2}$°.

Fahrenheit communicated his results to the Royal Society in 1724. Though he clearly never took water's boiling point as fixed, the Royal Society officially established it as such (1777), fixing the value at 212°, which occurred at a barometric pressure of 29.8 inches-Hg. Normal body temperature on this scale is 98.6°. The revised scale was quickly adopted in England and Holland and became the standard throughout the English-speaking world.

STEPHEN D. NORTON

Benjamin Franklin
1706-1790
American Scientist, Inventor and Statesman

Benjamin Franklin is one of the best known Americans of the eighteenth century. He made significant contributions to a number of fields, including playing a major role in the founding of the United States as an independent nation. In science and technology, he is remembered for his work in the theory of electricity and his many practical inventions.

Franklin was the eighth of ten children in a poor family in Boston. He educated himself through his extensive reading. Initially working for his brother on a Boston newspaper, at 17 he went to Philadelphia, where he worked as a printer, eventually buying his own press. He was so successful that he was able to retire at the age of 42. His financial security resulted from the success of his newspaper, *The Pennsylvania Gazette*, and the popularity of *Poor Richard's Almanack*, his book of common sense advice and commentary. Retirement allowed him to concentrate on his scientific interests and on political and diplomatic responsibilities.

His scientific studies centered on the new field of electricity, to which he made both theoretical and experimental contributions. He unified the disparate ideas and experimental information that had accumulated into a new theoretical base and provided ideas that proved to be fertile for subsequent research. In 1747 he undertook a series of ingenious experiments that provided him with a thorough understanding of electrical phenomena. He proposed that lightning was the same as electricity, leading to his famous experiment in which lightening was drawn from the sky via a kite into an apparatus that confirmed that it was electrical in nature. He proposed that electricity is a single electrical "fluid" (which we now know to be electrons) that may be transferred between bodies. He further proposed that the total quantity of this "fluid" is always conserved, a concept known as the conservation of electrical charge that is now regarded as one of the fundamental natural laws.

Benjamin Franklin. *(Library of Congress. Reproduced with permission.)*

He called a body "positive" when it contains an excess of electrical fluid, and "negative" when there is a deficiency of fluid. This and related terminology that Franklin developed form the basis of the terminology still in use today. He also noted the relationship between electricity and magnetism and attempted to magnetize an object using electricity. His scientific interests also included heat absorption, the Gulf Stream, ship design, and the relationship between sweating and body temperature.

Franklin's ingenuity was also evident in a number of practical inventions. The Franklin stove provides heat much more efficiently than a fireplace. Nearsighted and farsighted individuals should be grateful to him for the invention of bifocal eyeglasses. His work with lightning led to his invention of the lightning rod, credited with reducing the number of buildings destroyed by lightning.

The notoriety resulting from Franklin's work with electricity facilitated his political and diplomatic efforts. After serving the colony of Pennsylvania in a variety of ways, he lived in England for 15 years as the chief spokesman for the American colonies. He helped write the Declaration of Independence and then spent almost ten years in France obtaining support for the revolution of the colonies. He helped write the treaty with Britain that gave the colonies their independence. Subsequently he served as President of Pennsylvania and participated in the Second Continental Congress and the Constitutional Convention.

Benjamin Franklin was well known and appreciated in Europe as well as in America. He was the epitome of American independence, which served as a symbol for efforts to gain democratic government throughout the world. He was elected to England's Royal Society and the French Academy of Sciences and received honorary degrees from the University of St. Andrews and from Oxford University.

J. WILLIAM MONCRIEF

Sir Edmond Halley
1656-1742
English Astronomer and Physicist

Edmond Halley is best known for predicting the return of the comet that today bears his name and for the instrumental role he played in the publication of Isaac Newton's *Principia*.

Halley was born at Haggerton, near London, in late 1656 to a wealthy landowner and soapmaker. He attended St. Paul's School before matriculating at Queen's College, Oxford (1673). An experienced astronomical observer even before entering Oxford, Halley began assisting the Astronomer Royal John Flamsteed (1646-1719).

Leaving Oxford, Halley traveled to the island of St. Helena, where he established the southern hemisphere's first observatory and proceeded with the first telescopic mapping of the southern skies. Upon returning to England he published his *Catalogus Stellarum Australium* (1678) which gained him election to the Royal Society at age 22.

Also while at St. Helena he observed the transit of Mercury and realized that simultaneous measurements of a transit by astronomers at locations across Earth's surface could be used to determine the astronomical unit—distance from Earth to the Sun. He later concluded that the 1761 Venus transits would be more suitable (1679) and devised an appropriate observational method (1691,1694,1716). Measurements were made with a simplified version of Halley's method and a distance of 95 million miles was

Edmond Halley. *(Library of Congress. Reproduced with permission)*

computed, which compares well with the present value of 92.9 million.

Good friends with Isaac Newton (1642-1727), Halley encouraged him to publish his work on gravitation and motion through the auspices of the Royal Society (1684). However, due to the Society's financial difficulties, Halley assumed full financial responsibility. The *Principia* finally appeared in 1687.

In 1695 Halley undertook his now-famous study of comets. Compiling records of previous comets, he made calculations of their orbits. This led him to believe that the comets of 1531, 1607, and 1682 were the same object. In 1705 Halley predicted the object would reappear in December 1758. Confirmation of this prediction, 16 years after his death, was one of the three main supports for Newton's gravitational theory during the Enlightenment. The object is now referred to as Halley's comet in his honor. Halley was also the first to suggest that nebulae are clouds of interstellar gas (1715).

Halley's next great achievement was based on his study of Ptolemy's (c. 100-170) writings. His close scrutiny of Ptolemy's star catalog revealed discrepancies between positions measured in Ptolemy's day and those taken 1,500 years later. Allowing for observational errors and precession, Halley was left to conclude that stars have individual or proper motions. He detected such motions in 1718.

In 1720 Halley succeeded Flamsteed as Astronomer Royal. Correctly believing that accurate lunar measurements would improve methods for determining longitude at sea, he initiated observations of the Moon though its Saros cycle of 18 years—the period after which Sun, Moon, and Earth return to their same relative positions. Halley completed the measurements but died before their publication in 1749.

Halley's contributions outside astronomy include the first map of Earth's winds (1686), a mathematical relationship between height and pressure (1686), and research on tidal phenomena (1684-1701). His most significant geophysical contribution, though, was on terrestrial magnetism. Compass data indicated Earth's magnetic field was slowly drifting westward, which seemed impossible if Earth's interior were solid as was then generally believed. Halley proposed a core-fluid-crust model to explain the phenomenon (1692). According to the model, Earth's magnetic field is produced by a solid iron core. Earth's outer shell or crust is separated from the core by an effluvium-filled region. The westward drift of the magnetic field was the result of the core rotating eastward slightly slower than the crust. Halley further suggested that escaping effluvium, governed by Earth's magnetic field, produced the aurora borealis.

STEPHEN D. NORTON

James Hargreaves
1720?-1778
English Inventor

James Hargreaves, a relatively obscure figure in history, impacted the world with his influential invention. It was this innovation that made him a key contributor in the journey toward the Industrial Revolution, which began in Great Britain in the late 1700s. Hargreaves's *Spinning Jenny* changed the face of textile manufacturing.

Although little is known about his life, he was most likely born in Blackburn, Lancashire, England. Hargreaves was never formally educated and remained illiterate throughout his life.

Hargreaves learned the spinning trade as a boy, when he also worked as a carpenter. By the 1760s, Hargreaves had moved to the nearby village of Standhill. He worked as a spinner in his home, where he was proprietor of his own spinning wheel and loom.

During his time, Hargreaves's home of Lancashire was a center for England's production of cotton goods. However, as the Industrial Revolution had not yet taken a firm hold, it was common practice for most spinners to be relegated to their homes to complete the day's work. Cards, spinning wheels, and looms were all operated by hand.

Hargreaves began experimenting with a new machine that could spin two threads at once. His original design called for horizontally oriented spindles, which allowed threads to become tangled. It is said that Hargreaves's daughter, Jenny, accidentally knocked over her father's experiment and, as its wheel continued to spin freely with the spindles in a vertical position, Hargreaves had his vision. Thanks to his daughter, he also had a name for the new machine: the spinning jenny.

Hargreaves began building a spinning machine that would accommodate an ambitious eight threads. In 1764, Hargreaves had built the *Spinning Jenny*. The machine worked by threading eight spindles at once just by turning a single wheel. An operator could now easily spin eight threads at once. During this time, when it was common for children to work alongside adults, the spinning jenny was a friendly tool for their hands as well.

Although the machine was revolutionary, the yarn it produced was relatively weak and could only be used for weft, the horizontal threads that make up part of the mesh of fabric.

With its new weapon, the amount of yarn being produced by Hargreaves's household was unprecedented. But once Hargreaves began selling his invention to other spinners, the bulk of the local work force became alarmed. Fearing cheaper competition, Lancashire spinners broke into his home and destroyed his invention. Hargreaves responded to the attack by moving to Nottingham in 1768, where he set up a successful spinning mill.

It wasn't until 1770 that Hargreaves finally applied for a patent on the spinning jenny. However, by that time, having already sold several of his machines, the patent was declared invalid when challenged in court. Other spinners were now free to profit from his invention without paying Hargreaves royalties.

As with most inventions, others began improving upon Hargreaves's design. (Richard Arkwright and Samuel Crompton are credited with finally perfecting mechanical spinning.)

The machines evolved, eventually employing 18 to 120 spindles, and yarn production was greatly increased. The number of threads able to be produced at one time grew from eight to 80. When Hargreaves died in Nottinghamshire on April 22, 1778, more than 20,000 spinning jenny machines were producing yarn in Britain.

AMY MARQUIS

Sir Frederick William Herschel
1738-1822
German-English Musician and Astronomer

William Herschel is considered the founder of modern quantitative astronomy. His major accomplishments include his presentation of the first thorough and systematic study of celestial objects beyond the solar system, the discovery of the planet Uranus, and discovery of infrared radiation.

Herschel was born in Hanover, Germany. His father was an accomplished musician, oboist, and bandmaster of the Hanovarian Foot Guards. Young Herschel trained as a musician and became a member of his father's band. In 1757, during the Seven Years War, he escaped to England where he quickly established himself, first as a music copier, then as a performer and composer. In 1766 he became the organist of the chapel in Bath, a resort city in southeastern England. His sister Caroline joined him in 1772.

He was not satisfied with simply being a skilled musician. His innate inquisitiveness led him to try to understand musical theory as well. As a result, he read *Harmonics* by Robert Smith (1689-1768), which he found interesting enough to encourage him to also read Smith's *A Compleat System of Opticks*. The latter introduced Herschel to lenses and telescopes and resulted in a new hobby. He began to use telescopes to observe the planets and stars, becoming particularly interested in the celestial phenomena beyond the solar system. The telescopes that were available and affordable were not strong enough to observe objects at such distances, so he began to make his own. His skill grew until his telescopes were better than those at Greenwich Observatory, the center of astronomical observation in England.

He continued building telescopes and systematically surveying the heavens, making and recording accurate observations. Although these endeavors took a great deal of his time, as well as that of his sister Caroline Herschel (1750-1848) and brother Alexander who assisted him, he was

William Herschel. *(Library of Congress. Reproduced with permission)*

still an amateur astronomer. His study of the heavens remained a hobby, bringing in no income.

In 1781 he observed a faint object in the sky that behaved differently from the stars. He proved that this object was a planet that had not previously been identified, the first planet discovered since prehistoric times. He named the planet after the reigning English king, George III, but it is known today as Uranus. The announcement of his discovery brought him instant fame. He was awarded the Copley Medal by the Royal Society of London, which also elected him to membership. In 1782 he was granted a salaried position as the king's astronomer.

He was then able to devote himself totally to astronomy. He discovered moons associated with Uranus and Saturn, resolved nebulae into clusters of individual stars, and proposed a theory of the evolutionary origin of the universe. He measured the brightness of stars and cataloged thousands of nebulae, star clusters, and double stars. His method of systematically surveying the sky and accurately recording the positions of stars remains a basic method in astronomy today. Of particular importance to science in general was his discovery, in 1800, of "invisible light" from the stars. He showed that this "invisible light," now known as infrared radiation, displayed many of the properties of light even though it was not a part of the visible spectrum.

Infrared radiation has found extensive use in science and technology. As a result of his pioneering work, Herschel was knighted in 1816.

His sister Caroline, who assisted him in his work until his death, became a recognized astronomer herself. She discovered eight comets and three nebulae and received the gold medal from the Royal Astronomical Society in 1828.

Herschel was also assisted by his wife Mary, whom he married in 1788, and by his son John Frederick William Herschel (1792-1871) who continued his father's studies of nebulae and also became a pioneer in the development of photography.

J. WILLIAM MONCRIEF

James Hutton
1726-1797
Scottish Geologist and Chemist

James Hutton was born in Edinburgh, Scotland, in 1726. There was no early indication that this quiet, modest boy would achieve worldwide renown and secure his place in history as "the father of modern geology." With no corroborating data or earlier research to support his conclusions, his personal observations generated the concept of the "rock cycle," which—in three stages—shows that the matter of which rocks are made is never created or destroyed. It simply is redistributed and transformed from one type to another in an eternal recycling manner. This principle, called "uniformitarianism," is one of the foundations of modern geology.

This futuristic thinking matured over many years of study and field work that began when Hutton attended the University of Edinburgh, initially to study the law. He soon abandoned this pursuit for medicine and studied in both Paris, France, and Leiden, Holland. He earned his doctorate in Leiden but never actually practiced medicine.

He returned to Edinburgh to pursue an earlier interest in chemistry, which he had shared with his friend James Davie. The pair had investigated the possibilities of manufacturing sal ammoniac from coal soot in an inexpensive manner and decided to market the resulting product. The results were so successful that Hutton was soon a prosperous, independently wealthy man.

He used part of his fortune to become a gentleman farmer in Berwickshire, England. With this second good source of income at

Caroline Herschel. *(Library of Congress.)*

hand, he gave up active farming and returned to Edinburgh, where he planned to devote himself to scientific research—especially in the area of soil, rocks, and the natural processes that altered their appearance and locations.

Hutton was among the first researchers who made a cycle connection between the three known types of rock: igneous, sedimentary, and metamorphic. His observations were published in two papers in 1788 and later in the book *Theory of Earth* (1795). Unfortunately, Hutton was a far better scientist than author, and his 1795 book was not truly appreciated until it was explained, amplified, and supported by his friend John Playfair in 1802, when the latter published *Illustrations of the Huttonian Theory of the Earth*. In this work Playfair elaborated on the theme that eventually became fundamental to the world of geology: deep time. This was a phenomenal concept that centered on "Unconformity," the visual evidence of a loss of sedimentary layers that indicate a loss of geologic time.

Hutton described Earth as a self-renewing, eternal recycling machine that continues to renew itself in three distinct stages. Stage one is the inevitable decay resulting from the erosion caused by rivers, waves, and tides that wash the soils of the continents into the oceans, creating stage two: layers of sediment that build upon each other. The increasing weight of these layers precipitates stage three: the heating and melting of the sediments, thus producing magmas that generate intolerable heat forces. These forces in turn result in "uplifts": earthquakes, volcanoes, and other catastrophic actions that form new continents, islands, and other surface transformations.

This, of course, is a highly simplified version of Hutton's theories and observations. To this date, there are still thousands of pages he authored that have never been printed (or even translated, since he wrote in French much of the time). However, the impact of his work is still highly regarded and is part of the coursework in most geological curriculums. His contributions to the field receive high acclaim from notable authors and evolutionary biologists such as Stephen Jay Gould (1941-), whose *Time's Arrow, Time's Cycle* (1987) praises Hutton generously.

During the eighteenth century when Hutton came to his amazing conclusions, there was no way of proving the existence of the vast expanses of time in a geological sense. It was not until the twentieth century that scientists (specifically chemists) were able to use radioactive decay in estimating the ages of rocks and other ancient discoveries.

James Hutton died in 1797, leaving a staggering amount of written material to support his field work and studies. The last two sentences of one of his pioneering 1788 treatises have guaranteed him his rightful place at the top of the geological pyramid: "If the succession of worlds is established in the system of nature, it is in vain to look for anything higher in the origin of the earth. The result, therefore, of our present enquiry is, that we find no vestige of a beginning— no prospect of an end."

BROOK HALL

Pierre-Simon Laplace
1749-1827
French Astronomer and Mathematician

Pierre-Simon Laplace's *Celestial Mechanics* is generally regarded as the crowning achievement of Newtonian gravitation. Laplace also made fundamental contributions to mathematics introducing the Laplace transform method, Laplace coefficients, and establishing probability theory.

Laplace was born at Beaumont-en-Auge, Normandy, France, on March 23, 1749. From an

Pierre-Simon Laplace. *(Library of Congress. Reproduced with permission.)*

early age his education was directed towards an ecclesiastical vocation; he attended the Benedictine-run Collège in Beaumont-en-Auge (1756-66) before matriculating at the University of Caen (1766). At Caen his interest in mathematics was sparked, and in 1768 he left for Paris with a letter of introduction to Jean d'Alembert (1717-1773). Impressed, d'Alembert secured for Laplace an appointment at the Ecole Militaire.

Laplace quickly established his mathematical reputation and was elected to the Académie des Sciences (1773). Though he made many fundamental contributions to mathematics, Laplace was primarily interested in understanding the natural world. His collaborations with Antoine Lavoisier (1743-1794) initiated thermochemistry. They designed an ice calorimeter, measured the specific heats of selected substances, and determined that the heat required to decompose compounds into their elements equals the heat evolved when they are formed from their elements (1784). Laplace also investigated the cohesive properties of liquids, explained capillary action, and contributed to tidal theory by taking into account Earth's rotation. However, Laplace devoted most of his energies to celestial mechanics.

One of the most vexing problems for eighteenth-century astronomy was the solar system's stability. In addition to the cyclic variations of

lunar and planetary motions, there also exist cumulative orbital deviations. If these went uncor-
rected, it was generally believed the solar system
would fly apart. Isaac Newton (1642-1727) was
willing to accept Divine intervention to prevent
such a state of affairs. Laplace was intent on proving such anomalies were not cumulative but periodic effects governed by universal gravitation.

In five papers between 1785 and 1788,
Laplace demonstrated that these anomalies were
indeed periodic. He first showed that Jupiter's
orbital acceleration and Saturn's deceleration
over time resulted from their mutual, gravita-
tional-perturbing effect on each, which caused
their orbits to oscillate with a 929-year cycle. He
then showed that the Moon's increasing acceleration was also periodic, arguing that the gravitational influence of the Sun and planets on
Earth's eccentricity resulted in a lunar cyclic effect on the order of several hundred-thousand
years. Generalizing these results, Laplace
showed the total eccentricity of planetary orbits
had to remain constant, implying the solar system would remain stable indefinitely. Laplace
summarized his research on gravitational theory
in his five-volume *Celestial Mechanics* (1799-
1825). Presented with a copy of the volume,
Napoleon said he had been told there was no
mention of God, whereupon Laplace supposedly
replied, "I had no need of that hypothesis."

Laplace is best known for his version of the
nebular hypothesis (1796), according to which
the solar system originated from a rotating disk of
gas. As the gas contracted, centrifugal forces exceeded gravitational forces, causing rings of material to be shed. These rings then accumulated into
gaseous balls that later condensed into planets, all
rotating in the same direction about the core,
which condensed into the Sun. Accepted throughout the nineteenth century, the theory fell into dis-
repute when Venus's retrograde motion and the
eccentricity of Pluto's orbit were discovered.

During the French Revolution (1787-99)
Laplace served on the Commission on Weights
and Measures but was later dismissed. After
Napoleon became first consul, he appointed
Laplace, successively, as Minister of the Interior,
Senator, and Senate Chancellor. Laplace eventually turned on Napoleon and rallied to Louis
XVIII (1814). Elected to the French Academy
(1816), Laplace became president of that body
(1817). His last words were reported to have
been, "What we know is minute, what we are ignorant of is vast."

STEPHEN D. NORTON

Antoine Laurent Lavoisier
1743-1794
French Chemist and Economist

Antoine Lavoisier is regarded as the founder
of modern chemistry. Although he made
few discoveries of new substances or processes,
his work in chemical theory provided a synthesis of the discoveries of his contemporaries and a
framework upon which subsequent work could
be based. He is perhaps best known for his discovery of the role that oxygen plays in combustion, his statement of the conservation of matter
in chemical reactions, his clarification of the difference between elements and compounds (molecules), and his part in the development of the
modern system of chemical nomenclature.

The son of a prosperous lawyer, Lavoisier
was educated at the College Mazarin, where he
began his scientific studies after initially studying law. His early scientific publications led to
his election to the Royal Academy of Sciences in
1768, at age 25. His father bought a title of nobility for him in 1772, and he became a member
of the Farmers-General, a private company that
collected taxes for the royal government. His
personal wealth and political influence grew,
and as a member of the Gunpowder Commission, he lived in the Paris Arsenal, where he set
up a private laboratory to test the results of
chemical experiments performed by others and
to carry out his own. In 1791 he was appointed
Secretary of the Treasury.

Even though Lavoisier was involved in social reforms, such as old age pensions, and supported liberal political causes, after the revolution he was regarded with suspicion because of
his previous close connection with the royal
government. He was arrested and imprisoned in
1793, tried, and executed by guillotine in 1794.

Lavoisier's contribution to the founding of
modern chemistry was principally in the area of
theory. He confirmed, consolidated, extended,
and explained the many new discoveries made by
his contemporaries on the European continent
and in England, especially those of Joseph Black
(1728-1799), Henry Cavendish (1731-1810),
and Joseph Priestley (1733-1804). The result was
a new theoretical understanding of chemical
processes that provided the framework for the development of chemistry as a modern science.

Lavoisier discovered the part that oxygen
plays in combustion (burning) and developed a
theory that explained combustion, the oxidation

of metals, and respiration as all being similar reactions of chemical substances with oxygen gas. Additionally, his theory of combustion discredited the phlogiston theory, which had been a major detriment to scientific progress.

Although not the first to employ careful quantitative methods in the study of chemical processes, his endorsement and use of them was significant in the development of chemistry as a quantitative physical science. The use of a calorimeter by Lavoisier and Pierre Simon Laplace (1749-1827) to measure specific heats and heats of reaction was an important step in the founding of thermochemistry. Lavoisier was also the first to realize that all substances can exist in three states—gas, liquid, and solid. He played a significant part in the development of the metric system and in revolutionizing the nomenclature of chemical substances, both of which are still in use today in much the same form. In this new system of nomenclature, the name of a substance indicates the elements of which it is made.

Despite his brilliance and the enormous contributions he made to the founding of modern chemistry, Lavoisier was far from perfect. He was constantly enmeshed in disputes in which he claimed to be the first to make various discoveries, though his claims of priority had no basis in fact. He used the results of other scientists freely, often without acknowledging their work. Perhaps his contribution would have been even greater had he been more willing to work with his scientific contemporaries with greater cooperation and mutual appreciation.

J. WILLIAM MONCRIEF

Nicolas Leblanc
1742?-1806
French Surgeon and Chemist

Although trained as a physician and surgeon, Nicolas Leblanc is best known for his discoveries as an industrial chemist. In this role, he developed the Leblanc process of making soda ash (sodium carbonate) from common salt. Because of the wide variety of uses for soda ash, including making soap, glass, paper, and more, this became one of the most important chemical processing innovations of the eighteenth century.

Nicolas Leblanc was the son of an iron works director, probably born in 1742. He attended medical school, earning sufficient distinction to be named the private physician to the Duke d'Or-

leans at the age of 38. Intrigued by a competition sponsored by the French Academy in 1775, Leblanc became interested in the problem of making soda ash from common salt. At that time, the only source of this important chemical was by extracting it laboriously from wood ash or seaweed ash, or by mixing these ashes with whatever material required the soda ash for processing. One example of this is soap, which was often made by mixing lye (sodium hydroxide) or animal fat with wood ashes. However, the impurities present in the ash made this method less than desirable for many applications. The only source of high-purity soda ash was in some desert lakes, called soda lakes, or in some mineral deposits formed from soda lakes long ago.

French scientists suspected that, since both common salt (sodium chloride) and soda ash were both sodium compounds, it might be possible to change the one into the other cheaply and efficiently. However, five years after the competition began, this still had not been accomplished.

The Leblanc process consisted of treating salt with sulfuric acid to make salt cake (sodium sulfate). The salt cake was then mixed with limestone (calcium carbonate) and coal (primarily carbon) to form a black substance containing mostly sodium carbonate (soda ash) and calcium sulfide. Since the soda ash was soluble and the calcium sulfide was not, mixing this with water would dissolve the soda ash, which would then be recovered by boiling off the water or allowing the mixture to dry. Compared with previous methods, the Leblanc process was very simple and inexpensive.

The Leblanc process was a significant step forward in industrial chemistry because of the widespread use of soda ash. Purer soap became much less expensive and available to many more people. Glass also became cheaper and of higher quality because of the lesser quantity of impurities in soda ash produced by the Leblanc process. Soda ash is also used in paper manufacturing, ceramics production, petroleum refining, for water softeners, as a cleaning and degreasing agent, and as a process chemical in the manufacture of other chemicals containing sodium.

Although Leblanc won the award offered by the French Academy, he never collected on his prize because the French Revolution had begun by the time he completed his work. Nevertheless, he went on to construct a factory to manufacture large amounts of soda ash, only to have it seized by Revolutionary leaders in 1793. It was returned to him by Napoleon in 1802 but, lacking the money to resume his business, he was unable to run the factory. Depressed and disheartened by these setbacks, Leblanc committed suicide in 1806.

Although Leblanc's life ended tragically, his process was already widespread by the time of his death and continued to be used quite widely for nearly a century, until being replaced by the less expensive Solvay process. Today, approximately seven million tons of soda ash are used annually in the U.S. alone, giving an idea of the continuing importance of this chemical in modern life.

P. ANDREW KARAM

Johann Gottlob Lehmann
1719-1767
German Geologist

Johann Gottlob Lehmann established the foundations of stratigraphy, or the scientific study of sedimentary rocks with regard to their order and sequence. His *Versuche einer Geschichte von Flotz-Gebrugen* (1756), in which he classified mountains and established a theory of their origins, was the world's first geologic profile.

Lehmann was born on August 4, 1719, in Langenhennersdorf, then in the state of Saxony but now a part of Germany. He earned his M.D. at the University of Wittenberg in 1741, and went on to a medical practice in the city of Dresden. While working as a doctor, however, he discovered something that interested him more than medicine: rocks.

His amateur fascination with mines and mining soon turned into a vocation, and by 1750 the 31-year-old Lehmann had published a number of papers on ore deposits and their chemical makeup. These publications helped lead to an official commission from the Royal Prussian Academy of Sciences in Berlin to study mining as practiced in various parts of Prussia.

Six years later, in 1756, Lehmann published his classic, *Versuche einer Geschichte von Flotz-Gebrugen*. In it he observed that the placement of rocks on the Earth's surface or below it is not random; rather, it reflects a history (*Geschichte*) in geological terms. Obvious as this idea might seem now, it was far from apparent in Lehmann's time, and not only did his book establish a framework for stratigraphy, it also spurred on local geology, or the investigation of specific sites.

In 1761 Lehmann joined the ranks of the many German scientific minds recruited for the

Imperial Academy of Sciences in St. Petersburg, Russia. He became professor of chemistry, as well as director of the natural history collection at the Academy, but he remained interested in affairs back in Germany. Thanks to his efforts, a research institute called the Freiberg Bergakademie was established in his homeland in 1765.

While in Russia, Lehmann conducted a number of field investigations. For instance, in 1761 he traveled to the Beresof Mines on the eastern slopes of the Ural Mountains, and there gathered samples of an orange-red mineral. He called it Siberian red lead, and after studying it in St. Petersburg, described it as containing lead "mineralized with a selenitic spar and iron particles." This was in fact an early description of what would be identified in 1797 as chromium, a new element.

By that time, however, Lehmann was long dead, having passed away in St. Petersburg on January 22, 1767, at a mere 47 years old. His field studies in Russia later became an example for investigations of geologists who followed him.

JUDSON KNIGHT

Sir Isaac Newton
1642-1727
English Physicist and Mathematician

Sir Isaac Newton is regarded as one of the greatest scientists of all time. His work represents a major turning point in the history of science. He synthesized the work of his predecessors Nicolaus Copernicus (1473-1543), Johannes Kepler (1571-1630), Galileo Galilei (1564-1642), René Descartes (1596-1650), and others into a new understanding of the mathematical nature of the world, culminating in what has come to be known as the Scientific Revolution. Newton's major contributions include the discovery of the three laws of motion that form the basis of modern physics and the law of gravitation, the development of infinitesimal calculus, and the beginning of modern physical optics.

Newton had a troubled childhood. His father, an illiterate farmer, died before he was born, and he was raised by his grandmother after his mother married a man whom Isaac hated. These early circumstances may explain the fact that he was never psychologically stable. He suffered mental and emotional breakdowns in 1678 and 1693, and has been described as tyrannical, unstable, autocratic, suspicious, neurotic, and tortured.

Newton received a bachelor's degree at Trinity College, Cambridge University, in 1665. He was not a good student in the sense of doing required academic work well. Instead, he pursued his own interests, reading the works of scientists and philosophers such as Descartes, Thomas Hobbes (1588-1679), Kepler, and Galileo.

After graduation, he was forced to return home for two years because of the plague. It was during this period that his brilliance became apparent. By the time he returned to Cambridge in 1667 he had developed infinitesimal calculus, had made basic contributions to the theory of color, and had developed ideas concerning the motion of the planets. He was elected a fellow, received his masters degree in 1668, and was named the Lucasian Professor of Mathematics in 1669 when he was only 26 years old. He held this position for 32 years.

Newton published his first paper, on the theory of color, in 1672 and was elected to the Royal Society as a result of his invention of a reflecting telescope. His subsequent studies in optics resulted in the foundation of modern physical optics. He showed that white light is composed of a spectrum of colors, and he developed new theories of light and color based on the mathematical treatment of observations and experiments. He believed that light is made up of corpuscles instead of waves.

In the field of mechanics, his discovery of the three laws of motion that are obeyed by all objects provided the foundation upon which modern physics and, indeed, modern science are built. He also discovered the law of universal gravitation and explained the motion of the planets. The results of his work were published as *Philosophiae Naturalis Principia Mathematica*, often called simply *Principia*, in 1687.

In mathematics, Newton discovered the binominal theory in addition to infinitesimal calculus. He also studied and wrote in the fields of alchemy, history, music theory, chemistry, astronomy, and theology.

Newton never seemed eager to publish the results of his work. When, in the course of a conversation, the astronomer Edmond Halley (1656-1742) learned of Newton's theoretical ideas about mechanics, he encouraged their final development and publication, paying the printing costs of *Principia* himself.

Newton was widely recognized and honored for his groundbreaking work. He was elected to Parliament in 1689 and 1701, and was named master of the Royal Mint in 1696. In 1703 he was elected president of the Royal Society and was reelected each year until his death. He was knighted in 1705. When Newton died in 1727 his body lay in state in Westminster Abbey for a week, and his pallbearers included two dukes, three earls, and the Lord Chancellor—all honors reserved for the most highly respected individuals in Great Britain.

J. WILLIAM MONCRIEF

A KICK FOR SCIENCE

In eighteenth-century England Anglican Bishop George Berkeley was, for a time, an important and influential critic of English physicist Sir Isaac Newton's calculus and of a clockwork universe fully explainable by physics and mathematics. Berkeley attacked the lack of logical support for calculus—and explained away the fact that it worked so well in a wide variety of applications—by contending that calculus benefited from the mutual cancellation of logical errors. In Berkeley's view the growing reliance on science and mathematics was a diminution of religious scripture as the ultimate authority regarding the nature of the universe. In 1710 Berkeley published *Treatise Concerning the Principles of Human Knowledge,* in which he examined the nature and origin of ideas and sensations and attacked materialism based on the "unreality" of matter. According to Berkeley, the physical world was actually a manifestation of a human mind under the direct control of God. Akin to the question as to whether a tree falling in the forest makes a sound if no one is there to hear it, Berkeley asserted that there was no "real" matter—only perceived matter. Moreover, the world as man perceived it was simply an illusion impressed upon the human mind by divine will. Samuel Johnson, a Fellow of the Royal Society and contemporary of Newton, attended one of Berkeley's sermons and later, while standing with friends outside the church, participated in a debate concerning the merits of Berkeley's sophistry regarding the nonexistence of matter. When it was Johnson's turn to speak he sharply stubbed his toe on a large rock and proclaimed, "I refute it thus!"

K. LEE LERNER

Pierre Prévost
1751-1839
Swiss Physicist

Pierre Prévost is best known for his theory of exchanges, which he articulated in 1791.

Whereas scientists in his time postulated the existence of cold as a substance or phenomenon all its own, Prévost correctly stated that bodies radiate only heat. When a person's warm hand becomes cold because of touching snow, it is not because coldness passed from the snow to the hand, but because the difference in temperatures resulted in a transfer of heat from the hand to the snow. This would prove to be a crucial principle in the soon-to-emerge science of thermodynamics.

Prévost was born in Geneva—now the leading city of French-speaking Switzerland, but then an independent city-state—on March 3, 1751. His father was a Calvinist minister, and ensured that Prévost and his siblings received classical educations. As a college student, Prévost studied theology and later law, receiving his doctorate in 1773. Immediately he went to work as a teacher and tutor, a profession that took him to various parts of Holland and France.

His early career was consumed with the classics, and during his years as a tutor, Prévost focused on translating the works of Euripides. In 1778, he published *Orestes,* which won him acclaim among classical scholars, and later that year he was invited by King Frederick the Great of Prussia to join the Academy of Sciences and Belles-Lettres in Berlin. There he continued to pursue his already established interests, publishing works of moral philosophy and poetry; but he also met Joseph-Louis Lagrange (1736-1813), who encouraged him to engage in scientific studies.

Prévost published a number of scientific papers in Berlin, but was forced to return home when in 1784 his father died. He took the chair of literature in Geneva for a year, then moved to Paris, where he continued to occupy himself with the classics. He soon became involved in politics, serving as a member of Geneva's Council of Two Hundred upon his return to the city in 1786.

During the late 1780s and early 1790s, Prévost again became intrigued with scientific questions. This led to his publication of *De l'origine des forces magnétiques* (1788), a work on magnetism that won him attention among physicists. Soon he shifted his focus to the matter of heat phenomena, thanks to the 1790 publication of *Essai sur la feu* by Marc Auguste Pictet, also of Geneva. Pictet maintained that heat was a fluid in material form, and that radiations of heat were the result of expansions and contractions of this fluid in various bodies.

In "Sur l'équilibre du feu" (1791), Prévost maintained that heat was a "discrete" if material fluid composed of widely spaced particles that pass continuously between two bodies. His conviction of heat's materiality may have been incorrect, but his theory of exchanges—the idea that all temperature changes are a matter of heat loss and gain obtained through transfer between bodies of differing temperature—was revolutionary.

Prévost took the chair of philosophy and general physics at Geneva in 1793, and remained there until his retirement 30 years later. He wrote extensively, corresponded with a number of scholars throughout the continent, and translated a number of important works such as Adam Smith's *Wealth of Nations.* In 1796, he became a member of the Royal Society of Edinburgh, and in 1801 was invited to join the Royal Society of London and the Institute of France.

In his later decades, political involvements overshadowed much of Prévost's work. After the outbreak of the French Revolution, he argued strongly that Geneva should remain separate from France, and this led to a brief imprisonment by zealots in 1794. Ironically, when it became clear that Geneva was about to unite with France, as it briefly did in the Napoleonic era, Prévost was named to the commission that regulated this union. In 1814, with Napoleon all but defeated, Geneva returned to the status of a republic, and Prévost served on its representative council. He spent his last years studying the effects of aging, both on humans in general and on himself in particular, and died in Geneva on April 8, 1839.

JUDSON KNIGHT

Joseph Priestley
1733-1804
English Physical Scientist and Theologian

Joseph Priestley is best known for his discovery of oxygen, his fundamental studies of gases, and his contributions to the understanding of photosynthesis in plants.

Priestley was largely self-educated through his extensive reading. His formal studies were intended to prepare him for the ministry in one of the Calvinist nonconformist or dissenting churches that disagreed with the teachings of the Church of England. His growing liberal ideas in religion and politics later led him away from Calvinism. He would eventually become one of the chief spokesmen for Unitarianism in England.

Joseph Priestley.

Priestley's constant search for truth led him, in 1758, to begin scientific experiments. Although he was primarily a minister and theologian throughout his life and remained essentially an amateur in science, he was destined to make substantial contributions to the development of modern physical science.

He began teaching at the dissenting academy at Warrington in 1761 where, since the English universities were closed to dissenters, the emphasis was on practical rather than classical education. He wrote a number of textbooks in several subjects to facilitate this approach to education. Among these was *Rudiments of English Grammar,* which taught contemporary usage of the language rather than the idealized classical form normally taught in the English educational system.

Priestley was ordained in 1762 and received the LL.D. degree from the University of Edinburgh in 1765. His scientific interests resulted in his friendship with Benjamin Franklin (1706-1790), his election as a fellow of the Royal Society in 1766, and his publication of *The History of Electricity* in 1767 and *The History of Optics* in 1772.

Beginning in 1767 Priestley devoted most of his scientific attention to the study of pneumatic chemistry, that is, the study of gases and their chemical processes. It is said that his interest in this field resulted from living next to a brewery where he noticed that gases were emitted in the fermenta-

tion process. In 1773 he produced a publication on artificially carbonated water that won the Copley Medal of the Royal Society. During 1774-75 he carried out the experiments and observations that led to his discovery of oxygen. Since Priestley refused to give up the phlogiston theory of combustion, he could not fully understand his own discovery. Priestley actually called oxygen "dephlogisticated air." He discussed his work with Antoine Lavoisier (1743-1794) in 1774. It was Lavoisier who gave oxygen its current name and explained its role in oxygenation, respiration, and other chemical processes, and who pointed out that the discovery of oxygen and an understanding of its chemistry disproves the phlogiston theory.

In addition to oxygen, Priestley discovered eight other gases, including nitrogen, ammonia, nitrous oxide, hydrogen chloride, and sulfur dioxide. He also made significant contributions to the understanding of photosynthesis in plants. He discovered that plants take in air and purify it, producing dephlogisticated air (oxygen).

Throughout his life Priestley received significant financial support and encouragement from several influential individuals, such as Lord Shelbourne, who later became Prime Minister, and the industrial potter Josiah Wedgewood (1730-1795).

Priestley's nonscientific work was of significance as well. Twenty-five volumes of his theological writings were published after his death. He was among the originators of utilitarianism, the philosophy which advocates the pursuit of the greatest good for the greatest number, and influenced Jeremy Bentham (1748-1832), who subsequently developed utilitarianism into a full-blown philosophy.

Priestley's outspoken support of Unitarianism led to his growing unpopularity with the government, the Royal Society, and the citizens of England in general. In 1791 a mob destroyed his home and laboratory, and in 1794 he was forced to leave England for America. He settled in Pennsylvania where he continued both his theological and scientific studies until his death.

J. WILLIAM MONCRIEF

Karl Wilhelm Scheele
1742-1786
Swedish Chemist and Druggist

Swedish-born chemist whose record as a discoverer of new elements, compounds, and chemical reactions has long remained unequaled.

Karl Wilhelm Scheele was born in Stralsund, Germany (formerly the capital of Swedish Pomerania.) Scheele's interest in chemistry began during his experience as an apprentice to an apothecary in the town of Goteborg. He had no formal education. His teacher, Swedish chemist Torbern Bergman (1735-1784), obtained a small pension for Scheele from the Stockholm Academy of Science, which the young apprentice in turn used to fund his chemical experiments. Scheele spent most of life in poverty, serving as an apprentice in Malmö and Stockholm before settling in as a pharmacist in the small town of Koping. He preferred remaining an apothecary to becoming a university professor.

His lack of career advancement, however, in no way parallels his scientific achievements. While in Koping, Scheele discovered more new substances than any scientist of his time, and perhaps ever since. In 1773 he proposed that air was composed of two gases, oxygen and nitrogen. Scheele prepared oxygen from various oxides but is rarely credited with the discovery. While his research anticipated English chemist Joseph Priestley's (1733-1804) discovery of oxygen, Scheele's only book *Chemical Observations and Experiments on Air and Fire* did not appear until 1777, after Priestley announced his results. There is little doubt, however, that he obtained the gas about 1772, two years before Priestley. Both Scheele and Priestly, however, failed to recognize the significance of their findings and today Antoine-Laurent Lavoisier (1743-1794) takes credit for the discovery.

Scheele published his first scientific paper in 1770, documenting his isolation of tartaric acid from cream of tartar. The free acid is widely used in carbonated drinks, effervescent tablets, and some gelatin products.

The young chemist discovered an extraordinary number of elements: chlorine, barium, molybdenum, tungsten, and manganese. He was the first to isolate chlorine and show that it could bleach cloth. Manufacturers in England and France put his research to commercial use.

While Scheele is credited with identifying chlorine and barium, he believed they were compounds, not elements. In the early 1800s, British chemist Sir Humphry Davy (1778-1829) correctly identified chlorine as an element.

In 1776 Scheele discovered uric acid while analyzing a kidney stone. In the last years of his life, he proved it was lactic acid that made milk sour. He was also the first to prepare the chemi-

Karl Wilhelm Scheele. *(Library of Congress. Reproduced with permission.)*

cal compounds arsine and hydrogen sulfide. In 1783 he discovered the poisonous hydrocyanic acid, without realizing its toxic character.

Among his other discoveries were glycerine, copper arsenite (also referred to as the pigment Scheele's green), and the toxic gases hydrogen sulfide and hydrogen fluoride. Scheele also helped open doors to the modern world of photography by illustrating that sunlight removed salts from silver chloride, leaving only the metal behind.

Scheele died in 1786 at the age of 43. In 1931 the *Collected Papers of Carl Wilhelm Scheele* was published.

KELLI MILLER

Benjamin Thompson
1753-1814
British Physicist, Inventor, and Diplomat

As a scientist, Benjamin Thompson, also known as Count Rumford, is best remembered for his discoveries that provided the foundation for modern theories of heat and for a number of practical inventions that improved living and working conditions.

Thompson was a most intriguing individual. His life was characterized by ambition and

Benjamin Thompson, better known as Count Rumford. *(Corbis Corporation. Reproduced with permission.)*

opportunism. Throughout his career, his contributions to science and society were always interwoven with political schemes. It is fortunate that the significant results of his work were not ultimately harmed by the motives that led him to their accomplishment.

At the age of 19 he married a wealthy widow and, through her influence, was appointed a major in the New Hampshire militia. Because he spied for the royal governor against the colonists, he was forced to flee with the British from Boston in 1776. In London, he convinced the government that he was an expert on colonial matters, was appointed to assist the Colonial Secretary, and soon became Undersecretary of State. He returned to American briefly as the commander of royal troops in New York.

Soon after the end of the war, he moved on to Bavaria where he became aide-de-camp to the elector of Bavaria and a colonel in the Bavarian army. Thereafter, he was knighted by the English king and became the Bavarian minister of war, minister of police, major general, chamberlain of the court, and state councilor, holding all of these offices at the same time. In 1791 he was named Count Rumford of the Holy Roman Empire.

Subsequently, Thompson spent time in England. In 1798 he founded the Royal Institution of Great Britain. He also founded the Rumford Professorship at Harvard and successfully pro-

posed the establishment of the United States Military Academy at West Point. While in France, he became a friend of Napoleon and married the widow of the great French scientist Antoine Lavoisier (1743-1794).

Thompson followed the principle that a problem should be carefully and scientifically studied before a solution is attempted. Most of his scientific discoveries and inventions were the result of observations and experiments during attempts to solve military and political problems. His most important findings on the nature of heat came as a result of observations made while boring metal blocks to make cannons. From these observations he came to important conclusions concerning the relationship of work and heat and made some of the earliest measurements of the equivalence of work and heat. He proposed that heat is not a substance, but rather the result of the motion of particles.

His discoveries of convection currents and conductivity of heat resulted from studies he performed on various materials to be used for army uniforms, and his founding of the science of nutrition resulted from efforts to develop the cheapest way to feed the army. He pocketed the money saved.

Thompson's studies and writings dealing with improvements in social institutions are recognized as among the first in the field of sociology. He studied the economics of social organization, founded public schools, provided work for the beggars of Munich, and extended his findings to society in general. Again, his motives were not the best. He put beggars to work making army uniforms as a cheap alternative to professional tailors. The schools were provided so that the poor would be free to work without having to care for their children during the day.

These various efforts also led to practical inventions that improved life significantly at that time, and many are still in use today. Among these are the double boiler, the kitchen range, a portable field stove, the baking oven, a fireplace that does not smoke, the fireplace damper, table lamps, the drip coffeemaker, and steam heat using radiators.

J. WILLIAM MONCRIEF

Alessandro Giuseppi Volta
1745-1827
Italian Physicist

It is a mark of Alessandro Volta's influence on the world of science that the international unit of electrical potential difference or force is named the volt in his honor. His development of the first battery in 1800, along with his other experiments involving electrical current, made Volta a lasting name; however, he also conducted important studies with chemicals and gases.

Volta was born on February 18, 1745, in the town of Como in Italy's Lombardy region. He was one of nine children, and most of the others went on to pursue careers in the church, but after reading an essay by Joseph Priestley (1733-1804) at age 14, Volta decided to become a physicist.

By 1774, the 29-year-old Volta had been appointed professor of physics at the high school in Como. While working in this position, he developed his electrophorus, an early type of condenser for storing electrical charges. Volta's electrophorus made use of the earlier discovery by Charles-Augustine de Coulomb (1736-1806) that electrical charges are generated at the surface of a body rather than in its interior.

Also during the mid-1770s, Volta became involved in chemical experimentation. He discovered methane gas while studying marsh gas in 1776, and later made the first accurate measurements of the proportion of oxygen in the air. By this point his reputation had been growing, thanks in large part to the electrophorus, and in 1779 he became a professor at the University of Pavia. He soon set to work creating an electrometer for measuring electrical currents, and spent much of the 1780s in electrical experimentation.

In 1791, however, Volta was drawn into a curious controversy arising from claims made by his fellow Italian physicist Luigi Galvani (1737-1798). Galvani had noted that when he touched a frog's legs with probes of differing metallic composition, the muscles twitched, and this he attributed to what he called "animal electricity." Volta suspected these claims, and over a long series of experiments—some of which required him to be the subject—he showed that the electrical charge came from the two types of metal probes, not from anything inside the frog.

Continuing his earlier chemical experiments, Volta in 1796 found that the pressure of a chemical results from its temperature, not from pressure in the surrounding atmosphere. There was an ultimate purpose to his dual experiments in chemistry and physics, and this became clear in 1800, when he created the first battery. He did this by filling bowls with a saline solution and connecting them with a wire, one end of which was copper and the other zinc or tin.

Alessandro Volta. *(Corbis Corporation. Reproduced with permission.)*

Volta soon experimented with making a smaller battery, using stacks of copper, zinc, and cardboard soaked in a saline solution and connected by a wire. The stack was called a Voltaic pile. Later in 1800, English chemist William Nicholson (1753-1815) used a Voltaic pile to separate hydrogen from oxygen in water—the first use of electrolysis.

The work with batteries and Voltaic piles marked the zenith of Volta's career. He died on March 5, 1827, at the age of 82.

JUDSON KNIGHT

Abraham Gottlob Werner
1749-1817
German Naturalist

Abraham Gottlob Werner wrote the first modern textbook of descriptive mineralogy, *Vonden äusserlichen Kennzeichen der Fossilien* (1774; "On the External Characters of Fossils"). Taking issue with existing schools of thought regarding classification of minerals, he propounded his own theories of "geognosy" and "Neptunism." Ultimately Werner's views became insufferably dogmatic and impervious to conflicting evidence, but his early contributions were so great that he is remembered as one of the key figures in the establishment of mineralogy as a science.

Werner was born on September 25, 1749, in Wehrau, then part of Prussian Silesia and now part of eastern Germany. His family had worked in mining for many generations, and his father had a job as overseer at a metal foundry near Wehrau. By the age of 15, Werner had left school to work as his father's assistant, but he soon grew so interested in mining that he entered Freiberg's Mining Academy in 1769.

Having completed his course of studies at Freiberg, Werner in 1771 entered the University of Leipzig. There he became aware of problems in the existing systems for classifying minerals, of which there were many. Most, however, fell under two general headings: classification by chemical composition, and classification by physical characteristics. Displeased with this arrangement, Werner set to work, and within just a year's time produced *Vonden äusserlichen Kennzeichen der Fossilien.*

He was only 25 years old, and not only had he produced the first textbook in modern mineralogy, but Werner introduced a new means of classification that synthesized elements of the two existing systems. As brilliant as the book was, however, it also revealed the first traces of his dogmatism: in discussing crystalline forms, he revealed a prejudice against using mathematics in mineralogy, and treated crystallography as mere applied mathematics.

Soon after publishing his monumental book, Werner returned to Wehrau and went to work in the field, visiting mines and collecting minerals. But his work had so impressed the administration of the Mining Academy in Freiberg that they offered him a position as inspector and teacher of mining and mineralogy in 1775. As a teacher, Werner proved himself capable of stimulating curiosity and enthusiasm in his students, but he also imbued them with his zeal for an increasingly rigid account of the Earth's formation.

Defining his subject as "geognosy," or knowledge of the Earth, Werner claimed to offer an account of rock formations that avoided the wild speculations of many naturalists at the time. In fact his own system was at least as speculative as any of the ideas then being seriously considered by naturalists. Convinced that the Earth had once been completely covered by ocean—a school of thought dubbed "Neptunism"—Werner eventually succumbed to the vice of construing data to fit his theory.

He was convinced that the Earth was cold inside, and thus he dismissed volcanic rocks as the product of recent activity. Werner's conception of the Earth also forced him to ignore the potential of disturbances on the crust—folding or tilting that results in earthquakes—as evidence of seething energy within the planet's core.

Not surprisingly, Werner became more unshakable in his ideas as time passed and he grew older. He had no family to soften him: as a bachelor, his students took the place of a wife and children. He retired to Dresden, where on June 30, 1817, he died.

JUDSON KNIGHT

Biographical Mentions

Franz Ulrich Theodosius Aepinus
1724-1802

German scholar who produced the first detailed treatise on electricity and magnetism based on the principle of action-at-a-distance. Aepinus' research on the thermoelectric properties of tourmaline led him to consider the similarities between electric and magnetic effects. This insight was developed with great originality in his masterwork *Tentamen theoriae electricitatis et magnetismi* (1759). The *Tentamen* provided a model for applying mathematics to electric and magnetic phenomena and remained influential until the mid-nineteenth century.

Maria Angela Ardinghelli
fl. 1700s

Italian physicist and mathematician who is known for her translation into Italian of English physiologist Stephen Hales's book, *Vegetable Staticks* (1727), a classic text in plant physiology.

Laura Maria Caterina Bassi
1711-1778

Italian physicist who was not only the first woman to graduate from the University of Bologna, but received an official post at the university. Laura Bassi was born into wealth and home schooled with the assistance of her family physician. A child prodigy, she excelled at mathematics, philosophy, anatomy, natural history and several languages, which earned her entry to the University of Bologna, from which she received a Doctor of Philosophy degree. In 1745, she attained a monumental honor from Bologna's scientific community when she was awarded a post with the Benedictines of the Academy of Sciences established by the Pope. From 1745 to 1778, she gained respect for her research on mechanics, hydrometry, elasticity and other properties of gases, just as she earned widespread criticism for being a woman in a male-dominated field.

Antoine Baumé
1728-1804

French chemist who played an important role in the development of the chemical process industries in France. In 1752 he opened a pharmacy and chemical dispensary. His dispensary manufactured drugs in large quantities and distributed them in bulk to pharmacies and hospitals. Baumé also designed and built various instruments including an improved areometer. His production facilities supplied industrial and laboratory apparatus. Baumé remained committed to the phlogiston theory throughout his life.

Abraham Bennet
1750-1799

English physicist and inventor who invented the gold-leaf electroscope, an instrument used to detect electric charge. A gold-leaf electroscope contains two thin strips of gold that hang from either end of a rod. When the rod is electrically charged, the gold strips repel each other and spread apart. Bennet also invented a device called an electrical doubler. It was designed to multiply a small electrical charge until the charge was large enough to be detected by other instruments.

Claude-Louis Berthollet
1748-1822

French chemist who introduced chlorine into bleaching and established the composition of ammonia, prussic acid, and sulphuretted hydrogen. His masterwork, *Essai de Statique Chimique* (1803), contains the proposal that chemical reactivity depends on reactant masses (similar to the modern Law of Mass Action). Berthollet also proposed that compound composition can vary, contradicting Joseph Proust's now accepted Law of Constant Proportions according to which pure samples of a compound always contain the same elements in definite proportions.

John Bevis
1693-1771

English physician and astronomer who discovered the crab nebula (1731). In 1745 Bevis de-

cided to publish a new star atlas. His *Uranographia Britannica* was to have consisted of 52 large plates of the sky with accompanying explanations. However, just before the work was to go to press the publisher went bankrupt and the plates were sequestered. Bevis also suggested the feature retained by all modern capacitors of two conductors separated by an insulating, or dielectric, layer (1747).

Charles Blagden
1748-1820

English physician who is known for showing that normal body temperature is maintained by the process of perspiration; he noted this in his *Experiments and Observations in a Heated Room*, 1775. Blagden had a wide variety of interests ranging from vision to meteors, to the process of making liquor. He published the papers *On the Tides of Naples*, 1793 and *On the Heat of Water in the Gulf Stream*, 1781, which reported his observations on these and other subjects.

Ruggiero Guiseppe Boscovich
1711-1787

Croatian polymath who made significant contributions to the instrumental sciences of astronomy, optics, and geodesy as well as the theoretical disciplines of mathematics, mechanics, and natural philosophy. Boscovich attacked atomism, proposing instead that the universe was a plenum of thickening and thinning force concentrated in point-centers. Aggregation of point-centers accounted for all observable properties of matter. Boscovich's ideas were developed by Michael Faraday into the field concept, which has been an influential formative principle in physics.

Pierre Bouguer
1698-1758

French hydrographer primarily interested in navigation and ship design but best known as the father of photometry. He formulated Bouguer's law—in a medium of uniform transparency the intensity of a collimated light beam decreases exponentially with the path length through the medium—often incorrectly attributed to Johann Lambert. Bouguer also invented the heliometer (1748) and photometer. He was a member of Charles La Condamine's celebrated Peru expedition (1735-44), which helped verify Newton's theory of gravitation.

James Bradley
1693-1762

English astronomer renowned for discovering stellar aberration—apparent displacement of a star's position due to the combined velocity of Earth and starlight. Bradley exploited this effect to estimate the speed of light, thus providing the first direct evidence of Earth's motion. Correcting for aberration, his extensive astronomical observations were more accurate than those of his predecessors. Bradley discovered the nutation of Earth's axis, measured Jupiter's diameter, and succeeded Edmond Halley as the third Astronomer Royal.

Georg Brandt
1694-1768

Swedish chemist best known for discovering cobalt (1742). Brandt systematically investigated arsenic and its compounds (1733) and invented the classification scheme for the semimetals now called metalloids (1735). He published findings on how to produce sulfuric, nitric, and hydrochloric acids (1741, 1743) and demonstrated that gold dissolves in hot nitric acid and precipitates out when cooled. Brandt actively combated alchemical ideas and devoted his later years to exposing fraudulent transmutations of metals into gold.

Anton Friedrich Büsching
1724-1793

German geographer and major force in the development of contemporary geography. According to Büsching, the three foci of geography are the mathematical, the physical, and the human or cultural. He recognized the problems associated with data presentation and argued geography was more than just cartography. Büsching introduced statistical methods and census-based population density studies into geography. He also emphasized exact delimitation of areas using precise scales. His master work is *Neue Erdbeschreibung* ("New Earth Description").

John Canton
1718-1772

English schoolmaster best known for his electrostatic induction experiments and discovery that glass does not always charge positively by friction. Canton also developed a method for producing artificial magnets (1749) and correctly identified the cause of seawater luminosity. His observations led him to associate irregularities in compass needle diurnal orientations with unusual aurora borealis activity, today known as magnetic storms. He was awarded the Royal Society's 1765 Copley Medal for demonstrating that water is slightly compressible.

César-François Cassini de Thury
1714-1784

French cartographer who created the first modern map of France (1745). His project of producing an even more detailed map was almost completed before his death. Earlier in his career he sided with his father, Jacques, in defending René Descartes' prediction that Earth's polar axis is elongated against Isaac Newton's prediction that Earth is flattened at the poles. Cassini renounced the Cartesian theory after expeditions to Peru and Lapland called into question his family's earlier geodesic operations.

Jacques Cassini
1677-1756

Italian astronomer who fervently supported René Descartes' prediction that Earth is a prolate spheroid—elongated polar axis—against Isaac Newton's prediction that it is an oblate spheroid—bulging equator and flattened poles. Cassini adopted the Cartesian view after participating in his father's extension of the Paris meridian (1700-01). Results from this and other geodesic work seemed to support Descartes' hypothesis. Expeditions to Peru (1734-44) and Lapland (1736) later settled the debate decisively in favor of Newton.

Louis-Bertrand Castel
1688-1757

French physicist and leading opponent of Newtonianism. Most French scholars, with Castel at their head, viewed Isaac Newton's reliance on experimentation and observation as particularly undesirable, preferring instead the *a priori*, rationalistic approach of René Descartes. Castel published an anti-Newtonian theory that delayed acceptance of Newton's ideas and contributed greatly to the relative stagnation of French research. Castel is also known for his ocular harpsichord, which generated corresponding colors and musical tones.

Jacques Alexandre César Charles
1746-1823

French physicist who formulated Charles law—at constant pressure, a gas's volume is inversely proportional to temperature (1787). Unaware of Guillaume Amontons' original, unpublished discovery (1699), Charles communicated his results to Joseph Gay-Lussac, whose superior experimental techniques yielded more accurate measurements. Published by Gay-Lussac (1802), the relationship is often referred to as Gay-Lussac's law. Charles contributed significantly to ballooning, developing and ascending in the first hydrogen balloon (1783) and inventing the valve line, appendix, and nacelle.

George Cheyne
1671-1743

English physician and vigorous proponent of Newtonianism. His *Fluxionum methodus inversa* (1703) attempted to provide a mathematical treatment of the human body possessing the same rigor that Isaac Newton applied to celestial mechanics. However, it was of dubious mathematical quality. Addressing the theological implications of Newton's work, Cheyne argued that gravitation was an immediate proof of God's existence since attraction was not an essential property of matter. He was one of England's most widely read medical writers.

Adair Crawford
1748-1795

Irish physician and physicist who performed the first experiments to determine the specific heats of gases. Crawford believed the chemical changes in air brought about by respiration also altered air's heat capacity or specific heat. According to his application of William Irvine's theory of heat capacities, inhaled air had a greater specific heat than exhaled air, with the difference being transferred to and thus being the source of the body's heat. Irvine's theory of capacities was later shown to be untenable.

Axel Fredrik Cronstedt
1722-1765

Swedish chemist famous for discovering nickel (1751). Cronstedt established systematic blowpipe analysis, which he applied to the examination of minerals. This work led him to develop a rational classification scheme for minerals. First published in his "Essay on the New Mineralogy" (1758), the scheme divided minerals into four groups—earths, bitumens, salts, and metals—based upon chemical composition. Among Cronstedt's other researches are experiments with the new iron ore cerite, in which lanthanum was later discovered.

Georges Léopold Chrétien Frédéric Dagobert Cuvier
1769-1832

French anatomist considered the father of comparative anatomy. He extended Carolus Linnaeus's classification system by adding the phylum. Stressing similarities in internal structures rather than surface superficialities, he grouped Linnaean classes into four phyla: Vertebrata,

Mollusca, Articulata (jointed animals), and Radiata (all others). Cuvier's scheme continues to guide biological classification today. His pioneering work in classifying mammalian and reptilian fossils established vertebrate paleontology, and his studies of fish and mollusks inaugurated modern ichthyology.

Jean André de Luc
1727-1817

Swiss chemist, meteorologist, and geologist who made several firsts in scientific discovery. He first used the term "geologie" (1778), and interpreted the six days of Mosaic creation as epochs of geological time. He was first to provide correct measurement of the heights of mountains by the effects of heat and pressure on a thermometer and to publish the correct rules for equivalent heights to barometric pressure. Along with other meteorological instrument ideas, he invented a hygrometer using gut as the medium for measuring the humidity of the air and became involved with French scientist Horace de Saussure (1740-1799) in arguments over evaporative theory. He noted the independence of vapor pressure to atmospheric air pressure before John Dalton (1766-1844); described the chemical and electrical effects of the electric pile; and shared with Joseph Black (1728-1799) the discovery of latent heat.

Joseph-Nicolas Delisle
1688-1768

French astronomer best known for his efforts to determine the astronomical unit—the distance between the Sun and Earth—by observing transits of Venus across the Sun. The astronomical unit is crucial for determining the solar system size. Edmond Halley first proposed a method for determining this value from Venus transits. Delisle simplified Halley's method and produced a *mappemonde* indicating favorable observation sites for the 1761 and 1769 transits. He distributed the map and instructions through worldwide correspondence.

William Derham
1657-1735

English naturalist and theologian best known for his *Physico-Theology* (1713), which abounds in arguments from design to God, and *Astro-theology* (1715), which argues that Newtonian cosmology is ample evidence of God's existence. Derham also determined the speed of sound by timing the interval between the flash and roar of a cannon fired 12 miles (19.3 km) away (1705). His value of 1142 ft/sec (348.08 m/sec) is in good agreement with the presently accepted value of 1130 ft/sec (344.4 m/sec).

Nicolas Desmarest
1725-1815

French geologist who demonstrated the volcanic nature of columnar basalts in France's Auvergne region (1765), thus establishing that not all rocks are sedimentary. This was in opposition to Abraham Werner's neptunian theory. Desmarest's most original contribution to geology was a history of eruptions and lava flows from Auvergne's volcanoes (1775). His three-stage physiographical history extended historical geology to volcano studies. As an agent of the French department of commerce, Desmarest served as Inspector of Manufactures at Limoges (1788-91).

Jeremiah Dixon
1733-1779

Charles Mason
1728-1786

English astronomers famous for surveying the disputed boundary between Maryland and Pennsylvania (1763-68). The boundary they established, which has since come to be known as the Mason-Dixon line, settled the quarrel and remains the most famous boundary in the United States. As part of international efforts to accurately determine solar parallax, Mason and Dixon participated in a Royal Society expedition to observe the 1761 Venus transit. They separately observed the 1769 Venus transit.

John Dollond
1706-1761

English optician famous for inventing achromatic lenses. Objects viewed through lenses are fringed by interfering colors. This chromatic aberration is caused by lenses differently refracting the various wavelengths composing white light. The stronger the lens the more chromatically disturbed the image. Dollond succeeded in eliminating aberration by combining two lenses such that one reversed the effects of the other (1758). Chester Moor Hall (1703-1771) had been producing such lenses since 1733 but failed to publicize his work.

Henri-Louis Duhamel du Monceau
1700-1782

French agronomist, botanist, and chemist who first distinguished soda from potash (1736). Duhamel's explanation of the saffron plant blight in France brought him his first scientific recognition (1728). After conducting chemical experiments in the 1730s, his interests shifted to botany and agronomy. He published several works on wood's structural properties, a book on ship rigging (1747), and an exposition of Jethro Tull's agricultural writings (1750). Duhamel's pioneering contributions to agricultural practice proved influential in France.

Jeanne Dumée
fl. 1680

French astronomer remembered for her book *Entretiens sur l'opinion de Copernic touchant la mobilité de la terra* (1680). Dumée's expressed intent was not to argue for Copernicanism but rather to examine Copernicus's arguments. Nevertheless, she concluded that it was impossible to believe in the geocentric world view. The introduction to her critically acclaimed book challenged the widely held belief that women's intellectual abilities were inferior because their brains were smaller and lighter than men's.

Fausto d'Elhuyar
1755-1833

Juan José d'Elhuyar

Spanish chemists who first isolated the element tungsten in 1783. The brothers, who had studied mineralogy, were analyzing a mineral called wolframite when they made their discovery. For this reason, tungsten is sometimes called wolfram, and its chemical symbol is W. The brothers later went to South America to supervise the Spanish mining industry. Today, tungsten is used to produce very hard steels.

Antoine François de Fourcroy
1755-1809

French chemist whose most important contribution to chemistry was his advocacy of Antoine Lavoisier's views. His *Principles de chimie* (1787) was the first textbook based on antiphlogiston principles, and he collaborated with Lavoisier in revising chemical terminology. Fourcroy also collaborated with Louis Vauquelin. Together they discovered iridium (1803), produced the first satisfactory account of urea (1799), and prepared a relatively pure form of urea (1808). He was involved in educational reforms after the French Revolution.

Johan Gadolin
1760-1852

Finnish chemist best known for identifying the new earth yttria. During this period the term "earth" referred to any oxide resistant to heat and insoluble in water. Yttria was later shown to contain several new elements now called rare-earth elements. In 1886 Jean Marignac named the new rare-earth he had discovered gadolinium, the first element named after a person. Gadolin was active in politics and played an active role in Finland's secession from Sweden.

Johan Gottlieb Gahn
1745-1818

Swedish chemist and mineralogist who first isolated the pure metal manganese (1774). Gahn interacted closely with Karl Scheele, and conversations they had regarding Scheele's research led to Gahn's demonstration that bones contain phosphorus (1770). He also interacted with Torbern Bergman and contributed to Jönas Berzelius's discovery of selenium (1818). Gahn improved copper smelting methods and solved various other technical mining problems for which the Swedish College of Mining awarded him its gold medal (1780).

Thomas Godfrey
1700-1782

American mathematician and instrument maker who invented the double-reflecting quadrant (1730), which evolved into the modern sextant. John Hadley invented an almost identical instrument and published his work in 1730. A priority dispute arose, but the convenience of Hadley's design for ship-board use made his the instrument of choice for navigators. Godfrey published almanacs (1729-36), worked on mathematical and navigational problems, and took astronomical observations used to fix the longitude of Philadelphia.

John Goodricke
1764-1786

English astronomer who explained the nature of variable stars. Goodricke noticed the variability of ß-Persei's (more commonly known as Algol). Though not the first to observe these variations in magnitude, he was the first to establish their periodicity and accurately attributed their cause to a large orbiting body that regularly eclipsed ß-Persei. Goodricke was awarded the Royal Society's Copley Medal (1783). As the result of a severe illness in infancy, Goodricke was deaf and mute.

William Gregor
1761-1817

English mineralogist famous for discovering titanium. Gregor analyzed a black sand he found in Menacchan, Cornwall. The sand contained iron, manganese, and another substance that Gregor successfully extracted but could not identify. He published his findings in 1791 and proposed the name menacchanine for this new mineral. Martin Klaproth isolated the same element from a different source in 1795 and suggested the name titanium. Gregor was rector of Creed from 1793 until his death.

David Gregory
1659-1708

English mathematician and astronomer who was one of the first to lecture publicly on Newtonian science. Gregory defended Isaac Newton's views in his *Elements of Astronomy* (1702); however, he disagreed with Newton's position on chromatic aberration, suggesting it might be eliminated by combing two lenses such that one reversed the effects of the other. This was accomplished by Chester Moor Hall (1733), who failed to publicize his work, and then again by John Dollond (1758).

Jean-Etienne Guettard
1715-1786

French geologist remembered for discovering the volcanic nature of the Auvergne region of central France (1751). Controversy ensued over the nature of columnar basalt in the region. Guettard initially concluded that it was not volcanic but later became convinced otherwise. He prepared the first mineralogical map of France (1746) and identified trilobites in the slates of Anjou. Guettard also discovered sources of kaolin and petuntse, which made possible the manufacture of Sèvres porcelain.

Louis Bernard Guyton, baron de Morveau
1737-1816

French chemist who helped reform the system of naming chemicals. Guyton studied law at Dijon, where he practiced from 1756 to 1762. He served as an advocate-general in the Burgundy's provincial parliament, before retiring from the legal profession to devote himself to chemistry. In the 1780s, he proposed in his work *Méthode de Nomenclature Chimique* (Method of Chemical Nomenclature) that the existing method of naming chemicals be revised. At the time, chemicals were named by arbitrary means—for example the name of their discoverer, or their appearance. Morveau proposed that elements should be given simple names, while compound names should reflect their chemicals of origin. His recommendations were adopted throughout Europe. He also made great strides in the manufacture of gunpowder, and was the first to use chlorine as a disinfectant.

George Hadley
1685-1768

English natural philosopher who followed up Edmond Halley's (1656-1742) attempt at defining atmospheric circulation by correctly describing the trade wind circulation in relation to the rotation of Earth. A member of the early Royal Society, in 1735 Hadley presented a paper "Concerning the Cause of the General Trade Winds," theorizing from the observed data of trade wind motion. Halley stated that sun-heated air rose over the equator allowing air from the north and south to flow toward the equator, but the easterly orientation of this flow escaped him. Hadley realized that the rotation of Earth from west to east gave northerly or southerly flow an easterly trajectory, since motion from the north or south toward the equator in the reference of a rotating sphere appears slower. This anticipated the theory of Coriolis force, the conception of a force acting to the right on an object on a rotating body, in the next century.

John Hadley
1682-1744

English mathematician and instrument maker whose new grinding techniques made it possible for him to construct high quality reflecting telescopes. His first such instrument was constructed in 1721 and tested and praised by James Bradley and Edmond Halley. In 1730 he invented the reflecting quadrant. A priority dispute arose, but the convenience of Hadley's design for ship-board use made his the instrument of choice for navigators. His instrument evolved into the modern sextant.

Sir James Hall
1761-1832

English geochemist who defended James Hutton's theory that Earth's internal heat is responsible for geological change. Abraham Werner argued that Earth's surface and distribution of rocks were the result of a great deluge, noting—in opposition to Hutton—that basalts cool to glassy substances and limestone subjected to heat decomposes. Hall's experiments demonstrated that slowly cooled molten basalts yield crystalline structures and that limestone will not decompose at a high temperature if subjected to high pressure.

Francis Hauksbee
c. 1666-1713

English physicist remembered for his demonstrations while curator of experiments for the Royal Society of London. Hauksbee introduced the double-cylinder vacuum pump, which he employed extensively. He examined luminous, friction-induced effects in vacuums, confirmed the pressure dependence of sound's transmission in air, determined the density of water relative to air, and thoroughly investigated capillarity and surface forces. His researches were described in his *Physico-Mechanical Experiments* (1709, 1719) which was widely read during the eighteenth century.

René Just Haüy
1743-1822

French mineralogist considered the founder of modern crystallography. Haüy's mineralogical work was primarily morphological—seeking to describe and classify crystal structure through geometrical principles. He proposed that crystals of particular species are created by stacking constituent molecules (now known as unit cells) to form simple geometric shapes with faces having constant angles of inclination. Haüy also recognized that crystal species were limited because not all angles of inclinations are possible—the discontinuity principle.

Caroline Lucretia Herschel
1750-1848

German-English astronomer and first woman to discover a comet (1786). She discovered seven more comets and many nebulae. She was devoted to her brother William and submerged herself in his work, assisting in making instruments, taking observations, and preparing manuscripts for publication, leaving little time for her to make observations. She was awarded the Royal Astronomical Society's Gold Medal (1828) and was one of the first two women elected to honorary membership in the Society (1835).

Peter Jacob Hjelm
1746-1813

Swedish chemist who first isolated the element molybdenum in 1782. Today, molybdenum is used to strengthen steel and other metals for use at high temperatures. Hjelm also discovered that iron ore containing the element manganese produced steel that was harder and more resistant to corrosion than other steels. Hjelm was appointed master of assaying for the Royal Mint of Sweden. (Assaying is the process of determining the proportions of metals in ores or metal objects such as coins.)

Wilhelm Homberg
1652-1715

French chemist who introduced the new experimental chemistry to the Académie des Sciences. In *Essais de chimie* (1702-10) he concluded that salt, sulfur, and mercury were not present in all substances—an important move toward the modern concept of an element. Homberg's emphasis on analyzing substances into simple, stable chemical entities paved the way for eighteenth-century analytical chemistry, and his work on acid-alkali neutralization was essential for later work on the nature of salts.

Johann Juncker
1679-1759

German chemist whose most important contribution was the clarification of Georg Ernst Stahl's researches and theories. Juncker's systematic and coherent treatment of his teacher's work gave Stahlian ideas a wider audience, thus greatly increasing their influence on eighteenth-century chemistry and medical thought. Juncker adhered to Stahl's counsel that chemical and medical theories remain separate as the former had little to offer the latter. He also emphasized the importance of basing chemical theory on extensive experimental work.

Immanuel Kant
1724-1804

German philosopher whose master work, *The Critique of Pure Reason* (1781), initiated a "Copernican Revolution" in philosophy by treating many of the observed features of the world as constituted by the human knower. Considered by many the greatest modern philosopher, Kant's early work explicitly addressed scientific topics. His nebular hypothesis (1755) anticipated Pierre-Simon Laplace's more sophisticated formulation. Kant also correctly suggested the Milky Way is lens-shaped, that nebulous stars are galaxies, and that tidal friction slows Earth's rotation.

Maria Margarethe Winkelmann Kirch
1670-1720

German astronomer who married her teacher Gottfried Kirch in 1692 and worked as his assistant making observations and calendar calculations. She continued to research and publish after his death in 1710. She discovered the comet of 1702, published a pamphlet on the 1712 Sun-Venus-Saturn conjunction (1709), and wrote about the Jupiter-Saturn conjunction (1712). When her son Christfried was appointed astronomer of Berlin's observatory, she joined him and continued calculating calendars for various German cities.

Richard Kirwan
1733-1812

Irish chemist and mineralogist best known for his disagreement with James Hutton over the chemical composition of rocks and his support of phlogiston theory in *Essay on Phlogiston* (1787). Criticisms by Antoine Lavoisier and other caused Kirwin to reject phlogiston theory in 1791. He won the Royal Society's Copley Medal in 1780 for his work on chemical affinity, and his *Elements of Mineralogy* (1784) was one of the first systematic works on the subject in English.

Martin Heinrich Klaproth
1743-1817

German chemist remembered for his pioneering contributions to analytical chemistry and discovery of new elements. Klaproth deduced the presence of uranium in pitchblende (1789), zirconium in zircon (1789), and titanium in rutile (1795). He confirmed or independently discovered strontium (1793), titanium (1795), chromium (1798), tellurium (1798), and cerium (1803), always being careful to credit the original discoverers. Klaproth also pioneered the chemical analysis of antiquities, and his conversion to Antoine Lavoisier's new chemistry greatly influenced that theory's acceptance in Germany.

Ewald Georg von Kleist
1700?-1748

German physicist remembered for inventing the condenser or Kleist vial (1745). A similar device was later devised and better explained by Pieter von Musschenbroek in Leiden, Holland, thus its more common name, the Leyden jar. Attempts to explain the electric shocks Kleist and Musschenbroek received from these devices and subsequent experimental work of others indicated that the traditional view of electrical "atmospheres" was fundamentally mistaken. This led to increased efforts to quantify electrical phenomena.

Charles-Marie de La Condamine
1701-1774

French geographer whose celebrated Peru expedition (1735-44), along with Pierre Maupertuis' Lapland expedition (1736-37), helped verify Isaac Newton's theory of gravitation by demonstrating that Earth is flattened at the poles. Condamine was the first European to explore the Amazon region carefully. He shipped samples of the tree sap caoutchouc to France, thus introducing rubber to Europe, and he discovered curare. La Condamine was convinced that delicate scientific measurements were hampered by lack of standards and worked to establish an international unit of measure.

Nicolas-Louis de Lacaille
1713-1762

French astronomer whose redetermination of the Paris meridian revealed errors in Jacques Cassini's earlier measurements. This supported Isaac Newton's prediction that Earth is flattened at the poles. Lacaille observed nearly 10,000 stars during his Cape of Good Hope expedition (1750-1754) and included 1,942 of these observations in his *Coelium australe stelliferum* (1763). In conjunction with Joseph Lalande in Berlin, he measured the lunar parallax. Lacaille's 1761 Earth-Sun distance estimate was the first to treat Earth as other than a perfect sphere.

Nicole-Reine Lepaute
1723-1788

French astronomical computer who assisted Alexis Clairaut in calculating the gravitational effects of Jupiter and Saturn on Halley's comet and determining the exact time of its 1759 return. Lepaute also calculated the 1764 solar eclipse path for all of Europe and assisted Joseph Lalande with the Académie des Sciences' astronomical almanac for astronomers and navigators (1759-64). She investigated pendulum oscillations that her husband Jean-André Lepaute, the royal clockmaker, published in his *Traité d'horlogerie* (1755).

Anders Johan Lexell
1740-1784

Swedish astronomer and mathematician remembered for demonstrating that the object discovered by William Herschel on March 13, 1781, was in fact a new planet (Uranus). He further showed that perturbations of its motion could only be accounted for by another more remote planet. Lexell was proven correct when John Adams and Urbain Le Verrier's calculations led to the discovery of Neptune in 1846. Lexell also made important contributions to analysis, spherical geometry, and trigonometry.

Mikhail Vasilievich Lomonosov
1711-1765

Russian polymath and giant of eighteenth-century science. Lomonosov reputedly experimentally demonstrated mass conservation prior to

Antoine Lavoisier, proposed a kinetic theory of heat earlier than Count Rumford, developed a wave theory of light before Thomas Young, and formulated a kinetic theory of gases similar to Daniel Bernoulli's. He was first to record mercury freezing and observe Venus's atmosphere (1761). Lomonosov also produced a grammar that reformed and systematized Russian language and produced the first work on Russian history.

Pierre-Joseph Macquer
1718-1784

French chemist who published an influential chemistry textbook and was the first to publish a chemical dictionary that was organized along modern lines. Macquer's textbooks, *Elements of Theoretical Chemistry* and *Elements of Practical Chemistry* were important texts that helped mold the thinking of chemists for many years. In addition, his organization of chemicals into roughly modern categories helped to systematize the field, taking it yet another step from the practices of alchemy.

Jean Jacques d'Ortous de Mairan
1678-1771

French physicist who played a minor role in various important eighteenth-century scientific debates, typically seeking to reconcile Cartesian and Newtonian positions. Mairan was involved in controversies over Earth's shape, conservation of vis viva—the "living force" concept later called energy—and the nature of light. He correctly noted the cooling effect of evaporation (1749), and his climate studies led him to postulate a central fire within Earth as an important heat source. Mairan's *Physical and Historical Treatise on the Aurora Borealis* (1733) contains the first application of geophysical data to an astronomical problem.

Giovanni Domenico Maraldi
1709-1788

Italian-French astronomer known for his observations and improvements to the theory of Jovian satellite motions. Most of Maraldi's work was in positional astronomy, though he did participate in various geodesic operations, including a partial survey of the French-Atlantic coast (1735), a redetermination of the Paris meridian (1739-40), and Dominique Cassini III's map of France project. Maraldi also made regular meteorological observations and participated in an experiment to determine the velocity of sound (1738).

Nevil Maskelyne
1732-1811

English astronomer who established the celebrated *Nautical Almanac* (1766), whose tables and astronomical information were indispensable navigational aids. Maskelyne tested and devised better methods for determining longitude at sea; observed Venus's 1769 transit from which he computed the Sun's distance to within 1%; and calculated Earth's density, which compares well with presently accepted values, from plumb line deviation near Mt. Schiehallion, Scotland (1774). He succeeded Nathaniel Bliss as the Fifth Astronomer Royal (1765).

Pierre Louis Moreau de Maupertuis
1698-1759

French mathematician whose Lapland expedition (1736), in conjunction with Charles La Condamine's 1735 Peru expedition, confirmed Isaac Newton's prediction that Earth is an oblate spheroid—bulging at the equator and flattened at the poles—thus helping establish Newton's theory of gravitation. Maupertuis also formulated the principle of least action (1745), which was widely influential in eighteenth- and nineteenth-century physical thinking and was later incorporated into quantum mechanics and the biological principle of homeostasis.

Johann Tobias Mayer
1723-1762

German cartographer and astronomer remembered for his lunar tables, which allowed accurate determination of longitude at sea. Mayer claimed the British government's prize for best such method, and three years after his death the British Board of Longitude awarded his widow 3,000 pounds. Mayer also devised a method for fixing geographical coordinates independent of astronomical observations, invented a method for determining eclipses, investigated stellar proper motions, and produced an accurate map of the Moon's surface.

Charles Messier
1730-1817

French astronomer remembered for his catalog listing many of the most important nebulae and clusters. Messier was widely known by his contemporaries for discovering comets—15 by modern standards. Having been deceived numerous times by comet-like nebulosities, he compiled and published his Messier list of such objects. These are today known as Messier objects and designated M1, M2, etc, and include the Crab Nebula (M1), the Andromeda galaxy (M31), and the Pleiades (M45).

John Michell
1724?-1793

British geologist and astronomer who theorized about the origin of earthquakes, estimated the velocity of seismic waves in Earth's crust, and proposed various methods for determining earthquake epicenters (1760). His most significant achievement was his 1767 argument that double stars are binary systems because the number of doubles observed is too great to have resulted from random distribution alone. Michell is also remembered for making the first realistic estimate of stellar distances (1784).

Franz Joseph Müller von Reichenstein
1740-1825

Austrian mineralogist remembered for discovering tellurium. While working with gold ore in 1782, Müller isolated a substance he concluded was a new element. His work went unnoticed until Martin Klaproth asked for a sample of the ore and confirmed the discovery in 1798. Klaproth named the element tellurium. The Emperor Joseph II appointed Müller chief inspector of mines in Transylvania and in due time conferred upon him the title of Baron von Reichenstein.

Pieter van Musschenbroek
1692-1761

Dutch physicist who invented the Leyden jar, a device used for storing electric charges. Musschenbroek was a professor of mathematics and physics at the University of Leyden in Holland. Early in the eighteenth century, scientists were able to create electricity with friction, but not to store it. Musschenbroek was conducting an experiment in which he poured water into a glass jar and connected the jar by wire to a friction machine that produced static electricity. He discovered upon touching the wire that it created a shock, caused by electricity stored within the jar. The device, which was later named the Leyden jar, is said to be the prototype of capacitors, or electric conductors, which are used in radios, television sets and other electrical equipment.

William Nicholson
1753-1815

English chemist who discovered the electrolysis of water. Nicholson was a veritable jack-of-all-trades, working at times as a hydraulic engineer, inventor, translator and scientific publicist. In 1800, intrigued by the electric battery invented by Italian physicist Alessandro Volta, he built his own version. Nicholson became the first to produce a chemical reaction via the use of electricity, when he placed leads from the battery in water, which broke the water down into its separate elements, hydrogen and oxygen. His discovery was later used in chemical research and industry. In 1797, he founded the *Journal of Natural Philosophy, Chemistry and the Arts.*

Jean-Antoine Nollet
1700-1770

French scientist who made important scientific contributions to the field of electricity. He also aimed at teaching and popularizing physics with the use of instruments. To this end Nollet gave lectures in many schools of engineering and to Europe's *haute bourgeoisie.* His lectures gathered in the *Leçons de physique expérimentale* (6 vols., 1743-64) along with *L'Art des expériences* (3 vols., 1770)—explaining how to build instruments—were so successful they inspired self-taught scientists until the twentieth century.

Christopher Packe
1686-1749

English physician who produced the first map— *A New Philosophico-Chronological Chart of East Kent* (1743)—representing topographical features with scientific accuracy. He was also the first to use barometric readings to determine map-elevations and plot these as spot heights. Packe's anthropomorphization of the landscape contributed to geophysiological discourse—the view that Earth is a living organism—which in the twentieth-century produced James Lovelock's Gaia hypothesis. Packe also conducted chemical experiments and practiced medicine in Canterbury (1726-49).

Giuseppe Piazzi
1746-1826

Italian astronomer remembered for discovering the first asteroid, Ceres. Piazzi was founder and first director of the Palermo Observatory (1790). On January 1, 1801, while taking observations for his star catalog, he noticed a star-like object not on Nicolas de Lacaille's star list. Piazzi named the object Ceres and tracked its motions, which resembled those of a planet or comet. Carl Friedrich Gauss later showed that Ceres moved in the gap between Mars and Jupiter.

Madame du Pierry
1746-1789

French astronomer who was the first female astronomy professor appointed at the University of Paris. Pierry collected information about eclipses, which Joseph Jérôme Lalande consulted to study lunar movement. He dedicated his book *Astronomie des dames* (1790) to Pierry, declaring her a female role model because of her intellectual qualities. She gathered and computed eighteenth-century astronomical data to prepare charts of information about the duration of daylight at the Parisian latitude. Pierry taught a women's astronomy class in 1789.

Joseph Louis Proust
1754-1826

French chemist famous for discovering the law of constant proportions, or Proust's law, according to which pure samples of a compound always contain the same elements in definite proportions. His views contradicted Claude-Louis Berthollet's widely accepted claim that compound composition varies depending on how the compounds are prepared. Proust prevailed, but it was later shown that some compound compositions vary slightly—sometimes referred to as berthollides. Proust was the first to identify and study glucose.

Jeremias Benjamin Richter
1762-1807

German chemist famous for formulating Richter's rule: elements combine in fixed proportions to form compounds (1795). These proportions were later explained by John Dalton's atomic theory as relative molecular weights. Richter was obsessed with the mathematization of nature, believing it particularly relevant to chemistry, which is fundamentally concerned with determining the exact proportions of constituents present in compounds. He used the term "stoichiometry" to define the science of measuring the elements.

David Rittenhouse
1732-1796

American astronomer and instrument-maker known for the fine workmanship of his clocks and scientific devices. He produced high precision telescopes, thermometers, surveyor's compasses, and other instruments that were superior in quality to anything previously produced in America. His scientific interests were diverse, but astronomy was his primary interest. He observed the 1769 Venus transit and contributed to the best American calculations of solar parallax. Rittenhouse also played an important role in the American Revolution.

Ole Christensen Römer
1644-1710

Danish astronomer celebrated for demonstrating light's finite velocity. Römer noticed that the intervals between successive eclipses of Jupiter's satellites varied depending on Earth's positions—diminishing as Earth approached and increasing as it receded. He correctly attributed this to the time required by light to travel the Jupiter-Earth distance and in 1676 calculated light's velocity at 225,000 km/sec (139,808 mi/sec). Römer invented various scientific instruments including a micrometer, planetaria, and an alcohol thermometer that influenced Daniel Fahrenheit's thermometric researches.

Daniel Rutherford
1749-1819

Scottish chemist generally credited with dissevering nitrogen (1772). Joseph Black assigned him the task of investigating air incapable of supporting combustion. After carefully isolating and processing such air to remove carbon dioxide, Rutherford believed he had phlogisticated air—air having absorbed as much phlogiston as possible. According to phlogiston theory, respiration and combustion require the release of phlogiston, thus phlogisticated air could not support such processes. Antoine Lavoisier later described the true nature of nitrogen.

Horace Bénédict de Saussure
1740-1799

Swiss geologist remembered for his extensive alpine investigations chronicled in his *Voyages dans les Alpes* (1779-96) and theories regarding the formation of the Alps derived therefrom. Saussure made careful meteorological measurements, developed an improved thermometer and atmometer, and invented the hair hygrometer for measuring humidity (1781). He is recognized as the first experimental petrologist for his work on fusing granites and porphyries, and he helped popularize the term "geology," which replaced "geognosy" in the late eighteenth century.

Joseph Sauveur

French physicist who was the first to scientifically study acoustics. Not only did Sauveur coin the term "acoustics," he also studied it intensively, paying special attention to the relationships between notes and tones on the musical scale. Although of limited utility at the time, the science of acoustics has become very important in recent years in the physics laboratory, where it is responsible for sonoluminescence and other interesting phenomena. In addition, the science of acoustics led directly to the study of vibratory, oscillatory, and periodic motion, all of which are greatly important in physics, engineering, and other disciplines.

Willem Jakob 'sGravesande
1688-1742

Dutch physicist best known among his contemporaries as an exponent of Newtonianism, in particular for his *Mathematical Elements of Physics* (1720, 1721), which was easily the most influential semi-popular account of Newton's ideas prior to 1750. 'sGravesande elaborated on the methodology of Isaac Newton's work and helped to created a new school of experimental physics in Holland. 'sGravesande's 1722 free-fall experiments supported Gottfried Leibniz's interpretation of *vis viva*—"living force"—as the force of a body in motion.

Georg Ernst Stahl
1660-1734

German chemist and physician remembered for developing phlogiston theory—combustion and calcination consist of phlogiston loss. Stahl postulated that ash from burnt wood and calx from heated metals are devoid of phlogiston, and heating calx with phlogiston-rich charcoal yields metal by calx re-absorbing phlogiston. Accounting for observed weight changes required phlogiston to have negative weight. Though incorrect, Stahl's theory greatly facilitated the shift from alchemy to Antoine Lavoisier's new chemistry. Within medical thought, Stahl's vitalism provided an alternative to Hermann Boerhaave's mechanist theories.

Emanuel Swedenborg
1688-1772

Swedish natural philosopher and mystic widely remembered for founding a religious sect that maintains a loyal following to this day. Though his scientific interests were manifold, Swedenborg's most significant contributions were to geology and paleontology—marshaling evidence to show Scandinavia was once below water and arguing for the organic origin of fossils. A member of Sweden's Board of Mines, he assisted with technical mining and smelting projects and worked to improve copper ore refining.

Johann Daniel Titius
1729-1796

German astronomer famous for proposing a numerical rule describing the distance of the planets from the Sun (1766). Johann Bode uncovered Titius's work and popularized it (1772). Traditionally known as Bode's law but now referred to as Titius's law, the rule accurately accounted for the orbits of all known planets and the newly discovered Uranus (1781) and planetoids between Mars and Jupiter (1801). However, the law failed for both Neptune and Pluto.

Torbern Olaf Bergman
1735-1784

Swedish scientist best known for his work on elective affinity for which he compiled extensive tables listing relative chemical affinities of acids and bases. He provided the first comprehensive analysis of mineral waters (1778) and greatly improved qualitative analysis of minerals by introducing wet methods (1780). Bergman also discovered Venus's atmosphere during the planet's 1761 transit, investigated the pyroelectricity of tourmaline (1766), estimated the aurora borealis's height at 460 miles (740 km), and contributed to the development of crystallography.

Louis-Nicolas Vauquelin
1763-1829

French chemist who discovered the elements chromium and beryllium. Vauquelin was born in Normandy, and worked as a young boy in an apothecary, where he gained an interest in chemistry. He became close with Antoine-François Fourcroy, and was hired as his laboratory assistant. Vauquelin soon surpassed his mentor, becoming a prolific writer in the field of analytical chemistry. He was also a teacher, and was one of the first to use the laboratory method of instructing students. In 1798, he discovered beryllium in beryl and isolated chromium in lead ore obtained from Siberia. Vauquelin also discovered quinic acid, asparagines, camphoric acid, and several other organic substances.

William Watson
1715-1787

English physicist and botanist remembered primarily for his studies of electricity, especially those conducted with the Leyden jar. Watson improved this device both by lining it with lead foil (suggested by John Bevis) and thinning the glass. His experiments suggested, counter to effluvial theory, that electricity was a single fluid. Benjamin Franklin concurrently developed a similar theory, though in greater detail, which emerged as the standard view of electrical phenomena by the end of the eighteenth century.

Carl Friedrich Wenzel
1740-1793

German chemist whose most important work, *Lehre von der Verwandtschaft der Körper* (1777-82), addressed, among other issues, chemical affinity. Wenzel believed that substance concentration influenced the attractive forces between elements, and he was able to show that the rate at which a metal dissolves in an acid is proportional to the acid concentration. This was an early recognition that chemical reactivity depends on reactant masses, which was first clearly enunciated by Claude-Louis Berthollet (1803).

Johan Carl Wilcke
1732-1796

German physicist who is known for his work with electricity and magnetism. Wilcke used a Leyden jar, which is covered on the outside and inside with tin foil and has a metal rod through the lid connected to the inner foil, to experiment with the conduction of electricity. He also designed the first chart demonstrating magnetic inclination and developed a theory of specific heat independent of Joseph Black.

John Winthrop
1714-1779

American astronomer remembered for his work at Harvard during his tenure as the second Hollis professor of mathematics and natural philosophy (1738-79), including establishing America's first experimental physics laboratory and introducing calculus into the mathematics curriculum. His extensive astronomical observations include various Mercury transits and the Venus transits of 1761 and 1769, which contributed to international efforts to determine the astronomical unit. Winthrop was made a fellow of the Royal Society in 1766.

Christian Wolff
1679-1754

German philosopher whose systematic treatises were enormously important in diffusing Gottfried Leibniz's ideas and introducing Germany to recent scientific discoveries. He sought rapprochement between Scholasticism, modern science, and the new mathematics in a formal system where everything followed from self-evident axioms or preceding truths. This approach greatly influenced his contemporaries. Wolff was a confirmed corpuscularian, believed the physical world to be a deterministic machine obeying the laws of motion, and established German philosophical terminology.

**Thomas Wright
1711-1786**

English philosopher credited with originating the disk-shaped galaxy model. He constructed two geometric models producing the Milky Way's appearance. In one, Earth lies in a thin, spherical-shell of stars about a divine center with the universe full of such spheres. In the other, the universe is a series of concentric rings around the divine center. Wright later rejected both models, instead conceiving of man as living inside a cavity with the Milky Way a chain of burning mountains.

Bibliography of Primary Sources

Books

Banneker, Benjamin. *Benjamin Banneker's Pennsylvania, Delaware, Maryland and Virginia Almanack and Ephemeris* (1792-97). This work, which was updated annually between 1792-97, was the first scientific book published by an African American as well as the first almanac compiled by an African American. Readers did not seem concerned that the almanac's author was black. Recommended by abolitionist groups, the almanacs sold well throughout the United States, territories, and Europe, and Banneker acquired international acclaim. His astronomical information, tide calculations, and weather predictions were especially useful for farmers and sailors.

Bernoulli, Daniel. *Hydrodynamica* (1738). In this work Bernoulli explained both hydrostatics and dynamics via the mechanics of Isaac Newton with many examples of forces in fluid flow.

Buffon, Georges. *Histoire naturelle, générale et particulière* ("Natural History," 44 volumes, 1749-1804). This encyclopedic work offered a systematic account of natural history, geology, and anthropology.

Cronstedt, Axel Fredrik. *Essay on the New Mineralogy* (1758). Here Cronstedt offered a rational classification scheme for minerals. The scheme divided minerals into four groups—earths, bitumens, salts, and metals—based upon chemical composition.

Dalton, John. *Meteorological Observations and Essays* (1793). Dalton kept daily weather records from his childhood until his death and from this developed an acute interest in forecasting weather "with tolerable precision" as important "advantages" to humanity. He published his ideas about observing and making instruments in this book.

Derham, William. *Astro-theology* (1715). In this work Derham argued that Newtonian cosmology is ample evidence of God's existence.

Dumée, Jeanne. *Entretiens sur l'opinion de Copernic touchant la mobilité de la terra* (1680). Dumée's expressed intent in this work was not to argue for Copernicanism but rather to examine Copernicus's arguments. Nevertheless, she concluded that it was impossible to believe in the geocentric world view. The introduction to her critically acclaimed book challenged the widely held belief that women's intellectual abilities were inferior because their brains were smaller and lighter than men's.

Fourcroy, Antoine François de. *Principles de chimie* (1787). The first chemistry textbook based on antiphlogiston principles.

Gregory, David. *Elements of Astronomy* (1702). Gregory defended many of Isaac Newton's views in this work, but he disagreed with Newton's position on chromatic aberration, suggesting it might be eliminated by combing two lenses such that one reversed the effects of the other.

Halley, Edmond. *Catalogus Stellarum Australium* (1678). A star catalog based on Halley's observations on the island of St. Helena, where he established the Southern Hemisphere's first observatory and proceeded with the first telescopic mapping of the southern skies. This book gained Halley election to the Royal Society at age 22.

Homberg, Wilhelm. *Essais de chimie* (1702-10). Here Homberg concluded that salt, sulfur, and mercury were not present in all substances—an important move toward the modern concept of an element.

Hutton, James. *Theory of the Earth* (1795). In this work Hutton suggested that the weathering effects of water produced the sedimentary layers. This sediment was raised through volcanic action, and the erosion of wind, rain and rivers sculpted valleys and plains, starting the cycle again. However, based on observation and experimentation of river flow and mud content, Hutton realized this process would require much longer than 6,000 years. Hutton's theory attempted to explain the geological structure of Earth without resorting to catastrophic events such as worldwide flooding, comets, or massive earthquakes. This emphasis of slow, gradual change over time came to be known as Uniformitarianism.

Kant, Immanuel. *The Critique of Pure Reason* (1781). This work initiated a "Copernican Revolution" in philosophy by treating many of the observed features of the world as constituted by the human knower.

Kirwin, Richard. *Essay on Phlogiston* (1787). In this work Kirwin disagreed with James Hutton over the chemical composition of rocks and supported the phlogiston theory. Criticisms by Antoine Lavoisier and others caused Kirwin to reject phlogiston theory in 1791.

Kirwin, Richard. *Elements of Mineralogy* (1784). One of the first systematic works on mineralogy in English.

Lacaille, Nicolas-Louis de. *Coelium australe stelliferum* (1763). Lacaille observed nearly 10,000 stars during his Cape of Good Hope expedition (1750-54) and included 1,942 of these observations in this star catalog.

Lehmann, Johann Gottlob. *Versuche einer Geschichte von Flotz-Gebrugen* (1756). Here Lehmann classified mountains and established a theory of their origins.

Mairan, Jean Jacques. *Physical and Historical Treatise on the Aurora Borealis* (1733). Contains the first application of geophysical data to an astronomical problem.

Nollet, Jean-Antoine. *Leçons de physique expérimentale* (6 vols., 1743-64). A collection of lectures by Nollet that were aimed at teaching and popularizing physics with the use of instruments. To this end Nollet gave lectures in many schools of engineering and to Europe's *haute bourgeoisie*. His lectures were so successful that they inspired self-taught scientists until the twentieth century.

Nollet, Jean-Antoine. *L'Art des expériences* (3 vols., 1770). The lectures in this collection explained how to build scientific instruments.

Prévost, Pierre. *De l'origine des forces magnétiques* (1788). This groundbreaking work on magnetism won the author attention among physicists.

Priestley, Joseph. *The History and Present State of Electricity with Original Experiments* (1767). This book grew out of Priestley's friendship with Benjamin Franklin. After Franklin introduced Priestley to electricity, the Englishman's fascination led him to conduct many electrical experiments. At Franklin's behest, Priestley agreed to write a history of electricity, which resulted in this 1767 publication.

Saussure, Horace de. *Voyages dans les Alpes* (1779-96). In this book the Swiss geologist chronicled his extensive alpine investigations and theories regarding the formation of the Alps derived therefrom.

Scheele, Karl. *Chemical Observations and Experiments on Air and Fire* (1777). This book details Scheele's preparation of oxygen, which he obtained in about 1772. Although his discovery preceded Joseph Priestley's similar, independent act by about two years, Priestley published his findings before Scheele.

'sGravesande, Willem Jacob. *Mathematical Elements of Physics* (1720-21). This was easily the most influential semi-popular account of Isaac Netwon's ideas prior to 1750.

Wenzel, Carl Friedrich. *Lehre von der Verwandtschaft der Körper* (1777-82). This work addressed, among other issues, chemical affinity.

Werner, Abraham Gottlob. *Vonden äusserlichen Kennzeichen der Fossilien* ("On the External Characters of Fossils," 1774). This was the first modern textbook of descriptive mineralogy, Taking issue with existing schools of thought regarding classification of minerals, Werner propounded his own theories of "geognosy" and "Neptunism."

Articles

Hadley, George. "Concerning the Cause of the General Trade Winds" (1735). In this paper Hadley correctly described the trade wind circulation in relation to the rotation of Earth.

Prévost, Pierre. "Sur l'équilibre du feu" (1791). In this article Prévost maintained that heat was a "discrete" fluid composed of widely spaced particles that pass continuously between two bodies. His conviction of heat's materiality may have been incorrect, but his theory of exchanges—the idea that all temperature changes are a matter of heat loss and gain obtained through transfer between bodies of differing temperature—was revolutionary.

Priestley, Joseph. "Directions for Impregnating Water with Fixed Air" (1772). This famous booklet, which described how to make his soda water, set off a European craze for soda water.

NEIL SCHLAGER

Technology and Invention

Chronology

1701 Jethro Tull invents the seed drill, which saves seed and makes it easier to keep weeds down.

1709 Italian harpsichord-maker Bartolomeo di Francesco Cristofori invents the first piano.

1710 Jakob Christoph le Blon, a German-French painter and engraver, invents three-color printing.

1764 James Hargreaves builds the spinning jenny, the first practical application of multiple spinning by a machine, in England.

1769 James Watt obtains the first patent for his steam engine, which improves on ideas developed by Thomas Newcomen half a century earlier and is the first such engine to function as a "prime mover" rather than as a mere pump.

1778 English engineer Joseph Bramah introduces a flushing water closet—which, because it is connected directly to a cesspool, does not fully overcome the sanitary and aesthetic drawbacks of nonflushing toilets.

1783 Brothers Joseph-Michel and Jacques-Etienne Montgolfier give the first public demonstration of their hot-air balloon, sending up a large model made of linen lined with paper.

1784 Benjamin Franklin invents the first bifocal lenses for eyeglasses.

1785 Edmund Cartwright devises the first successful power loom in England.

1793 Eli Whitney invents the first cotton gin, which greatly stimulates the U.S. cotton industry—and helps perpetuate slavery in the process.

1796 Czech writer and inventor Aloys Senefelder invents lithography, which produces exceptionally faithful reproductions and is soon adopted by printers as a quick and effective method of mass-producing images.

1797 The modern parachute is born as André-Jacques Garnerin makes the first successful human parachute descent, jumping from a hydrogen balloon 2,300 feet (701 m) over Paris.

Overview:
Technology and Invention 1700-1799

Overview

The eighteenth century saw the transformation of technology from a small-scale, handcrafted activity to a mechanized industrial system. Building on improvements in agriculture, on established small-scale production (proto-industrialization), and on enhanced navigation and trade, this technological change relied on many new inventions, the increased use of steam power, the utilization of coal and iron, and labor-saving machinery. Taken together, these significant changes provided the foundation for an industrial revolution that was well in place by the century's end.

Agricultural Change

The ability to produce surplus foodstuffs with fewer agricultural laborers was essential for supporting an industrial work force. New crops, new tools, and new methods made that possible. For example, the introduction of the potato from the New World provided Western Europe with a new staple food that had a high caloric and vitamin content and the added advantage of being able to be grown on land less fertile than that needed for grains. In addition, developments such as better management of existing land, the cultivation of swamp land with improved drainage systems, the use of nitrogen-fixing and fodder crops such as alfalfa, clover and turnips, and the deliberate breeding of livestock and stall feeding increased food production. British agriculturists such as Charles Townshend (1674-1738) and Robert Bakewell (1725-1795) were especially influential through their use of new crops and innovations in animal husbandry. New mechanized agricultural technology appeared in the eighteenth century, including devices such as Jethro Tull's (1674-1740) seed drill, Andrew Meikle's (1719-1811) threshing machine, and Eli Whitney's (1765-1825) cotton gin. Taken together, these developments provided better diets, increased food and other agricultural production, and resulted in an increase in population. These were an essential and necessary foundation for industrialism.

The Age of Steam, Coal, and Iron

The Industrial Revolution was defined by steam as a new mechanical power source, by coal as a new energy source, and by iron as a new material. With the work of Thomas Savery (1650?-1715) and Thomas Newcomen (1664-1729) steam power met the increased need to drain water from mines with steam-powered pumping engines. It became even more valuable and widespread as an energy source when, in the later third of the eighteenth century, James Watt (1736-1819) incorporated several innovations into steam engine design, such as a governor, a separate condenser, and double acting piston motion. These innovations created a standard for factory use and, for more than a century, the steam engine was the primary power source for industrialization.

The increased use of coal as an alternative source of energy was the result of the overexploitation of wood in Western Europe. Providing higher temperatures than wood and widely available in Britain, coal became a fixture of industrial development. Coal's usefulness increased when Abraham Darby (1677-1717) used purified coal in the form of coke for iron smelting. Because of its high energy content and use for high temperature processes, coal fit the needs of the new technology well—so much so that black, coal ash smoke spewing from factory and locomotive smoke stacks became a hallmark of the industrial age.

The overuse of wood also created a shortage of timber as the traditional building material. That shortage, in addition to the increased industrial demands for stronger and more fireproof materials, made iron (and later, steel) an attractive new structural element. From pistons to pumps and from buildings to boilers, iron became the preferred and often necessary material for these devices. Its strength and durability made it advantageous compared with wood for the machinery and products of an industrial age. Replacing the easily shaped and manipulated wood, iron required a more complex process of extraction and refinement that depended on knowledge of both mining and metallurgy. The technology of industrialism required a higher level of technical knowledge and skill than the preindustrial era, which used natural materials and power sources as the basis of its production.

Mechanized Manufacturing

Mechanized manufacturing, so characteristic of industrialism, required precision machine tools to produce the standardized, interchangeable parts needed for large-scale production. Without carefully calibrated measuring instruments and the special tools these instruments made possible, such as lathes, planners, boring mills, drill presses, and milling machines, mass production was impossible. The contributions of Jesse Ramsden (1735-1800), with his dividing engine, Henry Maudslay (1771-1831), with his use of the lathe slide rest and pattern screw, John Wilkinson (1728-1808), with his precision boring mill, and, Joseph Bramah (1748-1814), with his hydraulic press, made possible the precision machining of metal and wood. In effect, these men and their devices provided the basis for the production of machines by machines—an unheralded but crucial foundation for mechanization.

In the last third of the eighteenth century the British combined steam power with mechanical equipment to transform textile production into the first widespread example of a highly mechanized process. Relying on John Kay's (1704-1764) flying shuttle loom, James Hargreaves's (1720-1778) spinning jenny, Richard Arkwright's (1732-1792) spinning frame, and Edmund Cartwright's (1743-1823) wool-combing power loom, textile production in Britain moved from a handcrafted, small-scale endeavor into a machine-centered, large-scale industrial technology. This transformation led to a centralized factory system with shift work, stringent worker discipline, and the wage system. Because it lent itself to mechanized production and because it had a ready marketplace as an inexpensive and comfortable cloth, cotton was the first successful large-scale product of the industrial process. The merging of steam power with special function machines and machine tools demonstrated the advantages of large-scale production at low unit cost. This kind of technology created a consumer culture with common products available at a modest cost for most buyers.

Mechanical Culture

The transformation of Western technology that occurred in the eighteenth century created a mechanical culture in which technology could thrive. Social attitudes toward invention, innovation and entrepreneurship tolerated and even encouraged deliberate technological change. To a degree not seen before, names of individuals were attached to several significant developments, and governments encouraged inventors and inventions with the patent system and with prizes for targeted technologies. Inventors and industrialists became symbols of progress and agents of positive change. Increasingly, technology was seen as advancing civilization—with materialism as a measure of improvement in a culture. This embrace of technological change permeated the whole culture. The middle class especially profited from and promoted mechanization, materialism, and held industrialism in high regard.

Conclusion

Technological methods in 1800 were quite different from those used in 1700. Industrialism transformed the way people performed technology. Special machine tools replaced the artisan's hand tools. Highly skilled workers, along with a much larger pool of unskilled laborers, superseded the artisans and craftsmen of a preindustrial era. Factory-based, large-scale production supplanted small-scale, home production. Simple machines gave way to special purpose, powered devices. Easily processed animal and vegetable substances diminished in importance and use as industrialism relied more heavily on minerals whose extraction and refinement required special knowledge and skills. Many people traded the subsistence self-sufficiency of an agrarian culture for the materialism of an interdependent industrial economy.

This transformation promoted technological change and rewarded those who created that change. Inventors, innovators, and entrepreneurs emerged as heroic figures whose work advanced the goals of a material world and made life more pleasant for the members of an industrial society. This ready social acceptance, along with the rewards of patent protection and other incentives for invention, stimulated the process of technological change with a resultant plethora of new devices and processes.

At the same time, the creation of this mechanical culture reshaped the work habits and work environments of countless laborers. The factory system, which took shape in this era and matured in the following century, mandated a rigorous work schedule, strict worker regulations, and the adoption of the wage system with almost no social net for most factory workers. Yet, in most cases workers accepted these changes in exchange for steady employment and the more diverse experiences of an urban, industrial culture.

The age of steam, coal, and iron created a new technological culture as well as the cornerstones of the industrial era. With the accelerated pace of technological change and production, the West began its embrace of industrialism as a hallmark of constant change, material comfort, and progress. Technology through industrialism became much more important to individuals and to society. From food production to cotton production, from power sources to energy sources and building materials, and from home to workplace the nature and degree of technology were transfigured so that invention and industrialism became synonymous with technology itself. For at least three centuries the industrial age dominated Western culture, provided the demarcation for classifying societies based on their level of industrialism, and provided a means for the creation of new wealth and prosperity for those who embraced this new means of manipulating the material world.

H. J. EISENMAN

The Social Impact of the Industrial Revolution

Overview

The Industrial Revolution increased the material wealth of the Western world. It also ended the dominance of agriculture and initiated significant social change. The everyday work environment also changed drastically, and the West became an urban civilization. Radical new schools of economic and philosophical thought began to replace the traditional ideas of Western civilization.

Background

The Industrial Revolution precipitated the world's second great increase in economic productivity. The first occurred 15,000-20,000 years ago during the Neolithic Revolution, when small communities became less nomadic and began to base their existence on animal husbandry and agriculture. The Industrial Revolution, which began in the mid-1700s and lasted into the mid-1800s, was similarly a revolutionary experience. It increased material wealth, extended life, and was a powerful force for social change. It undermined the centuries-old class structure in Europe and reorganized the economic and philosophical worldview of the West.

Preindustrial Europe was static and based upon privilege. The most powerful social group was the aristocracy. Its power came from the ownership of the means of production; this consisted of possessing the land and the mills that transformed the crops into material that could be processed into food. The class that labored to produce the agricultural wealth was the peasantry. They were at the bottom of society, and their lives were dictated by both the seasons and the direction of the landowner. They worked the noble's land and used his mills to process their grain. The lord also had the right to impose a tax demanding a certain number of days' labor from the peasants. The construction and repair of roads, dams, windmills, and canals were completed as a result of this tax.

Directly above the peasants were the artisans. These were highly skilled craftsmen who produced the utensils of the preindustrial world. They had far more control over their destiny than did the peasants. Most of their power rested in the right to form professional organizations known as guilds. These groups controlled standards, prices, and wages. The guilds were also a social welfare organization that had the responsibility of looking out for the craftsman's family if he should meet with an early death.

Finally, there was an emerging, vibrant, economic and politically powerful independent class known as merchants. This group made money by moving goods and services through the economic system of the preindustrial world. They were an urban class, acquiring charters from nobles that allowed them to incorporate towns. Many of these urban centers were guaranteed political autonomy and were run by a group of the most successful merchants, known as Burgers.

The family structure of preindustrial Europe was nuclear. The common belief that there were large extended families is an inaccurate description of life at this time. The average family con-

A yarn-spinning machine. *(Corbis Corporation. Reproduced by permission.)*

sisted of a husband, wife, and children. Everyone worked for the economic survival of the group. In an artisan household, the father practiced his trade and also trained the eldest boy to continue the business after he retired or died. His wife ran the shop that sold his products. The rest of the children had chores, usually determined by age and gender, all of which added to the economic success of the family. The husband and wife worked as a team, with the children supporting their efforts. Children usually left the household in their early teens. Boys of merchants and artisans usually went off for training or apprenticeship, while girls for the most part took positions as household servants. Since life was so precarious, couples usually did not enter into marriage before they had acquired the skills to insure an economically self-sufficient unit.

The Industrial Revolution was preceded by an agricultural revolution that increased the food supply while decreasing the amount of labor needed. Traditionally, the primary goal of agriculture was to produce enough food to prevent famine. This overwhelming fear of starvation made most farmers very conservative and highly skeptical of change. Poor harvests would lower the supply of food, which would result in increased prices. The basic effect of supply and demand was at the center of most of the class conflict in this preindustrial world. Both bad harvests and increased population affected the price of food. High prices increased the wealth of the aristocratic class and led to death and starvation among the peasants; therefore, the primary reason behind most peasant uprisings was the high price of food.

A multifaceted revolution in every aspect of agricultural production would eventually eliminate this ancient curse. The traditional mode of farm production that had existed for centuries was abolished. Small, disconnected plots of land were combined into large commercial operations that implemented both new crops and the latest farming methods. Practical research had identified a number of crops that would both provide feed for animals and increase the fertility of the land. Clover and turnips were two of the most widely used for this purpose. Information concerning these revolutionary changes was transmitted in the popular scientific literature of the time.

The greatest social and economic impact of the agricultural revolution came from the "enclosure movement," in which farmers were able to enclose their fields and grow different crops during different seasons. This drastically changed the centuries old model of the "open field system," in which farmland was used in a semi-public fashion that prevented farmers from growing patterns that differed from traditional ones. This method of farming had been established during a prior agricultural revolution that occurred during the early medieval period. Initially this led to a substantial increase in agricultural productivity. Gradually there was a Malthusian effect; the population increased and the threat of famine reemerged. This reality made the peasants question any proposed changes in the accepted method of farming. They were not risk takers, so they typically fought to maintain the status quo. Records show that by the middle of the seventeenth century, the price of grain, especially of wheat, increased considerably. The scientific literature of the time stated that production would increase if the land were used in a more structured fashion. Researchers insisted that small, individual peasant plots would become more productive if they were restructured into large agricultural farms. Labor, machinery, fertilizer, and seed would be used more efficiently, thus increasing the bushels per acre. Initially this was met with resistance on the part of the peasants. Eventually, the increase in the supply of food tempered much of the anxiety, and significant changes began to occur within preindustrial society.

The reliance on science and technology, the questioning of traditional methods of agriculture, and the centralization of factors of production set the stage for the onset of industrialization. Also, at this time the traditional peasant-lord relationship began to dissolve. The quest for large profits undermined the bond that once existed between the two classes, and at the same time the gradual acceptance of a market economy began to take root. This moved the aristocratic class to change its view of the peasant class. The highly religious and land-based society of the medieval world believed that social structure was ordained by God. The deep belief that all souls were equal in His eyes produced a social system where all classes had both rights and responsibilities. With the onset of a profit-oriented market economy, the wealthy landowners began to perceive the peasant as just a source of labor. It was the fruit of their labor that was of the greatest interest and importance in this entrepreneurial economy. This created an expectation of greater profits, which in turn increased the demand for greater material prosperity. This revolution in expectations would both stimulate and focus the drive toward industrialization. The new demand for consumer goods produced the first workshops and factories.

Impact

Industrialization increased material wealth, restructured society, and created important new schools of philosophy. The social impact of industrialization was profound. For the first time since the Neolithic Revolution, people worked outside of the local environment of their homes. They arose every morning and traveled to their place of employment. This was most often in a workplace known as a factory. The new machinery of the Industrial Revolution was very large and sometimes required acres of floor space to hold the number of machines needed to keep up with consumer demand.

As in all productive revolutions, skill greatly determined the quality of life. The most important aspect of this new economic order was the fact that the skills needed to succeed were in many ways different from those that had been needed in the earlier economy. Artisans had the easiest time transitioning to the new economic paradigm. The fact that they had highly developed manual skills enabled them to adapt to the new machinery much easier than their agricultural counterparts. This was also the case when it came to dealing with the new, enclosed work environment and strict schedules. The worker from the countryside had over the centuries constructed a cycle of labor that followed the seasons. There were times, especially during planting and harvesting, when he was expected to put in long hours, usually from sunrise to sunset. The term "harvest moon," which today is looked upon as a quaint metaphor for autumn celebrations, was in preindustrial Europe a much-needed astronomical occurrence that allowed the farmer extra time to harvest his crops. In turn, the long winter months were a relatively easy time. The lack of electricity and central heating kept most people in bed ten to twelve hours a day, affording them relief from the busy periods of planting and harvesting.

The industrial economy had a new set of rules and time schedules for the common laborer. The work environment not only moved indoors, but the pace of the work changed drastically. Instead of driving a horse that pulled a plow or wagon, the machines drove the worker. The seasons of the year were no longer relevant to the time spent at work. Adult males were now expected to labor twelve to fourteen hours a day, five-and-a-half days a week, all year long. This was a very hard transition to make. A great many people who had once been considered highly productive agricultural workers were unable to hold jobs because of their inability to adjust to this new regime.

In many ways women suffered more than men. In both the urban artisan economy and the rural agricultural world, women were traditionally regarded as playing an equally important role as men. They were full partners in the family's quest for economic success. Their status changed substantially as a result of the Industrial Revolution. Their labor became a commodity to be exploited. They were as a rule given the lowest-skilled, lowest-paying jobs. They were regularly bullied by both their bosses and their husbands. In many ways their labor and responsibilities doubled. They were not only responsible for their jobs in industry, but they were also expected to continue their traditional roles at home. They labored for ten hours in the factory and continued for untold hours once they arrived home. It must be remembered that by law men still controlled their families. Women had no political, social, or economic rights outside the home. They were forbidden to vote or own property. The "rule of thumb" was still supported by most courts in the Western world. This "rule of thumb" referred to the fact that a man could beat his wife with a stick, as long as it was not larger than the width of his thumb. Women did make some strides in their ability to choose a marriage partner; traditionally, marriages had been arranged for the most part to establish economic connections between families. When young women moved to the cities to work in the

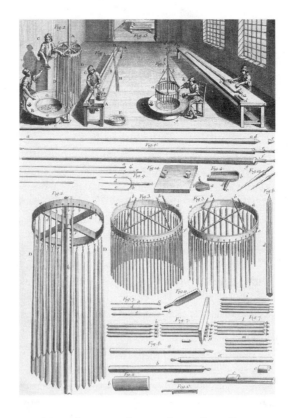

An eighteenth-century diagram of a candle-making factory. *(Corbis Corporation. Reproduced by permission.)*

factories, they most often chose a marriage partner from among the young men they came into contact with at their boarding houses or place of employment.

Child labor also changed as a result of the Industrial Revolution. Children were expected to help the family in the traditional economy, but usually they had been assigned tasks that were commensurate with their age. Not unlike their mothers, young children began to be exploited by their bosses. The most dangerous assignment for children in the factories was unjamming the great textile machines that wove cloth. Since their hands and arms were so small, they could reach into small spaces where the fabric tended to jam. The foreman would not turn the machine off but would insist the child reach in to dislodge the jam. If he were not quick enough, his hand or arm would become caught in the mechanism, and this could result in severe damage to the child. All laborers, male, female, and children, were eventually looked upon as interchangeable parts. As technology increased and machines became more sophisticated, the employer began to value machinery more than his work force. This would remain the case until the early 1830s, when legislation was passed to protect the workers.

The Industrial Revolution also accelerated the growth of the urban population. One of the more important consequences of urbanization was a rapid increase in crime. This was the result of three factors that dominated the urban landscape. The first two were poverty and unemployment. There was no job security or social security for the factory worker. If someone was injured on the job or laid off, he had little chance to replace his lost income. The few charitable organizations that were available were so over-taxed their aid never matched their good intentions. Overcrowding was the third important source of crime. Industrialization drew thousands of people to the urban areas in search of employment. Cities such as Manchester, England, were completely unprepared for the great influx of workers. This overcrowding fueled social dysfunction that resulted in a rapid increase in crimes against property and people.

One major attempt to deal with these problems was the creation of a professional, full-time police force whose members were trained in the latest techniques of crime prevention. Secondly, there was a vigorous attempt to reform the prison system. It was accepted among most intellectuals of the time that prisons should not be solely places of punishment. There was widespread acknowledgement of the belief that, through proper training and guidance, criminals could be reformed. Education would allow prisoners to find a productive place in the new urban industrial society.

The Industrial Revolution also accelerated change in the area of political and economic thought. The dominant economic model of the early industrial period was mercantilism, a command economy based upon the belief that there are a finite number of resources in the world. The primary economic goal of each nation was to control as many of these resources as possible. Its trade policies were a form of eighteenth-century protectionism. Great Britain not only forbade its colonies to develop any domestic industry, but the government also controlled colonial trade. Everything was done for the betterment of the "mother country."

The first major political economist to challenge this concept was Adam Smith (1723-1790). He was one of a number of Enlightenment thinkers who believed there were natural laws that governed the economic, political, and social relationships of men. These natural laws were discernible through the exercise of human reason. Smith believed that the natural law of economics revolved around the exercise of economic choice. He argued that the best way to bring about economic expansion was to allow people to make decisions regarding the products they wished to produce. The Industrial Revolution, because of its increased productivity, greatly expanded peoples' ability to choose. At the same time, it also raised expectations concerning issues of quality of life. Smith believed that through proper application of Enlightenment scientific principles, a nation could produce a society free of poverty. This revolutionary theory, that a government was obliged to create an environment in which these expectations could be met, would act as the foundation of the Declaration of Independence crafted by American revolutionaries in the 1770s.

By the beginning of the nineteenth century, economic thought became very pessimistic because of the inability of society to solve the conditions of the industrial working class. Over time it became widely accepted that the quality of life of the working class would remain forever wretched. Thomas Malthus (1766-1834), in fact, believed this condition was necessary to prevent widespread political and social unrest as the result of famine. Malthus's theory centered on the belief that if left unchecked, the population would outstrip agriculture's ability to feed it. Industrial wages were kept low in order to control the size of working families. This began the modern fear of the population bomb, which is still a hotly debated issue. This pessimism continued in the writing of David Ricardo (1772-1823). He formulated the Iron Law Of Wages, which mandated that monetary compensation be kept low. Ricardo believed that there were only a finite number of jobs available in industry. If a particular generation experienced too much economic security, they would marry earlier and have larger families. This would result in their children's generation experiencing the frustration of competing for a limited number of jobs. The resulting political and social upheaval would undermine the social stability of the nation. It is quite evident that both Malthus and Ricardo were still under the influence of a mercantilist world view in which humanity would be forever governed by the limits of its productive capacity.

The success of the Industrial Revolution in expanding both productivity and jobs eventually changed the economic pessimism of the early nineteenth century. The dismal view of the future of the working class would eventually be replaced by utilitarianism and socialism. Jeremy

La Nouvelle Yorck, by André Basset shows the bustling eighteenth-century New York City waterfront. *(Corbis Corporation. Reproduced by permission.)*

Bentham (1748-1832) helped create a philosophy based upon the concept that all social, political, and economic models should be concerned with creating the greatest happiness for the greatest number of people. This was based upon confidence in the correct application of reason. All social models would be judged on their ability to create utility. Thus, if properly implemented, the fruits of industrialization would be shared by all.

Socialism was an alternative theory, based upon the premise that true economic equality could only be attained if the workers controlled the means of production as well as the distribution of goods. This was a reaction against both the hardships of the working class and the economic inequality of capitalism. Basic socialist theory predicted that competition, the lifeblood of the free market, would eventually reduce to a small minority the number of capitalists controlling the economic system. Eventually, the large number of exploited workers would rise up and overthrow this small, very rich capitalist class. Economic, social, and political equality would then be achieved. The power of industrialization would be used to create a socialist utopia based upon the practice of the equal and rational distribution of goods.

In the end, utility carried the day. Over time reform legislation increased the basic social, economic, and political rights of the working class. They in turn realized it was in their own best interest to work for continued change through the existing political system.

The Industrial Revolution increased the material wealth of humanity, especially among the nations of the West. It increased longevity and accelerated the growth of the middle class. It helped to create the modern world view that through the proper use of science and technology, a more fruitful quality of life could be achieved.

RICHARD D. FITZGERALD

Further Reading

Clarkson, L.A. *Proto-Industrialization: The First Phase of Industrialization.* London: MacMillan, 1985.

Deane, Phyllis. *The First Industrial Revolution.* Cambridge: Cambridge University Press, 1979.

Kemp, Tom. *Industrialization in Nineteenth-Century Europe.* 2nd ed. London: Longman, 1985.

Rule, John. *The Vital Century: England's Developing Economy, 1714-1815.* London: Longman, 1992.

Thompson, E.P. *The Making of the English Working Class.* New York: Random House, 1966.

New Machine Tools Are a Catalyst for the Industrial Revolution

Overview

Machine tools were the instruments of industrialization. They enabled the first capitalists to mass-produce inexpensive, quality goods. They also allowed the industrial nations to eradicate many deadly diseases and to become the dominant economic and political forces in the world. These tools would also change forever the concept of human labor.

Background

Charles Darwin (1809-1882) shocked the nineteenth-century Victorians when he explained one of the most important evolutionary periods in humankind's history. In *The Descent of Man* Darwin described how our ancestors descended from the trees, our original home, to walk upright on the ground. This was the pivotal moment in the development of our species. Standing upright not only increased our locomotion, but it would eventually free our hands and allow us to become the premier tool-using species on the planet.

Our earliest *Homo sapiens* ancestors were not the only humanoid creatures competing for control of the environment, but by the end of the Paleolithic Age all the others had died off. This occurred because *Homo sapiens* had certain advantages over other humanoids. Most importantly, they had larger brains and free hands with opposable thumbs. This allowed for the development and use of tools and weapons. Scientists today know that there is an important cause and effect relationship between the use of tools and increased brain function. The process of creating and using these first tools stimulated the cortex of the brain, which increased the intellectual power of our early ancestors. Most anthropologists believe that language developed as a result of this increased brain function.

Tools were so important in humanity's development that social scientists actually classify different periods in prehistory by the material that was used to create many of these first implements. The Paleolithic or Old Stone Age is designated as the earliest period of humankind's development. These early people had two distinct characteristics in their technological evolution. They were the first to use existing tools to create new and more sophisticated instruments, and they were also the first to develop the idea of a toolbox. This is very significant because it shows that intellectual development had advanced to a point where there was a recognition that these tools would be used in the future to solve similar problems. This occurrence set the stage for the development of collective memory and the practice of passing on established knowledge and skills to the next generation. The impact of this was crucial: it not only insured that every generation had an established knowledge base, but it also provided a basis for the development of education and research. It thus created a model for increasing technological capabilities by developing the system of building upon prior knowledge and experience.

The most important invention of the Paleolithic Age was the ability to use fire. It was a multipurpose resource that went beyond cooking, heating, and light to allow early man to make more sophisticated tools. Ultimately, it was fire that allowed Paleolithic people to move out of Africa and into Asia and Europe at the end of the last ice age. Early humans used fire to harden their weapons and tools. This increased the effectiveness of hunters, which resulted in the adoption of a high-protein diet. The first effective tool was the hand axe, which allowed early humans to crack open the bones of animals to extract the protein-rich marrow inside. Archeologists believe these tools were constructed by skillfully using stones to fashion an implement that was both sharp and strong. Many anthropologists believe this was one of the original skills passed on from generation to generation. The hand axe eventually evolved into the spear and arrow points found among the artifacts of most Paleolithic sites. These tools allowed *Homo sapiens* to eventually dominate the food chain, which resulted in the extinction of other humanoids.

The Neolithic Revolution, which began between 15,000 and 20,000 years ago, witnessed the use of tools to drastically transform material culture and to create the conditions for the onset of civilization. This new way of life was based upon sedentary agriculture. These cradlelands of civilization were located along four great river systems: the Tigris/Euphrates in what is now modern Iraq, the Nile in Egypt, the Indus in

A modern woodworker uses a lathe to make furniture. *(The Stock Market. Reproduced by permission.)*

Pakistan, and the Huang He in China. Neolithic communities were able to take advantage of the dependable water and rich soil because they had the technological ability to do so. Digging sticks, axes, sickles, and eventually plows allowed the first farmers to successfully cultivate the land.

Man's tool-making ability continued to evolve during both the Classical and Medieval eras. Over the course of centuries, mathematical and scientific theory combined to create even more sophisticated tools. By the mid-eighteenth century many factors were in place to begin humanity's second great revolution in material productivity, and for this tools would become more important than ever before.

The basis of the Industrial Revolution (c. 1750-1900) was the movement away from the slow and more costly production of goods by hand to the use of machines and inanimate power. The most significant impact of machine tools was the reduction in the cost of production. Western manufacturing moved from the production of one product, which consisted of a certain number of handmade pieces, to the mass production of thousands of interchangeable parts from which hundreds of products could be constructed. This created the manufacturing model based upon the concept of the division of labor. For example, no longer was a highly skilled craftsman needed to create a rifle. Instead, a

semi-skilled group of workers could assemble one from a mass-produced set of interchangeable parts. This process increased the number of products that could be produced in a specified amount of time. This decreased the cost of manufacturing, which in turn lowered the cost of the item. The assembly line model would dominate manufacturing for almost 300 years.

The increase in production began with the development of a new generation of lathes that could hold a precision tool and increase the accuracy of the tool's performance. Increased precision helped create new inventions and designs, which consequently allowed for the development of mass production techniques.

The movement toward higher precision can be first seen in the development of highly accurate and affordable clocks. Antoine Thiout (1692-1767) and Robert Hooke (1635-1703) developed lathes that could uniformly cut the pieces of the internal mechanism of a clock. The two most important components of a timepiece are the clock wheels that accurately move the mechanism and the screws that hold the wheels in place. Once these two parts could be mass-produced, accurate time measurement would become a common characteristic of both manufacturing and scientific research. The availability of inexpensive, accurate timepieces also had a major impact on Western imperialism. One reason is that it allowed for the accurate measurement of longitude, which is the distance east or west of the Prime Meridian. Based on the precise knowledge of time, John Harrison (1693-1776) was able to use a chronometer and a quadrant to accurately pinpoint longitudinal position. Combined with the correct latitude, the distance north or south of the equator, it was now possible to develop highly accurate maps and charts. This enabled the industrialized nations to undertake voyages that would eventually spread Western civilization throughout the globe.

The early Industrial Revolution was powered by steam, and machine tools had a direct impact on the evolution of the steam engine. The theory behind this source of power was that boiling water created steam, which would exert pressure that could be harnessed to drive an engine or to operate a machine. The pressure would move a piston, which in turn drove the particular device. One of the major problems faced by both Thomas Newcomen (1663-1729) and James Watt (1736-1819) was to construct a cylinder large enough to hold a piston big enough to be useful in a steam engine. John Wilkinson (1728-1808) created a machine tool that could in fact bore a cylinder big enough to hold a large piston. This accelerated the use of the steam engine, which subsequently provided energy for the Industrial Revolution.

Steam radically changed the environment of the Western world. It produced a source of energy that could be used 365 days a year. Steam would eventually power the machines of the textile factories. These massive instruments would later replace thousands of spinners and weavers. Originally, these machines had been powered by falling water. The reliability of water, however, fluctuated with the seasons. When the water was high manufacturing could take place at a steady pace. When the water level was low or constricted by ice, its reliability was greatly reduced. Steam was steady and made production quotas far more predictable.

The mass production model also required the availability of affordable machines and tools. As with the production of clocks, most machines need gears and screws for their operation. Once again the model of the new power lathe was used to solve the problem. Henry Maudslay (1771-1831) and Joseph Whitworth (1803-1887) both perfected power lathes that could accurately mass-produce the cut screws and gears needed to run the new machinery of the Industrial Revolution. The impact on productiv-

ity was two-fold: they reduced the initial cost of manufacturing by making the machines less expensive, and they also decreased the "down time" for repairs. Instead of waiting for the services of a skilled craftsman, the defective part could be quickly replaced with an exact replica by the machine operator himself. The availability of inexpensive, interchangeable parts also allowed for greater control over production quotas and schedules.

Over time, improvements in machine tools were made. James Nasmyth (1808-1890) created the steam hammer. This powerful device enabled very large pieces of metal to be forged. Joseph Clement (1779-1844) invented a machine that could accurately plane metal. The ability to manufacture large, accurately planned pieces of metal helped give the industrial nations almost unlimited manufacturing and military power.

Impact

As has been the case throughout history, the effects of these machines were both positive and negative. Most of the important political, social, and economic reforms of the nineteenth century were the result of the new industrial environment created by these new tools. The refinements in the steam engine created the need for vast amounts of iron and coal. Iron was the material of the early Industrial Revolution. Initially, it was the only substance strong enough to contain the extreme pressure of the steam engine. Iron and coal are both found in nature in plentiful supply, but they are not easily extracted from it. The mining of these two substances created great hardship for the people involved, and the mines of the early Industrial Revolution were the most dangerous work environments. Water and bad air were major threats to the health of the miners. For instance, black lung disease was an illness that affected the majority of the miners. Dirt and dust from both coal and iron were common in the poorly ventilated shafts. Through the normal course of working, the miners ingested large quantities of iron and coal particles. Coal mines were especially dangerous because over time volatile coal gas built up and explosions were common. Children seemed to suffer the most because their bodies were still developing; the polluted air attacked their lungs more vigorously than those of adults. In many circumstances women and children were also used as draft animals. Belts and chains were placed around their bodies and attached to carts of ore. They were expected to pull these carts along mine shafts, sometimes getting down on all fours in order to complete the task. One of the first pieces of reform legislation was the result of hearings by the Ashley Mines Commission (1842), which found the conditions of miners intolerable.

The mass production model created by the use of machine tools also affected the quality of life of the common laborer. One of the reasons mass production was so profitable was that it greatly reduced production costs, and lower wages were a primary example of this. The precision work of these machine tools reduced the need for skilled labor. The tools became the most skilled part of the manufacturing process. A device that could cut screws and punch out gears to within one millionth of an inch was deemed far more valuable to the employer than the unskilled labor that tended the machine. These tools were in many ways perceived as more reliable than the worker. The machines never became tired or sick, and most problems could be alleviated by the simple replacement of a part. This was not the case for the unskilled laborer, whose work and home environment, poor diet, and long hours made him highly susceptible to disease.

The philosophy of socialism was an attempt to counteract this problem. Most early socialist theory focused on the new work and living environment in which the worker found himself. The squalor of the industrial city was attacked through utopian models that called for ideal communities. This new type of living space had a cleaner, more natural environment but also demanded a radical change in the basic structure and values of the capitalistic framework. Some called for the abolition of both the traditional structure of the family and accepted sexual practices. This was driven by the concept that the patriarchal, monogamous model of the family was just part of the greater oppressive paradigm of capitalism. This was especially aimed at the theory of private property, upon which the capitalist system rested.

This criticism did not entail a total rejection of the technological reality that was emerging in the late eighteenth century. The success of the new precision instruments was very appealing to many of the early socialists. They believed a society that could produce lathes and drills of a highly accurate nature could also solve the problems of industrial life. Henri de Saint-Simon (1760-1825) believed in the rational management of society. He wanted to create a techno-

An eighteenth-century smelting furnace in Maryland. *(Corbis Corporation. Reproduced by permission.)*

cratic state controlled by a board of directors who would, through the wise management of this new wealth, eradicate poverty. This governing body would consist of the most successful scientific and technological minds of the day. The very men who had created highly accurate productive machines would also establish a social system based upon the precise management of social resources.

Robert Owen (1771-1858) attempted to put this new theory into operation when he created the industrial community New Lanark in England. In opposition to the accepted economic theory of the classical economists, he paid his workers high wages and constructed quality housing. He also allowed his workers to take part in management decisions and instituted a model in which they were allowed to share in the profits. Owen was too far ahead of his time, and even though New Lanark was successful, other entrepreneurs never followed his example.

There were also many industrial motives for imperialism. Initially, it was the need for raw materials and new markets that drove the Europeans to expand around the globe. It was the West's advanced military technology that allowed Europe to dominate the world. Machine tools also played a major role in this enterprise. Alfred Krupp (1812-1887) was a leader in the manufacture of new weapons technology. The evolution of Nasmyth's steam hammer had a major impact on the development of both heavy and light artillery. Since the power of the hammer could be regulated, it allowed arms manufacturers to forge barrels to very precise standards, increasing the power and accuracy of the weapons. Artillery pieces were used mainly to support the infantry. Battles were won and lost when the infantry drove the opposing forces off the desired piece of territory. Because of the technological superiority of their weapons, the most important of which was the infantry's rifle, European armies were able to conquer and hold large tracts of land. Once again, interchangeable parts played a significant role. Eli Whitney (1765-1825) was the first to apply this concept to the manufacturing of rifles. His very limited success set the stage for a revolution in the area of light weapons. Eventually, the techniques of precision boring would lead to the development of rifled barrels. These advancements enabled the industrial nations to send relatively small numbers of men into battlefields around the world and continually dominate the tactical situation because of the reliability and accuracy of these mass-produced weapons. On another front, the impact of advancements made by Joseph Clement (1779-1844) in the planing of metal helped create a new generation of warships. Larger sheets of metal forged by steam hammers could now be cut to precise measurements by planing machines, which resulted in the construction of modern ships made from metal. These new tools also allowed for the construction of new and more powerful steam engines. Eventually, the navies of the industrial nations would control the sea-lanes of the world.

Finally, the evolution of machine tool technology would also have an impact on the health and living environment of the West. Precision tools that could drill, cut, grind, and polish accelerated the production of quality scientific instruments. These in turn were used to make a series of very important discoveries about the origin and transmission of disease. In 1714 Daniel Fahrenheit (1686-1736) created the mercury thermometer, which, along with the temperature scale that bears his name, allowed scientists and physicians to record detailed changes in body temperature. This was especially helpful in fighting disease. Doctors could now have a more accurate record of the impact of a particular dose of medicine by measuring how it reduced the fever of infection. Much of the knowledge concerning infection and germ theory was the result of the discovery and study of microorganisms. These advancements were made because of the development of the microscope. The renowned English naturalist Henry Baker (1698-1774) wrote a series of essays detailing the accuracy of this new scientific instrument. These were widely read by the scientific and medical community, which used the data the microscope provided to investigate the impact of microorganisms on the human environment.

From the earliest days of the human adventure, tools have played an important role in our development. Most importantly, the Industrial Revolution combined increased accuracy and power to reduce the cost of manufacturing and increase the material wealth of the human community.

RICHARD D. FITZGERALD

Further Reading

Ackerknect, Erwin H. *A Short History of Medicine*. Baltimore: Johns Hopkins University Press, 1999.

Burke, James and Robert Ornstein. *The Axemakers' Gift*. New York: Putnam's Sons, 1995.

McClellan, James E. III and Harold Dorn. *Science and Technology in World History*. Baltimore: Johns Hopkins University Press, 1999.

Pacey, Arnold. *Technology in World History*. Cambridge: MIT Press, 1993.

Smil, Vaclav. *Energy in World History*. Oxford: Westview Press, 1994.

The Industrialization of Agriculture

Overview

Agricultural technology changed more dramatically in the 1700s than at any time since the introduction of draft animals millennia before.

Mechanized planting and threshing made farms more efficient, threw workers off the farm, and altered the very shape of the countryside. Scientific approaches were applied to agriculture, and books helped spread new ideas and approaches. At the end of the century, cotton became a force for change: Whitney's gin made cotton profitable for the first time in the American South and helped support the continuation of slavery. Off the farm cotton mills led the way in industrialization. Farm mechanization made food supplies more stable and more plentiful, supporting a surge in population and leading to unprecedented growth in cities.

Background

In the eighteenth century, the world witnessed a revolution in agriculture led by three inventions—the seed drill, the threshing machine, and the cotton gin. Complementing these new tools were new ideas, set forth in books. The agricultural revolution paved the way for the Industrial Revolution, both by showing how the new ideas of science could be put to practical use and by freeing the manpower needed for factories.

Dramatic changes in agriculture were already in progress when the eighteenth century began. In 1700, 80% of the population was engaged in agriculture throughout most of western Europe. But this wasn't true for the Low Countries (the Netherlands, Luxembourg and Belgium). Pressed by the highest population density in Europe, these countries had made huge strides in farming efficiency by moving from the medieval practice of open-field farming to a field enclosure system. In addition to making agriculture more efficient, enclosure allowed landowners to raise sheep for the burgeoning wool trade.

Open-field farming was a communal activity. The land around a village was divided into rectangular plots called furlongs. Strips of about a morning's plowing were distributed within each furlong. This arrangement encouraged sharing of work and draft animals, and distributed good and poor soil equally among all the farmers. The open-field system also spread the ongoing burden of allowing one-third of the land to regenerate each year by lying fallow. Pasture and woodlands were held in common, so everyone was able to hunt, graze animals, and gather wood. The poor were granted the right of gleaning—they could go through fields after harvest and pick up any grain that had been left.

This type of farming had built-in inefficiencies, such as the need to move laborers and draft animals from field to field. It also discouraged innovation because the potential consequences were terrifying: Under the best of circumstances,

A French farmer sowing. *(Corbis Corporation. Reproduced by permission.)*

poor harvests could be expected every eight or nine years. Two crop failures in a row led to famine, so there was little margin for change without putting the entire community at risk. The tools of farming—the scythe, the wooden plow, and the hoe—remained as they had been for hundreds of years. People were even cautious about introducing new crops, such as potatoes and corn, that had been brought back from the Americas.

By contrast the enclosure system developed in the Low Countries, transformed farming into an efficient, pseudo-industrial endeavor. When plots of land were allocated to specific owners, the profitability of each tract became the responsibility of its farmer. This encouraged mechanization, reduced labor costs, and encouraged innovation. For the large landowners, who had sponsored enclosure laws, the result was an industrialization of farming, more arable land, and exceptionally high agricultural productivity. For smaller landowners, privatization was a disaster; farmers could no longer distribute risk or share resources, and they were even held responsible for the costs of fencing. Still, the overall effect was more food from less labor.

A key leader in the modernization of farming was Jethro Tull (1674-1741), a lawyer who never practiced law. Instead, he experimented with agricultural methods on his father's farm. In 1701, he introduced the seed drill. While devices of this sort had been used since Babylonian times, Tull's design matched the opportunities that enclosure provided. It set seeds into the ground in straight rows at uniform depth and automatically covered them after planting. This meant that far fewer seeds were needed per acre than broadcasting (scattering seeds onto the soil by hand), which spread the seeds unevenly and exposed them to wind, frost, birds, and animals. The neat rows allowed for more efficient cultivation, clearly separating weeds from crops, and allowed farmers to weed the ground between the rows. By the end of the eighteenth century, seed drills had gained broad acceptance.

In 1709, Tull observed the way that French and Italian farmers tilled their vineyards, breaking up the soil to improve yield. Thereafter, he became a strong advocate of the practice, and even introduced the horse-drawn hoe, another labor-saving device. Although Tull mistakenly believed that tilling provided food for the crops, the process's real benefits came from aerating the soil and allowing for more efficient use of water. Tull, a writer as well as an inventor, wrote *The Horse-hoeing Husbandry, or an Essay on the Principles of Tillage and Vegetation* (1733). The book aroused great controversy, but also spread the word on the benefits of tilling, mechanization, selective breeding of livestock, and using horses instead of oxen as draft animals.

While the Dutch rarely wrote about agriculture, the English became major publishers of manuals and books on farming. This began as a trickle with the founding of the Royal Society in 1662, but became a torrent in the 1700s. Richard Bradley wrote the first work on artificial cross-fertilization in 1739. Arthur Young published the first farming survey in 1771. These manuals and books had influence well beyond agriculture. The model for technical writing, reaching back to Greek and Roman times, had been an elegant and poetic style of prose. Tull and many of his contemporaries intended to make it easier for ordinary people to put their findings to work. While a few tried to convey facts and statistics in the old form, the new writers, often with apologetic references to classical works, produced a more direct and efficient form of prose. By doing so, they helped create a new and unadorned literature. At its best, this style lives on as clear, nonfiction prose. At its worst, it led to impenetrable technical writing.

Mechanization took another step forward with the 1788 invention of the threshing machine by a Scottish millwright, Andrew Meikle (1719-1811). His device, which removed the husks from grain, was the result of a 10-year effort. Meikle had built two previous machines, but his third try was successful. He had redesigned the machine's drum so that its operation beat rather than rubbed the grain (an idea he may have gotten from a flax-scutching machine). Meikle's machine wasn't successful commercially (he had to apply for relief in 1809), but by the 1830s mechanical threshing had been widely adopted in Europe.

There were many other important developments in farming in the 1700s: Charles Newbold invented a cast-iron plow that could dig more deeply into the soil in 1797. Joseph Boyce developed an early reaper (1799). New crops were introduced, and Viscount Charles Townsend helped end the practice of letting fields lie fallow by showing that rotating soil-enriching crops, such as turnips and clover, with traditional crops kept the soil fertile. This put 50% more land into use and increased the supply of available cattle feed.

Among all the agricultural inventions and innovations of the eighteenth century, Whitney's cotton gin (1793) stands alone as a direct link between the agricultural and Industrial Revolutions. As a young man, Eli Whitney (1765-1825) supported himself by inventing gadgets. While on a visit to the American South, he was approached by farmers who wanted a better way to separate cotton fibers from cottonseeds. Until this time, separating one pound of cotton lint from seed took ten hours of hand work. His southern hostess told her visitors, "Gentlemen, apply to my young friend, Mr. Whitney. He can make anything." Within six months Whitney devised a machine that pulled the fibers onto a coil of metal wires and could clean 50 lb (23 kg) of cotton in a day. The results were nearly instantaneous. For the first time, cotton became a profitable crop in the South, and, supported by slave labor, it soon became the region's most important agricultural product. Spinning machines and mechanical looms already existed, so the cotton gin quickly became an essential link in the manufacture of fabric.

Impact

One consequence of mechanization and other agricultural advances was that farms grew larger. Agriculture became a business and favored the formation of estates. By 1815, the majority of farms in Britain were owned by a minority of landowners (often absentee) who saw their holdings as financial properties, largely independent of tradition and community values. They invested in more agricultural innovations, changing agriculture even more. Larger farms were more profitable, and led to the dominance of plantation farming, which continues to this day with agribusiness. (The value of U.S. agricultural exports in 1999 exceeded $50 billion.)

From the beginning of the agricultural revolution, the small farmers who were most affected by the changes attacked new equipment and organized to stop mechanization and enclosure, with little political or economic impact. Even Thomas Jefferson, though he was an active innovator in agricultural science, imagined the U.S. as an agrarian democracy with independent and economically self-sufficient small-farm owners forming the backbone of the new nation. This was not to be. Today, about 100,000 people own half of all U.S. farmland, and farmers account for less than 1% of the population. One reaction to the growth of modern agribusiness has been the promotion of organic foods, grown without chemical fertilizers, pesticides, and herbicides. Other activists have organized protests against genetically engineered crops, which ironically, were developed to reduce the use of pesticides and herbicides.

A positive consequence of mechanization was that while life didn't necessarily get any easier for farmers, they were saved from much of the backbreaking labor that was common before machines came into widespread use. By the nineteenth century Cyrus McCormick's (1809-1884) reaper (which cut grain), invented in 1834, allowed two men to harvest 12 acres in a single day versus the one acre they would have harvested without the machine. John Deere's (1804-1886) self-cleaning steel plow and more advanced threshers both appeared in the 1800s. Charles and William Marsh invented the first true harvester (which cut and bound grain) in 1858. The mechanization of farms in the U.S. accelerated during the Civil War. Because of the labor shortages caused by mobilization, U.S. farmers invested an average of $200 for every hundred acres farmed during this period.

Although the most important farm machine, the tractor, did not exist until the very end of the nineteenth century, there was a prototype that existed in the eighteenth century. Nicolas Cugnot (1725-1804) invented a three-

Eli Whitney's cotton gin. *(Underwood & Underwood/Corbis. Reproduced with permission.)*

wheeled, steam-driven tractor in 1770. It was designed, however, to carry artillery, not to pull a plow. The tractor we know today had to wait for the invention of the internal combustion engine to become practical. Tractors have been used as a general farming device, not just to pull equipment, but to serve as portable engines to power other machinery.

Seed drills and automated tilling still dominate agriculture. In fact, the rotary mechanism in Tull's invention is the basis for all mechanical sowing devices. But tilling has become somewhat controversial. Loosening the soil might help get water to plant roots and limit weeds, but it also increases erosion, which reduces arable land, pollutes water, and clogs harbors and seaways. Over the last 30 years, no-till farming, which injects nutrients into the soil and controls weeds with herbicides, has been developed as an alternative. The fields are no longer plowed, which conserves soil and saves on labor and fuel costs.

The immediate outcome of new agricultural methods and tools was increased food supply. Between 1700 and 1870, English farmers quadrupled their productivity, and food prices dropped accordingly. The variety of food in the typical person's diet increased as well, partly because of the introduction of foods from other lands and partly because of the greater availability of meat, thanks to the production of feed crops (like turnips). Though there have been many famines since the close of the eighteenth century, most have been related to distribution problems rather than limited supply. For instance, during the Potato Famine of the mid-1800s, when millions of Irish were starving or emigrating, Ireland was producing more than enough food to feed its population.

In order for farmers to benefit from excess food production, they had to get their crops to market, so the agricultural revolution became an impetus for the development of roads, bridges, tunnels, and canals. The rapid growth of railroads was also strongly driven by the need to get a valuable, perishable product to consumers. To manage this market, commodities exchanges were created. The Chicago Board of Trade was established in 1848.

In addition to getting food to market faster, farmers could extend the value of surplus crops by making them last longer. Here they benefited from a prize offered by Napoleon in 1794 for a practical means of food preservation. In 1810, Nicolas Appert (1750-1841) invented canning. With a way to preserve their products indefinitely and deliver them to ever more remote destinations, markets once again expanded for farmers.

A natural consequence of more food was better diets for the general public and a rapid growth in population. In 1701, the combined population of England and Wales was 5.4 million people. By 1751, the population had risen to 6.2 million, and at the end of the century it was 9.3 million, nearly double what it was at the start of the century. Many of the members of this larger population were landless wage earners, a trend that was already beginning to emerge in 1700.

These workers were free to move around and become part of the successive waves of emigrants that settled in America, Australia, and New Zealand. They also began to drift into cities, particularly when the need for agricultural labor decreased and the demand for industrial labor was on the rise. As a consequence, urban populations grew quickly, providing concentrated markets for goods and services. (London, for example went from 600,000 to 1,000,000 inhabitants during the eighteenth century.) This labor force became an essential resource for industrialization, and the proportion of farmers in the population began to decrease inexorably as the Industrial Revolution progressed, a change that profoundly altered public attitudes and cultural norms. Industrialization also created a new

kind of poverty. In the cities, a large population of unemployed or underpaid workers became the urban poor we know today, plagued by crime, disease, and addiction.

Those workers who remained on the farm had to become more technical as machinery and modern methods took hold. Improvement societies, aimed at educating the ordinary farmer, were formed as early as the 1790s. These were the spiritual predecessors of agricultural colleges. The Morril Act of 1862 donated American public land to states so they could "provide colleges for the benefit of agriculture and mechanical arts." These land-grant universities have since gone on to become important centers of learning.

Wage-earning farmers provided the labor to physically change the landscape of the countryside by digging drains, clearing land, and erecting fences. The fences in particular helped to permanently end most common rights, including the practice of gleaning. By 1750, fully half of English farmland had become enclosed sheep fields. Extensive fencing led to new concepts of property rights and had natural consequences. Fenced land shut out wild animals and changed ecosystems, reducing biodiversity.

Even on the farm, the Industrial Revolution was beginning to be felt. As small-holding agriculture became less profitable, farmers who owned smaller fields were often forced to supplement their income to make ends meet. At the same time, higher efficiencies gave them the time for other activities. The result was the growth of cottage industries (originating in the 1600s) in which farm families did piecework for those who supplied the raw materials. The products, mostly crafts, were often sold in the cities and sometimes challenged established guilds. The competition, management, supply chain systems, and market shifts resulting from cottage industries provided experiences that were put to use during the start of the Industrial Revolution.

The cotton gin provided the biggest spur to the Industrial Revolution. The gin made mechanization profitable off the farm as well as on. It spurred the shift from muscle power to machine power, creating a wealth of new goods and services. The cotton gin also had its negative side. At this point in history, slavery in the U.S. had become uneconomical by most measures. Without Whitney's invention, slavery might have faded away, and, with it, the primary reason for the American Civil War (not to mention centuries of racial inequality and strife). With the gin, mills could be supplied with cotton and manufacturers could take advantage of the market demand for fabric—provided the cotton was grown and picked cheaply. Slaves provided the free labor needed to make this production system work.

Whitney did not profit from the cotton gin. A disastrous fire opened the door for imitations, and his opportunity for fortune was lost. However, a 1798 government contract gave him a second chance. The U.S. government paid him $134,000 to produce 10,000 muskets. Whitney developed what he called the "uniformity system" of using interchangeable parts. This allowed the guns to be manufactured by less specialized workings and permitted the use of standard replacement parts for repairs. Whitney had invented mass production, the system used to produce most commercial goods today. He had also sealed his reputation as an inventor without peer. He became the prototypical Yankee inventor and may have been the model for the hero in Mark Twain's *A Connecticut Yankee in King Arthur's Court.*

Because of the agricultural revolution, we now live in an urbanized world, filled with wage earners who are largely divorced from the traditions and rhythms of sowing and reaping. Many communal values have been replaced by appre-ciation of the individual. Hunger is still common (due, again, to a disruption in distribution), but starvation is not a regular experience for most of the world. The world population has risen continuously and is marching toward an expected 10 billion that will be absolutely dependent on the efficiencies of scientific farming for survival.

PETER J. ANDREWS

Further Reading

Books

Green, Constance McLaughlin. *Eli Whitney and the Birth of American Technology.* Reading, Massachusetts: Addison-Wesley Publishing Co., 1998.

Mathias, Peter and John A. Davis. *Agriculture and Industrialization: From the Eighteenth Century to the Present Day.* Malden, Massachusetts: Blackwell Publishers, 1996.

McClelland, Peter D. *Sowing Modernity: America's First Agricultural Revolution.* Ithaca, NY: Cornell University Press, 1997.

Mingay, Gordon. *Parliamentary Enclosure in England: An Introduction to Its Causes, Incidence and Impact, 1750-1850.* Reading, Massachusetts: Addison-Wesley Publishing Co., 1998.

Overton, Mark. *Agricultural Revolution in England: The Transformation of the Agrarian Economy 1500-1850.* New York: Cambridge University Press, 1996.

Music and the Mechanical Arts

Overview

The eighteenth century opened at the pinnacle of achievement for the master craftsmen of Western musical instruments. Fine string and keyboard instruments were produced by hand in small workshops using traditional methods and materials. In response to the changing desires of musicians and audiences and in keeping with the century's widespread interest in mechanical contrivances, these skilled artisans continued to innovate instrument design, leading to the invention and perfection of perhaps the most important instrument in the history of music, the piano.

Background

Through the seventeenth century, an assortment of keyboard instruments were in use throughout Europe. One of the most popular, the harpsichord, featured keys that plucked strings as the player pressed them; a rival for its dominance, especially in Germany, was the clavichord. The clavichord was similar in range to the harpsichord, but it used small hammers that struck the strings. These hammers would remain in contact with the string as long as the key was depressed, thus allowing the performer to control the volume of each note. Players could also, by rocking their finger while pressing down a key, produce a vibrated sound, similar to the vibrato produced by string players and singers.

Both of these instruments found favor with composers as well as performers, but by 1700 their shortcomings were well known. The harpsichord was a versatile instrument, capable of ensemble as well as solo playing and suited to either private or public performance. However, players could not modulate its volume at all, drastically limiting musical expression. While the clavichord provided the player both dynamic range and the expressive capability of vibrato, it was a delicate, intimate instrument poorly equipped for performance in public halls. An instrument that combined the power of the harpsichord with the expressiveness of the clavichord was desired to bring keyboard music into larger concert halls and to permit composers and performers greater creativity.

Interest in the design of a new keyboard instrument stirred across Europe. A craftsman in France presented four designs for "hammer-harpsichords" to the Académie des Sciences in 1716, and a number of makers from the German-speaking countries worked on examples of what came to be known as the pianoforte (the name combines the Italian words for "soft" and "loud" and was subsequently shortened to piano.) While German designers were largely responsible for the early development and evolution of the piano, credit for its initial invention must go to one man alone. Sometime between 1698 and 1700, Bartolomeo Cristofori (1655-1731), the keeper of instruments for the ruling family of Florence, the Medicis, produced the first working piano. Cristofori's piano had virtually all of the design elements of the modern instrument, but the complexity of its mechanism discouraged other makers from trying to produce similar instruments. In fact, Cristofori's design had more influence on pianos produced in the nineteenth century than on those made by his contemporaries.

In its final form (achieved in the mid-nineteenth century), the piano combines five mechanical components: strings (these vary in length, circumference, and number, depending on pitch), a frame to support the immense tension of the strings, a system of sounding boards and bridges to communicate the vibration of the strings, a mechanical "action" to bring the keyboard into contact with the strings, and a system of pedals that allow the performer to increase or diminish the damping of the strings. The action in particular posed a host of mechanical problems. Solving these satisfactorily was Cristofori's act of singular genius, and the chief conundrum for those who tried to copy or supplant his design.

The main challenge in designing the piano's action was constructing a system that allowed the hammer to strike the strings quickly and rebound immediately, leaving the strings free to vibrate. Cristofori achieved this with an intricate array of levers, springs, checks, and guides; his chief innovation was the design of a pivoting mechanism known as an "escapement" that allowed the hammer to strike the string from a short distance but still fall clear of it after making contact. He produced several pianos, but they failed to attract much interest in Italy, where harpsichords remained the favored keyboard instrument for years after Cristofori's death in 1731.

Interest in new keyboard instruments was greater elsewhere. Some German makers copied Cristofori's pianos directly; others worked on their own designs. While Cristofori's piano was essentially a modified harpsichord, other makers began from the clavichord and sought merely to make that instrument louder. Pianos based on the clavichord were usually rectangular in shape (Cristofori's piano and those like it had the wing shape similar to grand pianos of more recent times), and often featured a simplified action. Early German-made pianos came to the attention of important composers, including Johann Sebastian Bach in the 1740s and Wolfgang Amadeus Mozart in the 1770s. These musicians and others offered their opinions and suggestions about the new instrument and its capabilities, and composed pieces for it.

Beginning in 1756 the disruption of the Seven Years' War scattered German piano makers throughout Europe and helped to make England and France important centers for piano design and production throughout the second half of the century. By 1800, nearly 1,000 pianos were being manufactured each year in a dozen workshops in London, the largest center for piano production. Pianos became fashionable not only for public performances but also, gradually, in gracious homes. The instrument was far from standardized at that point; English and Continental pianos differed in significant ways, and all makers continued to experiment.

While no other technological innovation approached the significance of the piano, the eighteenth century saw other musical inventions as well. Many of the orchestral wind instruments, such as the clarinet, bassoon, and flute, achieved their modern form during this period; for the most part, these inventions were the result of gradual design improvements rather than new technologies or materials. String instruments had already reached their penultimate forms in the previous century, but instruments produced in Italian workshops in the early decades of the eighteenth century are widely believed to be the finest ever produced; scientists and craftsmen have studied them without success to try to determine the technological and scientific basis for their exceptional sound. So superior are these instruments that many are sought out and used by modern musicians.

A horizontal grand pianoforte. *(Corbis Corporation. Reproduced with permission.)*

Instruments for musical *reproduction* also evolved in this era. Mechanical musical instruments such as the musical clock and the barrel organ had been invented many centuries earlier, but these devices became more creative and more popular in the eighteenth century, carried along by widespread interest in automata and contrivances. The most significant mechanical music invention, the music box, came late in the century. This instrument consists of a revolving metal cylinder with properly spaced projecting pins to pluck tongues cut into a steel comb or plate. The comb resonates different pitches and overtones. A spring, clockwork, and fly regulator move the cylinder smoothly and at the appropriate speed to play a single tune. The music box first appeared in Switzerland in the 1770s, and became a popular instrument in many homes during the nineteenth century. Despite its limitations, it foreshadowed in importance and structure the more flexible music reproduction technologies of the player piano and the phonograph.

Impact

The impact of the piano on the subsequent development of Western music cannot be overstated. The piano quickly became an essential tool for musical composition, a kind of workbench at which most composers wrote their music. It became a vital ingredient in music instruction as it

provided a platform upon which to teach all of the ingredients of music theory. It was also an invaluable part of musical performance, as a virtuoso solo instrument and as a versatile accompaniment to other soloists or larger ensembles. Even the pitches used in musical performance have been determined by the conventional tuning of the piano. The place of the piano at the heart of the evolution of Western music has remained undisputed from the mid-eighteenth century into the twenty-first.

The piano was also a conduit through which music entered the homes and lives of thousands of people. Widespread enthusiasm for piano-playing at home was encouraged by the production, after 1811, of small "cottage" and other affordable upright pianos. Piano-making moved from workshop to factory, and by the 1830s pianos were being manufactured in America as well as Europe. It was an American-based maker, Henry Steinway, who produced the first fully "modern" piano in 1859, and ushered in the era of the piano's greatest popularity. At the time of the great Crystal Palace Exhibition in 1851, fewer than 50,000 pianos were manufactured worldwide annually; just before World War I, the real price of pianos had halved and annual production was around 600,000.

During its heyday from the 1860s to the 1920s, the piano was probably the most important source of entertainment in the home. But new musical technologies—most of which required considerably less effort—subsequently marginalized the piano. Automatic, or "player" pianos, introduced in the 1920s, ushered in other mechanical means of music reproduction, including the developing phonograph and radio. As these technologies improved, they provided easy ways to listen to a wide variety of music. The piano and other performance instruments lost their central role in domestic life, and became a subordinate, specialized form of entertainment. Ironically, the eighteenth-century's greatest advance in musical technology, the piano, was ultimately eclipsed in popularity by the descendants of the lowly music box as peo-

ple chose to enjoy musical reproduction rather than performance in their homes.

LOREN BUTLER FEFFER

Further Reading

Books

Ehrlich, Cyril. *The Piano: A History.* New York: Oxford University Press, 1990.

Gill, Dominic, ed. *The Book of the Piano.* Oxford: Phaidon, 1981.

Loesser, Arthur. *Men, Women, and Pianos: A Social History.* New York: Dover, 1990.

Pollens, Stewart. *The Early Pianoforte.* Cambridge & New York: Cambridge University Press, 1995.

Ripin, Edwin, et al., eds. *Piano: The Grove Musical Instrument Series.* New York: Norton, 1988.

Advances in Publishing and Bookmaking

Overview

Written language is one of the most significant inventions in human history because it allowed large amounts of information to be stored and conveyed from one person to another or between generations. Another significant step was the development of printing, which allowed the reproduction of this information easily and cheaply. In the eighteenth century, significant advances in the printing and bookmaking crafts made printing books even easier. Other advances, especially in lithography, also made it possible to reproduce drawings and, later, photographs along with the printed word. These advances helped make books less expensive and more appealing, resulting in ever-increasing book sales and readership.

Background

The written word goes back thousands of years and, indeed, we may never know exactly when writing was first developed. The first books were hand-written; in Europe, monks and commercial copyists produced virtually all books written until the middle of the fifteenth century. Although the first printing was developed in China nearly 2,000 years ago, it was the German Johannes Gutenberg (1398?-1468) who printed the first book using anything resembling "modern" techniques. Realizing that a great deal of time could be saved by using movable type, Gutenberg con-

structed the first printing press and in 1455, printed the now-famous Gutenberg Bible.

Printing existed before Gutenberg, but it was a laborious project that did not lend itself to mass production. Artisans would carve each page into a single block of wood or stone. These would then be inked and pressed onto paper, one sheet at a time. The next sheet would be carved and pressed onto the back of the paper, and then a new page would be made. By comparison, movable type allowed setting up a page by selecting from a selection of letters that were already formed. They would be set up and locked into a frame that would then be pressed into the paper. Afterwards, the letters could be removed from the frame and re-used for other pages or other books. This was a huge time-saver, helping to reduce the cost of books while improving the quality.

The next major innovations involved Jakob LeBlon's (1667-1741) development of three-color printing, Aloys Senefelder's (1771-1834) lithography process, and refinements in the manufacture of type, pioneered by Pierre-Simon Fournier (1712-1768). The first two developments helped bring high-quality drawings and artwork to printing while the third, better techniques for manufacture of type, helped make printing more reliable and simpler.

In lithography, a design is drawn onto a block of fine-grain limestone with grease-based

An eighteenth-century print shop. *(Corbis Corporation. Reproduced with permission.)*

ink or a grease pencil. A weak acid solution is then poured onto the surface and, repelled by the grease, collects on and etches away at the rest of the stone. This leaves the design standing above the rest of the stone, allowing it to be inked when the lithograph is used for printing. Lithography is one way of letting an image be printed repeatedly, allowing the reproduction of photographs, drawings, and other images in printed materials.

Along similar lines, stereotyping and stereography were introduced in Paris in the 1790s. Both of these procedures aim to make entire sheets of print that can be reproduced and run on numerous printing presses simultaneously. This is done by setting each page in type (as in then-conventional printing) and making an impression of the entire page in clay. When the clay hardens, lead or another metal can be poured into the mold. Each sheet that is made in this fashion can then be mounted on a separate printing press. By doing this, the most laborious part of typesetting, that is, setting the type for each page, needs to be done only once.

Impact

The importance of these innovations to the printing industry and their subsequent impact on society can hardly be overstated. Lithography gave printers the ability to make precise draw-

ings that, unlike many woodcuts, could be reproduced repeatedly without wearing out. Coupled with new techniques for color printing, printing began to expand into art, making it possible for anyone to bring quality art home with them, either as a framed print or between the covers of a book. Stereotyping gave the ability to print a book, newspaper, or anything else on multiple printing presses without having to go through the laborious and expensive process of typesetting multiple times. The development of new fonts gave printers more tools to make the printed word as distinctive as the written books of previous centuries, returning some of the artistry to printing. The improved techniques for making type helped reduce the cost of printing, in turn helping to bring the printed word into the reach of more people than ever before. These innovations improved the quality of printed images, reduced the cost of printing, and broadened the availability of both books and artwork in society.

By improving the quality of printed images, lithography and color separation helped to embellish books, newspapers, and other printed matters. Drawings or paintings could be added to books for artistic purposes, to illustrate particular points (such as, in a "how-to" book), or for any other reason. Before lithography, this was usually accomplished by carving a woodcut, which could become worn rapidly, reducing the quality of the image. Lithographic images, on the other hand, were as durable as the stone (or, later, metal) upon which they were etched, letting them be used repeatedly for years or decades (provided the plates weren't damaged through neglect or accident). In addition, lithographic plates could simply be re-ground once their image was no longer needed, helping to further reduce printing costs.

Color printing not only helped to enliven books (especially higher-quality ones), but could be combined with lithography to create color posters and color prints inexpensive enough for virtually anyone to own. This helped make artwork available to a wider number of people because it was no longer limited to those who could afford to commission paintings or purchase paintings from the masters. This has carried forward to the present, when just about anyone can buy a convincing print of famous paintings by Rembrandt, Leonardo da Vinci, and other masters. Color separations are also used in today's printing industry, making possible the beautiful color images found in *National Geographic*, *Smithsonian*, and other magazines. In fact, the fundamental principles behind today's color printing and lithography are not unlike those introduced by eighteenth-century printers.

Other processes, most notably stereotyping and improved methods of manufacturing type, helped to make printing more widespread and less expensive. As mentioned above, stereotyping made it possible to manufacture multiple copies of an entire sheet of type very easily. New methods of manufacturing type helped bring down its cost, again reducing the cost of the printed word. These developments made many things possible. For example, in the American colonies, it became possible for individuals to run small print shops, spreading information, ideas, and opinions throughout the colonies.

In fact, these two effects are intimately interrelated. By making it possible to print on multiple printing presses, books, pamphlets, and newspapers could be more widely distributed than ever before. At the same time, with a lower cost per item printed, the printed word was now affordable to more people than ever before. The Gutenberg Bible was relatively inexpensive compared to the hand-written books of the day, and it cost the equivalent of three years' wages for a clerical worker, and much more for the more typical person. This expense was largely due to the necessity of hand-set typing each page of the Bible, the need to illustrate by hand, and the inability to print on more than a single press. With fewer books printed, each volume had to be more expensive to offset the costs associated with their manufacture. Thus, by providing greater economies of scale, stereotyping helped bring down the cost of printing. Similarly, bringing down the cost of manufacturing type, one of Fournier's contributions, helped reduce the cost of printing in general.

These developments were important not only to the printing industry, but to readers as well. Printed materials were cheaper, encouraging wider distribution of books, newspapers, and the ideas they carried. Textbooks, works of fiction, subversive literature, revolutionary pamphlets, government propaganda, and all other words could be printed and widely distributed in a short period of time. While it would be an overstatement, for example, to claim that the American Revolution owed its existence to this, it would not be inaccurate to say that cheap printing helped the American colonies understand the issues at stake and come to a consensus about these issues and the way to resolve them. Similarly, French citizens were kept informed via printed pamphlets during the political upheaval of the French revolutions. In fact, even today, repressive regimes try to control printing as one way to control the spread of ideas dangerous to their control of the population.

In addition, by improving the quality of printed images, lithography and color printing techniques made printing more attractive and helped make it possible for the average person to own books and artwork of a quality previously reserved for the wealthy. This, in turn, helped encourage the appreciation of both books and artwork by people who may otherwise have remained ignorant about both.

P. ANDREW KARAM

Further Reading
Books
Olmert, Michael. *The Smithsonian Book of Books*. Smithsonian Books, 1992.

Inventions for Daily Life

Overview

Inventors in the eighteenth century worked to make life easier. Benjamin Franklin (1706-1790) introduced bifocal eyeglasses and a stove that warmed a whole room. The three-color print process and wood-pulp paper brought us newspapers, magazines, and wallpaper. The piano was invented and so was drip coffee. Joseph Priestley (1733-1804) gave us soda pop. A popular appetite for comfort, convenience, and amusement was created that has come to dominate the populations of developed countries, both culturally and economically.

Background

Daily life today is full of small comforts and conveniences that originated in the eighteenth century. Some of these, such as Franklin's bifocals, stemmed from personal needs. Others, like Joseph Priestley's soda pop, for example, came about by accident, as byproducts of research. Moreover, clever entrepreneurs such as Robert Barker, who invented the panorama, consciously pursued the market opportunities that emerged in rapidly growing cities.

Benjamin Franklin came up with a household improvement that ultimately changed the way houses (and buildings in general) were constructed. Franklin noticed that, while a heat source radiates in all directions, fireplaces were built into walls. Half the potential heat of the fire was thereby wasted. In 1740 Franklin solved this problem by inventing a stove that sat in the middle of the room. It had the added advantage of being made of cast iron, which absorbed heat so that, even when the fuel was exhausted, it continued to warm the room for awhile. David Rittenhouse (1732-1796) perfected the Franklin stove by adding an L-shaped chimney, which helped to vent the smoke and provided even more surface area to radiate the heat.

James Watt (1736-1819) developed a steam radiator in 1784 that provided uniform heat to his offices and eventually allowed heat, generated from a central source, to be distributed throughout a building. Along the same lines, various inventors, including Count Rumford (a.k.a. Benjamin Thompson, 1753-1814), developed cooking ranges. These devices allowed more efficient and controllable heating and cooking. Rumford devised other cooking conveniences as well, including the double boiler and drip coffee.

The science of the 1700s led to the development of measuring systems and devices that barely affected the daily lives of people then but are standard in most households today. Gabriel Fahrenheit (1686-1736) invented the mercury thermometer in 1714. Great Britain (and its North American colonies) adopted the Gregorian calendar in 1752. In addition, the metric system gained its first foothold in revolutionary France in 1795 and was eventually adopted by scientists and most nations.

An important step in commercial manufacturing occurred in 1790, when Nicolas LeBlanc (1742-1806), a surgeon, invented a process for producing sodium carbonate (soda ash or washing soda), an important ingredient in the manufacture of paper, glass, soap, and porcelain, from ordinary salt. At the time it was difficult to produce soda ash in industrial quantities, and the Académe des Sciences offered a prize for the development of a cheap and reliable method for producing the chemical. LeBlanc won the contest, and, by 1791 a plant was in operation producing 320 tons of soda ash per year. Unfortunately, politics worked against him and the French Revolution began before he could collect his prize. Although the National Assembly granted him a 15-year patent on the process, the government seized his plant a few years later. In 1802 Napoleon returned the plant to LeBlanc, but gave him little in the way of compensation. Discouraged, broke, and unable to reopen his factory, LeBlanc killed himself in 1806. Nevertheless, the process he discovered became the mainstay of the alkali industry and by 1885 it was being used to produce more than 400,000 tons of soda ash per year.

Hygiene received a much-needed advance with the first major improvement in the flush toilet since the invention of the "water closet" in 1596. In 1775 watchmaker Alexander Cummings added a water trap to the exiting pipe, stopping odors from backing up into the room and making indoor plumbing a much more desirable possibility, although toilets were still highly unsanitary affairs. Samuel Prosser invented the plunger closet two years later, and in 1778, Joseph Bramah put a hinged valve at the bottom of the bowl, and patented a float and valve flushing system that is still used today. Another century would elapse, however, before the modern toilet made its debut.

A number of inventions came together to create the newspapers and magazines we know today. Paper was made from wood pulp for the first time in 1719, and the first paper-making machine was invented in 1798 by Louis Robert. The three-color printing process was developed in 1710, and stereotyping, which permitted mass printing from durable plates, was invented in 1725. For reporters, the typewriter was invented in 1714, but it was not a practical alternative during the eighteenth century. Of direct benefit to reporters, however, were the replacement of quills with steel-pointed pens (1780) and the invention of the eraser (1770).

The new printing techniques also made homes more attractive by making wallpaper and decorative prints affordable. Another invention that was to become a fixture of middle-class living rooms, the piano, was developed in 1709 by Bartelomeo Cristofori (1655-1731). The most popular form of public art in the nineteenth century, the panorama, was created in 1789 by Robert Barker. It consisted of a series of paintings that fully encircled the viewer, offering a 360-degree view. Panoramas, which initially showed audiences views of the cities in which they lived, were an instant hit, drawing large crowds. When they became traveling exhibits, panoramas offered the general public its first visual experiences of foreign lands.

Impact

Products that add comfort and convenience to daily life are a major industry today. Some of the conveniences of the 1700s have been superceded: Rumford's double boilers may still be essential for gourmets, but most people now buy ready-made sauces and icings or use microwave ovens to melt butter and chocolate. Other inventions have continued to develop and improve. As early as 1802, the closed-top cast-iron cooking range began to replace earlier models. Mass production in the twentieth century lowered the price and extended the reach of conveniences, and electrical power made others, such as drip coffee, easier to enjoy. In the case of central heating and well designed toilets, eighteenth-century innovations came to be required by law. But the largest effect of the influx of new gadgets was the stimulation of an appetite for more. Today, some of us have central cooling as well as heating. Our

Benjamin Franklin wearing bifocals. *(Library of Congress. Reproduced with permission.)*

pens carry their own ink supplies and can be molded from soft materials that feel good in our hands. Bifocals, so common they have become an emblem of middle age, survive in stealth form as glasses with progressive lenses or as multifocal contact lenses. In our kitchens, electric or gas ranges are complemented by refrigerators and freezers. We also have minor gadgets we often feel we cannot live without, such as toasters, bread makers, food processors, waffle irons, and electric mixers.

Most of the forms of measurement developed in the eighteenth century are still with us. The Gregorian calendar, tuned to fractions of a second, is the worldwide standard for almost everything but religious or cultural activities. The thermometer, now a standard device, usually measures temperature in degrees Celsius, rather than Fahrenheit. Globally, the metric system today has only a few holdouts, most notably the United States. Many Americans, from auto mechanics to scientists, however, have been forced to become fluent in metric measurements.

With the new affordability of paper, newspapers and pamphlets became a force of revolution and popular government. In the Colonies, Thomas Paine's pamphlet "Common Sense" fired a revolution, and the Federalist Papers, published as weekly installments in newspapers, helped win public support for the U.S. Constitution. In 1807 English engineers improved Robert's papermaking machine, making it a practical reality. Inexpensive printing also increased general literacy and provided a vehicle for advertising that helped create markets for the products of the Industrial Revolution. Despite regular predictions of a "paperless office," both commercial and private use of paper increases each year. Indeed, computers are sold with printers, encouraging the use of paper as an essential adjunct to the "virtual world." While electronic publishing has made some inroads, paper's "form factors" (portability, "view-ability," ease of markup) make it indispensable. There are even pilot projects aimed at making paper and pens part of the electronic environment: the text of posters coated with electronic ink can be changed using pagers; and pens have been designed that broadcast their movements on paper to computers, allowing for the automatic capture of drawn images and text.

The role of images also grew. Today, even the humblest households are decorated with posters and art. The motion picture has its roots in the eighteenth-century panorama, which has been called the first mass medium. While people today watch *Titanic* at home on video, middle-class crowds once lined up to see panoramas of the Battle of Waterloo or the great earthquake of Lisbon. The panorama's popularity continued throughout the nineteenth century and, with the addition of movement and music, came to resemble movies more and more as time went on. Popular images of foreign lands, historical events, and myths were formed by the panoramas, which

traveled the world and earned their owners fortunes. While panoramas went out of vogue by 1900, they still exist in museums and art galleries. Walt Disney World, for example, has featured a popular multimedia panorama for more than a decade. The term "panorama" itself, coined by Barker, still survives, and a whole new genre of Internet panoramas has emerged in recent years.

Mainstays of popular culture, including soda, trading cards, printed T-shirts, ballpoint pens, greeting cards, coffee bars, and bumper stickers, all have their roots in the 1700s. To a time traveler from 1799, they might seem confusing but they would not be unrecognizable. In the eighteenth century, as the ranks of city dwellers and wage earners swelled, and the public sought the amusement and comforts that we expect today, their society began to resemble our own.

PETER J. ANDREWS

Further Reading

Books

Asimov, Isaac. *Isaac Asimov's Biographical Encyclopedia of Science & Technology*. New York: Doubleday, 1976.

Oetterman, Stephan. *The Panorama: History of a Mass Medium*. New York: Zone Books, 1997.

Thackeray, Frank W., and John E. Finley. *Events that Changed the World in the Eighteenth Century*. Westport, CT: Greenwood Publishing Group, 1998.

Wolf, Stephanie Grauman. *As Various as Their Land: The Everyday Lives of Eighteenth-Century Americans*. Little Rock: University of Arkansas Press, 2000.

Internet Sites

HyperHistory Online. http://www.hyperhistory.com/online_n2/History_n2/a.html

The Steam Engine Powers the Industrial Revolution

Overview

The invention of the steam engine in 1698 by Thomas Savery (1650?-1715) was among the most important steps toward the modern industrial age, in which machine power replaced human or animal muscle-power. Savery's 1698 patent of his steam engine—designed to help remove water that seeped into the bottom of coal mines—laid the foundation for a series of refinements and re-designs by Thomas Newcomen (1663-1729) and, most notably, James Watt

(1736-1819) that resulted in the transformation not only of work, but also of the entire society for whose support that work was done. While any number of other inventions and devices played major parts in the march toward industrialization, it was the steam engine above all that established the place and importance of the machine in the modern world, and made possible the creation of the large factories that were among the most significant undertakings of industrial civilization.

Background

By the late 1600s England had a fuel problem. Harsh winters and a growing population had resulted in the depletion of many of England's great forests, as trees were cut down and burned. What wood remained was more valuable as lumber than fuel. To solve the problem, England turned to its rich deposits of coal. Unfortunately, the mining of coal created its own set of problems, primarily the tendency of water to seep into the lowest reaches of coal mines, making it difficult if not impossible to extract the coal. Some attempts were made to remove the water by using hand-pumps to create vacuums in tubes: the water would then be sucked up the vacuum. Hand-pumping, however, was slow and ultimately ineffectual against the volume of water in the mineshafts.

Inventor and merchant Thomas Savery, of Devonshire, England, aware of experiments with steam and pressure, turned his technical abilities to the creation of a mechanical device for using those properties to raise water. Savery's device involved heating water in a boiler; the boiling water produced steam, which was routed through pipes to a receptacle, which was in turn connected to a suction pipe that extended down into the water to be raised. Once the receiving vessel was filled with steam, it was sprayed with cold water. The sudden cooling of the vessel quickly condensed the steam, creating a vacuum that resulted in a change in atmospheric pressure within the suction tube, forcing water up its length. Savery was granted a patent for the device in 1698.

Four years later he published a treatise describing his invention, *The Miners Friend; Or, An Engine to Raise Water by Fire*. His engines became known as Miners' Friends, although it is unclear whether any of them found actual use in coal mines.

There were some technical problems with the Miners' Friends, not least of which was the poor quality of metal fittings at the time. Because the device required control of internal pressure—technically, Savery's invention was an atmospheric engine rather than a true steam engine—it was vital that all of the pipes be tightly sealed. Such seals proved difficult to maintain, however, and Savery's devices suffered from constant failure of the pipe joints.

Another Devonshire inventor, blacksmith Thomas Newcomen, made refinements to Savery's device that turned it into a true steam engine. Where the Miners' Friend used an open fire, Newcomen's engine employed a sealed boiler. Above the boiler rested a piston cylinder; the piston was connected to a pump, whose weighted rod raised the piston to the top of the cylinder. With the piston at the top of the cylinder, the cylinder itself was filled with steam from the boiler. Cold water that was sprayed into the steam-filled cylinder condensed the steam, creating a vacuum, which drew the piston downward, completing the pump cycle. Each completed cycle raised the water in which the bottom of the pump rested. The first known working model of a Newcomen engine was deployed in 1712, although it is believed that many prototypes preceded it.

Newcomen's refinements greatly increased the effectiveness of the steam-driven water-lifting device, if not its efficiency. However, Newcomen's engine faced numerous legal problems as a result of Savery's patent and the belief that Newcomen infringed Savery's rights. After much negotiation Newcomen agreed to pay royalties to Savery and, later, to the syndicate that acquired Savery's patent after his death.

Efficiency issues were less easily resolved. While Newcomen's expertise as an ironworker contributed to improved joints and seals in the boiler and pipe system, there remained a severe flaw to his engine. Because the piston cylinder had to be cooled for every cycle and re-heated for each subsequent cycle, tremendous amounts of fuel were required simply to renew the steam. This mattered little at the coal mines, where there was, obviously, plenty of fuel. Indeed, for deeper mines, several Newcomen engines were used in sequence, raising water from one level to the next, each engine burning large amounts of coal to power each stroke of its pump. Elsewhere, though, the amount of fuel required became an obstacle to the engines' widespread use.

Newcomen engines did find wider use, however, with ongoing refinements and improvements, and the age of steam had truly begun. The full potential of steam, though, would await the attention of Scottish mechanical engineer James Watt.

Watt, an instrument maker, was asked in the 1750s to repair a Newcomen engine. Careful observation of the engine's workings inspired Watt. The inefficiencies were the result of combining the steam cycle and the condensing cycle in the same cylinder. Not only was the combination inefficient, it placed terrific and repetitive stress on the cylinder. During a Sunday stroll Watt had an insight: the condensation process

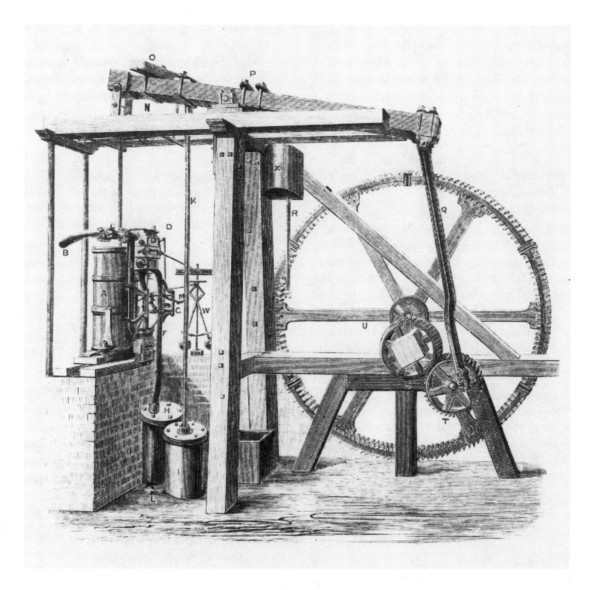

A diagram of James Watt's steam engine. *(Library of Congress. Reproduced with permission.)*

should be separated from the piston cylinder. He quickly designed a new steam engine, one in which the steam, having done its work, was guided away from the first cylinder and into a second, whose sole purpose was to serve as a vessel for condensation. Watt built a prototype—which still exists—in 1765 and was granted a patent in 1769.

The benefits of Watt's insight were dramatic. Improvements in thermal efficiency—the hot cylinder could remain hot, the cold cylinder cold—resulted in a great increase in the steam engine's economy of operation: less fuel was needed to accomplish more work. Backed by wealthy English manufacturer Matthew Boulton (1728-1809), Watt created in 1775 the Soho Foundry in Birmingham, England. From this

foundry Watt engines poured forth, as did ongoing improvements. Perhaps the most important improvement was Watt's 1782 introduction of the double-acting mechanism, which allowed the engine to drive both forward and backward piston strokes, essentially doubling its capacity for work. A further refinement adapted the steam engine for rotary motion—the engine could be used to power movement in any direction, making it ideal not only for up and down pumping actions but also for any repetitive cycle of movement. Soon, steam engines were driving cotton and wool mills and finding their way into other industries, notably the emerging transportation industry, which adapted steam engines for railroads and ships. Industry itself responded, finding ever newer and more ambitious uses for the now versatile engine.

Steam itself proved versatile under Watt's command of its properties. He heated his offices with steam. He devised instruments for measuring engine efficiency, and governors for controlling steam engines from improper or dangerous operation. Awareness of the potential dangers of steam engines, in fact, prompted one of Watt's few oversights: to the end of his life he remained opposed to the use of high-pressure steam, despite the dramatic increase in efficiency that higher pressure offered.

Nor were his inventions limited to the steam engine. As the success of his engine spread, so did the business demands Watt faced: deluged by paperwork, he invented a chemical process for copying documents, a process that remained in use until the arrival of the typewriter, long after Watt's death.

As the importance of the steam engine grew, Watt also applied himself—and his engineering skills—to documenting and measuring the engine's capacity for work. He created a unit of measurement based on comparing mechanical power with that of the previous universal power source, the horse. Watt's measurement unit, horsepower, equated the ability of a horse to lift 33,000 pounds (14,969 kg) 1 foot (.3 m) in 1 minute. Horsepower remains universally known as an index of mechanical energy.

But that was not Watt's only contribution to the language of energy measurement. When he died at the age 83 his refinement of the steam engine was already well on its way to transforming every aspect of human society. To this day his name is known not only as that of an engineer and inventor, but as perhaps the most common of all units of energy measurement: the watt.

Impact

The impact of the steam engine cannot be overstated—it belongs with the printing press, electric power, telegraphy and the telephone, and lately, the computer, as inventions that have exerted a dramatic and far-ranging influence on civilization.

From its beginnings as a device intended to serve a specific purpose—lifting water—the steam engine itself rose to lift all of society into the industrial age. The Industrial Revolution itself—the transformation of economy from a local mill-and-shop foundation to one based on huge central factories and wide, rapid distribution of goods—rests on a cloud of steam.

Virtually every industry was affected and altered by the steam engine. The coal mines for which Savery's Miners' Friends were first developed became more and more important as a source of fuel for the rapidly increasing number of steam engines; in turn, the increased efficiencies of the engines enabled the mines to be sunk ever deeper without flooding. Mines for other minerals quickly followed the route of coal mines, deeper and deeper into the earth. Metallurgy itself responded to the demand for ever tighter and more durable seals: improvements in metal-working made larger boilers possible, and larger boilers were able to drive larger pistons, which in turn powered larger and more powerful machines.

Those machines made possible the cornerstone of industrial society, the factory. Huge, centralized manufacturing operations grew up around the relatively cheap and increasingly efficient mechanical energy provided by steam engines. The textile industry was among the first beneficiaries of steam power, as engines were applied to driving great looms, tended by dozens of workers, producing vast quantities of fabric.

The ability to adapt steam engines, by way of mechanical linkages, to rotary drives enabled the creation of steam-powered transportation. Rail lines crossed nations and soon, continents; steamships traveled the oceans far faster than could sailing ships. Some models of early automobiles were steam-powered. Much of the freight carried by rail and steamships was produced in steam-driven factories. Steam heat, used for centralized heating, became more common.

As the importance of the factory increased, cities themselves increased in size as the population was drawn to work in the factories.

Industrialization proved a double-edged sword. Some workers, artisans, and craftsmen, seeing machines displace their skills, responded by following the lead of Ned Ludd and sabotaged factories. Over the course of the nineteenth century, conditions in factories prompted calls for social reforms and concerns for workers' rights. The growth of industrialized civilization, powered by steam (and later, electricity and petroleum) could be seen in skies that themselves grew dark with smoke from burning coal.

For good and ill, though, the inventions of Savery, Newcomen, and Watt altered the course of world civilization, providing a cheap and efficient source of energy and power that was adapted, and continues to be used in various forms, to drive civilization's engines.

KEITH FERRELL

Further Reading

Cardwell, Donald. *The Norton History of Technology*. New York: W.W. Norton & Company, 1994.

Carnegie, Andrew. *James Watt*. New York: Doubleday, Page & Company, 1905.

Lord, John. *Capital and Steam Power 1750-1800*. London: 1923.

Savery, Thos. *The Miners Friend; Or, An Engine To Raise Water by Fire, Described*. London: S. Crouch, 1702.

Thurston, Robert. *A History of the Growth of the Steam Engine*. New York: D. Appleton and Company, 1878.

Key Inventions in the Textile Industry Help Usher in the Industrial Revolution

Overview

While there is some debate over exactly when the Industrial Revolution's opening salvos were fired, there is no doubt that the introduction of new technologies for producing textiles were crucial. With the introduction of the flying shuttle loom in 1733, the invention of the spinning jenny (1764), the spinning frame (1768), and the power loom in 1785, Britain mechanized one of the world's most important industries. This, in turn, helped instigate social change, including unrest, raising issues that are still debated to this day. For better or worse, these inventions heralded changes that by now affect virtually every person on Earth, and will continue to do so for the foreseeable future.

Background

Clothing is one of the earliest and most fundamental inventions humanity has made. Stone tools have left more lasting records because they do not decay or rot, but there is evidence of clothing for nearly as long as for any other artifact. Clothing protects fragile skin from cuts and scrapes, provides protection from the cold or the sun, affords the wearer a degree of modesty, shows social status, and more. We are alone among animals in habitually covering ourselves. Part of the reason for this is that, as early hominids lost their body hair, they became more susceptible to weather changes. Then, as people moved into colder climates, the need for warmth and protection became even more important.

The earliest clothing was likely animal skins, which were supplemented by cloth of some sort by, at the latest, 4400 B.C., the date of the oldest known loom (a device for making cloth). Cloth was common throughout the classical world and elsewhere (especially China) by 2500 B.C., and had spread through virtually the entire civilized world by 1000 B.C. Also in use by this point was the spinner, a device used to combine relatively short fibers from animal hair or plants into long threads that could be used for weaving cloth. Early completely manual spinners were supplanted by spinning wheels, invented in India and brought to Europe in the Middle Ages. This made it possible to produce larger quantities of higher-quality yarn or thread than had previously been the case, although spinning wheels still required a person to operate each one.

At the beginning of the eighteenth century, then, this was the textile industry: a worker sitting at a spinning wheel spun fibers into thread or yarn, which was taken to another worker sitting at a loom, weaving cloth. The cloth, then, was collected on rolls, to later be made into clothing, linens, and other goods. This made the industry very labor-intensive, making cloth relatively uncommon and expensive.

This began to change in 1733, with the invention of the flying shuttle loom. In a loom, a set of horizontal bars hold two sets of threads that run the length of the loom. Each bar is attached to every other thread (collectively called the warp), and one bar moves up and down while the other remains stationary. Another thread (called the weft) is attached to a shuttle. With the movable bar, for example, in the up position, the shuttle is passed between the two sets of threads, drawing the weft behind it. This thread is packed tightly into the "V" formed by the warp, and the movable bar is lowered. This traps the weft thread, making one row of cloth. The shuttle is then passed in the other direction, the weft packed again, and the bar moved once more. By this process, cloth is formed.

In 1733, John Kay (1704-1764) realized that the shuttle could be manipulated more quickly by attaching a cord to it and jerking it

James Hargreaves's spinning jenny. *(Corbis Corporation. Reproduced with permission.*

through the warp threads rather than passing it through by hand. Not only did this speed up cloth-making, but it lent itself nicely to mechanization because a motor could pull the cord as easily as a person could, but more reliably, tirelessly, and more quickly. In spite of its advantages, however, Kay's invention was not readily accepted, in part because of fears it would lead to unemployment of weavers.

Just about 30 years later, spinning became mechanized by James Hargreaves (c. 1720-1778) (who invented the spinning jenny) and then by Richard Arkwright (1732-1792), with his spinning frame. These devices did for thread spinning what the flying shuttle had done for weaving; they made it possible for a single person to become significantly more productive. In fact, this was essential because, with the grudging acceptance of more mechanized looms, cloth making was outpacing thread production. Completing the mechanization of the weaving industry was the invention in 1785 of the wool-combing power loom by Edmund Cartwright (1743-1823). This device again mechanized activities previously performed by people; in this case, removing foreign materials from wool and aligning the fibers for more efficient weaving of higher-quality fabrics.

Impact

Each of these inventions, taken by itself, was significant. Collectively, they were, literally and fig-

uratively, revolutionary. Spinning and weaving had been intensely manual activities, and good cloth was neither plentiful nor cheap. These inventions helped to change that. At the same time, they helped contribute to a division of labor whereby individuals specialize in their work, making their wares available for purchase by those specializing in other areas. Both of these led, in turn, to widespread social, economic, and political changes that continue to this day.

To start with the most obvious, and simplest impacts to discuss, the quantity and quality of cloth available to the general public increased greatly with the introduction of these devices. In fact, machine-produced cloth became so common and so popular in England that Benjamin Franklin, wearing homespun fabric in his visits to England and France, made a conscious political statement about the values of the self-sufficient pioneers of the New World compared to those of pampered England.

This abundance of good cloth, however, was seen as a mixed blessing by spinners and weavers of the day. Fearful that advances in technology would destroy their jobs, they first refused to consider using the new technology, and later attempted to destroy it. In fact, Ned Ludd led the first such revolt in 1782, when, as legend has it, he destroyed a machine used to make stockings. Later, in riots that took place between 1812 and 1818, the Luddites (named, of course, after Ned) rampaged through parts of England, destroying many of the machines they feared were replacing craftsmen throughout the nation.

Many of these concerns persist to this day. It is not at all uncommon to hear news stories bemoaning the replacement of workers by automated equipment in the automobile, steel, and other industries. Workers fear replacement by robots, computers, and other new equipment. However, historically, automation does not seem to cost any society jobs, it merely reallocates them. For example, while England may have lost many jobs for spinners and weavers, people were needed to build, maintain, and operate the new machines. Greater production of cloth meant more cloth available for tailors to sew, and more shop keepers and sales people were needed to sell new products to new customers. Transporting these materials required more transportation capability, so jobs in the transportation sectors were formed, including those who built the carts and cared for the horses. Similarly, today automated factories open one set of jobs as they make other unnecessary. Unfortunately, both in the eighteenth century and today, those whose jobs are made obsolete are often those least able to master the skills needed for the jobs opening up. Because of this, any revolution of this sort will leave many able-bodied workers unemployed or underemployed while younger workers move into the new jobs that have been created.

These inventions led almost inevitably to a division of labor in which people increasingly became specialized in their professions. No longer could a single person do everything that was needed to be self-sufficient. In fact, some degree of specialization had been in effect for several centuries, because there had been bakers, weavers, farmers, soldiers, merchants, and so forth since the time of the Romans. However, many people could, for example, spin, weave, and make their own clothes and chose not to. By making the textile industry a place for skilled workers, the average person became unable to maintain even rough parity with manufactured goods, while textile workers had no other job skills.

All of these changes conspired to alter society irrevocably, especially when the marriage of steam power to machinery and the development of products with interchangeable parts led to mass production and ever more-efficient means of manufacture. In turn, industrialized nations steadily became more prosperous, economically and politically powerful, and began using ever-increasing amounts of natural resources to feed their mills, factories, and populations. In addition, with the economic and military power that come with high levels of industrialization, the developed world has repeatedly come under fire from the less-developed nations for having a disproportionate share of global wealth and for using an unfair share of global resources. Many of these questions remain unresolved and are likely to remain so for years to come. On a smaller scale, increasing industrialization and job specialization have also led to a wide gap between skilled, highly paid workers who design, manufacture, and tend the machines, and unskilled, low-paid workers who often see little opportunity for advancement in a technical world for which they are poorly trained. In this, the Luddites may find sympathy for their cause.

P. ANDREW KARAM

Further Reading

Books

Ashton, Thomas S. *The Industrial Revolution, 1760-1830.* Oxford University Press, 1998.

The Invention of the Chronometer

Overview

Locations on Earth are determined by a gridwork of lines, one set marking distance north or south of the equator and the other marking distance east and west of the Prime Meridian, running through Greenwich, England. For centuries, determining one's longitude, that is, one's position east or west of Greenwich, was nearly impossible, leading to the loss of life, ships, and property. This problem was finally solved by John Harrison (1693-1776), an Englishman, with his development of a highly precise clock called a chronometer. This invention revolutionized travel by sea, with repercussions that lasted until the 1990s.

Background

Today, for a few hundred dollars, virtually anyone can purchase a hand-held unit that, by detecting signals from an artificial constellation of satellites, will obligingly display one's latitude, longitude, and altitude with a precision of a few meters or less. The entire surface of the Earth has been mapped photographically, gravitationally, and recently, with space-borne radar to an unprecedented level of detail. It is difficult to remember that, until very recently, nobody on Earth knew where they were with this degree of precision. In fact, as late as the 1980s, before completion of the Global Positioning System (GPS), most ships at sea knew their positions to within only a few kilometers or more, and many still located themselves by celestial observations.

We live in a world that relies on GPS to tell us where we are at any time. Hikers carry GPS receivers in case they lose the trail, cars have GPS receivers to help navigate the streets of strange cities, and scientists use GPS to clock the uplift of Mount Everest. This accessibility of accurate geographic information is unprecedented in human history. Although the concepts of latitude and longitude have been with us for about two millennia (and we have been able to measure latitude for nearly that length of time), our ability to determine longitude is a recent accomplishment.

In fact, latitude measurements require only a simple instrument that can measure the distance above the horizon the Sun appears to be at noon, or the distance above the horizon a given star is seen to be. As we travel away from the Equator, the Sun drops lower in the sky while the pole stars climb higher. With but a smattering of math, this elevation can be turned into a latitude. Longitude is much more difficult and, for centuries, proved nearly intractable.

Because sailors could not accurately determine their location east or west on a map, early navigation often consisted of sailing north or south to a specific latitude and then striking out east or west along that imaginary line until one's destination was reached. Longitude was guessed, based on the captain's estimate of ship's speed, currents, wind speed, and other factors, but the captains were often wrong. In 1707, a wrong guess led to the sinking of four British man-of-war and the loss of 2,000 men when their fleet ran aground on islands just off the coast of Britain. In 1741, another wrong guess led the HMS *Centurion* to spend two additional weeks at sea on a ship struck by scurvy. By the time *Centurion* found port, 250 men had died. Eighty of those men died in the two weeks *Centurion* was delayed because her captain didn't know whether to turn east or west.

The longitude problem was recognized at the dawn of the Age of Discovery, in the early 1500s. Early suggestions were to use the Moon, the stars, or the position of Jupiter's satellites as clocks, but the practicality of making precise astronomical observations from the rolling deck of a ship proved insurmountable, and the positions of these bodies could not be predicted with a high degree of accuracy in that era. Other suggestions were to use clocks, if only they could be made sufficiently precise.

The Earth turns 360 degrees in 24 hours. In one hour, it will turn 15 degrees, and it will turn one degree in four minutes. One degree of longitude at the equator is about 60 nautical miles, or 70 "standard" mi (112 km). To measure longitude with any degree of certainty required a clock that was accurate to within a few seconds per day. And this degree of precision was required under ever-changing conditions of temperature, humidity, rolling of the vessel, air pressure, and more. Mechanical clocks were simply not up to the task, leading to the early interest in the heavens.

In the middle part of the eighteenth century, John Harrison (1693-1776) solved the longitude problem by constructing a series of chronometers that kept time more accurately than any previous such devices. Running without lubrication, changing temperatures could not cause grease or oil to thicken or thin. Neither could temperature cause expansion or contraction of moving parts, because Harrison coupled different metals to overcome such effects. On several test voyages, Harrison's chronometers kept nearly perfect time, especially his most famous clock, the H4 (built in 1759). In spite of this, many years were to pass before, in 1775, his chronometers were acknowledged to have solved the longitude problem—a problem that had stumped some of the greatest minds in Europe for nearly two centuries.

Impact

The impact of Harrison's chronometer on seafaring can hardly be overstated. However, it affected more than just ships at sea. National commerce benefited, and, somewhat less tangibly, conquering the longitude problem gave humanity a little more control over its world.

An eighteenth-century French ship. *(Corbis Corporation. Reproduced by permission.)*

The most obvious impact, of course, was on shipping. Captains were finally freed from dead reckoning and experience to determine their positions. Instead, they could do so scientifically, objectively, and with a high degree of precision. Because the captains now knew where they were with some degree of accuracy, they could better plan their landfalls and could ration their supplies more appropriately. The numbers of ships running aground did not, of course, drop to zero, if only because uncharted lands still existed. However, with more accurate maps of the world and a better idea of their ships' location on the globe, captains could now predict when they would reach port, determine the best course to take for repairs, or calculate their speed for the day. All of these, in turn, gave a much better idea of the amount of time left at sea, letting captains parcel out food to make it last until landfall.

In addition to these benefits, captains could now search for the most efficient routes across the oceans. Since they were no longer stuck with running along a line of latitude until they ran into their target, they could cut diagonals across the oceans, making for port in the shortest time following the most direct routes. Not only did this help speed up ocean travels, it also allowed a wider variety of shipping paths. This, in turn, helped reduce loss of shipping due to pirates or enemies because the shipping lanes were sud-

denly less predictable. Gone were the days when an enemy fleet would sit on top of the most widely used east-west parallel of latitude, waiting for a commercial or treasure fleet. Instead, with more routing options, there were simply too many paths to watch and attacks of this sort began to die off. This didn't happen immediately, of course, and attacks by privateers, warships, and pirates never ended entirely. However, their number dwindled and became less important as time went on.

This drop in shipping losses, of course, helped the treasuries of all sea-faring nations. Losing fewer ships, they had fewer ships to replace and fewer new crews to train. Faster landfall meant goods could reach market more quickly, giving a faster turn-around on money invested in a load of cargo from overseas. Fewer commercial shipping loses meant more money reaching the coffers of ship owners and investors, more tax revenue for the government, and more spending by everyone making a profit on the voyage.

Finally, this was but one more step in man's understanding and control of Earth. The longitude problem was of fundamental importance to the seafaring nations, and most of them offered large monetary prizes at various times for anyone who could solve the problem. News of possible breakthroughs caught the public's attention, and the public followed these events with some degree of interest. Those with family members or friends at sea had a more pressing interest, but in nations like Britain, the Netherlands, and Portugal, the entire country followed such issues because of the long and proud dependence on the sea for national sustenance and pride. Even nations like France and Spain were interested, having a proud seafaring tradition themselves. However, these land powers did not have as much at stake as did the smaller, sea faring countries, and their publics did not take the same interest in the problem.

When a solution to the longitude problem was finally announced, there were many who rejoiced because their loved ones, friends, or themselves were now at much less risk. At the same time, national voyages of discovery were continuing to set forth, intent on visiting and mapping all the lands of the globe. Armed with Harrison's chronometers and the imitations that quickly followed, these explorers could now chart islands, harbors, and other important features with unprecedented accuracy. This, in turn, gave their fellow sailors a much better idea of where to turn for supplies, repairs, or shelter, in addition to allowing the mapping of Earth in much better detail.

P. ANDREW KARAM

Further Reading

Books

Andrews, William. *The Quest for Longitude.* Cambridge: Harvard University Press, 1996.

Sobel, Dava and William Andrews. *The Illustrated Longitude.* Walker and Company, 1995.

Advances in Construction and Building Design during the Eighteenth Century

Overview

For building design and construction, the eighteenth century represents a period of transition. Engineers and architects, formerly lumped together, assumed separate identities and functions. The increasing complexity of buildings and bridges helped to spur training for architects, engineers, and builders, who until then had little formal instruction. Substituting design and calculation where rule of thumb and practical experience had been the order of the day meant that expert knowledge could be codified and formulated. These developments helped to give substance to Enlightenment notions of progress.

Background

In 1768 John Smeaton (1724-1792) described himself as a civil engineer and began a new vocation. From the earliest times up through the Middle Ages, public works such as the construction of aqueducts and harbors were carried out by armies and their captives. The technology was based on experience of trial and error, passed on by word of mouth. With time, interest in the underlying principles of construction grew, and by the eighteenth century engineering was a developing science in Europe.

In France in 1716, workers trained specifically to work on practical problems set by the government formed the Corps des Ponts et Chaussées. In 1746, a school was formed to supply recruits to the Corps. Toward the end of the century, polytechnic schools specializing in engineering were founded in Paris and Berlin, though it would be two decades before such schools appeared in Britain.

Bernard Forest de Bélidor (1697-1761) began his career as a professor of mathematics. After serving in the War of the Austrian Succession, Bélidor settled in Paris, where he took to writing books for artillery cadets and engineers. However, with *La science des ingénieurs* ("Engineering Science," 1729) and *Architecture hydraulique* ("Hydraulic Architecture," 1737-1739), he turned his attention to mechanical problems involving transport, shipbuilding, waterways, water supply, and ornamental fountains. Little in these books was original, but they served as a call to builders to base design and practice on the science of mechanics. Bélidor's volumes influenced generations of architects, builders, and engineers, among them the first two generations of engineers who could also be considered scientists.

The first cast-iron bridge, built in Coalbrookdale, England. *(Corbis Corporation. Reproduced by permission.)*

Since ancient times, canals have been used to transport passengers and goods inland, and for drainage and irrigation purposes. Locks were most likely invented in Italy, as early as 1440. Knowledge of canal building was brought to France by Italian engineers during the 1600s, and by the end of the 1700s, France had many hundreds of miles of canals for navigation. Some of the greatest French engineers of this period made their names in designing canals and improving rivers. In Sweden, canals made it possible to connect the North Sea with the Baltic by way of several inland lakes, bypassing a long and dangerous sea route.

Throughout the 1700s, road making as a profession was unknown. Daniel Defoe's (1660-1731) *A Tour through the Whole Island of Great Britain* (1724) does not stint on the sorry state of

British roads. France was probably the first country to realize the value of a good system of roads. The engineer Pierre Marie Jérôme Trésaguet (1716-1796) introduced innovations in construction, and revised the system under which French roads were built and maintained. The modern French highway system was completed during this period. The first turnpike in America, connecting Philadelphia to Lancaster, was built between 1792 and 1794.

The first road roller was built and tested in France on the eve of the French Revolution, in 1787, but the machine did not come into practical use until a generation later, and then first in England. Rock asphalt was discovered in Switzerland in 1712, but not used for pavements until the first decades of the nineteenth century.

Until the eighteenth century, bridges were built of stone or, less often, wood. Gradually iron began to be used in combination with wood. In 1739, the young Swiss-born engineer Charles Paul Dangeau Labelye (1705-1781) was the first to use caissons, which are watertight structures, in constructing the foundation for the first pier of Westminster bridge.

French bridges of the era were particularly noted for their elegance and increasing refinement in construction. Advances introduced by the engineer Jean Rodolphe Perronet (1708-1794) included reducing the thickness of piers and using flat arches. The building of the Pont de Neuilly over the Seine River in Paris required 872 workers and 167 horses. This bridge was the first to employ a horse-drawn mortar mixer.

American timber bridges combined arches and trusses in innovative ways. Timber bridges were also built in Europe. A major advance in bridge building was Abraham Darby III's (1750-1791) construction of the first iron bridge in England, between 1775 and 1779. Since the invention of the arch, the only innovation had been the use of timber frames. The bridge was built to replace the ferry over the Severn River, and had an arch of cast iron that spanned more than 100 ft (30 m).

Between 1759 and 1761, a self-taught mechanic named James Brindley (1716-1772) built the first modern canal in England and earned himself the title of father of the British inland waterways. The canal featured several innovations that were ridiculed at first, including an aqueduct bridge and a mile-long tunnel at one end that penetrated into the Duke of Bridgewater's coal mines at Worsley. In America, where by the end of the eighteenth century the science of civil engineering was still all but unknown, public improvements were slower to appear. A decade after the Revolution, the country could boast only three very short canals.

One of the first applications of the steam engine to supply water to dwellings was undertaken in 1787 in London. In France, until the nineteenth century, availability of water to the public was limited to fountains in public squares. In 1776, two French mechanics formed a company to supply Paris with water pumped from the Seine by steam. They obtained their engine from the English company that made it. Their intent was to supply water to public fountains for the poor and to private residences for those who could afford it. But the company failed in 1787. In America in 1796, water was brought to Boston through wooden pipes from Roxbury. Boston was the first city in America to have a public water supply. The first efforts to pump water for a public supply occurred at the waterworks in Bethlehem, Pennsylvania. The use of steam for pumping water and of cast iron for water mains were introduced in Philadelphia at the close of the century.

Aqueducts built in Lisbon, Montpellier, and Wales during this period were notable mostly for their beauty. Cast iron piping to carry water was used in England early in the eighteenth century, but not extensively. By the middle of the century, such pipes were in use in several French cities. Early innovations in cast-iron piping were introduced by British engineers. These included the first bell-and-spigot water pipes.

By the end of the eighteenth century, Paris had 16 mi (26 km) of sewers. The main beltline sewer was connected to a major aqueduct, which made possible the development of high-class residences and eventually modern Paris. In London, sewers laid in the mid-seventeenth century drained directly or indirectly into the Thames River, with undesirable consequences for public health.

The prevalence of waterways for travel and shipping up to the nineteenth century meant that ships had to be warned of dangerous coastlines and harbors protected from storms. John Smeaton's Eddystone Lighthouse, built between 1757 and 1759, featured interlocking masonry. In 1783, the French engineer Louis Alexandre de Cessart (1719-1806) commenced construction of a breakwater to protect Cherbourg Harbor. Dredges were used to deepen steam channels. In 1796, a four-horse-power Watt steam engine was used to operate a dredge in England for deepening Sunderland Harbor. The engine was designed to operate four "spoon" dredges that each raised a ton of earth 10 ft (3 m) high in one minute.

Up until the middle of the eighteenth century, lime was a standard material used in building. A problem in using it for foundations was that it would not always harden under water. Although various materials were used as hardeners in a process of trial and error, in 1756 John Smeaton found that adding clay in sufficient quantities would cause limestone to harden, although he did not understand why it worked. In 1796 it was discovered that water added to a fine powder of a clay called Sheppy stone generated a material that would harden very rapidly, even under water. The product was patented as "Roman" cement.

The production of cast iron in the sixteenth century led to widespread destruction of forests because of the need for charcoal used in smelting. In 1709, Abraham Darby II (1711-1763) was the first to use coke as fuel in the process, which made possible large-scale production of iron. Wrought iron made its appearance during the Renaissance, but the first rolling mill for iron dates from the early eighteenth century in England.

Impact

The closing years of the eighteenth century saw a move from an approximation of the world to what one scholar has called the "universe of precision." Up to that time, in the world of everyday reality, nature always got the better of man. Builders had exhausted the limits of geometrical reasoning and gauging dimensions by what they could see. There was no systematic body of knowledge to draw on. Technical engineering books were largely based on architectural treatises. Nature and people were at loggerheads.

The move in the 1700s to ground engineering in mathematics freed mathematics as a tool to be used in practical calculations. Moreover, investigations into the forces of nature, and the chemistry of materials, meant that engineers could begin to have a degree of technological control over their constructions. Just as improvements in surveying techniques led to better maps, which made it possible to plan cities, the scientific basis of engineering made it possible to model roads, bridges, and houses before building them. The full impact of these developments, however, was not realized until the nineteenth century.

GISELLE WEISS

Further Reading

Books

Forest de Bélidor, B. *Architecture hydraulique, ou l'art de conduire, d'élever et de ménager des eaux.* Paris: 1737-1753.

Perez Gomez, Alberto. *Architecture and the Crisis of Modern Science.* Cambridge, MA: MIT Press.

Picon, Antoine. *French Architects and Engineers in the Age of Enlightenment.* Translated by Martin Thom. Cambridge: Cambridge University Press, 1988.

Smeaton, John. *Narrative of the building...of the Eddystone Lighthouse.* London: 1791.

Balloons Carry Humans

Overview

With the balloon, the Montgolfier brothers brought flight to humans, initiating an era of fun and experimentation. Within a dozen years, there were scientific expeditions, competitions, recreational excursions, and military applications—a pattern that would be repeated when powered flight arrived a bit over a century later. Scientists took advantage of the new device to measure temperatures, wind patterns, and atmospheric composition. The military first used balloons to spot artillery and terrify the enemy. From the first, balloons attracted hundreds of thousands of spectators, and today, recreation is the main reason for ballooning, which continues to fascinate the public.

Background

Archimedes (c. 287-212 B.C.) established the principle of buoyancy in the third century B.C., but it was not until Joseph and Étienne Montgolfier began their experiments with balloons that the principle was put to use for flight. Francis Bacon (1561-1626) is sometimes given credit for the concept of the balloon, having written in 1250 about creating a flying machine by filling a hollow globe of copper with "ethereal air" or "liquid fire." In 1670 an Italian priest saw the air pump as a possible means for building Bacon's flying machine. His vehicle was based on scientific principles and looked good on paper. Unfortunately, when the copper was made thin enough to reduce its weight, the force of the vacuum collapsed the ball. A Brazilian priest had more success. In 1709 he created a small, working model of a balloon for the king of Portugal. He used hot air, just as the Montgolfier brothers would. When he lit the fire, the model drifted across the room and set the curtains on fire but failed to ignite anyone's imagination.

Independently of those predecessors, Joseph-Michel Montgolfier (1740-1810) conceived of the balloon in 1782 when he supposedly noticed the billowing of a shirt next to a dry-

Inflating a hydrogen balloon for flight. *(Bettmann/Corbis. Reproduced with permission.)*

ing fire. (A less romantic story holds that he was inspired by Joseph Priestley's experiments with gases.) Borrowing some rags from his landlady, he made a small model that took off into the air. He knew immediately that he had discovered something important. He sent for his brother Jacques-Etienne (1745-1799), and the two were soon experimenting by sending paper bags up a flue. On April 4, 1783, they held their first public demonstration. The bag was made of cloth and paper. (The Montgolfier family was in the paper business.) It was coated with alum to reduce its flammability—a good idea since the bag was filled with hot air created by burning straw. The contraption was held together with over 2,000 buttons. To the amazement and delight of the crowd, the balloon rose to a height of 2,000 feet (600 m) and traveled almost a mile (1.6 km).

The first balloon passengers were a sheep, a duck, and a rooster. The balloon that was used, however, was not a toy; it was 110 feet (34 m) in circumference and weighed 500 pounds (200 kg). The Montgolfiers set the balloon aloft in Versailles on September 19, 1783, before a crowd of 130,000. The animals traveled and landed safely. The next step was a tethered flight. Jean François Pilâtre de Rozier, a professor of physics, was the first human to fly, reaching a height of 85 ft (26 m) over Paris. It was now time for the first free flight. The king offered to provide a condemned prisoner as the test pilot, but Rozier wanted the honor for himself. He and François Laurent, an aristocrat and an infantry officer, took off on November 21, 1783. Theirs was not an uneventful journey. While in transit, the silk and paper balloon first sank to rooftop levels and then started to burn. The two men used wet sponges to keep their vehicle from burning up and landed safely after traveling about 7 miles (11 km). Local villagers were concerned about the strange device, but their worries disappeared when they were offered champagne.

Within weeks, Jacques Charles (1746-1823) and Nicolas Robert were aloft in a hydrogen-filled balloon. Theirs was a much longer journey, rising to 3,000 feet (900 m) and going 16 miles (26 km). Hydrogen, discovered in 1766 by Henry Cavendish (1731-1810), is significantly lighter than nitrogen, the main component of air. Apparently, Charles assumed the Montgolfiers were using hydrogen and was unaware of how innovative he was. When the balloon landed near a farming village, the villagers mistook it for a dragon and attacked it. After the balloon was totally demolished, they even had the local priest exorcise the demon. To protect future balloons, the king issued a royal proclamation against attacking balloons. Charles's new balloon was financed by charging some of the 400,000 spectators of the next flight (including Benjamin Franklin) a crown apiece.

Impact

The Montgolfiers continued their careers as inventors, developing the hydraulic ram (which uses the energy of flowing water to force a small portion of the water upward), the calorimeter, and a method for producing vellum, a fine-quality paper used for formal documents. Charles also continued to work as an inventor, improving many instruments and developing the hydrometer and the reflecting goniometer (used to measure angles). He is also famous for Charles's Law (different gases all expand the same amount with a given increase in temperature).

The dramatic and successful demonstrations by both the Montgolfiers and Charles created a ballooning craze across Europe. Within a year, there were flights in England, Scotland, and Italy. The balloonists worked to best each other, perfecting control of the craft, rising to higher and higher levels. Attempts were made to add propulsion, using oars, wings, and propellers. As early as 1784, a French general designed the first dirigible, which featured hand-cranked propellers and a cigar shape, but the first practical powered balloon was not invented until 1852, when Henri Giffard (1825-1882) flew a steam-powered dirigible.

The first trip across the English Channel by balloon was also the first instance of international airmail. Jean Pierre Blanchard (1753-1809), a French aeronaut, and John Jeffries, an American physician, took off from England in 1785 in a balloon laden with food, flags, and mementos. Unfortunately, their balloon began to sink as they traveled over colder water. They began to jettison everything to gain lift, including most of their clothing. When they landed in France, they had little more left than a bottle of brandy and a packet of letters. Blanchard continued as a pioneer. He invented the parachute and tested it by dropping a dog in a basket from a balloon in 1785. In 1793 he launched the first balloon in North America, from Washington Prison Yard in Philadelphia. But it was not until 1978 that the first transatlantic balloon trip was completed. The *Double Eagle II* was piloted by Americans Troy Bradley and Richard Abruzzo and went from Maine to France. In 1999 the *Breitling Orbiter 3* became the first balloon to travel nonstop around the world. It headed east from Switzerland and ended its journey in Egypt. It was piloted by Swiss psychiatrist Bertrand Piccard (1958-) and British pilot Brian Jones (1947-).

The French promoted the scientific use of balloons. In 1784 samples of air taken at high altitudes were brought back to Earth to be studied. Scientists also carried instruments up in balloons to measure temperature, pressure, and humidity. In 1804 Jean Baptiste Biot (1774-1862) and Joseph Louis Gay-Lussac (1778-1850) went up in a balloon left over from Napoleon's Egyptian campaign. They brought along small animals and scientific instruments and proved that Earth's magnetism was undiminished at altitudes up to 3 miles (5 km). Gay-Lussac later went up to a height of 4 miles (6 km).

More recently, balloons have been used to investigate radio rays and to establish the existence of cosmic rays. Balloons are still launched to determine the speed and direction of winds (a practice that gave rise to the phrase "send up a trial balloon"). Currently, 1,000 readings of wind, temperature, humidity, and pressure in the upper atmosphere are taken each day by weather balloons.

André Garnerin is the first to descend in a parachute. *(Corbis Corporation. Reproduced by permission.)*

It did not take long for the military potential of balloons to be realized. As early as 1793, Jean Pierre Coutelle used a balloon at the Battle of Maubeuge to spot artillery and demoralize the enemy. Napoleon used balloons and, during the Franco-Prussian War (1870-1871), balloons were dramatically used to carry 164 people and 20,000 lbs (9,000 kg) of mail out of the besieged city of Paris. Both sides in the American Civil war used them for reconnaissance. A Union volunteer, Count Ferdinand von Zeppelin (1838-1917), became fascinated with balloons and, beginning in 1891, dedicated his life to making practical airships. By adding propulsion and a lightweight frame, he invented the aircraft that bears his name. Zeppelins first flew in 1900. By 1910 they were providing regular flights between German cities and, later, they flew scheduled, commercial transatlantic flights, cruising at 68 miles (110 km) per hr. During Word War I, Zeppelins were used as bombers. At the same time, the British used traditional balloons for aerial photography. The military continued to find applications for balloons as late as World War II. Tethered by chains, they formed a part of antiaircraft defense, making attacks more difficult by forcing planes to fly above them. The British also used them for submarine detection and, after the war, balloons were used for defense as flying radar stations. Ultimately, the vulnerability of balloons to weather and enemy fire made their use in warfare obsolete, although in 1979, during the Cold War, a balloon was used by two families to fly from East to West Germany. Afterward, the sale of lightweight fabric was restricted in East Germany.

Today, while advertisement-covered, helium-filled blimps are the most visible application of balloons, recreational hot-air balloons actually dominate the lighter-than-air world. Through most of the early part of the twentieth century, ballooning was a rich person's sport, like yachting. The Gordon Bennett Balloon Trophy Races helped popularize the sport. In Europe, this tradition has continued but, in America, the invention of an inexpensive, highly controllable hot-air balloon popularized the sport. In the 1950s, Ed Yost, an employee at General Mills, put his

expertise in cellophane packaging to work for the U.S. military. He helped build giant polyethylene balloons for stratospheric research. Inspired by the new materials, he pursued the opportunity to design simple, relatively safe, and less-expensive balloons. Made of polyurethane-coated nylon fabric and powered by a propane burner, the *Raven* became the first modern hot-air balloon when it took to the skies in 1960.

PETER J. ANDREWS

Further Reading
Books
Asimov, Isaac. *Isaac Asimov's Biographical Encyclopedia of Science & Technology.* New York: Doubleday, 1976.

Kalakuka, C., and Brent Stockwell. *Hot Air Balloons.* New York: Metro Books, 1998.

Internet Sites
http://wings.ucdavis.edu/Book/History/instructor/balloon - 01.html Flight without Wings: Balloonists.

The Beginning of the Age of Canal Building in Great Britain

Overview

In 1760 the British Parliament approved construction of a canal to carry coal half a dozen miles (approximately 9.6 km) from the mines at Worsley to the city of Manchester. Until the eighteenth century, European canals were generally financed by the aristocracy and built for social or political reasons. But the agent who described the plans in the House of Commons argued that the new canal would make transporting goods easier and less expensive. The so-called Bridgewater Canal was the work of a self-taught engineer named James Brindley (1716-1772) whose name is synonymous with the early years of canal building. The success of Brindley's canal stimulated a new wave of construction that revolutionized Britain's transport system and contributed to the country's wealth over the next 50 years. Between 1760 and 1790, 25 new canal-building projects were begun. And in the short time between 1790 and 1794—a period known as canal mania—there were 46.

Background

Canals are artificial waterways that connect to rivers or other canals. First built in ancient times

in the Middle East to supply drinking water and irrigation, canals came to be used to enhance the navigability of natural waterways. The Romans constructed canal systems in Northern Europe and Britain for military transport and drainage. Although European waterway development went into decline with the fall of the Roman Empire, it revived in the twelfth century. In 1373 the Dutch invented the pound lock, a tightly closed chamber that could be flooded or drained as needed to allow a vessel to pass between bodies of water at different elevations. The modern era of canal building in Britain coincided with the beginning of the Industrial Revolution and lasted until the arrival of the railroads.

England's geography is diverse, and canals often pushed through well-populated areas. Thus, man-made obstacles as well as valleys, rugged hills, and changes in water levels at the junctions of rivers and canals tested the ingenuity of eighteenth-century engineers like Brindley and William Jessop (1745-1814), whose Grand Junction Canal sliced through the Chiltern hills. The innovations builders devised to meet the challenges of the landscape included locks, tunnels, and bridges and aqueducts.

Although the British did not invent locks, during the era of canal building in Britain locks became bigger and more complicated. Locks were an essential feature of canals wherever builders encountered a change in the level of the ground. So-called staircase locks, which shared gates, were used where the contour of the landscape changed very suddenly, and where closely spaced locks were required. Flight locks were a later innovation that consisted of a number of locks put together. Side-by-side, or duplicate, locks helped to relieve bottlenecks on busy canals, and stop locks were used to keep water owned by different companies separate.

Because opening and closing locks led to the loss of water, engineers were eager to try alternatives such as mechanical lifts. The most common lift was the inclined plane, which consisted of a railed track up which a boat could be hauled. The first inclined plane was constructed in 1788 on the Ketley Canal, a private canal serving the Shropshire iron works. Not many of these lifts were built, and none survive.

Engineers who did not want to go around hills (as Brindley preferred to do) or over them had the option of cutting or tunneling through it. The Trent and Mersey canal built by Brindley was the first large-scale effort to take a canal through a tunnel. The technology for tunneling was borrowed from mining, and early examples were little more than openings through hills. Tunnelers followed a straight line from one point to another, sinking shafts and aligning them using a telescope. But the science of geology was unknown at the time, and tunnelers had no way of predicting what lay beneath the surface. Thus, they might encounter underground water, quicksand, and difficult rock formations. Sometimes tunnels did not meet up correctly. To keep costs low, early tunnels lacked towpaths, and crews were required to "leg" their way through, moving the vessel forward by pushing with their feet against the sides or top of the tunnel. This work was so exhausting that it required professional "leggers." Later tunnels did include tow paths, and barges were pulled along the canal in the same way as above ground, by horses, mules, or donkeys. An alternative to tunneling was simply to cut a gap through the hill.

Canals often cut across private property and existing roads, and parliamentary law required that no one be inconvenienced. Thus bridges became an integral part of canal building, to preserve passages from one part of an owner's land to another when a canal cut through it. Bridges were commonly made of brick, masonry, timber, or cast iron. Additional features prevented the tow lines from getting caught up in the bridge. Although aqueducts were not popular with canal builders because of the problems posed by the weight of the water and the need to keep the aqueduct watertight, these structures were the ideal solution to certain kinds of problems. For example, crossing a valley meant bringing a canal down to the lowest level, which in turn meant going to the expense of building locks. In building the Bridgewater Canal, Brindley dodged the problem by constructing the canal on a viaduct carried on arches across the valley.

A dry canal would bring transport to a halt, and canal builders were perennially concerned about water supplies. Specially constructed reservoirs, feeder canals, and water pumped from wells and rivers were some of the ways builders used to ensure a supply of water to the canal system. In addition, means were devised to conserve water already in the system. For example, side ponds at locks helped to compensate water lost in operating locks and through leakage and evaporation.

New kinds of structures related to the operation of canals arose. Canal cottages provided accommodation for the lengthsmen, tollkeepers, and locksmen who worked the canals. Maintenance yards were constructed to allow area engineers and craftspeople to perform maintenance tasks such as dredging (removing mud from canals to keep them navigable) and lock repairs. Just as canals were developed to serve industry, so industrial buildings, such as potteries and mills, sprang up alongside the canals. Wharves and warehouses were constructed to handle goods that needed to be transshipped or stored for the short or long term.

Impact

In the late eighteenth century Britain was still a small country. People did not often travel, and the needs they had for materials were usually satisfied by those close at hand. Not that they had much choice—road transport was inefficient, expensive, and unreliable. The coming of the canals would change all that.

Canal building in Britain signaled a new age of engineering, and engineering on a massive scale. Although the American War for Independence (1775-1783) diverted the cash Britain needed for construction projects, in the peace that followed a growing number of people in newly expanding cities needed coal for lamps, for heat, and for power. That coal needed to be transported inexpensively. The Bridgewater Canal showed what was possible by cutting the cost of coal in Manchester in half.

The development of the country's inland waterway system to meet these demands was perhaps the most important factor behind the Industrial Revolution. Prosperity fed on prosperity, as industries scrambled to find canal-side sites that would gain them low transport costs for materials (such as pottery) and the promise of a less bumpy ride for delicate finished goods. Where canals met rivers, new towns sprang up to supply the infrastructure required to manage the canals and the commerce they brought. The city of Birmingham, for example, owed its growth to its position at the heart of a canal system connecting London, the Bristol Channel, and the Mersey and Humber rivers. Heading into canal mania, industrialists, developers, and investors were more than satisfied. Without canals, the Industrial Revolution would still have occurred, but it would have been slower.

Insofar as the stated purpose of the canals was to bring trade and affluence to the regions along their banks, they succeeded. But not every individual canal project was a success. Building a canal required an Act of Parliament and money in the form of investment capital. At first, the people putting money into canal building were those promoting the canals and local backers. But during the height of canal mania in 1793, canal projects attracted speculators as jam does flies. So intent were speculators on quick profits that they backed any plan that had the word "canal" attached to it. Problems posed by the landscape were often poorly thought through, and developers frequently underestimated costs. Many schemes were stillborn, and the number of failures helped to diminish the reputation of canals in the public mind. Moreover, hoping to keep costs low, engineers built canals too small, and they failed to envisage the canals as an interlinked transport system—there was, for example, no national standard for the size of canal locks.

In 1796 the specter of war with France brought an end to the free flow of funds from speculators. By 1797 engineers had taken note of the advantages of "roads with iron rails laid along them" over navigable canals. With the coming of the railways, the inland waterway system in England was abandoned.

GISELLE WEISS

Further Reading

Burton, Anthony. *The Canal Builders*. 2nd ed. London: David and Charles, 1981.

Burton, Anthony. *Canal Mania: 200 Years of Britain's Waterways*. London: Aurum Press, 1993.

Hadfield, Charles. *British Canals: An Illustrated History*. 7th ed. Newton, UK: David and Charles, 1984.

Britain and America Battle for Technological Prowess in the Eighteenth Century

Overview

America was founded to serve British commercial interests. British mercantilism (an economic system based on colonialism and a favorable balance of trade) aimed at orchestrating economic development in the colonies in the name of nation building. But by the late 1700s, America was agitating for a change from the old order. The debate over whether Americans should manufacture their own goods, and if so how they would do it, went to the heart of the colonies' desire for independence from Britain and their fear of succumbing to the excesses of a society based on manufacturing. While the debate went on, an unskilled America eagerly sought the technological know-how of a home country less than eager to give up its advantage.

Background

The Industrial Revolution began in Great Britain in the 1760s, and brought fundamental changes in the way people worked and where they lived. The core achievement of the revolution was to apply new sources of power to producing work and goods. Where sources of power had been humans and animals, the revolution substituted motors powered by fossil fuels. Manual tools such as sickles and foot-pedaled looms were replaced by power tools that needed less human guidance.

The Battle of Lexington (1775) marks the beginning of armed conflict between Britain and America. *(The Granger Collection, Ltd. Reproduced by permission.)*

Already in the 1730s, marketing opportunities for manufacturing production rose, owing to a number of factors. One of these factors was population growth, which doubled in Britain between 1750 and 1800. Increased demand for cloth for garments stimulated the development of mechanical yarn-spinning devices to replace inefficient yarn workers. In 1764, James Hargreaves (d. 1778) created the spinning jenny, a simple, inexpensive, hand-operated spinning machine that manufacturers were quick to adopt. In 1768, Richard Arkwright (1732-1792) built a water-frame and carding machine whose advantages over the spinning jenny were that it could spin several threads at once to make a strong, uniform yarn, and was powered by horses or water rather than by hand. Thus the British textile machine shop was a key site of industrial development, turning out such inventions as carding devices, power looms, and cloth printing machines. Short of adults to staff factories, factory masters turned to employing children. The abusive conditions under which both adults and children labored in industrial England were widely described.

In America, by contrast, the textile industry of the late 1700s was still primarily home-based. Although there was a market for mass-produced piece goods, sufficient amounts of quality yarn were in short supply. A similar problem had been solved in Britain with the institution of spinning factories. But transferring this technology to the United States proved problematic.

Before the 1840s, obstacles to the transfer of technologies between Britain and America included government prohibitions aimed at keeping the colonies agrarian, cultural barriers, and obstacles created by manufacturers themselves. In 1719, for example, Britain passed an act prohibiting metalworking in the colonies, although other handicraft operations such as breweries, glassworks, and printing continued. The Iron Act of 1750 was intended to reduce competition in manufacturing between the colonies and the home country by allowing America to export to Britain only basic iron and pig iron, not finished iron goods. In addition, Americans were forbidden, among other things, to establish furnaces to produce steel for tools, and to export colonial iron beyond the empire. The act was successful in suppressing the manufacture of finished iron goods, but it backfired in the sense that colonial production of basic iron thrived.

By the 1780s, responding to pressure from the woolen and cotton manufacturers, skilled British artisans were prohibited from emigrating for the purpose of carrying on their trade in America. Anyone attempting to export tools and machines related to the textile business and other trades risked being fined, having their equipment confiscated, and being subjected to 12 months' imprisonment.

In practice, these prohibitions proved difficult to enforce. For one thing, policing the coastline of the British Isles was difficult to do. Unskilled customs officers did not know how to recognize technology. And technologies were simple enough that artisans could simply memorize them. Moreover, machines could be disassembled into parts and carried on board. Restraints on emigration were lifted in the early 1800s, and on equipment somewhat later. Meanwhile, the illegal export trade business thrived.

Another obstacle to technology transfer was the disorganization of the British patent system, although the number of inventions being patented in England was so high that determined foreign investors could always find something of interest. This disorganization extended to conditions of work. Foreigners in search of new knowledge would get conflicting information from Britons in different textile districts. And manufacturers were protective of their technical secrets to greater or lesser degrees.

Still, most of these obstacles could be overcome. Until the 1780s, transfer of technology was one-way from Britain to the United States. Americans borrowed heavily from British skills, tools, and ideas, and because the American economy was hospitable to invention, many skilled British migrated there. Skilled emigrants to America included workers with textile and machine-making skills. In 1787 Tench Coxe (1755-1824) sent a secret agent to Britain to obtain the models for Arkwright's water frame. The agent was to send the models to France, where the then ambassador to France, Thomas Jefferson (1743-1826), would forward them to the United States. The scheme failed when customs officials, tipped off by British authorities, seized the models, and the agent was caught and fined. Still, within a year's time, Coxe had managed to obtain the models by other means. (Power loom weaving and mechanized calico printing were additional examples of technologies transferred illegally from Britain by British immigrants to the United States.)

Coxe had a robust zeal for the manufacturing cause. He and an influential group of like-minded Americans published pamphlets on the subject, picked the brains of new emigrants from Britain on the details of the latest technologies, and tried to import new mechanical inventions, prohibitions notwithstanding. In early 1791 Coxe created the Society to Establish Useful Manufacture. The society was accorded privileges by the New Jersey state legislature, and a great industrial works was planned to manufacture paper, shoes, pottery, and textiles. Although the enterprise collapsed before it could really get off the ground, its greater importance was in stimulating a dialogue that was ostensibly about the implications of industrial development but more broadly about the nature of American society.

Skilled as they were, British immigrants still needed the backing of American capitalists. In 1789, the Englishman Samuel Slater (1768-1835) arrived in New York City disguised as a farm laborer. Slater, who had trained in a cotton mill in Derbyshire, offered to help a Rhode Island merchant establish a textile works. In return for a good percentage of the profits, Slater would build and manage a spinning mill. Backed by the support of a group of American capitalists, Slater built dozens of cotton mills throughout Rhode Island and southeastern Massachusetts. He transferred his experience in the Derbyshire family system of production to solve the problem of staffing his mill buildings and machines. The result was called the mill village, and into the 1820s it proved to be a stable, harmonious community.

Not only did immigrants bring new technology with them as individuals, but they were instrumental in diffusing the technology internally in America. In 1793, a family of woolen clothiers from Yorkshire named Scholfield emigrated to Boston. With the support of a wealthy merchant, they built a woolen factory, and later the extended family had established carding and spinning mills across New England. They introduced the British woolen carding machine to the United States.

Impact

The struggle between Britain and America for technological prowess brought to the fore the question of what kind of a nation America should be. In 1775, the American philosopher and thinker Benjamin Rush (1745-1813) argued publicly that manufacture would save Americans money, provide employment for the poor and indigent, encourage immigration to solve the problem of underpopulation, and make the colonies less dependent on England. Moreover, science and industry would improve agriculture. Rush was supported by people like Tench Coxe and Secretary of the Treasury Alexander Hamilton (1755?-1804), who, in the aftermath of the

American War for Independence (1775-1778), argued that industry would help to stabilize the economy of the new nation and it would make women and children more useful.

Benjamin Franklin (1706-1790), who was himself no Luddite, feared all the same that manufacture would create in America the same disparities between rich and poor as existed in England. He was seconded by Thomas Jefferson, who argued that the happiness and permanence of government would outweigh the costs of having to import commodities from Europe.

Arguments aside, immigrants from Britain were instrumental in diffusing new technologies within the United States, although the technologies did not always fit American circumstances. For example, Americans were less interested in machines designed to produce high-quality products; they also relied more on wood than metal, which meant their machines were less durable than British models but could be improved at more frequent intervals; British equipment was labor-intensive, which meant it had to be modi-fied, considering the scarcity of American labor. Finally, technology did not only flow westward. Some American improvements to technology were quickly transferred back to Britain.

Despite the War for Independence, the United States was not fully free of its mercantile past until the War of 1812, which ended its entanglements in British affairs. The debate over manufacturing continued, although the country moved inexorably into the industrial age.

GISELLE WEISS

Further Reading

Books

Jeremy, David J., ed. *Technology Transfer: Europe, Japan and the USA, 1700-1914.* International Aldershot, UK: Edward Elgar, 1991.

Licht, Walter. *Industrializing America: The Nineteenth Century.* Baltimore: Johns Hopkins University Press, 1995.

Tucker, Barbara M. *Samuel Slater and the Origins of the American Textile Industry, 1790-1860.* Ithaca: Cornell University Press, 1984.

The Development of a Patent System to Protect Inventions

Overview

From the time when the very first man walked upright, invention was key to human development. The introduction of a spoken language paved the way for a true society; tools allowed for more efficient hunting and thus led to greater longevity; writing ushered in recorded history. On the American frontier, ingenuity was crucial for survival, yet when America finally gained its independence in 1776, all manufacturing and invention was tightly controlled by England. America's founding fathers quickly realized that in order to grow as a nation, they had to nurture the spirit of invention in their countrymen.

The first article of the Constitution provided for the creation of what would become the U.S. patent system. For the first time, inventors were rewarded for their ingenuity with true ownership of their work. From Alexander Graham Bell's telephone, to Thomas Edison's light bulb, the patent system helped make the United States one of the wealthiest, most powerful nations in the world.

Background

In the early days of the American frontier, self-reliance was essential. If something was broken, the frontiersmen were forced to fix it themselves. If something needed to be built for use around the house or in the field, chances were the frontiersmen had to build it. But while early America was forged on ingenuity and creativity, England still held the reigns when it came to manufacturing and invention.

Europeans were no strangers to the patent system. The earliest precursor was developed in 500 B.C. in the Greek colony of Sybaris, where it was a law that if a confectioner or cook invented a particular dish, no one else could create that same dish for a period of one year. The earliest English patent grant dates back to April 3, 1449. It allowed John of Utynam to be the sole practitioner of the art of making colored glass for a term of twenty years. Early patents granted in England were subject to stringent requirements. To qualify, an inventor not only had to finagle a

The patented Gymnasticon exercise machine (c. 1798).
(Corbis Corporation. Reproduced by permission.)

special act by Parliament, he needed to get in the good graces of the ruling monarch.

In the early colonies, the first exclusive right for use of an invention was granted to Joseph Jenks, Sr. of the Massachusetts Bay Colony in 1646. The General Court allowed him sole ownership for a term of 14 years of the water-mill engines he developed. The first known patent granted to a colonial representative was believed to have been in 1715, when Thomas Masters, a Pennsylvania Quaker, and his wife Sybilla applied for a patent for their water-powered engine used for grinding corn into corn meal. It was a milestone for the colonists, however the British still held the patent on their home soil.

The real interest in invention and ownership in America began in around 1775, coinciding with the drive for independence. After the Revolution, our founding fathers realized that for our nation to develop successfully without British rule, an incentive was needed to foster independent invention. For widespread manufacturing to take place, a monopoly must be granted to an inventor, not just in the area in which he lived, but one that spanned the entire nation, securing them the rights to their invention and its profits.

Impact

On May 14, 1787, delegates from the 13 states met in Philadelphia at the Constitutional Convention. Their goal was to draft a constitution that would replace the earlier Articles of Confederation. President George Washington presided over the Convention. Initially, there was no mention made of granting patent rights, but on August 18, James Madison of Virginia proposed a measure that would "secure to literary authors their copyrights for a limited time," and "encourage by premiums and provisions, the advancement of useful knowledge and discoveries." On the same day, Charles Pinckney of South Carolina also submitted a similar proposal, "to grant patents for useful inventions," and "to secure to authors exclusive rights for a certain time."

Both proposals were submitted to the convention. The committee, given some helpful coaxing with a presentation of John Fitch's brilliant new invention, the steamboat, gave their approval. In the first article of the Constitution, they provided for what was to become the foundation of the U.S. patent system.

Congress shall have the power to promote the progress of science and useful arts, by securing for limited times to authors and inventors the exclusive right to their respective writings and discoveries.

But this provision alone was not enough to secure patents for early inventors. Many, including Fitch, filed petitions with the first session of the first Congress in 1789 for patents and copyrights. None were granted, and inventors were understandably frustrated. On January 8, 1790, President Washington in his State of the Union Address called for the promotion of new and useful inventions at home and abroad. On January 25, 1790, the House called together a committee to draft a patent statute. It took nearly three months, but on April 10, 1790, the first Patent Act was signed into law by the President.

In the early days, there was no patent office. A board was established to administer the new patent system, led by the Secretary of State, Thomas Jefferson and including Secretary of War Henry Knox and Attorney General Edmund Randolph. There were stringent requirements regarding early patent applications. They had to contain a written description of the invention, along with drafts (for example, technical drawings) and a model. After the committee had approved the application, it was passed to the Attorney General for a legal review, then sent to

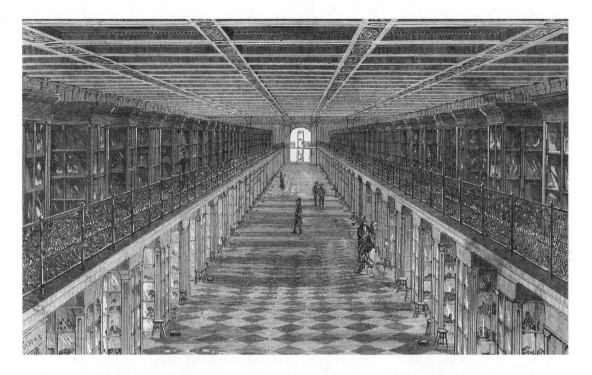

A drawing of the model room in the U. S. Patent Office. *(Corbis Corporation. Reproduced by permission.)*

the President of the United States for his signature, and finally returned to the Secretary of State to sign. If granted, a patent ran for a term of 14 years.

The first patent was signed on July 31, 1790, but only two other patents were granted in that year. This probably had to do with the rigorous procedure, as well as the fact that the Patent Board had to invent a system before they could even begin examining applications. Applicants were unsatisfied with the provisions of the 1790 Patent Act, and Patent Board members realized that they were unable to fulfill their responsibilities while carrying on with their full-time jobs. In response, the Act of 1793 was passed, abolishing the Patent Board. The State Department was now charged with registering patents, and the courts made the determination as to whether the patent was admissible.

The new Act allowed any inventor to receive a patent, provided they submitted an application complete with a model and had witnesses. While the 1790 Act had allowed both citizens and aliens to apply, now only citizens were given that right. The only drawback was the increased fee, from around five dollars to thirty dollars, which was a prohibitive amount for many farmers and laborers.

The first important invention patented under the new act was the now famous cotton gin, invented by schoolteacher Eli Whitney (1765-1825). Whitney applied for a patent in June of 1793 and was granted it in March of the next year. It has been said that his invention helped to develop mass production of cotton and even played a role in assuring the Northern victory in the Civil War.

In 1791, it was decided that the federal government move south from its present location in Philadelphia. A district was chosen near the settlements of Georgetown, Maryland and Alexandria, Virginia, in what is now Washington, D.C. On June 1, 1802, a clerk, known as the "Superintendent of Patents," was hired to oversee the issuance of patents, and the United States Patent Office was officially born. In the beginning, it consisted of a single crowded room in a building housing the State Department, as well as several other departments.

The 1793 Patent Act held firm until 1836, when a revision in patent laws charged a new Commissioner of Patents and Trademarks to oversee the office. The office itself remained in the State Department building until 1849, when it was moved to the Department of the Interior. In 1925 it was transferred again, this time to the Department of Commerce, where it remains today.

The same basic tenets that governed the office then, still govern the office today. Patents are issued to individuals for a period of 17 years.

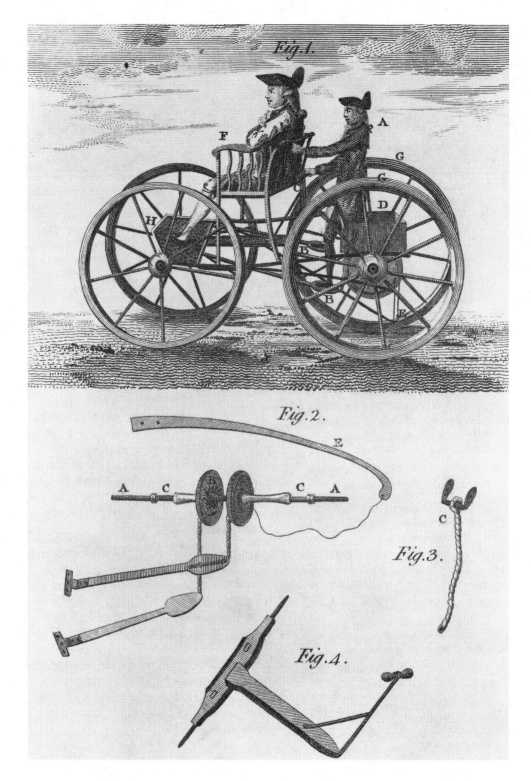

An eighteenth-century design for a horseless carriage. *(Corbis Corporation. Reproduced by permission.)*

After that, the invention is owned by the public. The 17-year term encourages the original inventor to improve upon his or her invention, in order to apply for a new patent. The Patent and Trademark Office administers the laws, examining applications to determine which inventions are entitled to patent protection. The office also maintains all records, enabling inventors to search for duplications of their own creation.

In the two centuries following the initiation of the Patent Office, the United States saw a technological explosion. Many notable inventors left their indelible stamp on history. Henry Ford built transportation that was easily affordable by the masses; Alexander Graham Bell made long distance communication possible; and the father of all invention, Thomas Edison, held an astounding 1,093 patents for everything from the phonograph to the electric pen. Each inventor was protected by the American patent system, and was allowed to prosper from their contributions. These inventors changed the face of American manufacturing, helping to make

the United States the richest, most powerful nation in the world.

STEPHANIE WATSON

Further Reading

Books

Dobyns, Kenneth W. *The Patent Office Pony: A History of the Early Patent Office.* Fredericksburg, VA: Sergeant Kirkland's Museum and Historical Society, Inc., 1997.

Videos

The American Documents: Patent Pending. Newsweek Productions, Republic Pictures Home Video, 1975.

Biographical Sketches

Sir Richard Arkwright
1732-1792
English Inventor

Sir Richard Arkwright was an English inventor and cotton manufacturer during the early years of the Industrial Revolution. He developed a mechanical machine for spinning cotton, a process that was up to this time done in small homes and farms. He also developed an early model for the factory system based on division of labor and a management system.

The youngest of 13 children, Richard Arkwright was born in Preston, England, on December 23, 1732. His parents were very poor and could not afford to send him to school. Instead, he was tutored by a cousin to read and write. After learning the basics, he was apprenticed to a barber and by 1750 was practicing his trade of wig-making. He acquired a secret method for dying hair and traveled about England purchasing human hair for the manufacture of wigs. These travels brought him into contact with people who were interested in designing and using machines for spinning and weaving. When the fashion for wearing wigs fell out of favor, he became interested in designing mechanical inventions to increase the speed of spinning the cotton thread used in making cloth textiles.

The industry most associated with the Industrial Revolution was the textile industry. Prior to the mid-1750s, the spinning of yarns and the weaving of cloth occurred primarily in the home, with most of the work done by people working alone or in conjunction with family members.

However, in Great Britain during the late 1700s, many specialized machines powered by water or steam appeared that eventually replaced the cottage system of textile production. With the aid of these machines, a single spinner or weaver now could turn out many times the volume of yarn or cloth that earlier workers had produced.

By 1767 a machine for carding cotton had been introduced into England, and James Hargreaves (1720?-1778) had invented the spinning jenny. These machines, however, required considerable labor as well as producing an inferior quality of cotton thread. This led Arkwright and colleagues John Kay (1704-1764) and Thomas Highs to design and develop a machine called the spinning frame or water frame. Arkwright's machine involved three sets of paired rollers that turned at different speeds and thereby applied the correct amount of tension to produce the thread. The rollers were able to produce a thread of the correct thickness, and the spindles twisted the fibers firmly together. Although the frame produced a coarse, hard thread, its strength was well suited as a warp thread.

As this first machine was put to use, the local hand-spinning weavers, concerned for their livelihood, forced Arkwright to relocate to another town. After moving the factory there, he went into partnership with Jedediah Strutt (1726-1797), the inventor of the stocking frame. The early frames were too large to be operated by people-power, and after a brief attempt to use horse-power, they began to experiment with the use of a water wheel. By 1771 Arkwright's invention became known as a water-frame. By 1790 his factories used the steam engine to pump water to the millrace of a waterwheel. Within a few years he was operating numerous factories equipped with machinery for carrying out all phases of textile manufacturing from carding to spinning.

These new machines required many workers to operate them. In one town of Cromford, he built a large number of cottages close to the factory. He imported workers from the neighboring countryside, preferring weavers with large families. While the women and children worked in the spinning-factory, the weavers worked at home turning the yarn into cloth. The factory employees worked from six in the morning to seven at night. Children, some as young as six, made up two-thirds of the 1,900-person work force.

Arkwright maintained his dominant position in the textile industry despite the loss of his comprehensive patent in 1785. Even though others copied his machinery, he was knighted by King George III in 1786 and accumulated a large fortune by the time of his death on August 3, 1792.

LESLIE HUTCHINSON

Bernard Forest de Belidor
1698-1761
French Engineer

Bernard Forest de Belidor was a French engineer and professor of artillery at La Fère military academy who wrote some of the best-known and comprehensive engineering manuals in both military and civil engineering. His work influenced the engineering practices in France and other European countries for nearly 100 years after their publication.

Belidor was born in Catalonia, Spain, where his father, Jean-Baptiste, was commanding a dragoon for the French army. Both his father and his mother died within five months of his birth, and his godfather's family raised him. As a child he had a natural aptitude for mathematics and inventing. This flair for the practical application of mathematics secured him an army post to survey the area from Paris to the English Channel. This project was completed in 1718 and covered in a book published by his colleagues Jacques Cassini (1677-1756) and Philippe de La Hire. After the publication of the book, Belidor's talents came to the attention of the Duc d'Orleans, who advised and influenced him to take an appointment as professor of mathematics at the new artillery school at La Fere.

As professor of artillery at La Fère, he became known as an author of textbooks and technical manuals. After serving a tour of active duty during the War of Austrian Succession, Belidor settled into the influential position of brigadier for the French Army. He was well known and regarded for his books, which were written with detailed instructions for the construction and firing of explosives and targeted at artillery cadets, engineers, and high-ranking officers in the French Army.

Two of his most important books, *La science des ingénieurs* (1729) and *Architecture hydraulique* (1737-53), addressed his ideas on the science of mechanics and fortifications for naval and military operations. These books proved to be invaluable resources to architects, builders, and engineers of the times. In addition to addressing common engineering problems encountered in shipbuilding, waterways, and road construction, his books also provided specifications for such things as foundations, arches, and ornamental fountains. Their many diagrams and detailed work plans for engineering and construction projects proved extremely popular.

Belidor's work unexpectedly ushered in a new science of mechanical engineering unknown at the time of the first editions of *Architecture hydraulique* and *La science des ingénieurs*. His practical ideas and solid engineering expertise influenced the proceeding two generations of scientists who became mechanical engineers after him. These unusual and innovative individuals became engineers of the "science of machines" and ushered in a new era that ultimately would make the Industrial Revolution possible.

Belidor was elected to the French Academy of Sciences in 1756. He died in Paris in 1761.

LESLIE HUTCHINSON

Joseph Bramah
1748-1814
English Inventor

Joseph Bramah was an English engineer and inventor during the early years of the Industrial Revolution. Considered one of the fathers of the tool industry, his inventions made a significant contribution to the emerging new industries of the early nineteenth century. His mechanical inventions included the burglar-proof lock and the hydraulic press.

Born Joe Brammer in Stainborough, Yorkshire, England, Bramah was incapacitated by a serious injury to his ankle when he was 16 years old and was unable to work with his family's agricultural business as planned. Instead, he became apprenticed to a carpenter. One of the jobs he was assigned was refitting a water closet (known today as a toilet). Frustrated with the poor mechanical design, he proposed a new one with better flushing capabilities and received a patent for it in 1778. At this time he also changed his name to Bramah, believing the new name would sell more of the redesigned water closets. Nearly 6,000 of these improved lavatories were sold, and the design was popular for nearly 100 years.

At the completion of his apprenticeship, Bramah set up his own carpentry and cabinet-making shop in London. During this time he became interested in the problem of designing a pick-proof lock. In 1784 he exhibited this lock in his shop window, offering a monetary reward for anyone capable of picking it. Despite many attempts, the Bramah lock was not opened for 67 years, until an American mechanic, Alfred Hobbs, successfully opened it after 51 hours of work. Bramah's lock design was successful because it was intricate; however, in order to manufacture it, he realized that he needed well-designed machinery capable of turning out precisely engineered machine tools. To assist him with establishing a machine shop, he hired a young blacksmith, Henry Maudslay (1771-1831), who became the head mechanic in Bramah's shop. This was the first machine shop established in London, and it was instrumental in designing and manufacturing many tools and parts necessary for the new manufacturing industries appearing during this period.

One of the first projects the two men took on was designing and producing a more efficient lathe known as the slide rest. Instead of a worker holding a cutting tool by hand against the metal to be cut, the iron fist of the slide rest held the tool firmly against the metal and moved the tool evenly along a carriage. This early prototype provided the groundwork for Bramah's success in the machine-tool industry. Without such machines it would not have been possible to manufacture the tools and parts that became the backbone of the industrial expansion of British manufacturing in the nineteenth century.

Another important Bramah invention was the hydraulic press. This innovation enabled machines to shape, extrude, or stamp materials under high pressure. Bramah was able to design and manufacture necessary tools and parts, such as nozzles, valves, pumps, and water turbines for

the new emerging industries. This opened up a tremendous new source of mechanical power to manufacturers during the Industrial Revolution.

Bramah was a talented and inventive individual. During the course of his life he secured nearly 18 patents. Some of his other inventions included a machine for numbering bank notes, a wood planing machine, a device to make quill tips for pens, a beer pump, a fire engine, and innovation in the manufacturing of paper. He was the first to suggest using a propeller to drive ships rather than using a paddle wheel.

Bramah died in 1814 in London and was well respected for his many innovations and inventions.

LESLIE HUTCHINSON

Edmund Cartwright
1743-1823
English Inventor

Edmund Cartwright was the inventor of a mechanical weaving loom that could be operated by horses, a waterwheel, or a steam engine. By 1791 this machine could be operated by an unskilled person (usually a child), who could weave three and half times the amount of material on a power loom in the time a skilled weaver using traditional methods could. This invention revolutionized the emerging textile industry in England.

The son of a wealthy landowner, Cartwright was born in Nottingham, England, in 1743. Because his family was rich, he was able to attend prestigious schools, eventually graduating from University College, Oxford. After completing his schooling he became rector of the church at Goadby Marwood in Leicestershire. Although he was employed in a profession he enjoyed, Cartwright had an interest in the inventions happening in the emerging textile industry.

In 1784 Cartwright visited a factory owned by Richard Arkwright (1732-1792). Arkwright had designed and built machinery capable of spinning yarn and thread, but the weaving was still done by independent household weavers. Cartwright, inspired by Arkwright's invention, began working on a powered weaving machine that would improve the speed and quantity of the actual weaving of cloth. Cartwright's first loom was clumsy and ineffective—primarily because he was unfamiliar with the construction and operation of the handlooms. In spite of the fact that the original machine worked poorly, he took out a patent. After employing a local black-

Edmund Cartwright. *(Library of Congress. Reproduced with permission.)*

smith and carpenter as consultants, he built two more prototypes and by 1790 had completed a loom capable of weaving wide widths of cloth with complicated patterns. All operations that could be done by the weaver's hands and feet could now be performed mechanically.

Cartwright had established a factory for his looms in 1786. Prior to this, the newly established factories manufactured only thread and yarn. Handloom weavers had been guaranteed a constant supply of thread, jobs, and high wages. The implementation of the power loom worried the local weavers who feared (correctly) that the machines would replace their services. During 1791 his factory was burned to the ground, possibly by a group of local unemployed weavers. (In 1799 a Manchester, England, company purchased 400 of Cartwright's power looms, but soon afterwards their factory was burnt down by another group of unhappy local weavers.)

The once-prosperous hand weavers eventually had great difficulty finding employment, and those who did were forced to accept far fewer wages than they had in the past. In 1807 nearly 130,000 individuals signed a petition in favor of a minimum wage in the factories. The local authorities replied by sending in the military; one weaver was killed and others seriously injured.

After the catastrophic fire and worker unrest, Cartwright found himself in financial difficulties, as many other business owners would not buy machinery from Cartwright. He attempted to offset these problems by inventing an ingenious wool-combing machine; however, local skilled workers opposed this also. Eventually he went bankrupt and was forced to sell his patents and factories. By the early 1800s, however, a large number of factory owners were using a modified version of Cartwright's power loom. Cartwright, having lost his patent, petitioned the House of Commons for compensation for others using his design. His claim was supported, and in 1809 he was awarded 10,000 pounds. He retired to a farm, where he applied his inventive abilities towards improving machinery used in agriculture. He invented a reaper and wrote pamphlets and essays on animal husbandry as well as using manure as fertilizer. He continued to develop new agricultural inventions until his death in 1823.

LESLIE HUTCHINSON

Henry Cort
1740-1800
English Inventor

Henry Cort developed two iron-making processes that, when combined, led to a fourfold increase in iron production throughout Britain within two decades. His patented grooved rollers and a puddling process for separating iron from carbon should have made him rich, but due to a disastrous business partnership, Cort was denied the fruits of his labors.

Cort was born in Lancaster, England, in 1740, and by the age of 25 was serving in the Royal Navy. In that capacity, he was responsible for improving ordnance made from wrought iron—primarily cannon balls. He soon became fascinated with the subject of ironworking, and using funds he had managed to accumulate, started his own iron business.

Cort's first major breakthrough was a process for creating iron bars with the use of grooved rollers, which replaced an older method of manually hammering the bars into shape. He patented the process in 1783. A year later he patented a second process, this one for separating out carbon by stirring molten pig iron in a reverberating furnace. As the iron decarburized, this had a purifying effect.

The two processes proved most useful when applied in tandem, and ironworking flourished throughout Britain as a result of Cort's two breakthroughs. Unfortunately, he was not able to enjoy his success. He had entered into a partnership with Samuel Jellicoe, whose father was a corrupt naval official involved in embezzlement of public funds. Because some of those funds wound up in the Cort-Jellicoe partnership, the British government punished Cort—though he had taken no part in the crime—by seizing the rights to his patents.

As a result of this misfortune, Cort's competitors were able to benefit from his efforts. Cort himself lived out his days on a small pension, and died at the age of 60 in Hampstead, London.

JUDSON KNIGHT

Samuel Crompton
1753-1827
English Inventor

Samuel Crompton. *(Library of Congress. Reproduced with permission.)*

The tale of Samuel Crompton is one of those unpleasant stories of an unschooled inventor who fails to protect the rights to his creation, and therefore dies in poverty. In Crompton's case, the invention was the spinning mule, which combined aspects of earlier devices to create a machine that would spin strong, smooth yarn efficiently. The spinning mule remained in use throughout the industrialized world for nearly two centuries, but because he never secured a proper patent for it, Crompton's benefits from his invention were modest.

Crompton was born in Hall-in-the-Wood, a village near Bolton, England, in 1753. His father died when he was five, and this forced his mother to take care of the family. Among her many tasks was her work with a spinning jenny, a then-recent invention of Englishman James Hargreaves (?-1778) for spinning yarn. But there was a problem with the jenny: yarn tended to break, and Crompton's mother—already pressed to her limits—would become increasingly angry with each breakage, which required her to restring the yarn. In his early adulthood, Crompton set out to build a spinning machine that would improve on the jenny.

Over the course of five years, young Crompton sunk all his money—he worked as a fiddler at a local theatre—and all his spare time into his invention. Finally in 1779, at the age of 27, he completed his spinning mule, so

named because like a mule, it was a hybrid. The mule borrowed from the spinning jenny the idea of a moving carriage that gently stretched the yarn, and from the water frame, developed by Richard Arkwright (1732-1792), it adapted the use of rollers to draw out the cotton fibers. However, unique to the spinning mule was a spindle carriage developed by Crompton. The carriage prevented tension from being applied to yarn before it had been completely spun, and thus the yarn produced by the spinning mule was strong, smooth and of the highest quality.

From the beginning, Crompton seemed not to understand how to make the most of his invention. At first his family simply sold their yarn at the local market, making a handy profit due to the fact that their product was better than any competitor's. But this only raised curiosity as to the invention behind it all, and eventually Crompton became the target of a campaign from all sides, as the textile industry pressured him to reveal his secret. He did not have the money to purchase a patent, so finally he agreed to sell the design to companies for a subscription of 70 pounds a year; but once the companies had the design for the spinning mule, they managed to wriggle out of the agreement.

For the first decade after he invented the machine, Crompton realized almost no proceeds at all from it. He was eventually able to shame enough companies into paying him a nominal fee that he collected 400 pounds—nothing compared to the untold fortunes the British textile industry had made on the machine, which had by then replaced both the spinning jenny and the water frame. In 1812 the House of Commons arranged a grant of 5,000 pounds for Crompton, but the inventor—nearly 60 years old at that point—was so deeply in debt that it mattered little. Twice he tried and failed to establish businesses. Finally, in 1824 a group of Crompton's friends anonymously donated an annuity of 63 pounds. This provided him with just enough to survive for the three years that remained before his death in Bolton in 1827.

JUDSON KNIGHT

Abraham Darby
1678?-1717
English Engineer

Though his name is far from a household word, Abraham Darby can be considered one of the individuals most responsible for the explosive growth of the iron industry—and thus for the success of the Industrial Revolution. His coke-burning blast furnace, introduced in 1709, spelled the end for charcoal as a means of heating iron, and replaced it with a much hotter, more efficient resource.

Born near Dudley, Worcestershire, England, in 1678, Darby went to work in the copper-smelting industry in the city of Bristol. At that time, the iron industry used charcoal for heating, but with charcoal it was difficult to produce a high, sustained level of heat. By contrast, the copper industry used coke, a derivative of coal produced by removing the sulfur and combustible impurities. Thus, coke never burst into flame, but rather delivered a very high, constant level of heat.

In 1708 Darby—then about 30 years old—opened the Bristol Iron Works Company near the village of Coalbrookdale in the upper Severn River Valley of western England, where supplies of coal and coke were plentiful. During the following year, he began producing small iron products such as cooking utensils with his coke-burning blast furnace. News of the Darby process was slow to catch on, but in time the inventor Thomas Newcomen (1663-1729) placed a large order for multi-ton cylinders to go in his steam-powered mine-pumping engines.

Darby died on March 8, 1717, before he was even 40 years old. Yet he had already managed to create a legacy, and his son Abraham II oversaw the production of the Newcomen cylinders for years to come. By 1758 the Darby foundry had produced 100 of the giant cylinders. The family business prospered even after the death of Abraham III in 1791, and in 1802 Richard Trevithick (1771-1833) commissioned the foundry to produce the first locomotive engine.

JUDSON KNIGHT

John Harrison
1693-1776
English Horologist

Perhaps the most famous clockmaker of all time, John Harrison solved the problem of reliably calculating a ship's longitude while at sea. By designing a highly accurate clock that allowed mariners to chart their position on Earth far more precisely, Harrison solved one of the most important scientific and technological problems of the eighteenth century.

John Harrison was the eldest of the five children of Henry and Elizabeth Harrison. He learned carpentry from his father and picked up excellent craftsmanship and a detailed knowledge of woods. John began building clocks quite early, enlisting his younger brother James to assist him. The first known Harrison clock dates from 1713, and the brothers built two more long case (grandfather) clocks in 1715 and 1717. How Harrison learned the complicated art of horology remains a mystery. At this time, clocks were very expensive and not particularly common, especially in isolated rural villages. No record exists of the young Harrison meeting with a clockmaker. Yet by 1720 he had built a clock for the tower of a local manor house. This clock elegantly employs Harrison's intimate knowledge of woods—the parts that would ordinarily need lubrication are crafted out of a tropical hardwood that naturally oozes its own oil.

In 1718 Harrison married Elizabeth Barrel. They had a son, also named John, in the summer of 1719. Elizabeth died in 1726, and John remarried within 6 months. He and his new wife (also named Elizabeth) spent the next 50 years together, producing two children: William in 1728, and a girl, again named Elizabeth, in 1732. While losing his first wife and remarrying, John had designed the two most accurate clocks in the world by 1727.

lyne (1732-1811), then both Astronomer Royal and a proponent of a competing astronomical solution, conducted these tests, which the watch, not surprisingly, failed. These new tests hounded Harrison until nearly the end of his life, ultimately denying him the full prize. A sympathetic King George III finally intervened on Harrison's behalf and Parliament made a special payment of more than 8,000 pounds in 1773.

John Harrison died three years later in 1776, a country carpenter who, despite active opposition from much of the scientific establishment, solved a key problem of eighteenth-century science and technology.

ROGER TURNER

Benjamin Huntsman
1704-1776
English Inventor

Benjamin Huntsman, an English inventor and businessman, is known for his 1740 invention of the crucible process for casting steel. This process resulted in steel with a more uniform composition and greater purity than any type of steel previously produced. Because of these properties, it made superior tools and cutlery. While this process is expensive and cumbersome compared to modern advances in steel making, it was the most significant development in steel production at that time.

Benjamin Huntsman was born in Lincolnshire, England in 1704. His parents emigrated from Germany just a few years prior to his birth. As a youngster, Benjamin showed a flair for mechanical crafts. He became extremely adept at repairing clocks and set up a clock-making business in Doncaster. He also practiced other general mechanical repairs and was a skilled surgeon. He became widely known for both these attributes.

In his craft, he set about improving the tools that he worked with. He became frustrated by the inferior quality of steel he obtained from Germany. This was especially true for the intricate springs and pendulums he needed to build for his clocks. In desperation, Huntsman began experimenting with his own steel production in order to upgrade the quality. After several unsuccessful attempts to fire up his furnace to a suitable temperature, he moved to Sheffield, England so that he would have a better fuel supply.

Huntsman was very secretive about his experiments. He finally perfected the crucible steel

John Harrison. *(Library of Congress. Reproduced with permission.)*

In 1714 the British Parliament established a 20,000-pound prize for any method or invention useful for accurately determining the longitude of a ship at sea. At the time, it seemed most likely that the solution to the "problem of longitude" would come from astronomy, and several scientists, including the Astronomer Royal, were named to the Board of Longitude as prize judges. By 1730 Harrison had traveled to London, met with a leading horologist, and obtained small grants to help him construct a seaworthy clock. Over the next 27 years, he built three large sea clocks, each weighing more than 60 pounds (27.2 kg).

Yet by the time the clock—called the H.3—was ready for testing in 1757, Harrison had imagined a radically new approach. Eschewing the hefty H.3, he designed a relatively tiny watch—just five inches (12.7 cm) across—endowed with an innovative new style of escapement. The H.4 proved as accurate as the H.3 and fit in a large pocket.

The H.4 cleanly passed the lengthy sea voyage stated as the test for the Longitude Prize. Surprised that a tinker's clock had passed the challenging test before an astronomical solution was discovered, the Board of Longitude then paid half of the 20,000-pound prize, but refused the other half until the watch passed a series of stringent, previously unstipulated, tests. Nevil Maske-

process and realized that it would be effective in making superior cutlery and tools. Unfortunately, there is no written record of the methods he adopted to overcome the difficulties he encountered during the development of the process. There must have been many problems because the process of manufacturing high quality cast-steel is an elaborate and delicate procedure. He had to discover the proper fuel, build an adequate furnace and make a crucible that could withstand more intense heat than anything previously known in metallurgy. Yet he was able to overcome all of these impediments and make excellent grade steel.

The crucible process technique that Huntsman developed for producing fine steel involved heating small pieces of carbon steel in a closed fireclay crucible placed in a coke fire. The temperature he was able to achieve with this method was high enough to melt steel, producing a homogeneous metal of uniform composition. Huntsman was extremely secretive about this process, but he did not patent it. In 1750, a competitor, Samuel Walker, discovered his method. While Walker profited considerably from the process, he was never able to usurp Huntsman's role as the producer of the best steel.

The demand for Huntsman's steel increased rapidly requiring him to move his factory to a new site in Attercliffe, England in 1770. This area later became the main location for the huge special-steel making industry of Sheffield. The Royal Society of London wanted to enroll Huntsman as a member in recognition of the merit of his invention of the crucible steel process. But he turned the honor down citing that it would conflict with his desire to work in seclusion and would also be against his principles as a member of the Society of Friends (Quakers).

Benjamin Huntsman died in 1776 at the age of 72. He was buried in the local churchyard in Attercliffe.

JAMES J. HOFFMANN

John Kay
1704-1764
English Inventor and Machinist

John Kay was an English machinist and inventor who patented the flying shuttle, a device that helped take an important step towards automatic weaving. When the flying shuttle was invented in 1733, it helped to increase the speed of the weaving operation and its use required

the development of more rapid spinning of yarns to feed the faster looms. Unfortunately, Kay did not reap the benefits of his invention and died in utter obscurity. In fact, so little is known about his later life, that historians cannot give a precise date of death. It is believed that he passed away in France sometime in 1764.

The early life of John Kay is also shrouded in obscurity. What little is known about his childhood places his birth in 1704 in Lancashire, England. He was the twelfth child of a farmer and woolen manufacturer. He was put in charge of his father's mill when he was still a youth. Kay made many improvements to the mill. This was especially true in the area of machinery and he even patented a machine for twisting and cording mohair and worsted wool to help the manufacturing process. However, his most important invention, the flying shuttle, would be patented in 1733.

Handloom weaving had been a slow, awkward process for centuries. The shuttle, which carried thread, was passed back and forth between cross threads by hand. Large fabrics required two weavers who were seated side-by-side to pass the shuttle back and forth between them. This made the manufacture of large fabrics especially expensive and cumbersome. Kay thought he could improve on this process. On May 26, 1733, he received a patent for a "New Engine or Machine for Opening and Dressing Wool" that incorporated his flying shuttle. The flying shuttle was mounted on wheels in a track and paddles were used to bat the shuttle from side to side when the weaver pulled a cord. Using the flying shuttle, one weaver could weave any width fabrics at a much greater speed than was previously known.

Some manufacturers used the flying shuttle, but they did not want to pay royalties on the invention. They in fact formed a union, which refused to pay any money to Kay. Because he was entitled to the money, Kay took these manufacturers to court, but the costs nearly ruined him and his business. Even worse, textile workers who feared that the new technology would make them obsolete ransacked Kay's house in 1753. These workers were concerned about their jobs and blamed Kay for their predicament. These events had a tremendous effect on Kay and he left England for France. There is no record of his life in France after the move and he was believed to have lived a meager existence before he died.

While Kay did not benefit from his invention, many people did. Besides the increased speed of weaving, the flying shuttle had other areas of impact as well. First, in order to meet the increased demands for yarn caused by the use of the flying shuttle, it was necessary to invent various spinning machines to increase yarn production. Second, it was a needed step in the development of power looms, where fabrics could be weaved without human manipulation. This greatly increased the speed of the weaving process and helped to keep costs extremely low.

JAMES J. HOFFMANN

Henry Maudslay
1771-1831
English Inventor and Engineer

Henry Maudslay was an English engineer and inventor who invented a variety of machines throughout his lifetime. Maudslay has been referred to as the father of the machine tool industry. The best known of these was a mechanical lathe that could manufacture threaded screws. Up until that time, threaded screws were built by skilled craftsmen, but did not have the exact consistency needed for some applications. Maudslay designed a lathe that could repeatedly cut the same threads on a screw resulting in identical sets. Maudslay recognized the importance of precision tools that produced identical parts and strived to manufacture them. He was considered a mechanical genius in his day and he shared his knowledge with others. Several other outstanding British engineers such as James Nasmyth (1808-1890) and Joseph Whitworth (1803-1887), learned their profession from Maudslay.

Henry Maudslay was born in Woolwich, England in 1771. His father was a workman at the Woolwich Arsenal and he encouraged his son to work hard and learn. At age twelve, Maudslay was working in the Arsenal filling cartridges. He was soon promoted to apprentice in the metal working shop and proved to be very useful. By the age of eighteen, he was considered to be one of the most promising young machinists, so he apprenticed with lock pioneer Joseph Bramah.

Bramah manufactured intricate locks that gave an unprecedented amount of security from a series of machines developed by Maudslay. These were among the first tools to be designed for mass production. The Bramah key was a small metal tube with slots cut into the end. Only when the key depressed the exact number of slides at the exact distance would the lock open. Bramah was so confident that the lock could not be picked, he offered a reward to the first person that could open it. It remained unopened for over 50 years until American locksmith A.C. Hobbs opened it. Despite their successes, Maudslay and Bramah quarreled over pay and Maudslay left to start his own company.

Maudslay began developing important machines for the Industrial Revolution. It was at this time when he developed his machine lathe for screw threads. He also designed the first micrometer, a device used to measure very small increments. He needed this device because he knew it was critically important that tools be standardized with interchangeable parts.

From 1801-1808 Maudslay went to work for Marc Brunel (1769-1849), a ship pulley manufacturer. One of his accomplishments was a series of machines that could manufacture wooden pulley blocks operated by ten unskilled laborers. Previously, it had taken 110 skilled craftsmen to perform that job. In 1807, he patented a table engine that served to be a useful, compact power source. He also manufactured steam engines and built England's first marine engine.

Maudslay made many other contributions to the Industrial Revolution. His list of accomplishments is long and includes methods for the purification of water and the invention of a punching machine for metal boilerplates. He even developed rounded corners on his designs that went against tradition, but proved to be stronger than the original designs. In addition, his legacy was passed on through many of the important inventors who had studied under him. Maudslay died in 1831 after traveling to France. He was buried in a cast iron tomb in the town of his birth.

JAMES J. HOFFMANN

Jacques Etienne de Montgolfier
1745-1799
French Aeronautic Inventor

Joseph Michel de Montgolfier
1740-1810
French Aeronautic Inventor

The Montgolfier brothers were eighteenth-century French businessmen who enjoyed conducting scientific experiments in their spare time. Becoming interested in the age-old dream of flying, they discovered an important aeronautical principle involving balloons and in 1783 were responsible for the first human flight in history. Since they worked closely together on this project and since neither accomplished anything else particularly noteworthy, their biographies are always combined.

Joseph and Étienne were the twelfth and fifteenth children of Pierre Montgolfier (1700-1793), a wealthy paper manufacturer whose factories were located near Lyon in southern France. Joseph was largely self-educated. He tried the paper manufacturing business but had little success. He was shy and extraordinarily absent-minded, two qualities that hampered his career. He was, however, very interested in technological advances. After the French Revolution (1787-1799) he retired from business and moved to Paris, where he worked at the Conservatoire des Arts et Métiers and helped establish the Société d'Encouragement pour l'Industrie Nationale in 1801.

Étienne was sent to Paris to study mathematics and became an architect. In 1772 his father decided that his older brothers were not capable of running the family business; Etienne was asked to give up architecture and become a paper manufacturer. He was a very successful businessman, especially after he and Joseph became famous. In 1784 King Louis XVI designated the firm as a manufacture royale. Paper manufacturing was Étienne's main pursuit for the rest of his life.

Joseph was the first to tinker with the problem of human flight. By 1782 he had stimulated Étienne's interest as well and the two men began conducting experiments in their leisure time. They initially tried to use hydrogen, discovered in 1766, to raise balloons, since it was 14 times lighter than air. But hydrogen was very combustible, expensive to make, and seeped through cloth as if it were a sieve. The brothers' experiments with silk and paper models led them to conclude that hydrogen could not be used to achieve flight.

Joseph then discovered that heated air had lifting power. Étienne used his mathematical background to work out the size and shape of a balloon to hold this heated air. They sealed the cloth with strips of paper, a natural idea given their family business. They erroneously believed that it was smoke that caused the balloon to rise. After numerous experiments, they settled on a mixture of damp straw and wool as the fuel providing smoke with the greatest lifting power. In one of history's fortunate accidents, these two scientific amateurs had inadvertently solved the problem of flight, because heated air becomes sufficiently rarefied (less dense) to lift a balloon.

On June 4, 1783, the Montgolfier brothers publicly demonstrated their discovery when they launched a 35 ft (10.67 m) balloon from a small town near Lyon. It ascended about 6,000 ft (1,829) and floated over a mile (1.61 km) away. Although it was unmanned, this was a giant step toward human flight. Joseph went to Lyon and Étienne to Paris to publicize their success. Étienne's mission was more significant because he succeeded in gaining the attention of Louis XVI, who loved dabbling with mechanical devices. In September, Étienne launched another montgolfière (the name given to hot-air balloons) for the royal court at Versailles. It carried several animals, all of which survived the ascent. Human flight was the next logical step.

Étienne, however, had competition. A popular Parisian science lecturer, Jacques Charles (1746-1823), had succeeded in constructing a hydrogen balloon, which he had flown unmanned. He and Étienne were now in a race to see who would get the first human aloft. Étienne constructed his balloon faster and won the race. On November 21, 1783, a montgolfière carrying two men floated over Paris for about 20 minutes. The two passengers were Jean François Pilâtre de Rozier (1757-1785) and François Laurent, marquis d'Arlandes (1742-1809). They were the first men in history to fly. Ironically, two years later de Rozier was the first human killed in an aerial accident.

Although the family received government grants and was ennobled in December 1783 (hence the "de" in their name), the brothers soon lost interest in aeronautics. Joseph made only one ascent, Étienne none at all (both men were middle-aged). Étienne returned to his paper manufacturing, while Joseph turned his attention to other inventions, the most important of which was a hydraulic device for raising the level of water. They had proven, however, that human flight was possible. Their success stimulated the development of the hydrogen balloon, which was far superior to their montgolfières in terms of range (both distance and height) and which soon supplanted their hot-air balloons.

ROBERT HENDRICK

William Murdock
1754-1839
Scottish Inventor

Scottish inventor William Murdock was a pioneer in the use of coal gas for lighting and made significant advances in the use of steam power. He was one of the first people in the world to use coal gas for lighting his home, and his invention of this technology was a very important step in helping people to light their homes, workplaces, and streets after dark.

William Murdock was born in Scotland in 1754. Little is known of his childhood, but it is known that, at age 23, he joined the engineering firm owned by James Watt (1736-1819) and Matthew Boulton (1728-1809). Two years later, in 1779, he helped supervise the construction of one of Watt's first steam engines, and by 1784 he was experimenting with an oscillating steam engine, the first of its kind built.

Murdock's experiments continued, and he built a steam-powered carriage in 1786, although this was ultimately unsuccessful. It did, however, help introduce ideas later used to develop the steam locomotive.

Murdock made other devices that eventually found their way into industry and transportation. Chief among them was the "sun and planet" gear, which allows a reciprocating steam engine (that is, one that simply moves a plunger back and forth) to turn a wheel in a circular motion. This was widely used for decades until a good rotary engine was developed to take its place. However, it still finds many uses and may be one of Murdock's most important inventions. Murdock also worked to find uses for compressed air and devised a gun powered by steam in the early 1800s.

In spite of his developments noted above, Murdock was best known for his discovery of a process to distill coal, forming "coal gas" that produced a fine white flame when burned. Murdock did much of the experimental work in this area on his own, carrying out many experiments in his home. Finally, in 1792 he succeeded in lighting his own home with coal gas. He was soon also able to light his offices.

Returning to Birmingham, where he had first worked with Boulton and Watt, Murdock continued his work with coal gas, primarily looking for more-effective and cost-effective ways to generate, store, and purify it. Although it took some time, Murdock's coal gas was eventually used for public and private lighting throughout Europe, America, and elsewhere in the world.

Murdock retired in 1830 at the age of 76. He had spent the majority of his life helping to design the machines and processes that helped

William Murdock. *(Library of Congress. Reproduced with permission.)*

power the Industrial Revolution. Thanks in part to his work, the stage was set for development of the locomotive, which was the most significant advance in land transportation since the time of the Romans, or earlier. And thanks to his pioneering work with coal gas, people began the process, still continuing, of claiming the night for more than simply sleeping. Before the advent of artificial lighting, little could be done by most people after dark because of the relative lack of lighting and the poor light produced by fires. Murdock helped to change that, and until the introduction of electric lights in 1879, a century later, no better technology was found.

William Murdock died in Birmingham, England, in 1839 at the age of 85.

P. ANDREW KARAM

Thomas Newcomen
1663-1729
English Inventor

Thomas Newcomen is often acclaimed as the inventor of the steam engine. An ironmonger by training, he converted Thomas Savery's primitive steam pump into a true, if inefficient, source of motive power. Originally developed to remove water from coal mines, the steam engine, as further refined by James Watt, provided the first reliable source of mechanical energy other than muscle, wind, or water power, and thus provided the key technical stimulus for revolutions in both transportation and industrial production.

Thomas Newcomen was born into a family of religious dissenters in Dartmouth, Devonshire, England. Nothing is known with certainty about his education or training, but it appears he entered into business in 1685 as an ironmonger (blacksmith and dealer in metals) with a partner, John Calley, a plumber and a fellow member of the Baptist sect. How Newcomen came to be interested in developing a steam engine, and how much contact Newcomen had with Thomas Savery (1650-1715), also a native of the Devonshire region and the inventor of the steam pump, is also a matter of speculation. Such contact is, however, fairly likely, as Newcomen visited the tin mines in the area and had some idea of their operation and problems.

Savery's steam pump had been developed for the rapidly growing coal mining industry, which had come into being to provide fuel in place of wood for heating, the forests of England having been alarmingly depleted. Coal was available in abundance, but mineshafts were likely to fill with water, which needed to be removed. Savery's invention, introduced around 1700, involved filling a pipeline to the water with steam, then cooling the source of the steam until it condensed to form liquid water, creating a vacuum that would pull the water upwards. Then, using additional steam applied from below, the water was pushed to the surface. Savery's pump, which required steam under pressure, was dangerous to operate as leaks and ruptures often occurred.

Newcomen introduced a cylinder and piston, which could be filled with steam pushing the piston one way and then cooled so that the steam condensed to form liquid water, leaving a near vacuum so that air pressure would push the piston back. Steam pressures only slightly greater than one atmosphere were then required, and atmospheric pressure did most of the work. At first, cooling was accomplished by passing water around the cylinder, but in 1704 or 1705, by a happy accident, cool water leaked into the cylinder, producing immediate condensation for a much faster power cycle. The first successful engine, with cool water injection into the cylinder at the appropriate times, was demonstrated in 1712. It was inefficient in that much of the heat energy was wasted heating up the cylinder after each cooling. The Scottish engineer James

Watt (1736-1819) would eliminate this problem by adding a separate condenser, which the steam entered after doing its work on the piston, so that the cylinder could remain hot.

It is now generally agreed that Newcomen was a skilled artisan who did not have a sound understanding of scientific principles but who worked by trial and error, combining bits of technology from earlier inventions. Nonetheless, it was the Newcomen engine, as improved by Watt, that led to the era of steamships and railroads and to factories built around powerful machinery. The steam engine would also serve as the basis for a new branch of physical science—thermodynamics—the study of the interconversion of heat and mechanical work, from which the modern laws of energy conservation and entropy increase would emerge in the nineteenth century.

DONALD R. FRANCESCHETTI

Johann Nepomunk Franz Aloys Senefelder
1771-1834
German Inventor

Though he remains an obscure figure, even among the annals of inventors, Aloys Senefelder had an enormous impact on the creation of the modern world. By inventing lithography in 1796, he made it possible for printers to mass-produce commercial images, thus influencing the spread not only of commerce, but of communication and literacy. Senefelder, who continued to improve on his invention, developed color lithography in 1826.

Senefelder was born on November 6, 1771, in Prague, now capital of the Czech Republic but then a provincial capital in the Hapsburg-controlled Austrian Empire. His family was German, and Senefelder's father worked as an actor in Prague's Theatre Royale. While Senefelder was a student at the University of Ingolstadt, his father died, and the young man tried to earn a living as an actor and writer. Failing in these endeavors, he turned to printing.

After learning the trade while working in a small printing office, Senefelder bought a press of his own and set up shop. Still striving to find a way to make money through his creative efforts, he initially intended the press as a means of publishing his plays. Since he could not afford to pay someone to engrave his printing plates for him, Senefelder decided to do the en-

dertake the experimentation that led to the development of color lithography in 1826. Printers were soon using metal plates, which were more efficient and more easily etched, but the name "lithography"—with its root word reflecting the use of stone in early plates—remained. Senefelder died on February 26, 1834, in Munich.

JUDSON KNIGHT

John Smeaton
1724-1792
Scottish-English Inventor

John Smeaton's impact on civil engineering in eighteenth-century England was so significant that the Institute of Civil Engineers—of which Smeaton had been a founding member in 1771—changed its name to the Smeatonian Society after his death in 1792. The results of his work, including mills, bridges, and harbors, were to be found throughout the landscape of England during the early industrial era.

Though Smeaton's family was Scottish, they lived in England, where he was born in 1724. His father, an attorney, expected him to follow in the legal profession, but Smeaton had no interest in law. His fascination lay entirely with machines, and he chose to pursue a career in civil engineering.

In 1753 Smeaton was elected to membership in the Royal Society, and two years later was selected to design a new lighthouse near Plymouth, England. For the lighthouse, completed in 1759, Smeaton developed a new variety of limestone-and-clay cement that would set under water.

Also in 1759, Smeaton published a paper on water mills, a principal source of power during the early Industrial Revolution. His paper dealt with the difference between undershot mills, or mills that derived their power from water flowing beneath them, and overshot mills. The latter is more widely known, thanks in part to Smeaton, who showed that an overshot mill is much more efficient than an undershot mill: due to gravity, the power produced is much greater when water flows over a wheel than when it flows under it. These findings won Smeaton the Copley Award in 1759, and he went on to build 43 mills throughout England.

From 1757 on, Smeaton designed a number of canal and bridge projects, and in the 1760s created a water-pressure engine for pumping water. The latter would be replaced a few years later, when James Watt (1736-1819) invented his

Aloys Senefelder. *(Corbis Corporation. Reproduced with permission.)*

graving himself, and went to work on a set of copper plates.

His efforts with copper met with little success, and Senefelder was at a loss for what to do when, almost by accident, he discovered the secret that would lead to his later success. One day he was writing a laundry list with a grease pencil on a piece of limestone, when suddenly he got the idea of etching away the surface around his markings, so that the lettering itself would be left in relief.

This incident occurred in 1796, and over the next two years, Senefelder conducted a number of experiments before developing a satisfactory means for flat-surface printing. In 1809 he became director of the royal printing office in Munich, capital of Bavaria. Nine years later, in 1818, he published *Vollständiges Lehrbuch der Steindruckerey,* later translated as *A Complete Course in Lithography,* wherein he explained the process he had created.

In later years a music publisher named Johann Anton André helped Senefelder establish a school in the town of Offenbach, where he taught the lithographic process. The printer also lived well from a pension granted by the king of Bavaria, and this gave him the freedom to un-

John Smeaton. *(Library of Congress. Reproduced with permission.)*

condensing steam engine. Smeaton also invented a tidal pump, installed at London Bridge in 1767, for supplying water to subscribers, and in 1769 invented a metal boring machine. In building Ramsgate harbor in 1774, Smeaton made several improvements to the existing design of the diving bell, adding an air pump. He spent his remaining years experimenting with steam engines and making improvements to them. He died in 1792 at Austhorpe, Leeds, England.

JUDSON KNIGHT

Jethro Tull
1674-1741
English Inventor

Jethro Tull invented the seed drill, which altered how land was used for agriculture. He also devised ways to aerate fields, eliminating the need for constant fertilization of crops. His writings were dispersed throughout Europe. Tull's innovative techniques and tools influenced agriculturists to adopt his methods and to improve implements. Some historians suggest that Tull was the catalyst for revolutionary changes in British agriculture, which resulted in more efficient land management and higher crop yields.

Tull was born in Basildon, England. Sources do not cite an exact birth date, but records indi-

cate that Tull was baptized on March 30, 1674. He grew up in a rural environment before enrolling at St. John's College at Oxford University. He graduated in 1691. Two years later, Tull began legal studies and was admitted to the bar in 1699. Tull never intended to practice law, envisioning his educational experiences as preparation for a political career. Because of ill health, he abandoned his political ambitions and became a farmer instead. Tull married Susanna Smith in 1699 and they lived on his father's land at Howberry.

Frustrated by his farm workers, who resisted his ideas about planting, Tull decided to create a machine that could operate more productively than human laborers. He adapted the "groove, tongue, and spring in the soundboard of the organ" and parts of other instruments "as foreign to the field as the organ is" and by 1701 had designed and built a seed drill. The name was derived from farmers' jargon; "drilling" referred to sowing beans and peas by hand in furrows.

Tull's device enabled him to plant seeds systematically in rows. As a result, the available soil in fields could be utilized more efficiently to plant more crops than were grown by using standard planting methods. Tull's tool encouraged increased development of fertile land. The space between rows could be cultivated to enhance yields. Sowing seeds with his drill minimized the fallowing requirements for preparing fields for production. The rotary mechanism Tull had developed to build his seed drill became the basis for subsequent sowing tools.

In 1709 Tull relocated to a farm he called "Prosperous." During the spring of 1711, he began traveling throughout Europe in an effort to improve his health. Tull studied Continental agricultural practices, which he experimented with when he returned to England in 1714. He had been especially impressed with French vineyards, where farmers plowed between rows instead of fertilizing the grapevines with manure. Duplicating French techniques on his English farm, Tull successfully grew turnips and potatoes without manure. He cultivated wheat for 13 years consecutively in nonmanured fields.

Tull advocated that farmers should crumble sod in order for air and water to reach plant roots. He invented a horse hoe specifically to achieve such soil pulverization. Visitors who came to Prosperous discussed agricultural methods with Tull and encouraged him to write an account of his work. Tull distributed *Horse-hoing Husbandry* in 1731 and a revised edition two years later. Members of the Private Society of Husbandmen and Planters, however, criticized Tull, saying that his pulverization ideas were without merit. Tull responded to these attacks with notes that were included in another edition, issued in 1743.

Tull also had to deal with the refusal of his own agricultural employees to learn how to use his tools and methods correctly. He died at his farm on February 21, 1741. He willed his property to his four daughters and sister-in-law, leaving his wastrel son only one shilling. His husbandry ideas inspired European farmers to experiment with "Tullian" methods. The noted French agriculturist Duhamel du Monceau annotated a translation of his work and followers of Tull's practices included such figures as Voltaire.

ELIZABETH D. SCHAFER

James Watt. *(Library of Congress. Reproduced with permission.)*

James Watt
1736-1819
Scottish Inventor and Scientific Instrument Maker

Although James Watt did not invent the steam engine, he made improvements to already existing engines that greatly increased their power. Watt also developed several other important inventions: a rotary engine to drive machinery; a double-action engine; a steam indicator to measure pressure in an engine; and a centrifugal governor to automatically regulate engine speed. He developed the concept of horsepower to describe the operating strength of engines. Watt also made important surveys of several canal routes and invented a telescope attachment to measure distance. In 1882 the British Association named a unit of electrical power measurement after Watt.

James Watt was born on January 19, 1736, in the village of Greenock in Renfrewshire, Scotland. Watt's grandfather, Thomas, had been a teacher of mathematics, surveying, and navigation; and his father James, the treasurer and magistrate for Greenock, ran a successful business building ships, houses, and mathematical instruments. Early in his life Watt demonstrated both manual dexterity and an aptitude for mathematics, and he spent much time in his father's workshop building models of such things as cranes and barrel organs. At the age of about 18 he was sent to live in the city of Glasgow with his mother's relatives, one of whom taught at the university. Watt soon moved to London to apprentice himself to a mathematical instrument maker. Watt was a sickly young man who suffered severe headache attacks all his life, and London did not suit him. By the age of 21 he had returned to Glasgow.

Through the connections of Andrew Anderson, an old school friend, Watt received an appointment as mathematical instrument maker for the University of Glasgow in 1757 and was allowed to establish a workshop on its property. He met important professors at the university, including chemist Joseph Black (1728-1799), whose studies of the heat properties of steam led him to develop the concept of latent heat. Black and Watt remained friends and corresponded until Black's death in 1799.

In 1710 Thomas Newcomen (1663-1729) and John Calley had developed a "fire" or steam engine, examples of which were being used by the mid-1700s to pump water from mines. The university had a model of one of these Newcomen engines that needed repair, and Watt undertook the task. During the job Watt noticed that the design of this engine caused large amounts of steam to be wasted and began thinking about an engine with a separate condenser that would solve the problem. At about this time Watt met John Roebuck, founder of the Carron Works factories, who urged the young inventor to build an engine that incorporated his ideas. With the help of money from Black, Watt built a small engine to test his ideas and then entered into a partnership with Roebuck in 1768. The following year Watt obtained his most famous patent, "A New Invented Method of Lessening the Consumption of Steam and Fuel in Fire Engines."

Watt had begun his surveys for canal routes all over Scotland in 1766, and this work kept him from devoting much time to his engine. In 1772 Roebuck's went bankrupt, and Matthew Boulton (1728-1809), who had inherited the Soho Works silver factories in Birmingham from his father, assumed a share of Watt's patent. Two years later Watt left Scotland and moved to Birmingham. The partnership seemed perfect; Watt needed Boulton's financial help, and Boulton's factories needed power. The effort nearly bankrupted Boulton, but by the early 1780s Watt's engine was used in copper and tin mines in Cornwall.

During that decade, often at Boulton's suggestion, Watt continued improving his engine. In 1781 he developed a "sun-and-planet" or rotary gear that replaced the back-and-forth motion of his engine with a circular one. The following year Watt patented a double-action engine in which the pistons both pushed and pulled. Since this engine required a new method to connect the pistons, Watt created a parallel motion apparatus in which connected rods drove the pistons in perpendicular motion. In 1788 he added a centrifugal governor that controlled the engine speed, and two years later Watt invented an automatic pressure gauge. Thus, by 1790 Watt's series of improvements and inventions had moved far beyond the scope of the Newcomen engine and had created a device that powered the Industrial Revolution. In 1800 over 500 Watt engines were installed in mines and factories all over Great Britain. Watt's patents also meant that he was a very wealthy man.

Watt's interests extended beyond business and industry. In the early 1790s Watt's teenage daughter Jesse suffered from tuberculosis. His friend Erasmus Darwin (1731-1802), a physician and grandfather of Charles Darwin, attempted treatment that did not help the girl; Darwin then suggested another physician, Thomas Beddoes (1760-1808). Beddoes had a practice in the seaport of Bristol and was attempting to use inhalation of various gases to treat tuberculosis and other diseases. Despite the efforts of Beddoes, Jesse died, but he and Watt developed a partnership to further investigate the gases. The inventor designed a breathing apparatus that was manufactured briefly by Boulton's firm and distributed to various physicians around Britain who were willing to try Beddoes's treatments. In 1798 Beddoes opened the Pneumatic Medical Institute in Bristol; this facility included a clinic, classrooms, and research laboratory. In 1799 the first human experiments with nitrous oxide, or "laughing gas," were completed here; Watt, his second wife, and two of his sons all participated. Unfortunately, the experiments at Bristol were not effective in treating diseases and ended the following year.

In 1763 Watt married his cousin Margaret Miller of Glasgow; she bore him several children but died in childbirth in 1773. Two years later Watt married Ann Macgregor; she had two children by Watt and outlived him by 13 years. By 1800 Watt was essentially retired. He received numerous honors during his life, including selection as a fellow of the Royal Society in 1785 and an honorary degree from the University of Glasgow in 1806. He spent much time in his final years travelling the European continent. Watt died on August 25, 1819, at his residence Heathfield Hall outside Birmingham. Two years earlier his son James Watt, Jr., had purchased the ship Caledonia and replaced her engines, making the vessel the first steamship to leave an English port.

A.J. WRIGHT

Eli Whitney
1765-1825
American Inventor

Whitney is credited with inventing the cotton gin. This mechanical device efficiently removed seeds from short-staple cotton bolls, resulting in that type of cotton being grown in more areas of the United States. Although the cotton gin was considered a labor-saving tool, it ironically caused the expansion of slavery in America. While the cotton gin relieved laborers from removing seeds from cotton fibers by hand, it created a demand for more raw cotton, thus increasing the need for workers to pluck the bolls from cotton fields.

Born on December 8, 1765, in Westboro, Massachusetts, Whitney was the son of Eli and Elizabeth Whitney. He financed his education at Yale by making nails and teaching. Graduating in 1792, Whitney accepted a position as a tutor on a Southern plantation. While visiting Catherine Greene (1755-1814), the widow of General Nathanael Greene, at her home Mulberry Grove,

Eli Whitney. *(Library of Congress. Reproduced with permission.)*

near Savannah, Georgia, he impressed her with his mechanical abilities and was invited to stay. (Scholars debate whether Whitney independently created the cotton gin or if Greene suggested the design. Other historians stress that slaves on Greene's plantation originally had the idea for the cotton gin, which Whitney patented as his own invention.)

Several of Greene's friends complained that short-staple cotton was unprofitable to grow because of the labor required to remove seeds. Greene suggested that agriculturists consult Whitney because of his technological talents. Whitney devoted several months to building a cotton gin on Greene's plantation. Nearby residents heard rumors about Whitney's work and thieves stole the prototype from his workshop. Agriculturists copied his design and built cotton gins prior to Whitney's receiving a patent in 1794. Whitney's cotton gin consisted of a long box with a revolving cylinder and saws that separated lint from seed. Underneath the saws, a brush (which some scholars say was Greene's idea) removed lint. The cotton gin enabled one person to clean fifty pounds of cotton per day instead of one pound processed by hand per day.

Whitney and Phineas Miller, Greene's plantation manager, established a partnership in 1793 to manufacture and sell the cotton gin, but they were plagued by patent-infringement trou-

bles. Whitney initiated at least 60 lawsuits against imitators and his patent was validated in 1807. He endured Congress's refusal to renew his patent in 1812 in addition to a factory fire and Miller's death. As use of the cotton gin increased, cotton cultivation became profitable and cotton exports increased from 189,500 pounds in 1791 to 60 million pounds in 1805. Southerners relied on even more slaves to plant and harvest more cotton. Approximately 657,000 slaves lived in the southern states in 1790 but, by 1810, the number had increased to 1.3 million. Many planters became wealthy from what was referred to as "white gold" and "King Cotton" came to dominate the Southern agricultural economy. More cotton was sold at lower prices, resulting in the textile industry's thriving in the South. Globally, Southern cotton became the favored material for fabric.

Whitney also applied his ingenuity to mass production. A 1798 government contract to manufacture 10,000 muskets resulted in Whitney's building an armory at Whitneyville, near New Haven, Connecticut. He proved that workers who were not skilled gunsmiths could use machine tools to create interchangeable, standardized parts. Whitney's factory was one of the first to demonstrate mass-production methods and division-of-labor strategies successfully. Unlike his cotton-gin experiences, however, Whitney profited greatly from this venture. Whitney's armory also inspired the construction of similar federal facilities. Dying at New Haven on January 8, 1825, Whitney was survived by his wife and four children. His cotton-gin design has been incorporated into more sophisticated, modern mechanical procedures to process as much as 15 tons of cotton per hour.

ELIZABETH D. SCHAFER

John Wilkinson
1728-1808
English Inventor

John Wilkinson is best remembered for developing the machine tools and techniques that helped make it possible to power the Industrial Revolution. By developing precision metalworking tools, Wilkinson was able to bore accurate, consistent cylinders for the steam engines under development by James Watt (1736-1819). These same techniques, it turned out, were also very useful in constructing cannons, pipes, and other similar devices.

Wilkinson was born in Clifton, England, in 1728, moving to Staffordshire at the age of 20. There, he helped build one of the first iron furnaces, making cast iron. Although cast iron is much stronger than native iron and is less brittle than wrought iron, it is difficult to work with. However, its superior material properties made the added effort worthwhile, while Wilkinson's furnace design helped make it affordable for a much larger group of people than had previously been the case.

Following this success, Wilkinson went to work at his father's factory in Wales. There, in 1775 he constructed his first cylinder boring machine, a device that could bore engine cylinders and cannon barrels more accurately than any previous such machine. In fact, existing boring machines were relatively crude, adapted as they were from wood-working tools and often used manually. By using better tools and mechanizing the process as much as was possible at the time, Wilkinson was able to produce borings that were remarkably precise, round, and consistent from one engine (or cannon) to the next.

Although this is a simple feat by today's standards, it was a phenomenal breakthrough in the eighteenth century. One of the major stumbling blocks in developing steam power was the fact that the cylinders leaked steam because they were not precisely round. The leaking steam, in turn, robbed the machines of efficiency, causing them to operate poorly. By making the borings precisely round and all the same diameter, Wilkinson's machine made it possible to coax more work out of the steam engines for the same cost in fuel. This, in turn, meant that more work could be gained for the same cost, a major improvement over previous practices.

As noted above, this same technology was also put to immediate use in boring cannon barrels. Like steam engines, cannons use the power of an expanding gas as a motive force; in this case, moving a cannonball out of the barrel at a high speed. Also like steam engines, a barrel that was not perfectly round threatened to cause the ball to bind, or resulted in gas leaking around it and robbing the shot of power. With more precise cannon barrels, balls (later shells) could be made to fit with closer tolerances in the knowledge that they would fit all cannon barrels, not bind while loading or being shot, and would get the most efficiency out of the load of powder. With more predictable characteristics from each cannon, artillery tables could now be drawn up that showed the distance a particular weight would be thrown for a given load of powder and elevation angle on the gun.

Other Wilkinson inventions included the world's first iron-hulled barge, used to transport the cannon barrels he manufactured, and the mating of Watt's steam engine to his boring machine to further automate the process of boring these pieces of equipment. At the same time, Wilkinson also worked with the French, teaching them to bore cannon barrels from solid iron and casting many miles of pipe and ironwork for the Paris waterworks.

Wilkinson died in 1808 at the age of 80. In accordance with his wishes, and very fittingly, he was buried in a cast-iron coffin he had designed some years earlier.

P. ANDREW KARAM

Biographical Mentions

Aime Argand
1750-1803

Swiss scientist who invented the first oil lamp. The Argand burner provided about 10 times the light of similar size lamps, and represented the first change to basic lamp design in thousands of years. Argand's device consisted of a cylindrical wick housed between two concentric metal tubes. The inner tube allowed air to rise into the center and support combustion, and a glass chimney increased airflow, allowing the oil to burn cleaner and more completely. Argand's concept was later adapted to gas burners, and the steady smokeless flame became the principle lighthouse illuminant for more than 100 years.

John Baskerville
1706-1775

British printer and inventor who, after beginning his career as a calligrapher and gravestone engraver, gained lasting recognition for developing a typeface in 1754 that is still used today. In the 1740s, John Baskerville became prosperous in the japanning trade before founding a printing press in 1750. In 1757, he published his first work, an edition of Virgil. Baskerville became a printer for Cambridge University in 1758, for whom he published his masterpiece, a folio Bible, which was printed in 1763 using his own typeface, ink and paper.

Thomas Bell

English inventor who invented a better way of applying prints to fabric. Using a cylindrical roller, Bell showed it was possible to print designs on fabric rapidly, consistently, and with little labor. Not only did this make it possible for everyone to enjoy more colorful and attractive clothing, but it also presaged the use of similar technologies in the print industry (e.g., ditto machines, rotary printing presses, etc.).

Johann F. Böttger
1682-1719

German alchemist and the ceramist of Augustus the Strong, King of Poland. As the first European to discover the secret of Chinese porcelain, Böttger is credited with the invention of European hard white porcelain, which has a higher kaolin content than the Asian type. His discovery led to the establishment of the first royal porcelain factory in Europe at Meissen near Dresden, Germany. Böttger died from alcoholism.

Thomas Boulsover
1706-1788

English cutler who invented fused plating, or the "old Sheffield plate," by which two metals could be fused to behave as a single metal. Boulsover made this discovery in 1743 while working with silver and copper. His finding paved the way for economical, mass commercial production of innumerable objects, from buttons to eating utensils. Coating a metal or other material with a hard, nonporous metallic surface improved durability. But previous practices were time-consuming, since plating a metal with another required craftsmen to fabricate a finished object then solder a thin sheet of plating on to it. Boulsover later invented a method for rolling saw-blade steel, previously made only by hand hammering.

Matthew Boulton
1728-1809

English engineer and industrialist whose financial backing and ability to raise funds from others enabled James Watt to perfect his steam engine. In 1759 when his father died, Boulton inherited his father's silver business and over the next several years expanded it into the great Soho works. By the early 1770s he had partnered with Watt, who needed financial help. Boulton needed greater power at his factory than nearby streams could provide. The effort nearly bankrupted Boulton, but by the late 1780s Watt's engine had reached profitable production. Ten years later Boulton's factory began minting coins for the British government and various colonies.

James Brindley
1716-1772

English engineer who constructed the first economically successful English canal. In 1759 the Duke of Bridgewater hired Brindley to build a canal to transport coal from the duke's mines to a textile manufacturing plant 10 miles (16 km) away. To accomplish this, Brindley designed a subterranean channel, which extended from the barge basin at the head of the canal into the mines, and the Barton Aqueduct, which carried the canal over the River Irwell. Brindley's self-made success eventually led to a network of canals totaling 360 miles (579 km), with all but one designed and constructed by Brindley. Since Brindley never took notes, the only record of his designs are the canals themselves.

David Bushnell
1740-1826

American inventor who designed the first military submarine. Before he graduated from Yale, Bushnell had learned how to detonate gunpowder in water and he planned how to construct a submarine. He secretly built a submersible wooden vessel, named the *American Turtle,* to use offensively against British naval forces during the American Revolution. Thwarted by alert British sentries, Bushnell then built mines that frightened the British into indiscriminately firing at Delaware River debris in an action called the Battle of the Kegs.

John Campbell
1720?-1790

Scottish navigator who invented the sextant. Apprenticed to a shipmaster, Campbell joined the English navy. He advanced in rank to become the commander of several ships. In 1757 Campbell recommended that John Hadley's double-reflecting quadrant be altered to represent one-sixth of a circle, attaining a range of 120 degrees. Renamed the "sextant," Campbell's revised tool was useful in measuring horizontal angles and distances involving the Moon and planets. In 1782 Campbell was named the governor of Newfoundland.

William Caslon
1692-1766

British engraver and type founder who made great contributions to the development of type design and, as a result, bookmaking. William Caslon began his career in London as a toolmaker and engraver of firearms who also cut brass letters for bookbinders. In 1720, he switched to type design and, in 1723, Caslon's foundry opened. The foundry's typefaces were very popular in England and in America, where they were used in 1776 for the printing of the *Declaration of Independence*.

Claude Chappe
1763-1805

French engineer and former priest who invented the first mechanical optical telegraph, or semaphore. During the French Revolution, Chappe proposed a visual signaling line between Paris and Lille, near the frontlines of battle. The idea was to transmit messages using light between distant points. Chappe's brother, a member of the Legislative Assembly, strongly supported the concept. With the assembly's backing, the two constructed a series of hilltop towers equipped with a pair of telescopes, each pointed in a different direction, and a semaphore with adjustable arms that could assume seven clearly visible angular positions. The device was capable of displaying 49 combinations that were assigned to the alphabet and a number of other symbols. Using this system, it only took 2 to 6 minutes to transfer a message, whereas riding couriers would have needed 30 hours.

Thomas Chippendale
1718-1799

British furniture-maker whose folio work, *The Gentleman and Cabinet-Maker's Director*, published in 1754, was the first comprehensive book on furniture. The book contained illustrations of every conceivable type of furniture, made Thomas Chippendale famous, and led to his becoming the best known of all such English craftsmen and designers. The book also resulted in Chippendale's name becoming synonymous with many types of eighteenth century English furniture.

Henry Clay

British inventor who, in 1772, introduced the technique of papier-mâché. This process, consisting of soaking paper strips in thin starch-based paste and applying them to a surface to form a shape upon hardening, is now used primarily for decorative objects. When first introduced, it was also used to create trays, moldings, and other objects.

William Congreve
1772-1828

English artillery officer and scientist who invented the Congreve rocket, which saw its first action in the Napoleonic Wars. This rocket was one of the first to be fired from a ship at shore targets and was used during attacks against the French in Boulogne and in the siege of Copenhagen. Unlike a cannon, the rocket carried its own propellant with it, causing less stress on the launcher while permitting it to reach higher speeds.

Nicolas Jacques Conté
1755-1805

French chemist and inventor famous for developing an improved pencil in 1795, and whose manufacturing method was still the basis of the twentieth-century pencil industry. Nicolas Jacques Conté's pencils, which combined clay with graphite in variable proportions to adjust for hardness and darkness, replaced pure graphite pencils whose contents were in high demand due to a supply blockade. Conté had a special talent for inventions, lending his skills to improving Napoleon Bonaparte's military balloons and building the first engraving machine in 1803.

Bartolomeo de Francesco Cristofori
1655-1731

Italian harpsichord maker credited with inventing the piano, which he called a "harpsichord that plays soft [piano] and loud [forte]" (c. 1709). In effect a grand piano, Cristofori's pianoforte was wing-shaped and featured an independent damper mechanism and strings struck with hammers. From his first design in the early 1700s, Cristofori made improvements to the pianoforte. By 1726 his instrument had all the elements of the modern twentieth-century piano.

Nicholas Joseph Cugnot
1725-1804

French military engineer who built the world's first true automobile—a large steam-driven tricycle. While serving in the Austro-Hungarian army during the Seven Years War, Cugnot began to plot out ways to haul artillery. After the war's end, he built two steam-propelled tractors, the second of which is now preserved in the National Conservatory of Arts and Crafts in Paris. This cumbersome carriage was a tricycle mounted with a single front wheel that performed both steering and driving functions. Although its two-piston steam engine operated for only 12 to 15 minutes before running out of steam, it proved the feasibility of steam-powered engines.

Claude de Jouffroy d'Abbans
1751-1832

French engineer who in 1783 launched the first experimentally successful steam-powered boat. Equipped with a double ratchet mechanism that produced continuous rotation of the paddle wheels, Jouffroy d'Abbans's vessel, the *Pyroscaphe,* ran against the current of the Saône River for 15 minutes. Jouffroy d'Abbans's success was marred by failure, however, when his patent was indefinitely delayed and he abandoned his experiments. In 1787 American Robert Fulton developed a more successful steamboat. Although he gave much credit to Jouffroy d'Abbans, the French inventor died bitter and forgotten.

John Adam Dagyr
?-1806

Welsh/American shoemaker known as "the father of American shoemaking." Dagyr was the first to operate a shoe shop set up like a factory, with several workers each specializing in one particular task. This approach helped to make higher-quality shoes at a more rapid pace, reducing costs significantly. Previously, shoemakers passed their trade down to children or apprentices, a much less efficient process.

Abraham Darby III
1750-1791

English inventor who built the world's first cast-iron bridge. Darby's bridge, which crossed the Severn River in Coalbrookdale in England, was a major improvement over the wooden or masonry bridges that preceded it. Unlike wood, iron would not rot (although it would rust if not properly maintained). Unlike masonry, the relatively high strength and low weight of cast iron allowed spanning larger rivers safely, without impeding navigation in the process. This made possible increasingly long bridges, culminating in such classics as the Golden Gate and Erasmus Bridges, among others.

Oliver Evans
1755-1819

American inventor who devised steam machinery. Apprenticed as a wheelwright, Evans developed mechanical processes to make industrial combs. He automated flour mills by using a rake that sifted and dried the flour, resulting in a higher-quality product. Many millers refused to reimburse Evans for his ideas. Frustrated by this patent infringement, Evans, interested in steam locomotion, worked on a high-pressure steam engine as well as a steam dredging machine called the *Orukter Amphibolos.*

John Fitch
1743-1798

American inventor who was one of the steamboat's developers. Self-educated, Fitch experimented with a steam-propelled boat, constructing a model in 1785 and a steamboat two years later. Plagued by a lack of money and public disinterest, Fitch also struggled to defend his invention's originality when James Rumsey simultaneously developed a jet-propulsion steamboat. Fitch received several state patents and the Patent Act of 1790 was partly a result of the Fitch-Rumsey conflict. Both men were awarded a patent on August 26, 1791.

Pierre Simon Fournier
1712-1768

French engraver and type founder who studied watercolor paintings before getting involved in type design and beginning a foundry (1736). Fournier, said to have cut 60,000 punches for 147 alphabets of his own design, also developed new type ornaments and improved methods for printing music, for which he invented a point system for standardizing music type. His two volumes entitled *Manuale Typographique,* published in 1764 and 1766, were the first books on punch-cutting and typefounding.

Benjamin Franklin
1706-1790

American writer, publisher, scientist, diplomat, and inventor who made significant contributions to a number of fields, including playing a major role in the founding of the United States as an independent nation. In science and technology, he is remembered for his work in the theory of electricity and his many practical inventions that include the Franklin stove, which provides heat much more efficiently than a fireplace; bifocal eyeglasses; the lightning rod; and daylight savings time.

André Jacques Garnerin
1769-1823

French balloonist who made the first public parachute demonstration in Paris in 1797. Garnerin was actually a balloon inspector for the French army, but in his spare time decided to develop a method to allow aeronauts to safely reach the ground if their balloon failed. His highest jump was from 8,000 feet (2,438 m) in England in 1802. The great majority of parachutes retained the fundamentals of his design until very recent advances were made.

William Ged
1690-1749

Scottish goldsmith who invented a process for making printing plates using molds that enabled future editions of a document by casting new plates from the original mold. The process, called stereotype, was patented by Ged in 1725. Although Ged himself only used the process to print a single work, he paved the way for other print shops to advance the techniques of stereotyping and stereography.

Pierre-Simon Girard

French inventor who, in 1775, invented the first water turbine. This device was the first to harness the power of water using a turbine instead of a water wheel. Although not significantly more efficient at the time than water wheels of the day, the turbine proved much more versatile over time. It was adapted for electrical power generation, ship propulsion, jet engines, and other uses, all of which continue to this day.

George Graham
1673?-1751

British instrument maker and inventor famed for his astronomical instruments as well as his clocks and watches. Apprenticed to a London clockmaker in 1688, George Graham took over the business in 1713 then perfected the first successful deadbeat (or recoilless) escapement for clocks in 1715. For nearly 200 years it was standard equipment in observatories. In 1721, he invented the mercury compensation pendulum and the cylinder escapement for watches. Graham also developed a 5 feet (1.52 m) mural quadrant with telescopic sights in 1742 as well as other astronomical instruments.

Catherine Greene
1755-1814

American plantation owner who financed Eli Whitney's cotton gin. Greene invited Whitney to Mulberry Grove, her Savannah, Georgia, home, and suggested that he invent a machine to remove seeds from cotton bolls. For six months, Greene provided Whitney with materials and money. She even placed a brush on the gin's teeth to sweep away lint. Greene insisted that Whitney file for a patent himself because of public disapproval of women inventors. The cotton gin dramatically increased Southern antebellum agricultural production.

John Greenwood
1760-1819

American dentist who, in 1790, invented the first dental drill. Greenwood was George Washington's dentist, an irony since Washington is famous for having had wooden dentures. Greenwood's invention of the dental drill helped dentistry to become a preventative practice because dentists could treat dental cavities by removing the dis-eased tooth, replacing it with silver, gold, or some other material. Previously, all that could be done was to pull the tooth when it became too painful or too rotten for the patient to tolerate.

Joseph-Ignace Guillotin
1738-1814

French physician who, during the French Revolution, proposed decapitation by a mechanical device as a more humanitarian means of executing criminals than the torture then in use. Such mechanisms were used in Europe until about 1700 to execute nobles; Guillotin simply advocated their revival and more egalitarian use. Adopted in 1792, the device quickly became the symbol of the "Reign of Terror." The public named it the guillotine, an "honor" Guillotin abhorred since he neither invented it nor approved of its use against political enemies.

Valentin Haüy
1745-1822

French abbot and inventor who is credited with the conception (1771) of an embossed letter system to educate persons with blindness. In 1783-84 Abbot Haüy founded the Institute for the Blind and began training his blind students to read with their fingers. By the end of 1784 he had helped develop a special printing press for the students to expand the Institute's library. Haüy's embossed system was used to educate the blind throughout Europe until the 1829 introduction of an adaptation of a military "night writing" system by Louis Braille, one of Haüy's students.

J. N. de la Hire

Inventor who developed the double-acting water pump in 1716. This seemingly simply invention produces a continuous stream of water by pumping on both the "up" stroke and the "down" stroke. Previous pumps, using a handle attached to a piston, would suck water into the pump while pulling a handle up, then discharge it into a pail while pushing the handle down. By discharging water on both strokes, de la Hire's pump greatly improved efficiency. Variants of this type of pump are still in use.

Jonathan Carter Hornblower
1753-1815

English engineer whose work lead to many advances in steam engine technology. Hornblower made a number of contributions to the growing steam engine industry, including development of new uses for Watt's engines and several of his own patented designs. Most of his work appears to have dealt with the use of steam engines for pumping water, but he was also influential in showing how steam engines could be used to replace work previously done by water, draft animals, and men.

Thomas Jefferson
1743-1826

American president, statesman and inventor who, among other things, is credited with helping to define the duties and regulations of the U.S. Patent Office. From 1790-93, Thomas Jefferson served as examiner of patents. Although he initially felt that patents would stifle national and international progress (he declined to patent any of his own inventions) he found that the availability of patents actually spurred inventions. To ensure that patents were warranted, he developed strict guidelines for patent approvals during his tenure in the office, and many of his ideas still flavor the patent process today.

Samuel Klingensteirna
1697-1665

Swedish mathematician best known for his work toward eliminating chromatic aberration in telescope lenses. Chromatic aberration occurs when, like a prism, light is split into its component colors as it passes through a lens. In addition to being irritating, this distorting effect interfered with scientific measurements. Klingensteirna's work was sufficiently important that he was awarded a prize by the Russian Academy of Sciences in 1761.

Charles Dangeau de Labelye
1705-1782

Swiss engineer who invented a pile-driving machine to help construct the Westminster Bridge in 1738. Labelye's machine, powered by three horses, repeatedly raised a weight and dropped it onto a piling, driving it into the soil to form a solid foundation for the bridge structure. The machine used a 1,698 lb (770 kg) weight that was raised 9.8 feet (3 m) and dropped 150 times an hour. Labelye's basic design remains in use today, although powered by engines instead of horses.

Benjamin Latrobe
1764-1820

American engineer who designed significant public buildings. Latrobe was a surveyor and architect in England prior to immigrating to the United States in 1795. He introduced the Greek and Gothic Revival styles, improved Philadelphia's waterworks, and served as engineer for canal projects. President Thomas Jefferson

named Latrobe surveyor of public buildings and together they improved the architecture of the White House and Capitol. The Basilica of the Assumption of the Blessed Virgin Mary in Baltimore, Maryland, is one of Latrobe's most outstanding works.

Nicolas Leblanc
1742-1806

French physician and inventor who is best known for inventing a process by which to make sodium carbonate. This material is widely used in making glass, soap, and other useful chemicals. In fact, Leblanc's process was widely used for over a century, although Leblanc did not benefit from this because his factory was seized during the French Revolution. Due to this and other problems, he committed suicide at age 64.

Jakob Christof LeBlon
1667-1741

German printer and engraver credited with being the first to print in four colors. A native of Frankfurt, LeBlon invented a color mezzotint process using three metal plates inked in red, blue, and yellow (1710). A few years later, he was the first to add a fourth plate for black, creating the four-color print process. In 1725 LeBlon published *Il Coloretto or The Harmony of Colouring in Painting*, containing nine full-page mezzotints in color, reproductions of paintings by the masters. This book was incorporated into his posthumously published *L'Art d'Imprimer les Tableaux* (1756).

Philippe Lebon
1767-1804

French engineer and chemist who pioneered experiments in the application of gas for light, heat, and power, and the recovery of by-products (patented in 1799). In 1801 Lebon headed the first public demonstration of gas for lighting and heating at the Hôtel Seignelay in Paris, using thermolamps he had developed in 1799. His experiments and exhibitions proved the practical application of inflammable gas. Lebon's career was cut short by his brutal murder on Paris' Champs Elysées in 1804.

Georges Louis Lesage
1724-1803

Swiss physicist and inventor who in 1774 developed one of the first electronic telegraph machines. Lesage's invention, unlike modern telegraphs, used a different wire for each letter of the alphabet. This device, although capable of sending a message for short distances, proved too cumbersome to use on a wide scale. In fact, the telegraph would not become popular until several decades later, when development of Morse Code permitted signals to be sent using only one wire.

Jacob Leupold
1674-1727

German engineer who collected, for the first time in print, the basic principles of mechanical engineering. Leupold's nine-volume treatise, *The General Theory of Machines,* helped systemize the current knowledge of the day. It also contained the first description of a noncondensing steam engine—that is, one in which the steam is released into the atmosphere after use in the engine. This was very similar to the first such engines actually built over 50 years later.

John Lombe
1690?-1722

English weaver and inventor who, with his brother Thomas, helped bring the silk industry to England. Understanding that the Italians held a monopoly on "silk-throwing" machines that spun silk into thread, Lombe traveled to Italy to learn their secrets. Upon his return, he and Thomas built the first English silk-throwing machine, sparking the birth of the English silk industry. A rumor at the time of Lombe's death suggested that he been poisoned by vindictive Italians.

Thomas Lombe
1685-1739

English weaver and inventor who, with his brother John, helped found England's silk industry. The Lombe brothers built England's first silk weaving engines, taking out patents on the process of turning silk filaments into thread. In fact, Thomas financed his brother's trip to Italy where, employed as a mechanic in a silk factory, John learned the secrets of the Italian machines. Upon his return, the two brothers constructed their first mill, launching what was to become an important English industry.

Sybilla Masters
?-1720

American inventor believed to have been the first colonist granted a patent. Masters invented a machine with mortars and pestles to crush and dry Indian corn without grinding it. Because of English-law prohibitions against women, Master's husband, Thomas, was named in the patent. They sold "Tuscarora Rice," the corn meal her machine produced. Masters recommended the corn meal as a health remedy, making it an early American patent medicine. She also patented a method to weave palmetto leaves into hats, baskets, and furniture.

Andrew Meikle
1719-1811

Scottish inventor who developed the first threshing machine, a device used for removing grain from the stalk. Meikle's invention of the drum thresher improved the efficiency of grain threshing, enabling farmers to produce more food from each acre of land. Though not portable, the drum thresher could be run by water, wind, or horse power.

Henry Mill
1683?-1771

English engineer who invented the first typewriter in 1714. Although most of the details of Mill's machine are unknown and it never became popular in its time, the typewriter and its successors later proved to be tremendously important devices. In the patent granted to Mill by Queen Anne, it was noted that Mill's machine made counterfeiting more difficult, and made reading and writing easier because any person could set down on paper letters that were "so neat and exact as not to be distinguished from print."

William Nicholson
1753-1815

British chemist who was the first to use electricity to decompose water into hydrogen and oxygen. Using the recently invented Voltaic battery, Nicholson placed electrodes into water and showed that it "broke apart" into hydrogen and oxygen gasses. This reversed earlier experiments in which these two gasses were combined to form water, thus showing that many chemical reactions could be reversed. Nicholson also founded the *Journal of Natural Philosophy, Chemistry, and the Arts* in 1797.

Jonas Norberg
1711-1783

Swedish inventor who improved lighthouse optics. In 1757 Norberg designed the first flashing light used in a lighthouse. At the Korso lighthouse, he arranged clockworks to rotate two curved mirrors in order to reflect light intermittently. He carefully calculated the most effective parabolic curve to achieve this success. Norberg also developed the first revolving light in 1781. He hung his rotating reflectors inside the Marstrand lighthouse to create the effect of light beams horizontally sweeping the surrounding sea.

Denis Papin
1647-1712

French physicist and inventor who developed several useful devices using steam as the source of power. These devices included a steam-powered version of Thomas Savery's pump used for de-watering mine shafts, the first engine to use steam as the motive force, and the "steam digester," a forerunner to today's pressure cooker, in which pressurized steam is used to speed cooking times.

Jean-Rodolphe Perronet
1708-1794

French civil engineer who designed and built many stone arch bridges throughout France. He is best known for his Pont de la Concorde bridge, which spans the River Seine in Paris. The original design, completed in 1772, was found too daring and was not constructed for nearly 15 years. Once started, Perronet let nothing keep it from completion, including the French Revolution. In fact, he even used some of the stone from the Bastille as a source of masonry for this bridge.

James Pickard

British inventor who in 1780 patented one of the first steam engines. Although others had developed variations on the steam engine, Pickard was the first to formalize this through the patenting process, in which he had to prove that the invention was new, useful, and practical. The steam engine was one of the primary forces ushering in and powering the Industrial Revolution.

Eliza Lucas Pinckney
1722-1793

American amateur horticulturist who initiated colonial indigo cultivation. Pinckney spent her childhood in Antigua. Her father, a British army officer, moved the family to South Carolina and left the teenaged Pinckney in charge of their plantation, called Wappoo. She experimented with planting West Indian agricultural crops, including *Indigofera tinctoria,* which was valued as a dye. Her efforts in cultivating seed and manufacturing indigo resulted in that crop's being grown throughout the colony and becoming a major commercial export to England.

Jean-Baptiste le Prince
1734-1781

French artist and engraver who developed the etching technique called "aquatint." This new method enabled engravers to duplicate some of the more delicate effects of watercolors and wash drawings. Le Prince, however, is best remem-bered as an artist rather than an inventor. He earned commissions from the Czar of Russia and was elected to the Acádamie as a history painter.

Robert Ransome
1753-1830

English inventor who developed chilled cast iron, which he used to produce improved plow-shares. Ransome noticed in his foundry that, when molten cast iron spilled onto the floor, the quickly cooled iron was harder than expected. He patented this process and used the "chilled cast iron" to make farm and other implements that were harder and more durable than had previously been the case. Today, Ransome's company remains an important firm, now renamed Ransomes, Sims, & Jefferies.

John Rennie
1761-1821

Scottish civil engineer who built three major bridges over the Thames River in London—the Waterloo Bridge, London Bridge, and South-wark Bridge. As one of the most productive civil engineers of his day, Rennie was also responsible for building many canals to help transport goods and food from their places of production to markets for purchase or export. In addition, he helped design and build several harbors, also important to the process of internal and international trade.

Nicolas Louis Robert
1761-1828

French engineer who invented the "continuous roll" process for making paper. This process, un-like earlier papermaking methods, enabled paper to be manufactured at a rapid rate, as it was continuously spun onto huge rolls as it is made. By so doing, papermaking became much more efficient, allowing large quantities to be turned out at very low cost. This was a vast improvement over previous manual methods that manufactured paper as individual sheets processed by hand.

John Roebuck
1718-1794

English inventor who developed the lead-chamber method of producing sulfuric acid and new ways of producing more malleable iron using a pit fire blasted with a forced draft of air. Sulfuric acid has become an important industrial chemical, used for storage batteries, papermaking, and many other industrial processes. Roebuck's method of iron-making not only helped lay the groundwork for later blast furnaces, but also helped turn brittle cast iron into more useful forms.

Thomas Savery
1650?-1715

English engineer who invented the "Miner's Friend," the first useful machine powered by steam. This device, a steam pump to remove water from mine shafts, was inefficient but worked well enough to inspire others to invent improved models. These progressively advanced steam engines helped to power the first phases of the Industrial Revolution, making Savery's invention, inefficient as it was, one of the most significant technological innovations of all time.

Johann Heinrich Schulze
1687-1744

German physician and anatomy professor who made a significant discovery in the development of photography when he observed that silver salts darkened when exposed to sunlight. In 1725, while attempting to create a phosphorescent material by combining chalk with nitric acid containing dissolved silver, Schulze noticed that sunlight turned the substance black. He tested words and shapes cut out of paper and placed against a bottle of the solution, but he never made a permanent image.

James Short
1710-1768

English astronomer who built some of the first telescopes to employ the design developed by French inventor Cassegrain. The Cassegrainian reflector used curved mirrors instead of lenses to concentrate light into a sharp and useable image for study. Although first designed in 1672, the technology to shape mirrors to the precise specifications needed for an effective instrument was lacking until 1740. The Cassegrainian reflector is still used today, as is its primary competitor, the Newtonian reflector.

Henry Shrapnel
1761-1842

English artillery officer who invented the shrapnel shell in 1793. This shell, designed as an anti-personnel device, exploded in mid-air, scattering shot, pieces of metal, and other materials designed to kill or wound nearby soldiers. This device was a great boon to the army because it greatly extended the lethal range of cannon balls, which no longer had to hit someone directly in order to injure them.

Samuel Slater
1768-1835

English-American engineer known as the "father of the American Industrial Revolution." Slater emigrated to America, bringing with him knowledge of the workings of England's most advanced textile machinery. Although the law forbade people with this knowledge from leaving England, he did so under an assumed name. Slater set up a mill in Rhode Island that became the basis for the American textile industry and helped launch the industrialization of the United States.

William Small
1734-1775

English-American scientist who helped found the "Lunar Society" in 1764, an organization that brought together some of England's finest engineers and inventors. The members of the Lunar Society (called Lunatics) included John Roebuck, James Watt, Joseph Priestly, Josiah Wedgewood, John Wilkinson, and corresponding participant Benjamin Franklin, among others. This important association facilitated exchanges and collaboration among scientists, leading to many inventions that made the Industrial Revolution possible. Small also had a profound influence on Thomas Jefferson as his teacher.

Jacques-Germain Soufflot
1713-1780

French architect best known as the designer of the church of Ste. Geneviève in Paris, which is more commonly known as the Panthéon. Trained in Rome (1734-38), Soufflot was known for his interest in Gothic architecture and for designs that incorporated Greek forms, such as his Greek Cross plan with a large dome, Corinthian columns, and portico on his 1757 plan for St. Geneviève, for which he also used a unique buried framework of iron bars, a precursor to reinforced concrete.

Jedediah Strutt
1726-1797

English inventor who developed a ribbing machine to produce stockings. While a seemingly minor item of clothing, stockings were nearly universally used and often expensive. Strutt's invention helped make higher quality stockings available to many who had not previously been able to afford such a luxury.

Emanuel Swedenborg
1688-1772

Swedish chemist, engineer, and mystic who published one of the first up-to-date accounts of mining and smelting techniques during the eighteenth-century. As with other influential publications of this era, Swedenborg's book was one of the first to systematically lay out the current principles and practices by which these ancient professions worked. By so doing, he helped to formalize the field, making it possible to teach standard, accepted techniques and practices to miners and smelters everywhere.

Robert Bailey Thomas
1766-1846

American publisher who, in 1792 during George Washington's second term as U.S. president, founded *The Old Farmer's Almanac*, the oldest continually published almanac of its kind. Thomas's *Almanac* presented farming features, astronomical information, and weather data that he calculated using a secret weather forecasting formula said to be 80% accurate. When Thomas died in 1846, he was rumored to be hard at work on his 55th edition of *The Old Farmer's Almanac*.

William Tiffin

British inventor who developed an improved form of shorthand in 1750 called phonetic stenography, which featured symbols that stood for sounds as opposed to earlier stenography, which employed symbols that stood for letters. Tiffin's innovation was vital to improving the speed of a stenographer because it allowed the stenographer to transcribe words using fewer symbols.

Ehrenfried Walther von Tschirnhaus
1651-1708

German mathematician, physicist, and chemist who made a number of technological advances in his search for the secret of true porcelain. As early as 1675 Tschirnhaus was experimenting with radiant heat and mirrors to discover the melting point of substances such as kaolin, a vital ingredient of the hard-paste porcelain eventually developed with the assistance of German alchemist Johann Friedrich Böttger (1682-1719) at the Meissen factory. The celebrated Meissen porcelain factory owed its inception to the association between Tschirnhaus and Böttger.

Jacques de Vaucanson
1709-1782

French inventor and engineer who developed a device that enabled silk to be woven into cloth. Previous automatic machines had been too clumsy to work with silk, a fragile thread that must be handled with great care. De Vaucanson's machine was a great improvement, and in 1770 he invented the chain drive to help run his machines. De Vaucanson's was the first chain drive in Europe, though such drives had been invented independently in China 800 years earlier.

Philip Vaughn

English inventor who developed radial ball bearings for use in carriages. Ball bearings help reduce friction between surfaces by giving a rolling, rather than a sliding, contact surface that is more efficient and prolongs the life of the contact surfaces. Vaughn used ball bearings in the axles of carriages, making carriages easier to pull. Ball bearings are now used in virtually all machines, from roller skates to ship's propellers to printing presses and conveyor belts.

Josiah Wedgwood
1730-1795

English potter, industrialist, and philanthropist who is recognized as one of the greatest potters of all time. He was among the first to successfully apply scientific and economic principles to industry and to combine art and industry. He invented the pyrometer for measuring high temperatures. Many of his techniques and designs are in use today. He was a leader in neoclassicism, the revival of classical styles in art and architecture. Charles Darwin was his grandson.

Bibliography of Primary Sources

Books

Bélidor, Bernard Forest de. *La science des ingénieurs* ("Engineering Science," 1729) and *Architecture hydraulique* ("Hydraulic Architecture," 1737-39). In these two works, Bélidor turned his attention to mechanical problems involving transport, shipbuilding, waterways, water supply, and ornamental fountains. Little in these books was original, but they served as a call to builders to base design and practice on the science of mechanics. Bélidor's volumes influenced generations of architects, builders, and engineers, among them the first two generations of engineers who could also be considered scientists.

Chippendale, Thomas. *The Gentleman and Cabinet-Maker's Director* (1754). This work was the first comprehensive book on furniture. The book contained illustrations of every conceivable type of furniture, made Thomas Chippendale famous, and led to his becoming the best known of all such English craftsmen and designers. The book also resulted in Chippendale's name becoming synonymous with many types of eighteenth-century English furniture.

Fournier, Pierre Simon. *Manuale Typographique* (2 vols., 1764, 1766). These two volumes were the first books published on punch-cutting and typefounding.

LeBlon, Jakob Christof. *Il Coloretto or The Harmony of Colouring in Painting* (1725). This work displayed

LeBlon's pioneering efforts as the first person to print in four colors. LeBlon invented a color mezzotint process using three metal plates inked in red, blue, and yellow (1710). A few years later, he was the first to add a fourth plate for black, creating the four-color print process. This 1725 work contained nine full-page mezzotints in color.

Leupold, Jacob. *The General Theory of Machines* (9 vols., 1724). This work represented the first time the basic principles of mechanical engineering appeared in printed form. Leupold's nine-volume treatise helped systemize the current knowledge of the day. It also contained the first description of a noncondensing steam engine—that is, one in which the steam is released into the atmosphere after use in the engine. This was very similar to the first such engines actually built over 50 years later.

Savery, Thomas. *The Miners Friend; Or, An Engine to Raise Water by Fire* (1702). Describes Savery's invention of an early steam engine to help pump water from the bottom of coal mines. His engines became known as Miners' Friends, although it is unclear whether any of them found actual use in coal mines.

Senefelder, Aloys. *Vollständiges Lehrbuch der Steindruckerey* (A Complete Course in Lithography, 1818). Described the technical process of lithographic printing, which Senefelder developed between 1796-98. He made it possible for printers to mass-produce commercial images, thus influencing the spread not only of commerce, but of communication and literacy.

Thomas, Robert Bailey. *The Old Farmer's Almanac* (1792). This was the first edition of the oldest continually published almanac of its kind. Thomas's *Almanac* presented farming features, astronomical information, and weather data that he calculated using a secret weather forecasting formula said to be 80% accurate.

Tull, Jethro. *Horse-hoeing Husbandry* (1731; 2nd edition, 1733). In this book Tull advocated that farmers should crumble sod in order for air and water to reach plant roots. He invented a horse hoe specifically to achieve such soil pulverization. Members of the Private Society of Husbandmen and Planters, however, criticized Tull, saying that his pulverization ideas were without merit. Tull responded to these attacks with notes that were included in a third edition, issued in 1743.

NEIL SCHLAGER

General Bibliography

Agassi, Joseph. *The Continuing Revolution: A History of Physics from the Greeks to Einstein.* New York: McGraw-Hill, 1968.

Anderson, E. W. *Man the Navigator.* London: Priory Press, 1973.

Anderson, E. W. *Man the Aviator.* London: Priory Press, 1973.

Asimov, Isaac. *Adding a Dimension: Seventeen Essays on the History of Science.* Garden City, NY: Doubleday, 1964.

Bahn, Paul G., editor. *The Cambridge Illustrated History of Archaeology.* New York: Cambridge University Press, 1996.

Barrow, Sir John. *Sketches of the Royal Society and Royal Society Club.* London: F. Cass, 1971.

Basalla, George. *The Evolution of Technology.* New York: Cambridge University Press, 1988.

Benson, Don S. *Man and the Wheel.* London: Priory Press, 1973.

Beretta, Marco. *The Enlightenment of Matter: The Definition of Chemistry from Agricola to Lavoisier.* Canton, MA: Science History Publications, 1993.

Boorstin, Daniel J. *The Discoverers.* New York: Random House, 1983.

Bowler, Peter J. *The Norton History of the Environmental Sciences.* New York: W. W. Norton, 1993.

Brock, W. H. *The Norton History of Chemistry.* New York: W. W. Norton, 1993.

Bruno, Leonard C. *Science and Technology Firsts.* Edited by Donna Olendorf, guest foreword by Daniel J. Boorstin. Detroit: Gale, 1997.

Bud, Robert, and Deborah Jean Warner, editors. *Instruments of Science: An Historical Encyclopedia.* New York: Garland, 1998.

Butterfield, Herbert. *The Origins of Modern Science, 1300- 1800.* New York: Macmillan, 1951.

Bynum, W. F., et al., editors. *Dictionary of the History of Science.* Princeton, NJ: Princeton University Press, 1981.

Carnegie Library of Pittsburgh. *Science and Technology Desk Reference: 1,500 Frequently Asked or Difficult-to-Answer Questions.* Washington, D.C.: Gale, 1993.

Crone, G. R. *Man the Explorer.* London: Priory Press, 1973.

Daston, Lorraine. *Classical Probability in the Enlightenment.* Princeton, NJ: Princeton University Press, 1988.

Dibner, Bern. *Ten Founding Fathers of the Electrical Science.* Norwalk, CT: Burndy Library, 1954.

Elliott, Clark A. *History of Science in the United States: A Chronology and Research Guide.* New York: Garland, 1996.

Ellis, Keith. *Man and Measurement.* London: Priory Press, 1973.

Erlen, Jonathan. *The History of the Health Care Sciences and Health Care, 1700-1980: A Selective Annotated Bibliography.* New York: Garland, 1984.

Frängsmyr, Tore, editor. *Linnaeus, the Man and His Work.* Canton, MA: Science History Publications, 1994.

Frängsmyr, Tore, et al., editors. *The Quantifying Spirit of the 18th Century.* Berkeley: University of California Press, 1990.

Galton, Sir Francis. *English Men of Science: Their Nature and Nurture.* London: Cass, 1970.

Gascoigne, Robert Mortimer. *A Chronology of the History of Science, 1450-1900.* New York: Garland, 1987.

Gasking, ELizabeth B. *The Rise of Experimental Biology.* New York: Random House, 1970.

Good, Gregory A., editor. *Sciences of the Earth: An Encyclopedia of Events, People, and Phenomena.* New York: Garland, 1998.

Grabiner, Judith V. *The Calculus as Algebra: J.-L. Lagrange, 1736-1813.* New York: Garland, 1990.

Grattan-Guiness, Ivor. *The Norton History of the Mathematical Sciences: The Rainbow of Mathematics.* New York: W. W. Norton, 1998.

Gregor, Arthur S. *A Short History of Science: Man's Conquest of Nature from Ancient Times to the Atomic Age.* New York: Macmillan, 1963.

Gullberg, Jan. *Mathematics: From the Birth of Numbers.* Technical illustrations by Pär Gullberg. New York: W. W. Norton, 1997.

Hankins, Thomas L. *Science and the Enlightenment.* New York: Cambridge University Press, 1985.

Hellemans, Alexander and Bryan Bunch. *The Timetables of Science: A Chronology of the Most Important People and Events in the History of Science.* New York: Simon and Schuster, 1988.

Hellyer, Brian. *Man the Timekeeper.* London: Priory Press, 1974.

Hindle, Brooke, editor. *Early American Science.* New York: Science History Publications, 1976.

Hodge, M. J. S. *Origins and Species: A Study of the Historical Sources of Darwinism and the Contexts of Some Other Accounts of Organic Diversity from Plato and Aristotle On.* New York: Garland, 1991.

Holmes, Edward and Christopher Maynard. *Great Men of Science.* Edited by Jennifer L. Justice. New York: Warwick Press, 1979.

Holmes, Frederic Lawrence. *Lavoisier and the Chemistry of Life: An Exploration of Scientific Creativity.* Madison: University of Wisconsin Press, 1985.

Hoskin, Michael. *The Cambridge Illustrated History of Astronomy.* New York: Cambridge University Press, 1997.

Knight, David M. *Sources for the History of Science, 1660- 1914.* Ithaca, NY: Cornell University Press, 1975.

Lankford, John, editor. *History of Astronomy: An Encyclopedia.* New York: Garland, 1997.

Lincoln, Roger J. and G. A. Boxshall. *The Cambridge Illustrated Dictionary of Natural History.* Illustrations by Roberta Smith. New York: Cambridge University Press, 1987.

Mayr, Otto, editor. *Philosophers and Machines.* New York: Science History Publications, 1976.

Porter, Roy. *The Cambridge Illustrated History of Medicine.* New York: Cambridge University Press, 1996.

Roger, Jacques. *Buffon: A Life in Natural History.* Translated by Sarah Lucille Bonnefoi, edited by L. Pearce Williams. Ithaca, NY: Cornell University Press, 1997.

Rothenberg, Marc. *The History of Science in the United States: An Encyclopedia.* New York: Garland, 2000.

Rudwick, M. J. S. *The Meaning of Fossils: Episodes in the History of Paleontology.* New York: American Elsevier, 1972.

Sarton, George. *Introduction to the History of Science.* Huntington, NY: R. E. Krieger Publishing Company, 1975.

Sarton, George. *The History of Science and the New Humanism.* New Brunswick, NJ: Transaction Books, 1987.

Scott, Wilson L. *The Conflict Between Atomism and Conservation Theory, 1644- 1860.* New York: American Elsevier, 1970.

Singer, Charles. *A History of Biology to About the Year 1900: A General Introduction to the Study of Living Things.* Ames: Iowa State University Press, 1989.

Singer, Charles. *A Short History of Science to the Nineteenth Century.* Mineola, NY: Dover Publications, 1997.

Smith, Roger. *The Norton History of the Human Sciences.* New York: W. W. Norton, 1997.

Spangenburg, Ray and Diane K. Moser. *The History of Science in the Eighteenth Century.* New York: Facts on File, 1993.

Stiffler, Lee Ann. *Science Rediscovered: A Daily Chronicle of Highlights in the History of Science.* Durham, NC: Carolina Academic Press, 1995.

Stwertka, Albert and Eve Stwertka. *Physics: From Newton to the Big Bang.* New York: F. Watts, 1986.

Travers, Bridget, editor. *The Gale Encyclopedia of Science.* Detroit: Gale, 1996.

Watkins, George and R. A. Buchanan. *Man and the Steam Engine.* London: Priory Press, 1975.

Whitehead, Alfred North. *Science and the Modern World: Lowell Lectures, 1925.* New York: The Free Press, 1953.

World of Scientific Discovery. Detroit: Gale, 1994.

Young, Robyn V., editor. *Notable Mathematicians: From Ancient Times to the Present.* Detroit: Gale, 1998.

JUDSON KNIGHT

Index

*Numbers in bold refer to
main biographical entries*

I